Biology

organisms and adaptations

For Susan, Karen, and Robert

For your unwavering support

For your commitment

For your love

In fond memory of Steve McEntee

June 19, 1968–August 26, 2012

Kate Nagle Photography

His inspiring artistry lives in this book.

Our memories of him will always live in our hearts.

Biology

organisms and adaptations

Robert K. Noyd

Jerome A. Krueger

Kendra M. Hill

BROOKS/COLE
CENGAGE Learning

Australia • Brazil • Japan • Korea • Mexico • Singapore • Spain • United Kingdom • United States

Biology: Organisms and Adaptations
Robert K. Noyd, Jerome A. Krueger,
Kendra M. Hill

Senior Acquisitions Editor: Peggy Williams

Publisher: Yolanda Cossio

Senior Developmental Editor: Nedah Rose

Assistant Editors: Shannon Holt,
  Suzannah Alexander

Editorial Assistant: Sean Cronin

Media Editor: Lauren Oliveira

Senior Market Development Manager:
  Tom Ziolkowski

Marketing Coordinator: Maureen Towle

Senior Brand Manager: Nicole Hamm

Art Director: John Walker

Content Project Manager: Hal Humphrey

Manufacturing Planner: Karen Hunt

Rights Acquisitions Specialist: Dean
  Dauphinais

Production Service: Dan Fitzgerald,
  Graphic World Inc.

Photo Researcher: Wendy Granger, Bill Smith
  Group

Text Researcher: Pablo D'Stair

Copy Editor: Graphic World Inc.

Art Management: Steve McEntee

Illustrators: Dragonfly Media Group,
  McEntee Art & Design, Graphic
  World Inc.

Text Designer: Jeanne Calabrese

Cover Designer: Jeanne Calabrese

Cover Image: © Ch'ien Lee/Minden Pictures

Compositor: Graphic World Inc.

Chapter Opener Illustrations:
  Steve McEntee: 1–9, 13
  Rob Duckwall, Dragonfly Media Group:
    10–12, 14, 17, 19, 20
  Mike Demaray, Dragonfly Media Group:
    15, 16, 18, 21

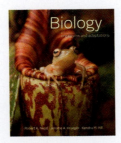

For product information and technology assistance, contact us at
Cengage Learning Customer & Sales Support, 1-800-354-9706
For permission to use material from this text or product,
submit all requests online at **www.cengage.com/permissions**
Further permissions questions can be e-mailed to
permissionrequest@cengage.com

Library of Congress Control Number: 2012950642

ISBN-13: 978-0-495-83020-7

ISBN-10: 0-495-83020-8

Brooks/Cole
20 Davis Drive
Belmont, CA 94002-3098
USA

Cengage Learning is a leading provider of customized learning solutions with office locations around the globe, including Singapore, the United Kingdom, Australia, Mexico, Brazil, and Japan. Locate your local office at **www.cengage.com/global**

Cengage Learning products are represented in Canada by Nelson Education, Ltd.

To learn more about Brooks/Cole, visit **www.cengage.com/brookscole**

Purchase any of our products at your local college store or at our preferred online store **www.CengageBrain.com**

About the Cover

Our cover image perfectly illustrates the approach of our text, *Biology: Organisms and Adaptations*, to the study of biology. It shows a tree frog perched atop the mouth of a pitcher plant in a mossy, upper-mountain forest in Sarawak, Malaysia. Carnivorous pitcher plants use fluid-filled leaves as traps to attract, capture, and aid in the absorption of nutrients from animal prey (see Chapter 13 for more about pitcher plants). The leaf trap creates a rich microhabitat for a community of organisms that exist in a mutualistic relationship with the pitcher plant. The plant relies on this community of bacteria, protozoans, and mosquito and midge larvae to help provide it with nutrients. Members of this miniature food web feed on one another and on the bodies of dead insects that drown in the pool. Although many amphibians fall victim to these traps, tree frogs (also called shrub frogs) apparently are unharmed by the fluid and use it to deposit their eggs for an unusual form of embryonic development that bypasses the tadpole stage and develops directly into small frogs, or froglets.

Printed in Canada
1 2 3 4 5 6 7 16 15 14 13 12

Brief Contents

Contents

15 The Biology of Fungi 468

Unit 4 Environmental Connections of Life

16 Physical and Chemical Cycles and the Biosphere 492

Unit 5 Genetic Connections of Life

About the Authors

Robert K. Noyd is a professor of biology at the U.S. Air Force Academy in Colorado Springs, where he teaches introductory biology and botany courses. He served as Director of Faculty Development for the academy, where he guided the professional development of over 500 faculty members. Bob actively writes and presents workshops throughout the country on course design and learning-focused teaching strategies. He earned a B.S. in biology education and an M.S. in plant biology from Central Connecticut State University and a Ph.D. from the Department of Plant Pathology at the University of Minnesota. As a native of New England, Bob started his career as a high school science teacher in 1978 and has been sharing his knowledge and excitement for the natural world with students at Bentley University, as well as several community colleges in Connecticut, Massachusetts, and Minnesota and now at the U.S. Air Force Academy.

Jerome A. Krueger grew up in South Dakota and received his B.S. and Ph.D. from the University of Minnesota and his M.S. from the University of Michigan. He has served on the biology faculty at the U.S. Air Force Academy and at South Dakota State University. His current academic affiliation is with Bitterroot College, a satellite campus of the University of Montana, and he is employed by the U.S. Forest Service as a Biological Scientist and Staff Officer with the Bitterroot National Forest. Years of teaching undergraduates in introductory biology courses led him to recognize the critical need for leading a change in the way science is presented to nonmajors. In addition to writing, he continues to consult on program and course development in the biological sciences.

Kendra M. Hill grew up in Maryland and received her undergraduate degree from the University of Delaware. She completed her M.S. in biotechnology from Johns Hopkins University and will soon complete her Ph.D. from South Dakota State University. She worked in biomedical research for several years before entering the biotech industry. For the past eight years she has been teaching Introductory Biology at South Dakota State University. In addition to teaching, she serves as Undergraduate Program Coordinator for the Biology and Microbiology Department. Through her experiences in the classroom and close interactions with her students, she is excited to present a new way to teach and serve students in introductory biology.

Preface

Welcome to an exciting new way to teach, learn, and experience introductory biology—the study of living things. The title of this book, *Biology: Organisms and Adaptations,* signals our approach to learning core biology concepts—organisms. Our approach was born out of a simple observation: Humans have an innate fascination with the lives and adaptations of living things. People love to explore, discover, and learn about the natural world. This observation provided us with a deep sense of purpose to build a biology book that translates the passions and interests of students and instructors into meaningful and enduring learning. It is our hope that you will find our approach motivating to yourself and your students.

An Organismal Approach

Our vision for a new and richer learning experience means that living things are out front doing the work—enhancing, engaging, connecting, providing applications, and driving a deeper level of understanding. Right from the start, each chapter opens with the story of an organism that illustrates the biological concepts to follow in the chapter. As the central organizing framework, organisms provide you with a powerful tool to:

1. **Interest and motivate your students to engage with biological concepts.**

   All instructors want their students to value their course and the learning it produces. This text directly acknowledges the great impact that positive feelings and emotions (appreciation, motivation, excitement, and curiosity) have on thinking and learning. The organisms have been carefully selected for their relevance and interest.

2. **Provide concreteness, familiarity, and a springboard for learning key concepts.**

   Key biological concepts are presented and learned in relation to familiar, recognizable organisms. A blue whale is classified; has an evolutionary history; lives in a particular habitat; has a body composed of organ systems, organs, tissues, cells, and subcellular components; feeds and metabolizes; mates, reproduces, and inherits traits. Thus the whale serves as a focal point to synthesize and connect the concepts that you have told us you want your students to learn.

   Educational research has shown that learning becomes more durable and transferable when it is constantly reinforced and repeated in different contexts. Traditional approaches often present a concept in the opening units and expect students to recall the details many chapters later in an entirely new context. The advantage of an organismal approach is the opportunity to reinforce processes and adaptations among different organisms throughout the book. We take every opportunity to reinforce the process of meiosis in animal, plant, and fungal systems. The same is true for other essential concepts, such as cell structure, natural selection, niche, gene expression, biomolecules, cell cycle, cellular respiration, and enzymes.

3. **Create a platform for inquiry, application, synthesis, and critical thinking.**

   We introduce a variety of organisms to provide engaging opportunities to develop question-asking skills, scientific thinking, data analysis, and the application and synthesis of biological concepts. Empowered with a deeper sense of anchored and connected understanding, students are then

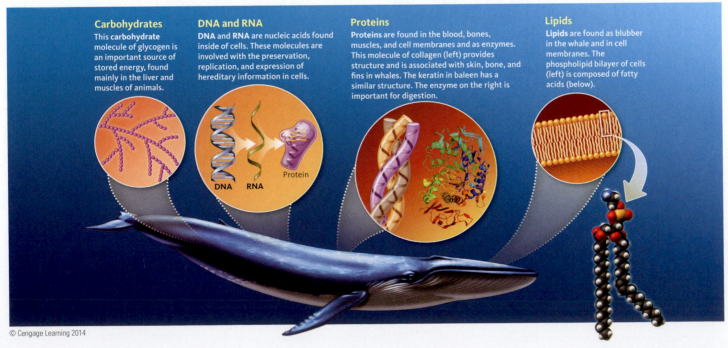

Carbohydrates
This **carbohydrate** molecule of glycogen is an important source of stored energy, found mainly in the liver and muscles of animals.

DNA and RNA
DNA and **RNA** are nucleic acids found inside of cells. These molecules are involved with the preservation, replication, and expression of hereditary information in cells.

Proteins
Proteins are found in the blood, bones, muscles, and cell membranes and as enzymes. This molecule of collagen (left) provides structure and is associated with skin, bone, and fins in whales. The keratin in baleen has a similar structure. The enzyme on the right is important for digestion.

Lipids
Lipids are found as blubber in the whale and in cell membranes. The phospholipid bilayer of cells (left) is composed of fatty acids (below).

DNA RNA Protein

capable of applying their knowledge and skills to other organisms that they encounter in their region of the country or the world. As an instructor, you have the flexibility of using the organisms provided in the text or using your own favorite organism.

The Organization of the Book

In terms of the biological hierarchy of scale and organization, our approach begins and focuses on the *organism* level. We believe that the most powerful frameworks to shape student understanding of living things are evolution and ecology. Therefore the first unit establishes these foundations and contexts within which the rest of the book is based.

Unit 1. Introduction to Organismal Biology—Establishes the Basic Foundation and the Framework.
Unit 1 lays the foundation, in a brief way, for the characteristics of life, hierarchy of scale and organization, cells, biological processes, and biomolecules. Our strategy is to give students enough information so that they can begin to apply it, and we can use it to go into greater depth in later units. For example, we present proteins as molecules that have a variety of shapes and perform many critical functions. We wait until Unit 2 to delve deeper into protein chemistry. This spares students from details at points of instruction where they do not need it. In addition to laying the groundwork, Unit 1 establishes two major themes or frameworks:

Evolutionary Biology—Evolutionary concepts provide a coherent explanation for the unity and diversity of living things. This central concept in biology is first introduced in Chapter 1 as a potent example of evidence-based scientific reasoning. We then follow a large-scale view (macroevolution) to a smaller-scale view (microevolution) in Chapters 2 and 3. Students love dinosaurs, and Chapter 2 uses *Archaeopteryx* and the evolution of birds to introduce large-scale evolutionary processes and concepts. These examples provide the frame to understand and launch our overview of biological diversity. Chapter 3 uses the mimicry of passion-vine butterflies to explore the processes of natural selection.

Ecological Approach—For nonmajors, understanding how the world works and the interdependence of organisms and the environment is really important. An ecological frame allows students to place adaptations in the context of where organisms live—their habitat and ecosystem. The coast redwood's narrow geographic range is used in Chapter 4 to present ecological concepts of habitat and the environmental factors that influence the activities and behaviors of organisms. Habitat is established and then used throughout the rest of the book.

© Cengage Learning 2014

For example, we present cellular respiration (Chapter 9) in terms of oxygen capture in both aquatic and terrestrial habitats. Photosynthesis (Chapter 13) is placed into an ecological context so that students can connect it to resources in their habitat—light, water, and carbon dioxide. We think this approach is much more motivating and concrete than placing it in a chapter devoted to abstract concepts of energy transformation, which emphasizes the detailed biochemical pathways.

Unit 2. Adaptations of Animals—The Most Familiar Organisms.
Most nonmajor students report that they find animals more interesting than other kinds of living things. Students see animals as dynamic and exciting because they move and behave in ways that they can relate to. In Chapter 6, the leatherback sea turtle is used to illustrate how animal structures reflect their function. The chapter starts with the most familiar level of animal structure—organ systems. We then use the digestive system of the leatherback turtle to present types of tissues and the cells that compose them. Cell structure and function are presented through an acid-producing cell of the leatherback stomach. Once structure is presented, the chapter shows how structures of the multicellular animal work together, communicate, and respond to change—in other words, homeostasis.

How these structures arise from a single fertilized cell is answered in Chapter 7, where the American lobster is used to show students patterns of growth and development, control, DNA structure and replication, and the cell cycle. Once again, the overarching context is a growing, developing organism that serves as a focal point to understanding the abstract mechanisms that produce that organism's body. Gene expression (transcription and translation) is not presented in abstract terms but rather as the driving process that results in the lobster's growth and differentiation.

The familiar dairy cow begins a two-chapter sequence that transforms food (Chapter 8) and oxygen (Chapter 9) into energy metabolism in the mitochondria. Cellular respiration is framed in the life of the clam that lives in the intertidal zone at the seashore. The battle between bats and their moth prey sets the stage for learning how nervous, muscular, and skeletal systems work together to enable the animal to eat, move, and find a mate (Chapter 10). Animal reproduction (Chapter 11) is a fitting way to pull Unit 2 together. Coho salmon display the interaction of animal populations in their habitat, their sensory systems in their drive to find a mate, their male and female organ systems, as well as sex cells and cellular processes. Finally, reproduction involves development and nutrient and gas exchange in the growing fetus.

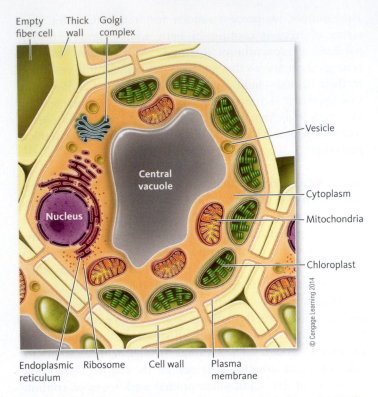

Empty Thick Golgi
fiber cell wall complex

Vesicle

Central
vacuole

Cytoplasm

Nucleus

Mitochondria

Chloroplast

© Cengage Learning 2014

Endoplasmic   Ribosome   Cell wall   Plasma
reticulum   membrane

Unit 3. Adaptations of Plants and Fungi—Moving from the Background to the Foreground. The lives of plants and students come together through coffee (Chapter 12); carnivorous pitcher plants (Chapter 13); and the symbol of the American Southwest, the iconic saguaro cactus (Chapter 14) in Unit 3. This unit explores how plants are constructed and presents plant cell structure through a photosynthetic cell in the leaf of a coffee plant. The reader is led to compare it to the animal cell presented in Chapter 6. The cell cycle and mitosis return as the cellular process that causes coffee buds to grow and flower. The complex reactions of photosynthesis are presented in context of the plant's light environment (Chapter 13). We address the long-held student misconception that plants don't respire by tracing sugar from its production in the chloroplast to its breakdown in the mitochondria. Plant activities and chemical cycles of the biosphere are understood through the absorption of water and minerals through roots. The reproduction of the flowering aspen and saguaro are the focal points for understanding asexual and sexual reproduction of flowering plants in Chapter 14. As in Unit 2, the sex lives of flowering plants pull together many core biology concepts, which include natural selection, gene flow, cell structure, mitosis, meiosis, and chemical signaling.

Although fungi are more closely related to animals than to plants, fungi are traditionally studied with plants and are placed in this unit. Fungi are framed in relation to the lives of college students through yeast infections, antibiotics, allergies, pizza baking, and beer brewing. Chapter 15 highlights how interesting organisms integrate key biology concepts. Cell structure is reinforced for the third time in the text as a hyphal tip cell of a mushroom. Other concepts such as mitosis, meiosis, enzyme activity, gene expression, and sexual reproduction are applied to these fascinating and ubiquitous organisms.

Unit 4. Environmental Connections of Life—Helping Understand Environmental Issues. A strong reason for students to take a nonmajor biology course is for them to become more informed and literate about issues confronting our society. Chapter 16 functions to help students apply organismal biology to global concerns such as clean water, carbon cycling, greenhouse gases, and climate change. Chapter 17 capitalizes on the fact that many students have visited Yellowstone National Park and will easily relate to environmental issues and the challenges faced by this ecosystem. Millions of people visit zoos and aquariums each year, and this chapter discusses their role in conserving global biodiversity.

Unit 5. Genetic Connections of Life—Helping Understand the Genetic Past and Future. The physical appearance and adaptations (phenotype) of animals, plants, and fungi serve as the basis to understanding the underlying genetic principles of inheritance (Chapter 18). This is why the unit is titled "Genetic Connections" and comes toward the latter half of the book. The familiar human is the lead organism used to get students to consider their own genetic history and future. Corn, the staple of American food products, is the organism used to introduce the application of genetics to genetic engineering of foods and ethical considerations (Chapter 19). Finally, this unit explains how genetic techniques help convict criminals, conserve biodiversity, and enhance our health care.

Unit 6. Applying Biological Concepts to Human Health and Disease—Promoting Health through Understanding. Students are very engaged when we discuss their personal health and welfare. This final unit applies concepts established throughout the book in the relationship between humans and their health. Chapter 20 begins with the story of the MRSA superbug and the issue of antibiotic resistance. Infectious diseases caused by bacteria, viruses, and protozoans take center stage to reinforce and apply concepts of cell structure, ecology, pathology, and immunology. While we acknowledge that the vast majority of bacteria are nonpathogenic, this chapter focuses on their structure and activities that cause disease. The last chapter (21) explains the pathology and public health issues relating to chronic disease—cancer, obesity, and heart disease. It is our sincere hope that our students will use this information to prevent the spread of disease and lead healthy, prosperous lives.

Chapter Features

- **Chapter Opening Vignettes with Enhanced Illustrations**— Each chapter opens with a story that profiles a fascinating organism. This organism may be familiar to some students, depending on their geography and personal experiences. The openers function to pique students' curiosity and motivate them to continue reading. They also illustrate and foreshadow the biology concepts to follow in the chapter.
- **Numbered Section Headings Present a Concept Statement**—This is a text-signaling device that telegraphs the

topic structure to students and helps them grasp the main point of the section. This feature allows instructors to refer to specific sections by number when assigning reading and preparation work.

- **Check & Apply Your Understanding**—At the conclusion of every section are bulleted key concepts followed by five questions that help students self-assess their understanding of the concepts presented. The first two questions assess basic knowledge, followed by questions of increasing difficulty. The fifth question asks students to transfer their conceptual understanding to other organisms or situations. These higher-level questions can also be used in lectures as one-minute papers or "clicker questions" asking students to make a specific choice regarding a scenario.

- **End of Chapter Review**—This section is designed to challenge and promote thinking and applying concepts.

  - **Self-Quiz Questions**—These questions are organized around each of the chapter's key concepts. For each concept, a series of matching and multiple-choice questions are posed. They are designed to assess semantic and conceptual knowledge and foster factual base building and independent learning.

  - **Applying the Concepts Questions**—These questions ask students to connect chapter concepts to each other and to other disciplines and to transfer their understanding to new organisms or situations. These questions can be easily incorporated into lectures to help students apply the material they have learned.

  - **Data Analysis Exercise**—This activity promotes higher thinking skills involving strategic knowledge (problem solving), application, and quantitative thinking skills. The *Data Analysis* feature challenges students to read graphs and tables from actual scientific papers. Here they analyze the freezing tolerance of lizards, photosynthesis in yucca plants, and the growth rate of giant pumpkins.

- **Question Generator**—One of the basic skills required for the development of scientific reasoning is asking questions. We intentionally work to develop students' question-asking skills through this feature. It's very important for students to realize that biologists don't pull their research questions out of thin air; rather, they arise from a base of knowledge and an understanding of how concepts are connected. Through the *Question Generator,* we present an intriguing aspect of the biology of an organism, which is related to chapter concepts. We then show students a diagram of concept relationships and use it to model how they can generate effective questions and develop this higher-level thinking skill.

Stunning Art Program. The biological hierarchy is a very effective schema to help students integrate their knowledge into understanding. As instructors, we move fluently among levels of the biological hierarchy. Students need time to locate the level at which they are working and how it connects upward and downward in scale. Our text's stunning art program takes students on a Google Earth–type visual journey from the organism level down to the tissues, cells, and molecules. The hydrogen bonds of a keratin molecule are visually connected to the feathers of a peregrine falcon. Through their connective thinking, students begin to reason and think like biologists.

Vocabulary Builds Language, and Language Drives Thinking. Biology, like all disciplines, includes many specialized terms that are networked together in specific ways to form its language. It is through this language that biologists and biology students understand and interact with the living world. And it is through biological language that they will construct a mental model, or learn biology.

We realize that the number of new technical terms in an introductory biology course can be overwhelming to students, and they often label biology as a memorization course. We recognize that the introduction and memorization of too many terms often impedes reading and learning. The language in this text has been carefully chosen, and terms are introduced when they are useful to learning. Before presenting and boldfacing a particular term, we asked ourselves several questions: How powerful is the term in explaining or organizing information? How often will the term be used again? Is it essential for furthering the understanding or application of a concept? A term that has been boldfaced has exceeded a threshold of importance. We also acknowledge that instructors using this book look for and are comfortable with specific terms. We boldface and define terms when they are first introduced in each chapter to inform the reader and develop conceptual understanding.

Instructor's Supplements

Instructor's Manual. A great planning tool that includes chapter objectives; key concepts; key terms; lecture outlines; ideas for further inquiry; websites, animations, and videos; suggestions for lecture enrichment and activities; possible answers to the text questions; and more. Content is mapped to learning objectives and includes alternate organisms to help reinforce ideas. Also included in Microsoft® Word format on the PowerLecture.

Test Bank. Test items ranked according to difficulty and Bloom's Taxonomy and correlated to the chapter learning objectives. Over 1,000 test items consisting of multiple-choice, matching, essay, and data analysis questions. Included in Microsoft Word format on the PowerLecture.

PowerLecture. The time-saving answer to course preparation, PowerLecture integrates all relevant resources into each chapter's Microsoft® PowerPoint® lecture so that there's no more hassling with multimedia files. Each chapter's lecture slides include the following resources, organized by chapter section: all chapter diagrams and photos; links to the animations, interactions, and videos; slides of book-specific questions that also appear on JoinIn on TurningPoint®; bulleted points listing key content; and many editing features, such as the ability to edit text on lecture slides, to present art in segments, and to edit labels, remove labels, or present one

label at a time. Available Online or on disc, PowerLecture also includes the Instructor's Manual and Test Bank.

ExamView® Access the Test Bank questions through this interactive testing software. ExamView allows you to easily edit questions, write your own questions, generate multiple versions of your test, and create your own online quizzing. ExamView can also be converted to work with your Black-Board or WebCT course. Included on the PowerLecture DVD.

Aplia with eBook. Aplia for Biology helps students learn and understand key concepts via focused assignments and active learning opportunities that include randomized, automatically graded questions; exceptional text interaction; and immediate feedback.

Biology CourseMate with eBook. Interested in a simple way to complement your text and course content with study and practice materials? Cengage Learning's Biology CourseMate brings course concepts to life with interactive learning, study, and exam preparation tools that support the printed textbook. With CourseMate, you can use the included Engagement Tracker to assess student preparation and engagement. In addition, CourseMate includes an interactive eBook where students can take notes, highlight, search, and interact with embedded media. Use it as a supplement to the printed text or as a substitute—the choice is your students' with CourseMate.

Student Supplements

Study Guide. Chapter summaries, key concepts, key terms, along with multiple-choice, fill-in-the-blank, matching, discussion questions, and more, all of which reinforce the text's learning objectives to help students with retention and achieve better test results.

Aplia with eBook. Aplia for Biology helps students learn and understand key concepts via focused assignments and active learning opportunities that include randomized, automatically graded questions; exceptional text interaction; and immediate feedback.

Biology CourseMate with eBook. Cengage Learning's Biology CourseMate brings course concepts to life with interactive learning, study, and exam preparation tools that support the printed textbook. In addition to interactive teaching and learning tools, CourseMate includes an interactive eBook. Students can take notes and highlight, search, and interact with embedded media specific to their book. Biology CourseMate goes beyond the book to deliver what students need!

Acknowledgments

We have been very fortunate to have the enthusiasm and support of so many talented professionals in developing this textbook with a new and different approach to teaching biology. Out front leading the charge was Senior Acquisitions Editor Peggy Williams, who initiated and managed this project over the years through her team of editors, art and media developers, designers, reviewers, and many other connections. Developmental Editor Nedah Rose kept this project on task and provided valuable insights and improvements to our writing and art program. Her perspective, suggestions, and advice were always helpful. Together, Peggy and Nedah saw our vision and made sure that it was translated into a package that others could use to inspire and improve biology education. Yolanda Cossio, Publisher, Life Sciences, provided consistent support and advocacy.

As you can imagine, introducing a new textbook into the product-saturated nonmajors market takes a lot of belief in the product and author team. Many thanks go to our Senior Market Development Manager, Tom Ziolkowski, as well as Senior Developmental Editor Joanne Butler, for making sure we had constant communication with biology educators around the country. Thank you for believing that *Biology: Organisms and Adaptations* can change biology education!

The art program was initially conceived and managed by Steve McEntee (McEntee Art & Design), who laid a beautiful and stunning foundation for us to build from. Steve worked closely with Jeanne Calabrese (Text and Cover Designer) to set the design for the textbook. Once Steve set the foundation, Rob Duckwall and other members from Dragonfly Media Group, McEntee Art & Design, and Graphic World were able to make remarkable contributions to the art program. Wendy Granger (Bill Smith Group) served as photo researcher and jumped through hoops tracking down high-quality photos and their associated credits. Along with good design and art, all good textbooks need a robust online media package; Lauren Oliveira was critical in developing digital instructor resources, and Shannon Holt and Suzannah Alexander guided the authors of the Instructor's Manual, Test Bank, and Student Study Guide. Sean Cronin performed dozens of tasks with efficiency and good humor. We are amazed at how each of these groups translated the vision inside of our heads and in our words to this book. You are very talented!

Content Project Manager Hal Humphrey (Cengage) and Dan Fitzgerald (Graphic World Inc.) somehow kept the authors, artists, and production on schedule. While the author team (and our families!) often thought this book would never come to fruition, the organization, communication, and management provided by Hal, Dan, Nedah, and Peggy kept us on track.

We would also like to thank our students and colleagues whom we have had the pleasure of working with over the years. Our experiences with students in the classroom have shaped our view on biology education and helped us to see what works and what doesn't work. Thanks to colleagues such as Dave Hale who reviewed a draft of the evolution chapter (Chapter 3), as well as John Putnam, Mark Pomerinke, and Tom Unangst who were always available to provide their expertise and resources on everything biological.

We could not have taken on this project without the sacrifices made by our families. Our deepest thanks to Susan, Karen, and Robert—your support allowed us the time and energy to think, write, and reflect. Thank you for your advice in the review of the manuscript, critique of art and photos, and endless conversations about presenting biology for the nonmajor.

In addition to the colleagues listed above, a number of professors participated throughout the development of this project—as reviewers, participants in focus groups, responders to surveys and telephone interviews, advisors, and contributors. You have helped us in many ways to shape and polish the content, sharpen the focus, clarify and improve the chapters, and create the resources for both students and instructors. We are grateful to every one of you.

Advisors and Contributors

Helene Engler, Interactive Concept Maps

Bob Harms, *St. Louis Community College at Meramec,* Study Guide author

Tara Jo Holmberg, *Northwestern Connecticut Community College,* Instructor's Manual author

Dubear Kroening, *University of Wisconsin-Fox Valley,* JoinIn

Fiorella Penaloza, *Briarcliffe College,* Online Quizzing author

Nicola Plowes, *Arizona State University/Mesa Community College,* PowerPoint Slides author

Laura H. Ritt, *Burlington County College,* ExamView and Test Bank author

Cara Shillington, *Eastern Michigan University,* ExamView and Test Bank author

Reviewers, Focus Group Participants, and Survey Respondents

Rebecca Abler, *University of Wisconsin–Manitowoc*

Julie Adams, *Elmhurst College*

Eddie Alford, *Arizona State University*

Troy D. Anderson, *The University of Texas at Tyler*

Enrique Aniceto, *Los Angeles Trade Technical College*

Senait Asmellash, *University of Advancing Technology*

Donna M. Becker, *Northern Michigan University*

Joressia Beyer, *John Tyler Community College*

Andrea Bixler, *Clarke University*

Lisa Ann Blankinship, *University of North Alabama*

Dennis Bogyo, *Valdosta State University*

Brenda D. Bourns, *Seattle University*

Richard Boyer, *Muscatine Community College*

James R. Bray Jr., *Blackburn College*

Peggy Brickman, *University of Georgia*

Mark Browning, *Purdue University*

Evelyn K. Bruce, *University of North Alabama*

Steven G. Brumbaugh, *Green River Community College*

Karen Campbell, *Albright College*

Geralyn M. Caplan, *Owensboro Community & Technical College*

Francesca Catalano, *American Public University*

Kerry Cheesman, *Capital University*

Thomas F. Chubb, *Villanova University*

Genevieve C. Chung, *Broward College*

Vickie Clouse, *Montana State University–Northern*

Reggie Cobb, *Nash Community College*

Lois V. Crichlow, *Valencia Community College*

Rhonda Crotty, *Tarrant County College–South*

Leslie Dafoe, *Sault College of Applied Arts & Technology*

Patrick Daydiff, *Arizona State University–West*

Paul Decelles, *Johnson County Community College*

Cynthia Lee Delaney-Tucker, *University of South Alabama*

Elizabeth A. Desy, *Southwest Minnesota State University*

Sondra Dubowsky, *McLennan Community College*

Jose Egremy, *Northwest Vista College*

Johnny El-Rady, *University of South Florida*

Randy Elvidge, *Georgia Military College–Augusta*

Kelly A. Fallon, *Waubonsee Community College/ McHenry County College/National University of Health Sciences*

Paul Farnsworth, *University of New Mexico*

Jim Fiedor, *Clarion University*

Michelle Finn, *Monroe Community College*

David Fitch, *New York University*

Rob Fitch, *Wenatchee Valley College*

Ted Fleming, *Bradley University*

Diana Fletcher, *Mount Royal University*

Margi Flood, *Gainesville State College*

April Ann Fong, *Portland Community College*

Edison R. Fowlks, *Hampton University*

Diane Fritz, *Gateway Community & Technical College*

Susannah B. Johnson Fulton, *Shasta College*

Edward Gabriel, *Lycoming College*

Kathy Gallucci, *Elon University*

Joseph Gar, *West Kentucky Community & Technical College*

Duyen Gauthier, *Westwood College*

Betsy Gerbec, *University of Wisconsin–River Falls*

Jennifer Gibbs, *Hinds Community College*

Beverly E. Glover, *Western Oklahoma State College*

Katie J. Goff, *Ferrum College*

Andrew Goliszek, *North Carolina A&T State University*

Larry Gomoll, *Stone Child College*

Jeanette Gore, *St. Petersburg College*

Scott Graham, *Colorado Mountain College–Aspen*

Mark Grobner, *California State University–Stanislaus*

Rhonda Gross, *Platt College*

Matt A. Haberkorn, *Phoenix College*

David Hale, *U.S. Air Force Academy*

Myra Hall, *Georgia Perimeter College*

Thomas Haner, *Illinois Central College*

April Harlin-Cognato, *Michigan State University*

Bob Harms, *St. Louis Community College at Meramec*

Steve Harris, *Clarion University*
Steve Heard, *University of New Brunswick*
Deena Hergert, *Rock Valley College*
Audrey Hernando, *El Paso Community College*
Gabriel Herrick, *Hillsborough Community College*
Tara Jo Holmberg, *Northwestern Connecticut Community College*
Jane E. Horlings, *Saddleback College*
Laura Houston, *Alamo Colleges–Northeast Lakeview Campus*
John Hunt, *University of Arkansas at Monticello*
Meshagae Hunte-Brown, *Drexel University*
Jessica Hutchison, *Cameron University*
Virginia Irintcheva, *Black Hawk College–Quad Cities*
Richard Jacobson, *Laredo Community College*
Robert Jonas, *Texas Lutheran University*
Katie Jordan, *Ferrum College*
Arnold J. Karpoff, *University of Louisville*
Karry Kazial, *State University of New York–Fredonia*
Leopold Keffler, *Marian University*
Ronald Keiper, *Valencia Community College*
Diane Kelly, *Broome Community College*
Rita Mary King, *The College of New Jersey*
Dennis Kingery, *Metropolitan Community College*
Dennis J. Kitz, *Southern Illinois University–Edwardsville*
Roger Klockziem, *Martin Luther College*
Todd A. Kostman, *University of Wisconsin–Oshkosh*
Holly Krahe, *Lynn University*
Dubear Kroening, *University of Wisconsin–Fox Valley*
Barbara Kuehner, *The University of Hawai'i Center at West Hawaii*
Rukmani Kuppuswami, *Laredo Community College*
Kim Lackey, *University of Alabama*
Thomas G. Lammers, *University of Wisconsin–Oshkosh*
Monica Lara, *St. Petersburg College*
Kaddee Lawrence, *Highline Community College*
Brenda Leady, *University of Toledo*
Maureen A. Leupold, *Genesee Community College–Batavia*
Cayle Lisenbee, *Arizona State University*
Michelle Lockett, *College of Southern Nevada*
Suzanne S. Long, *Monroe Community College*
David Loring, *Johnson County Community College*
Kathy Lowrey, *Jefferson Community & Technical College*
Ann S. Lumsden, *Florida State University*
Evelyn Lyles, *University of Maryland–Shady Grove*

John (Zhong) Ma, *Truman State University*
Mark Manteuffel, *St. Louis Community College*
Lisa Maranto, *Prince George's Community College*
Karen Marcus, *University of Alabama*
Jody Martin de Camillo, *St. Louis Community College*
Roy Mason, *Mt. San Jacinto College*
Juan Luis Mata, *University of Southern Alabama*
Elizabeth A. Mays, *Illinois Central College*
Vince McCracken, *Southern Illinois State University–Edwardsville*
Amanda McGowan-Attryde, *Arizona State University–West*
Linda J. McPheron, *Berkeley City College/San Francisco State University*
Michael McVay, *Green River Community College*
Jennifer A. Metzler, *Ball State University*
Sue Miller, *Bellevue College*
Jeanne M. Mitchell, *Truman State University*
Santiago Molina, *Tallahassee Community College*
Mark V. Mooney, *Bauder College*
Brenda Moore, *Truman State University*
Frances Moore, *Patrick Henry Community College*
Rachel Moreno, *Rock Valley College*
Syeda I. Munaim, *Schenectady County Community College*
Ann J. Murkowski, *North Seattle Community College*
Rosemary Nickerson, *Hagerstown Community College*
Han Chuan Ong, *Lyon College*
Chris Osovitz, *University of South Florida*
John C. Osterman, *University of Nebraska–Lincoln*
Kaleb Pauley, *Cedarville University*
Fiorella Penaloza, *Briarcliffe College*
Stacy Pfluger, *Angelina College*
Robert Pillsbury, *University of Wisconsin–Oshkosh*
David Pindel, *Corning Community College*
Nicola Plowes, *Arizona State University/Mesa Community College*
Narayanan "Raj" Rajendran, *Kentucky State University*
Beverly Ranney, *Western Oklahoma State College*
Jeffrey M. Ray, *University of North Alabama*
Laura H. Ritt, *Burlington County College*
R. Joseph Rodriguez, *The University of Texas at Austin, Center for Teaching & Learning*
Bill Rogers, *Ball State University*
William Ruf, *Indiana University–Bloomington*
Karina Rupert, *Cameron University*
Maysara Saadi, *Eldorado Emerson Private School*
Fred Schnee, *Loras College*

Jason F. Schreer, *State University of New York–Potsdam*

Erica Sharar, *Irvine Valley College*

Juanita Sharpe, *Chicago State University*

John Shiber, *Big Sandy Community & Technical College*

Cara Shillington, *Eastern Michigan University*

Brian R. Shmaefsky, *Lone Star College–Kingwood*

Marilyn Shopper, *Johnson County Community College*

Lakhbir Singh, *Chabot College*

Kelly Sjerven, *Rainy River Community College*

Dale Smoak, *Piedmont Technical College*

Phillip Snider, *Gadsden State Community College*

Bruce Stallsmith, *University of Alabama–Huntsville*

Robert D. Stark, *California State University–Bakersfield*

Lisa M. Strain, *Northeast Lakeview College*

Mark Sturtevant, *Oakland University*

Jean Stutz, *Arizona State University*

David Tapley, *Salem State College*

Sue Trammell, *John A. Logan College*

Linda C. Twining, *Truman State University*

Amie Voorhees, *Fresno City College*

Suzanne Wakim, *Butte College*

Charles M. Watson, *The University of Texas at Arlington*

Frances Weaver, *Widener University*

Gwen Wehbe, *Onondaga Community College*

Richard Weinstein, *University of Tennessee*

Lauren Wentz, *University of Wisconsin–Barron County*

Kelly J. Wessell, *Tompkins Cortland Community College*

Cheryl Wistrom, *Saint Joseph's College*

Edwin Wong, *Western Connecticut State University*

Corry Yeuroukis, *Cameron University*

Kerry L. Yurewicz, *Plymouth State University*

Craig A. Zoellner, *North Iowa Area Community College*

Robert K. Noyd

Jerome A. Krueger

Kendra M. Hill

Biology

organisms and adaptations

1 Biology: The Scientific Study of Life

1.1 The blue whale—the largest animal on earth

1.2 Organisms share many of the same characteristics

1.3 Organisms are complex interactive systems at all levels of organization

1.4 The unity and diversity of life are explained by evolution

1.5 Biologists use evidence to answer questions about the living world

1.1 The Blue Whale—The Largest Animal on Earth

The blue whale *(Balaenoptera musculus)* is a magnificent and impressive animal. It lives in all of the major oceans and is the largest animal on the planet—growing to over 100 feet in length and weighing nearly 400,000 pounds (**Fig. 1.1**). Blue whales are larger than any dinosaur that ever roamed Earth. They are so large that their tongues alone can weigh 6,000 pounds!

In spite of their size, their long, streamlined body is advantageous for swimming and traveling great distances in water. They have a fatty layer of blubber that is buoyant in water and helps them float. Without this buoyancy, the gravitational force of the whale's weight would crush its body organs. Blue whales are also some of the longest-lived animals, living up to 100 years.

Blue whales eat an enormous number of calories each day. To satisfy the daily needs of such a large animal, it eats as much as 4 tons (that's 8,000 pounds) of food. Whales, like other organisms, acquire their energy by ingesting important nutrients from their meal, using these nutrients as building blocks to build their own molecules. Think of the diverse ways animals catch and eat their food. Instead of teeth, the blue whale has evolved a set of over 400 baleen plates made of thick, bristly hairs to eat tiny shrimplike animals called krill (**Fig. 1.2**). When it opens its mouth to feed, a pouch under its jaw expands to take in the massive amounts of water needed to capture enough to make a mouthful. The whale then closes its mouth and uses its truck-sized tongue to push the water back through the baleen plates to filter out the krill. It may seem strange that the largest animal on Earth relies on one of the smallest animals for its survival. This unusual eating pattern is part of a particular food chain, the term used to describe the eating relationships and energy flow between interacting species.

Whales are considered highly social animals with complex communication patterns and intelligence. Whales rumble and grunt to communicate with mates, find offspring, and navigate about the ocean. In this area, the blue whale sets another record: it is the loudest animal and can communi-

Figure 1.1 The blue whale is an example of life.
© Cengage Learning 2014

Study of life through the blue whale

Blue whales have specialized baleen plates to help them acquire food

Whales navigate using echolocation

Like all living organisms, blue whales are made of cells and they reproduce

Krill is the food of choice for the blue whale

2 m

cate over 500 miles by emitting sound waves. In fact, blue whales have been recorded emitting sounds louder than those made by a jet taking off. As noise pollution increases in the ocean, there is great concern that important communication and interactions among whales will be interrupted.

The reproductive cycle of the blue whale is equally amazing. A female is pregnant for 10 to 12 months. She produces calves through sexual reproduction, a process involving the union of a sperm fertilizing an egg. The blue whale, like us, invests significant amounts of energy to find a mate, produce highly developed offspring, and care for its young for an extended period. Like many mammals, it takes a long time, between 8 and 10 years, for blue whales to reach sexual maturity. After the whales reach sexual maturity, their reproductive cycle is relatively slow, only producing one to two calves every three years. Blue whales migrate thousands of miles each year to reach warmer climates and mate in the late fall through the winter. During migration and mating, the whales don't eat and must rely on fat reserves for energy. The energy requirements of females are even greater after their calves are born. The calves consume about 100 gallons of milk a day and will nurse for seven to eight months. Female whales expend huge amounts of their energy reserves during this reproductive phase, which accounts for their huge appetites during their summer feeding seasons.

Blue whales have many important interactions with their environment. The study of these relationships is called *ecology* and is an important field of biology. One ecological relationship the whale has is with its food source— it has a predator–prey relationship with krill. Despite the vast numbers of krill removed from the food chain each year, the krill population remains stable. How can so many krill be removed from the food chain yet still have such a stable population from year to year? The reason is related to their life history and feeding habits. Female krill can lay 1,000 eggs at a time, and hatchlings mature very quickly. This combination means the krill population can grow and recover very quickly. In contrast, blue whales reproduce after they reach sexual maturity at 10 years of age and then only give birth to a single calf every other year. Because each species has different life histories, feeding patterns, reproductive rates, and ages, the study of these ecological relationships and life is both exciting and challenging for scientists.

So far we have introduced you to several aspects of the biology of the blue whale that include body structure, reproduction, and ecology. This information was collected, analyzed, and summarized by marine biologists over many years. Later in the chapter, we will elaborate on the processes that scientists use to collect and analyze data. As you can imagine, it's particularly challenging research because whales often live far from land and feed at more than one place.

The blue whale is among many organisms that can grab your interest and make you want to learn more about how organisms in the natural world are adapted to grow, survive, and reproduce. Redwood trees, leatherback turtles, bats, honey mushrooms, intestinal bacteria, and many other fascinating organisms each have amazing stories to tell you. Each organism can be used as a vehicle to present biological concepts for you to learn so you can apply your understanding to other organisms that live around you, on you, or inside you. In the book ahead, we open each chapter with an organism that "shows off" its adaptations, illustrating concepts and connections found in the chapter. We hope that it sparks your passion for the life around you and motivates you to ask more and deeper questions about it and its connection to your life. A study of biology through animals, plants, and microbes has enriched the lives of the authors, your instructor, and now, we hope, your life as well.

© National Geographic Image Collection/Alamy

Figure 1.2 Krill are the favored food of the blue whale. These tiny creatures are captured by the whale's baleen plates.

We will start your journey in biology, the study of life, with this marine animal illustrating the characteristics that make an organism a living thing. All organisms such as whales, bacteria, and trees are living systems that share many common characteristics. Living things are constructed of cells structured in an ordered manner, they acquire energy and raw materials, and they grow and reproduce. The traits an animal has are a result of the DNA it has inherited from its parents. Over many generations, this DNA can change, allowing organisms to adapt to their environment. In this first chapter, we introduce you to the characteristics of life and how it is studied. The chapter concludes with an introduction to evolution and evolutionary relationships that connect and unify all living things.

1.2 Organisms Share Many of the Same Characteristics

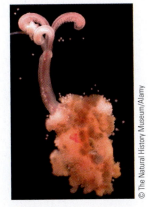

Figure 1.3 A boneworm. The boneworm feeds on the bones of dead whales.

© The Natural History Museum/Alamy

The boneworm has an important relationship with blue whales (**Fig. 1.3**). After a whale dies, its carcass drops to the bottom of the ocean where many organisms feed on it for months, breaking down the molecules and recycling them. The boneworm bores through the bones to tap into a rich energy source of fats and oils. What is amazing about the boneworm is that it does not have a mouth or stomach to break down the whale bones! So how could it possibly digest the decaying whale bones? This worm lives in close association with bacteria that are able to break down and digest the fats and oils found in dead whales. Once these nutrients are free, the worm uses its flower-like tail to absorb the nutrients as its food source.

The boneworm, its bacteria, and the blue whale share a surprising number of characteristics with each of us. Let's start by asking some big questions about living things. What characteristics make organisms different from nonliving things? What makes the blue whale alive or dead? What characteristics does it share with the krill that it eats or the boneworm that eats the fallen whale? In the section ahead, we will explore the characteristics of living organisms. To start, **Table 1.1** compares some components of the largest organism to one of the smallest organisms.

TABLE 1.1 Components of Large and Small Organisms

Component	Blue Whale	*E. coli* Bacterium
Atoms	C, H, O, N, Ca, Fe, Cu, S	C, H, O, N, Fe, Ca, Cu, S
Molecules—structural	Proteins, lipids	Proteins, lipids
Molecules—energy	Carbohydrates, lipids	Carbohydrates, lipids
Molecules—genetic information	Nucleic acids—DNA, RNA	Nucleic acids—DNA, RNA
Chemical reactions—energy generation	Cellular respiration requires oxygen	Cellular respiration requires oxygen
Mode of reproduction	Sexual	Sexual and asexual
Cell structure	Eukaryotic—complex compartmentalized	Prokaryotic—simple noncompart-mentalized
Levels of biological organization	Tissues, organs, organ systems	Cells

© Cengage Learning 2014

Life occurs at many levels of organization and scale

The biological world exhibits a ladder of organization, from atoms too small to see all the way up to global relationships among organisms and the environment. **Figure 1.4** illustrates life's organization as a series of circles, where each circle represents a particular unit of structure. From the figure, you can see that larger units are composed of smaller, nested units and that each level includes and builds on the level below it, creating the hierarchy of life. At the smallest, most basic level of organization are atoms, the fundamental building block of matter. All physical objects on Earth are composed of atoms. Atoms bond with one another to form molecules. The molecules that compose cells are large and complex, such as carbohydrates, lipids, proteins, and nucleic acids called DNA and RNA. Although these chemical components are critical to supporting life, atoms and molecules are not alive.

The cell is the smallest level of the hierarchy that operates as an adaptable living system, meaning it can survive and reproduce on its own. An organism is an individual that consists of one or more *cells*. A blue whale is made up of trillions of cells organized into *tissues* (such as muscle tissue and nervous tissue), *organs*, and *organ systems* that all communicate and interact in ways that keep the whole body alive. Although it may be almost impossible to imagine, an individual whale weighing 400,000 pounds started as a single cell.

At the next level of the hierarchy, organisms interact with each other in *populations* and *communities*. Individual populations consist of the same kind of organism, or species, in a specified area. A *community* is defined as all the populations of all species in a specified area. As an example, an ocean community may include the blue whale, krill, algae, seaweeds, fishes, corals, and many other species of organisms living in the vicinity.

The final level of organization includes the interactions of living and nonliving factors at the *ecosystem* level. The ecosystem includes interactions between organisms and their physical and chemical environment in a specified area. For instance, one of the ecosystems the blue whale lives in part of each year is the Antarctic Marine Ecosystem. The weather, sunlight, and ocean currents are examples of *nonliving* components that affect the *living* components such as algae, krill, and whales. The *biosphere*—the most inclusive level of life's organization—is one large ecosystem that encompasses all regions of Earth's surface, waters, and atmosphere in which organisms live.

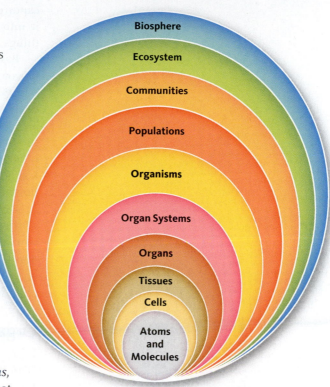

Figure 1.4 The hierarchical organization of life.
© Cengage Learning 2014

Organisms must constantly acquire energy

The blue whale, like all organisms, is a highly ordered system that requires energy to remain alive. The sun is the ultimate source of energy for living organisms, and nutrients are the source of chemical energy that organisms use for growth, development, and reproduction. Like the blue whale and the boneworm, each species has evolved with different adaptations to accomplish the same goal—acquiring energy from specific environments and food sources. Humans and other mammals have evolved specialized teeth adapted to breaking down food, jellyfish have tentacles to capture food, and frogs have long sticky tongue. Plants, algae, and some bacteria

Hierarchy (refers to the *hierarchy of life*)—larger units of life are composed of smaller, nested units. Each level includes and builds on the level below it. From the smallest to the largest, the organization is atoms, molecules, cells, tissues, organs, organ systems, populations, communities, ecosystem, and finally the biosphere.

capture carbon from the atmosphere and use the sun's energy to convert it into larger sugar molecules. Fungi break down living and nonliving things and then absorb the nutrients.

Based on how they process energy, organisms can be categorized into one of three groups: producers, consumers, or decomposers (Fig. 1.5). Producers, such as plants and algae, use photosynthesis to harvest the sun's energy and make sugar, a source of chemical energy immediately available for the plant. Consumers, such as animals, cannot make their own food and get energy and nutrients indirectly—by eating producers and other organisms. Decomposers, such as bacteria and fungi, use the chemical energy found in wastes or the remains of dead organisms, breaking down these remains into smaller molecules that can be used again by producers. As a result, producers, consumers, and decomposers recycle nutrients. In this way, nutrients serve two essential needs of organisms—they provide the *energy* that fuels the chemical reactions of life and the raw materials to produce new cells. The amount of available energy and raw *materials* is one reason that tropical rainforests have an abundance of organisms.

Organisms use DNA for the instructions to life

Every living system that biologists study contains and is the product of DNA. Deoxyribonucleic acid (DNA) is a long molecule that contains the information to direct life's activities and the instructions for each cell and organism to reproduce. The DNA molecule holds the information in the

Producers—use photosynthesis to harvest the sun's energy and make sugar, a source of chemical energy immediately available for the organism (such as plants, algae, or bacteria).

Consumers—cannot make their own food and get energy and nutrients indirectly by eating producers and other organisms. Animals are consumers.

Decomposers—use and obtain chemical energy from wastes or the remains of dead organisms, breaking down these remains into smaller molecules that can be used again by producers. Bacteria and fungi are decomposers.

Deoxyribonucleic acid (DNA)—a long double-stranded molecule that contains the genetic information to direct life's activities and the instructions for each cell and organism to reproduce. DNA contains a deoxyribose sugar-phosphate backbone and the nitrogenous bases adenine, guanine, cytosine, and thymine. In eukaryotic cells, DNA is found in the nucleus.

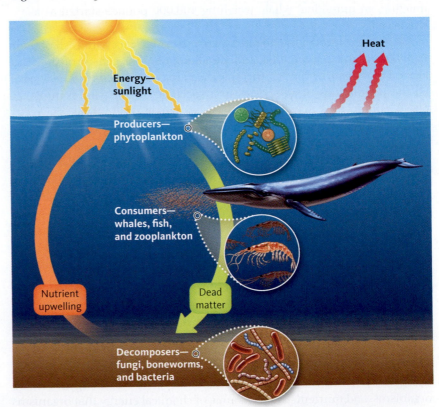

Figure 1.5 The one-way flow of energy and cycling of nutrients through an ecosystem. Energy from the sun flows through producers, then to consumers. All energy eventually flows out of the ecosystem, mainly as heat. Nutrients are concentrated in producers and consumers. Some nutrients released by decomposition are cycled back to producers.

© Cengage Learning 2014

specific sequence of molecules called *nucleotides* (**Fig. 1.6**). While each species has a different number of nucleotides comprising its genetic material, the amazing thing is that all living things use the same four nucleotides to build their genetic blueprint. Each organism's unique nucleotide sequence distinguishes it from all others. In a similar way, the English language uses 26 letters that can be combined into thousands of words. DNA is passed from adults to their offspring during reproduction. In humans, 99 percent of the DNA sequence is identical between individuals, yet we each have distinct features or traits. It is these small differences in our DNA that make each of us unique and provide the basis for broader evolutionary adaptations.

Since DNA is a universal molecule that all life forms use, it is important to understand some frequently used terms illustrated in **Figure 1.6**. All of the genetic content of the organism is referred to as the genome and is stored in structures called chromosomes. Individual segments of DNA coding for specific molecules, usually proteins, are called genes. Genes are the functional units of inheritance that pass on traits from generation to generation. In the human genome, there are 46 chromosomes (23 pairs) carrying a total of about 20,000 genes. Bacteria have a circular chromosome that on average carries a couple thousand genes.

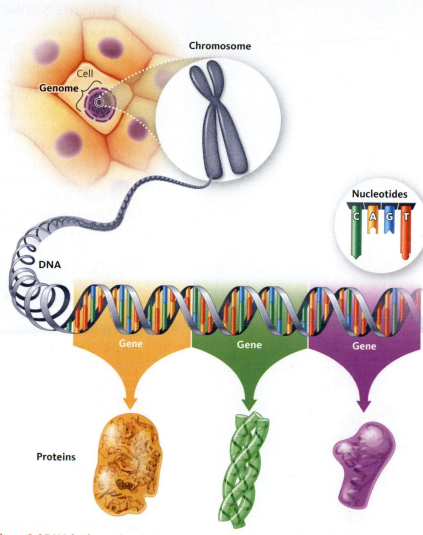

Figure 1.6 DNA is the molecule that provides the instructions for life. DNA is stored in chromosomes in the cell. Genes are the functional units that carry instructions for making proteins that perform cellular functions.
© Cengage Learning 2014

Organisms grow, develop, and reproduce

For the blue whale to reach its enormous size, it grows, or increases in size, for a long period of time. In boneworms, and other organisms with shorter life spans and smaller bodies, this growth period is much quicker. Multicellular organisms also progress through a series of stages of increasing complexity, called *development*, based on the information contained in the organism's genes. For instance, the whale must grow and develop for 8 to 10 years until it reaches sexual maturity and can reproduce.

All living things have the capacity to reproduce. Blue whales, like most animals, reproduce *sexually*, a process that involves a male and a female mating and passing their DNA to their offspring through sperm and eggs. Sexual reproduction results in offspring that appear both similar to and different from their parents and siblings. In contrast, some living things such as bacteria, plants, fungi, and even some animals can reproduce *asexually*, involving only one parent. Asexual reproduction results in offspring that are genetically identical to their parent. Regardless of the mode, in the process of reproducing, traits contained on chromosomes are inherited or passed by descent from one generation to the next.

Genome—all of the genetic content (DNA, genes, and chromosomes) of the organism.

Chromosomes—the structures in the cell that store the DNA.

Genes—the functional units of inheritance that pass on traits from generation to generation. Genes are individual segments of DNA coding for specific molecules, usually proteins.

Organisms sense and respond to change

The environment in which organisms live is constantly changing. By sensing and adjusting to change, each organism strives to maintain conditions in its internal environment within a range that favors survival and reproduction. This process is called homeostasis, and it is a defining feature of life. For instance, like many other large mammals, the blue whale must maintain its body temperature within a narrow range close to 98°F (about 37°C) to survive. During prolonged exercise or when swimming into warm tropical water, a whale's body temperature may exceed its range. The whale has specialized cells that sense this change and trigger a response. The whale adjusts its body temperature by increasing circulation near its outer body surface and decreasing it in the body's core. By redirecting blood flow, excess heat is lost to the surrounding water and the whale reduces its body temperature back within range. When the whale enters cold waters near Alaska, its blood is directed in the opposite direction—to the body core to preserve heat near vital organs and tissues.

All organisms sense and respond to change. Plants, fungi, and bacteria detect and respond to chemical and environmental signals, like changes in temperature and moisture levels. The leaves of many trees change color in the fall as a response to decreased sunlight and moisture. In addition to whales, bats and some other animals use *echolocation*, the transmission of sound waves to monitor their surroundings. Human adaptations for perceiving and responding to the environment include a highly complex sensory system consisting of sight, smell, taste, sound, and touch. Simple or complex, this ability to react to changes in the environment represents a significant evolutionary adaptation.

Populations of organisms adapt and evolve over time

Living things also respond to their changing environment over a longer period of time through adaptation. The temperature, moisture, and atmospheric conditions on Earth when life first originated 3.5 billion years ago were quite different than they are today. As environmental conditions have changed over time, populations of organisms that possessed adaptive traits survived, reproduced, and passed along their genes. The diversity of living things has been shaped by differential survival and reproduction in response to environmental changes. An adaptation is a change in a trait, behavior, or structure of an organism that allows the organism to be suited to its current environment. For example, whales have several critical adaptations for feeding and reproduction. They migrate thousands of miles between the cold polar waters, where they find most of their food, to warm tropical waters where they bear their young. They have the ability to send and receive vocal signals over long distances to locate mates. Whale calves are born with little blubber for insulation and require warmer waters to survive. While in warmer waters, whales fast for up to nine months of the year. Whales best adapted to migrate and fast for extended periods have a better rate of reproduction and pass on these favorable traits to future generations. Individuals that are best adapted to their environment tend to survive and have higher reproduction rates than those that are not well adapted. This ultimately creates genetic changes in the *population* as a whole. In section 1.4, you will learn more about how adaptations lead to evolution—the genetic changes in a population over generations.

Homeostasis—the process in which an organism senses, adjusts, and maintains conditions in its internal environment within a range that favors survival and reproduction.

Adaptation—a change in a trait, behavior, or structure of an organism that allows the organism to be suited to its current environment.

Evolution—the genetic changes in a population over generations.

1.2 check + apply YOUR UNDERSTANDING

- Organisms are unified through a set of shared key characteristics that define life.

- The biological world exhibits a ladder of organization, or hierarchy, from atoms all the way up to the biosphere.

- All organisms consist of one or more cells, which stay alive through ongoing inputs of energy and raw materials.

- Cells contain DNA inherited from their parents that encodes information needed for growth, survival, and reproduction.

- Living things sense and respond to changes in their external and internal environments.

- Organisms adapt and evolve over time.

1. Which level of hierarchy is the smallest that can operate as an adaptable living system?

2. The prime summer food for the blue whale in the Arctic is krill, a small aquatic animal. Based on your knowledge of how organisms process energy and referring to **Figure 1.5**, are krill producers, consumers, or decomposers?

3. Whale muscle contains an oxygen-storing protein called *myoglobin*. Which level of the hierarchy is whale muscle? Myoglobin?

4. When baleen whales dive, their heart rate slows, and blood is shunted toward their brain and lungs. This statement refers to which characteristic of living things?

5. Horses and zebras share many physical characteristics that would indicate a common ancestor. What evidence might confirm this?

1.3 Organisms Are Complex Interactive Systems at All Levels of Organization

Whale watchers may be lucky enough to observe the complex process of feeding by a humpback whale. Humpbacks, like blue whales, have comb-like plates of baleen that are used to filter krill and small fish from the water. To feed, humpbacks work in teams to blow bubbles and frighten fish into a large mass. Several members of the team dive under the fish and then swim rapidly toward the surface with their mouths open. With one large gulp the humpback takes in almost 16,000 gallons of seawater and fish! The mouth is then rapidly closed and the lower pouch contracts to expel water out of the mouth through the screenlike baleen (**Fig. 1.7**). Fish are trapped inside the mouth on the baleen before being swallowed.

Whales feeding at the *organism* and *population* level of the hierarchy ultimately depend on the structure and functioning of smaller levels of the hierarchy—the digestive and muscular *systems*, digestive *organs* of the mouth, and their *tissues, cells,* and *molecules*. The baleen material, for example, is made of tough keratin proteins that are made by special cells in the whale's upper gums. The keratin in whale baleen is the same material found in your hair and fingernails and in the horn of a rhinocerus. Success in acquiring nutrients occurs when all levels are integrated into a whole that works together—faulty structure or malfunctioning at any of these smaller levels means that the humpback will be less effective in capturing as much food energy as its competitors.

In the section ahead, we will introduce you to each level of the biological hierarchy that is responsible for the structure and function of organisms. These smaller, lower levels of the hierarchy allow you to understand

b.

a.

Figure 1.7 (a) Humpback whales work in groups to get food. They stun fish with a stream of air bubbles. The leader of the group sends a sound signal, and all the whales lunge upward, filtering mouthfuls of water with baleen plates to capture fish and other food. (b) Close-up of a baleen plate.

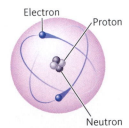

Electron
Proton
Neutron

Figure 1.8 An atom. Atoms are composed of electrons, protons, and neutrons.
© Cengage Learning 2014

Elements—substances that cannot be broken down into simpler substances. Examples include carbon, oxygen, and hydrogen.

Molecules—two or more atoms interacting, or bonding, together.

how organisms such as the whale, redwood trees, and bacteria function and are adapted to their environment.

Organisms are composed of atoms, molecules, and macromolecules

At the lowest level of the hierarchy, the atomic level, living things are composed of substances called *chemical elements*. Elements, such as carbon, oxygen, and hydrogen, are substances that cannot be broken down into simpler substances. Atoms of each element have a unique chemical identity based on the specific number and arrangement of their subatomic particles—protons and neutrons in their centers and electrons orbiting their outer regions (**Fig. 1.8**). Protons are positively charged particles, electrons are negatively charged, and neutrons do not carry a charge. If an atom has no net charge, then the number of protons equals the number of electrons. An atom with a charge imbalance (there are more or fewer electrons than protons) is called an *ion*. The sulfur atoms found in the keratin proteins that compose baleen or your hair, for example, have 16 protons, 16 neutrons, and 16 electrons and therefore do not carry a net charge.

The sodium in the saltshaker on your table is the same element as the sodium that causes your neurons to send messages in your brain. The salt's chlorine atoms are also found in the hydrochloric acid in your stomach. **Table 1.2** lists some important elements of the living world and describes some of their functions in the animal body. You may notice that many elements in the table are minerals found on food labels.

Atoms in living systems rarely exist alone; rather they interact, or bond, with other atoms to form molecules. The bonds that hold molecules together may be strong or weak depending on the strength of the attraction between the atoms. Strong bonds occur between atoms when a pair of electrons from one atom orbit both nuclei; in other words, the atoms share two electrons (**Fig. 1.9a**). One type of weak bond occurs when an electron is transferred from one atom to another. In this case the gaining atom becomes negatively charged, and the losing atom becomes positively charged; since opposite charges attract each other, they form a bond (see **Fig. 1.9b**). The way that atoms interact in a molecule gives the molecule a particular shape, which in turn influences its chemical properties and function in the body.

Water   Let's look at a molecule of water—a molecule that is essential for survival of all living things. Its chemical formula is H_2O, which means that it consists of two hydrogen (H) atoms bonded to an oxygen (O) atom (**Fig. 1.10**). The atomic interactions and bonds result in two important characteristics: (1) a bent molecular shape as shown in **Figure 1.10b** and (2) a negative charge at the oxygen

TABLE 1.2 Important Elements Found in Animals

Element	Chemical Symbol	Main Function(s) in the Animal Body
Carbon	C	Forms the backbone of all macromolecules—carbohydrates, lipids, proteins, and nucleic acids
Hydrogen	H	Part of the carbon backbone of all macromolecules and water
Oxygen	O	Component of all macromolecules and water
Nitrogen	N	Building block of amino acids, nucleic acids, and proteins
Calcium	Ca	Bones, muscles, and nerves
Phosphorus	P	Nucleic acids, membranes
Sodium	Na	Maintains water balance; involved in nerve conduction
Potassium	K	Electrical signals in nerves; major ion inside cells

© Cengage Learning 2014

a. Strong chemical bond

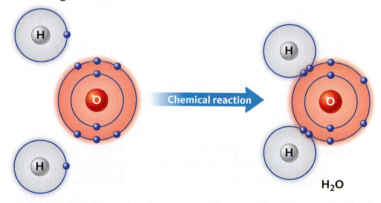

H_2O

b. Weak chemical bond

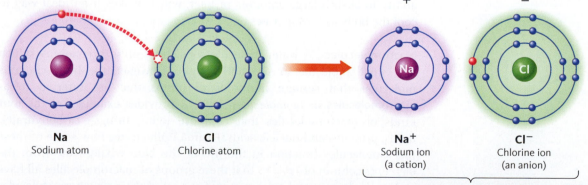

Na
Sodium atom

Cl
Chlorine atom

Na$^+$
Sodium ion
(a cation)

Cl$^-$
Chlorine ion
(an anion)

Sodium chloride (NaCl)

Figure 1.9 Strong and weak chemical bonds. (a) Two hydrogen atoms share electrons with an oxygen atom to form a water molecule, forming a strong chemical bond. (b) On the left side of the diagram, notice that sodium has one electron in its outer shell and that chlorine has seven. When the electron from sodium is transferred to chlorine, the sodium becomes positively charged and the chlorine negatively charged. These opposite charges attract, forming a sodium chloride molecule.
© Cengage Learning 2014

a.

b. H bond

Figure 1.10 Water molecules can stick together. (a) Water droplet. (b) Water contains two hydrogen atoms and one oxygen atom with a bent angle. The oxygen carries a slightly negative charge and the hydrogen atoms are slightly positive, giving water important properties for life.
© Cengage Learning 2014

end of the molecule and a positive charge at each of the hydrogen ends of the molecule. This is the result of unequal attractions for electrons. In a water molecule, electrons are attracted to the oxygen nucleus more than they are the hydrogen nucleus and thus "spend more of their time" at the oxygen end of the molecule, giving the oxygen a slightly negative charge and the hydrogen a slightly positive charge. The charged ends of the water molecule influence how water molecules interact with each other and other molecules in the body.

When two water molecules interact, the negatively charged oxygen atom of one molecule is attracted to the positively charged hydrogen atoms of another molecule (see **Fig. 1.10b**). This attraction among water molecules causes liquid water to be sticky and elastic and to form drops. Water's stickiness is an important chemical property that allows it to flow through blood vessels or move up the trunk of a tree. It also gives water its ability to absorb large amounts of heat, which makes it an ideal way to cool the body as it evaporates.

Macromolecules A water molecule consists of only three atoms, which is a small molecule in size and weight. However, the body contains molecules that contain hundreds of atoms that are referred to as **macromolecules** or *biomolecules*. **Table 1.3** provides examples of the four kinds of macromolecules that compose living things—carbohydrates, lipids, proteins, and nucleic acids. **Figure 1.11** illustrates how some of these macromolecules function in the life of the blue whale. Note from the elements column of **Table 1.3** that these groups of macromolecules all have carbon, hydrogen, and oxygen atoms as part of their structure. Molecules with carbon and hydrogen atoms as the backbone are also referred to as **organic molecules** (inorganic molecules do not contain carbon). As you progress through your biology course, you will learn more about how these important molecules function in the life of animals, plants, fungi, and bacteria, and thus they will reappear throughout this textbook.

Macromolecules—large, complex molecules that compose living things; also referred to as *biomolecules*. Examples include nucleic acids, proteins, carbohydrates, and lipids.

Organic molecules—molecules with carbon and hydrogen atoms as the backbone.

TABLE 1.3 The Four Macromolecules in Organisms

Macromolecule	Definition	Examples	Elements	General Function
Carbohydrates	Organic compounds made of smaller sugar units	Sugar, starch, glycogen, cellulose	C, H, O	Energy source; cell structure
Lipids	Fatty, oily, or waxy molecules with a hydrocarbon backbone; many contain chains of fatty acids	Fats, phospholipids, cholesterol, and waxes	C, H, O	Energy source, membranes, precursor of steroids and vitamin D
Proteins	Large molecules consisting of one or more chains of amino acids	Muscle fibers, collagen, enzymes, carriers, hormones	C, H, O, N May also include S (sulfur)	Animal structures, speed up chemical reactions, transport chemicals and molecules, cell-signaling molecules
Nucleic acids	Chains of nucleotides joined by sugar phosphate bonds	DNA, RNA	C, H, O, N	Store and carry genetic (hereditary) information

© Cengage Learning 2014

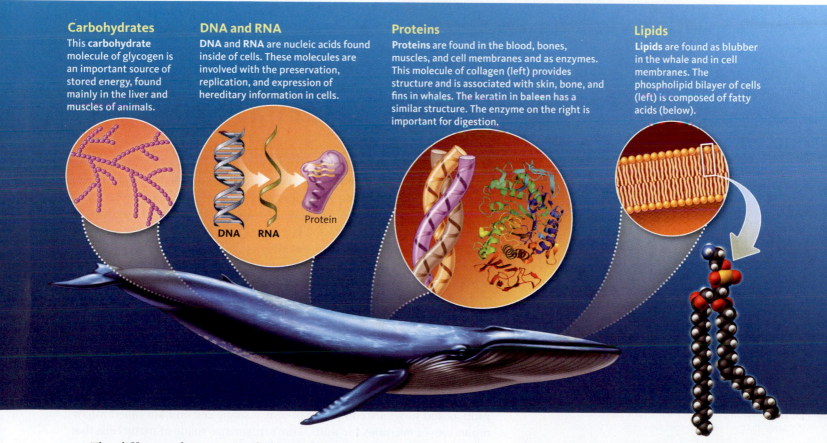

Carbohydrates

This **carbohydrate** molecule of glycogen is an important source of stored energy, found mainly in the liver and muscles of animals.

DNA and RNA

DNA and **RNA** are nucleic acids found inside of cells. These molecules are involved with the preservation, replication, and expression of hereditary information in cells.

Proteins

Proteins are found in the blood, bones, muscles, and cell membranes and as enzymes. This molecule of collagen (left) provides structure and is associated with skin, bone, and fins in whales. The keratin in baleen has a similar structure. The enzyme on the right is important for digestion.

Lipids

Lipids are found as blubber in the whale and in cell membranes. The phospholipid bilayer of cells (left) is composed of fatty acids (below).

DNA   RNA   Protein

The difference between nonliving and living systems lies not in the atoms themselves but rather in how they are bonded and arranged into these large, complex, three-dimensional macromolecules. Shape is so important for molecules, especially proteins, that it actually determines their biological function. Proteins consist of long chains of molecules that fold or twist into an immense number of shapes. Ultimately, the informational molecules DNA and RNA dictate the molecular building process and therefore determine the functioning in the cell. For example, the keratin composing a baleen plate is a very large protein that bonds together to form strong sheets, and the genes that direct the cells to produce the keratin are DNA molecules (see Fig. 1.11).

Figure 1.11 The four macromolecules of living things. The blue whale, like all living things, is composed of carbohydrates, proteins, lipids, and nucleic acids. This figure illustrates where a few of these molecules are found in the whale.
© Cengage Learning 2014

Proteins—long chains of molecules that fold or twist into an immense number of shapes.

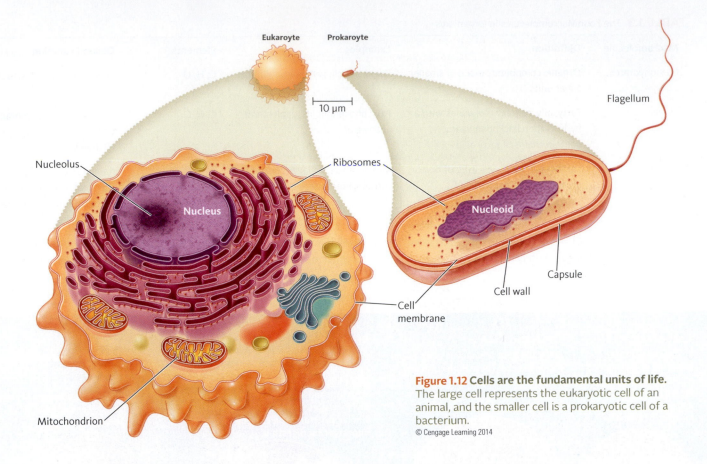

Eukaroyte   Prokaroyte

10 µm

Nucleolus

Nucleus

Ribosomes

Nucleoid

Flagellum

Capsule

Cell wall

Cell membrane

Mitochondrion

Figure 1.12 Cells are the fundamental units of life.
The large cell represents the eukaryotic cell of an animal, and the smaller cell is a prokaryotic cell of a bacterium.
© Cengage Learning 2014

The keratin, like all macromolecules, produced by whale cells is driven by an important group of molecules called *enzymes*. An enzyme is a protein (or sometimes an RNA molecule) that speeds up chemical reactions in the body. Most often, enzymes allow reactions to begin sooner than they would in their absence, thereby speeding up the chemical reactions of life. Without enzymes, an organism cannot transform energy to grow, survive, or reproduce. Life activities such as respiration, digestion, and photosynthesis would take too much time. The bottom line is that enzymes make life possible.

Cells are the structural and functional units of organisms

All living things are composed of cells. Organisms like bacteria and some protists and fungi are composed of only a single cell and are called *microorganisms*, or microbes for short. Those organisms made of many cells that work together are multicellular. Multicellular organisms can be only a dozen cells such as some algae or, like the blue whale, be composed of many trillions of cells organized at higher levels of the hierarchy—tissues, organs, and organ systems.

The cells that make up the millions of different types of organisms can be divided into two basic groups. Eukaryotic cells are larger, more complex, and consist of many specialized compartments (**Fig. 1.12**). Animals, plants, fungi, and protists are composed of these types of cells. On the other hand, prokaryotic cells are smaller, simpler, and lack the organized compartments of eukaryotic cells (see **Fig. 1.12**). Single-celled microbes such as bacteria are prokaryotic, whereas the amoeba is a eukaryotic microorganism.

Enzyme—a protein (or sometimes an RNA) that speeds up chemical reactions in living things.

Microbes—single-celled organisms, like bacteria and some protists and fungi.

Multicellular—organisms made of many cells that work together.

Eukaryotic cells—large, complex cells consisting of many specialized compartments. Animals, plants, fungi, and protists are composed of these types of cells.

Prokaryotic cells—small, simple cells that lack the organized compartments; includes bacteria and archaea.

Structurally, cells are separated from other cells and their environment by a thin **plasma membrane**. This membrane acts as a physical envelope and compartmentalizes the cell so that it can carry out its specialized activities such as storage, construction, or transport. Enclosed within this membrane is cytoplasm. Eukaryotic cells contain a large structure in the middle called the **nucleus** that houses the majority of the DNA in the cell (see **Fig. 1.12**). Prokaryotic cells do not have a compartmentalized nucleus but rather house their DNA within the cytoplasm. Cells that compose plants, fungi, and bacteria have an additional layer outside of the plasma membrane—a strong supportive cell wall.

Multicellular organisms form tissues, organs, and organ systems

The cells composing multicellular organisms are organized into the next level of the biological hierarchy—tissues. **Tissues** are groups of cells that perform a specialized function such as movement, protection, or reproduction. The long-distance migration and the playful acrobatics of the humpback whale are the result of nerve and muscle tissues that function together in structures at the next level of the hierarchy—organs.

Nerves and muscles are animal **organs** that are made up of two or more types of tissues. Plant organs include leaves, stems, and roots, all of which are composed of storage, support, and transport tissues. The familiar mushroom is an example of a fungal organ because it too consists of several types of fungal tissues.

To produce long, complex sounds and songs, humpback whales use several different organs that work together at the **organ system** level. Whales use their muscular diaphragm and lungs to force air through their windpipe and other organs of their respiratory system. Their sounds include grunts, moans, and rhythmic patterns, or singing, which is especially intense during the mating season.

Organisms are systems of complex interactions at all levels of the hierarchy. In a system all parts rely on other parts to function properly. As a living system, the specialized cells, tissues, organs, and organ systems all depend on the other parts of the system to function effectively. Therefore the organism works as a whole through the communication and interaction of its parts at all levels of the hierarchy.

Plasma membrane—a thin, structured lipid bilayer that separates cells from other cells and their environment. This membrane encloses the cytoplasm and acts as a physical envelope. It compartmentalizes the cell so that it can carry out its specialized activities such as storage, construction, or transport. Enclosed within this membrane is cytoplasm.

Nucleus—a large structure in eukaryotic cells that houses the majority of the DNA in the cell.

Tissues—groups of cells that perform a specialized function such as movement, protection, or reproduction.

Organs—a group of two or more types of tissues that work together to perform a specific function.

Organ system—a group of several different organs that work together to perform a specific task.

1.3 check + apply YOUR UNDERSTANDING

- The living world follows a hierarchy of scale that starts at the atomic level. Atoms interact to form bonds, forming the molecules and macromolecules that make up cells.

- The cell is the smallest level of the hierarchy that operates as an adaptable living system, meaning it can survive and reproduce on its own.

- In multicellular organisms, cells are organized into tissues, organs, and organ systems that all communicate and interact in ways that keep the whole body alive.

1. Distinguish between the following terms: *atom, proton, molecule,* and *element*.

2. Organisms are composed of four major categories of organic macromolecules. Name them and state the three elements common to all four groups.

3. **Figure 1.13** shows sweat droplets on skin. What chemical property

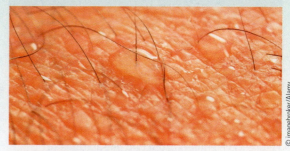

Figure 1.13 Sweat droplets on skin.

CONTINUED

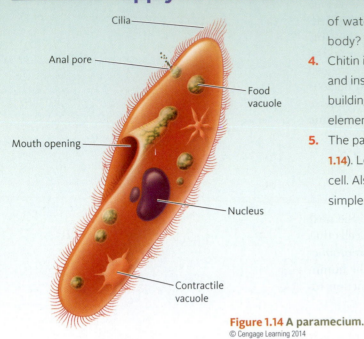

Cilia

Anal pore

Food vacuole

Mouth opening

Nucleus

Contractile vacuole

Figure 1.14 A paramecium.
© Cengage Learning 2014

of water accounts for the droplet and the ability of sweat to cool the body?

4. Chitin is a compound that is used to make the hard exoskeletons of crabs and insects. It is also found in cell walls of mushrooms and other fungi. Its building-block molecule has the chemical formula $C_8H_{12}NO_5$. How many elements and atoms compose a molecule of chitin?

5. The paramecium is a single-celled protist that lives in pond water (**Fig. 1.14**). Look at **Figure 1.14** and determine if it is a eukaryotic or prokaryotic cell. Also, compare the paramecium to a bacterium, explaining if it is simpler or more complex than a bacterium.

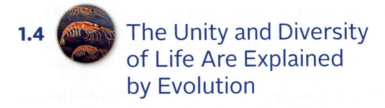

1.4 The Unity and Diversity of Life Are Explained by Evolution

The Mexican tetra fish provides an excellent example of adaptation and evolution. There are several populations of this particular species of freshwater fish. In their natural habitat, some populations live close to the water's surface and have normal eyesight, whereas others live in the total darkness of caves and lack eyes (**Fig. 1.15**). Because of this interesting characteristic, they are popular in household aquariums and often are referred to as *blind cavefish*. In addition to blindness, these cavefish lack pigmentation in their bodies, and they have modified mouth structures. They have a larger jaw and more teeth and taste buds than the surface fish that has normal eyesight. The blind cavefish illustrates what Charles Darwin referred to as *descent with modification*—this present-day fish is a descendant of the sighted surface fish with modified traits that reflect adaptations specialized to the environment where they live. Studies indicate that blindness and modified feeding traits are a result of changes in just a few genes that affect early development.

You may be curious as to why blindness and lack of pigmentation would be an adaptive advantage for the cavefish. One hypothesis is that since fish living in total darkness do not need to expend energy on eyesight or pigmentation, this energy could be diverted to other processes such as growth, reproduction, or acquiring food. The adaptations in feeding structures may have helped the fish efficiently acquire food in the environment where they lived. Scientists have also identified other gene mu-

Richard Borowsky

Figure 1.15 Members from two populations of Mexican tetra fish. The blind cavefish has less pigmentation than fish that live near the surface.

tations that have accumulated over thousands of years resulting in other trait differences between various tetra populations. Cavefish evolution is an example of how genetic changes that accumulate over time lead to adaptations in various populations. In this example, it is also interesting to note that evolution doesn't necessarily lead to an organism that is more complex but rather to one that is better suited to its environment.

Evolution occurs by natural selection over time

Take a moment and think about a population of organisms. Some individuals will have a close resemblance to their parents, whereas others are remarkably different. Like the rock pocket mice in **Figure 1.16**, each member of a population carries a different combination of genes, giving them different traits. This variation is the result of sexual reproduction and mutations in the DNA sequence. Sometimes these new variations of existing traits allow an individual of a population to be better suited to its environment. These new, more adaptive traits may improve chances of not only survival but also reproduction. Because reproduction passes on genetic information from one generation to the next, these modified traits are passed along from parents to their offspring. This differential in survival and reproduction among individuals of a population is called **natural selection**. You'll notice that the coloring of the mice camouflages them from hungry owls. Those mice with superior camouflage are more likely to survive and pass on the genes coding for effective camouflage. Over many generations, the effective camouflage trait becomes more common in the mouse population. When different forms of a trait are selected and become more or less common over successive generations, the overall genetic makeup of a *population* shifts, and evolution is under way.

Whether you consider blind cavefish, pocket mice, or blue whales, each of these species is influenced by evolutionary processes and events incrementally over each generation. Those organisms like bacteria and fruit flies that reproduce quickly have the ability to evolve more quickly, whereas organisms like the blue whale that reproduce more slowly will evolve

© Dr. Hopi Hoekstra, Museum of Comparative Zoology, Harvard University

Figure 1.16 Examples of protective coloration. The mice in the top two blocks blend into the environment and are difficult for owls and other predators to see. The mice in the lower half are easily preyed upon.

Natural selection—one mechanism of evolution that is the differential survival and reproduction among individuals of a population. Natural selection acts over time on variation, inheritance, selection, and adaptation.

Pakicetus
50 million years ago

Nostrils at front
of skull

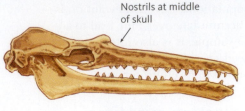

Aetiocetus
25 million years ago

Nostrils at middle
of skull

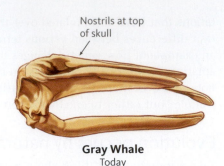

Gray Whale
Today

Nostrils at top
of skull

Figure 1.17 Transition of whale nostrils over 55 million years. They have moved from the front to the top of the head.
© Cengage Learning 2014

more slowly. In these more complex organisms, evolution is not an observable phenomenon, except over hundreds of thousands of years or even longer. The ancestors of blue whales, for example, once had blowholes (nostrils) in the front of their heads and walked on land. In an aquatic environment, natural selection favored the survival of those animals with blowholes closer to the top of their heads so that they could breathe while bringing less of their heads out of the water. As a result, over the last 55 million years the whale's blowholes moved from the front of the head to the top (**Fig. 1.17**).

Evolution is marked by common descent

Evolution is the process that successfully explains both the unity and the diversity of life. Charles Darwin carefully laid out the case for evolution in 1859 in his book *On the Origin of Species by Means of Natural Selection.* Darwin proposed that all organisms share a common ancestor and therefore they share important yet common characteristics. This proposal was based on his observations and analyses of many types of species over several decades. He noticed that humans artificially selected different breeds of animals and concluded that organisms in nature constantly and slowly changed with each generation. The specific traits that give an organism selective advantage become more common in each new generation. Darwin focused on these adaptations to formulate his concept of natural selection as the *mechanism* for evolutionary change. *Unity* refers to the fact that all living things share the same set of characteristics. This is explained by the fact that all living things descended from a common ancestor, meaning that all organisms, living or extinct, are related.

Natural selection is an explanation that accounts for life's *biological diversity*, or biodiversity. Natural selection acts over time on variation, inheritance, selection, and adaptation. Life is diverse because each species is uniquely adapted to its environment, and it is no surprise that an endless amount of unique environments exist for organisms to exploit. For instance, the blind cavefish lives in an environment that is different from that of the tetra fish, which lives at the surface. In Chapter 2 you will gain a greater appreciation for the biodiversity of Earth that we see in the millions of different species as a very impressive aspect of our world.

This textbook is titled *Biology: Organisms and Adaptations.* Throughout the text we will connect biological concepts to evolution. We will discuss adaptations of organisms, as well as evolutionary relationships. We use terms like *evolve, adaptations,* and *ancestors* with intention. This connection to evolution is the unifying theme of biology and will be developed in the next two chapters.

Biodiversity (or *biological diversity*)—the number and variety of organisms found within a given region.

- Evolution is a process that explains the unity and diversity of life. Populations of organisms respond to environmental factors. As they respond and adapt to their environment, populations evolve. Over time, these adaptations are passed on to their offspring.

- Evolution occurs by natural selection over time. Natural selection is the differential survival and reproduction of individuals in a population.

- The unity of life refers to organisms sharing a common set of characteristics. All living things descended from a common ancestor and are related.

1. What aspect of evolution did Charles Darwin propose in 1859?
2. What is the relationship between the time it takes for an organism to reproduce and the observation of evolutionary changes?
3. How do you respond to a friend who says cave animals like fish and crickets lost their eyes through evolution because they no longer needed them in a dark environment?
4. What is the relationship between DNA, genes, and proteins?
5. One misconception regarding evolution is that it leads to greater complexity or that it leads to life progressing up a ladder. Explain why this is a misconception (in other words, explain why this is a false statement).

1.5 Biologists Use Evidence to Answer Questions about the Living World

Let's return to the scientific study of blue whales. Much of the research on the blue whale is descriptive, or nonexperimental, because it is almost impossible to perform experiments on such a large animal. One descriptive inquiry you might pursue is estimating how many individual whales are in a population. The answer to this question would be of interest in monitoring their population numbers over time to conclude how well they are doing—are their numbers increasing or decreasing, and why? To answer these questions, in addition to being able to identify them, you have to know when and where to look for the blue whale. In spite of their large size, blue whales are hard to find. They live in most of the world's oceans, typically alone or in small groups, and they migrate vast distances each year.

The reality is that given the current state of science, we can never know exactly how many blue whales there are. However, marine biologists can collect enough observational data to make a reasonable estimate of the population. They can go on expeditions to locations where whales congregate, such as their Arctic and Antarctic feeding grounds where krill are abundant in the summer months and to the warmer regions where whales give birth. Scientists can also make observations along known migration routes, for example, the west coast of the United States. The difficulties involved in locating and observing underwater animals accounts for the large range in the estimate for the number of blue whales worldwide (between 5,000 and 12,000 based on estimates made in 2002). These observational studies not only enhance our understanding of the blue whale and its habitat, but population estimates provide critical data to help with whale conservation efforts. Establishing population trends is key for understanding the maintenance, loss, and restoration of biodiversity in various ecosystems. In this final section of this chapter, we will introduce you to the process of scientific investigation.

Scientific inquiry involves critical thinking

Science is the systematic investigation of the *natural* world. Biologists, for example, are engaged in understanding how life works and how it came to be that way. This work is challenging because the physical and living worlds are extremely complex and always changing. For instance, marine biologists who set out to study the lives and behavioral patterns of blue whales must also include krill, algae, water temperature, and all the other components of the whales' ecosystem.

Scientific study, like learning in general, does not start from a blank slate; it progresses and builds on the work of others. Thus biologists start by learning as much as they can about the organisms they are studying. They ask questions, make observations, evaluate the strengths and weaknesses of evidence, and then construct logical, well-reasoned explanations or hypotheses. They then test to see if the explanation works and, if so, under which conditions. Making connections through this exploration and analysis are all part of the scientific process. In this way, science tries to answer questions about the natural world.

Developing new knowledge involves thinking critically about our preconceptions and interpretations about our world. Clearly, there are many everyday ideas and customs that you accept without questioning their validity. Every day, however, we are bombarded with an overload of new information via the Internet, television, and print media. At a certain point, we are forced to think *critically* about the quality of information coming to us from so many sources. Thinking critically means discovering and carefully interpreting the facts before forming beliefs or accepting information. When you think critically, you evaluate the content of new information by assessing its authenticity and accuracy while generating alternatives. Critical thinking is not the same as scientific investigation, but it is an important piece of it. Scientists apply a pattern of critical thinking to help them evaluate their knowledge of nature. Scientists make every effort to put aside preconceived biases and design experiments that yield measurable, reproducible data that can be objectively analyzed.

It's important to note that scientific research also advances through a team effort. This is one reason why scientists gather at scientific meetings, where they present their data, share ideas, and discuss their work, as well as learn the most current information, make new connections between their research and the research of others, and ask new and deeper questions. Peer review by fellow scientists is an important step that provides critical feedback to ensure that any study methods or conclusions are sound.

Scientific investigation involves a variety of approaches

The word *science* is derived from the Latin word for knowledge, and the aim of every scientist is to produce a degree of understanding that is as close to reality as possible. In science, knowledge is obtained through the

Science—the systematic investigation of the natural world.

a.

b.

Figure 1.18 Scientists observe and measure nature. (a) Scientists studying fossils. (b) Scientist taking measurements of the natural world.

proper application of systematic investigation, sometimes called the scientific method. The investigation is conducted through a thorough, appropriate experimental design followed by a rigorous review by other scientists of the work and conclusions. The process is designed to be as objective as possible to reduce human bias in the perception and interpretation of data. You may be familiar with elements of this process from previous science classes.

Science works in two fundamental ways—nonexperimental and experimental. Nonexperimental methods seek to observe and measure nature as it occurs (**Fig. 1.18**). These methods are also called *descriptive* science and depend solely on observation and measurement by scientists. For instance, researchers in Australia tag blue whales to collect information on their feeding habits and movements. By monitoring the whales' feeding, scientists will better understand how the whale changes location to find food while also collecting baseline data on the impact of human disturbances on whale feeding patterns. Much of the information about the natural world included in this book was obtained in this way.

In contrast, experimental science attempts to manipulate systems and then observe and measure their response. One of the issues whale populations must contend with is a degree of disturbance from noise pollution from cargo ships and military sonar systems. Submarine- and ship-based sonar emits powerful high-frequency signals that might be damaging whales' ears or interfering with their navigation and communication. Experiments have been conducted that compare responses from whales exposed to this high-intensity sonar to those from whales not exposed. Analysis of data from current experiments indicates that whales are changing the way they communicate with each other. Whales exposed to this interference communicate less often and at a lower frequency than whales that are not exposed to this variable. **Figure 1.19** walks through the scientific process examining noise pollution and blue whale communication.

Let's summarize some terms that are critical to understanding the scientific process. An **experiment** is a carefully controlled test designed to make a discovery or to determine the validity of a hypothesis. A **hypothesis** is a testable, plausible explanation for a natural phenomenon. An example of a hypothesis for the example above is: *The high-intensity sonar from military systems interferes with the whale's ability to commu-*

Scientific method—the proper application of systematic investigation that includes a thorough, appropriate experimental design followed by a rigorous review by other scientists of the work and conclusions. The process is designed to be as objective as possible to reduce human bias in the perception and interpretation of data.

Experiment—a carefully controlled test designed to make a discovery or to determine the validity of a hypothesis.

Hypothesis—a testable, plausible explanation for a natural phenomenon.

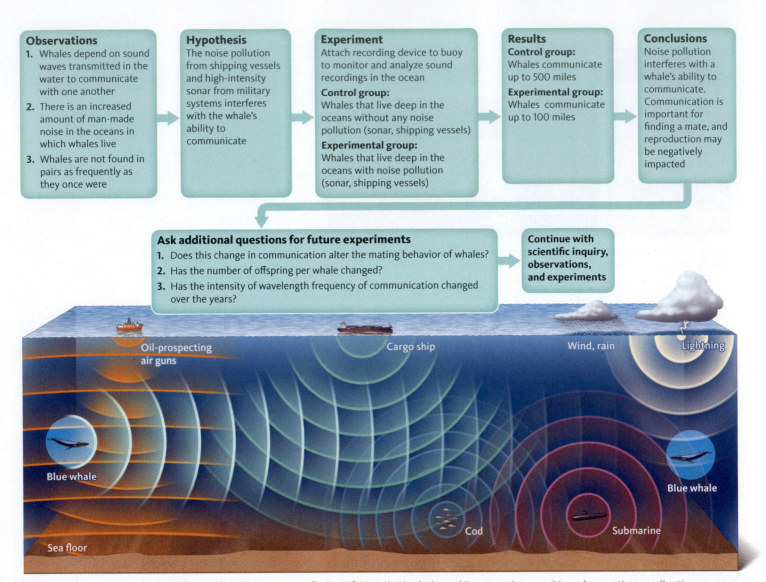

Observations
1. Whales depend on sound waves transmitted in the water to communicate with one another
2. There is an increased amount of man-made noise in the oceans in which whales live
3. Whales are not found in pairs as frequently as they once were

Hypothesis
The noise pollution from shipping vessels and high-intensity sonar from military systems interferes with the whale's ability to communicate

Experiment
Attach recording device to buoy to monitor and analyze sound recordings in the ocean

Control group:
Whales that live deep in the oceans without any noise pollution (sonar, shipping vessels)

Experimental group:
Whales that live deep in the oceans with noise pollution (sonar, shipping vessels)

Results
Control group:
Whales communicate up to 500 miles

Experimental group:
Whales communicate up to 100 miles

Conclusions
Noise pollution interferes with a whale's ability to communicate. Communication is important for finding a mate, and reproduction may be negatively impacted

Ask additional questions for future experiments
1. Does this change in communication alter the mating behavior of whales?
2. Has the number of offspring per whale changed?
3. Has the intensity of wavelength frequency of communication changed over the years?

Continue with scientific inquiry, observations, and experiments

Oil-prospecting air guns

Cargo ship

Wind, rain

Lightning

Blue whale

Sea floor

Cod

Submarine

Blue whale

Figure 1.19 The steps of scientific inquiry. The process of scientific inquiry includes asking questions, making observations, collecting evidence, analyzing results, and drawing conclusions. Often each step generates additional observations and questions.
© Cengage Learning 2014

nicate. Supporting evidence could come from examining whether there is physical damage to the whale's eardrum after a pulse of sonar. After scientists gather enough evidence, they draw their conclusions. This is one area in which scientists challenge one another—different scientists interpret the data and draw different, yet reasonable, conclusions. This disagreement then drives the development of further hypotheses and experiments.

Biologists and the public use the word *theory* in different ways. In everyday conversation, the word is used to express a guess or a hunch, but scientists use *theory* to mean a well-established explanation to a hypothesis or series of hypotheses. A **theory** is an explanation based on a large body of scientific observations, experiments, and reasoning—tested and confirmed as a general principle helping to explain and predict natural phenomena. Only after the evidence and results are scrutinized and tested by scientists over long periods of time does this explanation become a theory. Because of this very high standard, there are few theories in biology. The theory of evolution is based on thou-

Theory—an explanation based on a large body of scientific observations, experiments, and reasoning; it is tested and confirmed as a general principle helping to explain and predict natural phenomena.

sands of individual studies of the fossil record, DNA evidence, and the diversity of life. From these observations and studies that began in the mid-1800s, most scientists are convinced that evolution is a fact of life.

Science has its limitation in studying the natural world

It is important to realize that science is not a process that can solve all problems and answer all questions about nature. The realm of science is limited to solving problems about how the natural world works and not examining the philosophical meanings of life. Because science requires observable, testable, measurable, and repeatable data, it does not address the realm of values and ethics (questions about right and wrong), nor does it explore the supernatural aspects of life (questions about God or divine beings).

There are other limitations to scientific exploration. Critics try to discredit science because our understanding of the world changes, but this is not a fair assessment. Science is not a process that produces certainties, or absolute facts. The goal of science is to continually improve our understanding of the cause-and-effect realities driving the natural world. Truth is a relative concept, and science produces "possible" to "highly probable" explanations for natural phenomena; these are never certainties. With new information, tools, or approaches, earlier findings can be replaced by new findings, so what we know about the natural world today may be different 10 years from now.

Another limitation of science has to do with the people who are conducting and communicating the research. Science can be done poorly, just like any other human endeavor. Scientists often feel the pressure of public opinion or popular causes. Because the scientific process can be so convincing, there are those who apply the name of science to their efforts to "prove" their favorite cause, adding bias to their studies, taking shortcuts, or even changing the results of the data. For example, in 2004 and 2005, a Korean scientist published breakthrough results in the cloning of embryonic human cells. However, in 2006, after pressure from peers who questioned his findings, he admitted to fabricating his data. Scientific conclusions are subject to scrutiny and must be defensible with evidence and logical reasoning. As responsible members of society, scientists do their best to uphold the integrity of the discipline, and you, as a citizen, have the power to apply critical thinking skills to understanding the scientific news that is part of your world.

Despite its limitations, scientific knowledge is powerful in understanding and improving the natural world. Throughout history, basic scientific research has been applied to make great technological advances such as traveling to the moon or building an artificial heart. Science is the fuel that drives technology, and in turn, new technologies drive science. From the actual use of the knowledge of science to real-world problems, we have found that scientific knowledge is the most reliable information we have about the natural world.

The evidence of evolution illustrates scientific inquiry

The study of the reptiles from the Jurassic Period of over 200 million years ago is an excellent example of nonexperimental scientific inquiry. To improve our understanding of the relationships between reptiles

Baleen whale

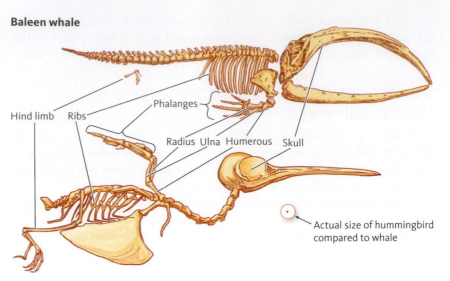

Hind limb   Ribs   Phalanges   Radius Ulna Humerous   Skull

Actual size of hummingbird compared to whale

Hummingbird

Figure 1.20 Comparative skeletal structures of baleen whales and hummingbirds.
Copyright © 2006 by The University of California Museum of Paleontology, Berkeley, and the Regents of the University of California. Used with permission.

and birds, scientists use findings from many branches of science such as geology, physics, chemistry, biology, mathematics, and computer science. Physical characteristics of skeletons and teeth help determine anatomical features, while molecular and chemical evidence from bone and feces helps establish evolutionary relationships, and the location of fossils in the rock helps determine geologic time references. Research of living species takes advantage of these same evidences as well. These pieces of evidence are woven together to form the larger picture and understanding of evolutionary relationships and the tree of life. Since evolution is such an integral part of biology, we will briefly walk through several lines of evidence that scientists use to establish relationships.

Observable Characteristics One of the most powerful pieces of evolutionary evidence comes from the physical characteristics of the organism. Just like members of your family tree, organisms that are closely related share similar anatomical traits. **Figure 1.20** compares the skeletons of a whale and a hummingbird, which share similar structures derived from a common ancestor. Nearly every bone in the whale corresponds to a similar bone in the hummingbird. This anatomical relationship can be observed in all types of animals, large and small. These similarities in anatomical traits result from common ancestry and are referred to as homologies or homologous structures.

Although much information can be gained from homologies and comparing anatomical structures, in many cases, comparative structures can be misleading. Therefore this method is only used as a first line of evidence when proposing evolutionary relationships. This information is important, but scientists exercise caution and look beyond superficial appearances to study the anatomy more deeply (**Fig. 1.21**).

Fossils The fossil record provides a rich picture of life in the past. Fossils are the physical evidence of organisms that lived in the past and include mineralized bones, teeth, casts, molds, shells, and wood. Trace fossils are evidence of an organism's activity such as footprints, burrows,

Homologous structures—the similarities in anatomical traits that result from common ancestry.

Fossils—the physical evidence of organisms that lived in the past and include mineralized bones, teeth, casts, molds, shells, and wood. Trace fossils are evidence of an organism's activity such as footprints, burrows, impressions, and preserved feces.

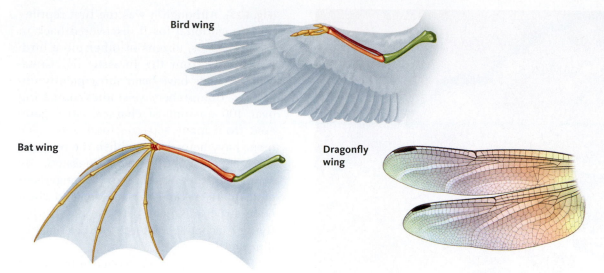

Bird wing

Bat wing

Dragonfly wing

Figure 1.21 Anatomical features may look similar but be different. When comparing the wings of a bat, bird, and dragonfly, you may conclude that all three have wings that are derived from a shared common ancestor. A closer examination will reveal that the underlying structure of a dragonfly wing is very different from that of a bat and a bird. In this case, three organisms share a similar functional adaptation of flight, but close examination of the internal wing structure provides evidence that the organisms are not closely related.
© Cengage Learning 2014

impressions, and preserved feces. **Figure 1.22** illustrates several different fossil formations. Scientists also use the fossil record to collect information about the habitat, diet, and behavior of organisms.

Researchers use the preserved observable structures found in fossil evidence to gather information about the anatomy and physiology of an extinct organism. Take the feathered *Archaeopteryx*, for example

Figure 1.22 Fossils are preserved in various ways. Casts are made; footprints, bones, and teeth are left behind; insects can be preserved in amber or frozen in glaciers.

Louie Psihoyos/Getty Images

Figure 1.23 Fossils are used to gather information about the anatomy and physiology of extinct organisms. *Archaeopteryx* shares characteristics with reptiles and birds.

(**Fig. 1.23**). Although it was the first reptile–bird transitional fossil discovered back in the mid-1800s, dozens of other more bird-like fossils from the Jurassic and Cretaceous Periods have been subsequently unearthed. Comprehensive studies comparing over 100 anatomical characteristics gathered from many separate fossilized specimens have helped to establish the relationship between birds and dinosaurs. By comparing the structure and arrangement of feathers, as well as shoulder and chest bone structures, scientists can deduce the flying mechanics of these extinct birdlike dinosaurs.

Establishing a time sequence by dating fossils and the rock record also reveals important information about ancestor-descendant relationships through time. Paleontologists are able to predict the time that particular features appeared within an organism's lineage if they know when the organism became a fossil in the rock formation. For instance, studying fossils laid down a few million years ago, scientists know that by 3.2 million years ago our human ancestors were walking upright and had a smaller brain capacity. Fossils from the first species we recognize as more modern humans date back about 2 million years ago. By studying fossils from this geologic time period from around the world, we see that humans began to migrate out of Africa shortly after this time. Paleontologists and other scientists continue to learn a great deal from the physical evidence fossils can provide.

Understanding the rock layers provides *relative* ages of fossils. Like placing newspapers on top of one another, each new layer of soil is deposited on top of the older underlying layer. The deepest layers of rocks are from earlier deposits of sediment and are therefore the oldest, while the top layers of sediment are more recent. This is one of the many ways that information on evolution over time is associated with the geologic record (**Fig. 1.24**). The geologic age of a fossil can be defined relative to other organisms, rocks, or events extracted from the rock layers. From this information scientists can determine the *order* in which species evolved but not the *absolute* ages of the fossils. To accurately determine the relative age of a particular fossil requires several different techniques of radiometric dating that take advantage of important chemical properties of atoms and elements (**Fig. 1.25**).

Developmental Patterns Sexually reproducing organisms, such as fungi, plants, and animals, develop from a fertilized egg to an embryo through a series of rapid developmental stages referred to as *embryonic development*. The scientific study of these developmental processes, known as *embryology*, is also used to establish evolutionary relationships. In the

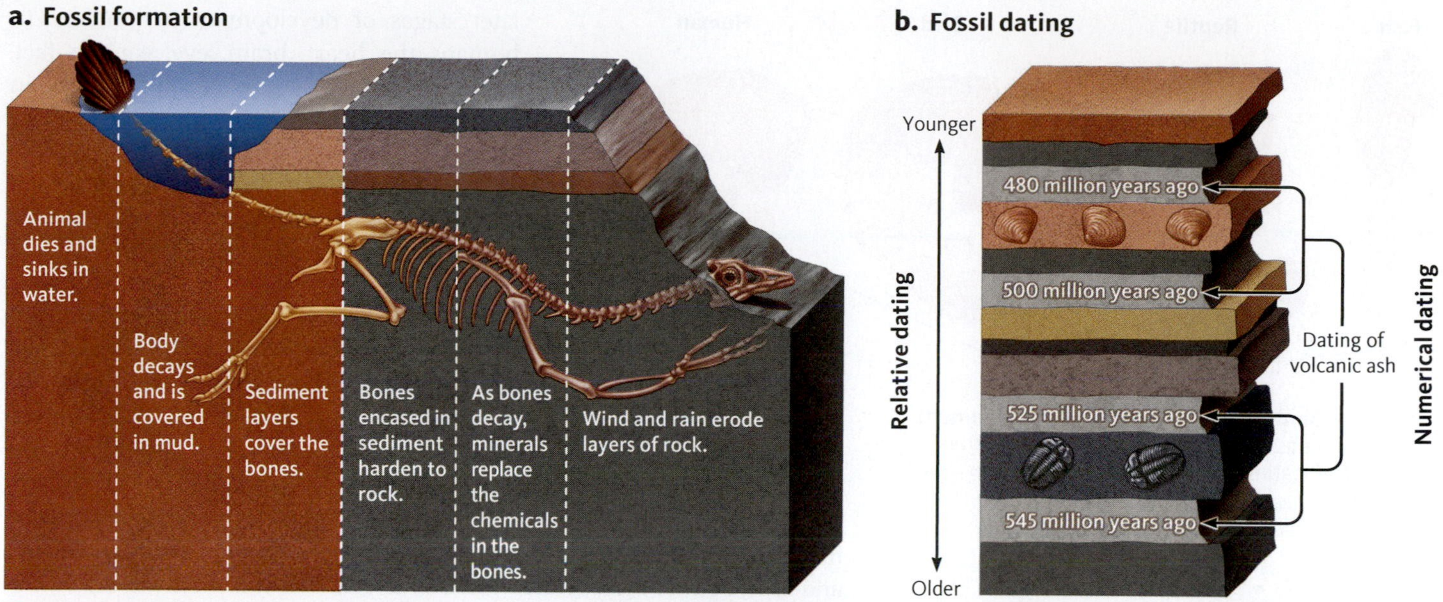

a. Fossil formation

Animal dies and sinks in water.

Body decays and is covered in mud.

Sediment layers cover the bones.

Bones encased in sediment harden to rock.

As bones decay, minerals replace the chemicals in the bones.

Wind and rain erode layers of rock.

b. Fossil dating

Younger

Relative dating

480 million years ago

500 million years ago

525 million years ago

545 million years ago

Older

Dating of volcanic ash

Numerical dating

Figure 1.24 Formation and dating of fossils. (a) Layers of gravel, sand, and mud accumulate, embedding and preserving fossils. (b) From these layers, relative ages of fossils can be determined. Radiometric dating can also be used to provide numerical data to associate with the geologic record.
© Cengage Learning 2014

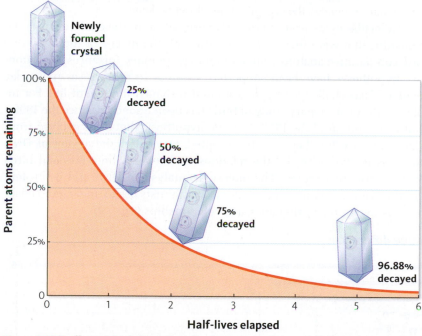

Newly formed crystal

100%

75%

50%

25%

0

Parent atoms remaining

25% decayed

50% decayed

75% decayed

96.88% decayed

0   1   2   3   4   5   6

Half-lives elapsed

Figure 1.25 Radiometric dating. Certain elements such as potassium, phosphorus, and carbon occur in multiple atomic forms and are naturally radioactive, with constant and measurable rates of radioactive decay. Based on the known rate of decay of these elements, radiometric dating determines when the "radiometric clock" started, and therefore the age of the rock layers associated with the fossil can be determined.
© Cengage Learning 2014

case of vertebrate animals, the embryos exhibit shared characteristics with ancestral embryos that provide evidence of their mutual evolutionary history. **Figure 1.26** shows several vertebrate animals (fishes, reptiles, amphibians, and mammals) that all share a common ancestor. Examination of their embryos reveals that, during corresponding stages of early development, the embryos appear very similar. All early vertebrate embryos have tails, which persist in some animals but are lost during the

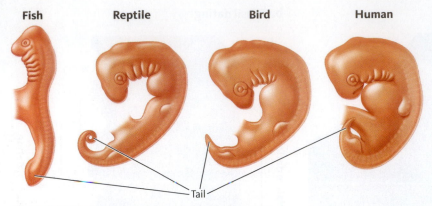

Fish   Reptile   Bird   Human

Tail

Figure 1.26 Similarities in embryonic development.
Vertebrates pass through similar developmental
patterns, indicating that they share a common
ancestor.
© Cengage Learning 2014

later stages of development in humans. In
humans the heart, brain, eye, ear—in fact,
all of our organs—pass through stages in
development that are similar to developmental
stages for fish and amphibians. Usually, the
longer embryos resemble each other through
the course of development, the more closely
related they are.

Molecular Data As you've learned from
observable structures, fossils, and developmental
patterns, the more characteristics that organisms
share, the closer their evolutionary relationship.
The evidence for relationships is also true when
comparing organisms on the molecular level. In fact,
one of the greatest similarities among all organisms is their shared genetic
material—DNA. Because all organisms utilize the same basic structures of DNA
and cells, both of these structures serve as evidence that all living organisms
are related through common ancestry. DNA codes for the production of
proteins and other cellular products that form every part of an organism's
body. This means that evolutionary change, or descent with modification,
occurs at the genetic level. It follows that if two organisms have very similar
molecular sequences, the organisms are closely related.

Molecular evidence is a powerful verification of the evolutionary relationships that were first established through the means of the fossil record, comparative anatomy, and embryology. In many instances, evolutionary hypotheses have been supported, whereas in other cases molecular
evidence has challenged hypotheses and reshaped the tree of life. For instance, the evolutionary study of birds has been underway since the 1800s,
but it was not until the 1970s that the hypothesis that birds evolved from
dinosaur ancestors began to be accepted. Only after the advent of DNA
analysis in the 1980s did the relationship between dinosaurs and birds
become *widely* accepted. The molecular analysis of DNA and its protein
products is another line of evidence that scientists use to understand adaptations, evolutionary timelines, and relationships (**Fig. 1.27**).

Number of DNA nucleotide differences

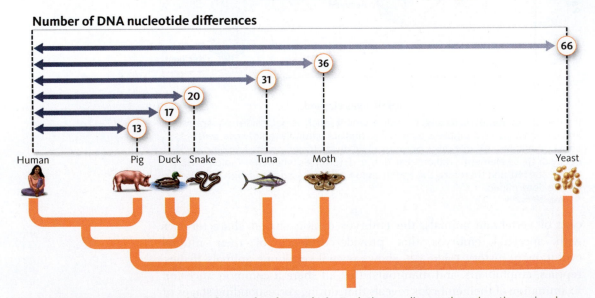

Figure 1.27 Evolutionary timeline based on molecular analysis. Evolutionary diagram based on the molecular
comparisons of the cytochrome c gene of several organisms.
© Cengage Learning 2014

- Biology, like all scientific disciplines, uses a dynamic process to gather evidence to answer questions about the natural world.

- Scientific studies include both experimental and nonexperimental approaches.

- Scientific inquiry involves critical thinking by asking questions, making observations and predictions, executing repeatable experiments, and interpreting data.

- Scientific evidence is used to study evolution. Observations of physical characteristics, fossils, developmental patterns, and molecular data are used to improve our understanding of evolutionary relationships and the tree of life.

1. Explain the two ways that scientists conduct research.
2. List four lines of evidence used to establish evolutionary relationships.
3. How would you respond to a friend who tells you that evolution is "only a theory"?
4. Imagine you have been invited to go on a fossil expedition on the cliffs of the Chesapeake Bay. Where will you find the oldest fossils—at the top of the cliff or the bottom? Explain your answer.
5. What makes conducting research into the development of plant roots more difficult than studying the development of plant leaves?

End of Chapter Review

Self-Quiz on Key Concepts

Life's Unity

KEY CONCEPTS: Organisms are unified through a set of shared key characteristics that define life. The living world follows a hierarchy of scale that starts at the atomic levels and builds in complexity through the organism to the biosphere. All organisms consist of one or more cells, which stay alive through ongoing inputs of energy and raw materials. They sense and respond to changes in their external and internal environments. Their cells contain DNA inherited from their parents that encodes information needed for growth, survival, and reproduction. Finally, organisms adapt and evolve over time.

1. Match each of the following key terms with the best definition. Select the best answer using each choice only one time.

   consumer   a. organism that acquires energy from breaking down other organisms
   producer   b. cell with membrane organelles
   decomposer   c. organism that acquires energy from the sun
   eukaryote   d. examples include bacteria and archaea
   prokaryote   e. organism that acquires energy from ingesting (eating) other organisms

2. Match each of the following key terms with the best definition.

   DNA   a. structure that holds an organism's genetic information
   genome   b. segment of genetic instructions
   chromosome   c. molecule containing an organism's genetic information
   gene   d. the entire set of genetic information of an organism

3. Match each of the following levels of biological hierarchy with the best definition

atom   a. comprised of many smaller chemical units bound together

molecule   b. the smallest adaptable living system

cell   c. the most inclusive level of life's organization

population   d organisms of the same species interacting with one another

biosphere   e. the fundamental unit of matter, makes up all physical objects on Earth

4. The fact that all living things use DNA as their information molecule provides evidence of:
   a. biodiversity.
   b. a hierarchy of scale.
   c. a shared common ancestor.
   d. ecological roles.

5. All organisms:
   a. have the capacity to sense their environment.
   b. have adapted strategies for obtaining energy from their environment.
   c. consist of one or more types of cells.
   d. all of the above.

The Study of Science

KEY CONCEPTS: Biologists strive to understand and make sense of the living world and to do this as objectively as possible. Biology, like all scientific disciplines, uses a dynamic process to answer questions about the natural world. It involves asking questions, making observations and predictions, executing repeatable experiments, and interpreting data.

6. Match each of the following key terms with the best definition.

theory   a. the dynamic process of studying natural phenomena

hypothesis   b. a controlled test of a hypothesis

experiment   c. an educated guess based on observation

scientific method   d. a repeatedly confirmed, tested, and repeatable explanation based on observation, experimentation, and reasoning

7. The two approaches of scientific research are experimental and:
   a. control.
   b. theoretical.
   c. hierarchical.
   d. descriptive.

8. The concept of critical thinking involves:
   a. evaluating information without tests.
   b. evaluating information before accepting it.
   c. relying on the opinions of others.
   d. experimental versus descriptive science.

Evolution—A Unifying Theory

KEY CONCEPTS: Evolution is a process that explains the unity and diversity of life. Populations of organisms respond to environmental factors. As they respond and adapt to their environment, populations evolve. Over time, these adaptations are passed on to their offspring. The theory of evolution explains life's diversity and unifies biology. Understanding evolutionary relationships is the foundation to understanding life.

9. Match each of the following key terms with the best definition.

natural selection   a. a change in a trait, behavior, or structure of an organism

evolution   b. changes to the genetics of a population over generations

adaptation   c. mechanism of adaption, survival, and reproduction

10. The fossil record provides us with:
    a. a historical record of adaptations.
    b. evidence for modern biodiversity.
    c. clues to past ecological relationships.
    d. all the above.

Applying the Concepts

11. In 2005, a South Korean scientist, Woo-suk Hwang, reported that he made immortal stem cells from 11 human patients. His research was hailed as a breakthrough for people affected by currently incurable degenerative diseases, because such stem cells might be used to repair a person's own damaged tissues. Hwang published his results in a respected scientific journal. In 2006, the journal retracted his paper after other scientists discovered that Hwang and his colleagues had faked their results. Some people think this incident shows that scientists are not telling the truth about the natural world. However, others think that the incident helps confirm the usefulness of a scientific approach, because other scientists quickly discovered and exposed the fraud. What do you think?

12. The diet industry is a multibillion-dollar industry built upon spokesperson claims of often incredible results using a particular product or diet plan. In spite of continual scrutiny from organizations like the U.S. Food and Drug Administration cautioning consumers about weight-loss products, consumers continue to flock to one diet after another. How would applying basic elements of critical thinking by the general public alter their diet plan selection?

13. Given your basic introduction to biodiversity in this chapter, how would you respond to someone challenging you about the often costly scientific research projects aimed at investigating ecological relationships among whales and krill? Could you make a strong argument for investing in research investigating less charismatic organisms like the European earthworms that are invading temperate forests in the northern United States?

14. In 2008, the *New York Times* published an article, "New Fossil Analysis Finds Species Walked Upright 6 Million Years Ago," that summarized a recent research article regarding human evolution. In this article, the information presented challenges the previously accepted hypotheses on (1) when human ancestors began to walk upright and (2) who might be our closest relatives. Do you think these findings invalidate other scientific findings? Explain your position.

Data Analysis

In 1986, the International Whaling Commission (IWC) passed a moratorium banning whale hunting. However, some countries such as Japan and Norway continue to hunt whales under a scientific research permit. In fact, the Japanese government subsidizes the whale hunts and has tried for years to overturn the international ban. Not only is whale hunting important to the Japanese cultural tradition, whale meat is also eaten as a protein source. Japan argues that existing whale population counts are not accurate and that the estimate of whale demographics underrepresents the age, sex, and distribution of whales.

The Makah Indian Nation, located on the Olympic Peninsula in Washington, is the only indigenous tribe in the United States that has guaranteed treaty rights to hunt whales for cultural purposes. The Makah tribe has approval from the IWC to take no more than five whales a year. Tribes in other countries have similar agreements with the IWC. These tribes follow their traditions while honoring the agreement.

You are attending an international conference on whales tasked with updating the moratorium that includes the Makah tribe rights to five whales per year. Do you support the agreement as originally written? Do you stop all hunting of whales?

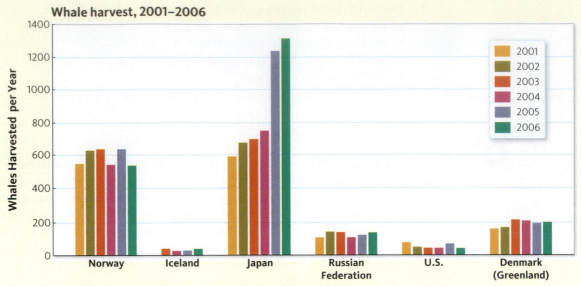

Whale harvest, 2001–2006

Legend: 2001, 2002, 2003, 2004, 2005, 2006

Y-axis: Whales Harvested per Year (0–1400)

X-axis: Norway, Iceland, Japan, Russian Federation, U.S., Denmark (Greenland)

Figure 1.28 Annual whale harvest from 2001–2006 (2006 figures are not complete).
Graph by mongabay.com, data derived from *Science*. Graph from http://news.mongabay.com/2007/1221-japan.html, accessed March 3, 2008.

Data Interpretation

15. According to the graph above, how many whales were harvested by Japan in 2006 (**Fig. 1.28**)?
16. Approximately what percent increase is this over Japan's 2002 harvest numbers?

Critical Thinking:

Use the Internet and the media guide to research and answer the following questions.

17. Research whale population estimates for humpback whales. How accurate do you think these estimates are?
18. Because Japan hunts whales under a scientific agreement with the IWC, what type of research do you think is being accomplished? Is it necessary to harvest over 1,000 whales a year to conduct this research?
19. As a U.S. representative to the IWC, do you support the proposal to maintain the moratorium on hunting whales? Defend your stance.
20. As a Japanese representative to the IWC, what is your opinion on the moratorium? Defend your stance.
21. As a Makah tribal representative to the IWC, what is your opinion on the moratorium? Defend your stance.

Question Generator

Hyraxes, Rodents, and Elephants

Background: The rock hyrax is a rabbit-sized mammal that lives in rocky outcrops in Africa and the Middle East (**Fig. 1.29**). The common name "hyrax" means "shrew mouse" because of its resemblance to rodents. Anatomically, rodents are classified as having a single pair of upper and lower incisors (specialized teeth for gnawing), and they have feet with five toes. Believe it or not, it has been suggested by some scientists that hyraxes are closely related to elephants.

Figure 1.29 Rock hyrax.

Below is a table that compares features of hyraxes, rodents, and elephants (Table 1.4). Use the background information along with the table to ask questions and generate hypotheses. We've translated the components of the table into a question and provided several more to get you started.

Here are a few questions to start you off.

1. What is the evolutionary relationship between hyraxes, rodents, and elephants?

2. Is the hyrax's common name appropriate?

3. How are characteristics weighed when making a decision about evolutionary relationships?

4. How important is the fossil record in inferring evolutionary relationships?

Use the information in the table to generate your own research question and frame a hypothesis.

TABLE 1.4 Characteristics of Hyraxes, Rodents, and Elephants

Characteristics	Hyraxes	Rodents	Elephants
Incisors (front teeth)	Upper incisors = 1 pair Lower incisors = 2 pair	Upper incisors = 1 pair Lower incisors = 1 pair	Upper incisors = tusks
Canine teeth	Absent	Absent	Absent
Toes	4 front; 3 back Short hooflike nail	5 toed	4 front; 3 back Short hooflike nail
Clavicle (collarbone)	Absent	Present	Absent
Digestion	Nonruminant; hindgut	Nonruminant; hindgut	Nonruminant; hindgut
Fossil record origin	40 million years ago	60 million years ago	40 million years ago

© Cengage Learning 2014

© Cengage Learning 2014

2 Evolution and the Diversity of Life

2.1 *Archaeopteryx* and the evolution of birds

2.2 The diversity of life is cataloged and classified by evolutionary relationships

2.3 Animals, plants, fungi, and protists are classified in the Domain Eukarya

2.4 Bacteria and archaea are prokaryotic microorganisms

2.1 *Archaeopteryx* and the Evolution of Birds

Imagine transporting yourself back in time 160 million years. You stand alone on a tropical beach of a vast ocean, the only human being walking the planet since humans have not yet evolved. Behind you, the lush rainforest is filled with ferns, palmlike cycads, ginkgo trees, and trees that look similar to modern redwoods. The continents have not yet separated and are clustered near the equator, so the climate is hot and muggy (**Fig. 2.1**). All around you is an amazingly diverse array of plants and animals. The ocean is filled with microscopic plankton, mollusks, fish, sharks, rays, giant marine crocodiles, and other marine reptiles (**Fig. 2.2**). The treetops are alive with flying reptiles and birdlike creatures gliding from branch to branch. In an open area far down the shoreline, you see herds of giant grazing dinosaurs that resemble the long-necked *Apatosaurus*. Walking into the rainforest, you catch fleeting glimpses of small rodentlike mammals scurrying under the brush as you swat at oversized insects buzzing about your head. You have just found yourself in the middle of the Jurassic Period!

A pigeon-sized primitive bird called *Archaeopteryx* runs, leaps, and glides through the forest (**Fig. 2.3**). Even with its feathered wings, you notice it is an awkward flyer. When you take a close look at the curious creature, it seems to be part dinosaur and part bird. It has many dinosaur characteristics that modern birds don't have, such as teeth, claws on its forelimbs, and a large, bony tail. On the other hand, although *Archaeopteryx* doesn't have a bill, it displays other features common to birds. The feathers are a giveaway, but it

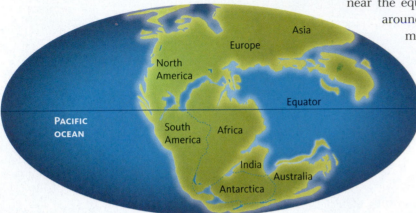

Figure 2.1 The continents 160 million years ago. At that time, the continents that we know today were one large land mass. Over millions of years the continents broke apart, drastically changing Earth's sea levels and climate.
© Cengage Learning 2014

Figure 2.2 Artist's image depicting the Jurassic Period 160 million years ago. At this time plant life was lush, the oceans were rich with life, and reptiles ruled the land.

also has a wishbone for flight, smaller claws and fingers, and on its foot there is a toe that is reversed for perching (see **Fig. 2.3**).

Now, all of these Jurassic organisms are extinct, but you can certainly see the similarities to their plant and animal descendants of today. You further contemplate the relationships of birds and dinosaurs, the fossil record, and the adaptation of flight. As you suspected from its appearance, the *Archaeopteryx lithographica* is now understood to be a transitional form between reptiles (dinosaurs) and birds. Transitional animal forms provide insight into the adaptations of organisms to changing conditions or new habitats—from water to land, land to water, or land to air. *Archaeopteryx* is especially interesting because it not only helps us define the relationship between dinosaurs and birds; it is also one of the earliest species to show adaptations to fly.

Flight gave reptiles a huge advantage over other organisms; they could escape predators, travel farther using less energy in search of food, and live where no other animals could. Pause and think about the many adaptations that needed to come together for a reptile to fly (**Fig. 2.4**). A reptile had to evolve light bone structure, an aerodynamic body shape, and wing bones and muscle structures to flap and provide lift. In the 1860s when the first fossils of *Archaeopteryx* were found, scientists were amazed

Wing with feathers (bird characteristic)

Wing claw (reptilian characteristic)

Toothed beak (reptilian characteristic)

Long tail with vertebrae (reptilian characteristic)

Foot with 3 toes pointing forward and 1 backwards (bird characteristic)

Figure 2.3 Fossils reveal the appearance of *Archaeopteryx*. They provide evidence on the relationship between dinosaurs and birds, and the adaptation of flight. The *Archaeopteryx* has both birdlike and reptilelike characteristics.
© Cengage Learning 2014

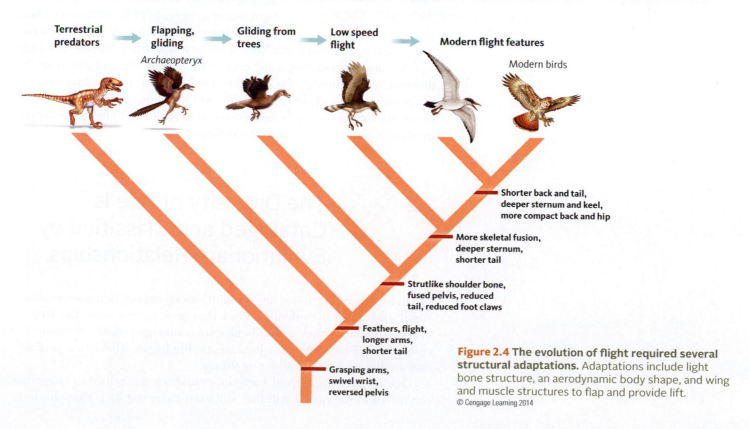

Terrestrial predators

Flapping, gliding

Archaeopteryx

Gliding from trees

Low speed flight

Modern flight features

Modern birds

Shorter back and tail, deeper sternum and keel, more compact back and hip

More skeletal fusion, deeper sternum, shorter tail

Strutlike shoulder bone, fused pelvis, reduced tail, reduced foot claws

Feathers, flight, longer arms, shorter tail

Grasping arms, swivel wrist, reversed pelvis

Figure 2.4 The evolution of flight required several structural adaptations. Adaptations include light bone structure, an aerodynamic body shape, and wing and muscle structures to flap and provide lift.
© Cengage Learning 2014

Figure 2.5 A transitional fossil. *Archaeopteryx* is considered a transitional fossil since it shares characteristics with reptiles and birds.

to see a fossil that had so many of the biological features of flight, including feathers arranged to allow takeoff, gliding, and maneuverability (**Fig. 2.5**).

Although most of the bird lineages that arose 100 million years ago during the Cretaceous Period became extinct, some survived and gave rise to the astonishing diversity of birds we see today. Today there are over 10,000 species of birds that have been identified. Small evolutionary changes here and there over millions of years altered each bird species enough to take the best advantage of a *niche*—a specific set of habitat conditions where it feeds, migrates, nests, and breeds in its own particular place and manner. When you look at the cardinal in your backyard or a sparrow on the street, you are seeing a descendant of the mighty dinosaurs that fascinate us so much!

The ability of many scientists in other fields to accurately present information from life 160 million years ago is amazing. Equally impressive is the immense amount of scientific evidence early naturalists and modern scientists have collected and analyzed to establish the evolutionary relationships. The lines of evidence include not just fossils, but also the observable characteristics, molecular data, and developmental patterns that you learned about in Chapter 1. In this chapter, we'll provide insight into the diversity of life and look at how evolutionary relationships allow us to chronicle the history of life on Earth. We'll show how scientific evidence is brought together and diagrammed using evolutionary trees. We will also trace the diversity of life through the domains and kingdoms of life. Knowing the characteristics of each of the domains and kingdoms of life will help you better place organisms into their positions in the tree of life and better understand their differences and relationships.

2.2  The Diversity of Life Is Cataloged and Classified by Evolutionary Relationships

When life began, Earth was hot, with little or no oxygen in the atmosphere. Fossil and mineral evidence shows that prokaryotes were the first to evolve 3.5 billion years ago and eukaryotes diverged from a common ancestor 1.7 million years after prokaryotic life began. These first prokaryotes are the ancestors to all living things.

One important group of bacteria, cyanobacteria, helped to shape the conditions of the early Earth that still exist today (**Fig. 2.6**). These bacteria

a. **b.**

Dr. Jeremy Burgess/SPL/Photo Researchers, Inc.

Tony Brian/SPL/Photo Researchers, Inc.

Figure 2.6 Cyanobacteria, one of the oldest known groups of prokaryotes. These bacteria altered the oxygen content of the atmosphere through photosynthesis. (a) Floating cyanobacteria. (b) Microscopic view of chains of cyanobacteria that use photosynthesis to acquire energy.

live in the water and produce oxygen through photosynthesis. The relatively oxygen-rich atmosphere in which life evolved was originally generated by these single-celled organisms billions of years ago.

Cyanobacteria also played an important role in the evolution of plant life. Plants photosynthesize using specialized organelles (chloroplasts) that originated from these bacteria living inside of a green alga ancestor of plants. So, although these bacteria are not directly related to plants, their chloroplasts are.

Other characteristics that help to identify cyanobacteria include their cell walls, lack of an internal nucleus, size, and the formation of round colonies. Biologists use this type of information to place organisms into a particular category—to slot them into their proper place in the catalog of life. Once identified, they are then able to use additional evidence from fossils and molecular studies to establish evolutionary relationships and to gain a better understanding of life. There are at least 3 million and perhaps as many as 30 million different species of living things on Earth today. Keep in mind that all of these forms of life—including cyanobacteria, algae, plants, reptiles, and birds—share common characteristics stemming from a common shared ancestor.

Classification organizes life's diversity

You may be wondering how biologists get their arms around the vast diversity of life on Earth. **Taxonomy** is a field of biology that classifies, identifies, and names organisms. Biologists who specialize in this field are referred to as *taxonomists*. They **classify**, or place organisms into categories, based on similar characteristics or traits. These include body form, anatomical structures, developmental events that occur, and the biochemistry of the organisms (**Fig. 2.7**). Relying on physical features is somewhat artificial because many organisms that appear similar on the surface really are evolutionarily unrelated to each other. For this reason, biologists place organisms into groups based on their evolutionary relationships.

Taxonomy—a field of biology that classifies, identifies, and names organisms. Biologists that specialize in this field are referred to as *taxonomists*.

Classify (or *classification*)—the process of placing organisms into categories based on similar characteristics or traits. This includes body form, anatomical structures, developmental events that occur, and the biochemistry of the organisms.

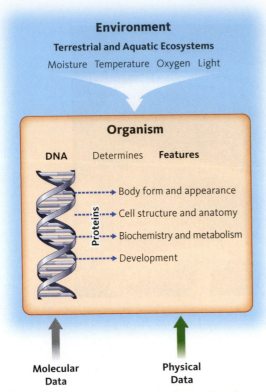

Environment

Terrestrial and Aquatic Ecosystems

Moisture Temperature Oxygen Light

Organism

DNA   Determines   **Features**

Proteins

- - - → Body form and appearance
- - - → Cell structure and anatomy
- - - → Biochemistry and metabolism
- - - → Development

Molecular Data   **Physical Data**

Figure 2.7 Classification of organisms. Organisms are classified based on body form, anatomical structures, developmental events that occur, and the biochemistry of the organisms. Physical and molecular data are analyzed to establish evolutionary relationships. Pressures from the environment greatly influence organism structure.
© Cengage Learning 2014

How do you determine the evolutionary relationships of many organisms? By compiling a wide range of evidence, biologists are able to establish strong evolutionary relationships and look at life's history as a whole. This process combines physical evidence with other lines of evidence such as fossils and molecular data. Today taxonomists have the tools to look deeply into the genes that produce the physical attributes of organisms. Molecular evidence involves sequences of bases on the DNA molecule and the protein molecules they encode. Other data involve comparing the molecular composition of cellular structures. When new research findings challenge the existing classification of an organism, taxonomists then reevaluate the evidence and data just like other fields of science. By understanding both taxonomy and evolutionary relationships, you can make important connections and understand a great deal about organisms.

The hierarchy of classification proceeds from the broadest, most inclusive group (domain) to the most specific and exclusive group (species). Perhaps you learned the mnemonic *Dashing King Philip Came Over for Great Spaghetti* in an earlier science class. The first letter of each word in this strange sentence refers to a specific level within the *taxonomic hierarchy*: domain, kingdom, phylum, class, order, family, genus, and species. **Table 2.1** shows the taxonomic classification of the blue whale and provides some characteristics of each of the taxonomic ranks. Keep in mind that domains include millions of species, whereas a single species stands alone as unique from all other species.

The power of classification comes from understanding how organisms are grouped into taxonomic levels such as the family or class. By using deductive reasoning, you can infer characteristics of the organism based on its classification. This type of reasoning takes a general fact and then assumes, or infers, more specific information about it. For example, knowing that the blue whale and the humpback whale belong to the same genus

TABLE 2.1 Taxonomic Classification of the Blue Whale

Taxonomic Classification		Familiar Characteristics	Other Members
Domain	Eukarya	Cells organized with complex structures enclosed by membranes (i.e., organelles such as a true nucleus)	Plants, animals, fungi, protists
Kingdom	Animalia	Multicellular, ingest food, lack cell walls, most reproduce sexually, distinct development stages	Mammals, reptiles, birds, amphibians, insects, worms, jellyfish, sponges
Phylum	Chordata	Notochord (which in many chordates becomes part of the backbone)	Vertebrates such as mammals, reptiles, birds, and amphibians
Class	Mammalia	Hair and sweat glands, some of which produce milk (mammary glands)	Mammals such as polar bears, whales, gray wolves, bats, dogs, and cats
Order	Cetacea	Latin for "large sea animal"	All whales
Family	Balaenopteridae	Includes all whales without teeth	Rorqual whales
Genus	*Balaenoptera*	Baleen plates used for feeding	Blue whales, fin whales, minke whales
Species	*Musculus*	Muscular (though its literal Latin translation can refer to muscle it may mean "little mouse," which might be a play on words!)	The blue whale

© Cengage Learning 2014

Balaenoptera, you can deduce traits about both whales—that both are large marine mammals without teeth, having instead a unique filter feeder mechanism that uses a structure called baleen plates. Classification provides the tools for you to make generalizations and predictions about organisms; it also establishes a common language understood by all biologists and students worldwide.

As our knowledge about the living world increases, organisms can be moved, or reclassified, into different groups. Today, most evolutionary biologists favor a classification system that organizes life by placing all organisms into one of three broad categories, called *domains,* based on an organism's specific DNA sequences, cell characteristics, and type. The domains are Domain Bacteria, Domain Archaea, and Domain Eukarya. Biologists assign the domain based on the two different types of cells composing organisms. Recall that living things are composed of either prokaryotic or eukaryotic cells.

Biologists consider both the archaea and bacteria domains to be the most primitive and the earliest of all life forms. Both domains comprise small prokaryotes that exist as a single cell without a distinct membrane-bound nucleus. Despite their size, prokaryotes in these two domains produce complex chemical compounds, some of which cause diseases in humans, some of which are useful as medicines, and some of which provide nutrient sources important for the growth of other organisms. They play ecological roles as producers or decomposers in nearly all of Earth's ecosystems. Bacteria include the cyanobacteria in the ocean, the *Lactobacillus* in your yogurt, the *E. coli* in your intestine, the *Staphylococcus* on your skin, and over 15,000 other species (**Fig. 2.8**). You actually have more bacte-

a.

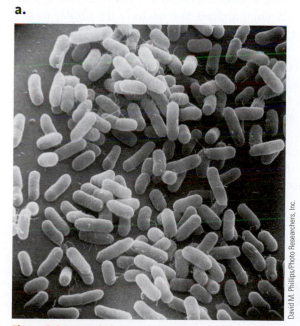

David M. Phillips/Photo Researchers, Inc.

b.

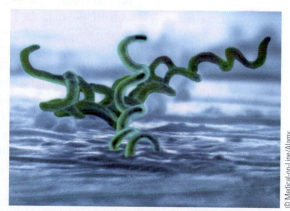

© Medical-on-Line/Alamy

c.

Science Source/Photo Researchers, Inc.

Figure 2.8 Bacteria have many different roles.
(a) *E. coli* are found in your digestive tract and help to break down foods, (b) *Treponema pallidum* causes the human disease syphilis, and (c) *Staphylococcus* is a decomposer commonly found on human skin but can cause disease if it enters the bloodstream.

a.

Photo by J. Krueger

Figure 2.9 Archaeans live in diverse ecosystems.
(a) Thermal pools at Yellowstone National Park.
(b) Hydrothermal vents deep in the ocean.

b.

NOAA

ria on and in your body than cells that make up your body. While archaea live in association with humans, they are also known to live in extreme environments—an indication of the ancient heritage of the domain. In fact, the word "archaea" translates to "ancient." The first archaeans studied were those that lived in places like hot springs in Yellowstone National Park and thermal vents on the floor of the ocean, sites once considered too hostile for life (**Fig. 2.9**). Although archaea and bacteria are similar in size and shape, they differ at the molecular level and in some of the habitats where they live.

You are most familiar with members of the Domain Eukarya, called eukaryotes. People, pets, plants, and mushrooms, for example, all have bodies composed of many cells. However, many yeasts (fungi) and the members of a group called protists are single-celled organisms. Life shows levels of organization that include a spectrum of organisms from the simple to the complex. As Earth changed over billions of years, more simple forms evolved into the more complex multicellular eukaryotes (**Fig. 2.10**). We will explore the fascinating eukaryotic organisms in more detail in Section 2.3.

Evolutionary trees summarize evolutionary relationships

Family genealogical trees chart the history and relationships within a family over many generations by showing the ancestors and their descendants. Family trees also illustrate who are siblings, cousins, uncles, and aunts. Using these diagrams, you can see all of the important family information by examining the tree. **Evolutionary trees** (**Fig. 2.11**, p. 46) provide a similar visual summary of a complex set of scientific data, linking taxonomy with evolutionary relationships. Because new information is continually being collected about organisms, evolutionary trees represent our best data about the pattern of relationships among groups of

Evolutionary tree—a tool that provides a visual summary of a complex set of scientific data, linking taxonomy with evolutionary relationships.

a. *Tatiana Belova/Shutterstock.com*

b. *© WildPictures/Alamy*

c. *Photo by J. Krueger*

d. *© MELBA PHOTO AGENCY/Alamy*

Figure 2.10 Eukaryotes are both familiar and diverse. Each of these organisms—(a) the polar bear, (b) the death cap mushroom, (c) a quaking aspen tree, and (d) an amoeba (a common protist)—represents some of the tremendous diversity of this domain.

species. These diagrams show who we think is most closely related to whom and illustrate pathways that follow descendants through hundreds, thousands, or millions of years of evolution. Let's return to the relationship between dinosaurs, *Archaeopteryx,* and modern birds. The evolutionary tree in **Figure 2.11a** reflects the current understanding of the evolution of birds. By examining the evolutionary tree, it tells you that (1) birds are the closest living relatives of dinosaurs, and (2) birds and dinosaurs descended from a common ancestor.

a.

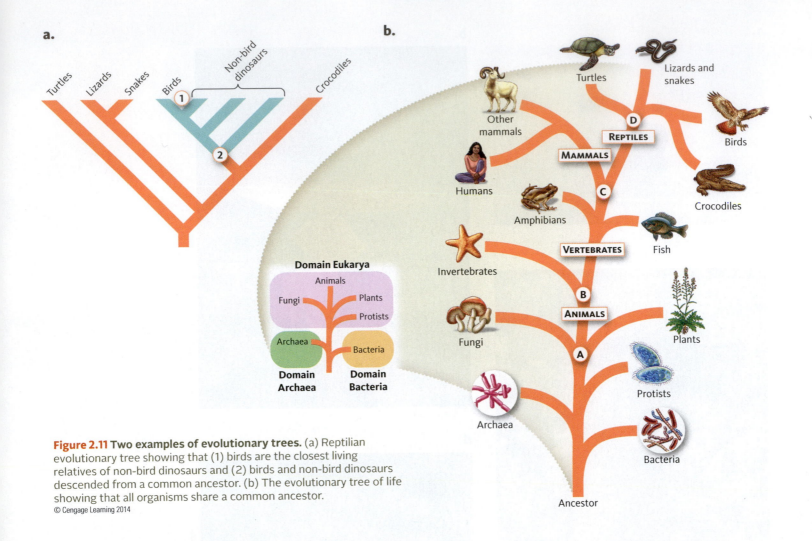

b.

Figure 2.11 Two examples of evolutionary trees. (a) Reptilian evolutionary tree showing that (1) birds are the closest living relatives of non-bird dinosaurs and (2) birds and non-bird dinosaurs descended from a common ancestor. (b) The evolutionary tree of life showing that all organisms share a common ancestor.
© Cengage Learning 2014

Let's use **Figure 2.11b** to show how some familiar animals are related. On the tree, each line or branch represents one evolutionary *lineage,* a line of descent from a common ancestor. You can see several places where two animal groups diverge (split) at a branch point known as a *node.* Note that the last common branching point for animals, plants, and fungi is node A, which makes protists their shared ancestor. Node B indicates that invertebrates like the starfish share a common ancestor with vertebrates such as fish, mammals, birds, lizards, and crocodiles. From node C you can conclude that mammals and reptiles all share an ancestor with animals. A common mistake made while interpreting evolutionary trees is focusing only on the branch tips rather than tracing the organism back to the node. By glancing at just the branch tips, you might incorrectly interpret that the turtle and other mammals are more closely related than the turtle and the lizards and snakes. The tree is correctly interpreted by looking at nodes C and D, concluding that turtles are more closely related to lizards and snakes. As you move closer to the base of the tree, you find that these familiar animals are all connected through shared characteristics.

Shared characteristics distinguish organisms and establish relationships

There are different types of evolutionary diagrams—some are similar to a family tree, others include a time frame, while still others illustrate the shared characteristics among the groups (**Fig. 2.12**). Shared characteristics are evolutionary novelties, like anatomical features, that groups of organisms have in common. These traits or characteristics can include physical attributes such as the presence of a bony skeleton or behavioral characteristics such as singing to attract a mate. These shared characteristics are used to separate organisms into groups and establish relationships.

Let's walk through the evolutionary diagram using the shared characteristics of the seven major animal groups to help you interpret the information that illustrates their relationships. First, notice that all of the animals are illustrated along the top of the diagram. If we compare a gorilla, a bird, and a frog, you can start to develop a list of similarities as indicated by the red bars, which indicate a common ancestor that developed that specific characteristic. For example, each one has vertebrae, a bony skeleton, and four limbs. If we look at more specific differences, you might note that for reproduction the gorilla and bird both use a watertight sac that they carry or an egg that they lay (called an *amniotic* egg or sac), but the frog uses a different type of egg. The gorilla and the bird share an evolutionary novelty. The more novelties two organisms share, the more closely related they are. In this example, biologists conclude that the gorilla and bird share a common ancestor that also used amniotic eggs or sacs for reproduction.

Let's now further examine the characteristics of reptiles and mammals shown in **Figure 2.12**. Look at the mammal branch of the diagram. What characteristic do mammals share that reptiles do not? One characteristic shared by all mammals is hair, a trait not found in reptiles. The development of hair distinguishes mammals from other groups while establishing relationships within the mammal group. Likewise, looking at the reptile branch, a shared specialized skull structure distinguishes this group of animals from all the others.

Evolutionary diagrams are valuable tools for learning in a glance about relationships and shared characteristics within a group of organisms. However, evolutionary diagrams are compilations of many lines of scientific evidence; and the evolutionary study of organisms, especially those that are extinct, is more complex than these diagrams indicate. Studying evolution and hypothesizing relationships require examining fossils, comparative anatomy, molecular data, and developmental patterns. With increasing knowledge and tools, evolutionary biologists continue to improve the hypotheses of how various organisms are related.

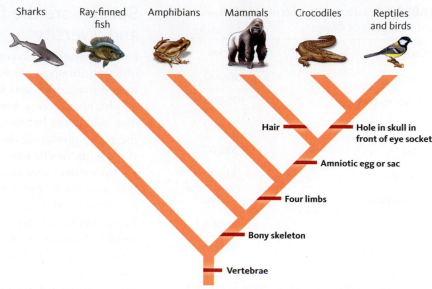

Figure 2.12 Shared characteristics of animals. This evolutionary diagram shows evolutionary relationships and shared characteristics (indicated by red bars). Animals above each of the red bars share the common trait. The point where the two branches meet below the trait is the point the animals split from their shared common ancestor.
© Cengage Learning 2014

Shared characteristics—evolutionary novelties, like anatomical features, that groups of organisms have in common. These traits or characteristics can include physical attributes such as the presence of a bony skeleton or behavioral characteristics such as singing to attract a mate. Shared characteristics are used to separate organisms into groups and establish relationships.

TABLE 2.2 Estimated Number of Identified Species on Earth from Various Groups

Category	Species
Insects	950,000
Flowering Plants	300,000
Worms	115,000
Fungi	100,000
Mollusks	81,000
Crustaceans	40,000
Algae	24,000
Ferns and Horsetails	20,000
Lichens	20,000
Mosses	15,000
Birds	10,000
Reptiles	8,250
Amphibians	6,200
Mammals	5,400
Corals	2,200
Conifers	980

Species—members of a population that actually or potentially interbreed and produce viable offspring in nature.

Species are the fundamental units of evolution and diversity

Earth's incredible diversity of insects, flowering plants, and other groups of living things is evident when you visit a garden, go to a zoo, or read nature magazines and field guides. These publications show you photographs, illustrations, and maps, along with information about the diverse diet, coloration, habitat, or geographic range of each species. Understanding and appreciating the huge number of Earth's life forms, or *biodiversity,* starts with the classification of life.

How many species currently inhabit Earth? This is not an easy question to answer. Estimates of the number of living species range from 3 million to 30 million species. This estimate varies widely because most living species are microorganisms and tiny invertebrates that live in inaccessible habitats or are difficult to find. Biologists have named and described over 1.8 million species of eukaryotes alone (**Table 2.2**), and thousands of new species are added each year. It is interesting to note that the mammals we are most familiar with make up less than 0.5 percent of the known species, whereas insects make up over 50 percent. It is also estimated that over 99 percent of all species that ever lived are now extinct.

When species are scientifically described, taxonomists place the species into their specific hierarchical groups beginning with the three domains of life (Domain Bacteria, Domain Archaea, and Domain Eukarya) and working their way down to the individual species of life. Whether we are discussing biodiversity, identifying organisms, understanding similarities between organisms, or tracing evolutionary history, it is important to understand what biologists mean when they use the word *species.* At a glance, do the spiders pictured in **Figure 2.13a** look like the same species to you? What about the birds shown in **Figure 2.13b**? Although the spiders look very different from each other, it may surprise you that they are actually the same species, whereas the birds are two different species of cardinals. It is not always easy to identify a species based only on its appearance, and today a combination of molecular, ecological, structural, and behavioral information is used when identifying different species.

So what makes two organisms members of the same species? At a distance, by looking at physical traits, you might think the two cardinals in **Figure 2.13b** are the same species—both are the same size, have a pointed head crest, and use identical songs and calls. The two species are distinguished from one another based on their coloration and the structure of their upper beak. Are these particular differences sufficient to call them different species? Or are the similarities sufficient enough to group them together? If you were an evolutionary biologist examining fossils of these two birds, could you distinguish them from one another? These questions have led biologists to look at the definition of species using another aspect of the organism—reproduction.

Sexual reproduction, although it produces genetic and physical variation, maintains the distinctive set of characteristics for a species from one generation to the next. Even though members of a species may appear different, they must be capable of breeding and passing their traits on to the next generation of their offspring. If members of a population actually or potentially interbreed and produce viable offspring in nature, they are considered a **species**. According to this biological definition, members of species belong to a *reproductive* community—they are reproductively compatible and, as a result, the entire population has a similar genetic and physical makeup. In the case of the two cardinal species, they do not interbreed in nature. Because this population does not breed with populations of other

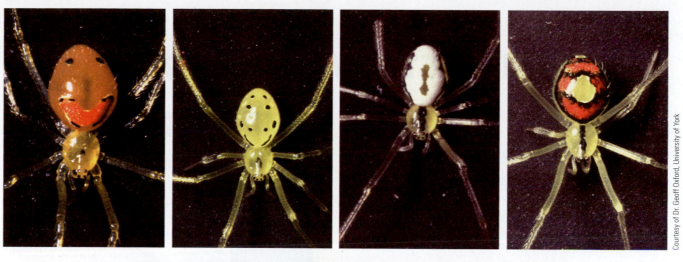

a.

Courtesy of Dr. Geoff Oxford, University of York

b.

© John Anderson/Alamy

Elliotte Rusty Harold/Shutterstock.com

Figure 2.13 **Members of a species may look different but can interbreed.** (a) Members of one species of happy face spiders displaying genetic variation. (b) The Northern cardinal (left) and the Pyrrhuloxia cardinal (right) may look similar, but they are two separate species of birds.

species, their molecular attributes, and thus their genes, are more similar in their own population. Therefore the cardinals are two distinct species.

To organize and understand information about species and life, biologists once classified organisms based only on their physical similarities. Today, biologists in all fields are classifying organisms based on their evolutionary history and relationships using physical, molecular, and reproductive data as evidence. From this classification, we gain a better understanding of life—we understand what it means to be a plant, an animal, or a bacterium; where these organisms came from; and how they are all connected.

Figures 2.14 and **2.15** summarize the history, relationships, and general characteristics of the kingdoms of life. With an appreciation of the history of life and an understanding of how scientists establish and trace life's diversity and evolutionary history, we will now shift focus to a brief introduction to the diversity of life. Because it is important to grasp a basic understanding of each kingdom's characteristics, our descriptions will be general in nature, but there are always exceptions. We hope that you appreciate the incredible diversity of life and how the millions of organisms are related to each other through immense periods of geologic time.

THE HISTORY OF LIFE ON EARTH

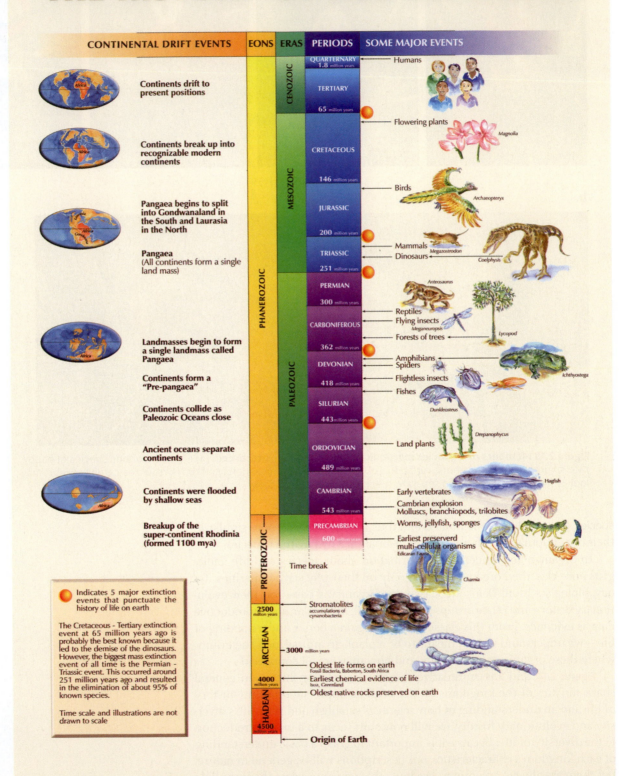

| CONTINENTAL DRIFT EVENTS | EONS | ERAS | PERIODS | SOME MAJOR EVENTS |

Continents drift to present positions

Continents break up into recognizable modern continents

Pangaea begins to split into Gondwanaland in the South and Laurasia in the North

Pangaea (All continents form a single land mass)

Landmasses begin to form a single landmass called Pangaea

Continents form a "Pre-pangaea"

Continents collide as Paleozoic Oceans close

Ancient oceans separate continents

Continents were flooded by shallow seas

Breakup of the super-continent Rhodinia (formed 1100 mya)

PHANEROZOIC
PROTEROZOIC
ARCHEAN
HADEAN

CENOZOIC
MESOZOIC
PALEOZOIC

QUARTERNARY 1.8 million years — Humans
TERTIARY 65 million years
— Flowering plants — Magnolia
CRETACEOUS 146 million years
— Birds — Archaeopteryx
JURASSIC 200 million years
— Mammals — Megazostrodon
TRIASSIC 251 million years
— Dinosaurs — Coelphysis
— Anteosaurus
PERMIAN 300 million years
— Reptiles
CARBONIFEROUS 362 million years
— Flying insects — Meganeuropsis
— Forests of trees — Lycopod
DEVONIAN 418 million years
— Amphibians
— Spiders
— Flightless insects
— Fishes — Ichthyostega
SILURIAN 443 million years
— Dunkleosteus
ORDOVICIAN 489 million years
— Land plants — Drepanophycus
CAMBRIAN 543 million years
— Early vertebrates — Hagfish
— Cambrian explosion Molluscs, branchiopods, trilobites
PRECAMBRIAN 600 million years
— Worms, jellyfish, sponges
— Earliest preserved multi-cellular organisms — Edicaran Fauna — Charnia

Time break

Indicates 5 major extinction events that punctuate the history of life on earth

The Cretaceous - Tertiary extinction event at 65 million years ago is probably the best known because it led to the demise of the dinosaurs. However, the biggest mass extinction event of all time is the Permian - Triassic event. This occurred around 251 million years ago and resulted in the elimination of about 95% of known species.

Time scale and illustrations are not drawn to scale

2500 million years — Stromatolites accumulations of cyanobacteria
3000 million years
— Oldest life forms on earth Fossil Bacteria, Baberton, South Africa
4000 million years — Earliest chemical evidence of life Isua, Greenland
— Oldest native rocks preserved on earth
4500 million years — Origin of Earth

Acknowledgements: Contents: Anusuya Chinsamy-Turan (UCT); Illustrations: Samantha van Riet; Design & layout: Featherline Art & Design cc

Figure 2.14 History of life on earth. Scientists recognize that Earth has changed since its origin 4.5 billion years ago. Continents have moved together and split apart, significantly changing sea levels and the climate. The levels of important gases, like carbon dioxide and oxygen, also drastically changed, allowing for life to evolve. Over the last 3.7 billion years, living things have diversified and adapted to many different environments. The oldest prokaryotic fossils date back to 3.5 billion years ago, and evidence for the first eukaryotic organisms dates back 1.7 billion years. Oxygen played a critical role in the evolution of eukaryotic organisms because it provided the metabolic machinery that allowed eukaryotic cells to become multicellular. For the first 3 billion years of Earth's history, life was confined to freshwater and marine environments. Eventually, plants and animals became more complex and made the transition from life in water to life on land.

Reprinted by permission of the South African Agency for Science and Technology Advancement (SAASTA)

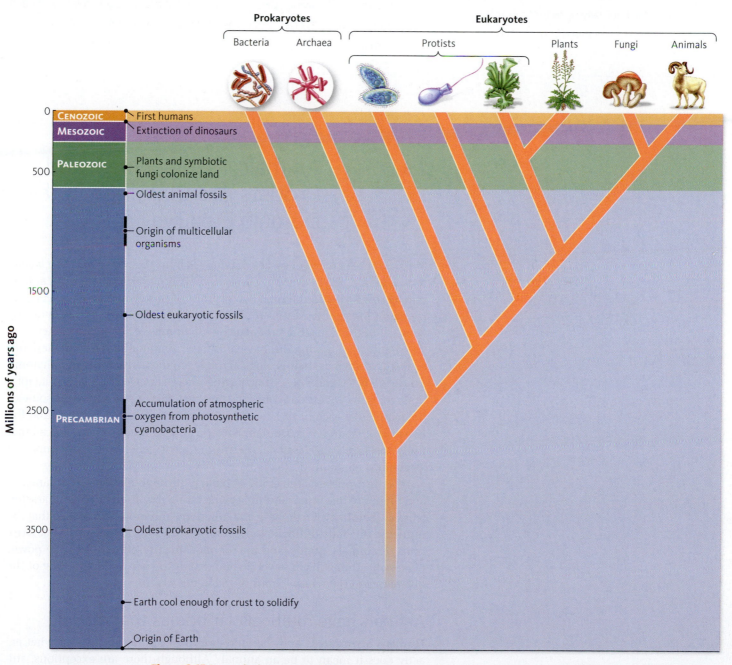

Figure 2.15 An evolutionary tree. This evolutionary tree shows the major groups of organisms with an evolutionary timeline noting milestones along the history of life.

© Cengage Learning 2014

- An understanding of evolutionary relationships is the foundation to understanding life and its history on Earth. The more closely related two organisms are, the more similar their characteristics.

- Evolutionary trees are one tool that biologists use to summarize evolutionary relationships.

- Species are the fundamental units of evolution and diversity and are defined as members of a population that actually or potentially interbreed and produce viable offspring in nature.

1. In your own words, what defines a species?

2. Looking at the evolutionary tree (see **Fig. 2.15**), which kingdom is most closely related to animals?

3. Two populations of great horned owls living in the same forest look very similar to one another, but they do not breed to produce offspring. Biologically, are they considered the same species? Explain your answer.

4. Look at the evolutionary tree in **Figure 2.11b**. Circle the common ancestor of (a) fungi and animals, and (b) protists and plants. Which common ancestor evolved first: that of animals or plants?

5. Which of the following pairs of organisms are *most* closely related to each other (see **Fig. 2.12**)? (A) Crocodiles and mammals. (B) Amphibians and mammals.

2.3  Animals, Plants, Fungi, and Protists Are Classified in the Domain Eukarya

So far, you have learned some cool things about the blue whale, *Archaeopteryx,* and the blind cavefish. These animals provide you a glimpse at how remarkable life is. In the final two sections of this chapter, we will generalize about the characteristics of the three domains of life, and provide examples of other amazing organisms with whom we share our planet.

Members of a kingdom have a unique combination of characteristics or features, such as their ecological roles, genetics, structural and anatomical differences, and ways of acquiring energy, as well as various adaptations for growth, reproduction, and survival. **Table 2.3** summarizes some of the features of each of the four kingdoms in the Domain Eukarya and the two microbial domains, Domain Bacteria and Domain Archaea. Recall that these top levels of classification are the most general and encompass a tremendous amount of diversity.

This text presents the study of biology through the lives of organisms, and therefore it is important to become familiar with their basic characteristics as members of a domain and kingdom. You must also realize that the immense diversity of life includes many exceptions to the general descriptions of animals, plants, and microorganisms presented here. The power of classification will allow you to add details to your understanding of the diversity and interconnections of the living world.

Animals have multicellular bodies that move

Humans are animals and therefore are most familiar to us, but what exactly does it mean to be an animal? Although there are exceptions, animals as a group have multicellular bodies that are on the move. They run

TABLE 2.3 Characteristics of the Domains and Kingdoms of Life

Domain	Kingdom	Ecological Roles	Structural Features	Acquiring Energy	Development and Reproduction	Examples
Eukarya	Animals	Consumers, predators, prey, parasites	Multicellular with complex tissues and organ systems. Most are mobile.	Ingesting other organisms	Complex development with mostly sexual reproduction	Whales, humans, penguins, frogs, ants, worms, jellyfish, and sponges
	Plants	Producers	Multicellular with cell walls (containing cellulose), chloroplasts, vacuoles, immobile. Many also have flowers or cones, roots, and vascular tissue.	Photosynthesis	Complex development with both asexual and sexual reproduction	Mosses, ferns, pines, and flowering plants
	Fungi	Consumers, decomposers	Multicellular with cell walls (containing chitin), immobile, use spores for reproduction, made of hyphae	Externally digesting and absorbing other organisms (dead or alive)	Complex development with both asexual and sexual reproduction	Yeast, puffballs, button mushrooms, powdery mildew, death cap, and honey mushrooms
	Protists	Decomposers, producers, parasites	Single-celled or multicellular	Photosynthesis, digestion, or ingestion	No development with asexual and/or sexual reproduction	Amoeba, paramecium, algae, and slime molds
Bacteria		Decomposers, producers, pathogens	Single-celled, cell walls	Photosynthesis, absorption, convert chemical nutrients to chemical energy	Asexual reproduction	*Escherichia coli*, *Streptococcus pneumoniae*, and *Salmonella enteritidis*
Archaea		Decomposers, producers, pathogens	Single-celled, cell walls (chemically distinct from plants, fungi, and bacteria)	Photosynthesis, absorption, convert chemical nutrients to energy	Asexual reproduction	Genus *Sulfolobus*

© Cengage Learning 2014

to catch prey or to escape being eaten. They fly, jump, dig, and swim, moving from one point to another. To move, many animals have muscles and sensory systems to perceive the world. Eyes, brains, and complex limbs evolved in the animal body plan. The evolution of animals was driven by the ability to move.

Animals, like plants, evolved in the sea from an aquatic single-celled ancestor similar to the amoeba. Animals are the most abundant of all eukaryotic species living today, with over 35 phyla that contain about 1,250,000 identified animal species. Although not a formal level of classification, animals are traditionally divided into invertebrates and vertebrates. Invertebrates, like insects, worms, and mollusks, lack a hard, bony backbone that houses the spinal cord. Mammals, birds, reptiles, amphibians, and fishes comprise the vertebrate phyla. **Figure 2.16** is an evolutionary tree showing the major groups of animals and some of the characteristics of each branch of the tree.

Figure 2.16 Brief evolutionary tree describing some of the major animal groups. A complete digestive system, segmentation, jointed legs, and a backbone are a few of the important characteristics that differentiate groups of animals.

© Cengage Learning 2014

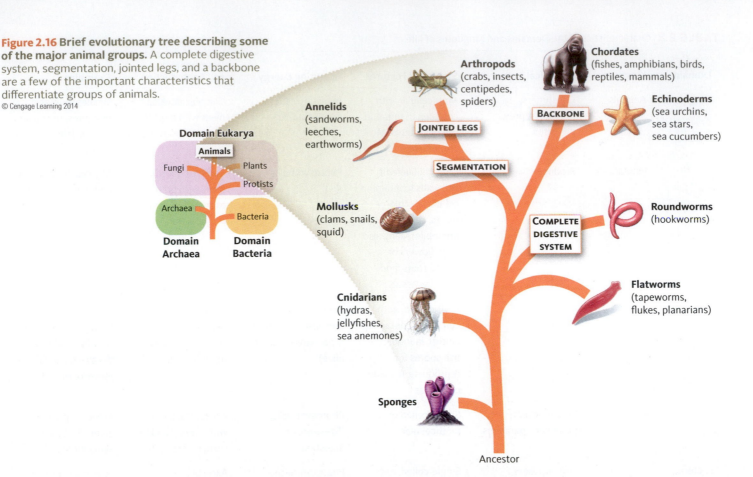

Ancestor

Figure 2.17 These animals share the characteristics of life, but each has different ways of fulfilling the requirements of life.

Animals are *multicellular* organisms with similar characteristics. The cells that compose their bodies are eukaryotic, complex, and like all cells are surrounded by an outer membrane. They differ from other organisms like plants and fungi in that they lack any type of supporting cell wall. Animals feed by actively ingesting the tissues, juices, or wastes of other organisms. Examples include whales, humans, penguins, frogs, ants, and the less familiar worms, jellyfish, and sponges (**Fig. 2.17**). Animals are identified, organized, and classified based on their body structure, form, and complexity. Most animals pass through a unique developmental stage before reaching maturity. Animals also have increasingly complex systems for maintaining their internal environment, acquiring and using nutrients, and reproducing.

Compared to other kingdoms, animals are distinguished from plants and many protists because of their way of obtaining energy from the ecosystem. Animals are **heterotrophs**, or *consumers,* that obtain nutrients from other organisms instead of photosynthesizing energy molecules. Animals can obtain nutrients exclusively from plants and be herbivores, they can eat other animals and be carnivores, or they can be omnivores with a mixed diet. Carnivores such as the gray wolf are predators that hunt and kill their prey. Many animals such as ants, ticks, and tapeworms are parasites and live on or in host animals, sometimes causing harm as they obtain nutrients. Earthworms, cockroaches, and turkey vultures serve as scavengers. These animals eat dead animals and plants, helping to recycle nutrients back into the ecosystem. Animals also use the oxygen that plants make and release carbon dioxide back into the ecosystem that plants use.

Plants produce food and oxygen for many living things

The Plant Kingdom, which includes mosses, ferns, pines, and flowering plants, is a major source of beauty and economic importance that touches each of our lives every day (**Fig. 2.18**). Most of your nutrition depends on

Figure 2.18 Diversity in the Plant Kingdom. (a) Mosses (on the rocks) are simple plants that lack specialized tissues. The larger ferns have evolved specialized vascular tissues for transporting water and nutrients but lack the adaptation to produce seeds. (b) Conifers have more elaborate structures for support and reproduction. (c) Flowering plants that dominate the landscape today include a common cattail found in wetlands across the United States, corn releasing its pollen, and a highly prized orchid flower.

Heterotrophs—consumers (like animals and fungi) that obtain nutrients from other organisms instead of photosynthesizing energy molecules.

Figure 2.19 Evolutionary tree of the plants. This diagram shows the relationships of the various plant groups that branched from a common algal ancestor. Plants are groups based on reproduction modes and on whether or not they have specialized vascular tissues for transporting water and nutrients.
© Cengage Learning 2014

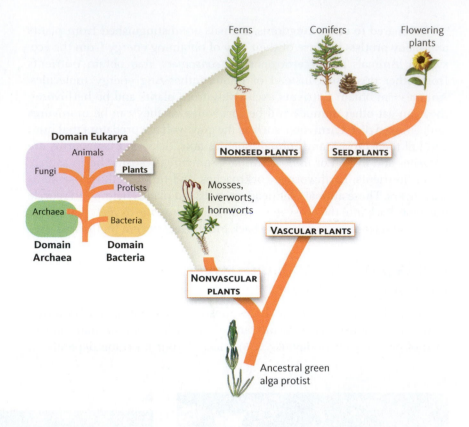

plants, either directly or indirectly. In fact, just three cereal crops—rice, wheat, and corn—form the basis for much of the human diet. Nonfood products such as wood and cotton fibers are also important as renewable resources that enrich our lives.

Plants are eukaryotic organisms that evolved from a green alga ancestor (**Fig. 2.19**). They are classified into twelve phyla primarily based on their reproductive structures, such as cones or flowers. There are over 300,000 named species of plants and over 90 percent of those are flowering plants, such as grasses, roses, geraniums, and oak trees.

One key feature of plants is their ability to transform sunlight energy and carbon dioxide into larger biological molecules, such as sugar, through the metabolic process of photosynthesis. Plants and other organisms that can "feed themselves" through photosynthesis are called autotrophs. In nearly all ecosystems on Earth, the flow of energy starts when plants intercept energy from the sun and convert it into chemical energy stored in biological molecules such as sugar and cellulose. Photosynthesis has had a dramatic impact on our planet—in fact, it has shaped life as we know it. One by-product of photosynthesis is the oxygen in our atmosphere that most living organisms rely on. Photosynthetic processes over millions of years helped to create and now maintain a breathable atmosphere (**Fig. 2.20**). Without plants, the ratio of carbon dioxide and oxygen in our atmosphere would be very different.

Plants grow in nearly all ecosystems, filling a variety of ecological roles. While you're most acquainted with plants that grow on land, many plants such as duckweed float in freshwater ponds and lakes and are entirely aquatic. Only a few groups of plants are adapted to live in salt water. Be-

Photosynthesis—the metabolic process that transforms sunlight energy and carbon dioxide into larger biological molecules, such as sugar.

Autotrophs—organisms, like plants, algae, and some microbes, that can "feed themselves" through photosynthesis.

Photosynthesis

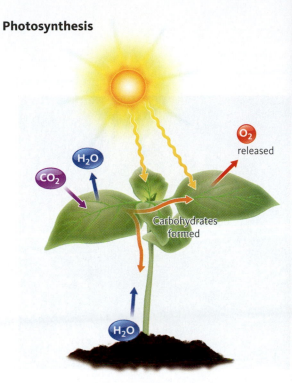

Photo by J. Krueger

Figure 2.20 A brief look at photosynthesis. Photosynthesis is the process of using the sun's energy to convert carbon dioxide to usable carbon molecules. Plants have evolved leaves, which are very elaborate structures for gathering sunlight. (a) Pine needle. (b) Oak leaf.

cause they photosynthesize energy from the sun and the air, plants act as *producers*. Other producers include some species of photosynthetic protists and bacteria. Positioned at the base of all food chains, plants provide energy to consumers. Plants also serve other functions that living organisms rely on. The roots of plants not only stabilize the plant and help it acquire resources, but they also help reduce soil erosion and aid in the development of soils. Plants also provide habitats and food for other organisms. Finally, plants play an important role in helping to balance water and carbon levels on Earth.

One of the largest living species on Earth is the giant redwood (**Fig. 2.21**). These magnificent evergreen trees can grow to over 300 feet tall, and they can live over 1,000 years. Giant redwoods grow best at altitudes above 2,000 feet and are found in a thin strip of land along the Pacific coast of North America. This area provides moderate year-round temperatures, and the winter rains and summer fog provide high levels of moisture. Redwoods, like all plants, rely on decomposers to gain access to nutrients and to recycle them back into the soil so they can be reused after the trees have died.

At the cellular level, plants and animals are different. We'll discuss animal and plant cells in greater detail in Chapters 6 and 12, but we introduce them here to help you understand the basic distinctions between organisms. Animal and plant cells are both eukaryotic and contain many of the same structures such as a nucleus, mitochondria, and an organized membrane system. However, plant cells have evolved some structures that

Figure 2.21 Gymnosperms produce seeds in woody cones. The seeds of the giant redwood are released from cones and are the size of sand.

animal cells don't have (**Fig. 2.22**). In addition to their cell membranes, plant cells have cell walls made of cellulose. This incredibly strong material supports the plant as it branches and grows. Like other living plant cells, each of the redwood's cells also has a structure called a *vacuole,* used for storing water and minerals. These vacuoles are relatively large and often occupy more than 80 percent of the cell's volume. To support photosynthesis, certain plant cells contain chloroplasts, specialized organelles where the solar energy conversion process takes place. In conifers such as redwoods, the chloroplasts are concentrated in the green needles (leaves) where they can intercept sunlight.

Trees have several remarkable adaptations to acquire and move water throughout their huge plant bodies. Mature redwoods move hundreds of gallons of water each day from the soil through their roots, stems, and leaves; much of this water eventually evaporates into the atmosphere. Redwood trees can also meet part of their water needs by converting heavy fog high up in their branches into their own rain showers. Another characteristic that sets plants apart from animals is their limited mobility. As adults, plants are rooted to a single location; therefore they need to grow toward resources such as water, minerals, and light. As the redwoods grow in height, their massive root systems must also continually grow to find and absorb the additional water the trees need. A plant's immobility makes it vulnerable to changing environmental conditions, attacks, or other danger. Over time, cells with unique attributes developed to help plants survive. Specialized cells produce the very thick bark on the outside of the redwood's trunk and help fend off insect attacks and fire damage. The cells inside the tree produce unusual amounts of chemicals, known as tannins, which aid in resisting rot and insects (which is why redwood is so highly prized for decks and fences).

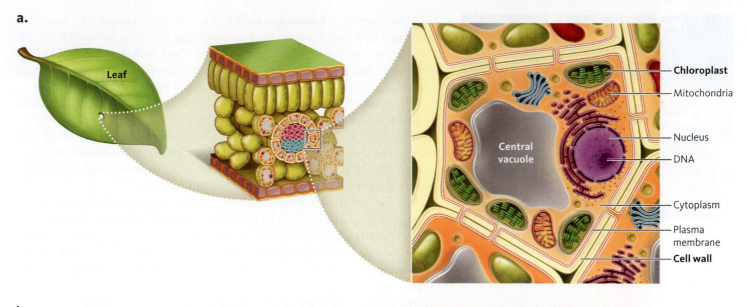

a.

Leaf

Chloroplast

Mitochondria

Nucleus

Central vacuole

DNA

Cytoplasm

Plasma membrane

Cell wall

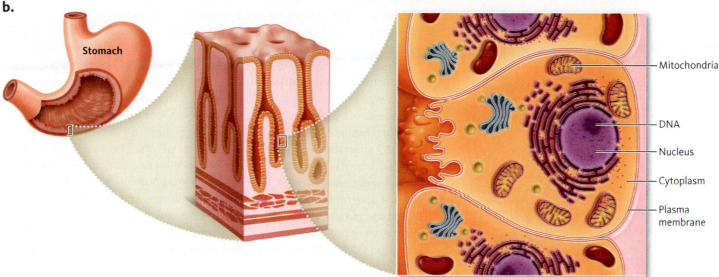

b.

Stomach

Mitochondria

DNA

Nucleus

Cytoplasm

Plasma membrane

Figure 2.22 General comparison of plant and animal cells. (a) Individual plant cell shown is a sugar-making cell surrounding the vein of a leaf. (b) Animal cell shown is an acid-secreting cell lining the human stomach.
© Cengage Learning 2014

Unlike most animals, most plants, including redwoods, can reproduce both sexually and asexually. In sexual reproduction, the male tree produces pollen (containing sperm) that lands on female cones (containing eggs) and begins the process of fertilization. Once fertilized, seeds are produced in woody cones. In total, redwood cones on a mature tree can produce over 6 million seeds in a year! Their asexual reproduction process is less complex and more common—the redwood sprouts a new shoot from its already established root system. The sprouts can grow very quickly by taking advantage of the existing root system to acquire water and nutrients.

Fungi break down living and nonliving materials

You probably don't think about fungi very often, but your relationship with fungi is more extensive than you may realize. Many fungi are used as food, such as the button mushrooms on your pizza or the shitake mushrooms in your stir-fry. Yeasts are used in the production of pastries, cheese, breads, wine, beer, and soy sauce. Some fungi, such as molds, grow on our food and spoil it. Scientists have learned to harness the products of some

Figure 2.23 The reproductive structures of the honey mushroom that we simply refer to as the mushroom.

fungi as sources of medicines, such as the antibiotic penicillin. Chemicals produced by fungi are used in detergents and for other industrial chemicals. There are species of fungi that live in fuel tanks, on camera lenses, and inside human bodies. Fungi also cause diseases in humans, such as irritating athlete's foot, ringworm, and vaginal yeast infections.

The diversity of Kingdom Fungi is comparable to that of the plant and animal kingdoms. Scientists have described over 100,000 species of fungi—with perhaps as many as 1.4 million species still unidentified. Fungi live throughout the world in soils; on dead materials; and as partners and parasites of plants, animals, and even other fungi. Kingdom Fungi is complex, and we'll discuss the details of fungi in Chapter 15.

You can see the mushroom growing in your backyard, but you may not be aware that, like an iceberg on the ocean, most of the fungus is out of sight below the ground. For instance, honey mushrooms live as parasites of woody trees, and because of their ability to grow to an immense size can devastate large areas of forests by depleting the environment of important nutrients. Several incredibly extensive honey mushroom individuals have been found in Washington and Oregon. The largest specimen in Oregon spans over 3.5 miles, crossing three county lines; it is estimated to be 2,400 years old. Honey mushrooms live three feet underground and are rarely seen except in the fall season when the reproductive structures called *fruiting bodies,* or *mushrooms* as we know them, emerge above ground to reproduce (**Fig. 2.23**).

Most fungi are multicellular, but one group, known as *yeast,* lives as single cells. Although plantlike in some characteristics, fungi are not able to photosynthesize or make their own food, nor do they ingest food like animals. Molecular evidence makes it clear that fungi are more closely related to animals on the tree of life than to plants. The Fungi Kingdom is divided into five phyla based on their reproductive structures (**Fig. 2.24**).

Whereas plants build large molecules through photosynthesis, fungi break these molecules down into simpler atoms and molecules. They are champion *decomposers* of the living world and therefore are critical to the cycling of matter in an ecosystem. Decomposers break down the wastes or remains of dead organisms, enabling other organisms to take up the released

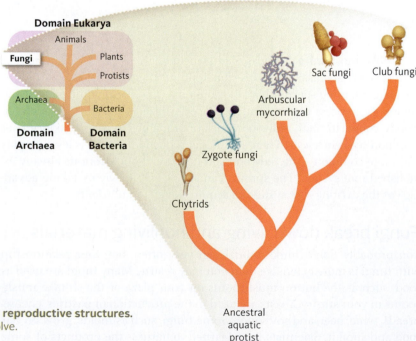

Figure 2.24 Evolutionary tree of fungi based primarily on reproductive structures.
Club fungi are one of the most recent fungal groups to evolve.
© Cengage Learning 2014

a.

b.

© Avico Ltd/Alamy

c.

© WildPictures/Alamy

Dana Richter/Visuals Unlimited, Inc.

Figure 2.25 Like plants and animals, fungi show tremendous diversity. Here are (a) the highly poisonous death cap; (b) the colorful decomposer, many-colored polypore; and (c) the underground structures of a mycorrhizal fungi critical to plant growth.

nutrients. You may be familiar with shelf fungi that grow on decaying wood using the components of wood cells as a source of energy and raw materials to build their bodies (**Fig. 2.25b**). In the process of extracting nutrients from the wood, they break it down and decompose it. The scarlet waxy cap fungus is an important decomposer that is often found living with the giant redwoods. Not only do organisms such as the redwood rely on fungi to recycle the nutrients in dead organic matter, but also without these efficient decomposers, our planet would have quickly filled up with the wastes of the living world long ago. Valuable nutrients would be unavailable for organisms, and our planet's living systems would ultimately grind to a halt.

The honey mushroom and other fungi share some characteristics with plants and in fact were originally classified with the Plant Kingdom. Fungi, like plants are confined to a fixed location and have a branching growth pattern used to seek out and acquire resources. Fungal cells also have vacuoles used for cellular storage, and they produce a diverse set of chemical compounds for defense and ecological adaptation. Like the cells of a redwood tree or corn plant, fungal cells all develop cell walls, although the main structural compound is chitin instead of the plant compound cellulose. Interestingly, the

Figure 2.26 Light and dark photos of *Mycena lucentipes*, a fungal species described in Brazil.

chitin produced by fungi is the same compound found in an insect's hard outer skeleton. Despite these similarities to plants, DNA analysis reveals that fungi are actually closer relatives of the Animal Kingdom.

Like plants, the honey mushroom requires water, oxygen, and a food source to live. As heterotrophs, fungi must obtain their nutrition from other sources. Some are decomposers, degrading and digesting dead plant and animal sources, and others are parasites feeding off living organisms. Fungi are very different from plants in the way they feed. They secrete specialized proteins called *digestive enzymes* into their surrounding environment. The enzymes break down their food externally, and the smaller molecules are absorbed into the fungus' body. Because of the diverse arsenal of digestive enzymes fungi produce, they have more versatility than plants in the great variety of food sources they can consume.

Fungi can reproduce both sexually and asexually. Many fungi release reproductive cells (spores) into the air as a strategy to maximize their success for reproduction. You have probably stepped on a puffball in your yard and observed a "puff" of white emerge. The white cloud contains millions of reproductive spores that germinate into new fungi when environmental conditions become favorable.

Honey fungus and 64 other species of fungi have another curious characteristic that distinguishes them from plants—they glow in the dark (**Fig. 2.26**). The glow is caused by chemical compounds in the fungi reacting with water, oxygen, and other components. The result is a pale yellow-green light. This phenomenon is sometimes called *foxfire,* and it has long been described in historical journals, folklore, and ancient texts. Why would a fungus spend energy to emit light—what could be the evolutionary advantage of this process? One hypothesis is that the chemical reaction may be a defense mechanism. The light attracts the predators of insects and these predators defend the fungus from other animals.

The protists are an assemblage of simple eukaryotic organisms

What do ice cream, marshmallows, cosmetics, and toothpaste all have in common? They all contain products of protists. The highly diverse Protist Kingdom includes organisms like paramecia, amoeba, and algae that live as single-celled or simple multicellular organisms filling a range of ecological roles (**Fig. 2.27**). In fact, this group is so diverse that many taxonomists now consider the Protist Kingdom to be several kingdoms. Historically, classifying protists has been difficult, but as molecular techniques improve, so does our understanding of this kingdom. Approximately 60,000 protist species have been identified, most of which are found in moist environments. Protists represent an important link in early evolution—they evolved from prokaryotes, eventually giving rise to each of the eukaryotic kingdoms.

From a classification standpoint, protists are eukaryotes that have characteristics that don't easily place them into one of the other kingdoms. Many, such as the amoeba, are microscopic single cells that are larger and far more complex than prokaryotes. Other protists grow into tree-sized, multicellular seaweeds such as kelp (see **Fig. 2.27**). In this kingdom, the deductive power of classification is weaker. Simply knowing an organism is a protist is not very informative because in the past protists have been grouped according to their similarities to other kingdoms. For example, there are animal-like protists, plantlike protists, and funguslike protists. Today biologists are using DNA analysis to classify protists and establish accurate evolutionary relationships (**Fig. 2.28**).

a.

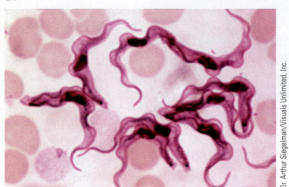

b.

c.

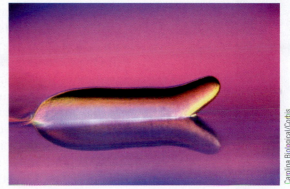

d.

Figure 2.27 Protists are varied. Protists are a "grab bag" of eukaryotic organisms that live in all environments. Some are single-celled, like diatoms (a) and the malaria-causing plasmodia (b); others are multicelled organisms like kelp (c) and slime molds (d).

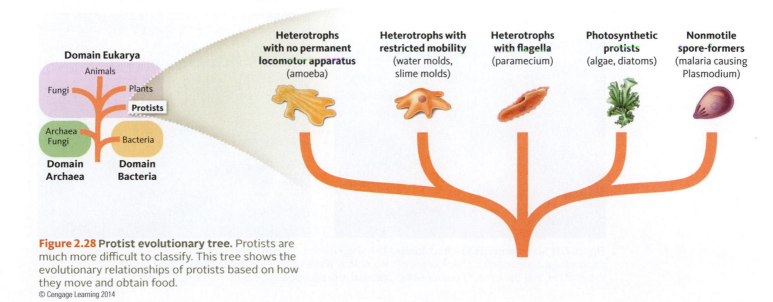

Figure 2.28 Protist evolutionary tree. Protists are much more difficult to classify. This tree shows the evolutionary relationships of protists based on how they move and obtain food.
© Cengage Learning 2014

Figure 2.29 Phytoplankton, a diatom. Diatoms are significant producers in aquatic habitats that serve as the basis of the marine food web.

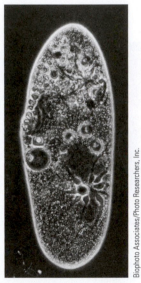

Figure 2.30 A paramecium. Paramecia are often found in a drop of pond water. They use cilia on the outside of their body to chase and sweep bacterial prey into their mouth before digesting it in a food vacuole.

Protists serve a variety of ecological roles as producers, consumers, decomposers, and parasites. Phytoplankton in the ocean is largely made of a group of photosynthetic, plantlike protists called *diatoms* (**Fig. 2.29**). The more than 100,000 species of diatoms are significant producers of organic compounds and oxygen. These microscopic creatures form the base for marine food webs. Interestingly, one particular species of diatom lives on the skin of the blue whale, giving it a yellow-green tint. This protist does not harm the whale, but living on the whale benefits the diatom. Diatoms also have many commercial and industrial uses. They have a glassy cell wall made of the hard mineral called *silica,* used in products such as toothpaste and scouring powders.

Funguslike protists serve as decomposers and parasites. The ones we're most familiar with cause disease. The potato famine in 1845 that affected much of the population of Ireland and Europe was caused by protists known as water molds. These water molds attacked the potatoes, ruined the crops for several years, and led to the Great Hunger that resulted in the Irish migration to the United States. Other water molds you may have heard of cause Sudden Oak Death, killing millions of trees in California, and downy mildew common to many ornamental plants.

Animal-like protists, such as the amoeba and paramecium, play an important ecological role by eating bacteria, plant cells, and algae (**Fig. 2.30**). Once these food sources are absorbed, the energy obtained may be transferred to the next level of the food chain when other organisms consume the amoeba and paramecium. Other protists are parasites to humans and other organisms. *Plasmodium* is a genus of animal-like protists that cause the devastating disease malaria. You will learn more about this disease-causing organism in Chapter 20.

Some protists live in colonies providing some level of specialization, with different cells performing unique functions. For example, *Volvox* is a colonial protist in which some cells are responsible for locomotion, while others are responsible for reproduction (**Fig. 2.31**). Individual cells in *Volvox* colonies work together to acquire food, grow, and reproduce. This adaptation of living in a colony was an important step in the evolution of multicellular organisms.

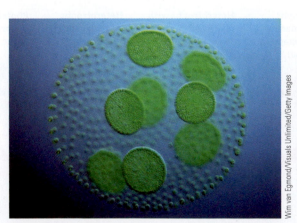

Figure 2.31 Some protists live in colonies. The single-celled *Chlamydomonas* (right) is closely related to the multicellular *Volvox* (left). The *Volvox* lives in a colony made of two cell types—one type is responsible for locomotion, and the other is responsible for sexual reproduction.

2.3 check + apply YOUR UNDERSTANDING

- Life on Earth shows great diversity. The Domain Eukarya includes animals, plants, fungi, and protists.

- Members of each of these kingdoms have a unique combination of characteristics or features, such as their ecological roles, genetics, structural and anatomical differences, and ways of acquiring energy, as well as various adaptations for growth, reproduction, and survival (see **Table 2.3**).

1. List five characteristics that define animals.
2. How do biologists establish evolutionary relationships among protists?
3. Some groups of fungi produce large, specialized aboveground structures (mushrooms). What function do they serve in the life of the fungus?
4. Looking at the evolutionary tree (see **Fig. 2.16**), which groups of animals have backbones?
5. The hydra in **Figure 2.32** is a multicellular organism that lives in an aquatic environment. It does not have cell walls, and it feeds as a consumer. Which kingdoms could it belong to? How could you determine to which kingdom it belongs?

© Natural Visions/Alamy

Figure 2.32 A hydra.

2.4 Bacteria and Archaea Are Prokaryotic Microorganisms

Microbiologists who study the relationships between humans and microbes estimate that we carry over 500 species of bacteria on our skin (even if you shower every day). They also estimate that in the adult human body, bacteria outnumber human cells 10 to 1; in fact, several pounds of your body weight is actually the hitchhiking microbes. Although bacteria that cause human diseases such as dental cavities, food poisoning, and strep throat often get the spotlight, most are harmless or beneficial.

The bacterial species *Escherichia coli (E. coli)* inhabits the intestines of humans, birds, and other mammals—hence the name *coli* for the large intestine, or colon. *E. coli* bacteria satisfy their nutrition requirements by helping us satisfy ours. In other words, the bacteria take some nutrients for themselves while helping us break down our food during digestion. In the process, they produce important vitamins for us.

Bacteria grow and divide rapidly when provided with plenty of nutrients. In fact, *E. coli* are great to use in the science laboratory because they divide every 20 minutes. To build large populations quickly, bacteria reproduce asexually using a process that produces two identical daughter cells. These bacteria remain tiny, enabling nutrients and wastes to efficiently move in and out of each individual cell across the outer cell membrane.

The section ahead gives you a sense of the diversity of prokaryotes, like *E. coli,* that literally live everywhere—from the darkness of the ocean depths to the polar deep freezes. They are found in the air, soil, water, and both on and in other living organisms (**Fig. 2.33**). They live in all ecosystems and play important roles in driving the global ecosystem.

Figure 2.33 Examples of bacteria. (a) A type of *Clostridium* found in the intestines of mammals and in the soil (decomposer). (b) *Rhizobium*, a soil bacteria (decomposer). (c) *Streptococcus* (soil decomposer and human pathogen).

a.

b.

c.

10 µm

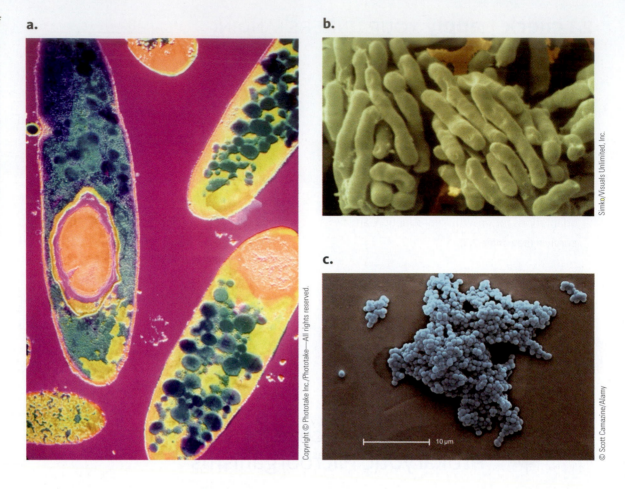

Bacteria and archaea share similar characteristics

All prokaryotic cells are classified into Domain Bacteria or Domain Archaea, and as prokaryotes they have similar structures (**Figs. 2.34** and **2.35**). Prokaryotes are microscopic, single-celled organisms that are smaller and simpler than eukaryotic cells. They are approximately one-tenth the size of a typical eukaryotic cell, and they lack internal membrane-bound organelles (like a nucleus). These unicellular organisms are found in three common shapes—some are ball shaped, others are rod-shaped, and some irregularly shaped like spiral springs and commas. Prokaryotic cells have an internal region called a *nucleoid* that contains their DNA, which is in a circle that forms a single chromosome. Like all cells, they have ribosomes for the production of proteins and an outer plasma membrane. On the outside of the cell membrane, prokaryotes are protected by a cell wall. Many also have a sticky outer capsule, an adaptation that helps them to attach to surfaces. **Figure 2.35** highlights these features.

Prokaryotes reproduce asexually through a process in which one cell grows and splits into two cells. As mentioned with *E. coli,* some species are capable of creating a new generation every 20 minutes. These rapid reproductive rates permit prokaryotes to quickly increase their population when environmental conditions are favorable. This is one reason why bacterial infections can grow and spread so rapidly.

Like all living things, prokaryotes sense their environment. This ability allows them to move toward areas where nutrients are more plentiful and

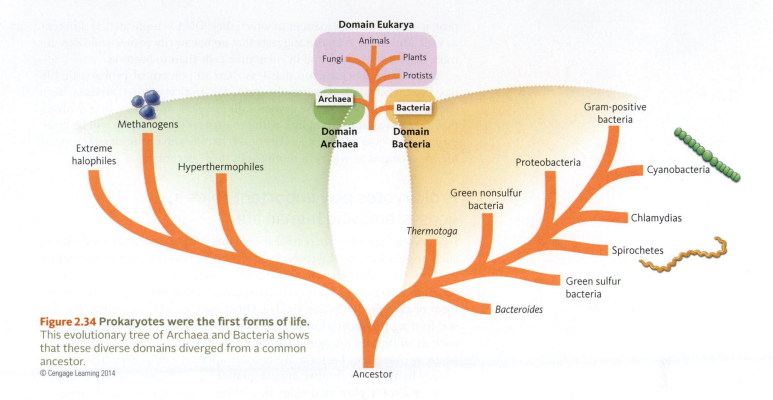

Figure 2.34 Prokaryotes were the first forms of life. This evolutionary tree of Archaea and Bacteria shows that these diverse domains diverged from a common ancestor.
© Cengage Learning 2014

where conditions favor growth. *E. coli* moves by way of numerous thin *flagella* located on its surface (see **Fig. 2.35**). Some prokaryotes require oxygen and navigate toward oxygen sources, whereas others are killed by oxygen and thus move away from it. An interesting feature of *E. coli* is that it can live in either anaerobic or aerobic environments (presence or absence of oxygen). It is able to adjust to either environment by using different metabolic pathways. Photosynthetic bacteria, like cyanobacteria, move toward light, but if the light is too intense they will move away. Some contain structures that act like an internal compass and respond to the Earth's magnetic field.

The first life forms were prokaryotes, and they were adapted to harsh conditions. Today we can find numerous species of bacteria and archaea that live under extremes of pH, temperature, and salt concentrations (see **Fig. 2.9**). Scientists refer to these organisms as "extremophiles," a reference to the fact that they are found in highly acidic and alkaline environments and also in very salty places such as the Dead Sea and the Great Salt Lake in Utah. *Sulfolobus* is an archaean that lives in the hot, muddy, acidic springs found in Yellowstone National Park. The bacteria *Thermus aquaticus* also can be found at Yellowstone. Both of these species thrive in geothermal pools at extremely hot temperatures as high as 180°F.

With so many similarities, you may be wondering why microbiologists place archaea and bacteria into separate domains. While their size, mode of reproduction, and cellular structure are similar, the macromolecules produced by these cells differ. In other words, at the *molecular level* archaea and bacteria are significantly different from one another. In fact, in archaea the proteins and carbohydrates that make up the plasma membrane and cell wall are distinctly different from both bacteria and eukaryotes. In addition, the processes and machinery that convert genes to functional

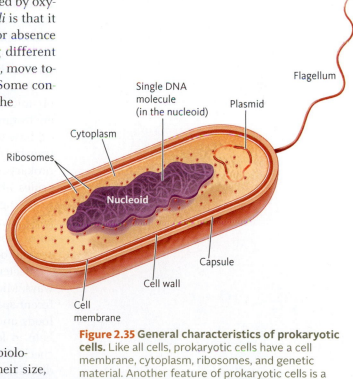

Figure 2.35 General characteristics of prokaryotic cells. Like all cells, prokaryotic cells have a cell membrane, cytoplasm, ribosomes, and genetic material. Another feature of prokaryotic cells is a cell wall.
© Cengage Learning 2014

proteins, as well as the manner in which their DNA is replicated, is different as well. Molecular evidence suggests that archaean ribosomes and DNA are more similar to those found in eukaryotic cells than to bacteria.

Microbiologists estimate that less than 10 percent of prokaryotic life has been classified. Because the evidence indicates that archaea have changed very little over the several billion years since they first evolved, they provide an important link with the earliest life forms. It will be interesting to see how current classification schemes and our understanding of life are changed as we learn more about these organisms.

Prokaryotes play important roles in ecosystems and human life

Prokaryotes are very successful organisms, in part because they exploit such a wide variety of energy and carbon sources for growth and reproduction. Cyanobacteria are a group of autotrophic producers and utilize the sun's energy to convert carbon dioxide to a usable energy source in the form of carbohydrates (see **Fig. 2.6**). Other species, like *Sulfolobus,* are also self-feeding producers, but instead of sunlight, they use chemical sources such as sulfur and nitrogen to acquire the molecules they need to survive. Many archaea and bacteria are heterotrophic decomposers that acquire energy by feeding off other organic matter.

Prokaryotes play vital roles in cycling important chemical elements between organisms and the environment. Both archaea and bacteria are particularly useful organisms in the cycling of carbon, nitrogen, and sulfur between the nonliving and living world by converting these elements into forms that organisms can use. For instance, photosynthetic cyanobacteria in the oceans play a small role in capturing carbon and producing organic compounds and oxygen for other organisms to use. Some prokaryotes, such as nitrogen fixers, live in plant roots and form a mutually beneficial relationship with them by converting atmospheric nitrogen into a chemical form that can be used by the plant. The prokaryotes that decompose organic material act as living recyclers by releasing nutrients back into the environment for use by other organisms. Decomposers also form the critical base of many food chains. At the bottom of the food chain, there is a transfer of molecules rather than a transfer of organisms. So, although prokaryotes do not eat other living organisms, they consume organic molecules like carbohydrates, proteins, and amino acids that cannot be directly ingested by other organisms. When other organisms consume prokaryotes, these organic compounds move up the food chain. As a result of all of these interactions, bacteria and archaea serve a wide variety of essential roles in maintaining global ecosystems.

Bacteria have a direct impact on humans in both helpful and harmful ways. Microbiologists think that in total there are as many as 100,000 different species of bacteria living inside the human body, helping us digest foods and make important vitamins. The bacteria that live on our skin help to lower the body's pH, keeping other disease-causing bacteria in check. Although most bacteria are harmless, some cause serious human diseases. Bacteria that metabolize food caught between our teeth cause the most common disease in humans, tooth decay. As a by-product of their metabolic activities, these bacteria secrete acids that break down our tooth enamel. Now you can think of brushing and flossing your teeth as depriving oral bacteria of a free meal! Other common bacterial diseases include pneumonia, tuberculosis, syphilis, *Chlamydia,* and Lyme disease. The role of archaea in humans is less understood.

Bacteria are essential to our modern lives, used in the commercial production of a variety of food products and medicines. They are used to make wine, cheese, yogurt, and soy sauce. As defense mechanisms, bacteria ward off other bacteria by releasing chemicals called *antibiotics*. Actinomycin, streptothricin, and neomycin are examples of antibiotics produced by a soil bacterium known as *Streptomyces*. The discovery of these bacterial antibiotics in the twentieth century has dramatically changed disease control for humans.

2.4 check + apply YOUR UNDERSTANDING

- The first life forms were prokaryotes. Prokaryotes include bacteria and archaea and are found in all ecosystems, playing important roles in driving the global ecosystem.

- Prokaryotic cells are classified into Domain Bacteria or Domain Archaea. Prokaryotes are microscopic, single-celled organisms that are smaller and simpler than eukaryotic cells. They lack internal membrane-bound organelles (like a nucleus) and reproduce asexually. At the molecular level, archaea and bacteria are significantly different from one another.

- Prokaryotes exploit a wide variety of energy and carbon sources for growth and reproduction. Prokaryotes play vital roles in cycling important chemical elements between organisms and the environment.

1. Prokaryotes play vital roles in cycling important chemical elements between organisms and the environment. Give an example from the text of how prokaryotes cycle carbon.

2. What important roles do bacteria play in our modern lives?

3. List four characteristics that archaeans share with bacteria and two that place them in a different kingdom.

4. Methanogens are archaeans that live in the stomachs of cows and help to digest cellulose. Would you expect to find archaeans living in the stomachs of other herbivores (like deer or sheep)? Or do you suspect the conditions are too extreme for archaeans to live in this environment?

5. How would you differentiate a single-celled plantlike protist, like algae, from cyanobacteria?

End of Chapter Review

Self-Quiz on Key Concepts

Evolutionary Relationships and the Tree of Life

KEY CONCEPTS: An understanding of evolutionary relationships is the foundation to understanding life and its history on Earth. Evolutionary trees summarize evolutionary relationships. The more closely related two organisms are, the more similar their characteristics.

1. Match each of the following terms related to **evolutionary trees** with its characteristic, description, or example.

    lineage
    species
    shared characteristics
    evolutionary tree

    a. evolutionary novelties or traits that groups of animals have in common
    b. ancestor–descendant relationship
    c. diagram showing evolutionary relationships
    d. group of organisms that are reproductively compatible

2. Which group of organisms shares the most recent common ancestor with plants, fungi, and animals?
   a. bacteria
   b. archaea
   c. protists
   d. viruses

3. The evolutionary history of an organism is most often determined through:
   a. anatomical structures.
   b. molecular data.
   c. developmental patterns.
   d. environmental adaptations.

Biodiversity and Organizing Life

KEY CONCEPTS: Life on Earth shows great diversity. Millions of different kinds of organisms, or species, have appeared and disappeared over the history of the Earth. Biologists use taxonomy to organize life based on similar characteristics. Evolutionary biologists use this information to establish evolutionary history. The domains and kingdoms of life group organisms together by structure, evolutionary history, ecological role, and behavioral patterns. Species are also grouped based on shared traits and evidence of descent from a common ancestor.

4. Match each of the following **groups of organisms** with their ecological roles on the right. Each term on the left may match more than one role on the right.

   protists
   fungi
   plants
   animals
   prokaryotes
   autotrophs
   heterotrophs

   a. producers that use photosynthesis to make organic compounds
   b. consumers that ingest nutrients from other organisms
   c. decomposers that break down and absorb living and nonliving things for nutrients

5. Protists have traditionally been subdivided according to their resemblance to which kingdoms?
   a. animals
   b. plants
   c. fungi
   d. animals and plants
   e. animals, plants, and fungi

6. Yeast, athlete's foot, blue cheese, and antibiotics are all associated with which kingdom?
   a. animal
   b. plant
   c. fungi
   d. bacteria

Applying the Concepts

7. The ongoing application of DNA comparisons among organisms is further refining our understanding of evolutionary relationships. Speculate what might happen with further investigations among such diverse groups as the protists over the next 20 years. Give reasons for your thinking.

8. **Figure 2.36** shows two wolf species. Do you think that these two wolves are closely related? Is looking alike the strongest evidence of relatedness? Why aren't anatomical characteristics the most reliable or accurate criteria for classifying an organism?

Figure 2.36 Two wolf species.

9. In learning about *Archaeopteryx,* we explored the importance of studying the characteristics of transitional organisms in the fossil record. Understanding that aquatic mammals like blue whales and dolphins evolved from animals that lived on land, what timeline within the fossil-bearing rocks would you search for transitional animals, and what transitional characteristics would you look for?

10. Exploration of the oceans' deepest layers is an ongoing and exciting area of research for marine scientists seeking new species of life. Recently, a new single-celled organism was discovered near hydrothermal vents at 12,500-foot depths where the vent temperatures neared the boiling point. What specific characteristics would you sample for to taxonomically place this new organism in one of the three domains of life?

Data Analysis

Today scientists accept that dinosaurs are a single group of animals that shared a unique set of anatomical features. For example, all dinosaurs had jaw muscles that extended onto the roof of the skull; a large muscle ridge on the thigh bone; and well-developed bony structures at the hip, knee, and ankle. Scientists use these shared characteristics to classify dinosaurs because they are *always* present in dinosaurs and they are *only* present in dinosaurs. Such shared characteristics are important for classifying organisms; they help to identify the development of novel structures in evolutionary history. In this case, these features arose in the ancestor of dinosaurs and were inherited by all its descendants, including modern birds. Interpret **Figure 2.37** to answer the following questions.

11. According to the evolutionary diagram, birds are most closely related to which group of dinosaurs?

12. To which group of dinosaurs are birds most distantly related?

Critical Thinking

13. Circle the most recent shared ancestor of sauropods and birds, and of horned and armored dinosaurs. What evolutionary novelty separates these two lineages?

14. What type of hip structure do modern birds have? What type of hips did armored dinosaurs such as the *Stegosaurus* have?

15. From this diagram, can you tell anything about the hip structure of crocodiles and their close relatives?

16. Are sauropods more closely related to armored dinosaurs or birds?

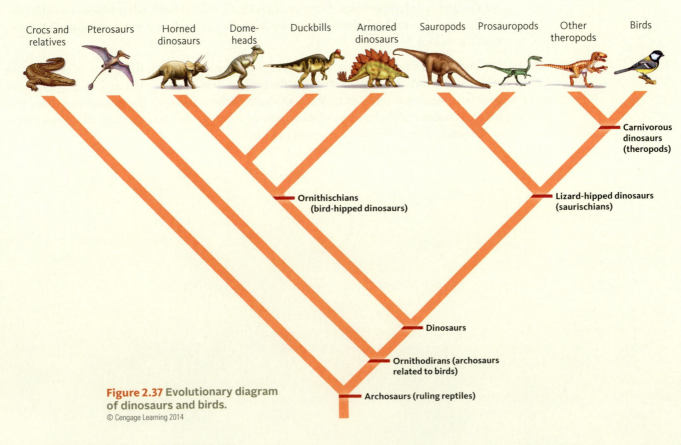

Figure 2.37 Evolutionary diagram of dinosaurs and birds.
© Cengage Learning 2014

Question Generator

Evolutionary Relationships of Carnivorous Plants

The aquatic waterwheel is a carnivorous plant whose leaves are arranged in a *whorl,* or a "wheel" around its central stem (**Fig. 2.38**). It has characteristics that are very similar to three other plants that also capture and digest small animals. First, it lives in freshwater lakes and ponds in Europe, Africa, India, and Australia. This habitat and geographical distribution are similar to those of the submerged carnivorous bladderworts. Second, the waterwheel has a set of modified and specialized leaves that act as a "snap-trap" mechanism that resembles the Venus flytrap. Third, waterwheel flower structures are remarkably similar to the flowers of sundew plants. Because of these characteristics, the waterwheel was placed in its own family and genus.

Below is a block diagram showing the three different characteristics and the other carnivorous plants they resemble (**Fig. 2.39**). Use the background information along with the block diagram to ask questions and generate hypotheses. We've translated the components and interaction arrows shown in the diagram into a question and provided a couple more to get you started.

Figure 2.38 Waterwheel plant. The carnivorous waterwheel plant uses its whorls of leaves to capture tadpoles, insect larvae, and snails.

Here are a couple of questions to start you off.

1. Does a common geographical distribution indicate a common center of origin where a common ancestor lived? (Arrow #1)
2. Does a similar snap-trap structure indicate a strong evolutionary relationship? Are they homologous structures? (Arrow #2)
3. To what degree are the flowers similar? Similar shape, numbers of parts, anatomy? (Arrow #3)

Use the block diagram to generate your own research question and frame a hypothesis.

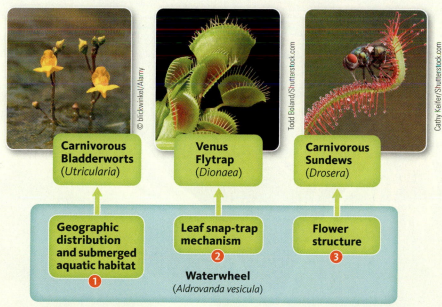

Figure 2.39 Factors relating to the characteristics of carnivorous plants.

3 Evolutionary Change and Adaptation

3.1 Warning coloration of passion-vine butterflies

3.2 Evolutionary change begins with genetic variation

3.3 Natural selection shapes evolutionary change, adaptation, and fitness

3.4 New species arise through the process of speciation

3.5 Each species has a unique evolutionary history

3.1 Warning Coloration of Passion-Vine Butterflies

Figure 3.1 Geographic range of *Heliconius sapho.*
This species is one of many species of passion-vine butterflies that display warning coloration.
© Cengage Learning 2014

Passion-vine butterflies that live in the understory of rainforests of Central and South America (**Fig. 3.1**) have a challenging life. In less than six months, they must navigate their way through the rainforest to find a jungle cucumber flower that is their food source, find a mate, and avoid predators. One such predator, the rufous-tailed jacamar, uses its long, sharp, forceps-like bill to capture the showy, slow-moving butterflies. Butterflies also are on the menu for tropical lizards, which forage on the ground during the day when butterflies are active. Whether they are in the air or on the ground, populations of butterflies are under tremendous pressure from predation—a strong selective force of evolutionary change.

From a predator's perspective, not all butterflies are equally desirable sources of food. In fact, some butterflies are to be avoided altogether because they are foul tasting or even harmful. Their bodies contain bitter or toxic chemical compounds that cause birds to quickly stop eating them. In some cases, the chemical compounds cause intense vomiting. Early biologists observed that the bitter-tasting butterflies also displayed distinctive, colorful wing patterns. This is the case with the passion-vine butterflies—all 45 species exhibit a spectacular diversity of eye-catching wing patterns. The phenomenon of "advertising" an animal's toxic quality to potential predators is called *warning coloration,* and it occurs across the animal kingdom from bright poison-dart frogs to coral snakes to dragonflies and other insects.

The evolutionary story of passion-vine butterflies begins with their highly specialized relationship with a single family of plants—the passion vines (**Fig. 3.2**). Female butterflies lay their eggs on a select few of the more than 500 species of passion vines. For example, a species of passion-vine butterfly called the Holstein butterfly *(Heliconius sapho),* lays its eggs on the young leaves of just one species of passion vine (**Fig. 3.2**). The eggs then hatch into caterpillars that feed on the young leaves. The leaves contain chemicals such as cyanide that are toxic to most species of caterpillars. However, a change in the butterfly's genetic information enabled passion-vine caterpillars to metabolize the toxin, complete their life cycle, and pass the cyanide-resistant genes to the next generation. In addition to cracking the defenses of the plant to exploit it as a source of food, the caterpillars retained the plant's toxins in their body as they became transformed into butterflies, making them bitter tasting to their predators. The caterpillars and butterflies with the genetic mutation survived in greater numbers and thus caused shifts in the relationship between the butterflies, their host plant, and their predators.

Let's now connect bitterness to the evolution of bright wing patterns and coloration. Which of the two traits evolved first? The bright coloration would make it easier to identify a mate of the same species, but it would have the unwelcome catch of making the butterflies easier to see by predators. Therefore it is likely that this trait evolved *after* the insect became toxic, in which case, having danger signs for wings could be the whole point. Biologists hypothesize that bright, colorful warning signals make the nauseating butterflies easier for their predators to remember than if they were camouflaged, or blended into the background. Evidence supporting this idea comes from experiments that show that birds and lizards learn from a few bad experiences to identify and avoid the noxious species of passion-vine butterflies based solely on their wing pattern. Possession

Passion-vine butterfly life cycle
These forest-dwelling butterflies are predominantly black with brilliant blue, red, orange, yellow, or white warning patterns that signal their distastefulness to predators.

Adult butterflies use their large compound eyes and antennae to find cucumber flowers where they, unlike many butterflies, eat protein-rich pollen. Adults have a distinctive fluttering flight pattern as they fly from cucumber flowers to lay their eggs on passion-vine leaves.

At maturity, caterpillars stop feeding and are transformed into a brown twisted **pupa**. Males of many passion-vine butterfly species mate with females by depositing a sperm packet as soon as they emerge from the pupa and become an adult butterfly.

Eggs are continuously laid on young leaves of the passion-vine plant and fertilized by the sperm packet carried by the female. Each egg has an opening through which sperm penetrate and fertilize the egg. The eggs mature at a slow steady pace.

Larvae (caterpillars) are protected by long spines. They have a mouth with strong jaws that easily chew nutritious leaves.

of a particular bright colorful wing pattern reduced the owners' chances of being devoured. Once predators learned to avoid specific wing patterns, passion-vine butterflies with other wing patterns were eaten more frequently and thus selected against in the population. As a result of natural selection, the bitter butterflies increased in the population, while nontoxic ones decreased—unless of course they adapted in other ways.

Another amazing phenomenon associated with warning coloration in butterflies occurs when two different species display the identical wing pattern, or when one butterfly mimics the other. In this case both species are toxic and bitter tasting. **Figure 3.3** shows the *mimicry* between two different passion-vine species that live in the same habitat in Ecuador. As you can see, except for a slight difference in shape, these two species take advantage of the identical protective pattern and its warning signal to poten-

Figure 3.2 Passion-vine butterflies are one of the longest living butterflies. Adults can live up to six months and sometimes longer, while most butterflies live only two weeks to two months. Adult butterflies rely on various species of cucumber flowers for pollen, and larvae depend on passion-vine leaves for food.
© Cengage Learning 2014

Heliconius sapho   **Heliconius cydno**

Figure 3.3 Wing coloration of *Heliconius cydno*. Another passion-vine butterfly species, *Heliconius cydno*, mimics the wing coloration and pattern of *Heliconius sapho*. These toxic species have overlapping geographic ranges in Central and South America.
© Cengage Learning 2014

tial predators. There are many cases where chance has favored more recently evolved nontoxic species with a warning pattern similar to that of a toxic species. Any predator with a memory would not want to risk eating anything it associated with a nasty experience.

In this chapter we will describe the forces of evolutionary change that have produced the immense diversity and adaptations of organisms. These forces act in a two-step process that first generates genetic variation and then eliminates those organisms that are less adapted to the living and physical parts of their environment. In the case of passion-vine butterflies, these forces play out in the ravenous caterpillars, plant toxins that become butterfly toxins, predator aversion to specific wing patterns, the deception of predators by mimicry, and, for the time being, evolutionary success for the butterflies.

3.2 Evolutionary Change Begins with Genetic Variation

Individual organisms change and acquire new abilities or characteristics throughout their lives. For example, humans can make major increases in their muscle mass and strength through lifting weights and exercise. They may also have laser surgery to correct their vision. Do these changes contribute to evolutionary changes in the population? Although exercise affects longevity and survival, these *acquired* changes in *body* cells (the bigger muscles) are not passed to the next generation and do not necessarily make the next generation more adapted to survive and reproduce. Only *inherited* abilities, or changes in the genetic information contained in cells that combine to make the next generation (sperm, eggs, spores), contributes to adaptation and evolutionary change. Organisms inherit the genetic information for adaptation, which is any feature that enables an organism to increase its ability to survive and reproduce in its environment.

Evolution is characterized by changes to the genetic composition of a population from generation to generation. The population of purple lupines shown in **Figure 3.4** is actively reproducing and evolving through time. Under favorable conditions, lupines reproduce rapidly to form meadows and fields blazing with a brilliant purple color. On a much smaller scale, viruses undergo genetic change within their host. The change has enabled some viruses to jump from a nonhuman host and infect humans. In the case of the bird and swine influenza viruses, the people who lived in close contact with birds or swine were the first to become infected with the new variants. Lupines and viruses exemplify the two-step process of evolution: (1) the generation of *genetic variation*, the processes that cause genetic changes, which then lead to greater variation in a population, and (2) the *selection* of those individuals that will survive and reproduce successfully, a process called natural selection. The section ahead introduces you to the processes at the genetic and cellular levels of the hierarchy that create a genetically diverse population—the first step in evolution and adaptation.

Figure 3.4 An example of genetic variation. The flowers displayed by this population of purple wildflowers produce genetic variation through sexual reproduction.

Shutterstock.com/Cucumber Images

Adaptation—feature of an organism that enables it to survive and reproduce in its environment.

Evolution—process that causes changes to the genetic composition of a population from generation to generation.

Natural selection—a process of favoring adapted individuals for survival and successful reproduction.

Populations contain individuals with different forms of genetic information

Let's look again at the purple lupines in **Figure 3.4**. Although they appear similar, each individual plant differs slightly in its height, the size of its leaves, and the number of flowers it produces. In other words, each purple lupine has a unique set of genetic information in its genome. If you compare the genome of one individual to another, you will find that some genes are identical in their sequences of bases. Genes that encode the assembly of cell walls or chloroplasts, for example, are the same. However, some genes may occur in different forms called alleles. The genes that encode information on flower color may exist in a form that causes a light purple color, whereas another allele will result in a darker purple flower. For the animals and plants around you on the hiking trail, each gene is located on two chromosomes and thus occurs as two alleles. The population, therefore, consists of individuals that all contain genomes with a different combination of alleles—the population has genetic variation. How does genetic variation arise?

Sexual reproduction is a powerful source of new genetic combinations

Organisms that reproduce more offspring more frequently transmit their adaptive gene combinations to their offspring in the next generation. Sexual reproduction generates genetic variation, whereas asexual reproduction has the opposite effect in a population—it reduces genetic variation. Let's look at how the sexual reproductive process produces variation as chromosomes are passed on by each parent to their offspring.

In eukaryotic organisms (animals, plants, fungi, and protists), sexual reproduction involves two important processes—meiosis and fertilization. In animals, meiosis is a reproductive process that produces sperm and egg cells. During meiosis, a cell in the testis of a male or the ovary of a female reduces the number of chromosomes in half, rearranges its information to create new combinations of alleles, and then splits twice to produce four cells (**Fig. 3.5**). This is analogous to shuffling a deck of cards before dealing each person a new hand. As a result of this process, each sperm and egg cell contains a unique set of genetic information. In plants and fungi, meiosis produces genetically unique spores that eventually develop into new offspring. The products of meiosis—spores, sperm, and egg cells—drive the evolutionary process.

Fertilization is the initiation of sexual reproduction, and it occurs when the male gamete (sperm) comes together with the female gamete (egg). When different combinations of alleles from each parent unite through fertilization, the resulting offspring are genetically unique combinations of traits. In terms of your family, you may appear similar to your brother or sister, but unless you are identical twins you are not physically or genetically identical. Even though you and your sibling possess the same genes as your parents, each of you has inherited different combinations of traits (alleles). The same is true of seeds produced from the flowers on a plant. Each seed is a product of fertilization and thus represents new variations on which natural selection can operate.

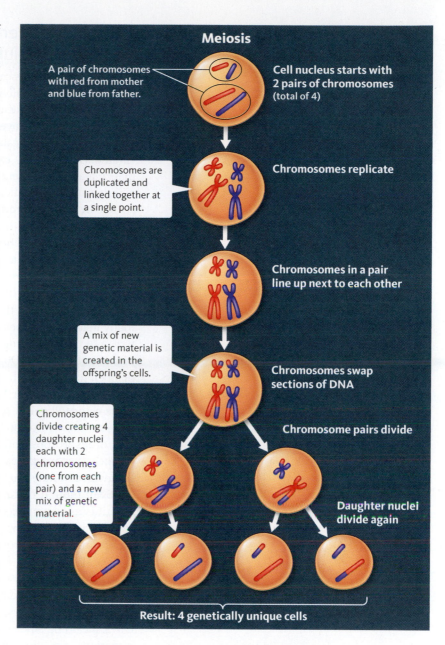

Meiosis

A pair of chromosomes with red from mother and blue from father.

Cell nucleus starts with 2 pairs of chromosomes (total of 4)

Chromosomes are duplicated and linked together at a single point.

Chromosomes replicate

Chromosomes in a pair line up next to each other

A mix of new genetic material is created in the offspring's cells.

Chromosomes swap sections of DNA

Chromosomes divide creating 4 daughter nuclei each with 2 chromosomes (one from each pair) and a new mix of genetic material.

Chromosome pairs divide

Daughter nuclei divide again

Result: 4 genetically unique cells

Figure 3.5 The stages of meiosis. Meiosis is the genetic process that first doubles the chromosomes, exchanges segments, and then assorts them into four different cells. The result is four genetically unique cells, which may become eggs, sperm, or spores.
© Cengage Learning 2014

Alleles—different forms of a gene that bring about particular traits.

Meiosis—reproductive process in specialized cells that reduces the number of chromosomes in half, rearranges its information to create new combinations of alleles, and then splits twice to produce four cells.

Fertilization—the event when the male gamete (sperm) unites with the female gamete (egg).

Mutations generate new variation and shape evolutionary change

During meiosis, genes can be mistakenly rearranged, moved from one place to another, or deleted entirely from the genome. Genes can also be mistakenly duplicated. Molecular analysis has shown that gene duplications are common and have played a major role in evolutionary change. When genes are rearranged, duplicated, or deleted, genetic combinations arise from the variation that already exists. However, there is a potent evolutionary force that creates new sequences of bases in the DNA of a cell and therefore creates new genetic information, new alleles, and new variation—collectively any change in the DNA sequence is called a **mutation**.

Small changes in the DNA can produce major changes that shape evolutionary success. Mutations can alter how genes function—a mutation may turn off master genetic "switches" to create entirely novel structures such as limbs, fingers, or fins. Mutations in passion-vine butterfly caterpillars allowed them to resist the toxins in passion-vine plants. Genetic mutations are often responsible for the activation or deactivation of genes. Humans have many inactivated genes in our DNA. For example, we carry about 900 genes that could help us recognize different odors, but only about half of them are active and working. Dogs, which are much more dependent on their sense of smell for survival than humans, have about the same number of olfactory genes as humans; far fewer are silenced by mutations, however. Some of these activated genes contribute to the ability of dogs to detect odor molecules at a much lower concentration. We have used this ability in dogs to find drugs, bombs, and even certain kinds of cancers. The hundreds of duplicate olfactory genes present in the human and the dog illustrate how some have mutated to become more or less functional through time.

Even though mutations have played a major role in evolutionary success, it is important to emphasize that mutations occur by accident and are not directed in any way toward making organisms more adapted to their environment. They are not produced to fulfill the needs of an organism.

Do all mutations have a positive effect on success? No, in fact most changes to DNA have no effect at all on the organism because they occur in regions of the DNA that do not code for proteins. Mutations may also have negative consequences; some cause a number of deadly human genetic diseases, including sickle cell anemia and Tay-Sachs disease. In most cases these mutations are eliminated by selection, but many persist in populations because of how the mutated genes interact with other alleles. For example, the majority of mammals can make vitamin C, but a mutation in primates, which include humans, has blocked our ability to do so. We must include sources of vitamin C in our diet. On the other hand, sometimes a mutation will be adaptive. If the adaptive mutation is passed to subsequent generations, its frequency will increase due to natural selection, and evolution will have occurred.

Not all genetic changes result in a change in phenotype or fitness

Variation is the raw material from which patterns of change emerge in populations of organisms. However, it is important to stress that not all changes at the genetic level translate into the outward appearance of a trait, or the **phenotype**, at larger levels of organization—the functioning organism.

Looking at the mutations that allow passion-vine caterpillars to resist plant toxins, it is easy to see how this change has positively influenced the insect's **fitness**, or ability to survive and reproduce in its environment. But in most

Mutation—a change in a DNA sequence.

Phenotype—the observable characteristics of a trait.

Fitness—the ability of an organism to survive and reproduce in its environment.

cases, it is difficult to predict how any particular genetic change will play out in nature. The fate of a particular genetic change depends on the type of change, the degree of change, where the change occurs, and the function of the gene that is altered. For instance, a large genetic alteration may occur in segments of DNA that don't control other genes or code for functional molecules such as proteins. This change would then not be translated into cellular, tissue, organ, and organismal change; it is considered neutral. On the other hand, small changes in genes that code for proteins may lead to large changes in the molecule's structure or function and affect fitness, like those for detecting a wide variety of odors. These changes are passed to offspring and descend with the lineage to successive generations. The accumulation of positive, neutral, and negative changes over an extremely long period of time creates the diverse collection of genes in an organism's genome.

Migration and dispersal add genetic variation to a population

The migration or movement of gametes, spores, or individuals into or out of a population is another source of genetic variation. These reproductive structures carry new alleles that in turn add variation to the receiving population. The movement of alleles among different populations is a process called **gene flow**. One of the most spectacular examples of gene flow occurred in the 1950s when beekeepers brought aggressive African honeybees to South America. The African bees interbred with the native bee population and passed on alleles for aggressiveness. The so-called Africanized bee was born. It is important to note that the bees already carried this gene, but the form of the gene that expressed aggression is what was introduced into the new population.

Since fungi and plants are confined to a single location, how does gene flow work for them? They move new alleles through the dispersal of their spores, pollen, and seeds. Pollinators such as the passion-vine butterflies move pollen between genetically different passionflowers, resulting in genetic variation in the new seeds. In some cases these structures move great distances by wind and water currents. Primates and other animals play a crucial role in dispersing the seeds of fruit trees in forests (**Fig. 3.6**). Research has found that individual monkeys and chimpanzees in an African forest move 140 seeds per acre in a single day—this adds up to thousands of seeds and their unique combinations of alleles moved each day.

Historically, as Northern Europeans migrated and mixed with other populations, they moved alleles for digesting lactose sugars in dairy products across the globe. Today, humans move our genomes great distances in short periods of time—we can fly through air travel and carry our alleles with us.

Figure 3.6 Animals are agents of seed dispersal. Blue monkeys gather fruits, eat the pulp, and spit out or eliminate seeds throughout the forest canopy. This makes them important agents for migration and gene flow in tree populations of the rainforest.

Bernard Castelein/ANP Photo/age fotostock

Gene flow—the movement (also called *migration* or *dispersal*) of alleles between different populations.

- Evolution is marked by changes to the genetic composition of a population from generation to generation.

- Evolution is a two-step process that begins with processes that generate genetic variation and the selection of those individuals that survive and reproduce successfully.

- Genetic variation arises through sexual reproduction (meiosis and fertilization), the mutation of genetic information, and the dispersal or migration of individuals in and out of populations.

1. How does genetic variation drive evolutionary change?
2. Describe three processes by which sexual reproduction produces genetic variation in offspring.
3. Give two reasons why genetic mutations are such a potent force in shaping evolutionary change.
4. Bottlenose dolphins live underwater and need to surface to breathe air. They are descended from air-breathing land mammals. Predict whether you would find inactivated genes to smell airborne odors in their genome. Defend your prediction.
5. Small spiders can travel from one location to another by a process called *ballooning*. To lift off of a surface, a spider climbs to a high point, casts out silk threads to catch a breeze, and "flies" away (**Fig. 3.7**). Spider ballooning illustrates which evolutionary process presented in this section? How does ballooning influence genetic variation in spider populations?

Courtesy of Rothamsted Research

Figure 3.7 Spiders "ballooning."

3.3 Natural Selection Shapes Evolutionary Change, Adaptation, and Fitness

Whether the organism is a butterfly, passion-vine plant, predatory bird, or human, individual organisms develop and change throughout their lifetimes—they respond and adjust to their environment. However, the changes taking place in an *indivdual* are caused by different processes than those that bring about *evolutionary* change. Evolutionary processes occur at the population level and are selective—individuals, often with slight differences, are selected for or against. The phrase "selected for" means that individuals with greater fitness have a particular set of abilities or characteristics that are more likely to reproduce and pass those abilities to the next generation than those indivuals that are "selected against," or have lower fitness. As a result, traits that are selected for are in higher frequency in the next generation than those selected against. As the genetic composition of the population changes, it evolves.

To illustrate how a population of wild mustard plants evolves to become better adapted to its environment, let's start with a single seed and its "struggle" to survive in an abandoned farm field in spring. The seed has been dispersed several years ago and has remained dormant all this time. As a product of sexual reproduction, it has a unique set of genetic traits that it inherited from its parents such as its rapid growth rate, ability to reproduce quickly, hard seed coat, food reserves, and bitter-tasting defensive compounds. Some of these traits are *adaptive* because they increase the mustard's ability to survive and reproduce in the field at this particular time. If the environment changes, the trait may no longer be adaptive.

Each of the thousands of mustard seeds varies in these capabilities. The first step in evolution, or the generation of genetic variation, sets the stage for the second step in the evolutionary process, one that is essentially a process of elimination by natural selection. How does it work?

The soil environment poses a collection of conditions, or selection pressures, on the seed—temperature, concentration of oxygen, amount of moisture, and light—that it requires to sprout (Fig. 3.8). Other selection pressures include competition with other plants for the same resources and predation from animals or penetration by fungi that will eat or decompose the seed. Will the seed germinate before its food reserves run out? Will it avoid detection, grow to the surface, and then develop into a seedling? Will it mature, flower, and attract pollinators? The answers to these questions depend on two things—the seed's luck and the interaction between the environment and the seed's inherited capabilities. To what extent is the mustard seed in the field adapted to the prevailing environment in which it finds itself? The section ahead examines how the environment eliminates some members of the population by the process of natural selection and therefore shapes adaptation. In the background is the process of chance—the genetically superior seed may just not be that lucky.

Selection pressures, requirements, and capabilities change with the life stage of the organism

Let's return to the wild mustard seed as it passes through the different stages of its life. At each stage, whether it's a seed, seedling, juvenile, or mature flowering plant, a whole new set of selection pressures comes into play (Fig. 3.9). This is evident by differences in the death rate at the various

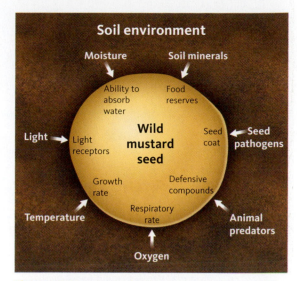

Figure 3.8 Genotypes determine traits and capabilities. A wild mustard seed inherits a genotype that gives it specific traits and capabilities. It is under strong selection pressures from the soil environment to absorb water, minerals, and oxygen. If successful, it will germinate, establish a root system, grow, and emerge from the soil environment into an environment with different selection pressures above the ground.
© Cengage Learning 2014

Selection pressures—conditions that exert natural selection on an organism at its particular stage of life.

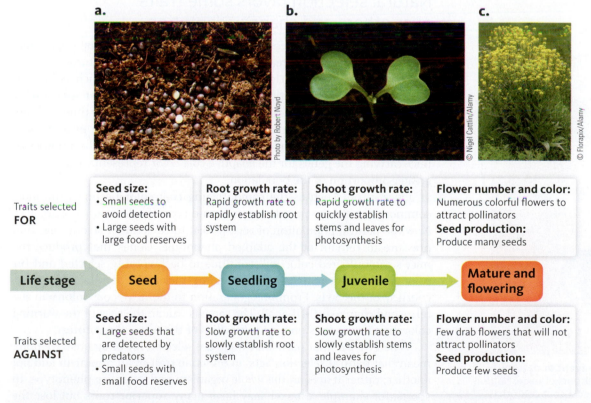

	Seed size:	Root growth rate:	Shoot growth rate:	Flower number and color:
Traits selected FOR	• Small seeds to avoid detection • Large seeds with large food reserves	Rapid growth rate to rapidly establish root system	Rapid growth rate to quickly establish stems and leaves for photosynthesis	Numerous colorful flowers to attract pollinators **Seed production:** Produce many seeds
Life stage	Seed	Seedling	Juvenile	Mature and flowering
Traits selected AGAINST	Seed size: • Large seeds that are detected by predators • Small seeds with small food reserves	Root growth rate: Slow growth rate to slowly establish root system	Shoot growth rate: Slow growth rate to slowly establish stems and leaves for photosynthesis	Flower number and color: Few drab flowers that will not attract pollinators **Seed production:** Produce few seeds

Figure 3.9 Life stages of a wild mustard plant. Each stage in the life of a wild mustard plant is marked by different and opposing selection pressures that shape characteristics such as seed size, growth rates, flower number, and color.
© Cengage Learning 2014

stages. For example, many seeds may germinate only to be eliminated at the seedling stage of life.

In the seed stage, large seeds would stand out and be detected by seed-eating birds more frequently than smaller seeds. However, larger seeds contain a larger amount of stored food that allow the seed to grow stronger and survive deeper burial in the soil.

In the seedling stage, rapid growth is favored so the plant can establish its root system and absorb water necessary for extending the shoot, growing leaves, and performing photosynthesis. The faster growing juveniles are more likely to become mature and flower. Finally, the flower is under strong selection pressures to be pollinated for the development of seeds, and the life cycle continues in a new environment.

Each life stage is marked by different processes and different requirements for light, water, and space. Seeds require less water than seedlings and thus are less susceptible to drought. Seeds that germinate later in the summer when conditions are drier have a lower chance of survival than those that get out of the starting gate early in the season. However, a seed that germinates early is more vulnerable to a late spring frost. On the population level, those mustard plants with traits that balance these pressures are selected for and are more likely to pass these traits to the next generation. Here natural selection acts in a very directed way, gradually eliminating less fit plants at every stage of their life cycle. Because those individuals that are most fit and best adapted to the current environment survive, the genetic composition of the population of wild mustards changes with each generation—the population becomes more adapted as it evolves.

Natural selection favors some traits over others and shapes adaptation

Now let's look at the process of natural selection in an animal population. Rock pocket mice illustrate the two common ways that natural selection places pressure on a trait (**Fig. 3.10**). First, an inherited trait, such as the coat color, may change in a particular direction through consistent selection for that trait. In this case, predation is the agent of selection. Over time, owls are more likely to detect and remove from the population those mice that are not effectively camouflaged against the rocks. Eventually, the number of mice carrying the genes for the adaptive coat color represents a greater and greater proportion of the population.

Once a trait such as the camouflaging coat color becomes the most common form in the population, a second type of selection pressure acts. Now the natural population of rock pocket mice is under strong selection pressure to remain at the adapted phenotype. Genes that produce the mice's camouflaged color will continue to be favorably selected and fur color that doesn't match the surrounding rocks will be selected against, courtesy of the owls. From one generation to the next, coat color will stabilize and change little if at all. The same is true in the case of the warning coloration in the passion-vine butterflies of the tropical rainforest.

When we say that particular traits are selected for or against, we don't mean that natural selection acts on only one trait of an organism and not another; rather it acts on the *whole* organism and its entire phenotype. In sports, for example, a boxer may have many superior traits but lose the fight because of one particular flaw, such as a lack of endurance. The lack of camouflaged coloration inherited by some rock pocket mice eliminated them from the game. Like athletes, organisms are constrained by the capa-

Figure 3.10 Predation is an agent of natural selection. Owls prey on rock pocket mice. Natural selection has strongly favored a coat color that matches the rock background. Once the coat color is adapted to the rock background, predation now selects against any change in the mice's coat color that will make them visible.

© Dr. Hopi Hoekstra, Museum of Comparative Zoology, Harvard University

Figure 3.11 Bighorn sheep butting heads. Male bighorn sheep fight for dominance by charging at each other head-on. Only the winning male has the right to mate with female sheep in a herd. Females also have horns, but they are much smaller.

bilities they bring to every situation—rarely is an organism superior in all areas of its biology, at all times, and in all environments.

Through the generally slow, gradual processes that produce and then select genetic variation, populations become adapted to their environments. It's important to emphasize that environmental conditions are constantly changing on the local and global levels, creating a moving target for natural selection to act on. In its "struggle for existence," an organism faces the current environmental conditions with the genetic endowment it received from its parents, who received it from their parents, and so on back in time. As a result, an organism confronts its *present* environment with a genetic composition and phenotype that reflect natural selections made in the *past*.

Sexual selection favors traits that increase reproductive success

Natural selection is often thought of in terms of survival—traits that allow a plant to compete for water, a butterfly to escape from predators, or a monkey to find fruits in the forest. However, an organism has another component to its fitness—traits that lead directly to *reproductive success*. For sexually reproducing organisms, success translates into passing their genes to the next generation. To do this, you need to find a suitable mate of the opposite sex. When particular traits that attract a mate are selected over others, it is called **sexual selection**.

Examples of sexual selection in animals generally involve males who compete with other males for females. Darwin observed that this type of selection takes on two forms. First, males can win over a female by intimidating other males or directly battling them for superiority or territory. This is the case when male bighorn sheep, with their massive horns, clash in spectacular head-on collisions with their rivals (**Fig. 3.11**). Gorillas, lions, and other social animals have a leader, the alpha-male, who has mating priority within the group. In these cases, evolutionary processes favor adaptations such as large body size, aggressive behavior, large antlers in the deer family, or the production and maintenance of weapons such as horns and spurs.

Sexual selection—pressures that favor particular traits that attract a mate.

Figure 3.12 Seahorses have elaborate courtship dances. They pair for the season and perhaps for life. They bond with an elaborate courtship dance that may last four days and include brightening their colors, quivering, pouch displays, and intertwining their tails.

Males can also win over females by being the most attractive, by being good providers, or by impressing them with their style. Favorable traits for this strategy include vivid, bright colors, elaborate courtship displays, and dances (**Fig. 3.12**). During summer evenings, male gray tree frogs can be heard serenading, or singing, to attract mates. The male hanging fly captures food, hangs from a leaf or twig, and releases chemical signals to attract a female. If the female accepts his gift of food, the pair mates while she eats. All of these strategies rely on the preferences of each female who chooses her mate. Whether it involves dressing up, singing a song, or providing dinner, the animal world is full of seemingly bizarre examples of traits favored to find a mate and produce offspring.

Animals are not the only organisms that mate and are under sexual selection pressures. In flowering plants, sexual selection occurs when flowers compete to attract bees and butterflies. Flowers vary in their color, size, shape, and odor to appeal to pollinators that carry pollen. Pollen grains then germinate, produce sperm, and literally race from the top of the flower to the eggs at the base through a passageway. The fastest sperm are selected and favored because they reach the eggs first, win the race, and successfully produce a seed. Thus every seed in a fruit represents an example of sexual selection.

The traits acquired for sexual selection can come at a cost, and sometimes that cost is decreased survival. On occasion, fights between males lead to injury and even the death of one of the rivals. Larger males have larger energy requirements and expend more energy than smaller males. Defending a harem of 50 female elephant seals requires a lot of energy—many bulls lose a third of their body weight during the 3-month breeding season. When frogs call out to find a mate, they also alert predators to their location. Producing pollen and colorful flowers is costly in terms of energy and sugar resources. Although some adaptations that are favored by sexual selection may reduce survival, adaptive alleles are passed to the next generation by way of increased reproductive success.

Coevolution occurs between interdependent species

When organisms interact and exert selection pressures on each other to produce adaptations, coevolution occurs. The life cycle of the passion-vine butterfly is an example of coevolution in two different ways. First, a biological "arms race" exists between the passion-vine plant and its herbivore. You'll recall that a series of mutations produced toxic compounds in its leaves. The compounds protected the plant from most herbivores but inadvertently selected for those passion-vine caterpillars that had mutations to tolerate or detoxify these toxins. The mutation benefited the caterpillar by providing a food source and reduced competition from other species that lacked the mutation. Second, the cucumber flowers were under different selection pressures to enter a mutually beneficial partnership—the cucumber flowers evolved a particular shape and pollen reward that favored the passion-vine butterfly's unique behavior and sensory characteristics over those of other butterflies. These specialized traits result in greater fitness for both species—the butterfly and the plant. In fact, many flowers have evolved together with their very specific animal pollinators (**Fig. 3.13**).

Throughout the book, you will find many examples of coevolution, whether they are a human and its pinworm parasite in a disease situation, or a plant and its growth-promoting root fungi. In all cases, the direction

Sunbirds are attracted to red colors and drink nectar with their long bills

Selection pressure

Selection pressure

Sticky-tube heather has long, tubular, red flowers; provides an abundance of sugary nectar

Figure 3.13 Coevolution. The long curved bill of a sunbird, who lives along the tip of South Africa, fits perfectly into the long tubular flowers of sticky-tube heather. This plant and its pollinator are an example of coevolution.
© Cengage Learning 2014

Coevolution—process that occurs when two or more organisms interact and exert selection pressures on each other to produce adaptations.

of *coevolutionary* change has been toward greater fitness and adaptation of those species involved.

Chance is a nonselective force of evolutionary change

Natural selection is only part of the evolution process; chance, or blind luck, also plays an important role in shaping the genetic composition of a population. Random genetic mutations produce variation in sperm and egg cells. These new genetic variations can create new traits leading to evolutionary adaptations. During sexual reproduction, the particular sperm that ultimately fertilizes the egg is a random event. Weather events are random acts of nature that can result in evolutionary change. For instance, winds may disperse mutated seeds or spores into a previously unfavorable soil environment where they thrive. Or the wind may blow these seeds onto the roof of a building. Even though the seeds may be highly adapted for success, random forces can mean that they never make it to the "starting line" of competition.

It's important to note that random events are *nonselective* and do *not* produce better-adapted organisms—they favor or eliminate individuals regardless of their fitness. The passion-vine butterflies, the wild mustard plants you see flowering in an abandoned field, or the rock pocket mice scurrying about the rocks are the result of many years of genetic variation, chance, coevolution, and natural selection.

3.3 check + apply YOUR UNDERSTANDING

- Organisms at every stage of their life history are under selection pressures from the physical environment and other organisms (for example, competition, mutualism, predation).

- Natural selection is a process of elimination that favors organisms that have characteristics that enable them to survive and reproduce. These adapted individuals have greater fitness.

- Sexual selection is a type of natural selection that acts on an organism's ability to reproduce sexually (that is, successfully find a mate, unite gametes, and produce offspring).

- Much of the evolutionary process occurs through coevolution, which results in adaptations of organisms that interact and depend on each other.

- Evolutionary change involves chance events that may retain less fit or eliminate genetically superior individuals in a nondirected, random way.

1. What are the two processes that act on genetic variation to produce evolutionary change?

2. Why doesn't chance result in a more fit and adapted population?

3. What is the relationship between an adaptive trait and an organism's fitness?

4. Female fiddler crabs choose male fiddler crabs based on the size of the male's enlarged claw. To attract the attention of a female, a male builds structures around the burrow entrance and waves or slams his enlarged claw to the ground. From this scenario, describe two specific evolutionary processes taking place.

5. The newly hatched birds of many songbirds such as robins, cardinals, and blue jays are naked, blind, and helpless (**Fig. 3.14**). Describe two selection pressures on birds at this stage of life. Name two traits that are selected for and two that are selected against.

Figure 3.14 Newly hatched robins.

New Species Arise through the Process of Speciation

Books, magazines, television shows, and our own observations of nature tell us of the rich diversity of the living world. For example, there are over a thousand species of bats in the world. Birds (10,000 species) are far more diverse than conifer trees (980 species). Butterflies are also a particularly diverse group, with almost 15,000 species worldwide. The principal characteristic used to distinguish them is wing structure and color pattern. Why are some groups like butterflies and birds more diverse than conifers? What processes drive speciation, or the rise of all these new species?

Biologists are gathering evidence to answer these questions. In particular, they are studying the 45 species of passion-vine butterflies living as neighbors in the tropical rainforests of Central and South America. Recall that a biological species is defined as members of a population that can interbreed in nature by sharing a similar set of genes. Members of a species form a reproductive community in which individuals mate within their own species. If this is true, then how do new species of passion-vine butterflies arise? In the case of passion-vine butterflies, biologists have discovered that two different species mate with each other about 10 percent of the time. They also found that these matings produced fertile offspring that can potentially breed with their parents. These matings create new combinations of characteristics for natural selection to act on. This breakdown in mating barriers is one piece of evidence of how new species begin. In the section ahead we will begin to look at the gradual processes that produce new species from their ancestors.

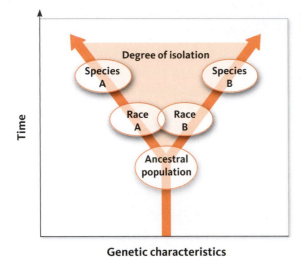

Figure 3.15 Divergence and speciation. Speciation occurs when an ancestral population diverges into new habitats under different selection pressures. As they diverge, they first become races and then species.

© Cengage Learning 2014

Populations diverge from an ancestral population

Populations diverge from the ancestral population for several reasons. Different populations may not be in contact with each other. Physical barriers may prevent the flow of genes among these populations and provide conditions in which each population evolves independently of the others. Genetic processes such as sexual reproduction, genetic rearrangements, duplications, and mutations constantly increase the genetic diversity in each population. Due to different environments and different selection pressures on each of the populations, some traits may be favored in one population, while different traits are favored in other populations. The end result is that the populations become different, or *diverge*.

Figure 3.15 shows the slow, gradual divergence of an ancestral population into two different races, which can interbreed but display different phenotypes and appearances. Ultimately, if a race, or population, accumulates enough phenotypic and genetic differences, their divergence leads to reproductive isolation. Reproductively isolated populations are unable to interbreed—they become separate species.

Reproductive barriers isolate populations and drive speciation

Speciation is the consequence of reproductive barriers and isolation, which prevent two organisms from breeding. Reproductive barriers fall into two broad categories—those that prevent fertilization, called pre-mating barri-

Speciation—the process that produces new species.

Species—members of a population that can interbreed in nature by sharing a similar set of genes.

Races—two different populations of the same species that can interbreed, but display different phenotypes and appearances.

Reproductive isolation—condition in which populations are unable to interbreed and can lead to separate species; drives speciation.

Reproductive barriers—conditions that prevent two organisms or populations from breeding.

ers, and those that prevent the fertilized egg, or zygote, from developing. These barriers are called post-mating barriers, which result from genetic incompatibility between the species.

Let's first look at four common pre-mating barriers—geographical, temporal, mechanical, and behavioral. Perhaps the most common barrier is *geographical,* where species can become separated either by large distances or by geographical barriers such as mountain ranges, large bodies of water, glaciers, or deserts. For example, the northern and Mexican spotted owls do not mate because they simply do not encounter each other in their ranges (**Fig. 3.16**). Human activities such as deforestation and development also physically separate species. During Earth's long history, continental movements separated landmasses (refer back to **Fig. 2.14**), which in turn physically separated and isolated populations and powered speciation. The same is true with the many unique species that inhabit the remote Hawaiian and Galapagos Islands. Once the species are separated geographically, they can no longer interbreed; they follow their own genetic pathway, leading to populations with unique traits. Geographical separation also involves different populations moving into different environments—different climates, resources, predators, or competitors. Speciation, therefore, is the consequence of adaptation to different environments.

Temporal barriers occur when species reproduce at different times of the day or different seasons of the year. Two species of evening primrose grow and bloom together in deserts of southwestern America. One species blooms before sunrise and attracts the early-rising bees, while the other blooms in the late afternoon and is pollinated by bees that fly later in the day. Even though both plants live side-by-side, they are subjected to a strong barrier to reproduction and rarely form hybrids.

Many insects have intricate male and female reproductive organs that must fit together exactly for successful copulation, or sex, to occur. Different species of insects have different sex organs to copulate in flight, or on the ground, or in the water. The *mechanical barriers* set up by these reproductive structures may also prevent any mating at all between closely related species. A number of plants with complex flower structures such as milkweed, orchids, mints, and snapdragons have mechanical barriers that prevent or interfere with pollination from other species.

Animals depend on communication to find a mate or time their mating behaviors. If these signals are not recognized by another species, a *behavioral barrier* isolates them. This is especially true of many insects, which communicate by way of visual, auditory, or chemical signals. One of the most dramatic examples of this type of barrier is the courtship display behavior of nocturnal fireflies. Male fireflies fly over open meadows and marshes emitting a characteristic flashing pattern designed to persuade receptive females to announce their position in the surrounding foliage. Each species has its own characteristic flashing pattern comprised of a signature color, rate of flashing, length of flash pulse, and intensity of the flash. Females flash back in response. On a summer night you may see several different species flashing and responding to their specific flash pattern, which behaviorally isolates them from each other.

If pre-mating barriers are overcome, as shown by the passion-vine butterflies, the two different species may successfully mate and mix their genetic information through fertilization. This results in a hybrid offspring. Hybrids are frequently encountered in plants such as birches, oaks, and willows. In animals, wolves and dogs have been crossed to produce wolf-

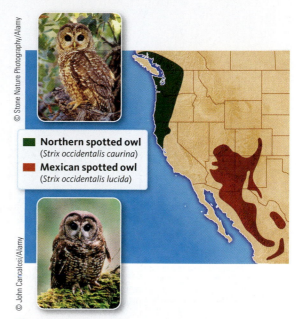

Northern spotted owl
(*Strix occidentalis caurina*)
Mexican spotted owl
(*Strix occidentalis lucida*)

© Stone Nature Photography/Alamy

© John Cancalosi/Alamy

Figure 3.16 Geographic separation contributes to speciation. The northern and Mexican spotted owls are geographically separated and are reproductively isolated.

© Cengage Learning 2014

Hybrid—the offspring of mating between two different species.

Figure 3.17 An example of hybrid sterility. A mule is the sterile offspring from a female horse and a male donkey.

dogs, which are often kept as pets. The domestic cow has been crossed with bison to produce a "beefalo." In fact, molecular data shows that over 5 percent of bison today have genes from domestic cattle in their DNA. Molecular data have also been used to support the hypothesis that red wolves are actually coyote–gray wolf hybrids.

Two types of post-mating barriers function *after* mating and fertilization have occurred. In these cases, the genetic program of one species is incompatible with that of the other. The first isolating barrier, called *hybrid inviability,* results in an early death of the embryo. Birds, frogs, and mammals such as domestic sheep and goats can mate but not produce offspring. The second barrier comes into play in cases where a mating successfully produces an offspring. However, the offspring are sterile, and this barrier is appropriately called *hybrid sterility.* Mules are sterile because they are the result of a cross between a female horse and a male donkey (**Fig. 3.17**). The offspring of lions and tigers, and zebras and horses, are also sterile. It's important to note that these mating barriers are not easily broken and require many matings and generations to produce them.

3.4 check + apply YOUR UNDERSTANDING

- Speciation results from reproductive barriers and isolation, which prevent two organisms from breeding and reduce gene flow among populations.

- Populations diverge into different races and ultimately become isolated into isolated reproductive communities called *species.*

- Reproductive isolation is caused by pre-mating barriers such as geographical, temporal, mechanical, and behavioral barriers, as well as post-mating barriers that render the hybrid inviable or sterile.

1. What is meant by the term *speciation,* and what processes drive it?
2. List four pre-mating barriers that isolate populations and keep species separate.
3. What is the relationship between reproductive isolation, gene flow, and genetic diversity?
4. Researchers have attempted to artificially fertilize eggs from bullfrogs with sperm from leopard frogs. The embryos did not develop. Which type of reproductive barrier keeps these two frog species separate?
5. After thousands of years, two isolated populations of green-eyed tree frogs in northern and southern Australia breed with each other to form a population of hybrid frogs. Each population has a different mating call (**Fig. 3.18**). In areas where all three populations overlap, research shows that female frogs respond and prefer calls from males of their own population. Is this preference selecting for or against the reproductive isolation of the hybrid frogs? Which kind of reproductive barrier keeps these species separate? Is sexual selection involved in this scenario? Explain.

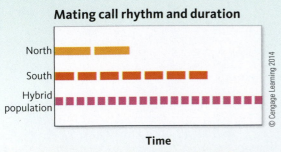

Mating call rhythm and duration

North
South
Hybrid population

Time

Figure 3.18 Mating calls in green-eyed tree frogs.

3.5 Each Species Has a Unique Evolutionary History

Every living thing is a modification of the genetic information inherited from its parents. This is what is meant by the phrase "descent with modification," which is the basis of understanding the family tree or the evolutionary history of an organism. Every organism around you, on you, or within you represents a living product of its species' evolutionary journey through time.

An interesting evolutionary history is shown by trilobites, which were a group of marine animals that looked similar to modern-day horseshoe crabs (**Fig. 3.19**). Trilobites first appear in the fossil record about 526 million years ago, and show a remarkably complex and diverse number of body plans. Over 17,000 species have been discovered, as they flourished for millions of years before they became extinct about 250 million years ago. Their history, as shown by their extensive fossil record, was marked by diversification into several different lineages.

Through the history of life, some species diversify extensively, whereas others survive millions of years with little change (**Fig. 3.20**). The fossil record indicates that species have a beginning and an end. As a general rule, most species die out within about 10 million years of their appearance. What events trigger these changes and patterns to emerge? What ultimately causes the death of all individuals comprising a species? In the section ahead we attempt to answer these questions and describe processes and conditions in the evolutionary history of several different organisms.

Some groups survive for millions of years with little change

The horseshoe crab could have been the inspiration for the old saying, "The more things change, the more they stay the same." Modern horseshoe crabs look almost identical to the *Mesolimulus* fossil found in limestone from the Jurassic Period of 150 million years ago (**Fig. 3.21**). Going even further into the past, the fossil record shows that horseshoe crabs are the surviving members of an ancient lineage that stretches back nearly 400 million years. The horseshoe crab is called a "living fossil" because it lived during ancient times and survived to live today. The fossil record contains many other examples of modern species that resemble fossils of organisms that lived millions of years ago.

Several factors probably combine and contribute to produce species that survive and remain unchanged through such an immense period of time. First, populations may have been under intense selection pressure to maintain a constant phenotype. Any variation was selected against and eliminated through natural selection. Second, an especially stable and constant environment may have driven selection against any genetic variation. Third, perhaps the horseshoe crab is constrained by development. In other words, maybe its unique embryonic development "backed itself into a corner" and left little room for change. Finally, reproductive barriers may have been especially strong and created many isolated populations. In the end, the best explanation

Figure 3.19 A trilobite fossil. Trilobites were a group of arthropods that lived in the ancient seas. They had hard shells and compound eyes like modern groups. Trilobites successfully radiated into nine major lineages and survived for over 300 million years.

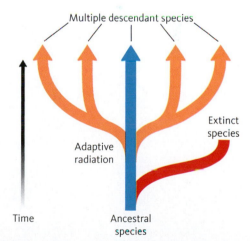

Figure 3.20 Evolutionary pathways. An ancestral species may remain unchanged for millions of years (blue), radiate into multiple species (orange), or become extinct (red).
© Cengage Learning 2014

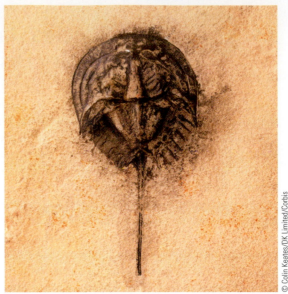

Figure 3.21 The horseshoe crab—a modern fossil. Modern horseshoe crabs (left) look almost identical to the *Mesolimulus* fossil (right) found in limestone from the Jurassic Period 150 million years ago.

probably involves some combination of these factors that permitted horseshoe crabs, like other living fossils, to adjust to changes of the environment without the need to change their phenotype.

Innovations and opportunities trigger adaptive radiation

While some species remain virtually unchanged, others have branched or radiated into numerous lineages, each with many species of their own—this is called **adaptive radiation**. Adaptive radiation often follows significant evolutionary innovations such as wings and feathers in reptiles, flowers in plants, and jaws in vertebrates. The ability to fly was an enormous advantage to reptiles that evolved this trait. Today there are thousands of species of birds that resulted from this evolutionary innovation. About 400 million years ago, adaptations to land environments provided new selection pressures on plants—unfiltered light and more rapid diffusion of gases meant greater rates of photosynthesis. The innovation of flowers allowed greater sexual reproduction, generated greater genetic diversity, and increased adaptive fitness in almost all habitats.

Although is it tempting to think of evolutionary innovations in the same way as innovative products such as the iPod or digital cameras, they are driven by very different processes. Market innovations are based on fulfilling consumer needs, whereas evolutionary innovations are driven by random genetic variation followed by natural selection. Breeding crop plants for seeds or livestock for milk production by artificial selection is also a "needs-driven" process. However, in nature evolution is not a needs-driven process where organisms evolve adaptive structures in *response* to environmental demands; rather, adaptations evolve because traits were selected for their fitness.

New opportunities trigger adaptive radiations. A well-known example of adaptive radiation involves a group of 14 finch-like species, which are native to the Galapagos Islands 600 miles off the coast of Ecuador in South America. On his visit to the islands in 1835, Charles Darwin observed feeding

Adaptive radiation—species have branched or radiated into numerous lineages each with many species.

Insects

Grubs

Tool-using finch

Leaves

Seeds

Buds/fruits

Figure 3.22 Darwin's finches. Darwin's finches are a group of 14 finch-like species that are all descended from a common ancestor that migrated to the Galapagos Islands from Central and South America. Over the past 2 million years, each of the species adapted to different food sources.
© Cengage Learning 2014

behaviors, nest building, and physical traits in these similar-looking birds. Darwin's observations led him to hypothesize that all 14 species of finches descended from a common ancestor. Several recent DNA studies strongly support this conclusion that "Darwin's finches" radiated from a single common ancestral species. It is estimated that the founding population of at least 30 individuals departed the Pacific Coast of South America approximately 2.3 million years ago, during a time of dramatic climate change. Once this founding population colonized and occupied the different island habitats, selection pressures to different food sources caused the finch populations to diverge. Some finches have bills that crush foods such as seeds, buds, and fruits, whereas other species have bills that grasp leaves or tools, or bills that probe plants for insects and grubs (**Fig. 3.22**). In addition to feeding, beaks also function in producing mating songs, and it has been hypothesized that changes in beak structure also formed behavioral reproductive barriers. These barriers further isolated finch populations and drove speciation. Darwin's finches show how evolutionary forces such as natural selection and speciation combine to produce diverse forms of life.

Extinction marks the death of a species

Environments and selection pressures are constantly changing—physical conditions and resources change, as well as the abilities of competitors, prey, and predators. Change may occur as a gradual shift or as a catastrophe, such as a hurricane or the introduction of an invasive species to the habitat. In the face of change, a species faces three possible outcomes: (1) genetic variation and natural selection will shape the future population to become better adapted, or (2) individuals will migrate and colonize new or more suitable habitats, or (3) the species will fail to adapt quickly enough, decline, and become extinct.

This chapter has explored the first two outcomes, so let's now look more closely at the third outcome—extinction. It is estimated that 99.9 percent of all species that have ever lived are now extinct. Extinction of a species can be caused in two different ways. First, some species are simply replaced along a lineage by species better adapted to change, as shown with the trilobite lineages in **Figure 3.19**. A second cause of extinction is through the elimination of all individuals. Given life's extremely long history, this makes sense. The fossil record reveals that at least five

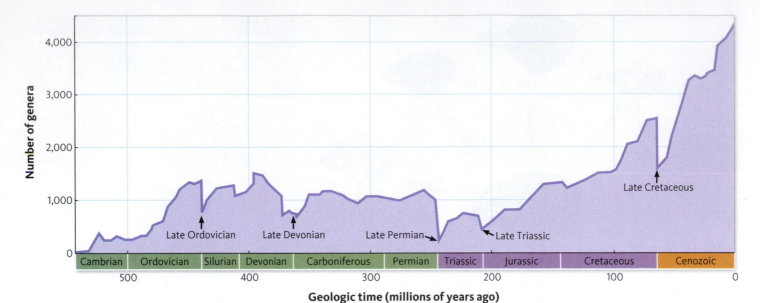

Figure 3.23 The record of mass extinctions. The fossil record indicates that there have been five mass extinction events in the history of Earth. Each time a vast majority of species became extinct in a relatively short period of geologic time.
© Cengage Learning 2014

separate **mass extinctions** have occurred in which a vast majority of species has disappeared in a relatively short period in geologic time. Almost all trilobites disappeared from the fossil record along with many marine organisms and insects in a mass extinction 250 million years ago (**Fig. 3.23**). It has been estimated that up to 96 percent of all marine species, 57 percent of insect species, and 70 percent of terrestrial vertebrate species went extinct. Scientists look for evidence that mass extinctions have been caused by gradual changes in the Earth's environment or a catastrophic event like an asteroid striking our planet. It is now generally accepted that the consequences of an asteroid or comet that struck the Earth triggered the most recent mass extinction 65 million years ago at the end of the Cretaceous Period. This event on the global level brought to a close the 140-million-year dominance of dinosaurs (see **Fig. 3.23**).

Local extinction occurs when a species has died out in a region but not globally. The extinction may be the result of a change in an ecological balance. The appearance of a stronger competing species that is able to capture resources can drive the native species to local extinction. Local climate, natural disaster, or diseases also have an impact. For example, fungal infections are causing a large population decline of at least six bat species in the northeast United States.

Human activities, including overhunting, habitat destruction, and the introduction of invasive species and predators, lead to a species' extinction. Before the arrival of humans, Australia and New Guinea had diverse populations of large animals—for example, 400-pound ostrich-like flightless birds and rhino-like marsupial, or pouched, mammals. Today these landmasses have no animals larger than a 100-pound kangaroo; humans hunted them to extinction about 40,000 years ago. Island populations are particularly vulnerable to introduced predators, which can prey on the local species long before the population can adapt effective defenses. You are probably aware that widespread habitat destruction is cited as a major cause of total extinction of species from Earth. It has been reported that about 18 million acres of the world's forests—an area larger than the state of West Virginia—are lost each year to deforestation. These areas are often

Mass extinctions—the death of all individuals comprising a species; elimination of a species in a relatively short period in geologic time.

converted to cropland, which establishes new ecosystems and new selection pressures on organisms that remain.

The population of a species may become so low that it cannot rebound. In this case a species' level of genetic diversity may be so low that any small change to its environment will push it over the edge to extinction. These particular species are often protected by law under the Endangered Species Act. This concept is developed further in Chapter 17 on conservation biology.

Throughout this chapter we have emphasized the evolutionary processes that generate genetic variation, drive natural selection, and shape adaptation. Adaptation is the ability to survive and reproduce in the current environment, but also includes the resiliency to absorb varying magnitudes of disturbance and change. All species adapt to constantly changing conditions in different ways, and this marks their evolutionary pathway through time. As you continue to learn about the remarkable unity and diversity of life in this book, be sure to remember that all organisms are part of their species' particular pathway—they confront their present environment with genes that have been selected and imprinted, or fixed, within the genes of their parents, grandparents, and all those in their past lineage.

3.5 check + apply YOUR UNDERSTANDING

- Every organism is a product of its species' evolutionary journey through time, which includes adaptation to constantly changing conditions.

- Evolutionary pathways include species that have changed very little, species that have branched into numerous lineages (adaptive radiation), and those that have failed to adapt and become extinct.

© Visuals Unlimited/Corbis

Figure 3.24 Invasive species can cause local extinction of nature species. Purple loosestrife is a beautiful but aggressive invasive plant that displaces native vegetation.

1. What is a living fossil?
2. Describe three factors that would contribute to an unchanging evolutionary lineage over a long period of time.
3. The process of forming hybrids has been called *reverse speciation*. Do you agree with this statement? Defend your position.
4. Which is more susceptible to extinction from a change in sea level: (A) a species with little genetic diversity, or (B) a species that has a large amount of variation in its members? Defend your answer.
5. Purple loosestrife is a beautiful but aggressive invasive plant from Europe that has taken over wetlands of eastern North America (**Fig. 3.24**). How can purple loosestrife drive the native wetland plants, fish, and other wildlife to local extinction?

End of Chapter Review

Self-Quiz on Key Concepts

Evolutionary Change, Genetic Variation, and Populations

KEY CONCEPTS: Evolutionary change of a population occurs in two separate steps: first, genetic variation is produced by sexual reproduction and mutation; and second, those individuals that will survive and reproduce successfully are selected. The result is a gradual evolution of the genetic composition of a population from one generation to the next. Small changes on the genetic level can lead to large changes at the molecular, cellular, organism, and population levels of scale and organization.

1. Match each of the following **processes** with its characteristic, description, or example. Each term on the left may match more than one description on the right.

   Fertilization
   Gene flow
   Meiosis
   Mutation

   a. a change in the DNA sequence
   b. a process that recombines genetic information in the production of sex cells (sperm, egg) or spores
   c. the union of a sperm and an egg
   d. the movement of genetic information into or out of a population

2. Match each of the following **genetic terms** with its characteristic, description, or example. Each term on the left may match more than one description on the right.

   Allele
   Genome
   Gene

   a. different forms of a gene
   b. a segment of DNA that codes for a particular trait
   c. the collection of genetic material in an individual
   d. the segment of DNA that codes for eye color
   e. the segment of DNA that codes for blue rather than brown eye color

3. Which of the following processes *decreases* the genetic diversity of a population?
   a. meiosis
   b. fertilization
   c. natural selection
   d. mutation

4. Which of the following changes will lead to *adaptive* change?
   a. a change acquired through physical training and exercise
   b. a change in a gene from the sun's ultraviolet rays that causes skin cancer
   c. a change in a gene that is considered neutral in its effect
   d. a mutation that leads to greater production of sperm

Natural Selection, Adaptation, and Fitness

KEY CONCEPTS: A population consists of a spectrum of differentially adapted individuals, all of which have traits that contribute to their fitness. The most fit or adapted organisms withstand a variety of selection pressures at each stage of their life cycle to live to reproductive age, find a suitable mate, and pass their genetic information to their offspring. The process of natural selection results in organisms that are better adapted to their living and physical environment. Sexual selection is a type of natural selection that favors phenotypes and behaviors that influence the chances of finding a suitable mate. Random events may also alter the genetics of the population, but unlike natural selection, do not produce adaptive change.

5. Match each of the following **evolutionary processes** with its characteristic, description, or example.

   Coevolution
   Natural selection
   Selection pressure
   Sexual selection

   a. a process of favoring adapted individuals for survival and successful reproduction
   b. conditions imposed on an individual that favor traits that are adaptive
   c. factors that favor particular traits that attract a mate
   d. two organisms that change together by exerting strong selection pressures on each other

6. Which of the following processes will *not* lead to adaptive change?
   a. selection pressure from predation
   b. competition for a mate
   c. random events such as a hailstorm
   d. strong selection pressures from freezing temperatures

7. Which of the following are components of an individual's *fitness*?
   a. the ability to survive and grow to reproductive age
   b. the ability to produce lots of offspring
   c. both a and b
   d. neither a nor b

Speciation, Adaptive Radiation, and Extinction

KEY CONCEPTS: New species branch from an ancestral species in the process of speciation. Populations diverge when reproductive barriers cause isolation. Isolated populations may exploit new habitats and branch into several lineages, remain unchanged through long periods of time, or become extinct.

8. Match each of the following **evolutionary processes** with its characteristic, description, or example.

   Adaptive radiation
   Local extinction
   Mass extinction
   Reproductive isolation
   Speciation

   a. the process that eliminates a majority of species over a relatively short period of geologic time
   b. the condition in which two populations are unable to interbreed
   c. a process where an ancestral population branches into many species or lineages
   d. a process that produces new species; the consequence of selection and adaptation
   e. the elimination of all individuals of a species in a specific area

9. Which of the following barriers prevent mating and reproductively isolate populations?
   a. geographical barriers
   b. temporal barriers
   c. behavioral barriers
   d. all of the above

10. Which of the following processes will *not* lead to population divergence and speciation?
    a. hybrid inviability
    b. hybridization
    c. natural selection
    d. all of the above

11. Which of the following is *not* a plausible reason for a species to remain unchanged for hundreds of millions of years?
    a. The species was strongly adapted to its environment.
    b. The development of new forms was constrained.
    c. Natural selection favored individuals that showed variation.
    d. The environment remained constant.

Applying the Concepts

12. The pileated woodpecker can strike up to 15 blows per second at an equivalent speed of 16 miles per hour (**Fig. 3.25**). Natural selection has selected for woodpeckers with a thick bony skull that can absorb the shock of the strike. Think of a reason why *very* thick skull bones would be a disadvantage or selected against in pileated woodpecker populations.

© First Light/Alamy

Figure 3.25 Pileated woodpecker.

Figure 3.26 Bdelloid rotifer.

13. Bdelloid rotifers are microscopic protists that live in freshwater and moist soil habitats (**Fig. 3.26**). Every member of the 370 species is female. Without males for sexual reproduction, these animals have surprisingly evolved into different species and inhabited different habitats. How do they generate the genetic variation for natural selection to act on?

14. Wood-decay fungi grow as filaments, or threads, within wood. When threads of one fungus encounter a *compatible* thread, they reproduce sexually by merging their bodies into a single thread. However, in some cases fungal threads encounter an *incompatible* fungus and initiate a killing reaction by secreting dark-colored chemicals into the wood. These protective chemicals sharply define the interaction zone between two incompatible fungi and result in *zone lines* in the wood (**Fig. 3.27**). Zone lines represent which type of reproductive barrier in these fungal species? Does this killing reaction promote or inhibit the generation of genetic variation? Does it promote or inhibit speciation?

Figure 3.27 Zone lines.

Data Analysis

Data Interpretation

Recent molecular studies estimate that humans diverged from our most recent relatives, the chimpanzees, no more than 6.5 million years ago. Humans and their ancestors are called *hominins,* and the graph below shows a large data set of published measurements of hominin skulls older than 10,000 years old (**Fig. 3.28**). It shows the cranial capacity, or the volume of the braincase, plotted against time in millions of years ago. Use this graph to answer the questions below.

15. Does the graph show an abrupt or a gradual change in the size of the braincase in humans?

16. Did cranial capacity increase for the first million years?

17. Is increased cranial capacity a trait that is being selected *for* or selected *against*? Explain.

18. How long ago were human ancestors' cranial capacities roughly half the size of living humans'?

19. Why do you think there is a range of capacities for each time period? For example, 2.5 million years ago cranial capacity measurements varied from about 300 to 600 ml.

20. Why do you think that the number of data points at each time period decreases as you go farther back in time?

Critical Thinking

21. Do you think cranial capacity is an indicator of intelligence or a more complex brain?

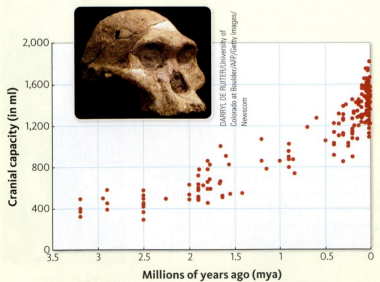

Figure 3.28 The cranial capacity of human ancestors.

Question Generator

Nectar-Robbing Bees

Many flowers have coevolved with their pollinators by offering energy-rich nectar to attract insects to transport pollen and effect pollination and seed production. Flower shape and structure are important cues to attract specific pollinators; for example, flowers with long tubes will place their nectar deep at the base of the flower and attract butterflies and hummingbirds with long tongues. This shape excludes pollinators such as bees that have short mouthparts. However, some species of bees skirt this process by making a hole into the flower and robbing it of its nectar (**Fig. 3.29**).

(From Inouye, D. 1983. The ecology of nectar robbing. In B. Bentley and T. Elias [eds.], *The Biology of Nectaries*. Columbia University Press, New York.)

Below is a block diagram (**Fig. 3.30**) that relates several aspects of the biology of nectar-robbing bees and the evolutionary fitness of their flower "victims." Use the background information along with the block diagram to ask questions and generate hypotheses. We've translated the components and interaction arrows shown in the map into a question and provided several more to get you started.

Figure 3.29 A nectar-robbing bee.

Figure 3.30 Factors relating to the effect of nectar-robbing bees on the fitness of their "victims."

Here are a few questions to start you off.

1. What is the relationship between nectar robbing, pollinator visitation, pollination, seed production, and plant reproductive success?

2. Are some flower shapes easier to rob than others? (Arrow #1)

3. Do nectar robbers reduce the number of visits made by pollinators? (Arrow #2)

4. What is the consequence of nectar robbing on the fitness of the plant? (Arrow #3)

5. How do flowers protect themselves against nectar robbing? (Arrow #1)

6. Does nectar robbing reduce the number of seeds made by the flower? (Arrow #4)

Organisms in Their Habitat

4.1 The habitat of the coast redwood

4.1 The Habitat of the Coast Redwood

The immense coast redwood tree *(Sequoia sempervirens)* is one of the most massive and longest living plants on Earth. It is not uncommon to find redwoods that are more than 200 feet tall, with some growing to nearly 400 feet—twice as tall as cell phone towers and five times higher than the tallest trees in a typical city park (**Fig. 4.1**). When environmental conditions such as temperature, nutrients, and moisture are favorable, redwood trees grow quickly and live more than 2,000 years. You might think that an organism that can survive so long would be flexible in its requirements for life, but the coast redwood is very sensitive to wind, salt, and other small differences in local environmental conditions.

The coast redwood grows naturally in only one area on Earth (**Fig. 4.2**). What does this small geographic region offer the coast redwood that no other place on Earth can? This narrow band of land provides ideal conditions, or habitat, for redwood trees to grow and reproduce. It has mild temperatures, heavy winter rains of more than 100 inches per year, a great deal of summer fog, and nutrient-rich soils. The rain and fog are important sources of moisture, and the valleys of coastal California provide large amounts of minerals to the trees.

Figure 4.1 Coast redwood trees. Coast redwood trees are large by any standard, dwarfing the people walking on the path. The redwood's first layer of branches starts about 100 feet over the walkers' heads. Mild winters, abundant rain, and mineral-rich soils support their growth.

Organic matter such as leaves, needles, and pieces of wood collects in the valley bottoms, where, with the help of decomposers, it is slowly broken down, releasing minerals back into the soil. These nutrients are quickly absorbed by the redwood's shallow but expansive root system. Like all living things, the redwood is adapted to the habitat in which it lives. We do not see redwoods growing in New York or Texas because the environmental conditions in those states do not fall within the specific range that supports the growth and reproduction of the redwoods. Although environmental conditions might be favorable in other locations, we do not find redwoods growing there because the redwood seeds have never reached those places.

The redwood's habitat includes hundreds of different plants, animals, mosses, lichens, fungi, and bacteria that live in the tree's shadow, or understory. Other tree species such as Douglas-fir and western redcedar live beside the redwoods, providing them protection from the coastal wind and salt and competing for many of the same resources. Banana slugs, insects, fungi, and bacteria that share the habitat serve an important role as decomposers, making minerals available to the redwood.

Redwoods not only live in their own specific habitat, but they also create habitats for other organisms. In the late 1980s, a group of biologists explored the redwood canopy and discovered entire communities of plants, lichens, fungi, reptiles, birds, and insects living high up in the trees (**Fig. 4.3**). They were amazed to find animals such as wandering salamanders spending their entire life cycle high in the canopy and small shrimp-like crustaceans living in pools of rainwater found in fern mats. The level of biodiversity discovered high in the redwoods rivals that of tropical rainforest canopies.

Figure 4.2 Geographic range of the coast redwood. This map shows the areas in northern California and southern Oregon where coast redwood trees grow.
© Cengage Learning 2014

Recall from Chapter 1 that each organism interacts with other organisms in unique ways that influence how communities and larger ecosystems operate. Wandering salamanders play an important role as predators, feeding on collections of small arthropods such as mites, spiders, and springtails. The arthropods are entirely dependent upon the habitat created in the mix of branches and decaying organic matter where they live and breed. The lichens on the branches shelter and collect moisture for these animals. Fungi and bacteria play a vital role by decomposing fallen branches that are caught in the massive canopy, providing other living organisms with needed nutrients. Similar unseen communities occur in the canopies of most trees.

With all of these interactions, it is easy to upset the balance within the redwood habitat, community, or ecosystem. In fact, because of human activities, redwoods are not always found where you would expect them in northern California and southern Oregon. In some places, an ideal redwood habitat is dominated by grasses and nonwoody plants because logging activities have greatly altered the habitat conditions. Nearly 70 years of unchecked logging has reduced the ancient redwood forests from more than 2,000,000 acres to about 275,000 acres. Redwoods are a desirable lumber because the rot-resistant wood makes it valuable for log cabin construction materials, decking, and outdoor furniture products. Each day we can see the human impact on our environment, but natural disturbances such as windstorms and droughts can also drastically alter habitats.

In this chapter, we'll examine the intimate relationship between organisms and the habitats in which they live. As we have seen with the redwoods, habitats provide resources necessary for survival and the favorable environmental conditions to which these trees are adapted. An organism's habitat consists of environmental or nonliving components, such as time, space, climate, and the resources it needs to survive, such as water, minerals, and gases. The habitat also includes living components—the other organisms that share that location. The collection of living organisms in a habitat all have an influence on the biology of an organism—for example, the human impacts to habitats, competitors, food, predators, and mates. An organism's habitat changes seasonally and unpredictably through disturbance events. Organisms respond to change in the short term, and populations adapt to change over the long term. We will explore how organisms interact with their habitat and the characteristics that all organisms share in their need for resources and their ability to respond to disturbances. Let's begin by examining why scale is an important factor when studying organisms in their habitat.

Figure 4.3 Examples of the incredible biodiversity found in the upper branches of coast redwood trees. Some animals such as the wandering salamander (left), shrimplike crustaceans (right), and jumping spiders (bottom) live out their entire lives in the canopy of this forest. The large clump of vegetation in the foreground is a mat of decaying vegetation with hundreds of ferns growing out of it. The decaying vegetation collects over years in the forks of branches, providing habitat for many plant and animal species.
© Cengage Learning 2014

Figure 4.4 **The western hemisphere from space.** View from space of the western hemisphere of Earth showing the white clouds, blue oceans, and greens and browns of land. Coloration differences in the continents are due, in part, to different ecosystem conditions.

Habitat—the environment in which an organism lives, which includes all of the living and nonliving factors that surround and influence its life processes. At a large scale the organism's *macrohabitat* includes regional environmental characteristics; its *microhabitat* is the smaller-scale environment—the conditions in its immediate vicinity that influence its physiology.

4.2 Habitats Can Be Described at Large and Small Scales of Size

If you were looking down on Earth from outer space, you would notice its blue oceans and brown landmasses enveloped by an atmosphere containing white, swirling clouds (**Fig. 4.4**). Earth's shape, distance to the sun, orientation of 23.5° angle of incline, and 24-hour rotation result in wide variations in climate around the world. The unequal distribution of solar radiation and heat drives ocean currents as well as wind and rain patterns. Variations in geography and climate create an immense variety of places for living things to inhabit—from the dark, near-freezing depths of the ocean floor to the rocky peaks of mountains.

Life is amazingly adaptive and is found in almost all locations on Earth, but no single species lives everywhere. Coast redwoods can be found only along the northern California and southern Oregon coasts, and emperor penguins live only in Antarctica. In contrast, blue whales live in almost all of the world's oceans (**Fig. 4.5**). Where organisms live is driven, in part, by their adaptations to the environmental conditions and the mix of living organisms found there. The environment, or habitat, of the blue whale includes all of the living and nonliving factors that surround and influence the whale. At a large scale the whale's *macrohabitat* includes regional environmental characteristics—the oceans this species lives in—whereas the whale's *microhabitat* is the smaller-scale environment—the conditions in its immediate vicinity that influence its physiology. Biologists analyze an organism's habitat at both large and small scales to answer a variety of questions. For example, why does an organism live where it does? What are the important components of an organism's habitat? How is a particular species adapted to live in locations where it is found? The answers to these questions help us understand an organism, its relationship to the living world around it, and its environment.

Figure 4.5 **Geographic range of the blue whale.** Blue whales can be found in most of the world's oceans. The darker blue area on the geographic range map (left) shows where blue whales might be found. For a sense of how large blue whales are, compare the sizes of the boat and the whale, which may grow up to 100 feet long.
© Cengage Learning 2014

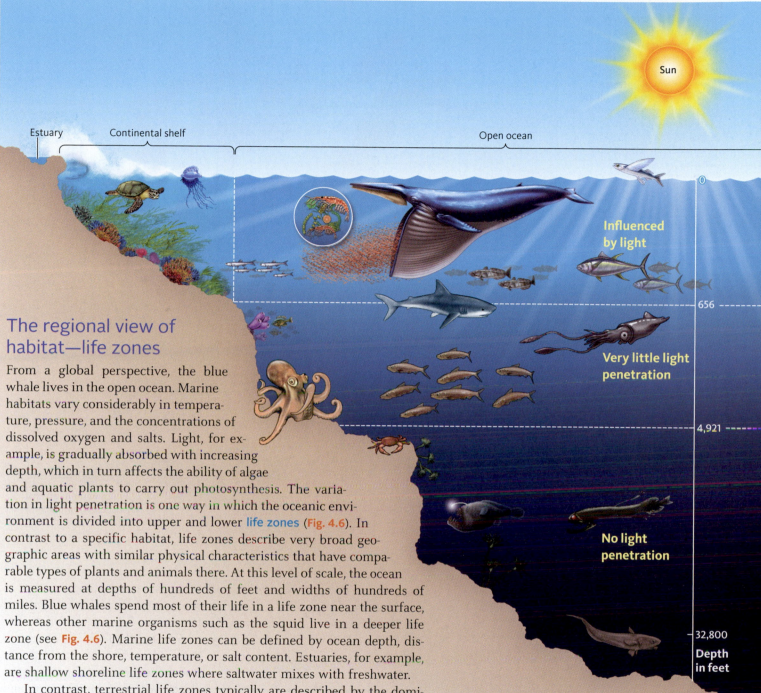

Estuary   Continental shelf   Open ocean

Sun

Influenced by light

0

656

Very little light penetration

4,921

No light penetration

32,800

Depth in feet

The regional view of habitat—life zones

From a global perspective, the blue whale lives in the open ocean. Marine habitats vary considerably in temperature, pressure, and the concentrations of dissolved oxygen and salts. Light, for example, is gradually absorbed with increasing depth, which in turn affects the ability of algae and aquatic plants to carry out photosynthesis. The variation in light penetration is one way in which the oceanic environment is divided into upper and lower life zones (Fig. 4.6). In contrast to a specific habitat, life zones describe very broad geographic areas with similar physical characteristics that have comparable types of plants and animals there. At this level of scale, the ocean is measured at depths of hundreds of feet and widths of hundreds of miles. Blue whales spend most of their life in a life zone near the surface, whereas other marine organisms such as the squid live in a deeper life zone (see Fig. 4.6). Marine life zones can be defined by ocean depth, distance from the shore, temperature, or salt content. Estuaries, for example, are shallow shoreline life zones where saltwater mixes with freshwater.

In contrast, terrestrial life zones typically are described by the dominant type of vegetation growing there. The collection of plants found in an area, whether they are grasses in grasslands or trees in forests, are linked to environmental factors such as patterns of precipitation and temperature. When you think of a tropical rainforest life zone, you think of a hot and humid habitat populated by many different species of trees. When you think of deserts, you think of a hot and dry life zone containing cacti and similar drought-resistant species. The major terrestrial life zones for North America are shown in Figure 4.7. The coast redwood grows where the winters are cool and rainy and the summers are dry. This region has annual rainfall between 78 and 118 inches and mild temperatures year-round. Factors such as latitude can play a role in vegetation patterns. For example, there are two different kinds of rainforests—temperate and tropical. A temperate rainforest exists in the Pacific Northwest, along a narrow

Figure 4.6 Ocean life zones. This chart of the ocean life zones shows where certain marine organisms live based on the depth of the water and distance from the shore. Many marine organisms are found close to shore in shallow water in the continental shelf life zone, while others inhabit deeper waters far from land. Light energy is an important resource that provides a basis for aquatic plants and algae, which in turn serve as a food source for other organisms.
© Cengage Learning 2014

Life zone—broad geographic areas with similar physical characteristics that have comparable types of plants and animals living there.

Figure 4.7 Distribution of major terrestrial life zones across North America. Each color represents an area dominated by groups of plants adapted to live in conditions with specific temperature and rainfall patterns. Temperature and rainfall are greatly influenced by mountain ranges, distance from the oceans, seasonal weather patterns, and distance from the equator (latitude).
© Cengage Learning 2014

ARCTIC OCEAN

PACIFIC OCEAN

Hudson Bay

ATLANTIC OCEAN

Gulf of Mexico

- ■ Arctic Ice Sheets
- ■ Tundra
- ■ Taiga
- ■ Hudson Plain
- ■ Northern Forests
- ■ Northwestern Forested Mountains
- ■ Temperate Rainforest
- ■ Eastern Temperate Forests
- ■ Great Plains
- ■ North American Deserts
- ■ Mediterranean California
- ■ Southern Semi-arid Highlands
- ■ Temperate Sierras
- ■ Tropical Dry Forests
- ■ Tropical Wet Forests

Scale

| 0 | 200 | 400 | 600 | 800 | mi |

| 0 | 400 | 800 | 1,200 | km |

Lambert Azimuthal
Equal Area Projection

coastal band that runs from western Washington north along the coast to southeastern Alaska. This region has cool temperatures year-round and can receive as much as 135 inches of rain per year. A tropical rainforest in the Amazon of South America receives an average of 108 inches of rainfall per year but has average temperatures higher than those of the temperate rainforest. The combination of rainfall and temperature characteristics helps define these life zones and the habitats within them. Moisture and temperature within a habitat also directly influence the types of animals, bacteria, and fungi found there.

Both elevation and latitude influence the development of life zones. For example, mountain life zones change with elevation (**Fig. 4.8**). As you climb the side of a mountain, you will notice a change in two environmental factors—temperature and wind. As a general rule in the Rocky Mountains of the western United States, the temperature drops 3° to 5°F for every 1,000 feet in elevation you climb. This drop in temperature results in a reduced growing season for the vegetation that characterizes each life zone. The direction the slope of the mountain faces also matters. In the northern hemisphere, south-facing slopes receive more solar radiation and are warmer and dryer than north-facing slopes; the opposite is true in the southern hemisphere. The timberline, where trees stop growing, begins at

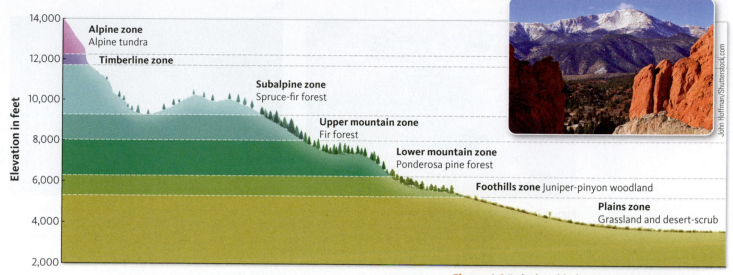

Figure legend labels (on the diagram):
Elevation in feet

- **Alpine zone** — Alpine tundra (14,000)
- **Timberline zone** (12,000)
- **Subalpine zone** — Spruce-fir forest (10,000)
- **Upper mountain zone** — Fir forest (8,000)
- **Lower mountain zone** — Ponderosa pine forest
- **Foothills zone** Juniper-pinyon woodland (6,000)
- **Plains zone** — Grassland and desert-scrub (4,000)
- (2,000)

John Hoffman/Shutterstock.com

Figure 4.8 **Relationship between elevation on a mountain and the distribution of life zones found on Pike's Peak, Colorado.** As you move up in elevation from the shrub-dominated foothills seen in the foreground in the picture at right, the dominant vegetation patterns change. As you gain elevation, the average temperatures decline and moisture conditions change, influencing the types of plant life growing there.
© Cengage Learning 2014

an elevation of about 11,000 to 12,000 feet in temperate mountains. With an increase in latitude (move further north or south of the equator), the timberline is found at lower altitudes based on the influence of general temperature patterns found there. Above the timberline, shorter plants dominate in the alpine life zone, where it is too cold and windy for trees to live. This life zone is similar to the Arctic tundra, where there are no trees and the vegetation is short and shrubby. Within the alpine life zone, as in each zone, there are a number of distinct habitats, such as wet meadows and rock fields. Animals such as the bighorn sheep live in the alpine zone along with marmots and other burrowing mammals. Other life zones include the subalpine, mountain, and foothill zones, where each has its characteristic collection of plants and animals.

The small-scale view of habitat—microhabitat

The blue whale's microhabitat is defined by the environmental factors and organisms in its immediate surroundings. At this level of scale, the whale directly experiences and adjusts to the water temperature and pressure on its skin, as well as the amount of light reaching its eyes. The microhabitat level of scale is most familiar to humans because this is the habitat in which we carry out our daily activities. The microhabitat of plants and fungi, which are rooted to one place, is smaller than that of animals, which move and experience larger areas of space.

An organism's microhabitat may be very different from its larger macrohabitat. Take an evergreen forest, for example: the microhabitat in the shaded area at the base of a tree will be cooler and more humid with less light and lighter winds than those conditions at the very top of the forest canopy. At an even smaller scale, the rough surfaces in the bark of the tree create countless numbers of microhabitats for small organisms such as insects or fungi to live in. Damage, disease, and birds such as woodpeckers create cavities in the trunks of trees (**Fig. 4.9**). These cavities form holes that collect rainwater, decaying organic matter, and airborne deposits of minerals, which combine to create conditions able to support a variety of organisms, including bacteria, mosquitoes, insects, and amphibians. These tree-hole microhabitats often have quite different physical characteristics and collections of living organisms than those in the macrohabitat of the forest.

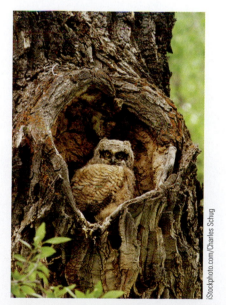

iStockphoto.com/Charles Schug

Figure 4.9 **Example of a microhabitat.** A tree hole like the one shown is created when damage occurs to the trunk of the tree. The hole collects rainwater, dust, and falling vegetation, creating a microhabitat where birds like owls and many species of plants and animals live.

a. David Hale

b. Photo by J. Krueger

Figure 4.10 Alpine habitat of the yellow-bellied marmot. (a) The rocky alpine habitat of the yellow-bellied marmot may seem harsh and uninviting to us, but it provides all the food and shelter this animal requires to thrive. (b) View from near the summit of Pike's Peak in the alpine life zone looking down toward the tree line. Marmots live in the exposed rocky terrain in the foreground.

TABLE 4.1 Marmot Habitat

Environmental Factors	Living Factors
Sunlight	Plants that it feeds on
Temperature extremes	Insects as food
Soil texture that it burrows in	Birds and their eggs for food
Oxygen availability	Presence of predators
Seasonal weather patterns	Presence of diseases
Wind and air currents	Parasites
Water availability	Other marmots
Spaces in rock crevices for hiding	Other animals that eat the same foods
Gravity	

© Cengage Learning 2014

Let's explore the habitat of the yellow-bellied marmot to understand the components of its habitat at several levels of scale. The marmot is a large ground squirrel that lives in the alpine life zones of western mountains in the United States and Canada (**Fig. 4.10**). Here in the open rocky areas near timberline forests, the marmot can easily spot predators and dig burrows to survive the long, cold winters. As the seasons change during the year, the marmot must adapt to a wide range of temperature extremes found in high mountain areas. Food supplies change along with the seasons, so the marmot has adapted to survive by eating a diverse array of plants, insects, and bird eggs based on their availability. Its skin plays host to parasites that live and feed on the marmot. What do you notice about this description? First, it specifies its large-scale life zone and then uses two general types of factors to describe the marmot's habitat—the environmental and physical factors, along with the living factors. It also includes a brief description of the marmot as a microhabitat for other creatures. **Table 4.1** lists several examples of these types of factors that you can use to characterize any habitat.

Knowing the habitat requirements for an organism provides you with a good deal of information about its adaptations and where you might expect to find it living. Habitats can be described at many different scales. Large-scale descriptions define the macrohabitat or life zones, which describe the living factors and environmental conditions across a broad geographic range, such as the alpine mountain habitat of the marmot or the ocean world of the blue whale. At the opposite end of the habitat scale are microhabitats, which describe the environmental conditions and living factors in the immediate vicinity of the organism. In the next section, we will explore how habitats provide resources necessary for survival and reproduction of organisms.

4.2 check + apply YOUR UNDERSTANDING

- Organisms are adapted to live in the habits where we find them.

- Where an organism lives can be described at many levels of scale, from the very small to very large.

- Life zones are one way to describe the physical conditions found in a habitat.

1. Distinguish between an organism's macrohabitat and its microhabitat.

2. How are life zones in aquatic habitats classified differently from those on land?

3. Describe environmental factors that comprise an organism's habitat.

4. Many plants live on the steep faces of cliffs in mountain canyons (**Fig. 4.11**), whereas others live on the level ground at the top of cliffs. Describe at least three environmental factors that differ between a vertical cliff habitat and a horizontal surface.

5. Douglas-fir is a type of tree adapted to live in cool, moist habitats. Predict whether it will be found predominantly on the north- or south-facing slopes of mountains.

Figure 4.11 Vertical cliff habitat.

Bruno Petriglia/Photo Researchers, Inc.

4.3  Organisms Acquire Resources from Their Habitat

Redwoods are massive organisms in nearly all respects, stretching up hundreds of feet and growing more than 20 feet in diameter. If you were able to look beneath the ground at the base of a coast redwood, you would see that the root system is equally impressive in size and extent, with hundreds of miles of small roots radiating out from the larger supporting roots near the center (**Fig. 4.12**). Earlier, we learned that the coast redwoods are adapted to a specific range of environmental conditions, but they also require resources such as water and many mineral nutrients they must obtain from the soil. Understanding how an organism interacts with and relies on resource availability in its environment is an important step in understanding the biology of that organism. The habitat provides food, water, shelter, and space to all the organisms that live there. Through millions of years of natural selection, each species has adapted to find, acquire, and use the resources in the habitat where it lives. In this section, we'll examine the key resources needed to support life and the relation-

Figure 4.12 **Root system of a redwood.** The root system of the enormous redwood is impressive, supporting the tree and reaching far out into the soil to obtain moisture and nutrients. The tree in this picture has toppled over, revealing a fraction of the original roots, which were torn away as the tree fell over.

ships between organisms and their habitats. Let's begin by looking at common resources organisms need.

Resources are necessary to support life

As organisms interact with their habitat, they acquire the necessary substances and energy they need to live—their **resources**. **Table 4.2** lists common sources for resources and examples of how organisms use them. All organisms require many of the same resources: water, carbon, nitrogen, and mineral elements. Organisms depend upon the same types of resources because they use them in similar ways to support their metabolism and build organic molecules such as proteins. In addition to these resources, plants and other photosynthetic organisms need light energy. Oxygen is a vital resource for plants, animals, and most fungi; however, some microscopic fungi, protists, and bacteria do not require oxygen. In fact, oxygen inhibits the metabolism of anaerobic organisms and can often kill them. Notice that producers and consumers require similar resources but obtain them from different sources in the environment. Fungi that decompose organic compounds in a dead tree on the forest floor require the same resources as other consumers.

Another resource that factors into all of life's activities is time. Time is an especially precious resource for alpine plants, such as moss campion, whose habitats are on mountain slopes above tree line. Here the environmental conditions are favorable for growth only 60 days per year. During this short time, moss campion needs to find and capture resources that will fuel the production of new roots and leaves and numerous pink flowers (**Fig. 4.13**). The resources the plant obtains must be sufficient to support growth during the short summer and also see the moss campion through the long alpine winter and permit new growth in the following spring. Many plants such as the common sunflower live an even more restricted life and must complete their entire life cycle from seed to flower and back to seed in only a single growing season. Insects, in general, also have short life spans in which they, like all living things, require the resource of time.

Figure 4.13 **Moss campion.** This short flowering plant found in the alpine life zone has a frost-free growing season of only 60 days.

Resources—the necessary substances required by an organism to survive that are provided by its habitat. **Table 4.2** lists common sources within habitats for resources and examples of how organisms use them. All organisms require many of the same resources: water, carbon, nitrogen, and mineral elements.

TABLE 4.2 Key Resources Needed by Living Things

Resources	Source in Habitat	Use
Water	Precipitation—rain, snow stored in the soil and bodies of water (lakes, ponds)	Photosynthesis Cellular respiration Growth Dissolving substances Transporting substances Entering chemical reactions Maintaining cell shape and form
Light	Sun—solar radiation	Photosynthesis Signals for biological rhythms
Oxygen	Atmosphere—O_2 Dissolved in water	Aerobic metabolism—cellular respiration and energy generation
Carbon	Producers use CO_2 in atmosphere for photosynthesis Consumers and decomposers consume organic compounds made by other organisms	Plants build sugars, starch, and cellulose All organisms use carbon in their body structure and life activities
Nitrogen	Producers derive from soil or dissolved in water Consumers and decomposers consume or decompose plant and animal matter	Builds proteins for body structures, membranes, receptors, enzymes, hormones, and immunity Heredity—nucleic acids (DNA/RNA) Plant pigments—chlorophyll
Minerals	Producers derive from soil or dissolved in water Consumers and decomposers consume or decompose plant and animal matter	Calcium—bones, shells, muscle, nerve Sodium—nerves, water balance Potassium—nerves, water balance Phosphorus—ATP, DNA, bones, shells Sulfur—protein structures Magnesium—chlorophyll, enzymes
Vitamins	Producers make vitamins Consumers and decomposers consume or decompose plant and animal matter	Metabolic reactions that result in energy generation or body structures
Heat	Sun Geothermal vents in deep ocean	Metabolic processes—photosynthesis and respiration
Shelter	Trees Burrows Caves	Protection from predators and adverse environmental factors

© Cengage Learning 2014

Species have different resource requirements

The alpine habitat of the moss campion provides limited water and minerals; few other plants can withstand these conditions (**Fig. 4.13**). Why? Most other plants require larger amounts of these two resources to support growth and development and therefore cannot survive in this habitat. In other words, each species has a specific minimum amount of any one resource it needs to survive.

Let's examine water requirements in humans as another example. Water is a critical resource necessary for our metabolism. By weight, adult males are 55% to 65% water, females are 45% to 55% water, and infants are 70% to 75% water. Who has the highest water requirements? Infants do; because a high proportion of their weight is water, they must drink proportionally larger amounts of fluid to maintain this condition. Humans

0–1%	Thirst
2%	Stronger thirst, vague discomfort
3%	Decreased blood volume, impaired physical activity
4%	Increased effort for physical work, nausea
5%	Difficulty concentrating
6%	Failure to regulate excess temperature
8%	Dizziness, labored breathing with exercise, increased weakness
9%	Muscle spasms, delirium, wakefulness
10%	Inability of decreased blood volume to circulate normally, failure in kidney function

Figure 4.14 Effects of dehydration on humans. Water is a critical resource for all life, and the physical effects of dehydration can occur rapidly. In humans, a loss of only 10% of water can cause internal organs to begin failing. While each organism has a unique need for resources, similar effects can result when other critical needs are not met.

© Cengage Learning 2014

Figure 4.15 Tropical rainforest canopy. The rich, dense collection of plants in the shade of the tropical rainforest canopy must cope with less light than plants in the upper canopy of the forest.

Andre Nantel/Shutterstock.com

Constraint—a resource available in less than optimal amounts that causes an organism to alter its behavior and/or metabolic processes in order to survive.

Tradeoff—a behavioral or metabolic shift that occurs in response to a resource constraint. All habitats have some resource limitations that influence the reproduction and survival of the organisms that live there.

are very sensitive to water loss; adults become noticeably dehydrated with as little as a 2% drop in body water (**Fig. 4.14**). As water loss increases, several physical changes occur: we breathe faster and our blood pressure and temperature changes. A 15% water loss in humans usually causes death.

Organisms that appear to be very similar may have very different needs for the same nutritional resource. Both men and women require iron for building and maintaining red blood cells. However, women 31 to 50 years old require about 18 mg of iron per day, whereas men require only 8 mg of iron per day. For women who are pregnant or nursing, iron requirements increase to as much as 27 mg per day. Extending this concept to other animals, the need for specific resources will be different based on age, growth rate, developmental stage, the environment in which an organism lives, how active it is, and its gender.

Resources limit or promote growth and reproduction

In any particular habitat, certain resources are abundant while others are scarce. A tropical rainforest habitat, for example, supplies plants with abundant amounts of water, yet light is less available for plants living near the forest floor (**Fig. 4.15**). How does a rainforest plant cope with a resource constraint, such as light that falls close to or below the plant's minimum requirement? Rainforest plants growing in low light compensate by shifting their growth pattern from roots to stems and leaves. This is one reason why vines are so prevalent in the rainforest—they typically allocate most of their resources into fast-growing stems and leaves that climb up and use the support of trees to find new sources of light. In general, leaves in light-limited habitats, such as rainforests, have larger leaf areas and come in a wide variety of shapes adapted to capture the lower levels of light at the forest floor. By redirecting its growth, the plant has evolved a tradeoff, exchanging the benefits of a larger root system for the more critical benefits of stems and large leaves necessary to obtain light. The larger leaves also serve an additional role of shading out competitors—competition for scarce resources is intense.

Animals compete in the same manner as plants for resources, with the added challenge of expending energy to find shelter and food. When an organism's daily energy needs are not met, it responds to this constraint by relying on stored energy reserves. American bison, for example, are large, plant-eating animals closely related to the domesticated cow (**Fig. 4.16**). In the winter when temperatures remain below zero for days, bison must eat about 2% of their body weight in grass each day to meet their energy needs. For a mature bull bison weighing nearly 2,000 lb, this can mean finding 40 lb of grass under several feet of snow each day! Bison compete to survive by having either enough energy reserves in fat to help them survive a long winter and/or the ability to find enough grass during the winter.

Just as constraints influence life by limiting growth and reproductive ability, at the opposite end of the spectrum excess resources also influence organisms and habitats; too much of any one nutrient can be toxic and actually kill. At larger scales, excess nutrients from fertilizers, farm fields, and even car and power plant exhaust promote rapid plant and algae growth in streams and large bodies of water such as the Chesapeake Bay and the Gulf of Mexico. This excess growth of algae has very negative environmental impacts when it blocks sunlight, creates additional sediment, and lowers the levels of oxygen and carbon dioxide in the water for local aquatic animals such as protozoans and fish.

Each species occupies its specific ecological niche

Although all organisms living in a specified geographic area share the same habitat, each species is adapted to a specific ecological role and range of microhabitat conditions that set it apart. This range of habitat conditions and its ecological role are called its ecological niche. Each organism's niche includes the environmental conditions, the resources necessary for survival and reproduction, and its ecological role or interactions such as decomposer, predator, or producer. Each fish species in a lake, for instance, can tolerate a range of salinity, temperature, pH, and food supply (**Fig. 4.17**). When the pH of the water is higher or lower than the tolerance levels shown by the cube in **Figure 4.17**, the fish cannot survive or reproduce. Likewise, if the temperature is above or below this range, the fish species shown in the figure cannot obtain enough oxygen.

The brown trout has a broad niche because it is adapted to a wide range of conditions. This ability to tolerate many different conditions means it can inhabit a broad geographic range. On the other hand, the closely related Atlantic salmon has a much narrower niche. It requires a higher water quality to live and reproduce, and it is more sensitive to environmental changes. In each case, the niches these fish species occupy represent the product of millions of years of natural selection that resulted in a unique series of adaptations to the environment.

Biologists are studying the niche of the endangered Pacific pocket mouse, a small omnivorous animal native to the sandy coastal dunes of southern California. Although biologists once thought the Pacific pocket mouse was extinct because of habitat destruction, they now are very interested in determining what factors affect the seeds and insects the mouse relies on for food, where the mouse makes its nest, and which animals prey on the mouse (**Fig. 4.18**). Understanding the role of these factors in the life of the mouse helps wildlife biologists preserve the mouse populations. The mouse has a narrow ecological niche and lives along the coast and similar areas where there are fine-grain, sandy soils; coastal dunes or river deposits; and sage scrub habitats near the ocean.

Figure 4.16 Bison feeding in winter. During the winter months, bison in Yellowstone National Park must be able to dig through several feet of snow to find enough grass to eat.

Jason Maehl/Shutterstock.com

Niche—an organism's range of habitat conditions and resources needed, along with the variety of interactions a species engages in for survival and reproduction.

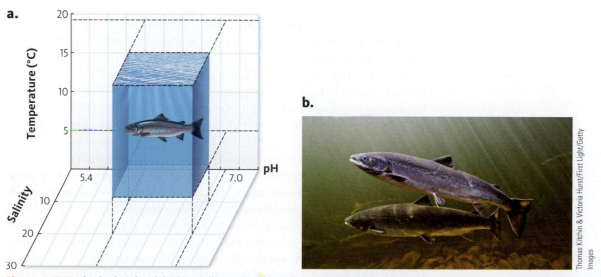

Figure 4.17 Ecological niche. (a) Three-dimensional graph illustrating a hypothetical niche for one species of salmon. The blue box indicates the niche tolerance range of this species for temperature, salinity, and pH. The fish can tolerate the conditions indicated within the shaded area but not in the area outside the box. (b) As the Atlantic salmon moves upstream to reproduce, physical conditions such as salinity and temperature change, but it remains within the bounds of its overall ecological niche shown in the three-dimensional graph at left.

© Cengage Learning 2014

Figure 4.18 Components of an organism's niche. An organism's niche (in red on map) includes many components needed for survival, growth, and reproduction. The niche for the Pacific pocket mouse includes its primary living locations, food sources, climatic characteristics, predators, and parasites or diseases.
© Cengage Learning 2014

In turn, each plant species the mouse feeds on has its own particular niche influenced by many of the same environmental factors as the mouse. Predators such as foxes and coyotes feed on the mice, and parasites such as ticks rely on the mouse for a blood meal. Along much of the mouse's habitat, there are continued concerns about human disturbance because of ongoing development. Destruction of these grasses and dunes eliminates or alters the mouse habitat and threatens the survival of this endangered species.

A significant interaction between an organism and its habitat is the requirement to obtain resources to support life. Whether it is a large redwood or a small mouse, an organism relies upon the resources available in its habitat. Although each species requires resources to build similar biomolecules, its specific requirements are unique and result from long-term evolutionary processes. Limiting resource availability constrains growth and reproductive potential, and checks the growth of populations. For each species, the relationship between an organism and the living and environmental factors in its habitat, along with its general ecological role, define its niche. Understanding an organism's niche provides a great deal of information about its general habits and ecological role. In the next section, we'll explore how organisms react to changes in their habitat.

- The niche describes the specific resources and habitat conditions an organism is adapted to.

- Habitats provide resources needed for organisms to survive.

- Resource needs are specific to different species and often limit growth and reproduction.

1. Define a resource for an organism and give an example.

2. Why do excess inputs into lakes of fertilizers that normally are of limited availability in nature cause populations of algae to explode?

3. What is meant by a species' ecological niche?

4. Two purple-flowered members of the aster family look alike except for their leaves (**Fig. 4.19**). Smooth blue aster has wide, large leaves, and tansy aster has narrow, short leaves. Predict which of these species lives in a shaded, moist habitat and which lives in a sunny, open habitat. Justify your answer.

5. Use the description below to decide whether bog spicebush shows a narrow or broad ecological niche (**Fig. 4.20**). Defend your answer. *Bog spicebush has a spotty distribution within its range from southeastern Virginia through the Carolinas west to southeastern Louisiana. It inhabits permanently moist to wet bogs where peat moss is abundant. Spicebush is found on very acidic soils that are high in organic matter.*

Figure 4.19 Two similar species of aster.

Figure 4.20 The geographic range of bog spicebush.
© Cengage Learning 2014

4.4 Organisms Respond to Changes in Environmental Factors and Resources

In nature, few environmental factors and resources found in a habitat remain constant. As conditions change over time, they influence how an organism lives and behaves. For example, in temperate regions of eastern Kansas, mature cottonwood trees produce new leaves each spring, followed by seeds in early summer (**Fig. 4.21**). The seeds have the ability to germinate quickly before temperature and moisture conditions become less favorable. As fall temperatures start to drop and the days become

Figure 4.21 Plains cottonwood. The plains cottonwood is a common sight throughout much of Kansas and the eastern half of the United States. As a deciduous tree, it produces new leaves each spring. And each fall the leaves turn color and are soon dropped, leaving the trees bare during the winter months.

shorter, the cottonwood leaves turn yellow and fall from the tree as it prepares for another winter. Light, temperatures, food sources, water availability, and shelter all are affected by the annual rhythms of seasonal changes or unpredictable catastrophic disturbance events such as fires or hurricanes.

Organisms as different as cottonwood trees and whales respond to seasonal environmental changes in many different ways, such as entering a dormant state, hibernating, reproducing, or migrating to new locations. Many species of bacteria, invertebrates, and plants have developed the ability to reduce their metabolism and resource requirements as they wait for more favorable conditions in the future. Disturbance events physically alter the habitat and result in a change in the availability of resources or environmental conditions. This type of change to a habitat creates new niche opportunities for those species that can quickly capture the new combination of resources.

Dormancy is a pause in an organism's activities

One common survival strategy in nature is found in all of the kingdoms of life. Many organisms respond to stressful environmental conditions by temporarily suspending growth, development, or reproduction. This period of an organism's life cycle is called dormancy. It occurs when environmental conditions exceed an organism's level of tolerance, such as when winter temperatures are too cold, summer temperatures are too warm, or water or other resources are too limited. Dormancy is a temporary condition that is reversed when conditions improve.

In temperate climates, trees such as the cottonwood and seeds of other plants predictably enter into a dormant period each fall. Deciduous trees such as cottonwood and maples reduce photosynthesis, shed their leaves, and survive for months on stored sugars until increased temperature, humidity, and moisture signal the trees to become active the following spring. Evergreen trees, including the coast redwood, retain their needles during the winter but curtail growth and other metabolic activities to conserve energy. Many plant species produce dry seeds that will remain dormant in the soil for years until environmental conditions become favorable for germination and growth.

Numerous species of bacteria, fungi, and protists also have dormancy mechanisms to lock water into their cells to survive periods of harsh conditions. The anthrax bacterium produces metabolically inactive cells that can remain dormant in the soil for decades. Under the appropriate environmental conditions, dormant cells become active again and can spawn many new generations of bacterial cells. Certain species of *Plasmodium,* an animal-like protist that causes the disease malaria in humans, goes in and out of dormancy based on changing environmental conditions. For example, when it is living within the mosquito host, the *Plasmodium* is inactive,

Dormancy—a survival-strategy adaptation of many organisms where they temporarily suspend growth, development, or reproduction in response to stressful environmental conditions. Dormancy is a temporary condition that is reversed when conditions improve.

but once inside a human blood cell, the protist becomes metabolically active and begins to reproduce.

One of the most interesting organisms that routinely enters a deep dormant state is the tardigrade, also called the water bear (**Fig. 4.22**). There are more than 1,000 species of these microscopic animals in the phylum Tardigrada. These small animals occupy many different niches in marine, freshwater, and terrestrial habitats throughout the world. When faced with adverse environmental conditions, they enter a dormant state where all metabolic processes cease. It seems that they can withstand any extreme condition that nature throws their way—freezing, dehydration, low oxygen, and excessive salts. Tardigrades in their dormant state are so robust they can even survive the inhospitable vacuum of space. Tardigrades can live indefinitely in their dormant state until conditions return to levels within their niche.

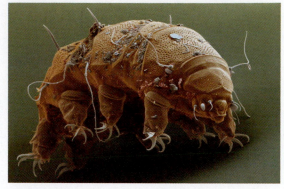

Figure 4.22 Tardigrade. The tardigrade, or water bear, is a common microscopic animal that lives in aquatic habitats. To survive unfavorable conditions, the water bear enters a dormant state, allowing it to survive extreme environments that may last for years.

Animals hibernate to survive unfavorable conditions

In the animal kingdom, maintaining a survivable body temperature and finding food become more difficult in habitats where winter temperatures are low for extended periods of time. One way that animals avoid these unfavorable conditions is by hibernating. **Hibernation** is a relatively long, seasonal period of inactivity during which the body alters its normal metabolic patterns by lowering temperatures and heart rates and slowing other bodily functions. Hibernation represents a tremendous physical advantage for many species, but it means a long time between meals.

Small rodents and many ground squirrels survive the harsh cold by creating deep burrows or well-protected nests. But how do mice last the winter without eating? Before settling in for several months of inactivity—or, in the case of Preble's jumping mice, up to eight months of inactivity—hibernators eat a large amount of food and store energy in the late summer and fall. For example, during summer seasons the weight for one species of jumping mice ranges from 11 to 25 g, but just prior to hibernation, the weight climbs as high as 35 g or more. By bulking up in the fall, the jumping mice have stored energy to survive during the hibernation period, and the animals then conserve their energy by reducing their metabolism. These metabolic changes can be very significant; the heartbeat of jumping mice slows from 420 beats per minute to as low as six beats per minute. Their body temperatures fall from summer norms of 95°F to about 40°F. Lowering their breathing rate lowers the amount of energy they need to survive and the amount of water they lose.

Not all hibernation lasts for extended periods of time or occurs just during winter months. Several species of frogs and tortoises and at least one species of lemur hibernate during summer months to avoid the extended hot, dry periods. In other cases, some animals lower their body temperature and are inactive for just a few hours each day. An excellent example is the ruby-throated hummingbird. It feeds on flower nectar during the daytime hours and hibernates at night by lowering its body temperature until it approaches that of the surrounding air. This temperature drop slows the bird's metabolic rate and conserves scarce resources. Each morning as the temperature climbs, the hummingbird's metabolism returns to normal levels and the bird becomes active again.

Hibernation—a survival strategy characterized by a relatively long, seasonal period of inactivity during which the organism's body alters its normal metabolic patterns by lowering temperatures and heart rates and overall slowing of other bodily functions.

Figure 4.23 Black bears have a modified hibernation. Although black bears are inactive for much of the winter and their heart rate slows dramatically, their body temperature does not drop as significantly. This modified hibernation allows them to be more alert to potential dangers. The red tags were attached to the ears of the bear cubs to allow wildlife biologists to track their activities over time.

Migration—the seasonal movement between habitats in response to changing habitat conditions—chiefly environmental extremes of temperature, moisture, or other resources. The physical expense involved must be exceeded by the availability of resources in the new habitat.

It may surprise you to learn that black bears are not true hibernators. During the winter denning months, black bears enter a modified state of hibernation. While their heart rates drop dramatically, their body temperature does not drop significantly. Their drowsy but ready condition allows the bears to defend themselves and their cubs should a predator disturb them. In addition, bears occasionally emerge from their dens in winter (**Fig. 4.23**).

Migration is the movement to more favorable habitats and resources

The Arctic National Wildlife Refuge in far northern Alaska is home to one of the largest seasonal animal movements in North America. The 1.5 million acre area on the coastal plain contains the greatest wildlife diversity of any protected area in the circumpolar north; it includes caribou, grizzly and polar bears, wolves, moose, and wolverines. Although the animals are well adapted to the harsh arctic habitats and are able to withstand a range of extreme environmental conditions, some species must relocate for parts of the year. This seasonal movement, or migration, is an important aspect of animal behavior related to changing habitat conditions. For example, each spring and fall, 30,000 caribou begin a seasonal movement between their summer and winter ranges in the refuge. They spend a great deal of time and expend scarce energy reserves to move from one area to the other (**Fig. 4.24**). The physical expense pays off; migration carries the herd to food sources that provide energy for their survival.

Migration is a strategy used by many mammals and other species that live in habitats that are difficult to survive in year-round. Just as the Alaskan caribou move during the spring and fall seasons, they survive by leaving the area for part of the year or part of their life and moving to habitats that are more hospitable. Animals migrate by flying, swimming, drifting, or walking. In some migrations, movement occurs in only a single direction; in others, only some members of a

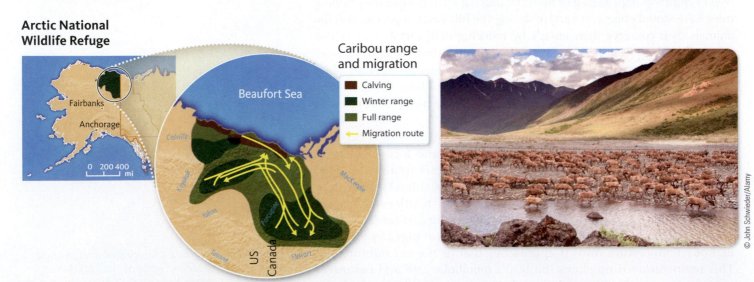

Figure 4.24 Migratory pattern of caribou. Migrating caribou create a dramatic sight in northern Alaska. Caribou migrate through parts of the Arctic National Wildlife Refuge in order to find better food sources and protection from weather extremes during different times of the year.
© Cengage Learning 2014

population migrate while others stay behind. In other cases, it may take several generations to complete a single seasonal migration.

What factors cause animals to make these movements? The trigger for this behavior can be changes in the local climate, the availability of food, the season of the year, or the signals tied to maturation processes and reproduction. The response is powerful, causing entire

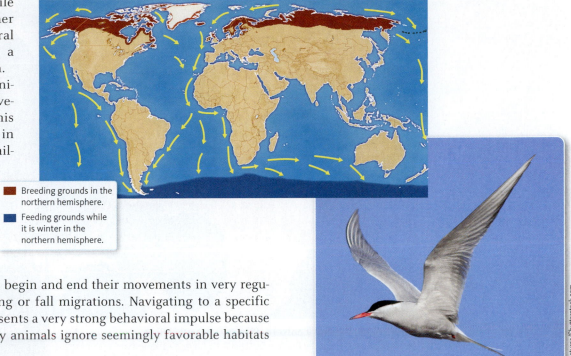

Breeding grounds in the northern hemisphere.

Feeding grounds while it is winter in the northern hemisphere.

populations of animals to begin and end their movements in very regular patterns, such as spring or fall migrations. Navigating to a specific geographic location represents a very strong behavioral impulse because many species of migratory animals ignore seemingly favorable habitats along the way.

The costs of migrating long distances include the expenditure of a considerable amount of energy, physical stress, and the risk of death. In evolutionary terms, the fitness benefits or reproductive advantages for the population must outweigh the costs. What are the benefits of migration? In some cases, such as that of monarch butterflies, migration is an adaptation to avoid unfavorable environmental conditions and food scarcity that occur during the colder months of the year in northern areas of the continent. Caribou move to take advantage of improved food availability. In other cases, such as that of the Arctic tern, it is a way for animals to find and exploit favorable habitats to support activities such as mating and reproduction.

Some animal migrations involve movement over tremendous distances between the summer and winter habitats. The champion migratory bird is the Arctic tern, which flies 12,000 miles between its northern breeding areas in the Arctic Circle south to the Antarctic continent each year (**Fig. 4.25**). During its lifetime, a tern may migrate as many as 500,000 miles—the equivalent of flying around the Earth 20 times! This long-distance movement places the birds in the most favorable habitats throughout the year. During the short summer period of the far north, the terns take advantage of the rich amount of available food to help them produce their offspring. When the feeding season ends in the north, the birds begin their long flight south.

Although relatively small in size, some insects also exhibit migratory behaviors. The monarch butterfly has a complex migratory pattern that includes a hibernation period. Because these insects cannot survive cold winters in the north, the butterflies migrate up to 2,800 miles to warmer winter hibernation spots where there is little threat of freezing temperatures (**Fig. 4.26**). To cover this distance, these seemingly fragile butterflies must fly between 50 and 100 miles each day. They stop at night and sleep in clusters to keep warm. Monarchs are the only insects that migrate these great distances and are the only insects with a two-way migration pattern. Their wing structure, metabolic endur-

Figure 4.25 Migratory pattern of the Arctic tern.
The Arctic tern is a champion migrator, moving 12,000 miles from one pole to another as the seasons change. The yellow lines show the tern's movement from its summer breeding grounds in the north to its feeding grounds in the south.
© Cengage Learning 2014

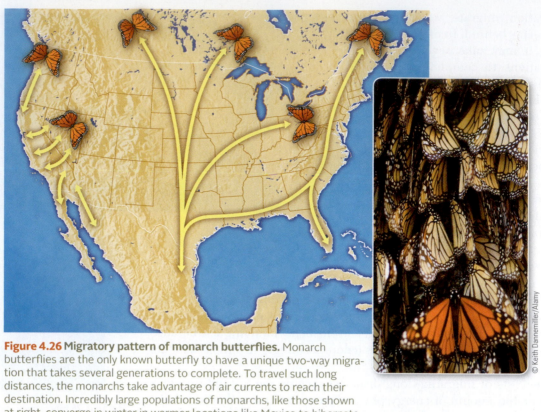

Figure 4.26 Migratory pattern of monarch butterflies. Monarch butterflies are the only known butterfly to have a unique two-way migration that takes several generations to complete. To travel such long distances, the monarchs take advantage of air currents to reach their destination. Incredibly large populations of monarchs, like those shown at right, converge in winter in warmer locations like Mexico to hibernate. In the spring, the monarchs return north, laying eggs on milkweed plants to produce the next generation.
© Cengage Learning 2014

ance, and annual reproductive cycle allow them to accomplish this phenomenal task.

Disturbance events select for opportunistic organisms

Disturbance is a strong agent of natural selection. Ecologically, disturbance events such as forest fires, tornadoes, hurricanes, and landslides can disrupt large areas of land, altering the physical conditions and changing the balance of available resources. In an ecological process called *succession,* disturbance often contributes to a shift in plant communities. After a forest fire, resources such as light and space suddenly become available on the forest floor where previously there was shade. The pioneering, opportunistic organisms that are able to disperse seeds into newly opened habitats germinate quickly and exploit the changed conditions. Pioneer species have the advantage over other plants and are the first to grow on disturbed sites (**Fig. 4.27**). Examine the ground surrounding a new construction site; those plants growing in the highly disturbed soils typically are pioneering weed species. Fires also select for seeds that can withstand the heat. In fact, many pine tree species actually need a fire to open their cones and release their seeds.

In time, as new plant species inhabit the area of disturbance, they further alter the habitat by creating shade and absorbing water and minerals as they grow. Animals and plants adapted to the altered conditions also soon migrate into the new habitat. Each succeeding group of plants and animals alters the habitat conditions, creating new niche conditions and opportunities for other species better adapted to them.

Disturbance—a range of physical events, such as a fire, hurricane, flood, or volcanic eruption, which alters the existing environmental conditions along with the ability of organisms to survive in that habitat. Disturbance includes biological events such as widespread disease or insect infestations that alter the plant and animal community structure.

Opportunistic organisms—pioneering organisms such as plants and animals that are able to quickly become established in recently disturbed habitats and exploit the changed conditions.

Figure 4.27 Disturbances create opportunities. With the overstory trees killed by fire, new plants begin to move in, taking advantage of light and water previously unavailable to pioneer plant species.

In this chapter, we have begun our exploration of the relationships between organisms and the world in which they live. Each species is adapted to live in a specific habitat, or area, with a range of environmental conditions and resource availability. Understanding these relationships allows you to apply your knowledge about an organism's tolerances and resource needs to the environment around you. For example, the eastern gray squirrel is a small animal that lives in forest habitat across much of the eastern United States (**Fig. 4.28**). It builds nests of twigs and leaves in tree branches, feeding throughout the year on tree bark and seeds. When its normal food source is limited, the eastern gray squirrel will eat insects, frogs, small rodents, small birds, and their eggs and young. In this regard, the eastern gray squirrel has broad niche adaptations, meaning it tolerates a wide range of environmental conditions, it can feed on a wide range of food, and its ecological role is that of a generalized omnivore. This species is well adapted to disturbance created by human development and takes full advantage of food sources such as bird feeders. In Chapter 5 we'll broaden our look at interactions among organisms in their habitats.

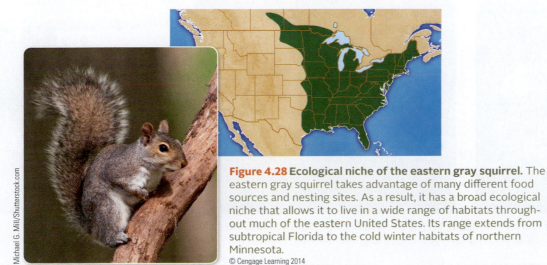

Figure 4.28 Ecological niche of the eastern gray squirrel. The eastern gray squirrel takes advantage of many different food sources and nesting sites. As a result, it has a broad ecological niche that allows it to live in a wide range of habitats throughout much of the eastern United States. Its range extends from subtropical Florida to the cold winter habitats of northern Minnesota.
© Cengage Learning 2014

- Organisms are adapted to respond to changes in environmental conditions.

- Organisms have evolved many different types of responses to cope with less than optimal environmental conditions.

1. Name three ways that animals respond to unfavorable changes to their environment.
2. List three conditions that cause tardigrades to enter dormancy.
3. Which hibernation characteristic distinguishes the ruby-throated hummingbird from the black bear?
4. Russian thistle is a major invasive weedy species located across much of the Great Plains. It rapidly invades newly disturbed sites such as plowed fields and construction sites. Explain what characteristics this successful opportunistic plant must possess (**Fig. 4.29**).
5. The volcanic eruption of Mount St. Helens in Washington State on May 18, 1980, was a major disturbance that destroyed dense evergreen forests and crystal-clear lakes and streams (**Fig. 4.30a, b**). **Figure 4.30c** was taken several years after the eruption. It shows red alder trees living along streams that cut through the blast zone still littered with downed trees. Would you expect the alder to be less or more tolerant of overstory shade than the tree species that occupied the site previously? Would the plants now growing under the mature alder share that characteristic?

Frank Awbrey/Visuals Unlimited, Inc.

Figure 4.29 Russian thistle.

a.

Photo by D.A. Swanson/USGS

b.

© 2005 Donald E. Hall, All Rights Reserved/Getty Images

c.

© Dan McShane

Figure 4.30 Eruption of Mt. St. Helens and its aftermath. (a) Ash cloud during eruption of Mount St. Helens in 1980. (b) Felled trees still are visible 25 years after the eruption. (c) Red alder trees have become established following the eruption of Mount St. Helens.

End of Chapter Review

Self-Quiz on Key Concepts

Habitat and Resource Needs

KEY CONCEPTS: Variations in Earth's geography and climate create an immense variety of habitats at various scales of size from large life zones to small microhabitats. Every species is adapted to its habitat and lives within its ecological niche. Furthermore, each species has a need for specific resources found in the habitat where it lives. Resource availability influences an organism's ability to grow and reproduce.

1. Match each of the following **ecosystem terms** with its description or definition.

niche	a. a resource that falls below the minimum needed
life zone	b. the habitat and specific range of conditions and ecological role that
habitat	an organism is adapted to
resource	c. all of the living and environmental factors that surround and
constraint	influence an organism
tradeoff	d. a broad-scale habitat designation, often based on the dominant
	vegetation
	e. using limited resources or energy for one activity instead of another
	f. necessary substances and energy used for living

2. A description of a habitat for a squirrel living in your backyard would include:
   a. the annual temperature cycles.
   b. the food it requires to survive.
   c. the tree that it builds a nest in.
   d. the ticks that live on the skin of the squirrel.
   e. all the above.

3. *Maggot* is a term used to describe the larval life stage for flies. Maggots hatch in the flesh of decaying animals and do best in warm temperatures up to 100°F. This would be an example of the maggot's:
   a. environmental requirements.
   b. biological requirements.
   c. niche.
   d. life zone.
   e. constraints.

4. What is the general relationship between environmental temperature and an organism's metabolic rate?
   a. As temperature increases, so does metabolic rate.
   b. As temperature increases, metabolic rates slow down.
   c. There is no general relationship between environmental temperature and metabolic rate.

5. The need for a specific resource for any one individual may differ based on:
   a. resource availability.
   b. activity level.
   c. constraints.
   d. geographic range.

Organisms Respond to Changes in Their Habitats

KEY CONCEPTS: Living and nonliving factors are important components of an organism's habitat that influence growth, survival, and reproduction. In nature, few environmental factors and resources found in a habitat remain constant, and over time these changes influence how an organism lives and behaves in its habitat. Disturbance events alter habitat conditions and create opportunities for new species. To maximize growth, survival, and reproduction, organisms respond to changes in environmental factors and resources through dormancy, hibernation, or migration.

6. Match each of the **behavioral adaptations** with its definition or description.

dormancy
hibernation
migration
opportunism

a. a temporary suspension of growth, development, and reproduction
b. a common response to disturbance
c. a relatively long, seasonal inactive period when metabolic rates and body temperatures are lowered
d. the seasonal movement toward more favorable habitat or resource conditions

7. In general, hibernation helps an organism conserve:
   a. body heat.
   b. time.
   c. energy.
   d. reproductive potential.

8. Migration refers to:
   a. a short-term alteration in resource adaptation.
   b. a change in behavior brought about by predator pressure on populations.
   c. the tradeoffs made between resource needs and environmental conditions.
   d. the deliberate movement of populations from one place to another.

9. Which of the following events might significantly disturb a habitat, causing new niches to be created and a new mix of living organisms to move into the area?
   a. a hurricane
   b. a flood
   c. a large-scale fire
   d. a landslide
   e. Each of the above qualifies as a disturbance event.

10. The period when monarch butterflies overwinter on trees, remaining stationary and lowering their metabolic rate, would be an example of:
    a. migration.
    b. dormancy.
    c. hibernation.
    d. none of the above.

Applying the Concepts

11. Habitats and life zones are described by researchers based on the living and nonliving characteristics found there. In this chapter we learned that patterns of environmental conditions such as rainfall and temperature play a major role in the distribution of life. Use **Figure 4.31** as a starting point to help you catalog the characteristics of the habitat and life zone where you live.

    Using the climograph in **Figure 4.32**, characterize the range of average temperatures and rainfall found in your life zone. Using **Figure 4.31**, if you were to travel straight north from your home, what life zones would you pass through on a direct line to the north pole?

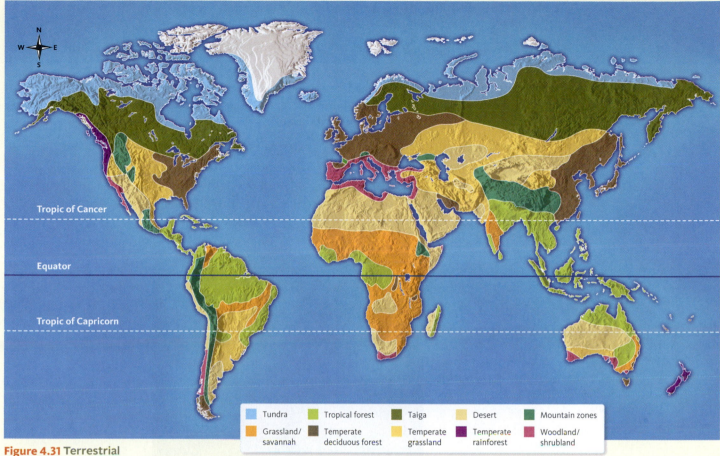

Figure 4.31 Terrestrial life zones.
© Cengage Learning 2014

Legend:
- Tundra
- Grassland/ savannah
- Tropical forest
- Temperate deciduous forest
- Taiga
- Temperate grassland
- Desert
- Temperate rainforest
- Mountain zones
- Woodland/ shrubland

12. The climograph in **Figure 4.32** shows the relationship between average moisture and temperature on the distribution of several life zones. Given the global concern over climate change and long-term climatic shifts, explain how a temperate rainforest might change if the average precipitation levels were cut in half.

Tundra life zones are found throughout much of the far north. They store a great deal of carbon in their soils, which remain frozen throughout most of the year. Examining the climograph in **Figure 4.32**, how would an average annual temperature increase of 5 degrees change these life zones?

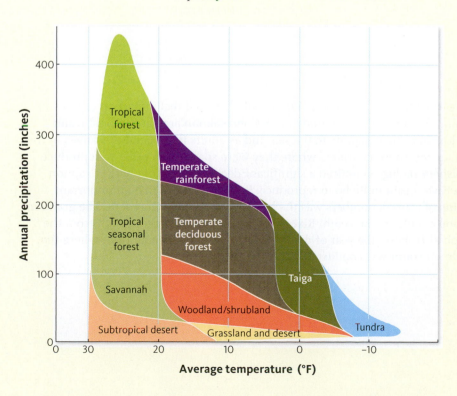

Figure 4.32 Climograph.
© Cengage Learning 2014

13. Why is the availability of certain elements such as nitrogen important when biologists study the ability of organisms to survive in a habitat?

14. Contrast the terms *habitat* and *niche*. Is it possible for multiple habitats to provide the same niche?

15. Acid rain is a type of chemical disturbance that lowers the pH of the soil, affecting the availability of minerals to plants. How would building a coal-fired power plant that promotes acid rain impact downwind plant communities over time?

16. How is it possible that organisms can live in the same habitat but not the same niche?

17. The big brown bat (**Fig. 4.33**) is a common insect-eating North American bat that often roosts and hibernates in lofts and buildings. How might the activities of the bat change prior to its entering hibernation? How would you describe changes in its cellular metabolism and core body temperature during hibernation?

Figure 4.33 Roosting site of big brown bats.

Data Analysis

The Migration of the Coho Salmon

In this chapter, we learned that habitat modification could have a direct impact on organisms that utilize that habitat for survival and reproduction. Coho salmon are born in freshwater rivers and streams. As juveniles they migrate to the sea, and as adults they spend their lives in the ocean. When mature, they return to the waters where they were spawned in order to reproduce. Starting in the 1950s, fishery biologists noticed a significant decline in the numbers of salmon returning to the Lemhi River basin in Idaho to reproduce. Ecologists noted that an important contributor to the decline of salmon numbers was the building of dams in the basin. The graph in **Figure 4.34** depicts salmon counts on the Lemhi River several hundred miles upstream from the Pacific Ocean. The graph also shows the year of dam development downstream from where the annual Lemhi River salmon count was conducted.

Coho Salmon Populations along the Lemhi River (1957–1997)

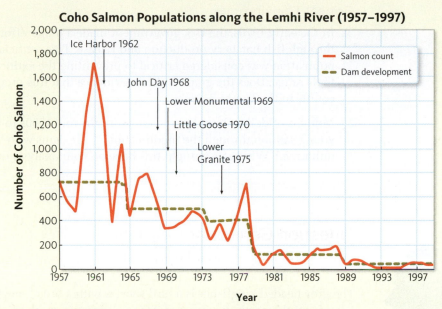

Figure 4.34 Salmon counts on the Lemhi River, 1957–1997.
© Cengage Learning 2014

Data Interpretation

18. Does the graph show an abrupt or gradual change in the size of the salmon population?

19. Did the peak salmon population on the Lemhi River occur before or after the installation of the first dam?

20. The federal government has mandated that each dam on the river be equipped with a fish ladder (**Fig. 4.35**) to assist salmon migrating upriver in getting around dams. Based on the data for the Lemhi River, have fish ladders been a successful investment since the 1960s?

Figure 4.35 Fish ladder next to a dam.

21. The Ice Harbor Dam completed in 1961 is situated on the Snake River far downstream from the Lemhi River. Speculate why this one dam seems to have had the greatest impact on salmon counts so far upstream in Idaho.

Critical Thinking:

22. In the early 1970s, the U.S. government made a major effort to increase salmon populations through fish hatchery production and release. This artificial means of increasing the fish population was considered critical to preventing the extinction of salmon throughout much of their range. Does the graph indicate evidence that this program has had any impact on salmon counts in the Lemhi River?

23. Millions of dollars have been spent since the mid-1980s to improve downstream migration of young salmon through the modification of the dams with structures that permit the fish to pass unharmed. What effect might this have on your analysis in the previous question?

Question Generator

Interpreting Migratory Patterns: The Grey-Headed Kingfisher

The grey-headed kingfisher is a bird species with a broad migratory range from the Cape Verde Islands off the northwest coast of Africa and across much of southern Africa and southern Arabia (**Fig. 4.36**). The grey-headed is a dry-country kingfisher, living in savannah and woodland habitats near water. Unlike most kingfishers, the grey-headed does not live in aquatic habitats; rather, it inhabits holes in steep riverbanks and aggressively protects its nest against predators. Environmental conditions related to seasonal shifts in rainfall are critical habitat cues that drive the birds' migratory and reproductive patterns. For example, as rainfall on the savannah increases, grasses grow and produce seeds that the kingfishers feed upon.

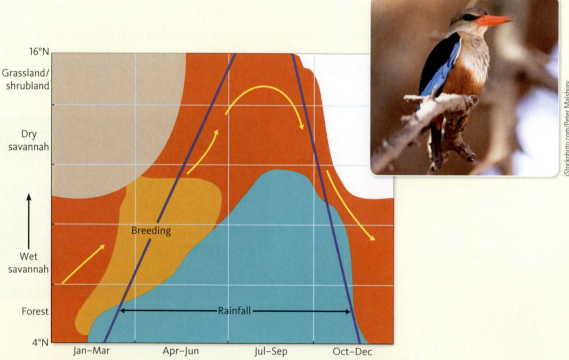

Figure 4.36 Migratory range of the grey-headed kingfisher. Grey-headed kingfisher resting on a branch. Relationship between the time of year (x-axis) and the bird's general location (y-axis). A blue line inside the graph indicates rainfall amount as a function of time of year. The bird's location shifts as a function of food availability, which is tied to rainfall, which occurs primarily between April and October.

Graph is adapted from Dingle, Hugh, *Migration: The Biology of Life on the Move*, Oxford University Press 1996, New York, p. 236, Elgood, J.H., Fry, C.H., Dowsett, R.J. 1973. African migrants in Nigeria. *Ibis* 115:1–45: 375–411 and Sinclair, A.R.E. 1983. The function of distance movements in vertebrates. In I.R. Swingland and P.J. Greenwood, eds. *The Ecology of Animal Movement*. Oxford, pp. 240–258.

Figure 4.37 is a block diagram that relates time of year, habitat type, and rainfall for the grey-headed kingfisher. Use the background information, the graph on page 128, and the block diagram to ask questions and generate hypotheses. We'll start you off by simply translating components of the block diagram into sample questions. Note that the arrows indicate the movement of the bird populations between habitats at different times of the year.

1. How does rainfall influence the movement of the kingfisher? (Arrow #1)

2. How does the impact of rainfall on food resources influence the movement of the kingfisher? (Arrow #2)

3. How much time of each year does the kingfisher spend in drier climates? Forest habitats? (Arrow #3)

4. Which habitats does the kingfisher prefer for breeding and why might that be? (Arrow #4)

Use the block diagram to generate your own research questions and frame a hypothesis.

Figure 4.37 Factors relating to kingfisher location.
© Cengage Learning 2014

5 Ecological Interactions among Organisms

5.1 Termites—a layering of communities and interactions

5.2 Organisms interact at different levels of scale

5.3 Populations of organisms interact in different ways

5.4 Energy flows through ecosystems

5.5 Interactions in social groups increase fitness

5.1 Termites—A Layering of Communities and Interactions

Figure 5.1 Damage caused by termites. Termites use their strong mouthparts and a partnership with microbes to feed off of the cellulose stored in wood fibers. As you can see in this picture of a log cabin destroyed by termites, the consequences of a termite infestation can be dramatic.

Imagine returning to your family cabin in the woods after several years away. You arrive with many fond memories only to find that the cabin has collapsed to the ground (**Fig. 5.1**)! In shock, you begin to investigate what happened. The foundation, along with metal hardware and pieces of glass, are intact and undamaged, but all the wooden beams in the cabin have collapsed. In fact, most of the boards and beams appear to have tunnels and tubes running through them. Termites *(Reticulitermes flavipes)* have destroyed your family cabin.

Although termites may look like ants that could really use some sun, they belong to a separate order of insects (**Fig. 5.2**). Most of the 2,500 different species of termites go unnoticed, living out of sight in decaying wood, soils, grasses, and forests—until they get our attention by destroying our houses. In spite of the damage caused by some termites, they play an important role in our ecosystems: recycling nutrients, enriching the soil, providing habitats for microbes, and serving as an important food source for predators.

One interesting fact about termites is their adaptation to a diet of cellulose-rich woody plant material. The cellulose molecule comprises long chains of sugar molecules bonded together in a way that is impossible for animals to break apart and digest. Termites not only possess powerful mouthparts able to cut apart tough wood fibers, but they also carry with them an arsenal of other organisms that enable them to digest this tough plant material.

Through the process of natural selection, a mutually beneficial relationship has evolved between termites and microbes such as bacteria and protists that live in the termite's digestive tract (see **Fig. 5.2d**). The termite–microbe relationship is so well established that each species of cellulose-eating termites has evolved with its own specific collection of microbe species living in its digestive tract. What makes the relationship so important is that the two very different types of organisms are completely dependent upon one another for survival.

The mutual survival of termites and microbes results from the specialized role each plays in the life of the other organism. The immense community of microbes in the termite's digestive tract secrete enzymes—specialized proteins that break down the large cellulose molecules into smaller organic molecules. The entire community, microbes and termites, benefits from the production of usable sugars that both termites and microbes can digest, and the microbes benefit from living in the low-oxygen, cellulose-rich habitat created by the termite body. The microbes rely completely on the food source provided by the termites, and the termites depend so heavily upon this relationship that if the microbes were removed from their digestive tract, the termites would starve.

Termites have other intriguing adaptations. For instance, they live in a highly ordered society with a well-defined division of labor and interactions. Termite societies can be very large—a single nest can contain over a million members. With so many termites under one mud roof, communication is important for their survival. They share information about food sources or dangers by touching one another or by leaving chemical signals that other termites recognize and act upon. Each termite belongs to a particular caste (or functional social group) that performs a specialized job in the colony (see **Fig. 5.2**). For instance, most termite nests contain several

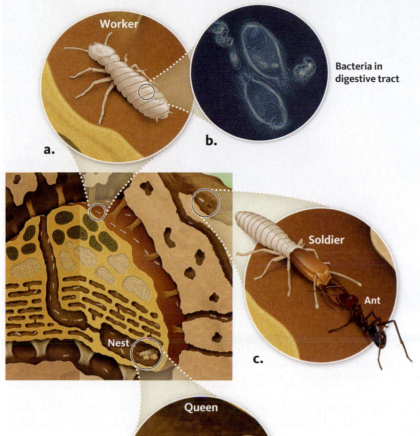

Figure 5.2 Termite castes. Termites have a distinct social order, with different termites performing different roles within the nest. The different roles, called castes, have very different physical forms that support their role. For example, the soldier termites have huge mouthparts for fighting, and the queen has an enlarged abdomen filled with eggs. (a) Worker termite. (b) Examples of the many species of bacteria and protists that live in the digestive tract of cellulose-eating termites. (c) Soldier termite defending nest against invading ant. (d) Termite queen.

reproductive pairs called *queens* and *kings*. The queens can live up to 10 years and produce 2,000 eggs each day, one reason that termite populations in a single nest can get so large. Queens can even reproduce without kings, producing new offspring asexually. Worker termites are the most numerous, and their activities dominate the termite nest. They care for the eggs and immature termites, find and deliver food to the colony, and build and maintain the nest. Soldiers defend the nest and food sites from invading insects and small animals. The soldiers are easy to spot in a termite colony because they have large mouthparts they use as weapons. Their mouthparts are so specialized for combat that the soldiers are unable to eat without the help of worker termites. Using overlapping generations and a cooperative system of care for the eggs and immature termites, members of the nest rely heavily upon one another for survival.

Back at the family cabin, if you cleared away the debris and began digging, you would find the underground termite nest filled with specialized chambers. Further exploration might reveal several termite nests in the surrounding oak–maple forest, where there is no shortage of wood for them to feed on. Each nest competes for similar resources and is part of a larger population of termites in the nearby woods. Not all termites make

Michel Lepage, Pascal Jouquet (Laboratoire Fonctionnement et Évolution des Systèmes Écologiques, UMR 7525-École Normale Supérieure) and Yves Le Goff (École Nationale Supérieure des Arts et Métiers, Laboratoire Mécanique des Fluides)

Figure 5.3 An above-ground termite nest. Some termites build part of their nests above ground. This example nest in Africa, built from moist soil and feces, is as hard as concrete. In this view the side has been cut way, revealing the many different internal chambers created by the termites—some for breeding, rearing young, and storing wastes, or used as vents to help keep the nest cool.

their nests underground; sometimes portions of the nest extend above ground, while other nests hang in trees and resemble large wasp nests. The worker termites move tons of earth to build nests and complex tunnels leading to the colony's food sources. Workers use their own feces, saliva, and mud to build protective structures as strong as concrete. The biggest colonies take years to build; they can get as large as 30 feet tall with tunnels 240 feet long (**Fig. 5.3**).

As soon as you expose the nest under the destroyed cabin, payback will come in the form of hungry raccoons, bears, and birds, which make a tasty meal of the termites. These feeding interactions represent pathways where energy is passed from one organism to another. In the forest, the flow of energy begins when photosynthetic plants use light energy to build organic molecules. Energy is stored in cellulose molecules that make up the bulk of wood used to build the cabin. The microbes and termites transform the stored chemical energy of wood into energy they can use for growth, development, and reproduction. The energy stored in the bodies of the termites is then passed along to termite predators that use their stored energy for similar activities.

In this chapter we will arrange individuals into their levels of organization and explore a variety of relationships that will allow you to more fully understand the natural ecosystems in which you live, work, or visit. Like the termites in their forest nest, each organism is both well adapted to the environmental conditions found in its habitat and able to compete for the resources needed for it to survive and reproduce. To be successful requires a diverse range of interactions of the organism with its environment and with other organisms, many of which involve specific behaviors that improve the population's evolutionary fitness. Within communities, organisms participate in many types of interactions that can have positive, negative, or neutral effects. For example, energy may enter an ecosystem and be transferred from one organism to the next through feeding relationships. Let's begin by examining various scales of interactions.

5.2 Organisms Interact at Different Levels of Scale

Cattle egrets are tropical and subtropical wading birds that nest in large colonies near bodies of water. Their diet consists mostly of insects, small frogs, and earthworms. The habitats around the egrets' range are alive with organisms large and small, which interact in many different ways and on many different levels of scale (**Fig. 5.4**). At the lowest level of scale, one organism interacts with another or with its environment; at larger levels of scale, the interaction may occur between large groups of organisms such as flocks of different species of birds. These interactions take on many forms and include competition, feeding relationships, and symbiotic relationships. For example, individual cattle egrets may land on the back of a large herbivore and feed on insects stirred up by its feeding and movement. Two egrets that land on the back of the same herbivore compete with one another for feeding territory.

Similar types of interactions can be found among organisms of all sizes. Within their gut, termites interact with a complex collection of specialized microbes that help digest their food. Externally, individuals interact with other termites in the nest as well as with other species, including those that prey on them. Finally, termites interact with the nonliving components of their environment that influence their daily lives and provide the resources necessary for survival. In this section, we'll examine organisms at many different levels of scale and their relationships with their habitats.

Awei/Shutterstock.com

Figure 5.4 Egrets develop relationships at all levels of scale. As their name implies, cattle egrets have a close relationship with cattle, feeding on insects and parasites found on the cows' skin. This white bird nests in large colonies, usually near bodies of water and often with other species of wading birds. The yellow tag on the cow's ear identifies who owns the animal, and the horns have been clipped for safety.

Donald M. Jones/Minden Pictures/Getty Images

Organisms of the same species form a population

In the opening section, you learned that each termite is host to many different species of microbes. Each species living in a particular geographic area with uniquely adapted individuals represents a separate population. For example, within a single termite's gut there are more than 100 different species of microbes, each representing a unique population confined to that physical area. Likewise, the forest habitat contains many populations of organisms, from microbial soil bacteria to larger, more recognizable populations of animals and plants (**Fig. 5.5**). Scientists and wildlife managers may also describe populations at larger geographic scales, such as state or regional levels.

Studying populations gives us information about which environmental or biological factors are important and cause change in the lives of organisms. Ecologists study interactions at the population level to learn more about the abundance of a species and how it is distributed across the landscape, and to better understand how the species and its ecosystem are interconnected. Recall from Chapter 1 that marine biologists collect data about populations of whales to make generalizations about the entire species. Ecologists also study populations to learn about what resources, such as food, may be limiting a population's growth. Understandably, populations are sensitive to changes in the availability of their food. Herbivores go hungry when droughts reduce plant growth significantly. Ultimately, they may perish from starvation.

Wildlife managers study populations to learn about the reasons behind changes in the population size of a species. White-tailed deer are common throughout much of the United States, and populations have

Population—a species living in a particular geographic area with uniquely adapted individuals.

Figure 5.5 Diverse populations can occupy the same habitat. Oak–maple forests contain many different ecological niches and a rich array of resources. You can see in this figure how many diverse populations of species occupy the same habitats, potentially interacting in diverse ways. Communities such as this support a high level of species diversity. Familiar examples include deer, worms, fungi, raccoons, and birds such as the American robin.

Figure 5.6 White tailed deer. Wildlife managers intensively manage white-tailed deer herds to ensure resource demand and resource availability are not too far out of alignment.

Carrying capacity—the ability of a habitat to sustain a particular population size based on the resources available over a specific period of time.

been closely studied for decades to document how they react to changes in their habitats. Throughout much of their range, deer herds lack natural predators, a situation that often causes their populations to explode. Similarly, mild winters or an abundance of food may allow populations to rapidly increase. Managers know that the available plant population used as their primary food source isn't enough to support the deer, which in turn will cause the deer population to suffer poor health or starvation. In place of natural predators, wildlife managers use hunting as a strategy to control deer populations (**Fig. 5.6**). However, if the deer population is considered too small, managers place stricter regulations on hunting to allow the population to increase. Biologists also monitor populations for the presence of diseases or other factors that may affect the entire species.

Ecologists use the term carrying capacity to describe the ability of a habitat to sustain a particular population size based on the resources available over a specific period of time. As populations rise and fall, competition for scarce resources can become more or less intense, ultimately influencing the size of those populations. The carrying capacity for large, migratory herbivores such as the American bison has been studied in Yellowstone National Park for decades (**Fig. 5.7**). Each summer, biologists count the bison in the park and estimate the amount of grass available to support that number of animals through the following winter. If the herd is too large to survive on the available grass, animals need to be removed or migratory corridors opened to support the movement of bison to food sources outside the park. Similarly, population levels below a habitat's carrying capacity provide an opportunity to increase population numbers to meet resource availability. The carrying capacity of a habitat varies for different species and changes over time with changing food availability, water supply, environmental conditions, and living space.

Communities support diverse groups of organisms

At the next higher level of organization are communities—collections of populations of different species located in the same habitat. Communities usually are labeled or described based on the dominant organisms found in a specific geographic area. For instance, the forest surrounding the cabin shown in **Figure 5.1** is referred to as an oak–maple forest community. The term community can also be used in a more general sense, such as the fish community in Lake Superior.

Communities rich in resources and different ecological niches generally support a large and diverse number of species (see **Fig. 5.5**). Recall from Chapter 4 that *ecological niche* describes a species' range of habitat conditions, resource needs, and ecological role. The oak–maple forest is an example of a resource- and niche-abundant community, having nutrient-rich soils, favorable annual rainfall and temperature patterns, and many potential microhabitats.

One index used by ecologists to describe communities is species diversity, a measure of the number and relative density of different species found in a geographic area. Species diversity represents a type of biological census, tallying the species present, their numbers, and their density in comparison to other species present. Species diversity can be measured at many different levels of scale, from within a single community to large regional areas. In contrast to the high species diversity of the oak–maple forest, communities such as deserts have limited resources and fewer niches and therefore support comparatively lower levels of species diversity. As with populations, changes in species diversity in a community provide biologists with clues about the importance of interactions between organisms and the influence of fluctuations in environmental conditions over time.

In Chapter 4, we learned that resource availability in a habitat has a direct impact on the organisms that live there. Changes in resource availability on a seasonal or year-to-year basis influence the species diversity in a habitat by altering the numbers of species. Similarly, an increase or decrease in population size changes the demand on resources and affects resource availability. The connections among diversity, population sizes, and resource availability are important to understanding how species relate to one another and to their habitat.

Biological and physical worlds interact in ecosystems

The next higher level of ecological scale is ecosystems, which contain collections of species, communities, and habitats all functioning in their environment. The oak–maple forest in the opening section represents a temperate deciduous forest ecosystem (see **Fig. 5.5**). Ecosystems, like communities, are often identified using dominant features such as vegetation types or terms that describe the environmental conditions. For example, the term *temperate* refers to the seasonal climate that is generally relatively moder-

Photo by Caitlin Krueger

Figure 5.7 American bison. Large migratory herbivores such as the American bison have huge food resource needs. Wildlife managers study the population size in comparison to grass availability to determine the carrying capacity of Yellowstone National Park each year.

Communities—collections of populations of different species located in the same habitat.

Species diversity—a measure of the number and relative density of different species found in a geographic area. Species diversity represents a type of biological census, tallying the species present, their numbers, and their density in comparison to other species present.

Ecosystems—geographic areas that contain collections of communities and habitats. Often defined by the dominant plant communities or environmental conditions found there.

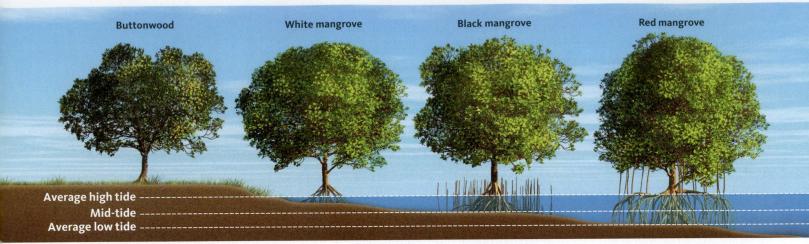

Buttonwood   White mangrove   Black mangrove   Red mangrove

Average high tide ----------------------------
Mid-tide ----------------------------
Average low tide ----------------------------

Figure 5.8 Different species of mangrove have slightly different niches. *Mangrove* is a term that refers to several closely related tree species, each growing in close proximity to the other but with slightly different niche adaptations. This diagram shows the transition from the red mangrove growing at the lowest elevation, which tolerates higher levels of water saturation, to the buttonwood growing at the highest shoreline elevation, which is adapted to dryer conditions. Note that while the diagram shows each species separated from the others, because of tidal variations, an overlap of mangrove habitats occurs.
© Cengage Learning 2014

Photo by J. Krueger

Photo by J. Krueger

Figure 5.9 The mangrove ecosystem. The mangrove ecosystem represents a transition between the ocean and terrestrial ecosystems, having characteristics of both. The top picture shows the prop roots of mangroves growing within the tidal zone, while the bottom picture shows the transition between the shallow estuary and the terrestrial mangrove ecosystem.

Transition zone—a location where the biological world and environmental conditions of the adjacent habitats blend together. The conditions of the blended habitat support species and contain ecological niches found in habitats on both sides of the zone.

ate, rather than extremely hot or cold, whereas *deciduous* describes the characteristic of trees that shed their leaves each fall. Other environmental factors, such as seasonal changes in light, climate, moisture, and nutrient levels of the soil, all affect the ecosystem. In some cases, ecosystems are named based on more general features such as being aquatic or marine.

To understand how an ecosystem works, let's examine the interactions between the biological and physical worlds in a mangrove ecosystem located along the coasts of south Florida. The term *mangrove* refers to both a specific tree species (for example, red mangrove *Rhizophora mangle*) and several closely related species that grow along the shore and have slightly different niche requirements (**Fig. 5.8**). Mangrove trees large and small dominate the shoreline to create a nearly solid green wall of vegetation. Mangrove ecosystems are found in the boundary area between the saltwater ocean and terrestrial habitats (**Fig. 5.9**). This boundary area is known as a **transition zone**, a location where the biological world and environmental conditions of the adjacent habitats blend together. The conditions of the blended habitat support species and contain ecological niches found in habitats on both sides of the zone.

The physical characteristics of this ecosystem are important to maintaining this diversity. The physical world of the mangrove forest ecosystem is heavily influenced by the ocean water's salinity and level, seasonal rainfall, and temperature patterns. Mangrove ecosystem requirements are different from those of the adjacent grass and sedge marshes found in the Everglades, which are dependent upon freshwater depths that change seasonally. Along the coastline, changes in water level driven by ocean tides result in exposure of certain mangrove species to the air part of the day and submersion at other times. While the landscape appears flat, there are small changes in height of a few inches that have a large impact on which mangrove grows there. Ocean water mixes with freshwater, supporting a range of species that are tolerant to changes in salinity. At high tide, saltwater fish may move into the submerged root zone, finding cover and food sources. At low tide when the water moves out, terrestrial species

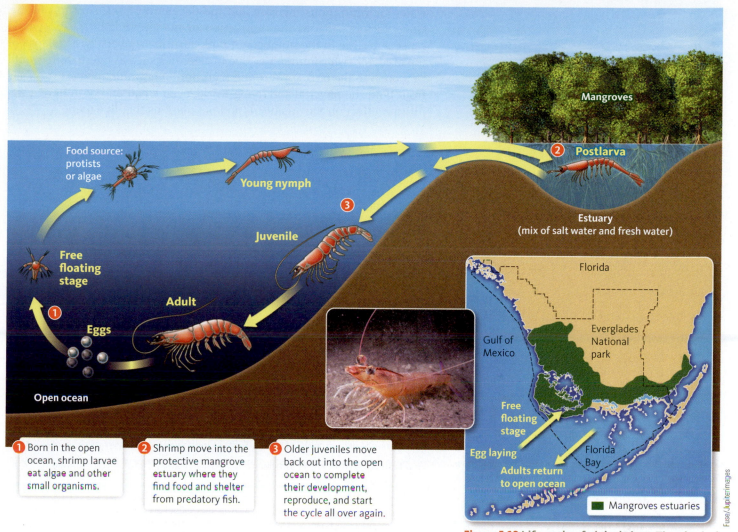

1 Born in the open ocean, shrimp larvae eat algae and other small organisms.

2 Shrimp move into the protective mangrove estuary where they find food and shelter from predatory fish.

3 Older juveniles move back out into the open ocean to complete their development, reproduce, and start the cycle all over again.

Figure 5.10 Life cycle of pink shrimp. The physical requirements for pink shrimp change as the shrimp develop. In response, they move among many different habitat locations, making them particularly sensitive to habitat disturbances in any one location that would alter their ability to develop and survive.

© Cengage Learning 2014

such as land crabs are present. Mangrove stems and roots create a tangled mass of potential sites for organisms to live and find refuge from predators. The abundant aquatic biodiversity attracts a wide range of wading birds, such as great egrets and roseate spoonbills, that feed and nest in or near mangroves. In this narrow geographic zone along the coastline, we find terrestrial, semiaquatic, freshwater, and saltwater communities coexisting alongside of one another.

One of the many aquatic species living part of its life in the mangrove forest ecosystem is the pink shrimp. Pink shrimp are an ecologically important species; they are predators of small invertebrates. In turn, the shrimp are a key food source for wading birds and larger predatory fish. The shrimp spend several months of their juvenile life living in the protected mangrove forests while they grow and develop. Shrimp are very sensitive to environmental factors such as temperature and the salinity of water. Juvenile pink shrimp prefer a habitat with moderate temperatures and salinity levels lower than that of seawater. The mangrove forests provide just the proper conditions unless something upsets the balance (**Fig. 5.10**).

During the summer rainy season, freshwater flows into the mangrove habitat from rivers. The flow of freshwater into the near shore habitat is critical to maintaining optimum water salinity and shrimp nursery conditions. If freshwater is diverted from the tidal zone by a canal or dam, saltwater concentrations increase and the developing pink shrimp can die. Other factors, such as the destruction of mangrove habitat, also would cause a decline in the numbers of pink shrimp. Conversely, a collapse of the predator fish communities due to activities such as overfishing would be a boon to the pink shrimp, and their numbers would increase. Impacts from both the physical and living world, even small ones, can have a significant effect on the organisms and populations that live there.

5.2 check + apply YOUR UNDERSTANDING

- Populations are made up of closely associated members of a single species.

- Communities are collections of different populations living within the same habitat or ecosystem.

- Biological and physical factors exert a significant influence of the types of organisms found in habitats.

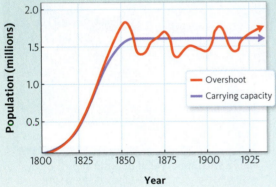

Figure 5.11 Changes in pheasant population.
© Cengage Learning 2014

1. In your own words, describe how the terms *population, community*, and *ecosystem* are related to each other.

2. Predict what might happen over time to species diversity in a forest community that experiences periodic severe forest fires.

3. Explain why transition zones often contain species found in adjacent habitats.

4. **Figure 5.11** shows 125 years of data illustrating changes in a fictional population of pheasants. Explain why the actual population of pheasants rarely meets the carrying capacity line after 1850.

5. Mountain ranges contain many ecological transition zones (**Fig. 5.12**). The slope and the altitude have an effect on the amount of solar radiation, rainfall, wind velocity, and temperature influencing the development of individual habitats. Which side of the mountain will have the greater number of transition zones, the windward or leeward side? Defend your answer.

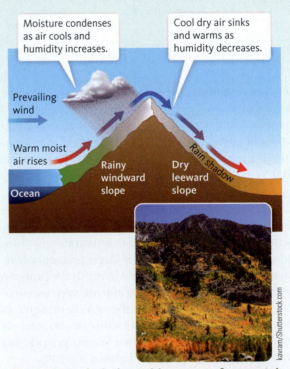

Figure 5.12 Ecological transition zones of a mountain.
© Cengage Learning 2014

5.3 Populations of Organisms Interact in Different Ways

The species that live in a community interact in a variety of ways. Some interactions are subtle, such as worker termites feeding their soldiers; other interactions are more dramatic, such as a wolf pack's coordinated attack on a moose. The interactions between organisms and their environment link them together at all levels of the hierarchy in an ecosystem. Consider the tiny house sparrow, a small seed- and insect-eating bird that weighs less than 1½ oz, which is about the weight of four quarters (**Fig. 5.13**). Although house sparrows aren't native to North America, they have successfully spread across the continent. House sparrows have thrived by being aggressive competitors. They are so aggressive, in fact, that they can displace larger birds such as bluebirds, forcing them from their territory. House sparrows bring their own way of interacting with the communities they overrun.

In many cases, the most visible interactions among community members are those involving feeding behaviors and the competition for resources (including territory or mates). Each interaction has an important influence on the types and numbers of a species you find in any community. As you read this section, consider both the interaction taking place and the range of potential outcomes that result from the relationship. With a better understanding of how and why organisms interact, you gain insight into the workings of the natural world around you. For ecologists, studying the interactions in a community allows them to make better-informed decisions concerning the management of wildlife and the resources they need to survive.

Organisms compete for resources in different ways

All species require many kinds of resources for survival, and their populations grow or decline in response to changes in resource availability. Within a community, organisms must compete with each other for similar resources such as food, nutrients, or living space. This struggle, called **resource competition**, occurs between members of the same species as well as between members of different species for the limited resources available in every habitat. Resource competition helps shape community structure by increasing the fitness of the superior competitors while reducing the fitness of or eliminating those members least able to compete. The best competitors get more food, better living sites, and access to mates, thus improving their survival and reproductive success.

Competition between members of *different* species, called **interspecific competition**, occurs when the species have some similarity or overlap in their ecological niche. The greater the *niche overlap*, the more intense the resource competition will be. For example, several species of squirrels and chipmunks living in a forest must compete for the same primary food resource—plant seeds (**Fig. 5.14**). The resources required by the animals are available in limited amounts, and niche overlap increases the degree of competition between the species

Figure 5.13 A pair of house sparrows. These small birds are mighty competitors, often forcing much larger birds from nests and food sources.

a. Chipmunk **b. Ground squirrel**

c. Abert squirrel

Figure 5.14 Three similar species compete for the same food source. Small animals such as the (a) chipmunk, (b) ground squirrel, and (c) Abert squirrel all inhabit Ponderosa pine forests in Colorado, and all compete for the same seeds as their primary food source.

Resource competition—a struggle, or competition, for resources such as food, nutrients, or living space that occurs within a community. This level of competition occurs between members of the same species as well as between members of different species for the limited resources available in every habitat.

Interspecific competition—competition between members of different species.

Figure 5.15 An example of niche overlap. Niche overlap between species A and B is far greater than with species C. Competition intensity increases with the amount of niche overlap.
© Cengage Learning 2014

(**Fig. 5.15**). In this graph, species A and B show greater overlap in terms of the way they use a resource than does species C. Due to the high amount of overlap for species A and B, the competition for resources is greater between them; each species will, therefore, try to exclude competitors from getting food resources by excluding other animals. Consider whether A or B even compete for resources with species C. With little or no overlap between A or B and C, little competition likely occurs. Given that resources are limited, it makes sense that the competition for food resources is a major component of territorial behavior in many animal species.

Species with similar resource requirements can coexist in the same area if they alternate their use of the shared resource during the year or obtain the resource from slightly different locations. This process of subdividing resources, called **resource partitioning**, reduces competition intensity. Interspecific competition is very common among plants (**Fig. 5.16**). Similar species of grasses and flowering plants growing in the same location compete for growing space, moisture, light, and soil nutrients. The species have a common, but not identical, need for these resources.

Songbirds that live in and around forest habitats provide an excellent example of resource partitioning between species with overlapping ecological niches. Although their primary food is insects, they usually feed on different types of insects, in different places, and at different times. The yellow-headed blackbird feeds on insects living in fields near the edge of marshes. The red-headed woodpecker feeds on a different set of insects found on the trunks of trees (**Fig. 5.17**). Larger birds such as the northern cardinal feed on larger insects, and smaller birds such as the chipping sparrow feed on small insects. Specialized beak shapes are among the adaptations that allow these similar species to successfully coexist in the same habitat.

Competition between members of the *same* species, called **intraspecific competition**, is very intense when there is complete niche overlap. Males competing for mates during the breeding season provide many dramatic examples of intraspecific competition. For instance, male Rocky Mountain goats fight each other for reproductive dominance (**Fig. 5.18**). This competitive process, called *dominance hierarchy,* eventually drives the younger or weaker male goats away from the females; fitness for the entire herd is increased. In other cases, animals of the same species subdivide habitats into territories, which contain sufficient food and den or nesting sites. Forming territories may ease direct resource competition between groups. Although the need for a specific resource varies with factors such as gender, development, and age, each member competes *intraspecifically* for the same types of resources.

Figure 5.16 An example of resource partitioning. Three plants (grass and two flower species) have similar but not identical resource needs. Slight variations in resource needs permit some degree of coexistence by partitioning the available resource, creating a differential ability to compete.

Adam Jones/Visuals Unlimited, Inc.

Resource partitioning—a process of subdividing the limited resources within a habitat that ultimately reduces competition intensity between species.

Intraspecific competition—an intense form of competition between members of the same species that occurs when there is complete niche overlap.

a.

Chas/Shutterstock.com

b.

Istockphoto.com/Rich Phalin

c.

Doug Lemke/Shutterstock.com

d.

Steve Byland/Shutterstock.com

Figure 5.17 Examples of food source partitioning. Songbirds living in the same habitat partition their primary food source of insects by feeding in slightly different locations, on different sized insects, or at different times. (a) red-headed woodpecker. (b) yellow-headed blackbird. (c) northern cardinal. (d) chipping sparrow.

William Church—Summit42.com/Flickr/Getty Images

Figure 5.18 Battling for mating rights is an example of intraspecific competition. Mountain goats engage in male-to-male fighting to establish dominance for mating purposes.

Predation is one organism eating another

All predators have diets made up of the available prey species in the community where they live. Predation refers to an interaction in which an animal kills and eats another to obtain its food energy. If you have ever watched a nature show, you are familiar with scenes of predators tracking and pursuing their prey. The blazing fast cheetah chasing the terrified gazelle across the African savanna is an example of predation in action. The unseen blue whale quietly eating thousands of pounds of krill each day is a less dramatic but equally important example of predation.

Some predators are *specialists,* feeding on just one or two types of prey species. Snail kites are birds of prey that live in the wetlands of the Everglades where they survive almost exclusively on apple snails (**Fig. 5.19**). The hooked portion of the snail kite's upper beak has evolved the exact degree of curvature necessary to reach inside the shell of adult apple snails and extract them for food. Specialization is a risky strategy that allows a predator to take advantage of a very specific food source unavailable to other animals. Specialists are very sensitive to changes in the ecosystem that alter food availability. In the case of the snail kite, habitat destruction and changing water levels in the Everglades ecosystem have significantly reduced the location and numbers of apple snails. As a result, the snail kite's population has fallen to critically low levels.

In contrast, many predators are *generalists,* feeding upon many species of animals as the opportunity arises. Coyotes are generalist feeders; their primary food sources are small mammals such as mice, ground squirrels, and rabbits. However, as the opportunity arises, coyotes will feed on large animals such as white-tailed deer or, if necessary, small beetle larvae and other insects. This high degree of dietary flexibility allows generalists to live in a wide variety of habitats. In addition, their ability to quickly switch food sources helps insulate them against changes in prey abundance or habitat disturbances.

Although all predators are dependent on their food source to sustain their population size, food availability is rarely balanced with demand. In some cases, this imbalance creates a time-lag response in year-to-year predator and prey populations. An example of this *population dynamic* occurs between wolves and moose living on remote Isle Royale in western Lake Superior (**Fig. 5.20**). The graph in **Figure 5.20** shows the repeated rise and fall in wolf and moose populations over 50 years. Compare the population changes between the two species. Notice that although they show similar patterns, the timing of the rise and fall is out of sync. As the availability of the wolves' primary food source, the moose, increases over time, the wolf numbers tend to rise in following years. The growing population of wolves eventually reduces the moose population to the point where the remaining numbers can no longer sustain the predator population, resulting in a steep decline in wolf numbers. When wolf numbers are low, moose populations start to rise again.

Even though they might be present in small numbers, certain organisms play a much larger role in communities. Ecologists use the term keystone species to describe organisms that have a disproportionally large effect on community structure by helping to regulate the types and numbers of various other species in a community (**Fig. 5.21**). So why are keystone species so important? Within a diverse community, a larger number of species are interacting and creating food webs that are increasingly complex. Removing a keystone species has a cascading effect on important characteristics such as diversity, food availability, and even

James Zipp/Photo Researchers/Getty Images

Figure 5.19 Snail kites are food specialists. Snail kites are the ultimate specialist feeders, surviving on a diet of native apple snails. Note the curved hook of the beak, which is perfectly adapted to reaching inside the shell to remove the snail.

Predation—an interaction in which one animal kills and eats another to obtain its food energy.

Keystone species—a term used to describe organisms that have a disproportionally large effect on community structure by helping to regulate the types and numbers of various other species in a community. An example would be a top-order predator.

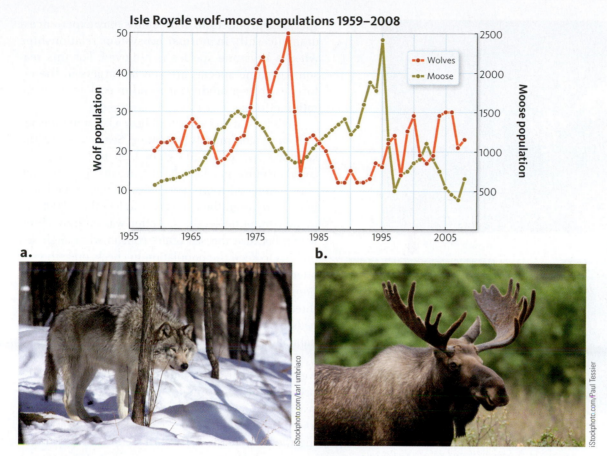

Isle Royale wolf-moose populations 1959–2008

Wolf population / Moose population

Wolves
Moose

Figure 5.20 Relationship between Isle Royale wolf and moose populations. On the remote island of Isle Royale, the gray wolf and moose populations are tied together—rising and falling slightly out of synch with one another. The graph shows the ever shifting relationships between (a) wolves and (b) moose over nearly 50 years.
© Cengage Learning 2014

Figure 5.21 Keystone species can be plants or animals. (a) Starfish. (b) American crocodile. (c) Whitebark pine.

Reproducing fish in kelp beds

Sea urchins

Figure 5.22 Sea otters are a keystone species. Sea otters feed on sea urchins, which keeps their populations in check. In places where sea otters have been removed, sea urchin populations have soared, which in turn reduced the kelp beds used by many fish species for reproduction.

species reproduction. Ecosystems may experience a dramatic shift in normal ecosystem relationships when a keystone species is removed. For this reason, keystone species are special targets in the efforts to protect biodiversity and in ecological restoration efforts.

Let's look at an example of how a keystone species influences community structure. Sea otters are the keystone species in the giant kelp beds along the coastline of the northern Pacific Ocean (**Fig. 5.22**). They once numbered more than 300,000, but unregulated hunting reduced their numbers to less than 2,000; in many communities, the sea otter was entirely eliminated. Because the sea otters consumed enough sea urchins to keep the population in check, once the sea otters were nearly eliminated, the sea urchin populations increased dramatically. As a result, sea urchins threatened their own primary food source, the kelp beds. The kelp beds represent a critical reproductive habitat for several species of fish, and as they declined the fish populations began to decline. When sea otters were protected by international law, their populations began to recover in many areas, and the number of sea urchins was reduced to normal levels. This allowed the kelp beds to become reestablished, improving the habitat for fish reproduction. Removal and recovery of this keystone species resulted in a cascading effect that significantly altered the species diversity for the entire coastal marine community.

Mutualism benefits both species in a relationship

Many organisms have evolved relationships in which both species benefit from the interaction. This type of relationship, called **mutualism**, is very common in nature. Mutualism develops because it results in an increased level of fitness. Recall the termites introduced in Section 4.1 and the collection of microbes that live in their gut. The benefits each obtains from this relationship ensure the others' survival. The wood-digesting microbes survive in the anaerobic, protected environment of the termites' gut and are continually fed chewed-up cellulose fibers. Without the digestive assistance of the microbe community, the energy contained in the cellulose-rich wood fibers would be unavailable to the termite.

The sea anemone and clownfish have evolved a mutualistic relationship based on protection and defense (**Fig. 5.23**). Anemones are stationary animals that attach to rocks or corals. They use their long, flexible tentacles to reach out into the water to feed. The tips of each tentacle contain toxins that the anemone can inject into its prey. Although the tentacles of anemones are dangerous to most fish, the clownfish actually takes refuge among them without experiencing harm. A slimy mucous coating secreted by the clownfish protects it from the anemone toxins. The sea anemone benefits from this relationship because the clownfish are aggressive and chase away the few predatory fish that can eat the anemone. In this relationship, both the anemone and clownfish have increased fitness based on the presence of and interaction between the two.

Mutualism—a very common relationship between different species in which both species benefit from the interaction, which develops because it results in an increased level of fitness for both species.

Figure 5.23 **Clownfish and sea anemones have a mutualistic relationship.** Clownfish hide in the protective cover provided by the sea anemone and at the same time chase away fish that feed on the anemone.

Commensalism benefits one organism in a relationship

Another category of interactions, called commensalism, occurs when only one of the two species involved receives a benefit from the relationship. The other organism involved is often a host that provides a home or transportation to another species but is neither harmed nor benefited by the interaction. For example, birds perching in the relative safety of a tree benefit, without benefit or harm to the tree.

Plants are often hosts to other organisms in commensal relationships. Many trees are home to plantlike organisms called lichens, which live on the outer surface of tree bark (**Fig. 5.24**). Lichens are an interesting organism—they are a mutualistic association between a fungus and either a photosynthetic alga or cyanobacterium. Tree bark provides a moist, shady habitat on which lichens can attach and grow. The lichens benefit from this relationship by having a favorable habitat, while mature maple trees with their thick, corky bark are unaffected from the interaction. You will learn more about these interesting organisms in Chapter 15.

Commensalism also occurs among a variety of aquatic organisms. The remora is a fish that has its top fins adapted into suction cups. Using these specialized fins, they temporarily attach themselves to larger fish such as sharks and hitchhike through the ocean (**Fig. 5.25**). The suction disks don't harm the host, and the streamlined remoras add little to no drag or energy expenditure to their host fish. Remoras also take advantage of the host's meals; their mouth projects forward and they are able to feed on scraps of food dropped by the host.

Parasitism benefits one organism in a relationship at the expense of another

Biologists use the term parasitism to describe the interaction between one organism, a parasite, that benefits at the expense of another organism, the host, which supplies resources. Perhaps you're familiar with an organism that fits the description, but in this case, we're referring to specialist feeders that live part or all of their lives in close association

Figure 5.24 **Lichen is a commensal organism.** Lichens grow in large numbers in the habitat provided by the bark of trees such as the sugar maple without causing the trees any harm. The lichen benefits from the space, moisture, and minerals provided by the maple tree trunk. The lichen organism itself represents a separate commensal interaction between green algae and fungi, seen as distinct layers in the cross section.
© Cengage Learning 2014

Figure 5.25 **Remora fish benefit from their relationship with sharks without harming their hosts.** Remora, like this one attached to the belly of the shark, hitchhike through the ocean attached to sharks or other fish. The remora gain access to food scraps from the shark but neither harm nor benefit their hosts.

Commensalism—a relationship between species where only one of the two species involved receives a benefit from the relationship. The other organism involved is often a host that provides a home or transportation to another species but is neither harmed nor benefited by the interaction.

Parasitism—an interaction between organisms where one organism, the parasite, benefits at the expense of another organism, the host, which supplies resources.

Figure 5.26 Bedbug. Bedbugs are small insect parasites that use their specialized mouthparts to draw blood from their sleeping host. In some cases, bedbugs go unnoticed by their hosts. In some cases, their feeding causes red welts similar to mosquito bites.

with a different species. All parasites specialize in drawing their nutrients directly from their host—some live and feed on the outside of the body while others make their way inside to places such as the digestive tract or muscles. Generally, most parasites such as fleas, lice, and tapeworms are much smaller than their host, and reproduce more quickly and in larger numbers. Over time, the loss of nutrients, the presence of the parasite itself, or infections that occur from the actions of the parasite harm the host. It's rare for a parasite to kill a host; death tends to happen when a new host with no defenses is infested or when huge populations of parasites overwhelm the host.

Humans serve as host to many types of parasites that have plagued us through the centuries. Bedbugs are found throughout the world. They are small (less than 1/4 inch long) insects that use their needlelike mouthparts to pierce skin and feed on the blood of human hosts or other warm-blooded animals (**Fig. 5.26**). Because bedbugs live in the cracks or crevices of the mattress and feed while you sleep, they are difficult to detect. Small red welts that itch or blotches on your skin similar to mosquito bites reveal locations where bedbugs have fed. Although bedbug bites are a nuisance and can become infected, they rarely cause long-term harm. Not being overly intrusive in their feeding impacts allows them to continue to feed night after night.

Plants also serve as hosts to parasites. Some of these parasites are bacteria, viruses, or fungi that cause disease. Some plants can be parasites on other plants. There are more than 4,000 species of parasitic plants that draw sugars, water, or minerals from their host. Mistletoe, a plant associated with the winter holidays, is actually a collection of damaging parasitic plant species. They are easy to spot, growing in large clumps on the stems or branches of larger plants. Although mistletoe plants have leaves for photosynthesis, they produce rootlike structures that penetrate the tissues of the host plant and siphon off water and nutrients (**Fig. 5.27**). Over time, the host plant may become weakened from the loss of nutrients and die.

In this section, we examined the role of interactions between organisms and how they shape the natural world around us. For example, predation is an important interaction that influences species diversity and distribution. In a similar manner, overlapping niche requirements create a

Figure 5.27 Mistletoe. Mistletoe are highly successful parasitic plants. The bushy, green clumps of foliage in these trees are collections of parasitic mistletoes and infected branches. These clumps are called witches broom for their unique appearance. Although they have green leaves, they draw most of their nutrition from host plants, sending feeder roots into the nutrient-bearing tissues of their hosts.
© Cengage Learning 2014

Sugar transport tissue

Water transport tissue

Parasitic mistletoe

Tree branch

competition for resources that also influences community structure. It is important to understand that as resource needs increase in similarity between different organisms, competition intensity increases, with the highest intensity occurring between members of the same species. You've also seen that certain plants or animals, such as the sea otter, play a keystone role in maintaining or shaping community structure. The next section explores the association between feeding relationships and energy within an ecosystem.

5.3 check + apply YOUR UNDERSTANDING

- Competition among and between species for resources takes on many forms.

- Interactions between different species include predation, mutualism, commensalism, and parasitism.

1. Define the term *commensalism*. How does commensalism differ from a mutualistic relationship?

2. Explain how competition intensity is related to the similarity of ecological niche.

3. If you were assigned to visit a natural community and determine its structure, what data would you collect?

4. Eyelash mites are microscopic animals that live in the hair follicles of human eyebrows and eyelashes. The mite has mouthparts adapted for eating skin cells and oils, which accumulate in the hair follicles. Usually, the mites cause no adverse symptoms and go unnoticed. However, in some cases the mite populations can dramatically increase, resulting in itching, inflammation, and other skin problems. What causes their populations to increase? Under normal conditions, what kind of relationship exists between eyelash mites and humans? Does this relationship change when skin problems occur?

5. In the mountains of Montana and Wyoming, two closely related birds—crows and the larger ravens—act as both predators and scavengers. They compete for the same territories and food resources. What kind of competition is described? Predict the intensity of this competition. Defend your answer using concepts in this section.

5.4  Energy Flows through Ecosystems

Every time you eat, whether it is an apple or a hamburger, you become part of an energy chain that can be traced directly back to the sun. The leaves of the apple tree (producer) converted the sunlight, carbon dioxide, and water into sugars in the apple, which you (consumer) in turn ate for a morning snack. Just two links in this chain lie between you and the sun. In the case of a hamburger, energy is transferred from the sun to the corn's leaves to its seed, and then to the cow and to you—three steps. This example illustrates the main concept of this section—energy flows from a source, such as the sun, throughout an ecosystem. This flow acts like an electrical current, which continually drives life's processes and activities.

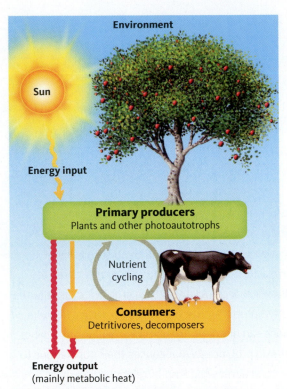

Figure 5.28 Energy flow through an ecosystem.
Energy inputs from the environment flow through producers, then through consumers. Some nutrients released by decomposition get cycled back to producers. All energy that entered this ecosystem eventually flows out of it, mainly as heat.
© Cengage Learning 2014

Primary producers—autotrophic organisms within an ecosystem, such as photosynthetic and chemosynthetic organisms, which are able to transform light or chemical energy into biological molecules directly.

Consumers—organisms such as animals and fungi that consume and convert the energy stored in producers, perhaps in the form of sugars, into fuel to run their life activities.

Trophic level—one position or level within a food chain or food web that describes the feeding or energy relationships in a particular ecosystem.

Food chain—an illustration that maps the specific feeding arrangements between organisms in an ecosystem.

When the microbes inside the termite's gut dismantle their host's cellulose meal, the process joins the termite and microbe together in a special kind of relationship—a mutualistic feeding relationship. If you expand your view outward from the termites to the large area of the forest, you will see that the flow of energy joins together all members of the community. The same is true on Isle Royale and the community of gray wolves, moose, and the plants that the moose eat. As organisms die or excrete wastes, the microbes living in the soil use the energy remaining to run their bodies. The sun's energy is channeled through the living world. By understanding how this happens, you will develop deeper insights into the interconnectedness of life in an ecosystem.

Food webs describe the feeding relationships in an ecosystem

The chain starts with a large source of energy, which for most ecosystems is the sun. However, in ecosystems at the bottom of deep dark oceans, heat and dissolved chemicals from geothermal vents drive and support life. A third source of energy can also originate from breaking down the chemical components of other organisms. Energy from the nonliving world is thereby captured and thus enters and drives the living world.

Let's work our way along the food chain starting with the first link—the producers. Plants, algae, and bacteria convert energy from sunlight or inorganic chemicals to create energy-rich carbon molecules. In an ecosystem, these autotrophs, or photosynthetic and chemosynthetic organisms, are referred to as **primary producers** (**Fig. 5.28**). Producers form the first link in the energy chain and therefore are critical to the functioning of all ecosystems.

Consumers occupy the next few links in the chain. Primary consumers, or organisms that eat producers, occupy the second link on the chain. This group includes animals and fungi that consume and convert the energy stored in producers, perhaps in the form of sugars, into fuel to run their life activities. Herbivores such as cows are one example. The third link of the chain includes organisms that feed on primary consumers—*secondary consumers.* These are the carnivores that eat herbivores. Carnivores such as wolves on the next link are called *tertiary consumers.* The leftovers, so to speak, are consumed by scavengers and *decomposers.* These heterotrophs, including many species of bacteria, archaeans, birds, insects, and fungi, feed on dead or dying organic matter, transferring the remaining energy and recycling the nutrients. Each of these positions in the chain is called a feeding level, or **trophic level**.

A more specific way to illustrate the feeding relationships in a particular ecosystem is a **food chain** diagram (**Fig. 5.29**). Food chains map specific feeding arrangements between organisms in an ecosystem. Producers, such as plants, algae, and phytoplankton, form the base of the food chain. From the primary producers, consumers are linked using a series of arrows that extend upward. The direction of the arrow indicates where one organism feeds upon another. In this way, food chain diagrams are simple ways to show feeding and energy-transfer relationships in an ecosystem.

The *trophic structure* of an ecosystem allows us to visualize the flow of energy, the network of links, and the hierarchy of feeding relationships among producers, consumers, and decomposers. Examine **Figure 5.30**, which illustrates the trophic structure for a terrestrial and an aquatic habitat. Notice that the grasshopper and zooplankton occupy the same trophic

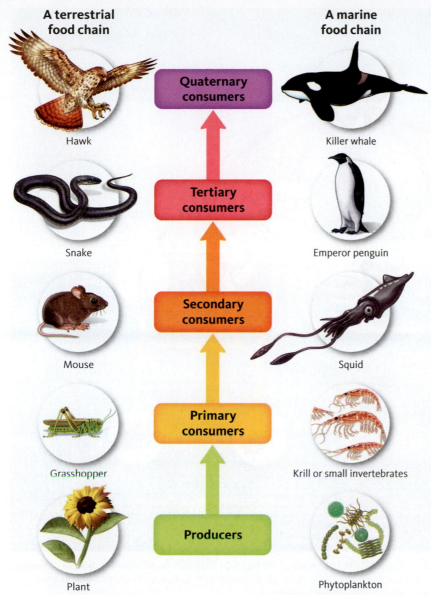

A terrestrial food chain

Hawk

Snake

Mouse

Grasshopper

Plant

A marine food chain

Killer whale

Emperor penguin

Squid

Krill or small invertebrates

Phytoplankton

Quaternary consumers

Tertiary consumers

Secondary consumers

Primary consumers

Producers

Figure 5.29 A food chain. A food chain diagram illustrates feeding relationships and energy flow in an ecosystem. Food chains can be drawn for any ecosystem, terrestrial or aquatic.
© Cengage Learning 2014

level as primary consumers in each ecosystem. Each is considered a primary consumer, feeding on the photosynthetic producers in their habitat.

Many consumers, especially generalists, feed on more than one source of food. Small rodents, ducks, shorebirds, and insect-eating birds can eat the same species of grasshopper (see **Fig. 5.30**). When several interlinking food chains are shown, the feeding relationships form a food web. As in the food chain, arrows show the flow of energy from one trophic level to the next. Note that arrows, and thus energy, move in only one direction—away from the source.

Not all organisms on each trophic level are consumed by organisms on the level above them. What happens to these organisms? Two groups—the detritivores and the decomposers—eventually "consume" the organisms by dismantling their organs, tissues, cells, and molecules. In the food

Food web—an illustration showing the range of potential feeding relationships among many organisms in an ecosystem.

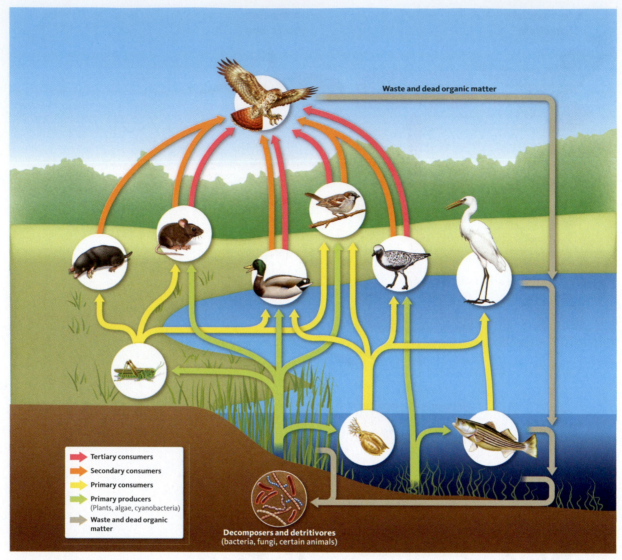

Waste and dead organic matter

Tertiary consumers
Secondary consumers
Primary consumers
Primary producers
(Plants, algae, cyanobacteria)
Waste and dead organic
matter

Decomposers and detritivores
(bacteria, fungi, certain animals)

Figure 5.30 A food web. Food webs show the multiple pathways of energy flow through an ecosystem. Not all pathways occur at the same time. Instead, which pathway energy flows through is dependent upon many factors such as the season, resource availability, species present, and disturbance patterns.
© Cengage Learning 2014

web, *detritivores* are scavengers such as earthworms, crabs, and vultures, which gain the energy and nutrients of decaying matter by feeding on the dead remains of plants and animals (**Fig. 5.31**). Many different species of bacteria, archaeans, and heterotrophic fungi function as **decomposers**, completing the process by breaking down the remaining organic matter and waste products and absorbing the remaining molecules into their bodies. The energy that started with the producers has flowed through the ecosystem, driving the organisms at each trophic level.

Energy transfer in food chains is limited

If you count the number of trophic levels in a food web diagram, you'll find that there are rarely more than five links between the producers and the top. Why are there so few links? The answer lies in the fact that or-

Decomposers—an organism such as a bacterium or fungus that feeds on dead organisms and breaks down their large biomolecules into simpler forms able to be recycled in an ecosystem.

ganisms do not incorporate 100 percent of the energy they consume into their body structures; energy is lost each step of the way (**Fig. 5.32**). In animals such as birds and mammals that maintain a constant body temperature, some of the energy they consume is lost as heat. Most predators eat only particular parts of their prey; for example, they do not eat the hooves or antlers. Even if they did, they are not able to completely digest, absorb, and use these body parts, because hooves, hair, and feathers are resistant to digestion. In fact, on average across many different food webs, only 10 percent of the energy transfers between each trophic level. This means the organism uses only 10 percent of the potential energy resource to produce biomass (growth); the remaining 90 percent is used to support metabolic reactions or is lost to the environment as heat.

This sequential reduction in energy transfer as you move up from producers to consumers forms an *energy pyramid* (see **Fig. 5.32**). In the pyramid, each trophic level narrows, containing less biomass and stored energy than the level below it. At the base of the pyramid is the energy available on Earth from the sun. Using photosynthetic metabolic pathways, primary producers convert about 1 percent of the total solar energy that strikes Earth's surface into chemical energy contained in molecules such as sugars. As the grasshopper (primary consumer) eats the plants, only about 10 percent of the energy is incorporated into its body (see **Fig. 5.32**). As the number of trophic levels increases, the amount of energy that transfers diminishes. This pyramid shows that fewer animals can live at the top of the pyramid on the food energy provided by their food source at the next lower trophic levels. This characteristic illustrates the application of an important law of physics to the biological world that limits the efficiency of energy transfer. Here lies the answer to the question of why there are only a small number of links in any food chain—at each level along the way, so much energy is lost as heat energy or waste that little is left to support more than five links in most ecosystems. In this respect, it is the inherent inefficiencies in energy transfer when one organism consumes another that limit the number of levels in a food chain.

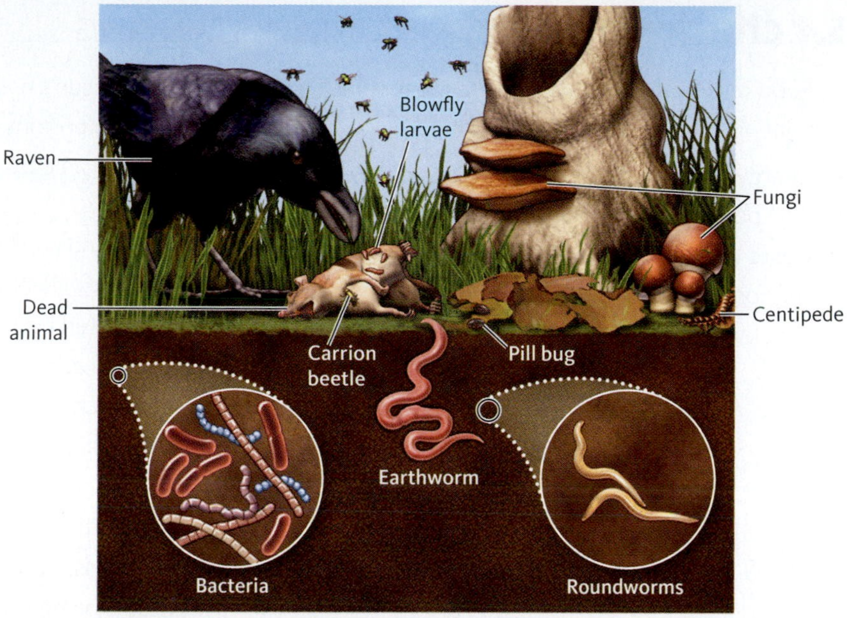

Figure 5.31 The role of detritivores and decomposers. Detritivores such as ravens and earthworms, along with decomposing bacteria and fungi, break down organic matter such as dead animals and plants into reusable products.
© Cengage Learning 2014

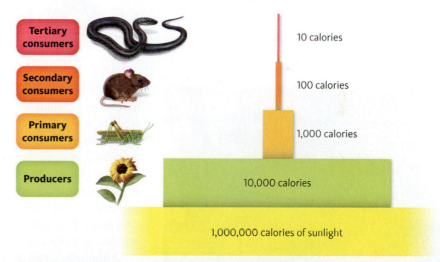

Figure 5.32 An energy pyramid. An energy pyramid illustrates the limitations on the energy flow between trophic levels. Plants convert a small percentage of sunlight into organic molecules, and only 10 percent of plant energy is transferred to growth in primary consumers. At each transfer step, this energy transfer limitation restricts the number of organisms that exist in the next higher trophic level. Calories are a unit of measure of energy.
© Cengage Learning 2014

- Food chains and food webs describe feeding relationships among organisms.

- Energy transfer in feeding relationships is not 100 percent efficient. In fact, much is lost as waste or heat while other energy is used to support life activities such as growth and development.

1. How is the sun's energy stored in the body of plant producers?

2. Why is it important to understand which trophic level an organism occupies in an ecosystem when discussing trophic levels and energy transfer?

3. Compare and contrast food chains and food webs. Are they different in the way they illustrate energy transfer within a community?

4. Domesticated herbivores such as cattle are often fed more nutritious grains such as corn. Although the grains are expensive, the cattle gain weight much faster than when they feed on pasture grasses. What characteristic of food chains and food webs is this meant to overcome?

5. For thousands of years, livestock breeding programs have selected animals with superior characteristics, including fast growth rates and large muscle mass, for reproduction. What factor in the energy-transfer process is being selected for by the farmers? How could a cow with these characteristics alter an energy pyramid?

5.5  Interactions in Social Groups Increase Fitness

Figure 5.33 Interactions within a social group. Each day you engage in a variety of interactions with numerous other people. Use of language, sounds, facial expressions, and gestures are among the ways we communicate with one another.

Behavioral ecology—a branch of biology that studies the adaptive behaviors of social animals to their environment that provide them a selective advantage over solitary animals.

Your life is filled with a diverse array of social interactions. Some are meaningful, such as a phone call from a loved one, whereas others, such as walking down a crowded hallway, pass nearly unnoticed. Compared to other animals large and small, humans use a variety of sounds, postures, and expressions to communicate information (**Fig. 5.33**). So far, this chapter has dealt with ecological interactions between species in populations, communities, and ecosystems. In this section, we'll explore interactions between individuals in groups below the population level. For instance, what's the advantage of belonging to a pack of wolves or a flock of birds?

Although not all animals live in groups, many species rely upon within-group social relationships for protection, finding food, and reproduction. At this level, individual animals interact and display behaviors designed to communicate very specific information to other group members. How these behaviors adapt animals to their environment and provide them with a selective advantage over solitary animals is the basis of a branch of biology called **behavioral ecology**.

Social interactions enhance reproductive success, defense, and hunting

Social groups increase the fitness of populations in three general ways: (1) care and nurturing of young, (2) protection and defense, or (3) obtaining resources such as food, territory, mates, or nesting sites. For example, animals that live in groups often have increased reproductive success because individual animals

Photo by Jason Forman for News21

are exposed to larger numbers of potential mates; the individual's success increases the population's reproduction success.

American robins show social interactions that enhance reproductive success through sexual selection and parental care of their offspring. In the spring, adult male and female robins gather in large groups to begin their mate selection process. The female robin selects her mate based on specific characteristics such as size, song, aggressive behavior, and bright coloration. Once selected, the male robin guards his mate while she is fertile to prevent other males from mating with her. Both robins help construct the nest, defend it, and care for the eggs. Following hatching, both parents feed, protect, and nurture the young until they are ready to leave the nest. After hatching, the close relationship ends, and the following year, mate selection and caring for the young begin all over again.

Parental care and nurturing of young is under genetic control in most mammals. In the common deer mouse, one gene controls nurturing behavior such as cleaning and nursing newborn pups (**Fig. 5.34**). The strength of this genetic trait is so important that mutant females that lack this gene fail to care for their young. The gene for parental care increases the survival of offspring and therefore the fitness of the deer mouse population. The process of natural selection has reinforced this trait in the population over time because of these positive outcomes. In very similar ways, specific genes have been linked to social behavior of insects such as termites, bees, and ants that form large, complex societies.

Birds provide many examples of social behaviors that promote mutual defense and improve group survival. Doves are common to most areas in the United States and often associate in large groups called flocks (**Fig. 5.35**). Flocking behavior gives each dove greater protection because the flock has more eyes available to detect dangerous predators. Once alerted to danger, the entire flock can fly away. Individual birds within the flock have a higher probability of survival because large groups that scatter in many directions confuse attacking predators. Finally, predators rarely catch and kill all members of a group, so while some members are caught, most escape.

Some social interactions increase fitness by favoring the acquisition of food. Predators that hunt as a group often are more successful than predators hunting alone. The gray wolf lives in harsh environments where obtaining food during winter months is very challenging. Cold temperatures and deep snow make it difficult to hunt and capture animals. Led by a mature alpha female, gray wolves live and hunt in social groups, consisting of 8 to 15 members, called packs. Cooperative hunting allows the much smaller wolves to kill animals as large as a 1,000-pound elk or moose. When a hunting pack attacks, the wolves take turns moving in to injure the moose while the others harass the prey. This tactic reduces the chance that any one member is injured or dies during a hunt. The pack hunting method is also more efficient—the energy expended per animal is lower as a pack than when individuals hunt alone.

Living in social groups has disadvantages as well as benefits. As numbers of a social group increase, there is more intraspecific competition for resources. This is evident when resources are severely limited. Large wolf packs experiencing food shortages will exclude older or weaker members from the pack or limit their ability to feed during difficult times. In this way, resources are reserved for members with a higher potential to help the entire group survive and reproduce.

Another disadvantage to living in social groups involves the transmission of disease, which can threaten the health of all members of the group.

Figure 5.34 Deer mouse mother and young. In many mammals, nurturing of young is an example of a behavioral trait controlled by certain genes.

Figure 5.35 Social behavior of doves. Doves group together in flocks. This behavior provides many eyes to look for potential predators.

You've seen how quickly a common cold is passed through high-density populations found in schools or dormitories. Similarly, rapid transmission of diseases is common in animal groups. A deadly example in canines is the highly contagious disease parvovirus, which is transmitted between dogs by contact with infected feces. Within members of the dog family, canine parvovirus causes high mortality among the young due to heart problems, intestinal tract damage, and dehydration.

Communication is essential to social behavior

For a social group to be successful, members need to communicate with each other. Communication involves a signal or stimulus transmitted by one animal and received by another. In some cases, signals are as simple as the visual cues of colorful feathers; in other cases, they are as complex as the hour-long song of whales. In general, as social complexity increases in a population, so does the complexity of the signals the group uses.

The method of communication animals use is a reflection of their ability to sense and process signals, as well as their patterns of daily activity. Animals that are active at night cannot rely completely on visual signals; they supplement their eyesight with combinations of smells or sounds. Wolves have a highly refined sense of smell; in fact, members of the dog family can sense odors at concentrations nearly 100 million times lower than humans can. When wolves are active at night, they use their powerful sense of smell in addition to their hearing to keep track of other pack members, detect potential threats, and find prey.

Visual signaling is an important means of communication for animals active during the day. Humans rely heavily on visual signals; our facial expressions can communicate emotions as different as happiness and anger. In the animal world, visual signals serve the same purposes. Mate selection often involves ceremonial visual signals that supplement coloration or auditory communication. For example, many bird species use physical posturing that includes highly exaggerated movements of the wings and body—all designed to attract a mate (**Fig. 5.36**). Another example of this behavior is the blinking-light visual communication used by fireflies. Fireflies, which actually are beetles and not flies, use species-specific flashing light patterns generated in abdominal organs to attract mates.

Animals that are active during the day also combine visual and auditory signals to increase fitness. Songbirds such as the northern cardinal use color and song to communicate with each other. You can see in **Figure 5.37** that the male shown on the right is more brightly colored than the female on the left. Brightly colored feathers are important signals indicating reproductive fitness. Females generally do not select males with damaged or dull feathers. Auditory signals are also important in attracting mates or defending territories. Male cardinals must sing a specific song and hit the right notes to attract females. Those that sing poorly or sing the incorrect notes are less likely to mate.

Physical touching, called tactile communication, is an effective tool to establish bonds between members or to send important signals. Humans use hand-holding or hugging to establish emotional bonding. In a similar way, animals may groom one another, lick, or nuzzle to communicate feelings. Bonding between mother and infants is strengthened by physical touch in combination with vocal communication.

Animals as diverse as dogs and insects use specialized chemical signals called pheromones to communicate information. Male members of

Figure 5.36 An albatross pair. Many species of birds have elaborate mating rituals that include wing and body movements by the male. The female albatross selects her mate based on the ability of the male to impress her with physical displays like this wing spread.

Doug Allan/The Image Bank/Getty Images

Figure 5.37 A male and female cardinal. Songbirds often have brightly colored males and more drably colored females. The brightness of the male coloration is an indicator of reproductive fitness and is a cue used by the females in selecting a mate.

Steve Byland/Shutterstock.com

Communication—animal behavior that involves a signal or stimulus transmitted by one animal and received by another. In general, as social complexity increases in a population, so does the complexity of the signals the group uses.

Tactile communication—physical touching between animals used for communication. Often used to establish bonds between members or to send important signals.

Pheromones—specialized chemical signals produced by animals used to communicate information.

the dog family release pheromones in their urine. Anyone who has taken a male dog for a walk understands the frequent urination on bushes, fences, or other structures. The urine leaves behind pheromone signals to communicate that dog's presence to other dogs in the area. Insects also have a powerful capacity to sense and recognize minute amounts of chemicals. Many species produce pheromones to attract mates, alarm members of their nest about danger, leave a trail for others to follow, or mark a territory.

Social groups often establish social orders and dominance hierarchies

Within social groups, conflicts often occur over leadership, access to resources, mate selection, and between juveniles and adults, an example of intraspecific competition introduced in Section 5.3. In many mammal species, social order is maintained through dominance hierarchy, a ranking of individuals based on social interactions. In some cases, the hierarchy is maintained by fighting for dominance. Social status associated with dominance hierarchy favors reproductive success for members that are the strongest and healthiest; their traits then are passed along to the next generation.

The term *pecking order* refers to the social hierarchy established in a flock of chickens. As in a wolf pack, an alpha female leads a chicken flock. When resources are abundant, the alpha female will mate and allow other females in the flock to mate as well. When resources are limited, the alpha female monopolizes the males and prevents subordinate females from mating. Pecking with the beak sends a physical message about behavior and establishes social position. Using this behavior, alpha female chickens will also defend food sources and regulate where the flock moves during different times of the day. Animal social groups led by the fittest members can improve survival rates for most members by maintaining order and ensuring that the most advantageous genes in terms of fitness are passed along to the next generation.

The purpose of communication in the animal world mirrors that in the human world—to convey information about resources, dangers, and reproductive fitness. Natural selection has resulted in the ability of signalers to transmit information and of receivers to recognize and react to these messages. In this way, evolutionary processes have favored social groups and their interactions.

We began this chapter by exploring the idea that organisms such as termites don't exist in isolation; in fact, each individual is an integral part of larger groups, populations, and communities linked together by a range of behaviors and relationships. Termites are involved in extensive relationships with microbes in their own bodies, other termites, plants that provide them shelter and food, and the world of predators that seek them out as food sources. The cumulative effect of interactions has important consequences for each species involved, often shaping species diversity and distribution in a community or ecosystem. Feeding relationships evolved as a means of fulfilling the energy resource needs for each species at each trophic level. Because the efficiency of energy transfer is very limited, the number of trophic levels in a food chain and the biomass of species within a trophic level are limited. In a similar manner, organisms that live in groups have evolved interesting and complex social behaviors, such as mating rituals and patterns of communication, which increase their overall evolutionary fitness. In the next chapter, we'll begin exploring how organisms are structured and adapted to live in their habitats.

Dominance hierarchy—a type of social interaction among members of a group that ranks members and creates a strict social order that often arises from the physical differences among individuals in relation to their access to resources based on certain characteristics.

5.5 check + apply YOUR UNDERSTANDING

- Many species have evolved interactions that enhance their survival and reproduction.

- Communication among and between species is often an essential form of interaction.

- Social orders and dominance relationships are used by animals that form groups to establish hierarchies of behavior for activities like feeding or mating.

1. Describe three ways in which animals communicate with one another.
2. Why is social behavior important to survival in many animal species?
3. How does the method of communication for animals that are active by day differ from that of those active primarily at night?
4. In the presence of danger to their nest, termite soldiers from the immediate areas are alerted within seconds and swarm to attack the invader. What method of communication common among insects might account for the rapid response of the soldier termites?
5. The black-tailed prairie dogs are highly social burrowing rodents. They are herbivores, feeding on the short-grass prairie plants of the northern Great Plains of North Dakota. They establish large colonies, up to several thousand animals within a very small area. What might be the benefits and costs associated with this social behavior?

End of Chapter Review

Self-Quiz on Key Concepts

Scale and Organization

KEY CONCEPTS: Individual organisms of a species live within larger groupings called populations. Collections of populations of different species living in one geographic area make up a community. Communities differ from one another in the types of species and their relative abundance. Communities then interact with the environmental factors within an ecosystem.

1. Match each of the following **ecological terms** with the best definition.

    transition zone
    population
    community
    ecosystem

    a. a single species located in one geographic area
    b. a collection of interacting species in one geographic area
    c. collections of communities and their environmental conditions
    d. a boundary area between adjacent habitats

2. Communities differ from one another in:
    a. the types of species found there.
    b. the abundance of each species found there.
    c. the resources that are available.
    d. all the above are correct.

3. Which of the following groups is not an example of a population?
    a. a group of three species of blackbirds living in a marsh
    b. two packs of wolves living in a mountain valley
    c. a group of manatees living in a mangrove lagoon
    d. the collection of killer whales living in Puget Sound

4. Communities that have abundant resources and a large number of ecological niches usually are:
    a. lacking in organisms that feed on decaying organic matter.
    b. characterized by shorter food chains.
    c. rich in biodiversity.
    d. restricted to transition zones.

Diversity of Interactions

KEY CONCEPTS: Interactions may occur between members of a single species or between species in a community. Interactions among populations in a community usually provide some benefit or harm to the organisms involved. Competition for resources occurs in many different ways influencing survival, reproduction, and community structure. In every community, feeding relationships called food chains, which illustrate patterns of energy transfer, link producers and consumers. Energy transfer in feeding relationships is not 100 percent efficient.

5. Match each of the following terms about **energy transfer** with the best description or example.

   keystone species
   trophic level
   food web
   resource partitioning

   a. an illustration that shows many potential feeding relationships in a community
   b. organisms that have a high influence on community structure
   c. dividing a resource in a way that allows two or more species to use it
   d. a feeding level of a food chain or food web; carnivore, herbivore

6. Match each of the following terms about **animal relationships** with the best description.

   parasitism
   commensalism
   predation
   mutualism

   a. a relationship where one organism benefits and the other is neither harmed nor helped
   b. a relationship where both organisms benefit
   c. a relationship where one organism benefits and the other is harmed
   d. an interaction where one animal eats another

7. Which of the following represents a mutualistic relationship?
   a. a turtle eating seagrass
   b. a virus that causes the human cold
   c. hummingbirds that pollinate flowers and get nectar in return
   d. birds that feed on nonharmful insects living on the back of a cow

8. Which of the following is a typical characteristic of parasites?
   a. They quickly kill their host.
   b. They typically are much smaller than their host.
   c. They exist in a win–win relationship with their host.
   d. They are part of the detrital food web.

9. Which trophic level forms the base of most food webs?
   a. predators
   b. herbivores
   c. producers
   d. heterotrophs

10. Energy transfer between species in a food chain typically is _____ percent efficient.
    a. 10
    b. 25
    c. 50
    d. 75

Behavioral Ecology

KEY CONCEPTS: The social behaviors and interactions between members of a population have important influences on survival and reproductive success. Many species rely upon social behaviors for protection, finding food and mates, and nurturing of offspring. Living in groups provides benefits to individuals, but there are also costs associated with this behavior. Communication is essential to successful social behavior. Chemical, acoustical, visual, and tactile signals transmit information between organisms. Natural selection has resulted in the ability of signalers to transmit information and receivers to recognize and react to the message being sent.

11. Match each of the following **animal behavior** terms with the best definition or explanation.

   dominance hierarchy a. specialized chemical signal produced by some animals

   pheromones b. a signal or stimulus transmitted by one animal and received by another

   communication c. a ranking of individuals based on social interactions

12. Which of the following represents a chemical signal used by many insect and animal species?
   a. a soil microbe that secretes digestive enzymes
   b. a pheromone
   c. physical rubbing or touching between members of a wolf pack
   d. the waggle dance of a honeybee

13. Which of the following represents a cost associated with living in a social group?
   a. greater energy expended by each individual in obtaining food
   b. increased mortality from predators
   c. lower success rates in finding food
   d. increased chances of transmitting harmful diseases

14. Which of the following is an example of tactile communication?
   a. the chemical trail left by worker termites that leads others to food
   b. the barking and growling signals sent by alpha wolves to members of the pack
   c. the physical touching and nurturing that happens between a parent and baby
   d. the interactions that occur between termites and their gut microbes

Applying the Concepts

15. White-tailed deer have as many as seven glands located on different parts of their body that are used primarily for communication. Gland secretions are made of fat-rich compounds that cling to the hair on the deer's leg and are rubbed off onto vegetation as the animal moves. Although both male and female deer produce the secretions, only the males use them as signals. Explain what type of signal these secretions represent and why they might be important tools for communication.

16. Why do ecologists prefer the term *food web* over *food chain*? What primary factor limits the number of trophic levels in a food web?

17. The big brown bat is a nocturnal mammal. Like most bats, it navigates at night using a form of echolocation. During the daylight hours, the bats roost in attics, hollow trees, or crevices of rocks. Big brown bats have good hearing, sight, and smell capabilities—similar to dogs. Which forms of communication are most likely used by big brown bats? Defend your answer.

18. **Figure 5.38** shows the relationship between six different species (y-axis) and intensity of competition for resources (x-axis) in a single habitat. Each letter on the y-axis of the graph represents the presence of a separate species and its presence or absence in response to competition. How does an increasing level of competition affect the number of species present in a habitat? Which species is least/most sensitive to competition?

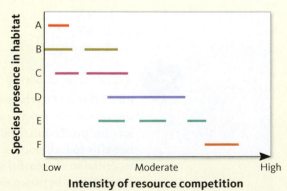

Figure 5.38 An example of resource competition.
Adapted from Figure 1. J. Vandermeer et al. 2002. Increased competition may promote species coexistence. J. Vandermeer et al. Proceedings of the National Academy of Sciences. V.99(13) 8731–8736.

Figure 5.39 Barnacles attached to a whale.

19. Barnacles are marine crustaceans related to the more familiar crabs and lobsters. As barnacles mature, they excrete a gluelike substance that permanently attaches them to surfaces including rocks and animals (Fig. 5.39). Most barnacles are suspension feeders and reach into the surrounding water with modified legs. These feathery legs beat rhythmically to draw plankton and detritus into their hard shell for consumption. Whales often have large collections of barnacles attached to their bodies that remain for the rest of the whales' lives. Based on this information, what is the relationship between whales and barnacles? Certain species of barnacles extend their feeding legs directly into their host and derive their nutrition that way. What type of relationship does this represent?

20. Examine the plant biomass/area production differences listed in Table 5.1. Marine estuaries, like mangrove forests, are highly productive centers of plant biomass production. Using this example data, would you expect the energy pyramid to have a broader base? Be taller?

TABLE 5.1 Biomass Production in Example Ecosystems

Ecosystem Type	Plant Biomass Production (Pounds of Production Per Square Yard)
Tropical forests and some marshlands	3.3–6.4
Temperate forests	2.4–3.4
Deserts	0.5
Marine estuaries	20+

© Cengage Learning 2014

Data Analysis

Communication Preferences in Chimpanzees

Chimpanzees share a good deal of genetic information and behavioral characteristics with humans. Internationally known scientists such as Dr. Jane Goodall have spent their careers studying the interactions among groups of chimpanzees and working to protect their populations (Fig. 5.40). Dr. Goodall's research on chimpanzee social and family life has revealed that chimpanzees have developed complex communication skills and behaviors. Scientists have learned that chimpanzees communicate in ways similar to humans, using a combination of nonverbal communications such as gestures and facial expressions, along with vocalizations.

Laboratory studies have revealed that chimpanzees use several different forms of signaling when trying to convey information about objects in their environment. These studies show that chimpanzees modify their method of communi-

Figure 5.40 Jane Goodall with wild chimps.

cation based on the attention of the researcher; they understand the importance of gaining an individual's attention to send a signal. The data presented in the histograms are from an experiment involving interactions between a human and chimpanzees that were attempting to get access to a tool needed to obtain food. In some cases, the scientist held a tool needed by the chimpanzee to obtain food. In other cases, no tool was present.

Data Interpretation

Use the histograms labeled a. and b. to answer the following set of questions. The columns labeled "Attention" refer to the researcher deliberately paying attention to the chimp attempting to communicate.

21. Examine **Figure 5.41a** to determine the following: in the presence of a tool, what was the chimpanzees' favorite mode of communication? Does this change in the absence of a tool?

22. Using **Figure 5.41b**, determine whether there is a communication method that demonstrates the greatest difference when a researcher faces toward or away from the chimpanzee. Why do you think there was this difference?

23. Based on this data set, how similar are chimpanzees to humans in their modes of communication?

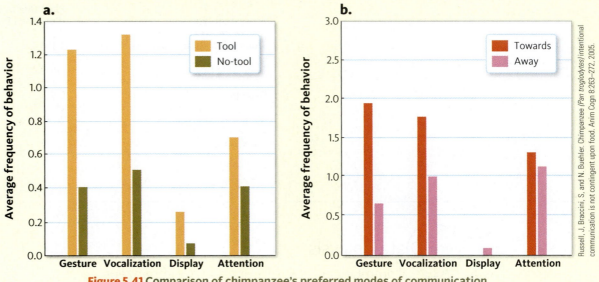

Figure 5.41 Comparison of chimpanzee's preferred modes of communication.

Russell, J, Braccini, S, and N. Buehler. Chimpanzee (Pan troglodytes) intentional communication is not contingent upon food. Anim Cogn 8:263–272, 2005.

Critical Thinking

24. Why do you believe chimps prefer vocal forms of communication with researchers whether or not a tool is present?

25. Would you expect different experimental results from "wild" chimpanzees?

Question Generator
Energy Transfer and the Nature of Food Webs

Energy transfer between organisms in a food chain is generally inefficient. Most ecologists estimate that on average only 10 percent of the energy contained in a consumed organism builds biomass or structure in the organism doing the eating.

Remember that in any community, the base of a food chain (or food web) consists of producers, organisms that convert light or chemical energy along with resources in the environment into biological molecules. For instance, most plants are able to convert only 1 to 2 percent of the light that falls on them into biomass. Because the upper trophic levels all depend upon the amount of organic energy created by producers, biomass production in the lowest trophic levels has a dramatic effect on the numbers of other organisms that a community can support.

Figure 5.42 is a block diagram that illustrates some of the relationships between energy flow and food webs. Use the background information, the graph, and the block diagram to ask questions and generate hypotheses. We'll start you off by simply translating several components of the block diagram into sample questions.

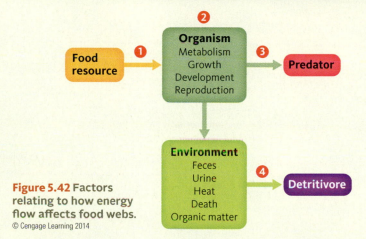

Figure 5.42 Factors relating to how energy flow affects food webs.
© Cengage Learning 2014

1. How does the efficiency of energy transfer affect the number of trophic levels in an ecosystem? (Arrow #1)

2. What is the relationship between amount of new growth by producers and the number of herbivores a habitat can support? (Arrow #2)

3. How does the position in the food chain (web) relate to population size? How does an organism's metabolic rate relate to its ability to accumulate biomass? (Arrow #3)

4. Why is a continuous accumulation of dead organic matter important to detritivores? (Arrow #4)

Use the block diagram to generate your own research questions and frame a hypothesis.

6 Animal Structure and Function

6.1 The leatherback turtle is on the move again

6.1 The Leatherback Turtle Is on the Move Again

Late one night on an Indonesian beach, a leatherback turtle *(Dermochelys coriacea)* laid the last of her eggs. She gently buried the eggs with sand, then crawled to the water (**Fig. 6.1a**) and began a 12,774-mile journey to her feeding grounds off the coast of Oregon. This large marine reptile migrated an average of almost 20 miles per day for 647 days. Her epic journey, tracked by a radio beacon on her back, was one of the longest recorded migrations of any vertebrate animal (**Fig. 6.2**).

The life of the leatherback is a continuous series of transoceanic migrations to reproduce and find food. From the very moment they hatch and emerge from their underground nest, these turtles orient themselves and head straight to the open ocean (**Fig. 6.1b**). During the early stages of their life, they live in areas that may be hundreds or thousands of miles from the beach where they hatched. Here they feed mostly on jellyfish, squid, and other slippery prey and grow faster than all other sea turtles. Adults range in length from four to six feet and weigh from 650 to 1,200 pounds, making them the largest living turtle species on earth. It is estimated that leatherbacks reach sexual maturity when they are 13 to 14 years old and that they live 30 years or more. Males spend their entire life in water. However, females leave their feeding grounds every two to four years and return to the same beaches from which they hatched to reproduce a new generation.

The leatherback's massive body is built for swimming. Looking at the turtle's body, you'll notice that its head and trunk are streamlined to minimize drag, or resistance to moving through the water (see **Fig. 6.2**). Unlike the land turtle, the leatherback does not have a hard shell. Instead, its upper shell, or carapace, is composed of thousands of small bones embedded in an oily, leathery skin. Water is channeled over this smooth waterproof surface and down seven ridges that run the length of the tapered carapace. This shape increases swimming efficiency and reduces the amount of energy needed by the turtle to swim long distances.

Figure 6.1 Leatherback turtles return to the beaches where they were hatched. (a) Above the high tide mark, a female turtle laboriously digs a deep nest chamber about three to five feet deep and lays about 85 eggs. The sand prevents the eggs from drying out and hides them from predators. She then buries the eggs and heads back to the water and into the open ocean. (b) After about 60 days the eggs hatch, and the hatchlings crawl out of their nest and head directly for the water.

a.

Kevin Schafer/Getty Images

b.

© Doug Perrine/Design Pics/Corbis

Anatomy of the Leatherback Turtle

Carapace

Lung

Esophagus

Heart

Stomach

Muscles

Oregon

Indonesia

Figure 6.2 The leatherback turtle is built for swimming. The leatherback turtle with its flippers, strong muscles, and digestive system is built for swimming. The leatherback feeds primarily on jellyfish and dives to great depths to hunt them. Its paddlelike flippers consists of five finger bones similar to those of humans. The map inset shows the migration path recorded for a leatherback that swam from its breeding ground in Indonesia to its feeding ground off the coast of Oregon in 647 days.
© Cengage Learning 2014

Stroke after stroke, a leatherback turtle propels her massive body forward. Large, strong shoulder muscles power her two large forelimbs, which have been modified into winglike flippers (see **Fig. 6.2**). The beating and rotating motion of these flippers provides thrust and further reduces drag through the water. In addition to using their front flippers for swimming, females use their front flippers to dig through sand and build their nests. Their hind limbs are shaped like paddles and act as rudders to maneuver and steer their direction in the water. The structure and functioning of the muscular and skeletal systems keep the leatherback swimming gracefully toward her destination.

Leatherbacks are one of the deepest diving of the air-breathing vertebrates, which enables them to search for and feed on jellyfish. Marine biologists observe that these turtles frequently dive to depths ranging from 100 yards to more than a half a mile while holding their breath for as long as 85 minutes. After taking a few short breaths of air at the surface, the turtle fills its wedge-shaped lungs and begins its dive. A three-chambered heart pumps and circulates red blood cells that contain a protein that picks up oxygen from the turtle's respiratory system and then circulates and releases it to body tissues and cells. Once the oxygen is delivered, leatherback muscles contain a second protein that stores the oxygen. Both blood and muscle tissue can store oxygen in large quantities, which helps the turtle to remain underwater for long periods of time. All of the structures coordinate and function on all levels of organization to make the deep dive possible.

Leatherback turtles are reptiles, a class of about 6,000 vertebrate species that have scaly skin, breathe air, and use sunlight to heat their bodies. Like all reptiles, turtles reproduce through internal fertilization and, like most reptiles, lay their soft-shelled eggs on land. A unique feature of turtles is the shell that protects their internal organs.

This chapter presents animal structure from a top–down perspective, first presenting the familiar organ systems and organs before examining the microscopic tissues, cells, and macromolecules that compose them. This chapter provides powerful insights into the inner workings of animals at the different levels of the hierarchy. Larger levels of organization provide the mental framework, while the smaller cell and molecular levels provide details of the mechanisms that run the animal "machine."

6.2  Animals Are Organized into Organ Systems, Organs, and Tissues

The leatherback turtle is an animal with a highly complex structure and organization compared to the simplest animal, the placozoan. Placozoans, whose name literally means "flat animal," have nearly transparent bodies consisting of two razor-thin sheets of cells a few millimeters long (**Fig. 6.3**). Their upper and lower surfaces have cells with hairlike cilia, which beat together and move the animal. Other types of cells are present but are not organized into more complex units of tissues, organs, or organ systems. Placozoans remind us that animals are heterotrophic multicellular organisms with different types of eukaryotic cells. They ingest their food and digest it internally. Animals as different as placozoans and leatherback turtles show us the wide range of structural complexity in the animal kingdom.

Vertebrates such as leatherback turtles, fishes, and humans are composed of hundreds of different types of cells. Each type of cell has a particular structure that performs a specialized function. In fact, a characteristic of complex animals is the division of labor among their different cells, tissues, organs, and organ systems. Recall from Chapter 1 that cells are the fundamental units and that they interact with each other. Different types of cells work together as a unit to create four types of **tissues**. Tissues associate and work together in different ways to form the next level of organization—the organs—each with its own specialized task. **Organs** work together as members of one or more organ systems, which ultimately make up the animal as an organism.

In the following section, we'll use this familiar vertebrate organization to develop general concepts of animal structure and function, starting with the largest and most familiar level of the biological hierarchy—organ systems. After establishing your understanding on this level, we will then link the organ and tissue levels to the structure and functioning of the animal in its habitat.

Organ systems are groups of organs that perform major functions

The animal body is organized into several organ systems that perform one or more of the basic functions of life, such as gas exchange, digestion, or excretion. **Table 6.1** lists each of the organ systems, the organs they contain, and the functions they perform. Note that many organ systems perform several different functions. For example, the skeletal

Figure 6.3 A placozoan. Placozoans consist of two thin sheets of cells, making them one of the simplest animals.

Light microscopy by Ana Signorovitch

0.2 mm

Tissues—a group of cells that work together to perform a specific function; they associate and work together to form body organs.

Organs—a group of tissues that work together to perform specialized functions; they form the structure and function of organ systems.

TABLE 6.1 Major Organ Systems

Organ System	Organs	Basic Function(s)
Integumentary system	Skin, hair, nails, claws Scales, feathers	Protection from infection, injury, and dehydration Regulation of body temperature Gas exchange (amphibians) Reception of sensory information
Skeletal system	Bones, joints, ligaments	Support Sites for muscle attachment Mineral storage Blood cell production
Muscular system	Muscles, tendons	Movement of body parts, production of heat
Nervous system	Brain, spinal cord, nerves, sense organs (eyes, ears, nose, tongue)	Sensory perception, communication, and coordination Integration of organ system activities
Endocrine system	Endocrine glands (pituitary, thyroid, adrenal, pancreas, testis, ovaries)	Regulation and control of development, metabolism, and other activities Integration of organ system activities
Circulatory system	Heart, blood vessels	Transport and distribution of materials and heat
Lymphatic system	Lymph nodes and vessels, spleen, tonsils	Immunity and defense of the body Return of extracellular fluids to blood
Respiratory system	Nasal cavity, larynx, trachea, lungs, gills, skin	Gas exchange—O_2 supply, CO_2 excretion Sound production
Digestive system	Mouth, tongue, esophagus, stomach, liver, pancreas, small and large intestines, anus	Ingestion of food and water Breaking down of foods for absorption Elimination of indigestible parts of foods
Urinary system	Kidneys, bladder, urethra	Regulation of body fluid composition and volume Excretion or disposal of metabolic wastes
Reproductive system	Male—testis, penis, seminal vesicles, prostate	Production of male gametes (sperm) Transfer of sperm to egg
	Female—mammary glands in breasts, ovaries, uterus, vagina	Production of female gametes (eggs) Protection and nutrition of developing embryo–fetus–infant

© Cengage Learning 2014

system supports the body and protects vital organs; however, it also makes blood cells and stores minerals. The division of labor among organ systems means that they are interdependent; each one depends on the functions provided by the other organ systems. **Figure 6.4** is a diagram that shows how the digestive system supports and is supported by the other organ systems of the body. In this case, the digestive system supplies the molecules to build and maintain the other systems, which in turn contribute to the specialized function of the digestive organs. No organ system works in isolation.

Organ systems are composed of body organs that work together to perform specific functions. The digestive system consists of the mouth, esophagus, stomach, and intestines, as well as the liver, gallbladder, and

Organ systems—a group of organs that work together to perform specialized functions, such as gas exchange, digestion, or excretion.

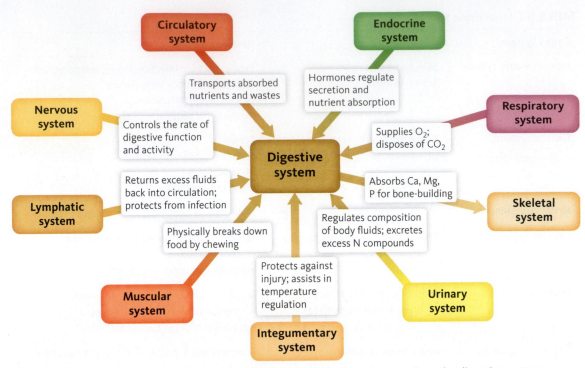

Figure 6.4 The body's systems are interdependent. This block diagram shows how the digestive system supports and is supported by the other organ systems.
© Cengage Learning 2014

pancreas. Each organ contributes to the overall functioning of the digestive system, whether it is the physical breakdown of food in the mouth or the storage of food in the stomach.

Organs are composed of tissues that work together

Body organs have a particular size, shape, and tissue structure to perform their function. Organs such as the urinary bladder and stomach are hollow saclike structures that store substances. Muscles are solid elongated organs that contract and pull on bones of the skeleton to bring about movements. Each organ is composed of different types of tissues organized to perform its specialized function. Organs have a network of blood vessels that delivers a constant supply of oxygen and nutrients for metabolism and energy, as well as a rich supply of nerves to communicate and interact with other organs.

Let's introduce you to the four types of tissues by examining the structure of the leatherback's esophagus, which extends from the mouth to the stomach (**Fig. 6.5**). As an organ in the digestive system, the esophagus receives food from the mouth and moves it to the stomach. The leatherback's esophagus has numerous backward-pointing spines that help the turtle trap and swallow its prey, slippery jellyfish (see **Fig. 6.5**).

The esophagus wall has three layers that contain the four types of tissue (**Fig. 6.6**). The innermost layer consists of **epithelial tissue**, which lines the passageway to the stomach. Epithelial tissues include specialized glands that excrete waste products or secrete substances, such as mucus, which lubricates the tube for the easy passage of food. Just beneath this lining is an extensive layer of **connective tissue**, which is rich in elastic fibers that stretch and expand the tube to accommodate large jellyfish. Connective tissue cells make and secrete these flexible fibers into a nonliving

Epithelial tissue—tissue that covers the inner and outer surfaces of the body; lines internal passageways of the body (digestive, urinary, reproductive); forms specialized glands that excrete waste products or secrete substances.

Connective tissue—tissue involved in support, storage, and protection; makes and secretes a nonliving layer, or matrix, in which cells are embedded; widely distributed throughout the body.

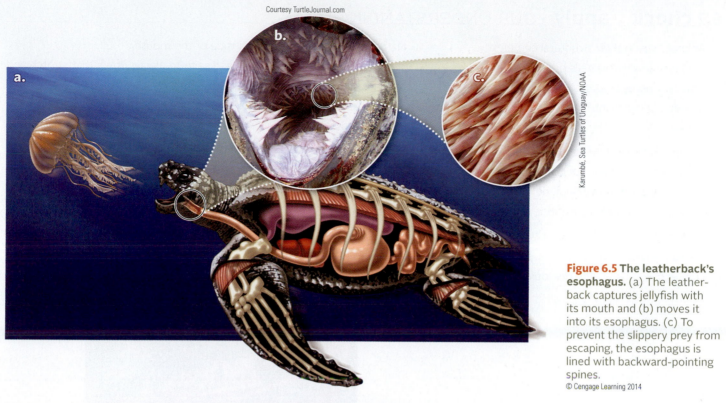

Courtesy TurtleJournal.com

Karumbé, Sea Turtles of Uruguay/NOAA

Figure 6.5 The leatherback's esophagus. (a) The leatherback captures jellyfish with its mouth and (b) moves it into its esophagus. (c) To prevent the slippery prey from escaping, the esophagus is lined with backward-pointing spines.
© Cengage Learning 2014

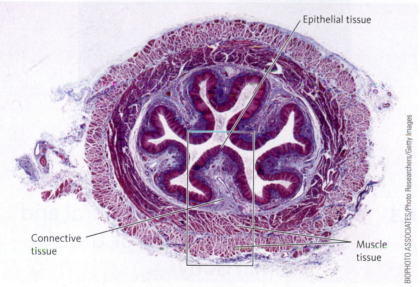

Epithelial tissue

Connective tissue

Muscle tissue

BIOPHOTO ASSOCIATES/Photo Researchers/Getty Images

Figure 6.6 Cross section of a leatherback's esophagus. As an organ, the esophagus consists of different tissues that work together. The innermost layer is epithelial tissue, which is shown at the top of the rectangle. Just beneath this layer is connective tissue, which provides passageways for nervous tissues and blood vessels. Muscle tissue contracts to move food along to the stomach, the next stop in the digestive system.

layer, or *matrix,* in which they are embedded. This connective tissue layer also contains blood vessels and nerves. Nerves are bundles of **nervous tissue** that stimulate the underlying layer of **muscle tissue** to contract and squeeze the esophagus closed. The combined action of nerve and muscle tissues expels seawater out the mouth or nose, traps the jellyfish, and propels it down the esophagus into the stomach. Thus, the four types of tissue combine in the leatherback's esophagus to fulfill its digestive function.

All body tissue can be classified as one of the four types we've described. For example, blood, bone, cartilage, tendons, and fat all are types of connective tissue. You can find three different types of muscle tissue in the body: skeletal, smooth, and cardiac. One type of nervous tissue senses changes in the environment, whereas another type stimulates muscles. Together, the four types of tissues combine to make a variety of organs and structures in animals.

Nervous tissue—tissue that senses and communicates information throughout the body; tissue that causes muscle tissue to contract.

Muscle tissue—tissue that contracts to produce movement; an effector in a feedback loop.

- Animals vary in their structural complexity. Simple animals have cells organized into tissues, whereas more complex animals have tissues organized into organs and organ systems.

- Complex animals have several different organ systems that perform specialized functions, such as support, movement, defense, gas exchange, and communication.

- Body organs are composed of two or more of the four major types of tissues: epithelial, connective, muscle, and nervous.

1. Name the tissue that lines the inside of the mouth.
2. Which type of tissue cells make and secrete elastic fibers that stretch and expand an organ?
3. True or false? Animal tissues are composed exclusively of living cells.
4. The thickness of epithelial tissue can range from a *single* cell layer to *many* cell layers, depending on where the lining is located and its function. Predict the thickness of epithelial tissue for the following locations and functions. Choose either single cell layer or many cell layers. Explain your reasoning. (A) Lining the lungs where gases move in and out of the blood. (B) Lining the esophagus. (C) Lining the small intestine where nutrients are absorbed into the bloodstream.
5. **Figure 6.7** shows a big-eyed tree frog trying to catch a fly with its tongue. Which of the four types of tissue (A) senses the fly, (B) makes the sticky solution on the tongue, and (C) flips the tongue out and pulls it back in?

Cathy Keifer/Shutterstock.com

Figure 6.7 Big-eyed tree frog.

6.3 Cells Are the Structural and Functional Units of the Animal Body

So far in this chapter, we have emphasized the hierarchy of biological organization as a framework to understand the animal body. We will now discuss the cellular level, which is arguably the most important in the organizational framework for two important reasons—structure and function. Cellular structures build body structures, and functionally cells perform the work of the body's organs and tissues.

To explore how cells operate as part of a tissue, let's zoom in on the epithelial lining of the leatherback's stomach and focus on one type of cell (**Fig. 6.8**). This specialized cell secretes hydrochloric acid (pH = 0.8) into the stomach and activates a powerful enzyme that breaks down proteins in the jellyfish that the leatherback has swallowed.

Figure 6.8c shows a highly magnified photograph of the epithelium. As you survey this tissue, you may notice that it is part of a single layer of

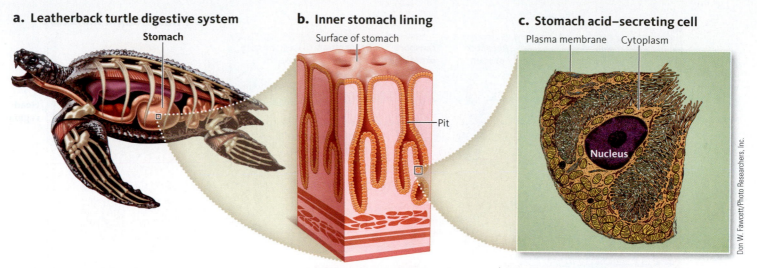

a. Leatherback turtle digestive system

Stomach

b. Inner stomach lining

Surface of stomach

Pit

c. Stomach acid–secreting cell

Plasma membrane   Cytoplasm

Nucleus

Don W. Fawcett/Photo Researchers, Inc.

Figure 6.8 The leatherback's stomach. (a) The leatherback's stomach stores and digests food. (b) The stomach is lined with pits, which contain a variety of cells that secrete digestive enzymes and acids. (c) The photomicrograph shows the cells responsible for secreting acids to break down the prey. (b) The leatherback's stomach cells secrete a digestive enzyme that begins to break down the proteins in the turtle's jellyfish prey.
© Cengage Learning 2014

cells that are in contact with each other. A triangle-shaped cell is magnified to show its three basic parts. First, a thin plasma membrane compartmentalizes the cell into a distinct unit so that it can carry out its activities. Even though it is only a few molecules thick, the membrane retains and supports the contents of the cell. If the membrane breaks, the contents will simply leak out into the environment and the cell will die. Second, enclosed within this membrane is the *cytoplasm,* which contains the basic substance of the cell. Third, in the middle of the cell is a dark structure called the nucleus. The nucleus houses the majority of the DNA and contains the genes that encode information on how to build the cell's structure and perform its specialized activities.

In the section ahead, we will zoom in for a more highly magnified view of this epithelial cell. We will start our tour of the cell with the outer plasma membrane and move inside the cytoplasm and nucleus, where we will show you how the different parts work together to perform their specialized functions. Along the way, we will highlight other cells where different cell structures play prominent roles in their performance.

The plasma membrane interfaces with the cell's environment

Every animal cell has a plasma membrane that functions as a boundary that interfaces with the cell's environment. The membrane must be strong enough to protect the cell from breaking open yet delicate enough to exchange materials with the cell's environment. Small oxygen molecules pass through the membrane and enter the cell to help fuel its activities. Going in the opposite direction are molecules of carbon dioxide, which is a waste by-product of cell metabolism. Even though these gases easily pass through the membrane, not all substances do. In fact, the plasma membrane has a structure that allows some substances to pass through easily while others need to be carried across; that is, the membrane is *selectively* permeable.

The structure of the plasma membrane consists of a foundation of phospholipid molecules with a variety of other molecules embedded

Plasma membrane—outer boundary of the cell; regulates movement of molecules in and out of the cell.

Nucleus—control center of the cell; houses the majority of the cell's DNA packaged with proteins into chromatin and chromosomes; contains the genes that encode information to build the cell's structure and perform its specialized activities.

Phospholipid—insoluble molecule that forms the membranes of the cell; an individual phospholipid molecule consists of a head (phosphate groups) and two fatty acid tails.

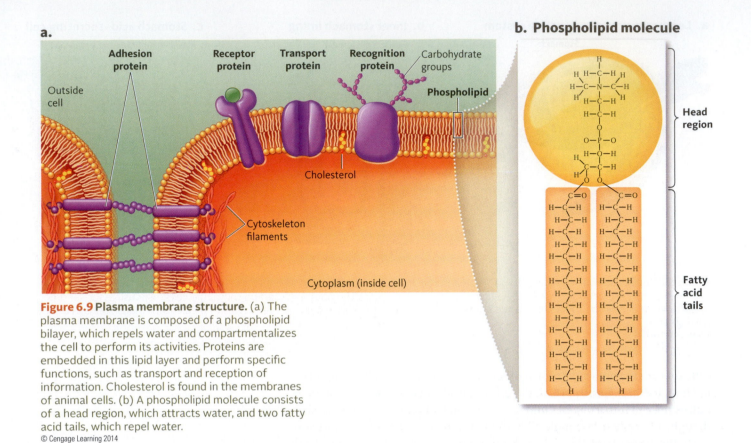

a.

Adhesion protein

Receptor protein

Transport protein

Recognition protein

Carbohydrate groups

Outside cell

Phospholipid

Cholesterol

Cytoskeleton filaments

Cytoplasm (inside cell)

b. Phospholipid molecule

Head region

Fatty acid tails

Figure 6.9 Plasma membrane structure. (a) The plasma membrane is composed of a phospholipid bilayer, which repels water and compartmentalizes the cell to perform its activities. Proteins are embedded in this lipid layer and perform specific functions, such as transport and reception of information. Cholesterol is found in the membranes of animal cells. (b) A phospholipid molecule consists of a head region, which attracts water, and two fatty acid tails, which repel water.
© Cengage Learning 2014

within it (**Fig. 6.9a**). Phospholipid molecules, like other lipids, do not dissolve in water. An individual phospholipid molecule consists of a head and two tails (**Fig. 6.9b**). Notice that the phospholipid molecules are organized into two layers—a bilayer—with the head regions oriented to the outside and inside of the membrane and the tails in the middle. This arrangement creates a membrane that repels water and separates the cell's contents from its watery environment. The abundance of phospholipids makes the membrane very flexible. Animal cell membranes also contain cholesterol, a lipid that stiffens the membrane and helps hold it together.

Proteins on the surface or embedded within the membrane play an important role in the life of the cell. **Proteins** are large, complex macromolecules that can form long fibers or fold into specific three-dimensional shapes. **Table 6.2** lists the different types of proteins and their functions, which range from pumping substances in or out of the cell, to receiving signals, to informing the immune system that the cell belongs to the body. Special adhesion proteins hold epithelial cells together and produce a multilayered tissue lining.

Proteins—large and complex macromolecules that can form long fibers or fold into specific three-dimensional shapes.

TABLE 6.2 Plasma Membrane Proteins and Their Functions

Protein	Function
Transport	Moves substances in or out of cell
Receptor	Transfers signals (nerve, hormones) from outside to inside of cell
Adhesion	Holds cell to plasma membranes of neighboring cells
Recognition	Informs immune system that cell belongs to the body

© Cengage Learning 2014

The cytoplasm contains many specialized organelles

The part of the animal cell between the plasma membrane and the nucleus is the cytoplasm. This complex mixture of water and dissolved substances contains protein tubules and filaments that form the cell's internal "skeleton," or cytoskeleton. This network of protein tubules, filaments, and rods supports and maintains the cell's shape, moves chromosomes during cell division, and moves different structures within the cell. In some cells, the cytoskeleton moves the entire cell. For example, the cytoskeleton in muscle fibers consists of a set of precisely aligned filaments that slide past each other and enable the entire cell to shorten. The "motorized" cytoskeleton extends into a *flagellum* that moves a sperm cell or a *cilium* on the surface of a lung cell that moves and sweeps dust particles out of the lungs.

The cytoplasm also contains many different specialized organelles (**Fig. 6.10**). Organelles are cellular structures that perform specialized functions for the cell, such as storage of energy, breakdown of sugars, or production of proteins. Some organelles form compartments that are enclosed by a membrane similar to the plasma membrane. The kinds and numbers of organelles in a particular kind of cell reflect the cell's specialization. For example, sperm cells have many organelles that power their whiplash swimming motions so that they can reach an egg. Let's continue the tour of the animal cell by describing the structure and function of the major cell organelles.

Mitochondrion. Many types of animal cells have hundreds or thousands of mitochondria in their cytoplasm. This sausage-shaped organelle (see **Fig. 6.10**) is the "power generator" of the cell, where the chemical energy in food nutrients is transformed into the fuel that drives the cell's many activities. Thus, the more active the cell, the more mitochondria it has. Epithelial cells in the stomach are actively making and secreting proteins. This intense activity requires large amounts of energy provided by its mitochondria. We'll cover the energy-producing processes in cells in more detail in Chapter 9. Unlike other organelles in the animal cell, a mitochondrion has a small circular molecule of DNA, and it even can reproduce itself by splitting in two.

Endomembrane system. The stomach cell has an extensive endomembrane system, which includes several organelles that work together to make, sort, and transport the digestive enzyme out of the cell into the stomach. This elaborate system begins with a continuous network of channels that wind their way throughout the cytoplasm. These channels are called the *endoplasmic reticulum.* In the typical animal cell, more than half of its volume may be occupied by these channels. Zooming in on these intracellular channels, you will notice that some channels have rounded structures attached to their outer surface. These structures, called ribosomes, have a critical function for the cell—they make proteins. Millions of ribosomes are also found free in the cytoplasm, where they make proteins wherever they are needed.

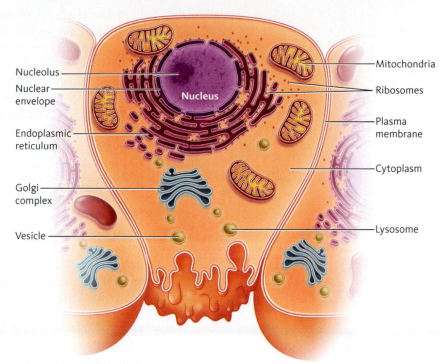

Figure 6.10 Animal cell structure. Cells that produce stomach acid are very active with many mitochondria and an extensive endomembrane network. The nucleus controls the cell's secretory activities.

© Cengage Learning 2014

Labels in figure: Nucleolus, Nuclear envelope, Endoplasmic reticulum, Golgi complex, Vesicle, Nucleus, Mitochondria, Ribosomes, Plasma membrane, Cytoplasm, Lysosome

---

Cytoplasm—complex mixture of water and dissolved substances that is located between the plasma membrane and the nucleus.

Cytoskeleton—forms the cell's internal "skeleton" through a network of protein tubules and filaments; supports and maintains the cell's shape, moves chromosomes during cell division, and moves different structures within the cell.

Organelles—cellular structures that perform specialized functions for the cell.

Mitochondrion—organelle that is the "power generator" of the cell where the chemical energy from food nutrients combines with oxygen to be transformed into fuel that drives the cell's many activities.

Endomembrane system—a system that includes endoplasmic reticulum, Golgi complex, vesicles that work together to make, sort, and transport substances within or out of the cell.

Ribosomes—organelles that make proteins; found free in the cytoplasm and along portions of endoplasmic reticulum channels.

Some channels of the endoplasmic reticulum have ribosomes, while other channels appear *smooth* because they lack these protein factories (see **Fig. 6.10**). These smooth channels contain enzymes that make lipids such as phospholipids and cholesterol. In liver cells, the smooth channels are also involved with drug detoxification. In reproductive cells, the smooth channels produce male and female sex hormones (for example, testosterone, estrogen, and progesterone). Another steroid hormone made in the smooth channels of endoplasmic reticulum is adrenaline, the "fight or flight" hormone. After these protein and lipid molecules are assembled in the endoplasmic reticulum, they are processed further in the next organelle of the endomembrane system, the Golgi complex.

The Golgi complex is an organelle that processes substances for secretion out of the cell. It appears as a discrete stack of channels, like a stack of pancakes (see **Fig. 6.10**). Proteins and lipids received from the endoplasmic reticulum are chemically modified and sorted as they move from one channel of the stack to the next. At the completion of their processing in the Golgi, these molecules are packaged and placed into small round vesicles. Vesicles pinch off of the Golgi complex and then move along the cytoskeleton within the cell or to the plasma membrane, where their contents are used to build or repair the membrane or the contents are moved out of the cell. The Golgi complex is an important part of the endomembrane system in cells lining the leatherback's stomach. Here, the digestive enzymes begin to break down the jellyfish swallowed by the turtle.

The endomembrane system may package powerful enzymes (proteins) into a special kind of vesicle called a lysosome. These enzymes are capable of engulfing and recycling worn-out organelles or, if they burst open, destroying the entire cell. During embryonic development, particular cells are genetically programmed to die when their lysosomes are activated to release their powerful enzymes to accomplish the mission. White blood cells moving throughout the blood and tissues defend the body by ingesting bacteria or foreign cells and then destroying them with their potent arsenal of enzymes in their lysosomes. A sperm cell has a lysosome at the tip of its head. When it contacts the outer surface of the egg, the sperm breaks open and dissolves the egg's cell membrane to facilitate fertilization. Therefore lysosomes are essential organelles in development, defense, cell maintenance, and sexual reproduction. Our tour of the stomach epithelial cell now takes us to the center of the cell—the nucleus.

The nucleus directs the cell's structure and activities

There are hundreds of kinds of animal cells—epithelial cells, muscle cells, and white blood cells to name a few. Despite their different shapes and functions, all of these cells contain the same genetic information. In animals, this set of genetic information, or genome, is stored in two locations in the cell—the mitochondria and the nucleus. Of these two organelles, the nucleus stores the majority of a cell's DNA. When an instruction from the genome is needed, the cell can activate a specific portion of the genome. In this sense, the nucleus is responsible for the construction and functioning of the entire cell.

The nucleus is enclosed by a nuclear envelope, which forms a boundary between the nucleus and cytoplasm. Unlike the plasma membrane, the nuclear envelope consists of two membranes with a number of large *nuclear pores* that control the passage of large molecules in and out of the nucleus (see **Fig. 6.10**). The nuclear pores enable the nucleus to exercise its "command and control" function. Messages are sent and received to direct

Golgi complex—organelle that functions to process substances for secretion out of the cell; receives and modifies proteins and lipids from the endoplasmic reticulum for transport.

Vesicles—small round organelles that transport substances within the cell or to the plasma membrane for secretion; originate from the Golgi complex and move along the cytoskeleton within the cell or to the plasma membrane where their contents are used.

Lysosome—special kind of vesicle that contains powerful enzymes capable of digesting the entire cell.

Nuclear envelope—membrane that encloses the nucleus and forms a boundary between the nucleus and cytoplasm.

and coordinate activities in the cytoplasm, endomembrane system, and plasma membrane. In this way, the nucleus expresses its genetic information to accomplish the cell's specific mission.

Our tour of the animal cell nucleus now takes us to a dark dense area (see **Fig. 6.10**). This region, called the nucleolus, contains several different chromosomes coiled together. The nucleolus contains instruction for the production of ribosomes, which you may recall are the cell's protein factories. Thus the function of the nucleolus is closely linked to the activity of the cell and its protein requirements. Actively growing cells have greater need for ribosomes and their proteins; therefore, they have more nucleoli (plural of nucleolus) in each cell or they have larger nucleoli.

Finally, it is important to emphasize the major role proteins play in the functioning of the nucleus. Proteins organize and package chromatin into chromosomes. They form part of the nuclear envelope and its pores. Lastly, as in the cytoplasm, proteins form the skeletal framework that mechanically supports and maintains the shape of the nucleus. The chemical structure of proteins will be presented in the next section.

Nucleolus—darkened region of the nucleus that contains instructions (genetic information) for making ribosomes.

6.3 check + apply YOUR UNDERSTANDING

- Animal cells are composed of three basic parts—outer plasma membrane, cytoplasm, and nucleus.

- The selectively permeable plasma membrane is composed of a phospholipid bilayer and embedded proteins that control which substances enter and leave the cell.

- The cytoplasm is a semifluid material that is supported by a protein scaffolding, or "skeleton." Specialized organelles such as mitochondria, endoplasmic reticulum, Golgi bodies, and vesicles work together to perform cellular functions.

- The nucleus contains the vast majority of the cell's genetic information and, through the expression of this information, controls the cell's activities.

1. Which cell organelle uses oxygen and nutrients to generate chemical energy to power the cell's activities?

2. Which of the two epithelial cells in the lining of the esophagus has a larger, more active nucleolus: (A) a younger dividing cell or (B) an older aging cell? Explain why.

3. Specialized cells called goblet cells line the inner stomach wall where they secrete the mucus protein. Name at least three cell organelles found in the cytoplasm that you predict would be abundant in goblet cells and explain why.

4. In the course of many types of cancers, epithelial cells detach from their neighbors and spread to other organs. Refer to **Table 6.2** to explain how a nonfunctioning membrane protein could cause cells to spread.

5. White blood cells (cells with the purple nucleus in **Figure 6.11**) move throughout the blood and tissues, where they change their shape so that they can squeeze through blood vessels into tissues. Here they patrol the body, looking for disease-causing microbes that they do not recognize as part of the body. From this description, predict (1) the plasma membrane proteins they use (*Hint:* refer to **Table 6.2**) and (2) the cell structures that allow them to change shape and move throughout the body.

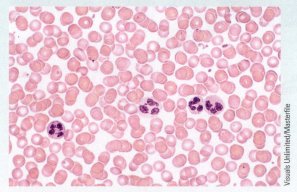

Visuals Unlimited/Masterfile

Figure 6.11 White and red blood cells.

6.4 Macromolecules Build Body Structures and Drive Life Processes

Let's continue down the digestive tract of the leatherback turtle and focus on a remarkable type of cell that lines its small intestine (**Fig. 6.12**). This epithelial cell absorbs nutrients such as glucose, fatty acids, and amino acids from the digestive tract and moves them into the bloodstream. It also imports molecules such as oxygen and glucose from the blood. Some of this molecular "trafficking" requires energy provided by its mitochondria. This cell's very active life is also a very short one; on average it lives for only three to five days before it has been entirely rebuilt by cell division and replaced by a brand new cell.

In the section ahead we explore how the different groups of macromolecules are used to build new cells. To accomplish this complex construction project, we need building block molecules, energy, and genetic information. From the previous section you know that different kinds of molecules compose the plasma membrane, membrane proteins, cytoskeleton, and organelles. On the chemical level, the particular shape of a molecule adapts it to perform a particular function, whether it is a phospholipid or a receptor protein. The information to build these parts is encoded on DNA, which is presented in more detail in Chapter 7.

Atoms bond together to form macromolecules with particular shapes and properties

Chapter 1 introduced you to the world of atoms, molecules, and elements. Recall that atoms bond with other atoms to form molecules. Bonds may be strong or weak, depending on the strength of the attraction between them.

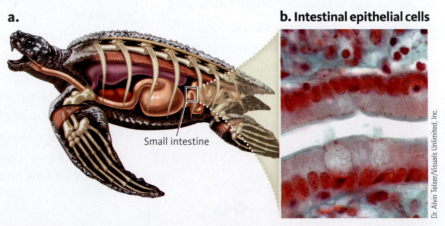

a.

b. Intestinal epithelial cells

Small intestine

Dr. Alvin Telser/Visuals Unlimited, Inc.

Figure 6.12 The leatherback's small intestine. (a) The small intestine of the leatherback completes the digestion of the jellyfish and is the principal site for nutrient absorption. (b) Cells that line the small intestine actively absorb nutrients from the digestive tract and shuttle them to the bloodstream for distribution throughout the body. Here you can see the column-shaped cells, each with a large, red-stained oval nucleus at its base.
© Cengage Learning 2014

Strong bonds occur between atoms when a pair of electrons orbits both nuclei; that is, the atoms share two electrons. This type of *strong* bond is called a covalent bond. Covalent bonds serve as the backbone of the macromolecules found in cells and determine a molecule's basic structure. However, many covalent bonds in molecules allow the atoms to rotate to form a large variety of shapes. This is where weaker bonds come in.

Weak bonds play important roles in determining the shape of a molecule and the function it plays in the cell or the body. One type of weak bond is called a hydrogen bond, where a hydrogen atom with a positive charge is attracted to another negatively charged atom, often an oxygen or nitrogen atom. This attraction does not involve sharing electrons, but it causes the backbone of the molecule to be attracted to other molecules or to fold back on itself to form a three-dimensional shape. Weak hydrogen bonds are very important in forming and stabilizing the helical shape of DNA and the coiled proteins that form hair, scales, beaks, feathers, and claws (Fig. 6.13). The proteins that physically glue an animal together owe their great strength to hydrogen bonding. Although an individual hydrogen bond may be *weak,* collectively many hydrogen bonds in a molecule exert a very *strong* force on its chemical properties. Water molecules also form hydrogen bonds with each other, which gives water its unique properties to support life and the living animal.

Proteins are the major structural and functional molecules in the animal body

After water, the most abundant molecule in the animal cell and body is protein. Proteins play critical roles at all levels of the biological hierarchy. Table 6.3 lists the many different types of proteins found in ani-

Figure 6.13 Keratin, a structural protein. (a) The beak, feathers, and claws of the peregrine falcon are composed of the tough structural protein called keratin. (b) Notice the numerous hydrogen bonds (red dotted lines) that hold this protein chain in a coil shape.
© Cengage Learning 2014

© Arco Images GmbH/Alamy

Covalent bond—strong bonds occur between atoms when a pair of electrons orbits both nuclei; serves as the backbone of macromolecules found in cells and determines a molecule's basic structure.

Hydrogen bond—a type of weak bond where a hydrogen atom with a positive charge is attracted to another negatively charged atom, often an oxygen or nitrogen atom.

TABLE 6.3 Major Types of Proteins and Their Functions

Type of Protein	Function
Enzymes	Speed up chemical reactions in cells; aid in digestion and metabolic processes
Structural proteins	Compose body parts (hair, nails, skin, scales, connective tissue, tendons, cartilage)
Contractile proteins	Muscles; fibers that pull chromosomes apart
Hormones	Send signals from one cell or organ to another (insulin, growth hormone)
Transport proteins	Carry substances (oxygen, fats) Store oxygen in blood and tissues
Antibodies	Protect body against infectious pathogens and disease
Nucleoproteins	Package DNA into chromosomes
Membrane proteins	Adhesion, transport, blood types
Toxins	Paralyze and hold prey; protect from predators
Cytoskeleton	Support cell's shape, move organelles

© Cengage Learning 2014

mals and their functions. Starting with the *organ systems,* the muscular system contracts when protein filaments slide past each other to cause muscles to shorten. *Organs* such as the outer layer of skin, hair, and claws are all composed of proteins. *Tissues* such as connective tissues have strong and flexible protein fibers, and a layer of proteins supports epithelial tissues. At the *cellular* level, membrane proteins act as receptors and signaling molecules, and the cytoskeleton in the cytoplasm and nucleus consists of protein threads. Ribosomes, which make proteins, are composed of dozens of different proteins. In the nucleus, proteins link DNA into chromosomes, which are moved apart during cell division by proteins. Sequences of DNA, called genes, code for and control the ribosome's production of proteins and therefore control the structure and function of the cells, tissues, organs, organ systems, and ultimately the organism.

A protein's function is determined by its molecular structure and shape. On the chemical level, proteins are large organic macromolecules composed of chains of hundreds and in some cases thousands of **amino acids**. Each amino acid consists of four parts (**Fig. 6.14a**): (1) a central carbon and hydrogen atoms, (2) a group of nitrogen and hydrogen atoms called an *amino* group on one end of the molecule, (3) an *acid* group on the other end of the molecule, and (4) a side group that attaches to the central carbon. The backbone of a protein consists of the central carbon along with the amino and acid groups.

The side group is what makes each of the 20 common amino acids unique with a different set of chemical properties (**Fig. 6.14b**). For example, some amino acids carry a negative or positive charge, whereas others differ in their solubility in water. Similar to the way the 26 letters of the alphabet are joined to form words; the 20 different amino acids are joined to form proteins. This sequence of amino acids is called the **primary structure** of the

Amino acids—building block molecules for proteins; each amino acid consists of four parts: (1) a central carbon and hydrogen atoms, (2) a group of nitrogen and hydrogen atoms called an *amino* group on one end of the molecule, (3) an *acid* group on the other end of the molecule, and (4) a side group that attaches to the central carbon.

Primary structure—sequence of amino acids that are linked together to form a protein molecule.

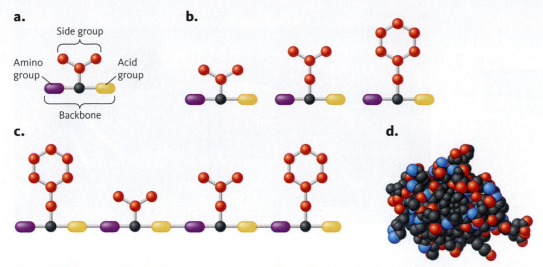

Figure 6.14 Proteins consist of chains of amino acids. (a) All amino acids consist of the same three parts bonded to a central carbon atom (gray): an amino group, an acid group, and a side group. (b) Three different amino acids are shown, each containing a different side group (red). (c) A chain of four amino acids bonded together to form a protein. (d) The side groups interact and form a globular-shaped protein called insulin. Insulin is a protein that regulates blood sugar and consists of two chains (21 and 30 amino acids, respectively), which are connected by a sulfur bridge.
© Cengage Learning 2014

protein molecule. The amino acids are covalently bonded together; the amino group on one end is bonded to the acid group on the next amino acid in the sequence (**Fig. 6.14c**).

Along the protein molecule, the side groups interact by forming hydrogen bonds with each other to twist, fold, or pull the chain into a particular shape. The shape of the molecule is critical to it performing its function in the same way that a key's shape is critical to opening a particular lock in a door. Some proteins such as insulin (**Fig. 6.14d**) form a round, *globular* shape, whereas others such as keratin (see **Fig. 6.13b**) form a *helical* shape. Hemoglobin is another example of a globular blood protein that consists of four separate chains of amino acids bonded together. If only one of the 578 amino acids is replaced by an amino acid with different chemical properties, the entire protein's shape may be altered and the protein's function changed or lost. This is the case with the disease sickle cell anemia, where a single change in an amino acid causes a drastic change in the shape of hemoglobin and inhibits its ability to bind to oxygen.

During the extremely long time that animals have evolved, changes to amino acids have led to adaptations. For example, changes in protein structure and shape have enhanced the ability of some fish to prevent their blood and tissues from freezing in frigid waters. Armed with these helical antifreeze proteins, these fish were able to extend their range into habitats such as the Antarctic Ocean.

Fats are lipids that insulate, store energy, and cushion delicate parts

Lipids are a large class of organic macromolecules that do not dissolve in water. They include phospholipids, cholesterol, and fats. Recall that each of the trillions of cells has a plasma membrane that is constructed with a

Lipids—a class of organic macromolecules that do not dissolve in water; include phospholipids, cholesterol, and fats.

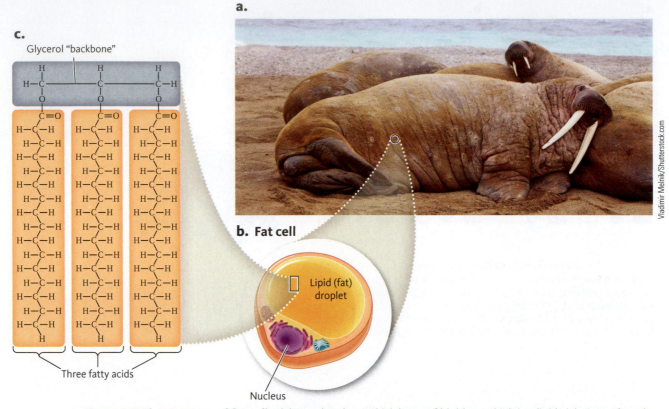

c.

Glycerol "backbone"

Three fatty acids

b. Fat cell

Lipid (fat) droplet

Nucleus

Vladimir Melnik/Shutterstock.com

Figure 6.15 The structure of fat cells. (a) A walrus has a thick layer of blubber, which is a lipid-rich tissue found under the skin of many marine mammals, such as elephant seals, whales, and dolphins. (b) Fat cells are round and filled with lipid droplets that store energy. (c) Animal fat is mostly composed of triglycerides, which consist of three fatty acid molecules (tan) attached to a three-carbon backbone called glycerol (gray).
© Cengage Learning 2014

Triglyceride—type of fat that contains three fatty acid molecules attached to a three-carbon backbone called glycerol; macromolecule that stores energy in fat cells and tissues.

Fatty acids—insoluble in water; building block molecules of fats (triglycerides); long chains of carbon, hydrogen, and oxygen.

phospholipid bilayer and *cholesterol.* Fats play a variety of functions in the animal body. A layer of fat underneath the skin in many mammals insulates the body against heat loss and stores large amounts of energy. Blue whales, elephant seals, walruses, and other marine mammals have lipid-rich *blubber* under their skin (**Fig. 6.15a**). In some large whales the fat layer may be a foot thick and constitute 40 percent or more of the whale's body weight. Blubber also streamlines the body shape and provides buoyancy. Whales can migrate for thousands of miles without eating. They use the fuel in their blubber as they move from their feeding grounds in cold polar waters to temperate breeding grounds. The leatherback turtle's leathery shell is saturated in oil that protects it against the oceanic environment.

Fat tissues are composed of clumps of fat cells, which specialize in storing large droplets of lipid molecules within their cytoplasm (**Fig. 6.15b**). The single large lipid droplet is characteristic of *white fat* cells. This type of fat cushions eyeballs and muscles, and it forms the bulk of breast tissue. *Brown fat* is found only in newborns and hibernating mammals, in which it generates heat. Brown fat cells contain numerous smaller fat droplets and many mitochondria.

Chemically, the lipid droplet inside the fat cells is classified as a **triglyceride**. Triglycerides contain three fatty acid molecules attached to a three-carbon backbone called glycerol (**Fig. 6.15c**). **Fatty acids** are long chains composed of the familiar elements of organic molecules—carbon, hydrogen,

a. Sugar (Glucose)   **b. Complex carbohydrate (Glycogen)**

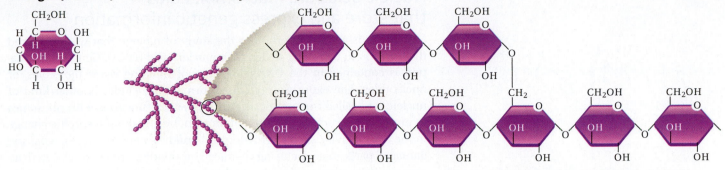

Figure 6.16 Simple and complex carbohydrates.
Carbohydrates are in the form of sugars or complex carbohydrates. (a) Glucose is a six-carbon ring bonded to hydrogen and oxygen atoms in the ratio $C(H_2O)$. These bonds carry a large amount of energy, which is released when they are broken during metabolism. (b) Complex carbohydrates such as glycogen are long chains composed of thousands of glucose molecules bonded together.
© Cengage Learning 2014

and oxygen. The carbon and hydrogen form a strong covalent bond. Covalent bonds contain a large amount of energy, and a triglyceride contains many bonds (see **Fig. 6.15c**). This is why fats contain more calories than other macromolecules and why animals use triglycerides to store large amounts of energy in small spaces.

Carbohydrates fuel the building and activities of organisms

The wide variety of carbohydrates can be placed into two categories depending on their size. The first category includes **sugars**, which are relatively small molecules that contain five or six carbon atoms that form a ring (**Fig. 6.16a**). Some sugars contain a single ring, whereas other sugars, such as table sugar, consist of two rings bonded together. Sugars are easily digested and are the primary nutrients in fruits such as apples and oranges. Many mammalian species include fruit in their diet as a rich, concentrated source of energy. The five-carbon sugars are essential to building the hereditary molecules DNA and RNA. The six-carbon molecule called **glucose** is a ring-shaped molecule with hydrogen and oxygen atoms (see **Fig. 6.16a**). It is a "hydrated carbon molecule" or a carbo (hydrate). Glucose molecules have several strong covalent bonds that carry a large amount of energy that is released when the bonds are broken apart during metabolism. Therefore glucose is the primary fuel that powers cellular activities such as contracting muscles, moving substances across membranes, and driving electrical impulses in the brain. Glucose, along with fats, fuels the leatherback turtle's 12,700 mile migration across the Pacific Ocean.

The second category consists of **complex carbohydrates**, which are long chains of glucose sugars bonded together (**Fig. 6.16b**). Animals store energy by making the complex carbohydrate called **glycogen**. Whereas fats store enough energy to last for days and weeks, glycogen stores energy that lasts only for several hours. Whenever an animal eats an excess amount of glucose and other sugars, cells of the liver and muscles bond thousands of glucose molecules together into long branched chains of glycogen. When needed, the glucose rings are quickly broken off to fuel the muscles, brain, and other active tissues.

Although carbohydrates are mostly used for fuel, one kind of complex carbohydrate called **chitin** forms the structure of the outer shell, or exoskeleton, of insects, crabs, and other arthropods. Chitin consists of long chains of glucose, but unlike other carbohydrates it contains nitrogen. This modified complex carbohydrate is a very strong and durable structural molecule that supports and protects the vital organs of the arthropod body.

Sugars—small carbohydrate molecules that contain five or six carbon atoms that form a single or double rings; building block molecules for complex carbohydrates.

Glucose—a six-carbon ring-shaped molecule; sugar that fuels life activities.

Complex carbohydrates—large macromolecules that consist of long chains of sugars bonded together.

Glycogen—complex carbohydrate used to store energy in animals.

Chitin—a modified (contains nitrogen) complex carbohydrate composed of long chains of glucose; a very strong and durable structural molecule that forms the outer shell, or exoskeleton, of insects, crabs, and other arthropods.

Nucleic acids are macromolecules that store and express genetic information

Nucleic acids are so important to the study of biology that we introduced them to you in Chapter 1 as deoxyribonucleic acid (DNA). Genetic information is contained in the DNA's sequence of building block molecules, or nucleotides. The expression of this information includes another kind of nucleic acid called ribonucleic acid (RNA), which functions with ribosomes to produce a particular protein that, in turn, brings about a specific change in the cell's activity, such as growing and building new structures, repairing damaged parts, manufacturing chemicals, or dividing into two cells. Activating specific portions of the DNA molecule orchestrates the cell's functioning in the tissue, organ, organ system, and ultimately the organism.

Within the nucleus, long strands of DNA are wrapped around proteins to form long fibers called chromatin. Chromatin occurs as separate "genetic packages" called chromosomes. Chromosomes become visible only when the cell is about to divide; otherwise, they are stretched out and dispersed within the nucleus. Each animal species has a characteristic number of chromosomes in the cells of its body. Human cells have 46 chromosomes, orangutans have 48, and several turtle species have 22. Male stinging jack jumper ants contain only a single chromosome, the lowest number known for any animal. More detailed information about DNA and RNA is presented in Chapter 7 and then applied to various organisms throughout the book.

Atoms and molecules move passively or actively across cell membranes

Intestinal cells lining the small intestine are one of the most active cells in the animal body. The digestive system breaks down large macromolecules (proteins, fats, complex carbohydrates) into their respective building block molecules (amino acids, fatty acids, sugars), which must be absorbed across cell membranes before they are moved into the bloodstream. Without this intense and coordinated movement of molecules, an animal will starve to death.

To accomplish this remarkable feat of absorption, an intestinal cell faces two different environments with which it exchanges materials. One side faces the digestive tract, with its watery mixture of amino acids, glucose, fatty acids, and salt molecules. The other side faces a watery fluid compartment called tissue fluid, which exchanges materials with the bloodstream (Fig. 6.17). The tissue fluid is the cell's "internal environment," which brings oxygen, building block molecules, and minerals into the cell where they are used or assembled, while waste products and secretions are moved out of the cell. How atoms and molecules move across the plasma membrane depends on whether the molecules require energy to move "uphill" or move passively "downhill."

Passive transport moves oxygen, water, and glucose molecules across the membrane "downhill," or down its concentration gradient. That is, these molecules are moving from areas of *higher* concentration to areas of *lower* concentration. No inputs of energy are required for passive transport because the molecules themselves have energy. Molecules are always moving and colliding with each other, and as a result of these collisions, the molecules move away from each other to areas where there is more space. The higher the temperature, the faster the molecules move and the faster they diffuse away from each other. Also, the steeper the gradient, the faster the molecules move. This process is called simple diffusion. Oxygen

Chromatin—a macromolecule found in the nucleus; composed of long strands of DNA wrapped around proteins; packaged into chromosomes.

Chromosomes—"genetic packages" in the nucleus of a cell; consist of long strands of chromatin (DNA and protein) that store and express genetic information.

Tissue fluid—fluid surrounding cells and tissues that exchanges materials with the bloodstream; referred to as the cell's "internal environment," which brings oxygen, building block molecules, and minerals into the cell.

Passive transport—movement of substances "downhill," or down its concentration gradient.

Simple diffusion—movement of substances such as gases down their concentration gradient from an area of higher concentration to an area of lower concentration.

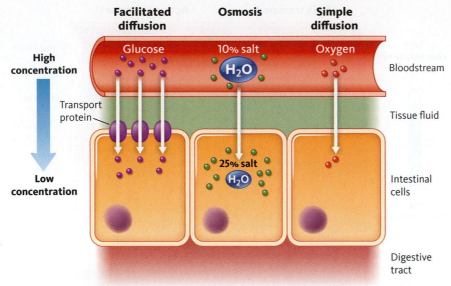

Facilitated diffusion **Osmosis** **Simple diffusion**

Glucose   10% salt   Oxygen
           H_2O

High concentration

Transport protein

Low concentration

25% salt
H_2O

Bloodstream

Tissue fluid

Intestinal cells

Digestive tract

Figure 6.17 Passive transport. Passive transport is an important process in moving substances across membranes from high concentrations in the bloodstream to lower concentrations inside cells. Gases such as oxygen and carbon dioxide move by simple diffusion. Osmosis is the passive transport of water, and facilitated diffusion requires a transport protein. It is important to note that substances move passively in both directions—into cells and out of cells.
© Cengage Learning 2014

moves by simple diffusion into cells, where it is used in the mitochondria (see **Fig. 6.17**). Carbon dioxide is produced in high concentrations by the mitochondria and thus moves in the opposite direction, out of the cell into the bloodstream. These gases and their movement will take center stage in Chapter 9, where we present animal breathing and gas exchange.

Water is the most abundant substance to move through the cell membrane. It has been estimated that about a billion water molecules move in and out of the cell each second. Water, like oxygen, moves "downhill" according to its concentration gradient. It is important to note that the greater the concentration of dissolved materials in water, the lower the concentration of water. Thus, a 25 percent salt solution has a lower concentration of water than a 10 percent salt solution because the salt molecules occupy space. In this situation, water will passively move by **osmosis** across the membrane from the 10 percent salt side to the 25 percent salt side of the membrane (see **Fig. 6.17**). To maintain water balance, the body regulates the concentration of dissolved materials in the blood and body fluids. For example, to draw water into a cell, the cell maintains a high concentration of dissolved material relative to the surrounding fluids.

Lastly, glucose is in higher concentration in the blood than inside cells, so a continuous gradient exists for glucose to move downhill into cells. However, glucose is a relatively *large* molecule that needs some help, which is provided by special membrane proteins that grab a glucose molecule on the outside and flip it to the inside of the cell (see **Fig. 6.17**). Although a "helper" protein is used, the process is still considered passive diffusion because glucose moves downhill into cells without the input of energy. This type of diffusion is called *facilitated diffusion*. Interestingly, this helper, or transport, protein is under the influence of the hormone insulin, which causes glucose to move from the blood into cells where it generates energy by mitochondria.

Active transport, in contrast to passive transport, is a process that moves molecules with the expenditure of energy. In most cases, molecules move uphill against the concentration gradient. Amino acids and many minerals are actively transported through the membrane by transport proteins (**Fig. 6.18**). Another type of active transport involves the movement of large macromolecules such as proteins or fats. **Bulk transport** processes involve enclosing these macromolecules in a vesicle before moving them into the cell or

Osmosis—passive movement of water across a cell membrane; water moves down its concentration gradient toward the compartment with the highest concentration of dissolved materials.

Active transport—a process that moves molecules uphill against the concentration gradient with the expenditure of energy.

Bulk transport—a type of active transport that moves large macromolecules by enclosing them in a vesicle before moving them into the cell or secreting them out of the cell.

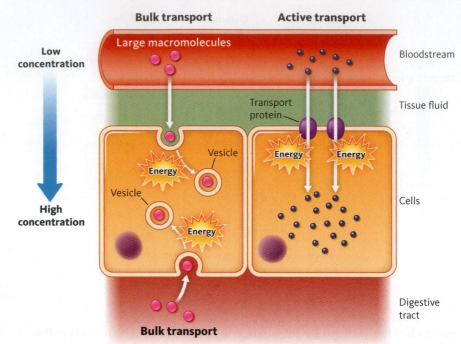

Bulk transport

Active transport

Low concentration

High concentration

Large macromolecules

Bloodstream

Transport protein

Tissue fluid

Energy

Vesicle

Energy

Energy

Vesicle

Cells

Energy

Digestive tract

Bulk transport

Figure 6.18 Active transport. Active transport uses transport proteins and energy to move substances against their concentration gradient. Bulk transport moves large macromolecules such as proteins and fats. Note that active and bulk transport moves substances in both directions—into cells and out of cells.
© Cengage Learning 2014

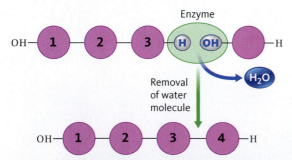

Enzyme

OH— 1 — 2 — 3 —(H)(OH)— —H

Removal of water molecule

H_2O

OH— 1 — 2 — 3 — 4 —H

Figure 6.19 Building macromolecules. Enzymes remove water between two small building block molecules, thus building large molecules. The circles could represent several different kinds of building block molecules—glucose for building glycogen, amino acids for building a protein, or fatty acids for bonding to a glycerol molecule to build a fat. Different enzymes are used to build the different types of macromolecules.
© Cengage Learning 2014

secreting them out of the cell. Large long-chain fatty acid molecules are often wrapped in a membrane vesicle and moved this way (see **Fig. 6.18**). The endomembrane system plays an important role in bulk transport.

Building block molecules are assembled into macromolecules

Now that you can visualize what the macromolecules look like and how they enter and exit cells, we will look at how they are put together, or synthesized, inside cells. Like all building processes, the building of large macromolecules such as glycogen, lipids, and proteins requires energy. Specific enzymes accomplish this macromolecular building process, or *biosynthesis,* by forming a covalent bond between two building block molecules. How do enzymes do this?

Let's first look at the molecular structure of the building blocks—glucose, fatty acids, glycerol, and amino acids. Notice that they all have an OH, or hydroxyl group, on one end of the molecule. Through the action of an enzyme, two glucose molecules are joined together when the OH group from one glucose is combined with an H atom from the other to form H_2O, or water (**Fig. 6.19**). By removing water from the two molecules, the enzyme strongly bonds the two glucose molecules together with a covalent bond. The same reaction causes three fatty acids to form covalent bonds with a glycerol molecule. Amino acids are coupled together into a long protein chain by the same process—the removal of water by enzymes. Thus water is a by-product of the cell building process. As you will see in Chapter 8, the reverse of this process *breaks down* macromolecules into their building block molecules. Here, digestive enzymes *add* water to the bonds to break them apart. As you will see in Chapter 15, many kinds of fungi specialize in this molecular demolition process.

- Molecular shape is determined by a backbone of strong covalent bonds and a number of weak hydrogen bonds.

- Molecular shape determines the function of animal structures.

- Passive transport processes (simple diffusion, osmosis, and facilitated diffusion) move substances down a concentration gradient from a higher to a lower concentration.

- Active transport processes require energy to move molecules from a lower concentration to a higher concentration.

- Enzymes that remove water between building block molecules build macromolecules.

1. Strong bonds require lots of energy to be built and then release lots of energy when they are broken. Which kind of bond serves as the strong backbone of macromolecules?

2. What are the four parts of an amino acid? Which two parts join to form a protein?

3. How does the cell build a glycogen molecule to store energy?

4. Consider two sides of a plasma membrane. Side 1 contains a 20 percent salt solution and Side 2 contains a 5 percent salt solution. Assume the salt does not pass through the membrane. In which direction will there be a net movement of water down its concentration gradient: (a) from Side 1 to Side 2, or (b) from Side 2 to Side 1? Explain why.

5. During lactation, mammary gland cells make and secrete very large immune system proteins called antibodies into a system of ducts within the breast (**Fig. 6.20**). Newborn infants then gain some of this protection by drinking their mother's breast milk. (A) Predict the process that is used by the mammary gland cell to transport antibodies from within the cell out into the milk duct? (B) Is energy required to do this?

Figure 6.20 Mammary glands.
© Cengage Learning 2014

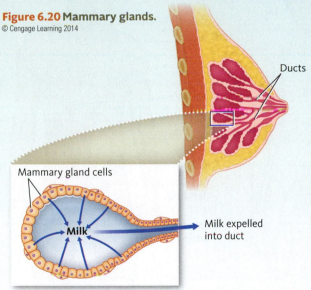

6.5 Body Structures Work Together to Respond to Change

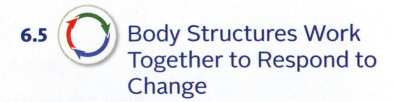

Change and responding to change are key characteristics of life. From one moment to the next, our bodies lose water through our skin and lungs. Changes in body posture alter blood pressure in different parts of our body. How does the body know whether to make a small adjustment or a dramatic response to the situation? How do all of these body parts on all levels work together as an integrated system called an organism? The one-word answer is *communication*.

The two organ systems involved in communication and body integration are the nervous system and the endocrine system (**Fig. 6.21**). Fast-acting nervous tissue cells, or **neurons**, receive information and communicate with other neurons in a complex network of connections. The generally slower-acting hormones are chemicals that signal cells to change their activities over a longer time frame of hours, months, or years. In both cases, nerves and hormones are involved in one part of the body communicating with another part.

To successfully compete and survive, organisms have evolved elaborate communication networks to process incoming signals and the information they carry. Some information concerns *internal* conditions of the body, which include temperature, water, and nutrient levels. Other signals

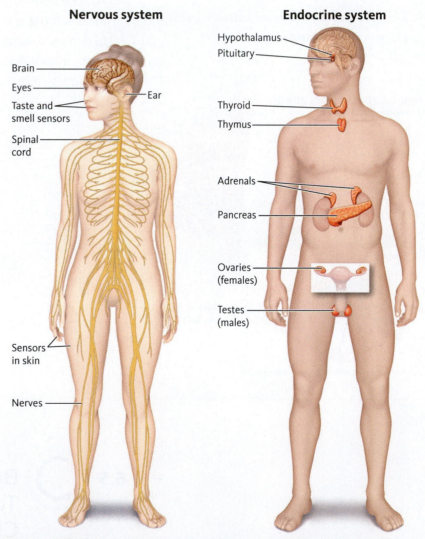

Nervous system

Brain
Eyes
Ear
Taste and smell sensors
Spinal cord
Sensors in skin
Nerves

Endocrine system

Hypothalamus
Pituitary
Thyroid
Thymus
Adrenals
Pancreas
Ovaries (females)
Testes (males)

Figure 6.21 The nervous and endocrine systems. The nervous and endocrine organ systems integrate the body's activities through communication among organs, tissues, and cells. The nervous system communicates information through electrical impulses, whereas the endocrine system communicates messages through chemical signal molecules called hormones.
© Cengage Learning 2014

Neurons—cells composing nervous tissue; receive information and communicate with other neurons in a complex network of connections.

arise from the *external environment,* involving conditions such as temperature, daylight, or the presence of a predator, prey, or mate. The consequences of failing to communicate and rapidly respond to internal and external signals are decreased performance, lower competitive ability, sickness, and maybe death. Natural selection is very intense in this area of an organism's biology.

In the section ahead, we explore how the trillions of cells in the human body work together, communicate, and coordinate their activities. Although we highlight the human system, the same processes occur in other animals such as jellyfish and the leatherback turtle that preys on it. Communication among parts allows a system to adjust to continually changing conditions.

Internal conditions are allowed to fluctuate within a normal range

As the human body continually exchanges energy and materials with the environment, it copes with constant change. Normal activities such as eating, drinking, breathing, and excreting wastes cause changes in the composition of blood and body fluids (the cell's environment). Inhaling air increases the concentrations of oxygen in the blood; eating adds nutrients to the blood; and excretion removes waste products from the blood. Running to class means an increase in body temperature. In the face of this constant change, body cells require a fairly constant internal environment that supplies nutrients and oxygen as well as disposes of waste products. The temperature and pH of the blood increase and decrease. How much change is too much?

The nervous and endocrine systems function to keep conditions within a particular range that is optimal for body functioning. That is, these systems keep the body in *a steady state.* Each factor has its own characteristic normal range with upper and lower limits. For example, the concentration of sodium in the blood is strictly controlled within a narrow range, whereas the concentration of sugar in the blood has a wider range of normal values and is less strictly controlled. When blood sugar levels rise above or fall below the range, a series of actions restores the levels (**Fig. 6.22**). The process that keeps all of these conditions, such as blood sugar level, within their normal ranges is called **homeostasis**. Essentially all organs and tissues of the body perform functions that maintain their own homeostasis. We'll see how they work together to contribute to the homeostasis of the whole body next.

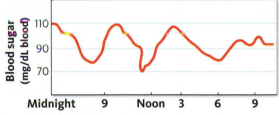

Figure 6.22 Normal range of blood-sugar levels.
Blood sugar level rises and falls throughout the day within a range of values. Notice that blood sugar level rises after a meal and then gradually declines. During sleep the level is maintained by the breakdown of glycogen stores in the body.
© Cengage Learning 2014

Feedback control systems continually adjust internal conditions

At this moment your body, like that of all multicellular animals, is keeping hundreds of varying factors within their normal ranges. In most cases, the body is using a system called **negative feedback** control, which detects changes in a factor and *counteracts* that change to maintain the steady state. Blood sugar, blood pressure, body temperature, and acid–base balance all are controlled through negative feedback systems.

Let's use the example of body temperature to explain how three components of the feedback loop interact with each other. Body temperature varies depending on where it is taken (rectal, oral, armpit), the time of day (lowest during sleep and early morning), and the individual's overall rate of metabolism. As a general rule, the normal range, or *set point,* for

Homeostasis—process that keeps conditions within the internal body environment within their normal ranges.

Negative feedback—a control system that detects changes in a factor and *counteracts* those changes to maintain the steady state.

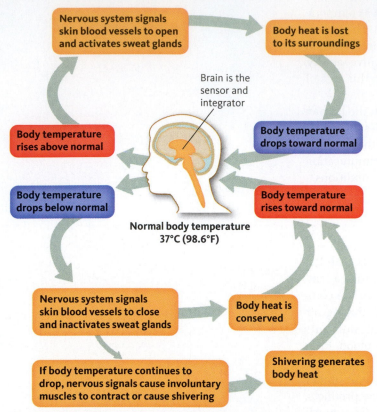

Figure 6.23 Negative feedback control loop. Body temperature is maintained within its normal range by the activities of feedback control. The upper loop is activated when the temperature exceeds its upper set point; the lower loop is activated when body temperature falls below its lower set point limit.

© Cengage Learning 2014

Positive feedback—control system that increases, or *amplifies,* the signal in the direction of the stimulus.

humans is 98.2° to 99.9° F. The body uses feedback control when the temperature exceeds or falls below its normal range (**Fig. 6.23**). Cells in the brain act as both *sensors* and *integrators* in regulating body temperature. In one area of the brain, temperature sensors are specialized neurons that continually send signals to an integration area of the brain. Here, the integrator part of the feedback system compares the information it receives to the set point. When the body temperature exceeds the set point, the integrator activates the third component, the *effectors,* which includes the sweat glands and blood vessels of the skin that release heat to the environment. However, if the body temperature falls below the normal range, the integrator activates other types of effectors—muscles—to generate heat through shivering. The effectors are acting to counter the change. This negative feedback system controls body temperature, which makes it possible for body cells to continue to live and function properly.

A **positive feedback** system uses the same three components. However, the effector increases, or *amplifies,* the signal in the direction of the stimulus. Using the body temperature example, if the temperature exceeded the set point, a positive feedback system would set into motion processes to further increase the body temperature rather than decrease it back toward the set point. Therefore the body rarely uses this type of control system because it does not lead to balance and stability. However, a few body functions do use positive feedback control. One example is the birthing process. When muscles of the uterus begin to contract to push the baby's head through the cervix, the stretch of the cervix sends stronger signals back to the uterus to further increase the muscular contractions. With the birth of the baby, the feedback loop is broken and the body returns to normal.

This chapter presented the structure and function of the animal body using the incredible leatherback turtle as the main example. We started with the levels of organization most familiar to you—organ systems and organs—and then worked our way down to the microscopic tissues and cells and then to the macromolecules that compose them. We saw that all life activities ultimately are cell activities, which depend on the movement of molecules across the plasma membrane. Finally, we focused on how all of the different organs and organ systems work together through feedback loops. An animal depends on its feedback control systems to keep all of its different structures communicating and functioning with each other to maintain its internal environment and keep it alive.

- Normal daily activities cause the body to continually respond and adjust to changes in the internal and external environments.

- Factors in the cell's internal environment fluctuate within a range or a set point.

- The nervous and endocrine systems monitor and respond to changes in the internal and external environments. The body's system of responses that maintain a factor within its normal range is called homeostasis.

- Feedback loops involve a sensor that monitors body conditions, an integrator that compares the conditions against a set point, and effectors that take action. Negative feedback loops operate to counteract the direction of a stimulus and are most often used to maintain homeostasis.

1. When a factor such as sodium concentration is said to be "strictly controlled," would you expect it to have a *narrow* or a *wide* normal range? Explain.

2. List several environmental factors that are balanced or regulated to maintain the steady state of an organism's internal environment. Explain.

3. What are the three parts of a feedback loop and what do they do?

4. Compare a negative and positive feedback control system and explain why a positive feedback system does not maintain the steady state of the animal.

5. When blood sugar level rises above its set point, receptor cells in the pancreas perceive and communicate this information to other specialized cells in the pancreas that secrete the hormone insulin. This causes the liver to reduce blood sugar levels by moving sugar from the blood into storage as glycogen. Which organs are the three components of the feedback control system in this scenario?

End of Chapter Review

Self-Quiz on Key Concepts

The Level of Structural Organization in Animals Varies from Tissues to Organ Systems

KEY CONCEPTS: Animals vary in their structural complexity. Simple animals consist of cells and tissues; more complex animals are further organized into organs and organ systems. Organ systems are composed of four different types of tissues—epithelial, connective, muscle, and nervous tissues. The animal system is functionally integrated into a whole through the communication and interaction of its parts at all levels of organization.

1. Match each of the following **levels of the biological hierarchy** with its characteristic, description, or example.

    Atom
    Cell
    Molecule
    Organ
    Organ system
    Tissue

    a. group of cells that perform specific functions
    b. heart, lungs, kidney, and stomach are at this level of organization
    c. highest level of organization in the animal body
    d. formed when electrons are shared and bonds are made
    e. fundamental unit of structure and function of the animal body
    f. fundamental unit of matter; consists of protons, neutrons, and electrons

2. Match each of the following **tissues** with its characteristic, description, or example.

    Connective tissue
    Epithelial tissue
    Muscle tissue
    Nervous tissue

    a. capable of shortening to produce movement
    b. fat, blood, and bone are types of this tissue
    c. functions to transmit signals
    d. lines the inside and outside of body

3. Which type of tissue directly opens and closes the iris of the eye to focus light?
   a. connective
   b. epithelial
   c. nervous
   d. muscle

4. Which type of tissue is specialized to make and secrete sweat from sweat glands?
   a. connective tissue
   b. epithelial tissue
   c. muscle tissue
   d. nerve tissue

 ## Cells are the Fundamental Units of Animal Structure and Function

KEY CONCEPTS: An animal is an assembly of cells that perform fundamental life processes. Animal cells are eukaryotic cells composed of an outer plasma membrane, cytoplasm, and a nucleus. The plasma membrane interacts with the cell's environment by receiving stimuli, transporting materials, and attaching to other cells. The cytoplasm includes a variety of organelles that perform specialized functions to break down and build new cell parts. The nucleus contains the majority of the genetic information because it houses DNA, which controls the activities of the cell.

5. Match each of the following **cell parts** with its characteristic, description, or example.

Golgi complex	a. contains majority of DNA; control center of cell
Lysosome	b. sorts and packages products for export
Mitochondria	c. transports large molecules to the plasma membrane
Nucleolus	d. region of chromatin that codes for the production of ribosomes
Nucleus	e. assembles proteins
Endoplasmic	f. channels and transports proteins and lipids
reticulum	g. transforms energy in food nutrients into cellular fuel
Ribosome	h. contains powerful digestive enzymes that break down a variety of
Vesicle	macromolecules within the cell

6. Match each of the following **membrane proteins** with its characteristic, description, or example.

Adhesion protein	a. pumps substances across the membrane
Recognition protein	b. informs the immune system that the cell belongs to the body
Transport protein	c. sticks to neighboring cells, joining them together

7. Epithelial cells that line the digestive system contain many vesicles. These vesicles contain digestive enzymes that will be exported from the cell into the gut. From what organelle are these vesicles derived?
   a. Golgi complex
   b. lysosome
   c. mitochondria
   d. endoplasmic reticulum

8. Red blood cells pick up and deliver oxygen to every cell in the body. Which body cell organelle uses the oxygen to break down nutrients and generate energy?
   a. Golgi complex
   b. lysosome
   c. mitochondria
   d. endoplasmic reticulum

9. Which of the following is the most abundant component of an animal cell's plasma membrane?
   a. cholesterol
   b. phospholipid
   c. protein
   d. all of the above

The Chemical Level of Organization includes Ions, Water, and Macromolecules

KEY CONCEPTS: Animal cells use chemicals and molecules to build structures and speed up metabolic reactions. Cells move ions, molecules, and macromolecules across their membranes to build new parts and drive their activities. Proteins are versatile molecules that form cellular structures and function as enzymes, hormones, defensive compounds, and muscle. Lipids are compact sources of energy and function as storage molecules. Carbohydrates are the primary source of energy for animal cells.

10. Match each of the following **classes of macromolecule** with its characteristic, description, or example.

    Carbohydrate a. composed of chains of amino acids
    Lipids   b. glucose and other sugars
    Nucleic acids c. fatty acids attached to glycerol backbone; insoluble in water
    Proteins   d. DNA and RNA that store and express genetic information

11. Match each of the following **types of transport** with its characteristic, description, or example.

    Active transport   a. transport process that requires the input of energy
    Bulk transport   b. movement from area of low to higher concentration with assistance of
    Facilitated diffusion   a protein
    Osmosis   c. water moving across a membrane from low to higher concentration
    Simple diffusion   d. movement from an area of higher to lower concentration
                           e. active transport of large macromolecules into or out of the cell

12. Which of the following large macromolecules is mismatched with its building block molecule?
    a. chitin; glucose
    b. protein; fatty acids
    c. glycogen; glucose
    d. triglyceride; fatty acids and glycerol

13. Which group of chemical elements must be available to build a general amino acid?
    a. carbon, hydrogen, oxygen, nitrogen
    b. carbon, oxygen, nitrogen, sulfur
    c. calcium, hydrogen, oxygen, nitrogen
    d. calcium, hydrogen, oxygen, phosphorus

14. Animals store short-term energy as ____ and long-term energy as ____.
    a. fat; starch
    b. fat; glycogen
    c. glycogen; fat
    d. starch; fat

Animals Respond to Change and Function as a Whole through Homeostasis

KEY CONCEPTS: Communication among parts, at all levels of organization, allows an animal to respond and adjust to changing conditions. Internal and external conditions of the body are continually monitored through communication networks, which receive, process, and respond appropriately to information through feedback control systems.

15. Match each of the following **components of a feedback loop** with its characteristic, description, or example. Each term on the left may match more than one description on the right.

    Effector   a. component of feedback loop that secretes a hormone or moves a
    Integrator   muscle
    Sensor/receptor   b. component of a feedback loop that detects change in a factor
                      c. component of a feedback loop that compares input to a set point and
                          coordinates output
                      d. neurons in the brain
                      e. pain nerve endings in the skin

16. Match the following **type of feedback loops** with its characteristic, description, or example.

Negative feedback
Positive feedback

    a. counteracts the signal in the direction of the stimulus; maintains a steady state
    b. amplifies the signal in the direction of the stimulus

17. Most factors are maintained within a range of normal conditions through ____ feedback control systems.
    a. negative
    b. positive
    c. neutral
    d. automatic

Applying the Concepts

18. Geckos are a group of 850 species of lizards (reptiles) that have two amazing adaptations. First, geckos can rapidly run up walls and along ceilings, leaves, and other horizontal surfaces. The secret lies in the structure of their toe pads, which are covered with millions of microscopic branched elastic hairs (**Fig. 6.24a**). In order to run, the toe pads must be able to attach and detach within milliseconds and not stick to each other.

    a. Predict which kind of *tissue* and biological *macromolecule*—carbohydrates, lipids, or proteins—compose the hairs on the bottom of the gecko toe pad.

    b. Briefly describe how biologists would study these remarkable foot hairs at the molecular, cellular, tissue, organ system, and organism levels of the hierarchy to understand this amazing adaptation.

    Second, when confronted or attacked by a predator, a gecko can actually break off its tail (**Fig. 6.24b**). The detached tail can wiggle and move to distract the predator just long enough for the gecko to make a quick getaway. The tailless lizard will grow a new tail over a period of weeks.

    c. Predict how each of the four types of tissues functions in this defense adaptation. What tissues compose the detached tail so that it can move? Explain.

Figure 6.24 Gecko adaptations.

a.

b.

Dennis Kunkel Microscopy, Inc./Visuals Unlimited, Inc.

iStockphoto.com/Stephan Hoerold

JOEL SARTORE/National Geographic Stock

19. The garden snail is a type of mollusk that feeds on vegetation. When the snail feeds, its mouth protrudes a tonguelike structure covered with a set of razor sharp teeth that move like a chain saw to cut and scrape food apart (**Fig. 6.25**). These teeth are lubricated by the mucus of the salivary gland. This sticky mucus traps food particles, and the beating of cilia lining the esophagus drives food to the stomach for digestion. From this scenario, (A) determine which organ system and organs are described, and (B) describe how the four types of tissues are specifically involved in the snail's feeding process.

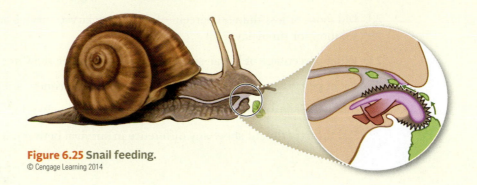

Figure 6.25 Snail feeding.
© Cengage Learning 2014

Data Analysis

One of the only insects found on the continent of Antarctica is a type of wingless fly called the Antarctic midge (**Fig. 6.26a**). During the summer, adults live for a little more than a week, during which time they mate and lay eggs. Eggs develop into wormlike larvae (**Fig. 6.26b**), which live in this larval stage for two years to gain enough energy to develop into adults. The midge lives in terrestrial microhabitats near the ocean, where it is drenched in both saltwater from tidal spray and freshwater from melting snow or rain. Zoologists investigated the ability of midge larvae to survive different concentrations of salt over a 10-day period. The concentration of salts in freshwater and inside midge larval cells is about 400 mOsm kg^{-1}. Pure water is close to 0 mOsm kg^{-1}, and pure seawater has a concentration about 1,000 mOsm kg^{-1}.

(From Elnitsky MA, Benoit JB, Lopez-Martinez G, Denlinger DL, Lee RE Jr. Osmoregulation and salinity tolerance in the Antarctic midge, *Belgica Antarctica*: seawater exposure confers enhanced tolerance to freezing and dehydration J Exp Biol 212, 2864–2871, 2009.)

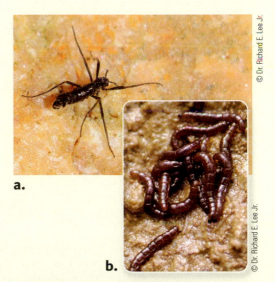

a.

b.

© Dr. Richard E. Lee Jr.

Figure 6.26 The Antarctic midge.

Data Interpretation

Refer to the graph of Antarctic midge survival in four different concentrations of salt (**Fig. 6.27**) to answer the following questions.

20. What is the relationship between survival, salt concentration, and submergence time? Explain.

21. What is the relationship between survival and concentration of seawater?

22. How long could midge larvae tolerate the two highest concentrations of seawater?

23. Did more or less than 50 percent of the midge larvae survive at the most concentrated salt solution for three days?

24. What percentage of larvae survived exposure to pure seawater for six days?

25. What percentage of larvae survived the 10-day experiment submerged in pure seawater?

26. Can midge larvae survive being submerged in freshwater?

27. Did the experiment show any difference in survival between 0 and 400 mOsm kg^{-1}?

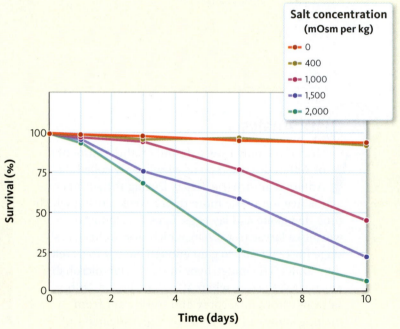

Figure 6.27 Midge survival rates in varying salt concentrations.
© Cengage Learning 2014

Critical Thinking

28. Apply the principle of osmosis to explain why being submerged in high concentrations of saltwater is detrimental to midge larvae.

29. Even though it did not affect survival, what would be the net direction of water movement when larval cells were submerged in pure water (0 mOsm kg^{-1})?

30. Assess the tolerance of Antarctic midge larvae to various solutions to which they are subjected. Your assessment should connect survival time and salt concentrations. Are they well adapted to tolerating a wide range of salt spray?

31. Think of an advantage to the dark color of the midge larvae (see **Fig. 6.26b**).

Question Generator
Fueling White-Throated Sparrow Migration

White-throated sparrows are medium-sized songbirds that feed on seeds, insects, and berries (**Fig. 6.28**). Each fall they migrate from their summer breeding grounds in Canada to their feeding grounds near the Gulf of Mexico. The birds do not feed during their migration, which takes several days and covers a distance of thousands of miles.

(From McFarlan JT, Bonen A, Guglielmo CG. Seasonal upregulation of fatty acid transporters in flight muscles of migratory whitethroated sparrows [*Zonotrichia albicollis*]. J Exp Biol 212, 2934–2940, 2009.)

Figure 6.29 is a block diagram relating several aspects of the biology of white-throated sparrows and their long-distance migration. Use the background information along with the block diagram to ask questions and generate hypotheses. We've translated the concept blocks and linking interaction arrows shown in the map into questions and provided several more questions to get you started.

Figure 6.28 White-throated sparrow.

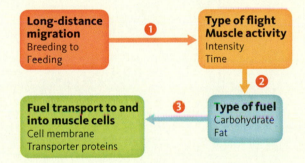

Figure 6.29 Factors relating to long-range migration of white-throated sparrows.
© Cengage Learning 2014

1. What type of muscle activity is required for the birds to make this journey? (Arrow #1)

2. How long would this activity need to be sustained? (Arrow #1)

3. How do fuels vary in their ability to sustain muscle activity? (Arrow #2)

4. How do fuels vary in their rate of transport from storage to muscle? (Arrow #3)

5. What type of fuel(s) would sustain this muscle activity over several days? (Arrow #2)

6. At the muscle cell level, do sparrows have a large number of transporter proteins to speed up fuel uptake? (Green block on diagram)

Use the block diagram to generate your own research question and frame a hypothesis.

7 Animal Growth and Development

7.1 Growth and development of the American lobster

7.2 Animal groups show characteristic patterns of growth and development

7.3 External and internal factors control growth and development

7.4 The cell cycle carries out growth at the cellular level

7.5 Growth and development are highly regulated processes at the molecular level

7.1 Growth and Development of the American Lobster

You have tied on your bib and are ready to begin the delicious business of eating a lavish lobster dinner. You may not realize it, but it took six to eight years for that lobster to grow large enough to fill your dinner plate (**Fig. 7.1**). What is involved in lobster growth and development that takes so long?

Let's look at the developmental journey the American lobster *(Homarus americanus)* took to arrive on your dinner plate. The lobster began as a tiny fertilized egg glued to the underside of its mother, along with tens of thousands of siblings (**Fig. 7.2**). Depending on environmental factors, female lobsters carry their eggs for nine to twelve months. Lobster embryos develop through stages (**Fig. 7.3**) in which they form the familiar body plan, which includes a head, midsection, and abdomen. During this time, organs such as the heart and eyes develop, and the embryos change into small larvae ready for hatching. Your future feast hatched at night between May and September. After hatching, it immediately swam upward to the warm water's surface, where it began feeding on microscopic plankton, and continued its development to the postlarval stage. The larval stage is the *free-swimming phase* of its life cycle and a time when the lobster is extremely vulnerable to predation by fish and birds. It is estimated that only one of every thousand lobsters survives the larval period. Your dinner was lucky—until it found the lobster trap, of course.

During the postlarval stage, a metamorphosis, or transformation, occurred. The lobster developed new structures, such as a long second pair of antennae and the tail fan, which make it look more and more like the lobster you are looking at now. As sensory organs developed, changes in behavior also occurred. For instance, new responses to light and gravity caused the lobster to descend to the ocean floor to start its *bottom-dwelling*

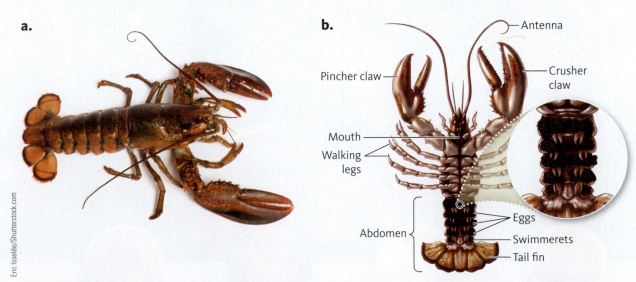

a.

Eric Isselée/Shutterstock.com

b.

- Antenna
- Crusher claw

Pincher claw

Mouth

Walking legs

Eggs

Abdomen

Swimmerets

Tail fin

Figure 7.1 The American lobster. (a) Lobsters are a type of arthropod called a crustacean, a group with hard shells that includes crayfishes, shrimps, and crabs. This American lobster is 6–8 years old and at commercial weight of 1 pound. (b) Lobsters are well adapted for moving, capturing, and eating prey. One structure is their large antennae that are used in detecting motion.
© Cengage Learning 2014

a.

b.

Peter Demmen/Aurora/Getty Images

Courtesy of Prue Talbot, PhD

Figure 7.2 Fertilized eggs of a lobster. (a) Female lobsters carry tens of thousands of fertilized eggs on their underside up to 12 months before they release the developed embryos into the ocean. (b) A close-up view of lobster embryos.

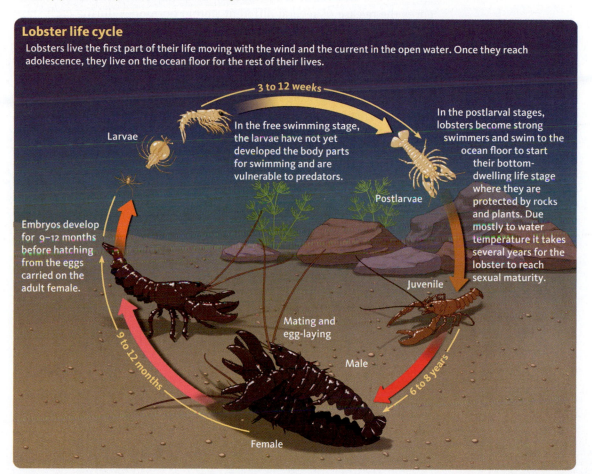

Lobster life cycle

Lobsters live the first part of their life moving with the wind and the current in the open water. Once they reach adolescence, they live on the ocean floor for the rest of their lives.

3 to 12 weeks

Larvae

In the free swimming stage, the larvae have not yet developed the body parts for swimming and are vulnerable to predators.

In the postlarval stages, lobsters become strong swimmers and swim to the ocean floor to start their bottom-dwelling life stage where they are protected by rocks and plants. Due mostly to water temperature it takes several years for the lobster to reach sexual maturity.

Postlarvae

Embryos develop for 9–12 months before hatching from the eggs carried on the adult female.

Juvenile

Mating and egg-laying

9 to 12 months

Male

6 to 8 years

Female

Figure 7.3 Life cycle of a lobster. The lobster starts its life cycle as an embryo attached to its mother (orange arrow), where it grows and develops into a small larva until it is ready for hatching. The hatchlings go through a metamorphosis and develop into sexually mature adults in 6–8 years.
© Cengage Learning 2014

phase. Here the lobster led its solitary life, preying on shellfish, sea urchins, and marine worms. It burrowed and built shelters in rock crevices as it grew larger and finally reached sexual maturity. Even though development was complete, the lobster continued to grow in size during its life.

Throughout its life cycle (as a larva and an adult), the lobster's body was supported and protected by a hard, rigid shell. As it grew larger, the lobster shed its shell, or molted. The newly emerged lobster was pale, with a soft, rubbery texture and a wrinkled appearance. At this stage the lobster absorbed water and expanded considerably, its weight increasing by 40 to 50 percent. The lobster then ate most of its shell to recycle many of the minerals. Over the next month, your vulnerable lobster grew larger and built a new shell. It also might have taken this time to regenerate any missing limbs, a feature not found in more complex animals like mammals. Growth and production of a new shell involve cell division, which generates layers of epidermal cells that secrete and build the shell. Altogether, your lobster molted more than 20 times, developed through several dramatically different stages, and grew for about six to eight years before it reached its commercial size of one pound.

You probably never thought about lobsters in this light. The lobster's complex life cycle is controlled by environmental factors. Temperature is the major factor that determines the well-timed events of egg production, hatching, larval development, molting, and growth rate. Larval development occurs above 50°F (10°C). Lobsters molt more often and grow more rapidly in the warm water of summer.

Temperature drives development in many animals, including insects, amphibians, and reptiles. Specifically in the lobster, temperature influences the release of several different steroid hormones during each stage of the life cycle. Many steroid hormones found in lobsters are identical to hormones that regulate metamorphosis in insects and reproductive cycles in mammals. These steroid hormones act directly on specific genes in the DNA in cells directing development. The timing, rate, and completion of the complex and intricate life cycle of the American lobster are the result of the interplay of other factors in addition to temperature—day length, food supply, stress, and pollution also play a role.

In this chapter, we'll examine the characteristics of animal growth and development. Like the American lobster, each animal species has a characteristic pattern of growth and development that includes passing through a number of life stages. Each life stage involves the growth and development of cells under the influence of environmental and hormonal factors that initiate and modify processes of the cell cycle and cell and tissue differentiation.

7.2 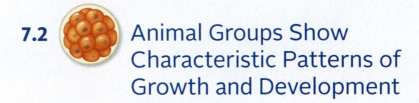 Animal Groups Show Characteristic Patterns of Growth and Development

We have already outlined the many features of lobster growth and development. One aspect we have not discussed is the complexity of the processes involved in patterns of growth. As with many animal species, male lobsters grow faster than females. In the wild, the average life span of a lobster is 15 years. Because they continue to grow throughout their lifetimes, they can reach large sizes. Bubba, one of the largest known lobsters, weighs in at 23 pounds and measures 21 inches long (**Fig. 7.4**). The lobster's increase in size, or **growth**, is the result of three different but closely related processes: increasing cell numbers, increasing individual cell size, and secretion of various proteins and other materials around the cells. Each process, in combination or alone, contributes to growth. From the beginning, lobsters, like all animals, pass through a number of developmental stages in which they progressively increase in size and complexity.

In Chapter 1, you learned that cells contain DNA, which encodes the information needed for growth, development, survival, and reproduction. As an animal grows in size, it passes through a series of developmental stages driven by its unique genetic code. **Development** refers to the changes in shape, form, function, and structure animals undergo as they mature from

Growth—an increase in size as a result of three different but closely related processes: increasing cell numbers, increasing individual cell size, and secretion of various proteins and other materials around the cells.

Development—the changes in shape, form, function, and structure organisms undergo as they mature from embryos to adults.

Figure 7.4 **Bubba (shown on the left).** One of the largest known lobsters, Bubba is 23 pounds, compared to the average 1.5-pound lobster (shown on the right).

AP Photo/Keith Srakocic

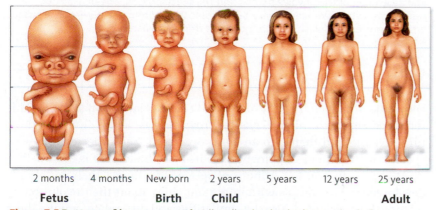

| 2 months | 4 months | New born | 2 years | 5 years | 12 years | 25 years |
| **Fetus** | | **Birth** | **Child** | | | **Adult** |

Figure 7.5 **Pattern of human growth.** Like all animals, the human body form, structure, and shape change as the person grows and develops to maturity.
© Cengage Learning 2014

embryos to adults. For instance, the developmental stages of the lobster include the larval, juvenile, and adult, and each of the stages has a distinguished body form (see **Fig. 7.3**). Another more familiar example is humans. At birth the human is 50 percent body trunk, 32 percent head, and 16 percent legs, whereas an adult human is 50 percent body trunk, 10 percent head, and 29 percent legs (**Fig. 7.5**). In addition to changes in body form, the cells of the body become organized into tissues, organs, and organ systems, which establish the form and function of a complete organism. Even though different organs and regions grow at different rates, the general body plan of most animals is established very early during embryonic development.

Each animal species has a characteristic pattern of development

Like all animals, humans go through various *life stages* that involve growth and development. Following the life cycle, you see that after the sperm fertilizes the egg, a special form of cell division called mito-

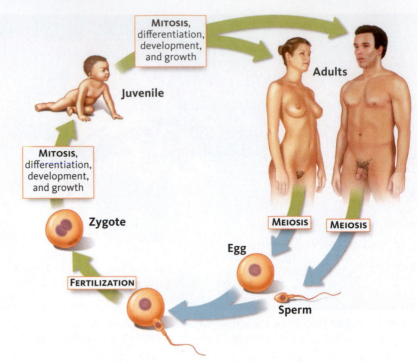

Figure 7.6 The human life cycle. The human life cycle includes stages of growth and development that involve mitosis and cell specialization. The blue arrow represents a stage with only one set of chromosomes from one parent. The yellow arrow represents a full set of chromosomes from both parents.
© Cengage Learning 2014

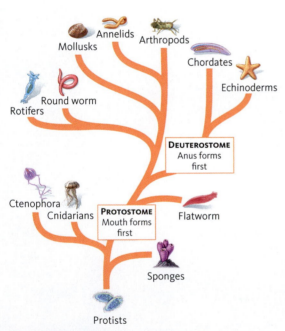

Figure 7.7 Embryonic development of protostomes and deuterostomes. Animals are divided into two groups based on embryonic development patterns. Protostomes are animals in which the mouth develops first, and in deuterostomes the anus forms first.
© Cengage Learning 2014

Zygote—a fertilized egg that grows and develops first into an immature juvenile and then into an adult organism.

sis occurs, followed by cell growth and cell specialization (**Fig. 7.6**). This fertilized egg, also called a zygote, grows and develops into an immature juvenile and then into an adult organism. Each species follows a specific pattern of development once its basic body plan is established.

When we look at the millions of species of animals, their embryonic development provides us with important clues about their evolutionary relationships. Scientists use the similarities and differences among these patterns of early development as a characteristic to classify animals. What is most striking is that most animals fall into one of two large taxonomic groups based on which of the openings to their gut forms first. **Figure 7.7** shows an evolutionary tree with the division of animals. Mollusks, worms, and arthropods are on the branch where the mouth forms first, and echinoderms and chordates are on the other branch where the anus forms first. Another example of animal groups that show similar developmental patterns is vertebrates. Recall from Chapter 2 that all vertebrates—mammals, fish, amphibians, reptiles, and birds—share characteristic anatomical traits such as a backbone. There are many other examples of how developmental patterns provide insight into life's history and relationships.

Although each species has a different pattern, they have some overall similarities in their development. For example, animal development starts with the head region and progresses to the tail end of the embryo. **Figure 7.8** shows a comparison of the four basic stages of development of the sea urchin, frog, and human. Shortly after fertilization, the zygote in all three animals undergoes rapid cell division but does not grow in overall size. This growth pattern results in a large increase in the number of cells. With each set of cell divisions, each cell generation gets smaller as the resources are

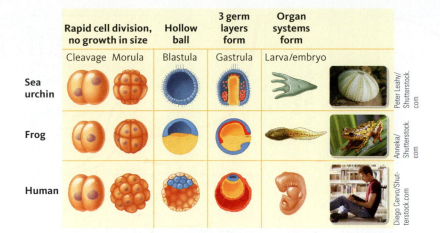

	Rapid cell division, no growth in size		Hollow ball	3 germ layers form	Organ systems form	
	Cleavage	Morula	Blastula	Gastrula	Larva/embryo	
Sea urchin						Peter Leahy/Shutterstock.com
Frog						Anneka/Shutterstock.com
Human						Diego Cervo/Shutterstock.com

Figure 7.8 Embryonic development of sea urchins, frogs, and humans. Different animals show similar patterns of embryonic development.
© Cengage Learning 2014

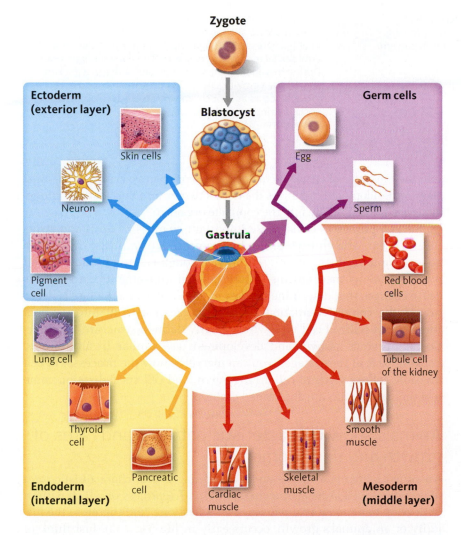

Figure 7.9 Formation of tissues. During embryonic development, the germ layers give rise to different tissues in the body. This diagram shows examples of specific cells originating from each of the germ layers.
© Cengage Learning 2014

divided among a larger number of cells. When the number of cells reaches about 128, the cells form a hollow ball with a cavity in the middle. and three distinct germ layers of cells begin to develop. These layers are called **germ layers** because they will give rise to the organism's different types of tissues (**Fig. 7.9**). The outer layer, the ectoderm, becomes the epithelial tissues of the body surfaces, while the inner layer, the endoderm, becomes the lining of the digestive system. The mesoderm is the middle germ layer that develops

Germ layer—any of the three cellular layers (ectoderm, endoderm, or mesoderm) that give rise to the various tissues and organs of the body.

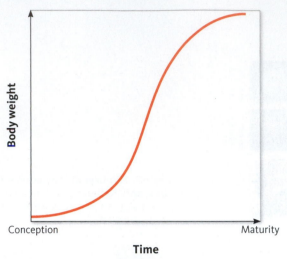

Figure 7.10 Determinate growth pattern. Many animals reach an adult size and then stop growing. This is referred to as a determinate growth pattern.
© Cengage Learning 2014

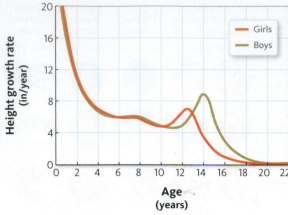

Figure 7.11 Human growth rates over time. The human life cycle starts at conception, progresses to an 8-week embryo stage, then a 7-month fetal growth stage, followed by birth into an infant stage. During these stages, growth occurs at a rapid rate. Over the next 12 years, the infant progresses through toddler and young child stages until sexual maturity is reached in the early teenage years. During this adolescent stage, humans undergo a growth spurt with a growth rate of 25 percent. After adolescence, growth occurs at a much slower rate as development continues into the adult stages.
© Cengage Learning 2014

into the muscular, skeletal, circulatory, and reproductive systems. The final stage of development organizes the body's tissues into organs, then into organ systems, and finally into the complete organism.

Several processes cause these important changes of form and structure to occur. The embryo grows by the process of mitosis. Cells move about, communicating with each other through cell-to-cell contact and other signaling mechanisms. These cellular movements and signals cause changes that determine the fate of each cell. Once the fate is determined, the **cell differentiation** process changes each cell into its final form—a specialized cell type with a clearly defined structure and function. Embryonic growth and development are similar in all animals. Looking back at these stages, you can marvel at how the processes of cell division, cell movement, cell arrangement, and cell specialization come together to produce a fully formed embryo.

Animals show characteristic patterns and rates of growth and development

An animal's life cycle can be divided into stages with characteristic *rates* of growth as well as developmental patterns. Regardless of which pattern of development an animal follows, the highest rate, and therefore the majority of an animal's growth, occurs early in life (**Fig. 7.10**). Just think of yourself. You grew at a faster rate during the beginning stages of your life compared to your adult life stage (**Fig. 7.11**). You may be familiar with growth charts used by doctors and parents to track the progress of a child's early development by comparing an infant's weight versus her length (**Fig. 7.12**). Another chart compares the infant's rate of growth of head circumference to normal rates of growth to determine if the infant's brain is developing at the expected rate.

Cell differentiation—the process that changes each cell into its final form—a specialized cell type with a clearly defined structure and function.

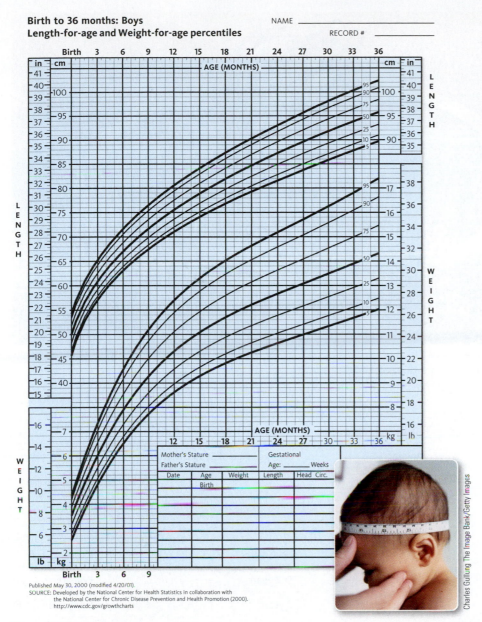

Published May 30, 2000 (modified 4/20/01).
SOURCE: Developed by the National Center for Health Statistics in collaboration with the National Center for Chronic Disease Prevention and Health Promotion (2000).
http://www.cdc.gov/growthcharts

Figure 7.12 Growth chart. Growth charts measure the progress of early childhood development. Head circumference is another indicator of healthy growth and development.
© Cengage Learning 2014

Animals show two basic *patterns* of growth. Animals as different as humans and birds show the first pattern called *determinate growth*. These animals grow to a particular size and then stop growing when they reach maturity (see **Fig. 7.10**). Other animals, such as many lobsters, fish, and amphibians, continue to grow throughout their lifetimes as long as adequate resources are available. These animals show the second pattern of growth, and because this type of growth is indefinite, it is referred to as *indeterminate growth*.

In general, animals (such as humans) that have a long life span take more time to grow and develop than do animals with a short life span. In some cases, development occurs at specific points in time. For example, animals such as the lobster, butterfly, and frog undergo dramatic changes

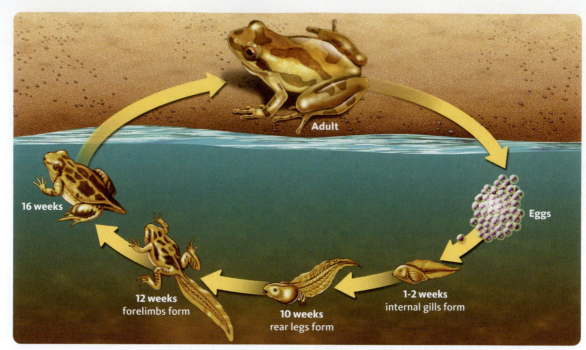

Figure 7.13 Life cycle of a frog. Amphibians, such as this frog, undergo metamorphosis with characteristic growth and development at specific milestones.
© Cengage Learning 2014

TABLE 7.1 Various Animal Development and Life Spans

Animal	Average Age of Sexual Maturity	Life span
Queen termite	2–7 weeks	50 years
Soldier termite	Remains sexually immature	1–2 years
Monarch butterfly	1 month	6–8 weeks
Leatherback turtle	12–15 years	45 years
American lobster	5–8 years	15 years
Red sea urchin	2 years	100 years
Poison dart frog	1.5–2.5 years	5 years
Human	12–15 years	75 years

© Cengage Learning 2014

in body form called metamorphosis (**Fig. 7.13**). **Table 7.1** compares the life spans of several animals to their age of maturity.

Not only do whole organisms show characteristic growth rates and patterns, but each of the four animal tissues—epithelial, connective, muscle and nervous tissue—also shows distinct patterns of growth. Epithelial tissue mainly grows through the process of cell division, which creates new cells. Bone, a type of connective tissue, grows both by cell division and by secreting an extracellular matrix of calcium that enlarges the tissue. Nerve and muscle tissues grow by enlarging cells rather than by increasing the number of actual cells. Muscle growth is triggered by repeated stimulation and hormone signals.

7.2 check + apply YOUR UNDERSTANDING

- Different animal groups show specific patterns of growth and development, which are controlled by the expression of different genes. Cell division, cell movement, cell arrangement, and cell specialization each is an important process in embryonic development.

- Animals advance through various life stages that involve growth and development of the zygote into a mature adult organism.

- Each species follows a specific pattern of development based on its basic body plan.

- An animal's life cycle can be divided into stages with characteristic *rates* of growth. Growth rates are especially high early in life. Some animals stop growing at maturity, whereas other species continue to grow their entire life.

1. Cell differentiation is an important process of animal development that begins with a single cell. What is cell differentiation?
2. Various tissues grow in different ways. Name the three ways tissues grow.
3. Cell division, cell movement, cell arrangement, and cell specialization are associated with animal growth and development. Which of these cellular processes contribute to growth and which contribute to development?
4. Describe the difference in the pattern of growth in lobsters and in humans. Do they follow a determinate or an indeterminate pattern?
5. The tiger salamander is an amphibian with a life cycle similar to that of the frog (see **Fig. 9.2**). (You can see a photo of a tiger salamander in **Fig. 7.14**.) For each set of genes, state whether they are turned on or off at the particular stage given: (1) yolk synthesis in the tadpole stage, (2) gill development in the adult stage, (3) limb development at the tadpole stage.

7.3 External and Internal Factors Control Growth and Development

The California tiger salamander undergoes metamorphosis and transforms from a larva, to a legless aquatic juvenile with gills, to a four-limbed terrestrial adult with lungs during its life cycle (**Fig. 7.14**). In order for the salamander to transform into its mature adult body shape—a four-legged animal with a rounded snout and long tail—the growth and development of the various tissues, organs, and organ systems must occur in an orderly fashion. As in all animals, the salamander's development is tailored to the growth rates of cells ensuring appropriate organ and organism growth. Cells ensure proper growth by communicating with each other. If specific cues are missed, the cells do not grow at the correct rates and tissues may not grow proportionally.

Animal growth is the result of a complex interaction of external and internal factors that include genetics, environment, nutrition, and hormones. You may be familiar with livestock production that occurs on farms. Breeders control the temperature and lighting in their housing systems, the nutrition levels of the herd, and the genetics of their stock, and they administer various hormones to enhance and control growth and development. Through strict control over environment, diet, and genetics, livestock producers have increased the growth rate of farm animals as well as the overall quality of the products derived from them.

Figure 7.14 Two stages of tiger salamander growth and development. (a) Aquatic juvenile tiger salamander with external gills. (b) Terrestrial adult tiger salamander with internal lungs.

Environmental factors trigger growth and development

Although growth and development through the various life stages of an organism are "programmed" in the organism's DNA, several factors in an organism's habitat influence these processes. The transformation of an organism from an embryo to an adult requires building block molecules and energy that are obtained from the environment. For instance, the nutrient levels in the water drive the continued growth of the tiger salamander and of many other amphibians and fish, whereas the nutrients in the human diet affect our development. Food supply, oxygen levels, space, and temperature are among the factors that influence growth and development.

Temperature and availability of food drive the timing, duration, and success of the life cycle stage of many animals. Recall that water temperature of 50° F is a key factor that triggers lobster larval development. The development of salamanders, frogs, and lobsters also corresponds to seasonal changes in their habitats. Many other animals that live in temperate climates, such as rabbits, deer, and squirrels, give birth in the springtime because the higher temperatures increase survival and food sources are readily at hand. In addition to food supply, increased temperature has a direct impact on an organism's metabolism. Temperature affects the rate of the chemical metabolic reactions that generate building block molecules and energy. Within a certain range and as a generalization, the higher the temperature, the faster the rates at which animals metabolize molecules and build new cells and tissues.

Another factor in animal growth and development is the length of the day. Longer days trigger foraging activities in many wild animals. Livestock breeders have taken advantage of this natural trait in animal production. Increasing the amount of light exposure to cattle, sheep, and poultry increases the amount of food that they eat, making them grow big and fat. Some factors, such as nutrient levels, directly influence the rate of growth, whereas other factors, such as light, work in indirect ways. So, how does light cause cells to divide and grow?

Hormones play a major role in controlling growth and development

Light, temperature, and other external factors often cause specific cells to produce hormones. As highly specialized chemical messengers, **hormones** bind to cellular receptors on specific target tissues. The receptors in turn

Hormones—highly specialized chemical messengers that bind to cellular receptors on specific target tissues, causing metabolic, physiological, or behavioral responses in organisms.

cause various genes that initiate a growth response to turn on or off. In the case of the lobster, hormones made in its eyestalks regulate both metamorphosis and molting.

Let's examine the role of hormones in controlling development, using yourself as a familiar example. Triggered by the production of sex hormones between 10 and 13 years of age, adolescent humans enter a developmental stage of sexual maturation called puberty. In males, increased levels of the hormone testosterone not only trigger the maturation of the sex organs but also act on connective muscle and bone tissue, enhancing strength, deepening the voice, and changing the shape of the face and skeleton. In a secondary reaction, testosterone acts on epithelial tissue, accelerating facial and body hair growth. In females, estrogen hormone produced in the ovaries triggers breast enlargement, widening of the pelvis, and an increase in the amount of body fat in hips, thighs, buttocks, and breasts. Increased estrogen levels also induce growth of the uterus and start a female's monthly menstruation cycle. Think of all of the different parts of your body that changed during puberty. These body parts changed because they had cells with specific receptors that bound and responded to the hormones circulating throughout the body.

Testosterone is a good example of the complexity of hormone interactions and regulation. **Figure 7.15** illustrates the interactions of several hormones and organs that are involved with testosterone production in human males. The hypothalamus stimulates the pituitary gland to secrete two hormones: follicle-stimulating hormone (FSH) and luteinizing hormone (LH), which are present in both males and females. In males, FSH promotes the production of sperm and LH controls the production of testosterone. In Chapter 11, we will cover in more detail the hormones of both the male and female reproductive systems.

The next two sections will explain how cell-signaling molecules, such as hormones, work at the cellular level. The release of one molecule triggers or inhibits the release of other molecules. Cells communicate with each other by making and transporting these various signaling molecules. This information will allow you to apply your understanding to the many different hormones that can cause dramatic changes in animals.

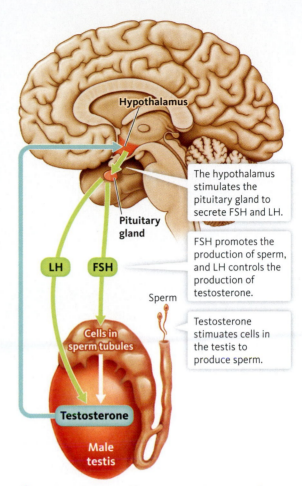

Figure 7.15 The role of hormones in human male sexual development. Several hormones interact with specific tissues to trigger sexual development in males. FSH, follicle-stimulating hormone; LH, luteinizing hormone.
© Cengage Learning 2014

7.3 check + apply YOUR UNDERSTANDING

- Animal growth and development are the result of complex interactions of many external and internal factors, which include genetics, environment, nutrition, and hormones.

- Temperature, availability of food, and length of day drive the timing, duration, and success of the life cycle stage of many animals.

- Light, temperature, and other external factors often cause specific cells to produce a hormone that binds to cellular receptors on specific target tissues, triggering gene expression.

1. List three external factors that influence growth and development.
2. Describe how a hormone interacts with a cell to cause it to divide.
3. In addition to altering daylight conditions, livestock breeders provide growth hormones to their animals. These hormones increase muscle mass by 20 percent. What target cells do these growth hormones act on?
4. In males, what would happen if the epithelial cells lacked receptors for testosterone?
5. The Arctic krill (see **Fig. 1.2**) is an arthropod like the lobster. Krill release about 1,000 eggs at one time each summer. The eggs develop into larva over a two-week period. Predict what environmental factors cue krill to begin the reproduction phase of their lives.

7.4 The Cell Cycle Carries Out Growth at the Cellular Level

In the previous sections you learned about growth and development from the perspective of hormones, tissues, organs systems, and the complete organism. Now we will look more closely at the cellular processes that underlie growth on all levels of organization in the animal body. Growth is an increase in body mass, often from an increase in cell number. Different types of cells divide at different rates. Some cells constantly divide, like those that produce your hair, fingernails, and blood. Nerve cells in the brain and muscle cells of the heart go through several rounds of cell division during the early stages of their development and never divide again. Some liver cells stop dividing at maturity but can be induced to divide to repair an injury.

Like organisms, cells of the body proceed through a number of distinct stages during their lifetimes. This life cycle of the cell is referred to as the **cell cycle**. The cell cycle is the foundation for growth and development in an organism—from a single-celled fertilized egg (or zygote) into a mature organism. The cell cycle is also the process by which cells are replaced and renewed. Some cells live for only a few days, others for months, and some for decades. Cells that divide continually complete the cell cycle, whereas others remain in one phase of the cycle.

It is important to note that sexually reproducing organisms use two types of cell division. In this chapter we will focus on *mitosis,* and in Chapter 11 we will cover *meiosis,* the process that produces reproductive cells in sexually reproducing organisms.

Overview of the cell cycle

The cell cycle of a cell varies depending on the cell type and location. Some developing embryo cells can complete the cycle in less than ten minutes. Some types of liver cells spend a year in the cell cycle, whereas the epithelial cells of your mouth and other high turnover cells spend about 24 hours in the cell cycle. Another example of the cell cycle in action is your skin. A mammal's skin, or epidermis, is divided into layers of epithelial cells. Cells are formed through cell division at the innermost layer of the epidermis (**Fig. 7.16**). Each

> **Cell cycle**—the distinct stages that a cell goes through to duplicate and divide. The cell cycle provides growth and development in organisms, as well as replacement cells. It includes interphase, mitosis, and cytokinesis.

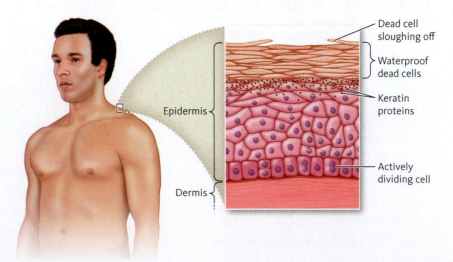

Figure 7.16 The skin is constantly renewed through the process of cell division.
© Cengage Learning 2014

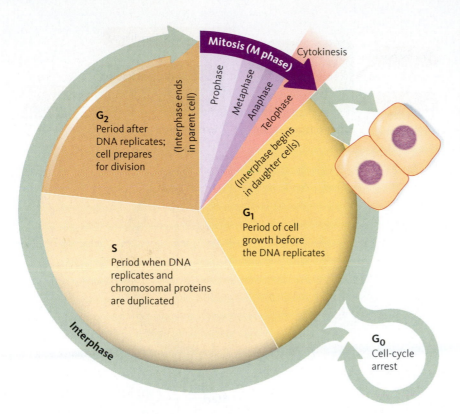

Figure 7.17 The cell cycle is divided into interphase and mitosis. Interphase is the period when the cell performs its job, grows its cytoplasm, replicates its organelles, and duplicates its chromosomes, located in the nucleus. During the mitotic phase (mitosis and cytokinesis), the cell splits into two identical daughter cells. Mitotic cell division is important for growth, wound repair, asexual reproduction, and the replacement of worn out cells.

© Cengage Learning 2014

newly formed layer of cells pushes up the previous layer. The older cells change shape and composition as they differentiate and become filled with waterproof keratin proteins. Eventually, the cells are cut off from the blood supply, die, and slough off. In humans, there are more than 25 layers of dead cells at the skin's surface. The bottom layer of the skin is continually renewing itself through the process of cell division. The life span of the epithelial cells in the skin is about three weeks, whereas the life span of the epithelial cells that line the intestine is approximately three days.

In eukaryotes, cell division is a complex and highly controlled process that is divided into two broad stages: interphase and the mitotic phase (**Fig. 7.17**). **Interphase** is the period when the cell performs its job, grows its cytoplasm, replicates its organelles, and duplicates its chromosomes located in the nucleus. The *mitotic phase,* which follows interphase, is the stage when the cell separates the duplicated nuclear material and then splits into two separate cells, called daughter cells. Looking at the cell cycle diagram you see that a typical dividing cell spends about 90 percent of its life in interphase and 10 percent in the mitotic phase. You can also see that the mitotic phase is split into mitosis and cytokinesis. **Mitosis** is the process that separates the chromosomes and produces two new identical nuclei in one cell. After mitosis, most eukaryotic cells then split the organelles, cell membrane, and cytoplasm into two identical daughter cells through the process of **cytokinesis**.

First we will briefly introduce the overall process of cell division, then we will explore the steps in greater detail throughout this section. Interphase is triggered by hormones and other signaling molecules that enter the cell and bind to receptors. Once the cell has the necessary components to split into two cells, the mitotic phase begins, in which the duplicated nuclear material splits in two through a series of steps (see **Fig. 7.17**). The cell pinches in two, and the cell membrane re-forms to divide the cytoplasmic material into two new daughter cells.

Interphase—the phase of cell division in which the cell spends most of its time. During this phase the cell performs its job, grows its cytoplasm, replicates its organelles, and duplicates its chromosomes located in the nucleus.

Mitosis—the cell division process that separates the chromosomes and produces two new identical nuclei in one cell. This process is common for tissue growth or repair, or for asexual reproduction.

Cytokinesis—the process that occurs after mitosis that splits the organelles, cell membrane, and cytoplasm into two identical daughter cells.

Figure 7.18 The cell cycle. Cell division is highly regulated by various checkpoints throughout the cell cycle. In order for the cell to progress to the next phase, specific criteria must be met. For instance, to progress into the mitotic phase, DNA replication and cell growth must have occurred and environmental conditions must be favorable.

© Cengage Learning 2014

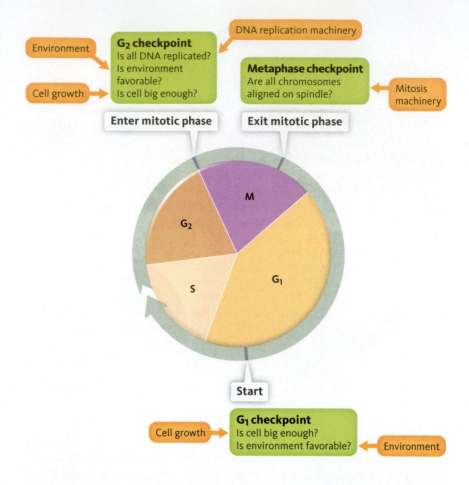

Returning to the epithelial cells in your skin—like all eukaryotic cells, these cells must also pass checkpoints to progress through the cycle. The checkpoints ensure not only that the cell faces favorable environmental conditions but also that cell growth and DNA replication have properly occurred during interphase (**Fig. 7.18**). If the internal conditions and cues of the cell are not correct, the cell does not progress into the next stage of the cell cycle. Certain cell types, such as cardiac and some nerve cells, do not divide when they are mature. Once their development is complete, these cells enter a holding state during which the cell is active but its cycle is arrested. For this reason, tissue damage from a heart attack or injury to the nervous system can be deadly.

A closer look at interphase

Let's look at the details of interphase using the skin epithelial cells as an example. Interphase is divided into three distinct phases during which specific cellular processes occur (see **Fig. 7.17**). Following the cell through interphase, the first stage is G_1, which is the growth or gap phase 1. During this time, biochemical processes of the skin cell operate at a high rate, producing enzymes and other proteins needed to support DNA replication and other activities of cell division. With the cell geared up with the necessary enzymes and proteins for DNA replication, the skin cell enters the S, or DNA synthesis, phase. During this critical phase, the cell's DNA code is replicated. For instance, at the end of the S phase, each of the human skin cell's 46 chromosomes has been replicated, producing identical copies of the DNA strands called **sister chromatids**. With its DNA duplicated, the cell enters the G_2 phase and makes additional structural proteins, such as microtubules, de-

Sister chromatids—two identical copies of the DNA strands produced during chromosome replication at the end of the S phase during cell division.

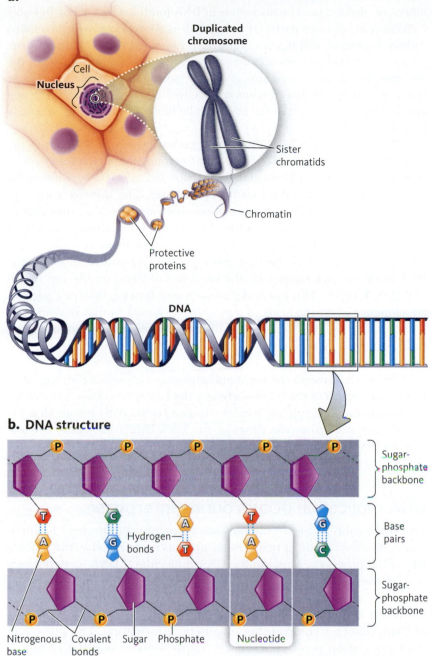

a.

Duplicated chromosome

Cell

Nucleus

Sister chromatids

Chromatin

Protective proteins

DNA

b. DNA structure

Sugar-phosphate backbone

Base pairs

Hydrogen bonds

Sugar-phosphate backbone

Nitrogenous base Covalent bonds Sugar Phosphate Nucleotide

Figure 7.19 The structure of DNA. (a) Chromosomes are associated with proteins and packed tightly in the nucleus. (b) The DNA molecule features a double helix, which consists of relatively weak hydrogen bonds holding two strands together, and specific complementary base pairing. The chemistry of this helix allows for DNA to unzip and copy itself.
© Cengage Learning 2014

signed to help carry out mitosis. The cell also continues to grow and replicate its organelles. The G_2 phase lasts until the cell begins mitosis.

Revisiting DNA structure

DNA replication occurs during interphase, so we will discuss this topic next. However, before we dive into DNA replication, it is helpful to recall the structure of the DNA molecule and chromosome. The DNA molecule is carried in the cell's chromosomes, which are made of two long chains of DNA wrapped in a double helix surrounded by protective proteins. Because DNA molecules are so long, they are condensed and packaged with structural and regulatory proteins into compact complexes, called chromatin, which can be stored in the cell's nucleus (Fig. 7.19a).

Recall from Chapter 1 that DNA is constructed from smaller building blocks, or nucleotides. The nucleotide in DNA (outlined by the box in **Figure 7.19b**) consists of three parts: (1) a sugar called *deoxyribose,* which contains carbon, hydrogen, and oxygen atoms in a five-sided ring, (2) a *phosphate* group, which contains phosphorus and oxygen atoms, and (3) one of the four *nitrogenous bases*—adenine, guanine, cytosine, and thymine. Each base is a ringed molecule that contains carbon, hydrogen, oxygen, and nitrogen atoms. The entire length of the DNA can be built by joining together individual nucleotides.

DNA has a simple ladderlike arrangement (see **Fig. 7.19b**). Along the outside of the molecule are two strands of phosphate and sugar molecules that are strongly bonded together. Bonded to each sugar along the sides of the ladder are the four different bases. The middle "rungs" of the ladder are formed by weaker bonds between bases on one side of the ladder with bases on the other side. Adenine (A) always pairs with thymine (T) and guanine (G) always pairs with cytosine (C); these pairings are referred to as the base pairing rules. As you read across the DNA molecule (see **Figure 7.19**), the sequence of bases on the top strand is T, C, A, T, and G. This particular sequence of bases codes for a particular sequence of amino acids in a protein. A change in this sequence, or a *mutation,* may lead to a change in the sequence of amino acids and can result in a change in the sequence, shape, and function of the protein for which it codes.

The bonds between the sugar and phosphate groups are strong covalent bonds, whereas the bonds between the base pairs down the middle of this narrow molecule are weak hydrogen bonds. This means that the strands are strong but the rungs of the DNA ladder are easily broken. The molecular structure of the DNA molecule helps DNA to replicate during cell division.

DNA replication occurs during interphase

The replication of each chromosome is an important stage of interphase, so we will look at this cellular process more closely. **DNA replication** is the process of copying the DNA molecule. If DNA were not replicated, then each time the cell divided, the genetic information would be cut in half, eliminating much of the genetic information necessary to survive. Replication ensures that each new cell has a *full complement* of DNA, that is, a complete set of genetic information. The process of DNA replication involves three key ingredients: a number of helper molecules, building block nucleotides to form the new strands, and energy input to power the replication process. The cell uses all of these ingredients to replicate its entire set of chromosomes in less than an hour.

When you were in the womb, each of the trillions of cells in your body was dividing at a very rapid rate. Each cell replicated its DNA with very few errors so that you could grow and develop as an embryo. When you think about it, the lack of errors is amazing considering that each of your cells has 23 pairs of (or 46) chromosomes containing more than 20,000 genes totaling about six billion base pairs of DNA that it needs to replicate. The nearly error-free copying is accomplished through the *semiconservative* nature of DNA replication.

The steps of DNA replication occur with the aid of structural proteins and several key enzymes (**Fig. 7.20**). First, the compact and twisted

DNA replication—the process of copying the DNA molecule from an existing DNA strand as a template.

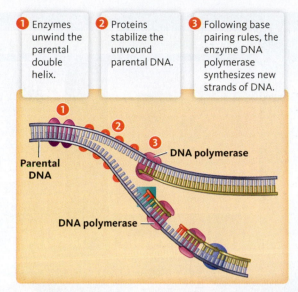

1 Enzymes unwind the parental double helix.

2 Proteins stabilize the unwound parental DNA.

3 Following base pairing rules, the enzyme DNA polymerase synthesizes new strands of DNA.

Parental DNA

DNA polymerase

DNA polymerase

Figure 7.20 Enzymes play a major role in DNA replication. DNA replication occurs during interphase prior to cell division. A specialized enzyme unwinds the double helix. Using base pairing rules, another enzyme (DNA polymerase) attaches free nucleotides to synthesize a new DNA molecule. When the chromosome is replicated in its entirety, the enzymes fall off and the DNA winds into the double helix. The duplicated chromosomes stay attached until cell division occurs.
© Cengage Learning 2014

chromosome is unwound and "zipped open." Each half of the DNA double helix (blue strands in **Figure 7.21**) serves as a template for the formation of new DNA strands (yellow strands in **Figure 7.21**). Notice that one strand is kept (conserved) in its original condition, and the other is newly made from the original template strand. A specialized enzyme called DNA polymerase adds bases to the new DNA molecule (see **Fig. 7.20**). Because each new molecule contains the same genetic information and uses the base pairing rules, replication is highly accurate and produces the exact copies needed. Errors could be harmful, so DNA polymerase and other enzymes proofread and repair any mismatched nucleotides.

A closer look at mitosis

After the chromosomes of the cell are replicated and checked for errors, the sister chromatids stay attached to each other until they are set to divide in the final stage of cell division—mitosis. The mitotic phase of the cell cycle splits up the duplicated genome and interior materials and moves them to the daughter cells. This continuous process is divided into four subphases: prophase, metaphase, anaphase, and telophase (**Fig. 7.22**). During the mitotic phase, the sister chromatids condense, DNA and cell organelles are arranged, and the duplicated chromatids are pulled apart by microtubules (structural protein fibers) to opposite sides of the dividing cell. At the end of telophase, the cytoplasm divides in two, producing two identical daughter cells.

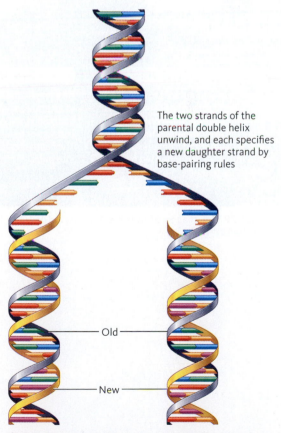

The two strands of the parental double helix unwind, and each specifies a new daughter strand by base-pairing rules

Old

New

Figure 7.21 The double helix "unzipped." In DNA replication, one strand of the DNA molecule is used as a template (blue strand) for making a new strand (yellow strand).
© Cengage Learning 2014

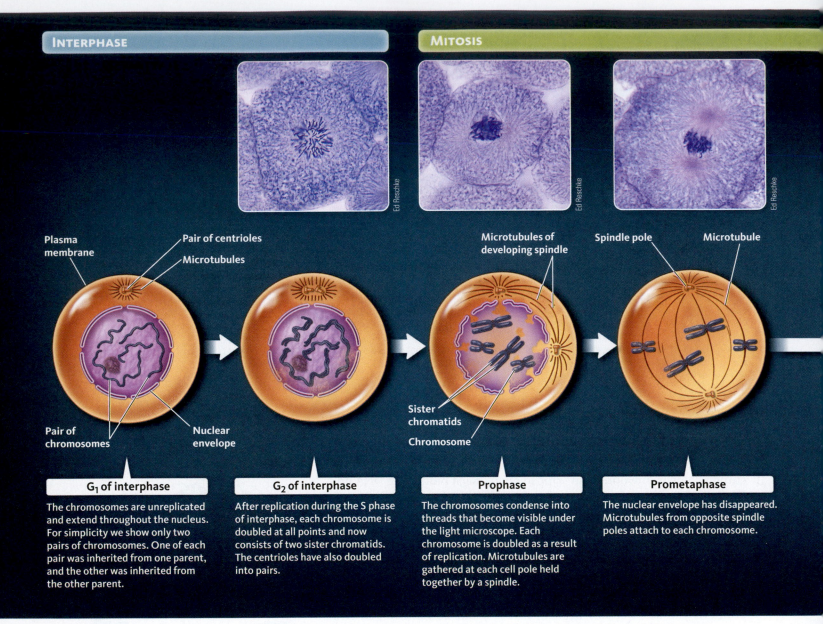

Plasma membrane
Pair of centrioles
Microtubules
Microtubules of developing spindle
Spindle pole
Microtubule

Pair of chromosomes
Nuclear envelope
Sister chromatids
Chromosome

G₁ of interphase	G₂ of interphase	Prophase	Prometaphase

G₁ of interphase

The chromosomes are unreplicated and extend throughout the nucleus. For simplicity we show only two pairs of chromosomes. One of each pair was inherited from one parent, and the other was inherited from the other parent.

G₂ of interphase

After replication during the S phase of interphase, each chromosome is doubled at all points and now consists of two sister chromatids. The centrioles have also doubled into pairs.

Prophase

The chromosomes condense into threads that become visible under the light microscope. Each chromosome is doubled as a result of replication. Microtubules are gathered at each cell pole held together by a spindle.

Prometaphase

The nuclear envelope has disappeared. Microtubules from opposite spindle poles attach to each chromosome.

Figure 7.22 The stages of mitosis. Mitosis in the whitefish embryo, drawn here with two pairs of chromosomes, follows distinct stages.
© Cengage Learning 2014

We have discussed mitosis as a cellular and genetic process that underlies embryonic development, but it also serves other important functions in animals. Mitotic cell division is involved in wound repair and asexual reproduction in some animals, such as corals, jellyfish, and starfish. Stepping back and thinking of the epithelial cell in your skin, you now know that during embryonic development the cells in your skin tissue were dividing and growing rapidly through mitosis. Now think of what it takes to maintain your skin cells. If you cut your skin, the damaged tissue quickly sends signals to repair the damaged tissue. By understanding that each cell has a specific life span, you know your skin cells are constantly dividing to form new cells that maintain the skin's function and keep it intact.

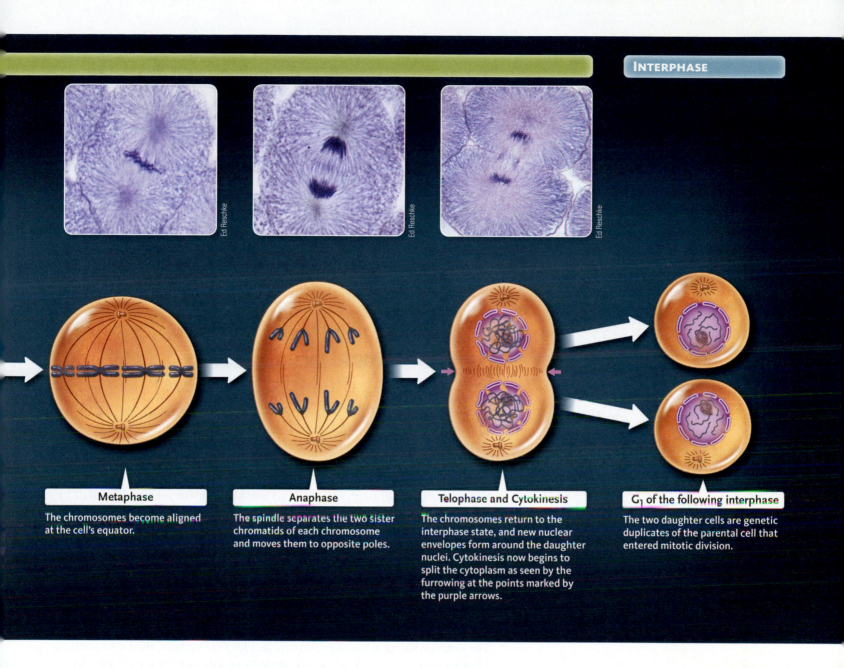

Metaphase

The chromosomes become aligned at the cell's equator.

Anaphase

The spindle separates the two sister chromatids of each chromosome and moves them to opposite poles.

Telophase and Cytokinesis

The chromosomes return to the interphase state, and new nuclear envelopes form around the daughter nuclei. Cytokinesis now begins to split the cytoplasm as seen by the furrowing at the points marked by the purple arrows.

G$_1$ of the following interphase

The two daughter cells are genetic duplicates of the parental cell that entered mitotic division.

7.4 check + apply YOUR UNDERSTANDING

- The cell cycle consists of distinct phases with specific checkpoints that ensure proper cell division.

- Growth is accomplished by the cell cycle and is highly regulated by hormones and external factors.

- Interphase is the period when the cell performs its job, grows its cytoplasm, replicates its organelles, and duplicates its chromosomes located in the nucleus.

1. Distinguish between interphase and mitosis in the life of a dividing cell. Mitosis serves several functions in animals. List four functions of mitosis.

2. In cells that continually divide throughout their life span, in what phase does the cell spend the majority of its time and what is it doing during that phase?

3. In DNA replication, if the original DNA sequence is ACCGTCG, what is the sequence of the newly synthesized strand?

4. How does the cell cycle of a nerve cell compare to that of an epithelial cell that secretes mucus in the intestine?

CONTINUED

- The structure of DNA allows it to replicate in a semiconservative fashion and thus maintains the integrity of the genetic code.

- During the mitotic phase (mitosis and cytokinesis) of the cell cycle, the cell splits into two identical daughter cells. Mitotic cell division is important for growth, wound repair, asexual reproduction, and replacing worn out cells.

5. The drug taxol is used in chemotherapy for treatment of cancer. It blocks the assembly of microtubules. Explain how taxol inhibits cancer cells from dividing. Would it also stop normal cells from dividing?

7.5 Growth and Development Are Highly Regulated Processes at the Molecular Level

Growth from cell division is fundamental to all living things. Although single-celled organisms such as bacteria undergo a cell division process, it is different from that of animal cells. Precise control in response to external and internal cues is critical to development, tissue renewal, and cell differentiation. Although each cell has mechanisms for regulating cell division and differentiation, sometimes mutations lead to unregulated cell division, which results in diseases such as cancer.

We now can build on your understanding of animal growth and development at the organism, tissue, and cellular levels by going deeper into processes at the molecular level. At the molecular level, genes control cell function by controlling what proteins the cell makes. For instance, in lobster, insect, and frog metamorphosis, environmental cues trigger genes to produce and release hormones that act as cell signals resulting in change. The gene products act as signaling molecules that turn on the expression of other genes, which leads to the growth or differentiation of cells. By regulating gene expression, the cell controls the steps that switch on genes and regulates cell cycle, cell division, and growth. Complex interactions between various proteins control the progression of the cell cycle through the various checkpoints (see **Fig. 7.18**).

Cells respond to a variety of signals

All cells need a way to communicate with each other to carry out their highly specialized, controlled, and coordinated functions. They also need to be able to sense and correctly respond to their environment. To accomplish all this, cells have intricate processes that support cell communication. Cell communication occurs through direct contact between adjacent cells and through intercellular chemical messengers, such as ions, hormones, and other small molecules. The most common way that cells communicate is through intercellular messengers. Changes during puberty, the daily regulation of blood sugar levels, and bone metabolism are just a

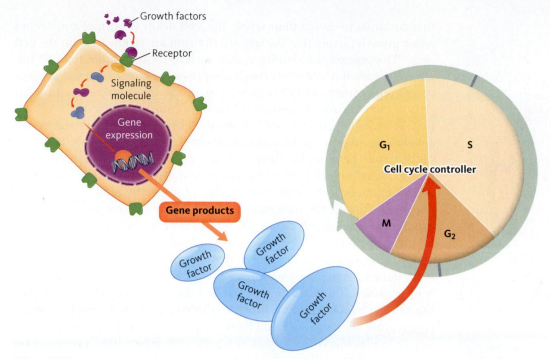

Figure 7.23 The role of growth factors in gene expression. Growth factors act as signaling molecules that initiate gene expression. The gene products produced interact with cell cycle control points that often trigger the cell cycle. For simplicity only one growth factor is shown in this diagram, however, several growth factors work together to control the cell cycle.
© Cengage Learning 2014

few examples of pathways that rely on messenger signaling molecules to trigger a cellular response.

The cell uses different signaling molecules, receptor molecules, and pathways to achieve many different outcomes. Cells are primarily stimulated to divide by a group of proteins or hormones called **growth factors**. Growth factors signal the cell to remain in the cell cycle, to exit the cycle, or to continue to differentiate and become specialized (**Fig. 7.23**). Some growth factors stimulate cells to grow, leading them from the gap phases in interphase to mitosis; others stimulate DNA replication; and still others stop the cycle, leading to a cell's programmed death.

Timing of the cell cycle is critical to growth and development. The two major timing events occur when the DNA is replicated and when the cell enters into the mitotic phase of cell division. Several internal and external factors influence this timing through the interactions of growth factors. For mitosis to begin, the cell must first have enough cellular mass and be growing at a specific rate. These factors give the cell the necessary cues that it is ready to divide. Some genes regulate the timing of events in embryonic cells, keeping them on track for the overall growth and development of the organism. Lastly, in order for the cell to enter into mitosis, the cell checks to make sure that the chromosomal DNA has replicated. At any point, if the cell "decides" these factors are not in order, signals are sent to induce the cell into "suicide," or cell *apoptosis*.

How long a cell lives is highly variable and controlled by a number of factors, such as cell type, environmental conditions, nutrient and resource availability, and genetic controls. For cells such as epithelial cells

Growth factors—a group of proteins or hormones that stimulate cells to divide by signaling the cell to remain in the cell cycle, to exit the cycle, or to continue to differentiate and become specialized.

a. General signaling pathway

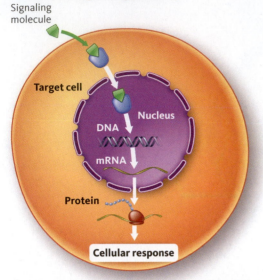

Signaling molecule

Target cell

Nucleus

DNA

mRNA

Protein

Cellular response

b.

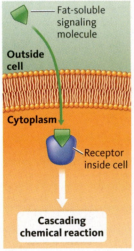

Fat-soluble signaling molecule

Outside cell

Cytoplasm

Receptor inside cell

Cascading chemical reaction

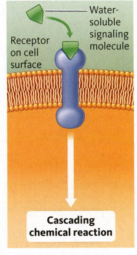

Receptor on cell surface

Water-soluble signaling molecule

Cascading chemical reaction

Figure 7.24 The cellular communication process. (a) Steroid hormones cross the cell membrane and bind to a specific receptor molecule inside the cell to form a complex. This complex triggers a cell response by turning genes on or off. (b) Some receptors are located inside the cell, and others are located on the cell membrane. In both cases, activated receptors trigger specific cellular responses.
© Cengage Learning 2014

Transcription—the gene expression process that transfers the code in the DNA to intermediate messenger RNA (mRNA) molecules.

Translation—the gene expression process that converts the messenger RNA (mRNA) code to a protein sequence.

RNA (ribonucleic acid)—a single-stranded nucleic acid found in both the nucleus and the cytoplasm of the cell that is intimately involved in the production of proteins. RNA contains a ribose sugar–phosphate backbone and the nitrogenous bases adenine, guanine, cytosine, and uracil. The three classes of RNA are messenger RNA (mRNA), transfer RNA (tRNA), and ribosomal RNA (rRNA).

that continue to divide their whole life, cell death is normal and occurs when growth factors that initiate cell division are removed from the cell cycle. The process is so common that each of us loses more than 50 billion cells each day, mostly through apoptosis, a preprogrammed process of cell replacement. Other cells are lost in routine ways—the epithelial cells of our skin are scraped away when we brush our hair or scratch an itch. Why is preprogrammed cell death necessary? Consider the newly developing fingers of a human embryo in the mother's womb. The cells of the tissues between the fingers die in a preprogrammed manner, separating the individual digits of our hands and feet. Without the process of preprogrammed cell death during this process, the skin webbing between our fingers and toes would remain. Many scientists consider preprogrammed cell death a positive control over the cell division process. Limiting the number of divisions an individual cell may undergo prevents uncontrolled, cancerlike cell division.

A general signaling pathway is used in cells to create a cellular response, such as the regulation of cell division (**Fig. 7.24a**). The first step in the cellular communication process is initiated when a signaling molecule reaches the target cell and binds to a specific receptor. Some signaling molecules bind the receptor inside the cell; others bind a receptor on the surface of the cell membrane (**Fig. 7.24b**). Both examples are common and result in similar outcomes of cell communication. When the signal binds a receptor inside the cell, a second step of cell signaling occurs. Binding of the signaling molecule with its receptor activates a series of cascading chemical reactions that turn on or off genes in the receiving cell. Together, these signaling molecules and their receptors change the activity of the cell by turning on specific genes to produce a molecule such as a protein. The gene product or protein may trigger cell division, cell differentiation, or cell death.

Now let's apply the signaling pathway to what you learned earlier about human sexual maturation. Prior to puberty, the sex hormones (primarily estrogen and testosterone) are produced at a low level, but at the onset of puberty, their level begins to increase. Sexual maturation is initiated by hormone signals released by the brain and received by cells in the ovaries and testes. These signaling molecules are transported throughout the body in the blood. When they reach the cells in the target tissue, the hormones pass through the cell membranes (target cells) and bind to their specific receptors inside the cell (see **Fig. 7.24a**). The target cells respond to these signals many times by turning on specific genes that alter cell division and gene expression.

The sex hormones along with the thyroid hormones are fat-soluble and can diffuse through the cell membrane and follow this pathway. Most hormones are water-soluble and act on receptors on the cell surface. When these hormones bind to their receptor, the receptor protein produces a secondary messenger signal that acts on the inside of the cell, generating cellular changes. The system that controls growth and molting in lobsters utilizes both water- and fat-soluble hormones and, therefore, follows a combination of these two pathways.

Different cells express different genes

The human body consists of more than 200 cell types, and it is amazing to think that each of these cell types is unique when they all carry the identical genetic information. Each cell is different because each expresses a unique combination of genes that influences cell specialization. One cell can interact with a nerve growth factor that causes it to differentiate into a neuron (nerve cell), while the other cell interacts with an epithelial growth factor, resulting in an epithelial cell (**Fig. 7.25**).

Different types of cells produce different collections of proteins during different phases of the cell cycle. These different molecules give each cell its structure and function. The production of these proteins involves acquiring information from the nucleotide sequence in the DNA, transferring this information to the site of protein synthesis (ribosome) outside the nucleus, and arranging the protein building blocks in the correct order. Gene expression, or the molecular production of a functional protein, includes two processes (**Fig. 7.26**). **Transcription** is the process that transfers the code in the DNA to intermediate RNA molecules, and **translation** converts this RNA code to a protein sequence.

We haven't discussed **RNA (ribonucleic acid)** yet, so we will explain it here. RNA is a second kind of nucleic acid, and, like DNA, is intimately involved in the production of proteins. Structurally, RNA differs from DNA in four ways (**Fig. 7.27**). First, RNA functions in both the nucleus and the cytoplasm. One kind of RNA shuttles genetic information from DNA in the nucleus to ribosomes in the cytoplasm and rough endoplasmic reticulum. Second, RNA is primarily a *single*-stranded molecule with only one sugar–phosphate backbone (in some viruses RNA exists as a double-stranded molecule). Third, the sugar in the backbone is *ribose* rather than deoxyribose, which is a slightly different molecule. Fourth, RNA contains adenine, guanine, cytosine, and *uracil* (rather than thymine). Therefore, when RNA functions with DNA, an adenine base pairs with uracil, which is similar to thymine but is not identical.

Transcription: the process where DNA produces RNA

In order for a cell's genes to express proteins, several molecular interactions take place. At the molecular level, the first step in protein synthesis is transcription—the process that rewrites or transcribes specific gene sequences in the DNA to create a messenger RNA (mRNA) strand (**Fig. 7.28**). Let's look at just one of the thousands of genes expressed in a goblet cell that makes mucus in the small intestine's epithelial tissue (**Fig. 7.29**). Mucus is a slippery secretion of water, salts, and a protein called mucin. The goblet cell makes mucus when the gene for making mucin is transcribed to an RNA sequence, which is then translated to its protein form.

In eukaryotes, transcription takes place inside the nucleus and requires the specific gene segment of the DNA helix containing the information to unwind and be exposed (see **Fig. 7.28**, step 1). Transcription of the mucin

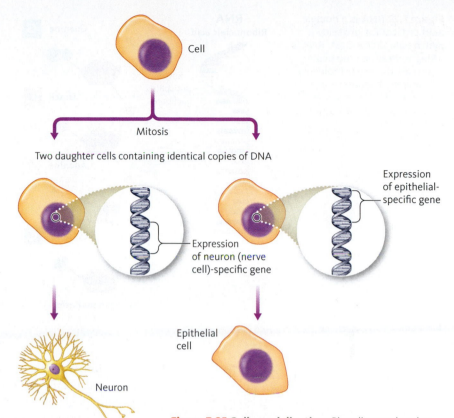

Figure 7.25 Cell specialization. Signaling molecules trigger the differential expression of genes, which leads to cell specialization. In this case, when the single cell is exposed to one set of factors, it differentiates into a neuron; however, under other influences, it becomes an epithelial cell.
© Cengage Learning 2014

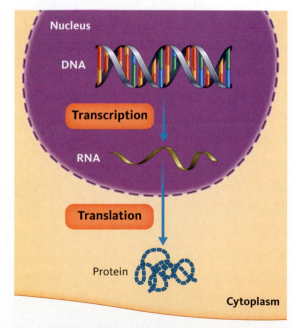

Figure 7.26 Gene expression is a two-step process. The path from the DNA gene sequence to a protein involves transcription of the DNA code in the nucleus, followed by translation of the mRNA to a sequence of amino acids in the cytoplasm.
© Cengage Learning 2014

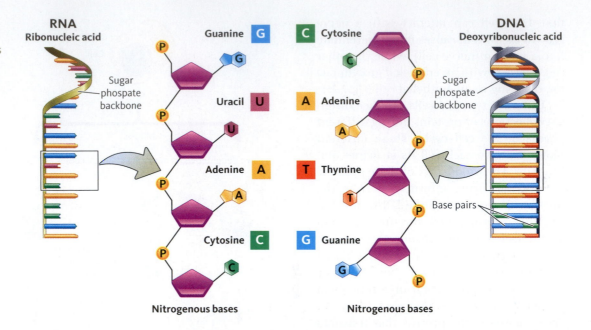

Figure 7.27 RNA is a nucleic acid critical for protein synthesis. Unlike DNA, RNA is a single-stranded molecule that uses the sugar molecule ribose in its backbone. It contains adenine, guanine, cytosine, and uracil (rather than thymine).
© Cengage Learning 2014

gene is initiated when a special enzyme binds the mucin gene **promoter**, which is a specific start sequence for the gene. Once transcription begins, the exposed gene sequence provides the template that is transcribed into an mRNA strand through complementary base pairing (step 2). This procedure is similar to the DNA replication process discussed earlier. Transcription of the mRNA continues until a "stop" at the end of the gene is reached. This "stop" sequence signals the enzyme to separate from the DNA strand, the mucin mRNA to detach, and the DNA strands to bind back together, re-forming the double helix. Like many mRNA strands, the mRNA strand containing the code for the mucin protein goes through some minor modifications before it leaves the nucleus to be translated into the protein. The next step in gene expression is the interpretation of the genetic code to produce the protein (steps 3 and 4).

The genetic code is universal among all living things, from bacteria to humans, once again illustrating our common ancestry (**Fig. 7.30**). The genetic code consists of sets of three nucleotides called **codons**. One or more codons codes for each of the 20 amino acids used in protein synthesis. Because there are four nucleotides, 64 codons are possible. Many amino acids have more than one codon, but each codon is specific to only one amino acid. Most codons code for amino acids, but some signal for the start or stop of translation. Interactions between various types of RNA, amino acids, and ribosomes allow the code to be translated into proteins.

Translation: the process where RNA is interpreted to protein

Sticking with the mucin protein example (see **Fig. 7.28**), the modified mRNA strand carrying the code for the protein leaves the nucleus. The strand next attaches to a ribosome, where it is interpreted or *translated* into the amino acid sequence of the protein (see **Fig. 7.28**, step 4). *Translation* refers to the chemical bonding of amino acids in the correct order to make a protein. Amino acids located in the cytoplasm of the cell are brought to the ribosome by another type of RNA called transfer

Promoter—a specific start sequence for a gene.

Codons—sets of three nucleotides that comprise the genetic code. Each of the 20 amino acids used in protein synthesis is coded for by one or more codons. The codon base pairs with an anticodon on the tRNA that carries a specific amino acid.

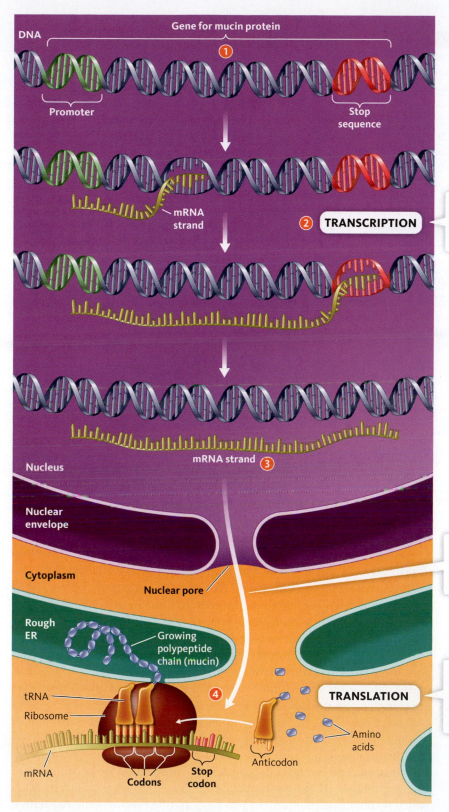

Figure 7.28 Transcription and translation. Gene expression involves two processes (transcription and translation) that occur in four steps. The DNA code is rewritten inside the nucleus into an mRNA code during transcription (steps 1 and 2). The mRNA leaves the nucleus and binds to a ribosome, where translation occurs (step 3). During translation (step 4) tRNA molecules carry amino acids to the ribosome, where they are incorporated in the growing peptide chain.
© Cengage Learning 2014

The DNA code is rewritten inside the nucleus into an mRNA code during **transcription**.

The mRNA leaves the nucleus and binds to a ribosome in the cytoplasm, where **translation** occurs.

During **translation**, tRNA molecules carry amino acids to the ribosome, where they are incorporated into the growing polypeptide chain.

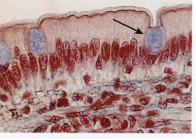

Figure 7.29 Goblet cell. Goblet cell of epithelial tissue that lines the small intestine, releasing newly synthesized mucus.

Dr. Donald Fawcett/Visuals Unlimited, Inc.

Second base of codon

		U		C		A		G		
U	UUU	Phenylalanine	UCU	Serine	UAU	Tyrosine	UGU	Cysteine	U	
	UUC	Phenylalanine	UCC	Serine	UAC	Tyrosine	UGC	Cysteine	C	
	UUA	Leucine	UCA	Serine	UAA	Stop codon	UGA	Stop codon	A	
	UUG	Leucine	UCG	Serine	UAG	Stop codon	UGG	Tryptophan	G	
C	CUU	Leucine	CCU	Proline	CAU	Histidine	CGU	Arginine	U	
	CUC	Leucine	CCC	Proline	CAC	Histidine	CGC	Arginine	C	
	CUA	Leucine	CCA	Proline	CAA	Glutamine	CGA	Arginine	A	
	CUG	Leucine	CCG	Proline	CAG	Glutamine	CGG	Arginine	G	
A	AUU	Isoleucine	ACU	Threonine	AAU	Asparagine	AGU	Serine	U	
	AUC	Isoleucine	ACC	Threonine	AAC	Asparagine	AGC	Serine	C	
	AUA	Isoleucine	ACA	Threonine	AAA	Lysine	AGA	Arginine	A	
	AUG	Methionine; start codon	ACG	Threonine	AAG	Lysine	AGG	Arginine	G	
G	GUU	Valine	GCU	Alanine	GAU	Aspartic acid	GGU	Glycine	U	
	GUC	Valine	GCC	Alanine	GAC	Aspartic acid	GGC	Glycine	C	
	GUA	Valine	GCA	Alanine	GAA	Glutamic acid	GGA	Glycine	A	
	GUG	Valine	GCG	Alanine	GAG	Glutamic acid	GGG	Glycine	G	

First base of codon (left axis) — *Third base of codon* (right axis)

Figure 7.30 The genetic code. The genetic code in mRNA consists of codons that code for specific amino acids. The start codon is in green and stop codons are in red.
© Cengage Learning 2014

RNA (tRNA). The tRNA is structured with a three-base sequence on one end, called an anticodon, and an amino acid on the other end. The amino acid carried by the tRNA is specific to the anticodon that it carries. At the ribosome the mRNA codon for a specific amino acid is exposed. Through base pairing rules, tRNA binds to the complementary codon and releases its amino acid, which then is covalently bound to the growing protein strand. One by one, amino acids are delivered to the growing polypeptide chain until a specific stop codon at the end of the mRNA is reached.

Once the amino acid sequence is assembled, the protein detaches from the ribosome so that it can be further processed to serve its function. Following translation, the mRNA degrades and the nucleotides return to the nucleus to be reused. The newly synthesized mucin protein is packaged with other chemical parts to form mucus. The cell then releases the finished mucus to perform its job (see **Fig. 7.29**).

We started out this chapter describing the transformation a lobster goes through as it matures from a single-celled fertilized egg to the one-pound lobster on your dinner plate. The intricacies of these processes are complex and fascinating. The formation of the lobster's body plan is a result of very specific patterns of gene expression that are influenced by the lobster's genetic programming, the environment in which it lives, and the availability of nutrients. Lobster growth and cell specialization are accomplished by the cell cycle, which is regulated by hormones and external factors. At the molecular level, genes are transcribed to RNA, which is translated into protein products. In these terms, gene regulation and expression are highly ordered chemical interactions that regulate cell differentiation, growth, and development of the lobster. It is incredible to think of the complex interactions between the environment, each organ system in the lobster, its tissue and cells, its DNA, and various proteins that all work together in order for the lobster to live its life. Later in Chapter 21, we will explore further influences on gene expression, growth, development, and human health.

- Cells signal other cells to initiate processes of growth, development, reproduction, and survival. Cells communicate commonly through intracellular chemical messengers.

- Cells are specialized to perform specific functions through the expression of genes.

- In the nucleus, genes are transcribed to an RNA template in a process called transcription. Transcription produces an mRNA that carries the universal code of life in codons.

- At the ribosome, tRNA interprets the code in the mRNA into protein products through the process of translation.

- Gene regulation and expression are highly ordered chemical interactions that regulate cell differentiation, growth, and development.

1. In early stages of development, the lobster goes through rapid cell division, changing from a fertilized egg to a larva. List the two ways cells communicate with each other during this transformation.

2. Many growth factors are signaling molecules that bind to the promoter region on DNA. What is the promoter's function?

3. Why is apoptosis (pre-programmed cell death) important as a part of embryonic development? Give an example where it operates.

4. How do the functions of transcription and translation differ?

5. During embryonic development, epidermal growth factor (EGF) triggers the growth differentiation of many cell types. Refer to **Figures 7.23** and **7.25** to explain the general mechanism of how EGF triggers cell growth and differentiation.

End of Chapter Review

Self-Quiz on Key Concepts

Animal Growth and Development

KEY CONCEPTS: Different animal groups show specific patterns of growth and development, which are controlled by the expression of different genes. Very specific patterns of gene expression control the formation of body plans in animals.

1. Match each of the following **processes** with the best definition.

   Growth   a. processes in which the cell becomes specialized in form and function
   Cell differentiation   b. overall increase in body size
   Development   c. series of changes an animal undergoes as it matures from an embryo to an adult

2. Look at **Figure 7.7** and choose from the list below the group of animals that follows a different developmental pattern from the others.
   a. segmented worms
   b. mollusks
   c. chordates
   d. arthropods

3. Which statement best describes the function of the outer germ layer in complex animals?
   a. It gives rise to organs such as the brain and skin.
   b. It gives rise to the internal organs such as the lungs and pancreas.
   c. It gives rise to the muscle and skeletal systems.
   d. It gives rise to the cardiovascular system.

External and Internal Factors Control Growth and Development

KEY CONCEPTS: Animal growth and development are the results of complex interactions of many external and internal factors, which include genetics, environment, nutrition, and hormones.

4. The environment includes external factors such as _____.
   a. daylight
   b. temperature
   c. nutrients levels
   d. all of the above

5. In lobsters, which internal factor controls growth and development?
   a. temperature
   b. food availability
   c. genetics
   d. metabolic rate

The Cell Cycle

KEY CONCEPTS: The cell cycle consists of distinct phases with specific checkpoints that ensure proper cell division. Growth is accomplished by the cell cycle and is highly regulated by hormones and external factors.

6. Match each **stage of cell divisions** with the best definition or description.

   Interphase
   Mitosis
   Cytokinesis

   a. part of the cell cycle in which the nuclear material splits into two cells
   b. part of the cell cycle that forms two identical daughter cells
   c. part of the cell cycle in which the DNA replicates

7. Following mitosis, the daughter cell has _____ as the parent cell.
   a. half as many chromosomes
   b. the same number of chromosomes
   c. twice as many chromosomes
   d. none of the above

8. During interphase of the cell cycle, a cell is _____.
   a. dividing nuclear material
   b. dividing into two cells
   c. doing its job
   d. It depends; many cell types do not enter interphase.

Cell Communication

KEY CONCEPTS: Cells signal other cells to initiate processes of growth, development, reproduction, and survival. Cells communicate commonly through intracellular chemical messengers.

9. Which of the following statements regarding hormones is false?
   a. Hormones bind to cellular receptors on specific target tissues.
   b. Hormones trigger molting and metamorphosis in many arthropods.
   c. Hormones are produced in small quantities by glands.
   d. Hormones elicit the identical response in every cell type at all stages of life.

Gene Expression

KEY CONCEPTS: Cells are specialized to perform specific functions through the expression of genes. At the molecular level, genes are transcribed to an RNA template, which then is translated into protein products inside the cell. Gene regulation and expression are highly ordered chemical interactions that regulate cell differentiation, growth, and development.

10. Match each of the following **processes** with the best definition or description.

   Transcription
   Translation
   DNA replication

   a. process in which chromosomes are duplicated
   b. process in which RNA is converted to protein
   c. process in which DNA is converted to RNA

11. mRNA molecules play an important role in translation by _____.
    a. carrying the code in the form of codons
    b. interpreting the code
    d. forming the ribosomal unit
    c. assembling the amino acids of the polypeptide chain

12. In gene expression, _____ takes place in the nucleus and _____ at the ribosome.
    a. transcription; protein modification
    b. transcription; translation
    c. protein modification; transcription
    d. translation; transcription

Applying the Concepts

13. You have just adopted a basset hound puppy (**Fig. 7.31**). Apply the following concepts—patterns of growth, germ layers, life stages, and life span—to the growth and development of this puppy by answering these questions:
    a. List three areas of the puppy's body that are currently *undergoing cell division.*
    b. Describe how the puppy's body shape changes as it develops into an adult. How do its head, eyes, ears, and legs change through time?
    c. Refer to **Figure 7.9**, what germ layer gives rise to the dog's muscles? Lungs?
    d. Hounds are known for their great sense of smell. From what germ layer do these olfactory neurons arise?

14. Mitosis occurs for three basic reasons in multicellular organisms. List these three reasons and describe one in detail. Describe how mitosis relates to the cell cycle.

15. Many cancer drugs stop cell division. List several mechanisms that could hinder cell division.

16. In humans the gene that codes for insulin is located on chromosome 11. This gene is about 150 base pairs long. The sequence AAACTCCAC is a small portion of the insulin gene. Take this DNA sequence and transcribe it to an mRNA sequence. Describe how this mRNA is translated into a protein.

a.

iStockphoto.com/Christopher McDowell Mullins

b.

pixel-pets/Shutterstock.com

Figure 7.31 Basset hound. (a) Puppy. (b) Adult.

Data Analysis
Alcohol Exposure and Prenatal Development

Exposure to alcohol affects prenatal development in several ways. One in 100 babies is born with fetal alcohol spectrum disorders (FASDs), one of the leading causes of preventable birth defects. Fetal alcohol syndrome (FAS) is a severe form of FASD characterized by facial malformations, impaired growth, and central nervous system abnormalities. In the United States it is estimated that 1 in 8 women drinks while pregnant, and 40,000 babies with FAS are born each year.

Data Interpretation

17. **Figure 7.32a** shows a plot of the percentage of alcohol use among women who are pregnant versus those who are not pregnant. Use the 2005 data to create a summary table comparing both "any alcohol use" and binge drinking among pregnant and nonpregnant women.

18. Examining the developmental chart (**Fig. 7.32b**), what two major structural birth defects are most common in babies born to mothers who drink alcohol?

a. Alcohol use among women aged 18–44, 1991–2005*

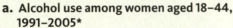

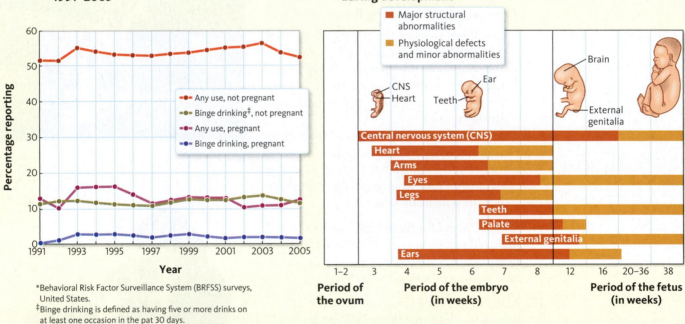

b. Vulnerability to alcohol-induced birth defects during development

*Behavioral Risk Factor Surveillance System (BRFSS) surveys, United States.
‡Binge drinking is defined as having five or more drinks on at least one occasion in the pat 30 days.

Figure 7.32 Alcohol use among pregnant women.
© Cengage Learning 2014

Critical Thinking

19. In 2009, there were about 4.2 million births in the United States. If the same trends from the graph shown in **Figure 7.32a** hold true, how many babies were born from mothers who used alcohol during pregnancy? How many were born from binge-drinking mothers?

20. In the United States, 50 percent of pregnancies are unplanned. This means that women who are not planning on getting pregnant usually are not aware they are pregnant until 4 to 8 weeks into the baby's term. Use your understanding of development to explain why this is a concern.

21. Human brains go through a rapid growth spurt during weeks 26–38 (the third trimester), so they are particularly vulnerable to the effects of alcohol during this period. Developing neurons are impaired by the interactions with alcohol. Signals indicating that development is not progressing normally are sent. The end result is that the neurons are sent into apoptosis. How do you think this ultimately affects the brain?

Question Generator

Hormones and Monarch Butterfly Metamorphosis

Insect development is regulated by several growth hormones. In response to environmental cues, the brain releases the first hormone (prothoracicotropic hormone [PTTH]), which in turn releases another hormone (ecdysone). This second hormone crosses the cell membranes in each of the tissues that will undergo metamorphosis. Inside these cells, the hormone molecules bind to their receptors and trigger gene expression. A third hormone released by the brain, called juvenile hormone, is also a factor in insect development. (**Figure 7.33** shows the stages of a Monarch butterfly's metamorphosis.) Whereas one set of hormones induces metamorphosis (orange blocks in **Fig. 7.34**), juvenile hormone inhibits metamorphosis (green blocks in **Fig. 7.34**). Acting together, these hormones trigger the metamorphosis response in the butterfly and other arthropods such as lobsters.

Figure 7.34 shows a block diagram of the interactions of hormones on cells involved in Monarch butterfly metamorphosis. Use the background information along with the diagram to ask questions and generate hypotheses. We've translated the components and interaction arrows shown in the map into a couple of questions and provided several more questions to get you started.

Figure 7.33 Stages in Monarch butterfly metamorphosis.
© Cengage Learning 2014

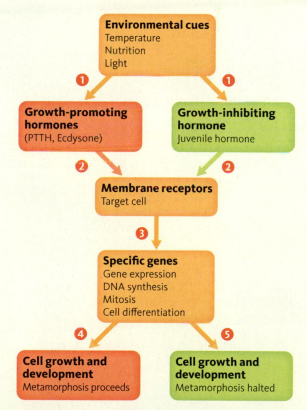

Figure 7.34 Factors relating to butterfly metamorphosis.
© Cengage Learning 2014

1. Where are the environmental cues received by the insect? (Arrow #1)

2. At which range of temperatures are these hormones released? (Arrow #1)

3. What is the relationship between insect growth and development and hormones? (Arrow #2)

4. How are the ecdysone hormone and the butterfly related? (Arrow #2)

5. Does juvenile hormone suppress the expression of specific genes? (Arrow #3)

Use the block diagram and background to generate your own research question and frame a hypothesis.

8 Animal Feeding and Digestion

8.1 Cattle partner with microorganisms to unlock the energy in grass

8.2 Animals display a wide spectrum of adaptations for acquiring food

8.3 Foods are composed of nutrients and fiber

8.4 The digestive system breaks down food to small molecules for absorption

8.1 Cattle Partner with Microorganisms to Unlock the Energy in Grass

Cattle *(Bos taurus)* are a familiar sight in much of rural America, where they live in herds in close association with humans. In the United States, more than 100 million cattle worth $6 trillion supply us with meat, milk, leather products, and fertilizer. Before the invention of the modern tractor, cattle known as oxen pulled carts, wagons, and plows.

Cattle feed exclusively on plants such as grasses and other short-growing plants, which classifies them as herbivores (**Fig. 8.1**). They consume very large quantities of these plants. During the summer, each cow may eat 30 pounds or more of grass each day. Why do cattle eat so much food? One reason is that they have large bodies to maintain. A female cow may weigh 1,000 pounds, and a bull may reach 1,500 pounds or more. Another reason is that they eat plant stems and leaves, which provide less energy, or fewer calories, as well as lower amounts of nutrients than similar amounts of fruits, seeds, or meat. Imagine how much lettuce you would have to eat to provide the 1,500 to 2,000 calories you require each day.

The nutritious plant food grazed by cattle is found inside cells that are encased in walls composed of cellulose, otherwise known as fiber, a complex form of carbohydrate (see **Fig. 8.1**). Like the termites we discussed in Chapter 5, vertebrates such as cattle and humans do not produce the specific chemicals necessary to digest cellulose. Therefore, a large proportion of the plants eaten by cattle provide no nutrients or energy. So how does a 1,000-pound cow fuel its activities and obtain the raw materials it needs to build such a massive body? The answer lies in the diverse populations of microbes that, as with termites, live in its digestive system. These microbes break down

> **Herbivores**—animals that feed exclusively on plants or parts of plants (roots, stems, leaves, flowers, fruits, and seeds).
>
> **Fiber**—the indigestible portion of plant foods; cellulose, a complex carbohydrate that composes plant cell walls.

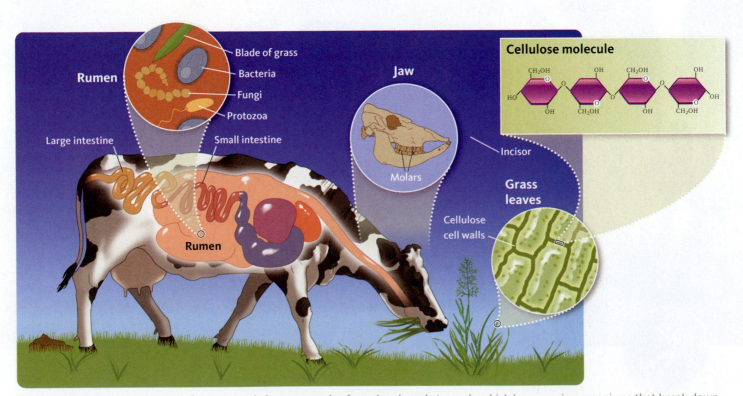

Figure 8.1 The digestive system of a cow. Cattle have a complex four-chambered stomach, which houses microorganisms that break down the tough cell walls of plants. Their efficient digestive system extracts nutrients and energy from grass.
© Cengage Learning 2014

and digest the massive amounts of plants that cattle eat to provide the nutrition they need to meet their large daily energy requirements.

It may surprise you that all animals, including humans, use mutualistic microbes to varying degrees to assist them in their digestion. Koalas, which live in the forest canopy, have a diet comprised exclusively of plant leaves, just like cattle. Their intestines have a large saclike pouch that contains large populations of plant-digesting microbes. Grasshoppers, rabbits, elephants, and horses all are herbivorous animals that have digestive systems specifically adapted to enhance microbial activity. Humans also have large populations of microbes living in their intestines. In fact, we have more bacteria in our intestines than cells in our body. For humans, the benefit of the partnership is relatively small compared to that in leaf-eating monkeys and grazing cattle, which derive a large proportion of their energy and nutrients from the chemical actions of microorganisms. What makes the cow's digestive system so well adapted to provide the environment for microbes to live?

Cattle, along with sheep, giraffes, and deer, belong to a special class of herbivores called *ruminants*. After a small amount of chewing, they swallow the foliage, passing it into the first chamber, or rumen, of their large four-chambered stomach (see **Fig. 8.1**). In the rumen, microbes use their special set of chemical enzymes to begin breaking down the fibrous plant cell walls. To continually grind the tough plant material into smaller and smaller bits, ruminants regurgitate the partially chewed food, rechew it, and swallow it several times. This process is known as chewing their cud, which continuously exposes new surfaces of cellulose to rumen microbes and is an important part of the cattle's digestive process.

In many ways, when you feed cattle, you feed their microbes, which include a diverse number of prokaryotes such as bacteria and archaeans, as well as protozoans and microscopic fungi (see **Fig. 8.1**). All of these microbes have the ability to thrive without oxygen in the rumen environment. Within this warm, moist, nutrient-rich chamber, microbes continually reproduce to form populations that are staggering in size: 20 drops of fluid (about 1 mL) contains more than 16 billion bacteria and more than four million protozoa. The cow's rumen holds more than 50 gallons of fluid that contains trillions of microbes! As they actively break down plant materials, the microbes release nutrients that can be absorbed into the circulatory system of the cow. As they die, the rumen microbes themselves are digested and absorbed by the stomach, thus making up a significant portion of the protein in the cow's diet.

Despite the colossal scale of microbial activity in the digestive system of cattle, much of the plant material cannot be digested and is eliminated as feces. Plant cell walls are very tough to digest, and some plants are much easier to digest than others. In addition to the kind of plant eaten, digestion time plays an important role in the amount of nutrients a cow extracts from each mouthful. The longer the cellulose is in contact with the cellulose-degrading microbes, the more thoroughly the cellulose is digested. Large ruminants such as cattle and bison take up to 80 hours to digest and process plant fiber, whereas smaller ruminants such as white-tailed deer have a shorter processing time of 45 hours. Humans, who eat both plant and animal tissues, have even shorter processing times: about 24 hours from feeding to elimination.

In this chapter we will explore how animals acquire energy and raw materials through feeding and digestion. We start with the wide diversity of adaptations for capturing food and then look at the chemical level, or the nutrients, in foods. The last section takes you through each compartment of the human digestive system (**Fig. 8.2**), which functions to take

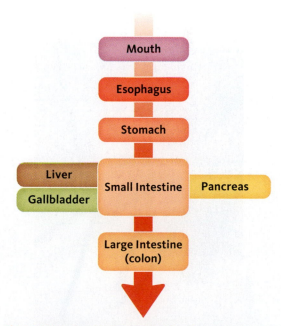

Figure 8.2 A simple diagram of the digestive tract. Food moves through the digestive tract, which is a series of compartments (organs) each having a specialized structure that performs particular digestive processes. The liver, gallbladder, and pancreas secrete digestive substances into the first part of the small intestine, where most of the digestion and nutrient absorption occurs.
© Cengage Learning 2014

Figure 8.3 Koalas have a specialized diet. They eat only the leaves, stems, and bark of eucalyptus trees. These trees form forests in the eastern and southeastern regions of Australia.

large, complex macromolecules and break them down to small molecules that are absorbed across cell membranes into circulation. In the case of cattle, their success is tied to the success of the microbes they "farm" in their stomach—their ability to survive and fuel their lives is entwined together. However, this is only part of the process; animal cells require oxygen to completely break down and metabolize the nutrients they obtain. Breathing and the acquisition of oxygen will be presented in Chapter 9.

8.2 Animals Display a Wide Spectrum of Adaptations for Acquiring Food

Animals are adapted to eat a wide variety of foods. The American black bear commonly eats fish, insects, and honey, as well as fruits, nuts, and seeds. Bears are *generalists* and are able to switch their food sources depending on which foods are available. On the other end of the spectrum are the familiar koalas of Australia, which are *specialists* that eat a highly selective diet— they only eat the leaves, stems, and bark of *Eucalyptus* trees (**Fig. 8.3**). The fibrous leaves are low in nutritional value, so koalas consume over a pound of foliage each day. The koala is highly adapted to eat these leaves, which are toxic to most other animals. They have two agile thumbs that allow them to better grip branches and leaves. Koalas also have powerful jaws that are outfitted with sharp front teeth that clip leaves and grinding teeth that reduce the leaves to a fine paste. These herbivores are not ruminants like cattle, which maintain microbes in their large stomachs. Instead, koalas have large populations of microbes in an extremely long pouch between their small and large intestines. Here microbes break down the cellulose and inactivate toxins in the leaves. Koalas spend almost their entire life eating, resting, and sleeping (20 hours per day) in the treetops; they only come to the ground to move between trees. Their inactive lifestyle means that they have low energy requirements and don't need to feed very often.

The koala's specialized diet illustrates the main theme of this section: how different animals are adapted to eat particular *food items* in a particular *way.* Some animals, such as mosquitos, pierce and suck the blood or body fluids of animals, whereas blue whales filter plankton from the water. Many familiar animals, such as insect-eating birds and frogs, target, capture, and eat their prey whole. The feeding process provides many examples of how an animal's mouthparts are specialized for capturing and eating food and acquiring the energy they need for reproduction and other life activities. As one of the most important activities of animal survival, feeding mechanisms of animals are under powerful forces of natural selection.

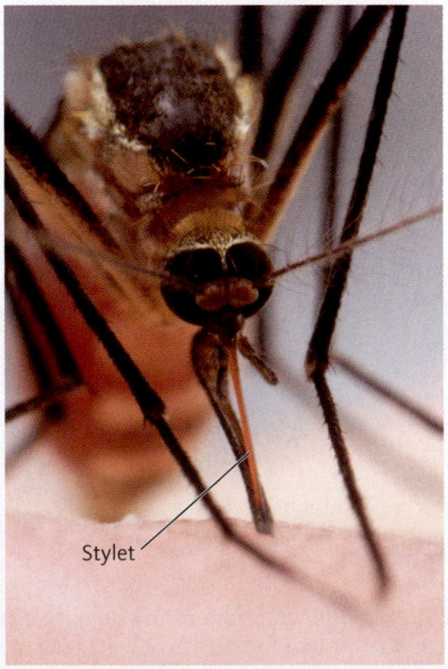

Fluid feeders ingest nutrient-rich animal blood or plant sugars

Many animals obtain nutrients by feeding on the nutrient-rich body fluids of animals or plants. The common mosquito, along with fleas, lice, bedbugs, and biting flies, sucks blood from amphibians, birds, and mammals. Technically, mosquitoes don't bite you; rather, they pierce your skin with a long, thin, needlelike tube through which blood can be sucked directly from your blood vessel (**Fig. 8.4**). This species of mosquito is an example of

Stylet

Figure 8.4 A fluid feeder. The female mosquito pierces the skin with scissorlike mouthparts before drawing blood up through her stylet. The saliva injected when the mosquito feeds causes an allergic reaction by the body. This reaction is shown by a small red bump and itching sensation.

a *fluid feeder*. Only female mosquitoes ingest blood, because the blood proteins are essential for the proper production and development of eggs. Surprisingly, she does not use the nutrient-rich blood to fuel her own metabolism or flying activities. For this, she, like her male counterparts, feeds on sugary fluids, or nectar, from flowers. In fact, most of the 2,500 species of mosquitoes feed on sugar and do not require blood for their reproduction.

Stealing blood from a vertebrate is not easy because the host's blood cells clump together, or clot, and pain receptors sound the alarm. To counter these defenses, blood-sucking animals commonly produce saliva with special blood-thinning chemicals called *anticoagulants* to stop blood from clotting or keep the blood flowing through their mouthparts. Bedbugs use their mouthparts to pinch and pierce the skin. They then inject saliva that has an *anesthetic* chemical that numbs nerves in the skin to prevent its sleeping prey from being awakened for the 3 to 10 minutes it feeds. Saliva is also a source of pathogens that can cause serious human diseases. For example, certain species of ticks transmit bacteria that cause Lyme disease, and fleas carry the pathogen that causes the plague. Vampire bats are known to transmit the rabies virus to humans.

Bulk feeders often use teeth and jaws to attack and feed on individual organisms

Vertebrates feed upon other organisms by either tearing off pieces of their body or eating them whole, a process called *bulk feeding*. The notoriously voracious piranha fish of the Amazon in South America feed this way, as do many familiar animals such as cows, owls, and frogs.

One of the most important adaptations for bulk feeding is teeth. Structurally, teeth are covered by enamel, which is the hardest biological substance known. This characteristic also makes teeth ideal structures for fossilization or use in identifying human remains in forensic investigations. Unlike modern-day birds, *Archaeopteryx* fossils show that it had small teeth similar to reptiles.

Teeth have characteristic structures that are linked to feeding. Each species has its own pattern of teeth in the mouth as well as individual tooth shape and cutting surfaces. Tooth shapes have evolved for particular functions. For example, piranha fish have razor-sharp teeth shaped like triangles (**Fig. 8.5**). Piranha teeth interlock like a saw blade and clip off small pieces of flesh from larger animals that move in the water. **Figure 8.6** shows a set of chimpanzee teeth typical of many mammals, including humans. At the front of the mouth are two pairs of **incisors**, which are used for cutting, clipping, or gnawing food. The incisors determine the size and shape of the bite and thus the food available to an animal. A narrow, delicate set of incisors enables grazing mammals, such as sheep and kangaroos, to clip grass plants closer to the ground. The incisors in elephants are modified into a pair of long *tusks*. **Canine** teeth occur on either side of the incisors and are specialized for piercing and tearing food (see **Fig. 8.6**). In **carnivorous**, or meat eating, mammals, the canines typically are enlarged, pointed, and long. Premolars and **molars** are located toward the back of the mouth. They have a more complex pattern of surfaces associated with the organism's particular food. These teeth are used for chopping up and grinding food, and they are especially well developed in herbivores such as cattle (see **Fig. 8.1**) and koalas. To chew and crush the tough exoskeletons of insects, small mammals have a continuous row of molars with sharp ridges.

Figure 8.5 Piranha teeth are suited for tearing flesh. Their sharp, triangular teeth work like a saw blade, shearing off pieces of flesh from their prey. Piranha are notorious for their aggressive feeding habits; however, serious attacks to bathers and swimmers are rare and the threat to humans has been largely exaggerated.

iStockphoto.com/Peter Jochems

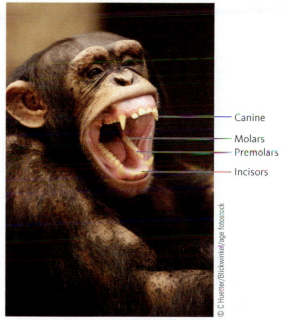

Canine
Molars
Premolars
Incisors

© C Huetter/Blickwinkel/age fotostock

Figure 8.6 Mammal teeth. From the front to back are the incisors, canines, premolars, and molars. Each type of tooth is specialized for a different kind of action.

Incisors—type of teeth at the front of the mouth, which are used for cutting, clipping, or gnawing food; incisors determine the size and shape of the bite and thus the food available to an animal.

Canine—type of tooth that occurs on either side of the incisors and is specialized for piercing and tearing food; well developed in carnivorous mammals.

Carnivore—animal that feeds on other animals or their eggs.

Molars—type of teeth used for chopping up and grinding food; are especially well developed in herbivores; located toward the back of the mouth.

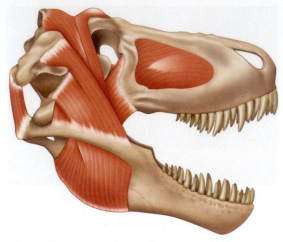

Figure 8.7 The jaws of *T. rex*. *Tyrannosaurus rex* had massive jawbones and powerful muscles that allowed it to generate extremely large bite forces to seize and kill prey. *Tyrannosaurus rex* and other carnivores do not chew their food extensively, as humans do.
© Cengage Learning 2014

TABLE 8.1 Animal Bite Pressures

Animal	Bite Pressure (psi)
Humans	55–280 Average 162
Dogs	3–300 Average 58
Lions	900
White sharks	600
Hyenas	Average 56–337 Max 1,010
Dusky shark	325
Alligators	2,200
Tyrannosaurus rex	3,000

© Cengage Learning 2014

Teeth may also be used to prevent the escape of prey rather than for biting and chewing. For example, rattlesnakes have modified hollow teeth, called fangs, which they use to inject venom into their prey. This helps immobilize and subdue the prey before it is swallowed. Finally, not all vertebrates have teeth. Turtles, birds, and a few mammals lack teeth altogether and swallow their prey whole (see baleen of blue whale in Chapter 1).

Vertebrate animals also evolved strong jawbones and powerful muscles for biting down on their prey with tremendous pressure. **Table 8.1** lists the bite pressures of several different animals. Humans have a very powerful bite pressure, especially for a short jaw. The jawbone is one of the strongest bones in the body. However, our bite pressure of about 162 pounds per square inch is weak in comparison to that of the *Tyrannosaurus rex,* which is estimated at 3,000 pounds per square inch (psi). This is equal to over a ton of weight, or the weight of a small car, pressing down on an area of a square inch. Bones easily crack under these pressures, which are generated by long, thick, powerful jaw muscles that are attached across the hinged joint between the jawbone and the sides of the skull (**Fig. 8.7**). The mouth and jaw structure of *T. rex* and other carnivores allows these animals to rapidly close their mouth and jaw with enough force to seize and kill prey.

Filter feeding is common in aquatic animals

Many animals that live in aquatic environments use filter feeding to capture food. Animals as large as the blue whale and as small as sponges have adaptations for trapping or filtering organisms suspended in water. Animals that use this mode of feeding vary in the type of filtration system they use. The largest living fish, the whale shark, is an example of an active filter feeder that swims along and opens its huge mouth to take in water (**Fig. 8.8**). It then expels the water through its gills, trapping plankton, algae, krill, and even small squid and fish on its gill plates. These organisms are then swallowed and digested.

Smaller marine organisms, such as feather duster worms and barnacles, stay fixed in one place, attached to shells, rocks, or other hard sur-

Warren Baverstock/ZUMA Press/Newscom

Figure 8.8 The whale shark is a filter feeder. It opens its large mouth to swim through and filter microscopic plankton and small fish from the water. This feeding adaptation fuels the growth of some whale sharks to be over 40 feet long and weigh over 20 tons.

faces. They extend specialized structures that trap floating particles in the water and then pull their appendages into their mouths to feed. Sponges create water currents that flow in from the side and out the top of their body, collecting and ingesting bits of bacteria and small microscopic animals called plankton.

8.2 check + apply YOUR UNDERSTANDING

- An animal's diet may include a wide variety of foods or it may be specialized to a narrow range of foods.

- An animal's feeding mechanisms are adapted for feeding on fluids, trapping suspended particles in water, or eating prey whole or in pieces.

- Bulk feeding animals have skulls, teeth, and jaws that reflect their diets.

1. Why is the black bear considered a generalist and the koala a specialist feeder?

2. What role do chemicals such as venoms, anesthetics, and anticoagulants play in animal feeding?

3. After subduing their prey (mostly with poisons), many spiders bite their prey to make a small wound. They then pump digestive fluids in and out through this hole and suck out the digested tissue. Classify this type of feeding mechanism as (A) fluid feeder, (B) bulk feeder, or (C) filter feeder.

4. What kind of teeth shapes would you expect to find in the following bats? (A) A fruit-eating bat, (B) a carnivorous bat that eats delicate moths, and (C) a vampire bat that cuts the skin and licks the blood of mammals.

5. Compare the teeth of a horse and those of a coyote (**Fig. 8.9**). Which of the figures shows the herbivorous horse and which shows the carnivorous coyote?

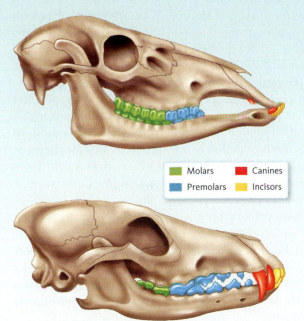

| 🟩 Molars | 🟥 Canines |
| 🟦 Premolars | 🟨 Incisors |

Figure 8.9 Two sets of teeth.
© Cengage Learning 2014

8.3 Foods Are Composed of Nutrients and Fiber

a.

Lori Sparkia/Shutterstock.com

b. Body composition of white-tailed deer

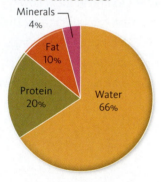

Minerals 4%
Fat 10%
Protein 20%
Water 66%

Figure 8.10 Deer graze on high-protein food sources. (a) Deer graze on tender shoots, a high-protein food, to fuel their growth. (b) Greater than 95 percent of a deer's body is composed of water, protein, and fat.
© Cengage Learning 2014

It's springtime, and white-tailed deer throughout North America hungrily graze on newly sprouting vegetation in woodlands and farm fields (**Fig. 8.10a**). They survived the food-scarce winter by living off of their fat reserves and what little plant materials they could find. Now, the deer's increased activity coincides with the new plant growth that provides them with nutrients such as water, vitamins, minerals, carbohydrates, lipids, and proteins. Each food item they select provides a different mix of these nutrients. Deer prefer plants such as clover and alfalfa, which are rich in protein. Why? The answer lies in the deer's body composition.

A deer's diet needs to support the building of its body. Deer, like most animals, are made up mostly of water and protein (see **Fig. 8.10b**). Proteins are the main biomolecules that form the deer's fur, skin, hooves, antlers, and muscles. The deer's blood is mostly water and protein. All of its cell membranes are made of lipids, and it has a layer of insulating fat beneath its skin. The hard bones of the deer skeleton contain calcium and phosphorus minerals. Successfully building a deer body requires sufficient *quantities of food* as well as a particular *quality of food.* Adequate amounts of food provide the necessary energy, measured in calories, to fuel growth and development. The quality of food depends on the kinds of nutrients provided by the food. Low-quality foods supply fewer nutrients and more fiber compared to higher-quality foods. Fiber is the indigestible portion of plant foods, and deer, like cattle, have microbes to assist them.

A doe prefers to eat protein-rich sprouts in the spring to support her growing fetus. Deer breed in the late fall and give birth in the spring. A high-quality diet supports the burst of growth during the final month of development as well as the production of protein-rich milk for the newborn fawn. A malnourished doe will give birth to an underweight fawn that will be too weak to stand and nurse and to resist unfavorable weather. The fawn simply does not survive.

In the section ahead, we will explore the foods that animals eat—their diet and nutrition. In addition to the study of the chemicals that compose food, **nutrition** is a specialization in biology that examines how nutrients are digested, broken down, absorbed, and used in the body. *Foraging,* or searching for food, is critical for survival and is closely associated with an animal's behavior, growth, development, body composition, and ultimately its fitness in its environment.

Foods provide the body with raw materials and energy for life's activities

Foods are composed of nutrients. **Nutrients** are chemical substances essential for survival because they provide energy and raw materials, and/or they support body processes such as the growth, maintenance, or repair of tissues. The major nutrients were introduced in Chapter 1 as the molecules of life, or biomolecules. The nutrients in foods are (1) carbohydrates, (2) lipids (fats and oils), (3) proteins, (4) vitamins, (5) minerals, and (6) water. On the atomic level, organic nutrients (see **Table 8.2**) are composed of carbon (C), hydrogen (H), oxygen (O), and nitrogen (N) atoms along with several other elements. In fact, greater than 96 percent of all the chemicals

Nutrition—a specialization in biology that examines how nutrients are digested, broken down, absorbed, and used in the body.

Nutrients—chemical substances essential for survival because they provide energy and raw materials and/or support body processes such as growth, maintenance, or repair of tissues.

TABLE 8.2 Animal Nutrients

Class of Nutrient	Chemical Composition	Energy (calories per gram)	Recommended Amount in Human Diet (per day)	Main Functions
Carbohydrates	*Organic* Carbon, hydrogen, oxygen	~4	~130 grams	Primary fuel source
Lipids	*Organic* Carbon, hydrogen, oxygen, phosphorus (in phospholipids)	~9	~20–35 grams	Source of energy at rest Fuel source and energy storage Body structure (membranes) Communication (steroid hormones)
Proteins	*Organic* Carbon, hydrogen, oxygen, nitrogen	~4	~52 grams	Structure (hair, muscle) Enzymes, hormones Energy
Vitamins	*Organic* Carbon, hydrogen, oxygen	0	Mostly required in milligrams or microgram amounts	Help regulate body processes
Minerals	*Inorganic* Na, K, Cl, Ca, P, S, Fe, Mg	0	Ca, P: 3–4 grams Others: milligrams or micrograms amounts	Help regulate body processes
Water	*Inorganic* Hydrogen, oxygen	0	~3–4 L = ~1 gallon 13 cups	Supports all body functions

© Cengage Learning 2014

in an animal's body are made from some combination of just these four elements. **Table 8.2** distinguishes the different nutrients in various foods.

Each food provides a unique mix of nutrients. Let's look at the nutrient profiles of three common foods: a hamburger, its bun, and a glass of skim milk (**Fig. 8.11**). The ground beef in a hamburger is animal tissue composed of muscle protein and fat; notice that it does not contain carbohydrates. The bun is made from plant tissue found in wheat seeds and is composed mostly of carbohydrates with small amounts of fat and protein. The skim milk, another animal product, is composed of carbohydrates and proteins. Milk produced by mammals in nature also provides fat. These three nutrient groups—carbohydrates, fats, and proteins—are called **macronutrients** because they are organic molecules required in large amounts by the body (see **Table 8.2**). These large, complex molecules contain *energy* stored in their chemical bonds. When these bonds are broken, they release energy measured as **calories**. The energy provided by different nutrients is based on calories per gram. If you think of the weight of a pen cap, for example, you'll have a good idea of how much a gram weighs. Fats consist of long carbon chains with many bonds. They provide the most energy at about 9 calories per gram, followed by carbohydrates and proteins at about 4 calories per gram. Which of the three foods provides the most energy? If you answered the hamburger, you are correct. It supplies nearly 200 calories, the bun 117 calories, and the skim milk 83 calories.

In addition to energy, macronutrients provide the *raw materials* to build or replace body structures. During digestion, large macronutrient molecules are broken down, or digested, to their smaller molecular units, which in turn serve as the building blocks for new cellular products. The

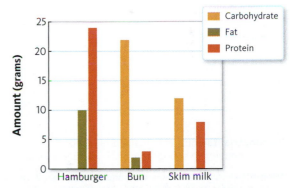

Figure 8.11 Nutrient profiles for hamburger, a bun, and an 8-ounce glass of skim milk are different. The hamburger supplies proteins and fats, whereas the bun provides mostly carbohydrates. With fat removed, skim milk is a good source of carbohydrates (in the form of sugars) and proteins.
© Cengage Learning 2014

Macronutrients—organic molecules required in large amounts by the body; large, complex molecules containing *energy* stored in their chemical bonds that, when broken, release energy.

Calorie—a measure of energy contained in food.

a.

b.

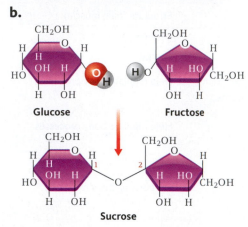

Glucose   Fructose

Sucrose

Figure 8.12 Fruit bats eat fruit and nectar. About 30 percent of bat species feed on nectar and fruit, which are rich sources of simple sugars. In plants, the sweet-tasting sucrose is formed by the removal of water from the ends of glucose and fructose molecules.

© Cengage Learning 2014

Micronutrients—nutrients that are required in smaller amounts to support normal body functioning.

Minerals—inorganic substances; animals need a variety of minerals, which usually are ingested as salts.

Metabolism—chemical reactions that release and use energy found in carbohydrates, lipids, and proteins.

Essential nutrients—nutrients the body cells can't make and *must* be obtained through the diet.

Simple carbohydrates—commonly called sugars; provide the major fuel source to power the animal body, especially the nervous system; type of carbohydrate that contains one or two sugar units bonded together.

Glucose—the most abundant sugar molecule in human diets; the main source of energy for cells; the sugar flowing through human blood vessels (blood sugar).

Sucrose—chemical name for table sugar; composed of two sugar units (glucose and fructose molecules) bonded together.

saying "you are what you eat" truly illustrates this fact. The atoms and molecules we eat are transformed through a wide variety of metabolic functions into our body structures. The carbon in the milk, as well as the bun, lettuce, and beef, become the carbon found in each of the trillions of cells in our bodies. Some ends up in the carbon dioxide molecules we exhale. Ultimately, all nutrients in the animal body can be traced back to nutrients captured by plants in photosynthesis or minerals absorbed from the soil.

The hamburger and milk also contains smaller amounts of vitamins and minerals. These two groups of nutrients are called **micronutrients**, which are required in smaller amounts to support normal body functioning. Vitamins are organic compounds, whereas **minerals** do not contain carbon and are inorganic substances (see **Table 8.2**). Micronutrients play important roles in cellular chemical reactions, or **metabolism**, that release and use energy found in carbohydrates, lipids, and proteins. For example, the beef provides minerals such as iron, which is used to produce hemoglobin in blood cells. A glass of skim milk delivers calcium, a mineral essential for muscle contraction in animals as diverse as humans, insects, snails, and kangaroos.

The animal cell's metabolic machinery can transform some kinds of nutrients into other forms of nutrients by rearranging their atomic components and chemical bonding. A familiar example occurs after you eat an excess amount of carbohydrate in the form of sugars. Your body cells can transform these sugars into fat. Likewise, excess proteins are also converted into fat. However, the body's cells can't make all nutrients needed by the body—some nutrients *must* be obtained through the diet. These nutrients are called **essential nutrients**. Because nutrients are chemical compounds, a basic understanding of their chemistry will allow you to connect foods that animals eat to an understanding of the biology of animals—structure, growth, and reproduction.

Plants are rich sources of carbohydrates

Plants, along with insects, are the most abundant foods on Earth. Herbivores obtain carbohydrates that are formed in plant parts—fruits, seeds, stems, leaves, and roots. Our diet is filled with carbohydrates such as fruit juices, starches from potatoes, and breads made from wheat, corn, and rice seeds. Chemically, carbohydrates fall into two broad categories based on the number of sugar units: simple and complex.

Simple carbohydrates are commonly called sugars and provide the major source of fuel to power the animal body, especially the nervous system. This type of carbohydrate contains one or two sugar units bonded together (**Fig. 8.12**). There are three slightly different types of sugar units. Each forms a ringlike structure consisting of the same number of atoms: 6 carbon atoms, 12 hydrogen atoms, and 6 oxygen atoms. One of these sugar units is called glucose, another is called fructose, and a third is called galactose. Slight differences in the arrangement of the atoms in these units cause large differences in the way they function and appear in nature. For example, the carbohydrate in skim milk is lactose, a double sugar that contains galactose units. **Glucose** is the most abundant sugar molecule in our diet and is the main source of energy for cells. The sugar flowing through our blood vessels is glucose. Table sugar, or **sucrose**, is composed of two sugar units, glucose and fructose molecules, bonded together (see **Fig. 8.12**). Sucrose is the main sugar flowing throughout plants. Sugars are easily dissolved in water and are the primary nutrients in fruits such as apples and oranges. Thus, animals

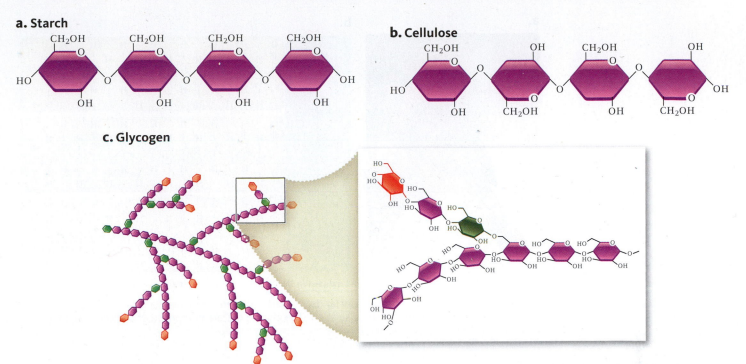

a. Starch

b. Cellulose

c. Glycogen

Figure 8.13 Complex carbohydrates. Plant leaves contain both starch (a) and cellulose (b), which are composed of chains of glucose molecules with different types of bonds. Notice the linkages between glucose units are oriented in different directions. (c) Glycogen is a branched chain of glucose units that store energy in animal tissues.
© Cengage Learning 2014

like primates and fruit bats that feed on fruits get rich, concentrated sources of energy from this type of food (see **Fig 8.12**).

Complex carbohydrates function as storage carbohydrates as well as forming cell wall structures in plants. Chemically, complex carbohydrates consist of long chains of glucose molecules. The key difference between these carbohydrates is the structure of their chains and the way the rings are bonded together. At first glance, these differences may appear small, but they have important functional consequences for the animal.

Figure 8.13 shows the structure of the two complex carbohydrates made by plants—starch and cellulose. **Starch** is the energy-storage carbohydrate found in plants. This is why starch is so abundant in plant roots, bulbs, and seeds. Many rodents and birds store and feed year round on seeds from wheat, corn, or rice, taking advantage of the stored carbohydrate energy provided by plants. **Cellulose** is a structural carbohydrate used by plants to build their cell walls and seed coats. Starch and cellulose are very similar and differ only in the bonds between the glucose rings (see **Fig. 8.1**). Interestingly, during digestion, animals do not have the ability to break the bonds of cellulose, but they are capable of breaking the bonds of a starch molecule. Therefore, animals can power their bodies with starch, but they cannot digest or use cellulose as a supply of energy unless they partner with microorganisms able to break this bond. Because cellulose is not digestible by humans, it is also called dietary fiber. Fiber adds bulk to foods and exercises the muscles of the digestive system, helping them maintain their ability to contract. Functioning intestinal muscles play an important role in maintaining a healthy digestive system.

The storage carbohydrate made by animals is **glycogen** (see **Fig. 8.13c**). Like starch, glycogen consists of a long chain of thousands of glucose molecules linked by bonds that animals can make or break. Unlike starch, glycogen is a highly branched molecule that makes it more compact than starch (see **Fig 8.13c**). When the body needs energy, individual glucose molecules are broken off and fuel the cell's energy metabolism. Glycogen is built from excess glucose in our diet and stored along with

Complex carbohydrate—functions as energy-storage molecules; in plants, structural molecules that compose cell walls; chemically, consists of long chains of glucose molecules.

Starch—the energy-storage carbohydrate made by plants; abundant in plant roots, bulbs, and seeds; a long chain of thousands of glucose molecules bonded together.

Cellulose—a structural carbohydrate used by plants to build their cell walls and seed coats.

Glycogen—the energy-storage carbohydrate made by animals; a long, branched chain of thousands of glucose molecules.

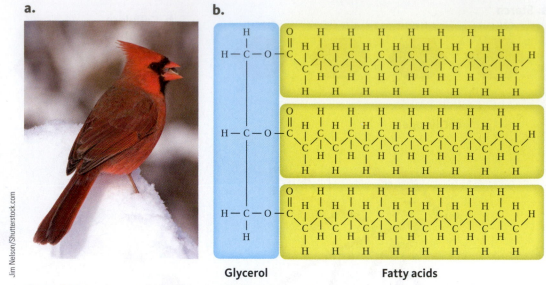

a.

b.

Glycerol   Fatty acids

Figure 8.14 Seeds contain high-energy lipids. (a) Birds that remain in northern areas during the winter feed on oils in seeds for energy and insulation. (b) Fats and oils are composed of triglycerides that consist of a glycerol backbone and three long-chained fatty acids.
© Cengage Learning 2014

Jim Nelson/Shutterstock.com

water in the liver and muscles. The average person stores approximately 2,400 calories in the form of muscle and liver glycogen. How much energy is this? The body's glycogen stores will sustain a 145-pound man running on a treadmill and expending 500 calories per hour for almost five hours.

Meat, eggs, milk, nuts, and seeds contain fats

Lipids, such as fats, oils and cholesterol, are another diverse class of organic nutrients that are familiar to us as greasy, oily substances that do not mix with water. Examples from our diet include butter, margarine, and cooking oils. Animals that eat bird eggs and meat and drink milk obtain significant amounts of fat from their diet. Fats are also stored in many plant seeds, such as nuts, which provide a dense and mobile nutrient package for birds and small mammals (**Fig 8.14a**).

The major groups of lipids are fats, which chemically are called **triglycerides,** named for the three fatty acid chains attached to a small glycerol molecule (**Fig. 8.14b**). When enzymes in the digestive system chemically digest lipids, the fatty acids are often broken off of their glycerol backbone before they are absorbed into circulation. Within body cells, these fatty acids have many chemical bonds that provide energy for the body. They also supply the basic components for cell membranes (phospholipids), steroid hormones (testosterone and estrogen), and some vitamins. Fats in the diet are necessary to absorb some vitamins (A, E, and K). Fats also function to cushion and support organs in the abdominal cavity and insulate the body against heat loss.

The main function of fats and oils is energy storage. In the winter, birds eat oily seeds as their primary energy source. Birds have a rapid metabolic rate for their body size, and seeds are well suited to meet their large energy demands. Gophers and other hibernating animals rely on the breakdown of stored fats to provide energy during their hibernation periods. The blue whale does not eat during its long migration and relies on its thick layer of fat (blubber) for energy. About 10 percent of a deer's body is composed of fat.

Lipids—organic nutrients that are greasy, oily; do not mix with water; fats, oils and cholesterol.

Triglycerides—chemical name for fats and oils; chemically composed of three fatty acid chains attached to a small glycerol molecule.

TABLE 8.3 Percentage of Body Fat in Humans

	Male	Female
Minimum body fat	3%–5%	12%–14%
Average college-aged adult	12%–15%	22%–27%
Average older adult	~18%–27%	26%–35%

© Cengage Learning 2014

Figure 8.15 Egg white (albumin) is composed almost exclusively of protein. Protein serves as a source of nutrition for the growing chick embryo. The yolk is an excellent source of lipids.
© Cengage Learning 2014

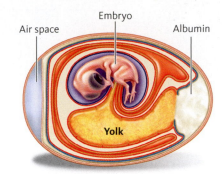

Humans store large amounts of energy as fat. Females have almost twice as much body fat as males of the same weight and, therefore, will be able to sustain their activity longer than males (**Table 8.3**). How long will the body fat sustain a male who weighs 145 pounds with 15 percent body fat (21 pounds of fat) running on a treadmill and expending 500 calories per hour? His body fat will provide more than 88,000 calories and theoretically will sustain his running activity for more than 176 hours! Recall that, by comparison, the glycogen in his liver and muscles would sustain less than 5 hours of treadmill work.

Animal muscle, milk, eggs, insects, and seeds are sources of proteins

An animal's dry weight consists mainly of protein. Skin, hair, connective tissues, bone, and muscle (the largest component of many animals) are mainly constructed from structural forms of proteins. Insects and other invertebrates are other good sources of protein. For example, the dry weight of grasshoppers consists of roughly 50 to 75 percent protein, which forms a large part of the rigid exoskeleton. Insects are a highly digestible source of food for many small mammals. The egg white, or albumin, serves as a source of protein for development of the chick embryo (**Fig. 8.15**). Mammals produce milk with protein to provide the raw materials for growth. To build and maintain their bodies, carnivores acquire large amounts of protein by feeding on other animals or their eggs.

In addition to their structural function, proteins play many critical roles in animal systems. Proteins in the form of hormones trigger growth and development. **Enzymes** are proteins that speed up chemical reactions, such as the digestion of food, and the conversion of these smaller molecules to energy. Proteins bind materials together, act as signaling molecules throughout the body, and provide a range of important molecular services. Proteins are actively involved in cell division and growth. They also defend the body against pathogens.

Plants also have proteins in their cells that are used in photosynthesis, cellular respiration, cell division, and reproduction. For these reasons, green growing shoots and leaves are preferred sources of protein for many grazing herbivores. Underground, the rapidly growing tips of roots are eaten by soil animals to acquire protein. We have already seen that seeds are excellent sources of storage carbohydrates and lipids, and now you can add proteins to the list of nutrients found in seeds.

Chemically, proteins are made from building block molecules called **amino acids**. In plant and animal cells, there are about 20 different types of nitrogen-containing amino acids. Of the three macronutrients, proteins are the only one that contains nitrogen. Recall from Chapters 6 and 7 that amino acids are linked together into long chains; the average protein is about 200 amino acids long (**Fig. 8.16**). This arrangement allows some side

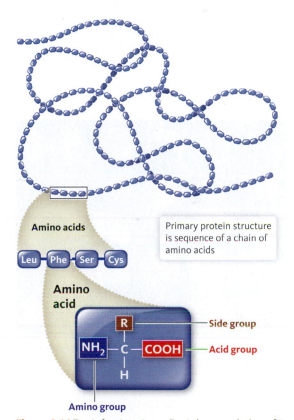

Amino acids

Primary protein structure is sequence of a chain of amino acids

Leu – Phe – Ser – Cys

Amino acid

R — Side group
NH₂ – C – COOH — Acid group
H
Amino group

Figure 8.16 Protein structure. Proteins are chains of amino acids. Each amino acid has a nitrogen portion, or the amine group, and a carboxyl (acid) group. An amino acid's chemical identity is due to its side group represented by the letter R. A protein's structure and identity are determined by the kinds and number of amino acids in its chain. In this diagram, four different amino acids are highlighted. Leu is an abbreviation for leucine, Phe, for phenylalanine, Ser for serine, and Cys for the sulfur-containing amino acid cysteine. Each of the 20 different amino acids have a unique side group.
© Cengage Learning 2014

Enzyme—protein that speeds up a chemical reaction, such as digestion of food and conversion of macronutrient molecules to energy.

Amino acid—building block molecules of proteins that contain nitrogen; there are about 20 different types of amino acids.

groups of atoms or attached molecules to attract or repel each other like magnets, resulting in a highly specific three-dimensional shape. The different side groups also give each protein specific chemical properties. The shape and chemical properties account for the large variation in the function of the many protein molecules. When proteins are digested to their final end products—amino acids—they provide the body with the necessary building blocks to build new proteins.

Most foods contain vitamins and minerals

In the human diet, 14 vitamins and 12 minerals have been identified as *essential,* which means they must be supplied in the diet because the body cannot make them. For instance, calcium and phosphorus are typically found in dairy products. Vitamin A is abundant in orange–yellow fruits and vegetables such as carrots and cantaloupe. When a cow eats grass, it is obtaining a wide variety of vitamins and minerals made by plant cells to support their metabolism. Unlike macronutrients, vitamins and minerals are not broken down by the digestive system. Digestive action releases them from the cells in the food before they are absorbed into circulation.

Minerals   Animals need a variety of minerals, which usually are ingested as salts. Salts that include sodium, chloride, sulfates, and phosphates are important components of body fluids. Calcium phosphate is an important material that makes bones, teeth, and mollusk shells rigid and hard (**Fig. 8.17**). Transport proteins, such as hemoglobin, contain iron for oxygen transport. Other metals, such as copper and zinc, assist enzymes to speed up chemical reactions.

Vitamins   Vitamins are a diverse group of organic chemicals that are generally required in small amounts in the body. They are classified as either fat- or water soluble, based on their chemical properties. Fat-soluble vitamins are stored in body fat, whereas water-soluble vitamins are not stored and must be continually provided in the diet. Excess water-soluble vitamins, such as vitamin C, are excreted in the urine. Vitamins serve a number of important functions (**Table 8.4**), from maintaining the integrity of body tissues to blood clotting and absorbing calcium across the intestinal lining. The B vitamins are active in all body cells, where they work with cellular enzymes to accomplish vital chemical reactions.

Fedorov Oleksiy/Shutterstock.com

Figure 8.17 Mollusk shells are composed of calcium.

Vitamins—a diverse group of organic chemicals that are generally required in small amounts in the body; play vital roles in the body by assisting enzymes in metabolic reactions.

TABLE 8.4 Vitamins and Their Main Functions

Class of Vitamin	Vitamins	Main Functions
Fat-soluble	A	Maintains epithelial tissue, visual pigments
	D	Assists calcium absorption, bone and tooth development
	E	Maintains red blood cells
	K	Essential for blood clotting
Water-soluble	B vitamins Folic acid Biotin	Assist cellular enzymes for energy metabolism, synthesis of biomolecules
	C	Collagen and connective tissues

© Cengage Learning 2014

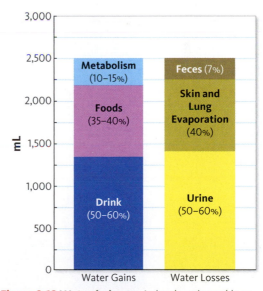

Figure 8.18 An African watering hole. In dry habitats such as the African plain, water holes are critical features of an animal's environment.

Let's look at a familiar vitamin—vitamin C—which is found in broccoli, citrus fruits, and strawberries. Vitamin C is a water-soluble vitamin that plays a critical role in the synthesis of collagen, a protein that acts as the body's glue to bind teeth, bones, skin, and blood vessels together. Without adequate amounts of vitamin C, the body cannot make collagen, which leads to a painful disease called scurvy. Scurvy posed the single greatest danger to the health of men sailing the globe during the Age of Discovery in the 15th and 16th centuries. At times, the crew went months without fresh food because much of the fruits containing vitamin C were eaten early in the voyage or spoiled on long sea journeys. The lack of these vitamins caused the sailors to suffer from swollen and bleeding gums, wobbly teeth, and a sense of exhaustion. The skin seemed to be falling from their bones. Their bodies were literally coming apart without this essential vitamin, and death was a common outcome.

Water is an essential nutrient for body functioning

Life evolved in water, and animals must acquire it, like other nutrients, to survive. Virtually all animal processes are directly or indirectly connected to water. Water is essential for chemical reactions, metabolic processes, and regulation of body temperature. It is a chief component of digestive secretions, as well as blood and body fluids, which transport nutrients, hormones, and waste products throughout the body. Animals must find water to survive and often risk being killed by predators to obtain it from water holes (Fig. 8.18).

Animals are composed mostly of water and can tolerate only small losses in their water content. For example, mammals undergo severe stress when they have a 10 to 15 percent loss of body water. For animals to maintain water balance and homeostasis, water that is lost must be balanced by an equal gain. Although the number varies from person to person, in general, 3 days without water is a serious threat to human survival (3 weeks without food). In the daily intake and output for humans, water is added to the body as liquids or in foods (Fig. 8.19). Interestingly, koalas don't drink water; thus, the water in the leaves they eat must be efficiently extracted and absorbed by their digestive systems to meet their needs. Energy metabolism also produces about 10 percent of human water needs each day. Water is lost through evaporation from the body surface and the lungs. It is also lost in urine and feces. In total, the human body exchanges more than 5,000 mL, or about one gallon, of water with the environment each day (2,500 mL in and 2,500 mL out).

Figure 8.19 Water balance. Animals gain and lose water daily. This bar graph shows the relative daily water gains and losses of humans. Typically, 90% of our normal daily water loss is through urination and evaporation through the skin and lungs. In dry environments, animals have adapted by maximizing gains and minimizing water losses. Adaptations include dry feces, concentrated urine, and thick skin to reduce evaporation.
© Cengage Learning 2014

- Foods are composed of nutrients that are chemical substances essential for survival.

- Macronutrients include proteins, lipids, and carbohydrates, which are nutrients needed in large amounts by organisms. They provide energy and raw materials for growth and development.

- Micronutrients include vitamins and minerals, which are needed in small amounts. They play essential roles in metabolism and formation of body structures.

- Water is a nutrient that is essential for chemical reactions to occur within cells, for transport of nutrients and hormones throughout the body, and for removal of wastes.

1. What two things do nutrients supply to organisms?
2. Water is an essential nutrient required in large amounts by the body. List at least three essential functions of water in animals.
3. Name the top two macronutrients that dominate the nutrient profile of the following foods: (A) animal muscle, (B) growing plant shoots (sprouts), (C) bird eggs. Which of these three food items supplies nitrogen?
4. Collagen is an important structural protein that supports the skin and imparts great strength to connective tissues that pull on muscles, such as tendons and ligaments. What are the building block molecules for this important molecule? Predict its shape as either (A) round and globular, or (B) a long, twisted chain. Explain your prediction.
5. During metamorphosis, a caterpillar uses its stored energy reserves to fuel its change into a butterfly (**Fig. 8.20**). Which two macromolecules does it use to fuel this process? Explain their role.

Figure 8.20 **Zebra longwing caterpillar.**

Leroy Simon/Visuals Unlimited, Inc.

8.4 The Digestive System Breaks Down Food to Small Molecules for Absorption

Animals rely on their digestive systems to break down the food they eat into the chemical molecules their bodies need. The digestive system progressively breaks down large, complex food molecules into molecular pieces that are small enough to be absorbed across cell membranes. The mammal digestive tract acts like a conveyor belt, moving food into a series of specialized compartments beginning with the mouth, progressing through the esophagus to the stomach, small intestines, and large intestines, and ending with the elimination of feces at the anus.

Let's compare two grazing mammals, cattle and horses, to show how their different digestive systems handle the breakdown of cellulose in the plants that make up their diets. Recall from the opening of the chapter that cattle are ruminants that swallow grasses and herbs with little chewing before passing the material into a large four-chambered stomach (see **Fig. 8.1**). Here, immense populations of microbes digest cellulose, converting it to sugars and starches. The larger food particles are recycled from the stomach back to the mouth and chewed several times for physical breakdown of the plant material before it is swallowed again. This type of digestive system is very efficient because microbes break down plant materials *before* the materials reach the small intestine, where the proteins from the microbes and other nutrients are absorbed. This system allows cattle and other ruminants, such as antelope, deer, and sheep, to graze on low-quality,

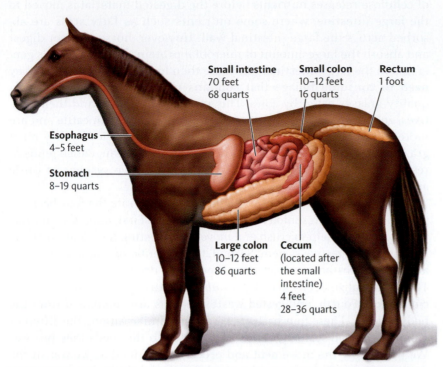

Small intestine
70 feet
68 quarts

Small colon
10–12 feet
16 quarts

Rectum
1 foot

Esophagus
4–5 feet

Stomach
8–19 quarts

Large colon
10–12 feet
86 quarts

Cecum
(located after
the small
intestine)
4 feet
28–36 quarts

Figure 8.21 Horse digestive system. Horses have a digestive system that includes a single stomach followed by the small intestine and a large pouch, the cecum, which houses dense populations of cellulose-degrading microorganisms.
© Cengage Learning 2014

high-cellulose foods to meet their nutritional requirements. Ecologically, this means that these animals can graze in areas that have a dry season when grasses lose much of their nutrition.

Horses have a different digestive system with a single stomach. They move and process plant materials more rapidly than do cattle (**Fig. 8.21**). Food is chewed only once before it is moved to the stomach for digestion and then on to the small intestine for nutrient absorption into the bloodstream. Like cattle, horses partner with microbes to break down cellulose, but, unlike cattle, the microbes do their work in a large pouch called the **cecum**, which is located between the small intestine and the large intestine (see **Fig. 8.21**). In the cecum, microbial digestion

TABLE 8.5 Comparison of Cattle and Horse Digestion

Digestion	Cattle	Horse
Food	Grasses and herbs	Grasses and herbs
Amount of chewing	Several times	Once
Breakdown of cellulose	Microorganisms	Microorganisms
Main site of microbial action	Rumen (stomach)	Cecum and large intestine
Main site of nutrient absorption	Small intestine	Small intestine
Percent of cellulose used	60%–70%	45%–55%
Passage time	Long (80–100 hours)	Short (30–45 hours)

© Cengage Learning 2014

Cecum—a pouch located between the small intestine and large intestines in many cellulose-eating animals; houses large populations of microbes that further digest cellulose and release nutrients. In humans, the cecum is the first part of the large intestine.

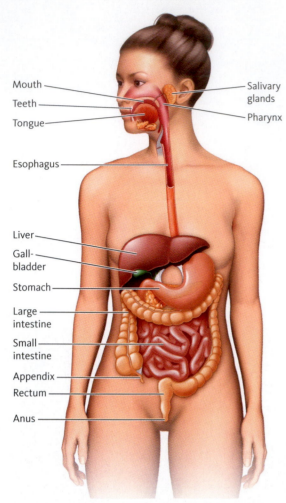

Mouth
Teeth
Tongue

Salivary glands
Pharynx

Esophagus

Liver
Gall-bladder
Stomach
Large intestine
Small intestine
Appendix
Rectum
Anus

Figure 8.22 The human digestive system. Humans have a series of compartments that progressively mix and break down large, complex food molecules to their simple building blocks our body is able to absorb.
© Cengage Learning 2014

Salivary glands—glands that open into the mouth; produce mucus and enzymes that begin the digestion of starch.

Mucus—a thick secretion composed mainly of water, complex carbohydrates, and proteins; binds food particles together and lubricates their passage through the digestive tract; protects the stomach from digesting itself.

Digestive enzyme—speeds up chemical reactions that break down large macronutrient molecules into smaller molecules outside of cells; breaks a specific chemical bond of a specific molecule under specific ranges of temperature and pH.

Substrate—the molecule that is acted upon by an enzyme; fits into the active site of an enzyme.

of cellulose releases nutrients before the digested material is moved to the large intestine, where some nutrients such as fatty acids are absorbed across the large intestinal wall. However, horses cannot digest and absorb the large amount of microbial protein and amino acids generated in the large intestine, which are then lost in the feces. Thus, they need to consume foliage that contains more protein than the low-quality grasses eaten by cows. This means that horses and their relatives, zebras and rhinoceroses, eat more frequently than cattle and are more selective in the kinds of plants they eat. Ecologically, they must graze in areas that provide lots of nutritious vegetation, which tends to restrict their range. **Table 8.5** compares the digestive systems of cattle and horses.

Herbivores such as cattle and horses help illustrate the four basic digestive processes presented in the section ahead. First, muscular contractions *physically* help to crush and *move* the digesting food and nutrients through the digestive system. Second, the *secretion of digestive enzymes* and other substances assists the chemically driven digestive process. Third, specialized cells and tissues *absorb* nutrients into the blood for circulation. Fourth, undigested waste products are *eliminated* from the body as feces. These four basic processes are similar among the different animal digestive systems, which are adapted to the foods they process. We will follow the movement and processing of food as we transit the digestive tract of a human with a single stomach like the horse (**Fig. 8.22**). The journey starts at the mouth, where the first digestive enzymes begin their work.

The mouth begins the physical and chemical digestion of food

The mouth and its parts (such as the lips, teeth, and tongue) acquire, hold, rip, and shred food into smaller pieces. The act of chewing physically digests food into smaller pieces so that food not only moves more easily through the tract but also has a larger surface area for chemical digestion to occur.

The mouths of most animals have several **salivary glands** that produce mucus and digestive enzymes. **Mucus** is a thick secretion composed mainly of water, complex carbohydrates, and proteins. It binds food particles together and lubricates their passage through the digestive tract. Mucus also protects the lining of the mouth against chemical damage to the epithelium.

Digestive enzymes speed up chemical reactions that break down large macronutrient molecules into smaller molecules. They are unique types of proteins that break a specific chemical bond of a specific molecule under specific ranges of temperature and pH (**Fig. 8.23**). Specialized cells lining the digestive tract secrete these enzymes where they work outside of cells.

Human saliva contains the digestive enzyme *amylase,* which breaks down long chains of starch into simple sugars. Salivary amylase is a large protein consisting of about 500 amino acids and calcium and chlorine atoms that hold it together in a precise three-dimensional shape (**Fig 8.24**). Amylase, like all enzymes, forms a kind of pocket called the *active site,* which interacts with a starch molecule. The molecule that fits into the active site of an enzyme is called the **substrate**. To break apart the starch substrate, the enzyme chemically adds a water molecule to the bonds between glucose units. This addition of water is called **hydrolysis**. Several amylase enzymes break apart the starch chain into double sugars with its glucose building blocks (**Fig. 8.25**).

Amylase's specific shape and active site act like a key in a lock; amylase does not break down other molecules such as glycogen, fatty acids or cellulose. Other kinds of enzymes are specialized to break these molecules apart. The shape of amylase is sensitive to environmental conditions. Amylase operates most effectively in the neutral pH range (7) and at the warm body temperature (37°C) of the mouth. At higher temperatures the proteins bonds weaken and amylase physically falls apart so that it is unable to operate. Recall that animals can't digest cellulose because they don't produce the necessary enzyme. In other words, they can't make the key to fit the cellulose "lock." Animals only succeed in unlocking cellulose by partnering with specific microorganisms that make this special "key" enzyme.

Food that has begun to be physically and chemically digested by amylase is pushed to the back of the mouth, and the tongue moves up to the roof of the mouth during swallowing. Three openings are closed: the mouth itself, the passageway up to the nasal cavity, and the opening to the trachea. The ball of food moves into the next region of the digestive tract, the *pharynx,* before it enters the **esophagus** (see **Fig. 8.22**). The esophagus is a tube lined with many glands that secrete mucus. These mucus-secreting glands are especially important near the junction with the stomach. Anyone who has felt the pain of heartburn knows that the esophagus can be damaged from acids that reflux back from the stomach.

The stomach stores and mixes food and begins the digestion of proteins

The stomach is an expandable muscular sac that stores food and begins the chemical digestion of proteins. It further breaks down large food particles by churning and mixing them with stomach secretions to create a

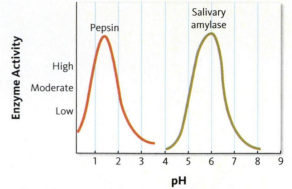

Figure 8.23 Digestive enzymes and pH. Each digestive enzyme operates at a particular range of pH. Stomach acid activates the enzyme pepsin, which is inactivated as it passes into the basic pH of the small intestine.
© Cengage Learning 2014

Figure 8.24 The digestive enzyme in saliva is amylase. This large, specialized protein breaks the bonds between glucose units on the starch molecule. In addition to being produced by the salivary glands, amylase is also made by the pancreas.
© Cengage Learning 2014

Hydrolysis—chemical process that breaks down large molecules through the addition of water by enzymes.

Esophagus—the digestive tube that moves food from the back of the mouth to the stomach; lined with many glands that secrete mucus.

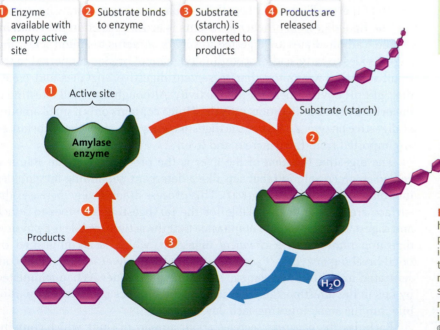

① Enzyme available with empty active site

② Substrate binds to enzyme

③ Substrate (starch) is converted to products

④ Products are released

Active site

Amylase enzyme

Substrate (starch)

Products

H₂O

Figure 8.25 How digestive enzymes work. Enzymes have a three-dimensional structure that forms a pocket called the active site (1). The substrate fits into the active site (2), where the enzyme adds water to break the bond between glucose units (3) and releases double sugars as products (4). The double sugars will be further digested to individual glucose molecules as they pass into cells lining the small intestine.
© Cengage Learning 2014

creamy, semifluid mixture. The gastric juice found in the stomach is highly acidic and creates a very hostile environment for most microbes, forming a barrier to most microbial pathogens that may have been swallowed with food or drink. The epithelial cells lining the stomach are tightly bound together, forming a physical barrier to microbes. The lining also prevents many molecules from being absorbed across the stomach wall directly into the blood. Only a few fat-soluble substances, such as alcohol and some drugs such as aspirin, are absorbed through the stomach wall.

The acidic pH (1.0–3.5) results from the secretion of hydrochloric acid, which activates the digestive enzyme *pepsin.* Pepsin specializes in breaking bonds that link amino acids together and therefore chemically digests proteins. Like amylase, pepsin is an enzyme that adds water, or *hydrolyzes,* bonds between amino acids. Pepsin not only digests proteins in foods but also proteins of cells lining the stomach. For protection, stomach cells are separated from these protein-cutting enzymes by a thick layer of mucus.

The stomach also controls the rate at which its contents empty into the small intestine. Forensic scientists apply the rate of digestion to the stomach contents of a victim to help estimate the time of death. Depending on the type of food, about half of the stomach contents is emptied into the small intestine in about two and half to three hours, and the stomach contents are totally emptied after four to five hours. Food then enters the small intestines, where it spends the next three to four hours being further digested into molecules small enough to be absorbed into circulation (see **Fig. 8.22**).

The small intestine is the main site for absorption of water and small nutrient molecules

The acidic, creamy mixture entering the small intestine from the stomach begins to signal the secretion of substances from the small intestine, pancreas, liver, and gallbladder (**Fig. 8.26**). The juices made by these organs would fill almost a gallon-size jug each day. Virtually all of this fluid, along with the quart of gastric juice, is absorbed back into circulation along the 20-foot length of the small intestine. Less than a quart reaches the large intestine.

The final digestion of macronutrients is accomplished in the small intestine: carbohydrates are digested to sugars, proteins to amino acids, and triglycerides to fatty acids. A coordinated team of organs connected to the small intestine works together to secrete digestive enzymes and create the right pH environment for their activity. Although the small intestine is the primary site for digestion, intestinal juice contains only a small amount of digestive enzymes. Most of the digestive enzymes and other substances are imported from the pancreas and liver.

The digestive function of the liver is the production of bile. **Bile** is a yellowish green solution that acts like a detergent, separating fat globules into many small droplets (**Fig. 8.27**). This process dramatically increases the surface area of the fats available for the fat-digesting enzymes to attack and digest. It's important to emphasize that the activity of bile is a *physical* digestion rather than a *chemical* digestion, which is accomplished by bond-breaking enzymes. Excess bile made by the liver is stored and concentrated in a thin muscular sac called the **gallbladder**. When stimulated by fats in the intestine, the sac contracts and empties about half a quart of bile into the small intestine each day.

Each day the **pancreas** secretes about a quart of fluid, which contains digestive enzymes that act on all three macronutrients, as well as sub-

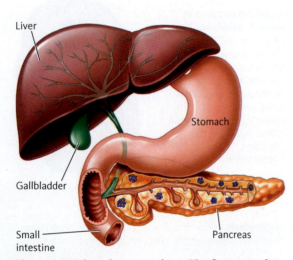

Liver

Stomach

Gallbladder

Small intestine

Pancreas

Figure 8.26 Digestive secretions. The first part of the small intestine receives digestive secretions from the pancreas, liver, and gallbladder.
© Cengage Learning 2014

Bile—a yellowish green solution made by the liver that acts like a detergent, separating fat globules into many small droplets; the liver makes half a quart of bile each day.

Gallbladder—muscular digestive sac that stores and concentrates excess bile; contracts and empties bile into small intestine when stimulated by fats in the intestine.

Pancreas—digestive organ that secretes enzymes that act on all three macronutrients (carbohydrates, lipids, and proteins); secretes substances that neutralize stomach acids and adjust the pH to slightly alkaline (7.5–8) levels.

stances that neutralize stomach acids and adjust the pH to slightly alkaline (7.5–8) levels. The pancreas, like the stomach, produces enzymes that digest proteins and, like the stomach, is protected from self-digestion by a layer of mucus. Fats are digested for the first time in the small intestine, where enzymes from the pancreas break bonds to release long fatty acids from their glycerin backbone molecule. Amylases and other enzymes break down complex carbohydrates into simple sugar units.

Enzymes embedded on the membrane of epithelial cells lining the small intestine accomplish the final step in the digestion of carbohydrates and proteins. Membrane transport proteins then shuttle amino acids and simple sugars into circulation. Fatty acids and other nutrients are absorbed into circulation through the enormous surface area of the small intestine. A large surface area is important because a large amount of nutrients can be absorbed and delivered to cells quickly. How does the small intestine's structure create such a large surface area?

The inner surface of the small intestine contains many folds with fingerlike projections called villi (Fig. 8.28). Villi increase the absorptive surface area 10 times over that of a smooth surface. A further increase in surface area comes from minute projections on each intestinal cell called *microvilli*. Taken together, folds, villi, and microvilli produce a surface area of almost 300 square yards—an area more than 50 feet long by 50 feet wide, or about the area of a tennis court.

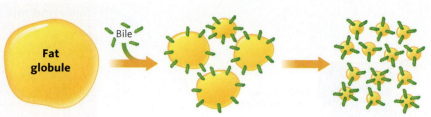

Figure 8.27 How bile works. Bile is produced by the liver and stored in the gallbladder. When signaled that fat is present in the small intestine, bile is released by the gallbladder into the small intestine. Here, bile physically acts on fat globules to make them into smaller fragments. Fat-digesting enzymes now have a greater surface area to chemically break fat globules down into fatty acids.
© Cengage Learning 2014

The large intestine consolidates undigested wastes and eliminates them from the body

By the time food passes from the small into the large intestine, virtually all of the available nutrients have been absorbed into circulation. What remains is undigested solid matter, or the stool, mixed with about a quart of watery digestive secretions. In the large intestine, also called the colon, the last process of digestion takes place—the consolidation of indigestible remains for elimination from the body. This consolidation occurs over several hours, when muscular contractions compact and push the stool through the six feet of intestine. Along the way, water and mineral ions are absorbed from the stool, making it a solid mass of feces. Of the quart of watery fluid that initially passes into the large intestine each day, less than a quarter cup is lost in feces. For desert animals, the small amount of precious water lost in feces is significant. Animals such as pocket mice, kangaroo rats, and rabbits actually reabsorb almost all of the water in the colon and therefore produce nearly dry feces.

Human feces are normally about 75 percent water and 25 percent solid matter. The solid portion consists of one third fiber from plants, one third dead bacteria, and one third a combination of fat, inorganic matter, and very small amounts of protein. The brown color of feces comes from a pigment in bile that is produced by the metabolism of a portion of hemoglobin, the oxygen-carrying red pigment found in red blood cells. If a gallstone blocks bile from entering the small intestine, then feces become grayish white.

Humans house microbial populations in the colon, where bacteria are capable of breaking down small amounts of cellulose from plants in our meals. This process provides a few calories, although the amount is

Villi—fingerlike projections that greatly extend the absorptive surface area of the small intestine; contain *microvilli* projections that further increase surface area.

Large intestine—also called the colon; the last segment of the digestive system that functions to absorb water and consolidate feces for elimination through the anus.

Feces—mass of undigested solid matter, or the stool, which is eliminated from the body; substantial amounts of water may be lost from the body as a component of feces.

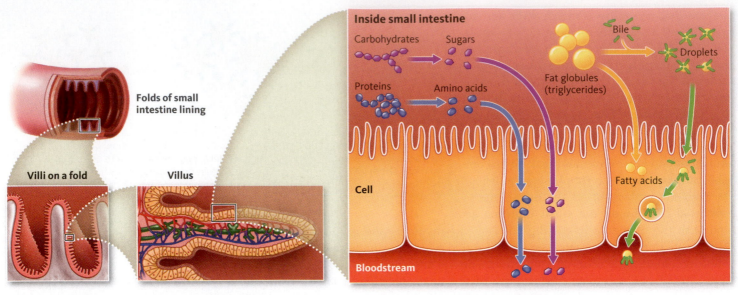

Figure 8.28 The surface area of the small intestine. Folds, villi, and microvilli greatly increase the surface area of the small intestine for the absorption of simple sugars, amino acids, vitamins, and minerals. Bile increases the surface area of fat globules and helps make the fatty acids more soluble in water. Attached to bile, fatty acids enter cells lining the small intestine. Virtually all nutrient absorption occurs across the cells lining the small intestine.
© Cengage Learning 2014

Figure 8.29 Rabbit feces. (a) Rabbits along with some rodents and shrews recover significant amounts of proteins and vitamins by eating their wet, grape-like feces. This behavior enhances nutrient extraction by passing food through their guts twice. (b) They also pass wastes in the form of hard, dry fecal pellets.

relatively insignificant compared to that in herbivorous mammals. These bacteria also produce vitamin K, which can be absorbed into the bloodstream, as well as intestinal gas, or *flatus,* which includes carbon dioxide, methane, and the smelly compound hydrogen sulfide. Medical researchers are now investigating the important role that colon microbes play in human health.

Many cellulose-eating animals, including humans, horses, and koalas, have a cecum, where large populations of microbes help further digest cellulose and release nutrients. These nutrients, with the exception of some vitamin and mineral micronutrients, are absorbed across the lining of the cecum into the bloodstream. Horses lose these micronutrients, but many small mammals, such as rabbits, shrews, and most rodents, recapture these essential nutrients by eating their own feces. Rabbits actually make two kinds of feces (**Figure 8.29**). The grapelike cluster (see **Fig. 8.29a**) is moist and vitamin rich and is eaten directly from the anus. It passes directly from the cecum to the anus. The dry fecal pellets (see **Fig. 8.29b**) are not consumed; they are passed through the large intestine where water was absorbed. Thus, feces represent an important source of nutrition for these animals.

Up to this point, an animal has invested a considerable amount of valuable energy and time into finding, capturing, ingesting, and digesting food. Nutrient molecules such as glucose, fatty acids, amino acids, vitamins, and minerals have been absorbed into body fluids and transported throughout the body, where they pass into cells. What happens to these molecules next? How will they be used? Where is the payoff for all of this investment? The answers to these questions depend on the energy requirements of the animal and the supply of oxygen to the cells' metabolic machinery by the respiratory and circulatory systems. The delivery of oxygen is the main topic in Chapter 9.

8.4 check + apply YOUR UNDERSTANDING

- The digestive system accomplishes four digestive processes: physically breaking down foods and moving them along, secreting digestive enzymes and other substances, absorbing nutrient molecules into circulation, and eliminating undigested waste products from the body.

- Digestive processes occur along the digestive tract, starting in the mouth and moving to the esophagus, stomach, small intestine, cecum in some animals, large intestine, and out of the body through the anus.

- The liver, gallbladder, and pancreas add secretions to the small intestine to aid digestive processes.

1. What are the four basic processes that digest food?

2. Why is the protein's molecular shape so critical to the functioning of a digestive enzyme?

3. Digestive organs that secrete protein digestive enzymes also secrete mucus. Why?

4. Constipation is medically defined as having fewer than three bowel movements per week. Feces, or stools, usually are hard, dry, small in size, and difficult to eliminate. How is this condition related to each of the following? (A) A low-fiber diet, (B) the rate the stool moves through the colon, and (C) the amount of water absorbed from the stool back to the bloodstream.

5. Two different animal digestive systems are shown in **Figure 8.30b and c**. One belongs to an omnivorous fox and the other to an herbivorous pika (**Fig. 8.30a**). Select which system is the fox's and which is the pika's. Explain your reasoning.

a.

visceralimage/Shutterstock.com

b. **c.**

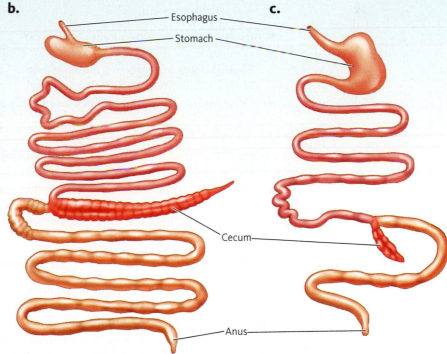

Esophagus

Stomach

Cecum

Anus

Figure 8.30 Two digestive systems.
© Cengage Learning 2014

Self-Quiz on Key Concepts

Animals Use a Variety of Structures, Mechanisms, and Chemicals to Acquire Food

KEY CONCEPTS: Animals obtain nutrients from eating whole organisms, their parts, or their fluids. Animals have evolved adaptations for feeding: capturing, chewing, and ingesting food items. These adaptations include mouthparts to pierce skin and suck blood, different types of teeth, jawbones, powerful jaw muscles, and elaborate filtering devices.

1. Match each of the following **feeding mechanisms** with its characteristic, description, or example. Each term on the left may match more than one description on the right.

    Bulk feeder a. method of feeding that involves collecting food particles from water

    Filter feeder b. obtains nutrients from whole or large pieces of plant and animals

    Fluid feeder c. animal that feeds on animal blood and plant sap

    d. blue whale

    e. cattle

    f. anteater

2. Which of the following pairs are mismatched between teeth and their action?
    a. incisors—cutting food
    b. molars—chopping up and grinding food
    c. canines—piercing and tearing food
    d. premolars—gnawing food

3. Which is a true statement about carnivores?
    a. Their incisors are small and delicate for clipping grass.
    b. Their incisors are large for piercing food.
    c. Their canines are large for piercing food.
    d. Their canines are small and delicate for clipping grass.

Foods Provide Nutrients and Energy

KEY CONCEPTS: Different parts of plants and animals provide different amounts of nutrients and energy for herbivores and carnivores. Foods such as eggs, meat, milk, seeds, and leaves supply calories through the carbohydrates, fats, and proteins they contain. Nutrients in food serve as the source of raw materials for building new molecules and as a source of energy to drive cellular work. Foods also provide the essential vitamins, minerals, and water needed to keep cells and the body functioning.

4. Match each of the following **nutrients** with its characteristic, description, or example. Each term on the left may match more than one description on the right.

    Carbohydrates a. composed of amino acids

    Lipids b. fats, oils, and cholesterol

    Minerals c. simple sugars such as glucose, lactose, and sucrose

    Proteins d. supplies the most energy per gram of all nutrients

    Vitamins e. dominates the nutrient profile of fruits

    f. classified as either fat- or water-soluble

    g. enzymes, most hormones

    h. salt (NaCl), calcium, iron, phosphates, potassium

    i. form long chains such as glycogen, cellulose, and starch

    j. form animal structures such as hair, hooves, and antlers

    k. primary source of energy for cell metabolism

    l. calcium and phosphates that make a snail's shell rigid

5. Match each of the following **nutrient combinations** with its food source. Each term on the left may match more than one description on the right.

Carbohydrates and proteins
Fats and proteins
Carbohydrates, fats, and proteins

   a. meat or animal muscle
   b. bird eggs
   c. seeds and grains
   d. animal milk
   e. shoot tips (sprouts)

6. The largest energy reserves in humans are:
   a. stored proteins.
   b. stored carbohydrates.
   c. stored fats.
   d. nutrients stored in your blood.

7. A nutrient that cannot be made by the body is called:
   a. simple.
   b. complex.
   c. essential.
   d. catalytic.

8. Which of the following is NOT a function of lipids in the human body?
   a. cushioning of internal organs
   b. insulation against heat loss
   c. source of energy reserves
   d. synthesis of steroid hormones
   e. component of enzymes that speed up chemical reactions

Digestion Breaks Down Large, Complex Molecules for Absorption into Circulation

KEY CONCEPTS: Large, complex molecules are broken down into their smaller building blocks by digestion. The digestive tract acts like a conveyor belt that moves food into a series of compartments, beginning with ingestion of food at the mouth and ending with elimination of feces at the anus. Along the way, digestive enzymes, acids, bases, and other chemicals are secreted and mixed with food. The end products of digestion are small molecules that are absorbed into circulation before they are delivered to cells.

9. Match each of the following **digestive system organs** with its characteristic or description. Each term on the left may match more than one description on the right.

Gallbladder
Pancreas
Large intestine
Liver
Mouth
Small intestine
Stomach

   a. begins the chemical digestion of starch
   b. produces bile that emulsifies fats into small droplets
   c. produces hydrochloric acid
   d. folds and villi create an enormous surface area
   e. stores and concentrates bile
   f. starts the digestion of protein by secreting the digestive enzymes
   g. main site where nutrients are absorbed into the bloodstream
   h. secretes digestive enzymes that break down all macronutrients
   i. consolidates undigested solid matter for elimination
   j. secretes substances that neutralize stomach acids

10. Digestive enzymes break bonds of proteins, complex carbohydrates, and triglycerides by:
   a. the addition of water.
   b. the removal of water.
   c. changing the pH of the substrate.
   d. raising the temperature of the substrate.

11. Large populations of microorganisms help break down _____ in the _____ of horses and rabbits.
    a. proteins . . . stomach
    b. proteins . . . cecum
    c. cellulose . . . stomach
    d. cellulose . . . cecum

12. Which of the following pairs are mismatched between digestive secretion and the nutrient it acts upon?
    a. pepsin—proteins
    b. amylase—starch
    c. bile—lipids
    d. hydrochloric acid—complex carbohydrates

Applying the Concepts

13. Blood Alcohol Content (BAC) is the percentage of alcohol (ethanol) in the blood by volume (volume of ethanol divided by volume of blood). Blood is mostly water, so when a person drinks, the alcohol is readily dissolved into the blood and body fluids. *Task:* Two college-aged people, a male and female of the same age, weight, and body build, drink the same amount of alcohol in the same amount of time (**Fig. 8.31**). Refer to **Table 8.3** to help you predict which of these people will have the higher BAC. Why? What about a college-aged male and an elderly male of the same weight and body build? Which will have the higher BAC? Explain.

One mixed drink with 1.5 fl oz (44 mL) of 80-proof liquor (such as vodka, gin, scotch, bourbon, brandy, or rum)

5 fl oz (148 mL) of wine

12 fl oz (355 mL) of beer or wine cooler

Volodymyr Krasyuk/Shutterstock.com

Julian Rovagnati/Shutterstock.com

Suto Norbert Zsolt/Shutterstock.com

Figure 8.31 Alcohol content in wine, beer, and cocktails.

14. Springboks are fast-running animals that are closely related to gazelles and impalas. They inhabit dry, open areas in southwestern Africa (**Fig. 8.32**). Their diet includes young tender grasses, flowers, seeds, and leaves of succulent shrubs. *Task:* Draw conclusions regarding the following characteristics of springboks. (A) Which of their teeth are most prominent? (B) Do you think they harbor a rich or poor flora of microorganisms? (C) Do you think they defecate moist or dry feces?

Mitrofanov Alexander/Shutterstock.com

Figure 8.32 Springbok, a gazelle-like animal.

15. The black rhino is an endangered species and is often part of captive breeding programs (**Fig. 8.33**). Rearing rhinoceros calves requires knowledge of their nutritional requirements such as the composition of their milk, which varies greatly among mammalian species, depending on habitat. The rhinoceros, for example, lives in hot African grasslands. In contrast, reindeer live in freezing habitats near the Arctic Circle. *Task:* Compare the milk of these two mammals

(**Table 8.6**) and predict which milk is made by the rhinoceros and which is made by the reindeer. Which milk provides the most calories? Which animal drinks a greater amount of milk daily? Explain your answer by connecting milk composition to habitat.

Figure 8.33 Feeding an orphaned black rhinoceros.

TABLE 8.6 Milk Composition

Animal	Fat (%)	Protein (%)	Water (%)
A	22.5	10.3	64%
B	0.3	1.2	90%

© Cengage Learning 2014

Figure 8.34. Kangaroo rat.

16. The kangaroo rat is an animal that shows a remarkable ability to remain in water balance despite living in the dry, arid deserts of North America (**Fig. 8.34**). It can live indefinitely on dry seeds and plant material, yet it never drinks. Even though its body is 66 percent water, how is it able to survive?

17. The diet of many carnivorous animals is determined by analyzing their stomach contents. Recently, a team of marine biologists examined the stomach contents of the porbeagle shark. The team found that more than half of porbeagle stomachs were empty. When filled, stomach contents included a wide diversity of prey. *Task:* What does this observation possibly indicate about each of the following: (A) Frequency of feeding, (B) length of feeding period (C) speed of digestion, (D) volume of food consumed, (E) type of feeding—*generalist* (eats anything) or *specialist* (eats only certain prey)?

Data Analysis

Feeding Behavior and the Human–Dog Relationship

Increasing rates of obesity is a problem for pets as well as humans (**Fig. 8.35**). Pet obesity carries similar health risks and challenges; for example, heart disease and diabetes are similar for obese humans and pets. Obesity results when the long-term energy intake exceeds the energy needed to maintain metabolic activities. Excess calories are converted and stored as fat by the body. How severe is this problem? Research estimates that 34 million dogs and 54 million cats are overweight, up by 2 percent and 5 percent, respectively, between 2007 and 2009. Studies suggest that both pet and owner behaviors play a significant role in these alarming trends.

Figure 8.35 The obesity epidemic affects pets as well as humans.

In a German study of 60 normal-weight and 60 overweight dogs, researchers examined the relationship between behavior and obesity in the human–dog relationship. Information about the dog's behavior patterns and the owner's lifestyle were collected. Use this background information and the data presented in **Table 8.7** to answer the questions below.

Data Interpretation

18. How many meals per day do most obese dogs receive?
19. What percentage of owners (obese and normal dogs) express a strong or very strong interest in their dog's nutrition?
20. How does the number of snacks per day influence the obesity rates of dogs? Is the trend consistent?
21. How do obesity rates differ for pet owners?
22. In looking over **Table 8.7**, decide which factors have a direct impact on the occurrence of obesity in pet dogs.

Critical Thinking

23. Dog size was not reported in this study, but it is well established that small dogs have a higher resting metabolic rate than do larger dogs. Understanding this, would you expect this factor to influence the obesity rates reported for these types of dogs?
24. Examine the group of data related to preventive health for dog owners. Is there a strong relationship between the behavior patterns and health of the dog owner and the weight of the dog? Describe that relationship.
25. Whether for a human or dog, effective weight control requires knowledge about the foods you eat and regular exercise. In response, veterinarians and pet food producers have developed public information campaigns aimed at improving pet owners' knowledge about weight control in dogs and cats. Have these campaigns been an effective tool in raising the concern of all pet owners?
26. Storage of excess calories as fat is considered an adaptive strategy of animals to maximize energy storage during times of food abundance for use later when there are shortages. Would you consider this a favorable adaptation for modern domesticated dogs and cats?

TABLE 8.7 Factors Contributing to Dog Obesity

	Overweight Dogs	Normal Dogs		Overweight Dogs	Normal Dogs		Overweight Dogs	Normal Dogs
Work Exercise and Training			**Feeding of food scraps to dog**	25.4	13.3	**Preventative Health for Dog Owners**		
Scale 1–7 (1 = important, 7 = not important)			**Interest in Dog Nutrition**			Regular meals	80	63.3
						Regular exercise	21.4	36.7
My dog would trade its food for a walk.	3.9	2.8	Strong or very strong	27.1	39.0	Occasional exercise	25	35
			Medium	32.2	47.5	No sports	53.6	28.3
Note: Numbers below are given as percentages			Low or nonexistent	40.7	13.5	Overweight	23.7	8.3
						Normal weight	76.3	91.7
Feeding the Dog			**Weight Control in the Dog**			Daily consumption of cigarettes		
Number of meals per day			Weekly	0.0	8.8			
One	44.8	59.3	Monthly	17.9	38.3	Up to 10	29.4	36.8
Two	27.6	30.5	Yearly	14.3	2.9	11 to 20	29.4	57.9
Three	27.6	10.2	Irregular intervals	67.8	50.0	21 to 30	41.2	5.3
Number of snacks per day								
Up to one	30.3	62.5						
Two	26.8	21.4						
More than two	42.9	16.1						

Adapted from Kienzle E, Reinhold B, Mandernach A. A comparison of the feeding behavior and the human-animal relationship in owners of normal and obese dogs. J Nutr 128, 2779S–2782S, 1998.

Question Generator

Feeding Efficiency in Baleen Whales

Baleen whales such as blue, humpback, and fin whales have evolved baleen filters that allow them to feed on vast quantities of small prey. This form of feeding is effective and efficient for two reasons. First, it supports rapid growth and the building of very large bodies. These whales are the largest living animals on Earth. Second, whales store large amounts of fat during their summer feeding to fuel their long-distance migration to breeding grounds that may be thousands of miles away.

Figure 8.36 Lunging baleen whales.

Baleen whales feed by engulfing large volumes of water containing dense populations of plankton or small swimming animals. They find their prey at the sea surface (**Fig. 8.36**) or at depths that exceed 1,500 feet deep. When they feed, they speed up or lunge forward toward a patch of prey with their mouths open wide. Water flows into the whale's mouth and stretches its grooved belly to accommodate the thousands of gallons that enter. Whales then close their mouth, and the engulfed water is filtered through the baleen plates, thereby leaving prey inside the mouth where it is swallowed.

Shown below is a block diagram (**Fig. 8.37**) that relates several aspects of the biology of filter-feeding baleen whales. Use the background information along with the diagram to ask questions and generate hypotheses. We've translated the components and interaction arrows shown in the map into a question and provided several more questions to get you started.

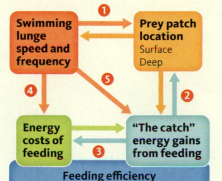

Figure 8.37 Factors relating to baleen whale feeding behavior.
© Cengage Learning 2014

1. Do baleen whales swim faster or lunge more often when they feed at the surface or in deeper waters? (Arrow #1)
2. Do whales filter more prey when they feed at the surface or when they dive deeply? (Arrow #2)
3. How does their energy gain compare to their energy costs to swim and feed? (Arrow #3)
4. How much energy does a baleen whale spend swimming and lunging for prey? (Arrow #4)

Use the diagram to generate your own research questions and hypotheses.

Goldbogen JA, Calambokidis J, Oleson E, Potvin J, Pyenson ND, Schorr G, Shadwick RE. Mechanics, hydrodynamics, and energetics of blue whale lunge feeding: efficiency dependence on krill density. J Exp Biol 214, 131–146, 2011.

9 Animal Respiration, Circulation, and Metabolism

9.1 Breathing and gas exchange in the tiger salamander

9.2 Animals obtain oxygen from their environment

9.3 Respiratory systems exchange gases between the environment and the body

9.4 Circulatory systems transport oxygen to cells

9.5 Oxygen enters cells to meet an animal's energy demands

9.6 Animal cells use aerobic and anaerobic metabolism to produce ATP

9.1 Breathing and Gas Exchange in the Tiger Salamander

After five months of winter, the melted snow and recent rains have made for soggy hiking trails. Along the trail, excess water collects in shallow pools or small ponds (**Fig. 9.1**). These temporary spring pools, called vernal pools, have no source of water other than rain and snowmelt and will often dry up and disappear before summer. Although they exist for only a short time, vernal pools play an essential role in the life cycle of the Eastern tiger salamander (*Ambystoma tigrinum*) (**Fig. 9.2**).

Adult tiger salamanders live most of their lives on land and usually only return to the water to breed. In the early spring, usually after a warm rain thaws the ground, adult tiger salamanders migrate to vernal pools to mate. A day or two after mating, females dive to the bottom of the pool to deposit a large mass of 20 to 80 eggs. Unlike many frogs that lay their eggs on the surface of water, salamanders attach the egg mass to sticks, stems, and leaves on the bottom of the pool (see **Fig. 9.2**). Tiger salamander eggs contain a large amount of yolk to supply nutrients and are covered by a jelly-like capsule composed of carbohydrates and proteins. The capsule protects the developing egg from injury, infection, and predation. After they are deposited, the eggs take up water, swell, and begin their development. Time is of the essence because the salamander eggs must grow and develop into adults before the vernal pool evaporates.

Salamander growth and development are fueled by the breakdown of nutrients in the yolk and oxygen from the surrounding water (**Fig. 9.3**); thus the rate of growth is closely linked to the rate at which oxygen diffuses into the eggs. Oxygen dissolved in the water diffuses across the permeable gel capsule and egg membranes and into the cells of the growing embryo. Within each cell, oxygen enters the mitochondria to produce a large amount of chemical energy, which drives cell division, cell differentiation, and the

Figure 9.1 Vernal pool. Salamanders lay their eggs in temporary wetland areas that do not support fish populations, which prey on frog and salamander eggs.

Michael P. Gadomski/Photo Researchers, Inc.

Figure 9.2 Life cycle of the tiger salamander. The three-month life cycle of the tiger salamander starts underwater with the fertilized egg and larval stages. It is completed on land with a mature adult stage. Adults may live 10 years or more.

© Cengage Learning 2014

building of new body structures. Tiger salamander eggs may contain green algae, which
will use light to produce additional oxygen
and sugars through photosynthesis. There is
evidence that eggs with algae develop at a
faster rate than eggs without algae. In addition to oxygen uptake, the developing salamander excretes carbon dioxide into the water. The living, growing egg mass is actively
exchanging gases with its aquatic habitat.

After about 20 days, an egg has used up
most of its yolk and has changed into a larva.
The larva hatches, emerging from its jelly-like
capsule, and swims freely in the water, where it
will spend the next 60 days. Salamander larvae are carnivores that feed on
small insects and crustaceans. Under crowded conditions, some larvae become cannibals and attack and eat their own kind.

With a larger body size and greater growth rate, larvae require a greater
amount of oxygen than when they were smaller eggs. Salamander larvae
are adapted to absorb oxygen dissolved in the water with three pairs of
external gills (see **Fig. 9.2**). These respiratory organs are composed of numerous branched filaments that give them a bushy appearance. The outer
tissue of these filaments is very thin and contains a rich supply of blood
vessels. As the larvae swim through the water, oxygenated water flows
over their gills, which increases the amount of oxygen that diffuses into
the blood. In addition to gills, larval tiger salamanders swim to the water
surface to breathe air in through their mouth. They contract body muscles
to pump fresh air in and out of their developing lungs. Using both their
gills and lungs, rapidly growing larvae are well adapted to meet their increasing demand for oxygen.

During their two-month larval period, tiger salamander larvae mature
and leave the water as adults. They have lost their larval structures such as
gills and flattened tail (see **Fig. 9.2**). Their organ systems have become modified for life on land. With the exception of one specialized group, salamanders develop a pair of simple lungs to breathe oxygen from the atmosphere.

Adult salamanders can also breathe oxygen directly through their
moist skin. The skin of amphibians is extremely thin and permeable to
gases that cross into an extensive network of blood vessels. Water from the
salamander body also moves across the skin to the environment, posing
the constant danger of dehydration. Thus salamanders need to limit the
time they are away from moist habitats to hunt and feed.

This chapter and Chapter 8 work together to present how animals have
evolved an amazing diversity of adaptations to acquire the food, water,
and oxygen they need to live. Chapter 8 traced the path of nutrients from
the environment into the animal body, where they are digested and absorbed into circulation. In this chapter we will follow a similar strategy by
tracing the path of oxygen from an animal's environment all the
way through the respiratory and circulatory systems to its ultimate
destination—the mitochondria in each body cell—where the oxygen plays
a key role in converting nutrients to energy. The final section of the chapter takes you to the molecular level to understand the essential role of oxygen in energy production. Together, these chapters present powerful biological concepts that you can apply to your own life to maintain a healthy
diet and lifestyle. Understanding these concepts can also bring deeper insights into the adaptations of animals living in habitats around you.

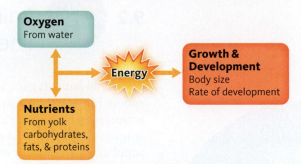

**Figure 9.3 Nutrients and oxygen fuel growth and
development.** Nutrients contained in the yolk and
oxygen from the water fuel the growth and development of the salamander eggs.
© Cengage Learning 2014

9.2 Animals Obtain Oxygen from Their Environment

Terrestrial animals such as the adult salamander are immersed in an atmosphere filled with oxygen. In fact, about one out of five molecules in the atmosphere is oxygen. Compared to animals living in terrestrial habitats, animals such as fish that live in water must cope with an environment that supplies much lower amounts of oxygen. Depending on physical factors, water contains about 20 to 40 times less oxygen by volume than air. Also oxygen is not very soluble in water, and gases diffuse 10,000 times more slowly in water than in air. To obtain an amount of oxygen equal to that obtained by air-breathing animals, fish must move much larger amounts of water over their respiratory surfaces, or gills. Water is a more dense medium than air and thus requires more energy to move it in the attempt to obtain oxygen. Together these chemical and physical constraints make meeting their oxygen needs a challenge—unless the creature is a water spider that literally brings the atmosphere underwater.

Interestingly, a water spider physically surrounds its body in a large air bubble (**Fig. 9.4**). The spider builds a unique air bell out of silk, collects air from the surface, and transports it underwater. The bells are usually tethered to water plants or stones. Within its air bell, it breathes a rich supply of oxygen that allows the spider to eat and digest its prey, have sex, and raise offspring underwater. It has brought one oxygen environment into another.

In the section ahead we compare the oxygen in the environment of animals living on land, in underground burrows, and in the water. Animals, like all organisms, live under the laws of their physical environment, and this environment includes the behavior and movement of gases. We begin at the source of oxygen in the environment—photosynthetic organisms or, in other words, plants, algae, and cyanobacteria.

Figure 9.4 Water spider. Water spiders are the only spiders that spend their entire life underwater. They weave and build underwater air bells to keep an air bubble around their bodies.

© WILDLIFE GmbH/Alamy

Photosynthesis is the source of oxygen in the environment

The early atmosphere of Earth contained only a trace of oxygen. Oxygen began to accumulate in the atmosphere with the rise of the first photosynthetic organisms, the cyanobacteria. Although a more precise timeline is still being investigated, fossil evidence shows that cyanobacteria arose about 2.7 billion years ago and became so abundant by 2.4 billion years ago that oxygen accumulated in significant amounts in Earth's atmosphere. The increase in oxygen had a profound effect on the direction of life in two ways. First, it formed an atmospheric layer of ozone (O_3) that shielded early living organisms from harmful ultraviolet rays from the sun. Second, it provided the oxygen gas (O_2) for new life forms, like animals, to evolve.

Today, all animals depend on photosynthetic organisms such as plants, algae, and cyanobacteria to produce oxygen. In the chemical reactions of photosynthesis, water, carbon dioxide, and light are used to produce sugars and oxygen. In addition to *producing* oxygen, plants also *use* oxygen to generate energy in their mitochondria. In fact, plant roots depend on a supply of oxygen from the soil. Since plants and other photosynthetic organisms produce much more oxygen than they use, it accumulates in the atmosphere or the surrounding water (**Fig. 9.5**).

Figure 9.5 Oxygen gas bubbles. Through the process of photosynthesis, algae produce oxygen gas bubbles in aquatic habitats.

Stéphane Bidouze/Shutterstock.com

TABLE 9.1 The Gas Composition of the Atmosphere and an Underground Burrow

Gas	Percent in Atmosphere (Dry Air)	Percent in Underground Burrow*
Nitrogen (N$_2$)	78.09	78.09
Oxygen (O$_2$)	20.95	12–18
Other gases	0.93	0.93
Carbon dioxide (CO$_2$)	0.03	2–7
Total	**100.00**	**100.00**

*Highly Variable with Depth and Soil Properties

© Cengage Learning 2014

Terrestrial animals obtain oxygen directly from the atmosphere

Animals that live on land breathe in air, which is composed almost entirely of two gases—nitrogen and oxygen. Other gases such as carbon dioxide, helium, and argon are found in very small quantities in the atmosphere (**Table 9.1**). Nitrogen is the most abundant gas in the atmosphere and does not chemically react in our cells—it just moves in and out of our body with each breath. On the other hand, oxygen enters into the chemical reactions of each body cell to convert the energy in nutrients into a form of energy that can be used to accomplish life activities.

Animals like the mountain goat that live at high altitudes must obtain oxygen from the "thin" air they breathe (**Fig. 9.6**). How is living at high altitude different from living at sea level, and how does it affect the mountain goat? The oxygen molecules in "thin" air are more dispersed because the *atmospheric pressure* is lower. At 12,000 feet, there is simply less air pushing down on the mountain goat than on animals living at sea level. Lower atmospheric pressure means that there is less oxygen in a given volume of air. Each breath the mountain goat takes delivers only about 60 percent of the oxygen in a breath taken at sea level. It is unclear exactly how the mountain goat is adapted to this decreased oxygen availability; its adaptations may include a faster breathing rate, more efficient extraction of oxygen from the air, or changes in its heart and circulation to maintain adequate oxygen delivery to its tissues.

Animals that live in underground tunnels, like the pocket gopher, face a more variable atmosphere than those living on the surface (**Fig. 9.7**). This terrestrial habitat has limited airflow, which limits the exchange of gases with the atmosphere. Also, soil organisms living adjacent to its burrows and tunnels, as well as the animal itself, use oxygen and produce carbon dioxide. Together these two factors cause the underground atmosphere to have less oxygen (12–18 percent) and more carbon dioxide (2–7 percent) than the air at the surface. Like the mountain goat, the pocket gopher is adapted to an atmosphere low in oxygen through its respiratory and circulatory systems.

Terrestrial animals, whether they live at high altitude, on the surface, or in underground burrows, acquire oxygen by diffusion. **Diffusion** is the physical process by which oxygen molecules move from an

Figure 9.6 Mountain goat in its high-altitude habitat. Mountain goats are adapted to extracting oxygen from air under a lower atmospheric pressure than animals living at sea level.

Figure 9.7 Pocket gopher in its underground habitat. Animals that live in burrows underground face lower oxygen and higher carbon dioxide levels than animals that live aboveground.

Diffusion—the movement of chemical substance from an area of high concentration (or pressure) to an area of lower concentration (or pressure).

Figure 9.8 Diffusion of oxygen and carbon dioxide between the atmosphere and the body's cells. Oxygen moves from an area of high concentration in the atmosphere to lower and lower concentrations in the lungs, blood, and body cells. Carbon dioxide moves in the opposite direction, starting in high concentrations in the mitochondria, where it is produced by aerobic metabolism.
© Cengage Learning 2014

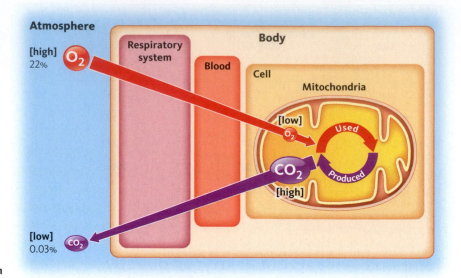

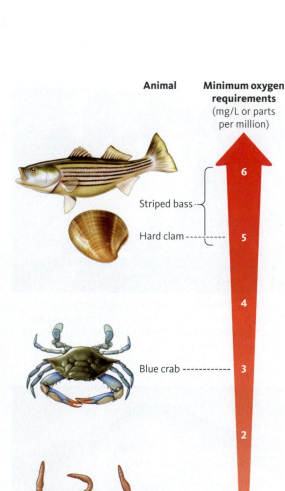

Figure 9.9 Dissolved oxygen requirements for marine organisms. The minimum dissolved oxygen requirements for key organisms living in Chesapeake Bay off the coast of Maryland. Active swimming organisms such as striped bass generally require higher levels of dissolved oxygen than blue crabs and marine worms.
© Cengage Learning 2014

Dissolved oxygen—the amount of oxygen dissolved in water; the oxygen content of water.

area of high concentration to an area of lower concentration, or down its concentration gradient. This movement occurs because all molecules are in a constant state of motion and always colliding with other molecules. These collisions cause the molecules to spread out into the space around them, or to areas of lower concentration. Biologically, oxygen molecules move from the atmosphere into the body and then into the blood and eventually cross the cell membrane and into the mitochondria of body cells (**Fig. 9.8**). Carbon dioxide also follows its *specific* concentration gradient in exactly the opposite direction: its production within the mitochondria creates the highest concentration, the next highest in the blood, then the lungs, and finally the lowest concentration is in the atmosphere, which contains only 0.03% CO_2.

Aquatic organisms depend on dissolved oxygen in water

An important part of an aquatic animal's ecological niche is the oxygen content, or dissolved oxygen, of the water in which it lives. Fish and other animals must move to locations in water that meet their requirements, or they do not survive. In fact, the dissolved oxygen is considered an indicator of the health of a body of water and its ability to sustain aquatic life. High-quality water contains higher levels of dissolved oxygen and is able to meet the oxygen requirements of a greater number of fish, clams, crabs, and worms that live there (**Fig. 9.9**).

Several biological and physical factors determine the dissolved oxygen level in water. Biological factors include inputs from photosynthetic sources and outputs from organisms consuming oxygen such as microscopic decomposers. Physically, the temperature of the water, the amount of salts, and the physical turbulence of the water influence the amount of dissolved oxygen available to animals. Let's examine each of these factors as they affect oxygen and a specific fish species, Coho salmon.

Coho salmon, like many active, fast-swimming fish species, require water with a high dissolved oxygen level for their eggs to hatch and for their migration. During their long-distance migration through fast-moving streams, slow-running pools, and the salt waters of the Pacific Ocean, they have to adjust to differences in the dissolved oxygen in the waters in which they swim.

Salmon begin life in a shallow pool of water in a freshwater stream in the Pacific Northwest. Here, as in all waters they inhabit, the primary sources of oxygen include photosynthesis of aquatic plants and diffusion of atmospheric oxygen across the air–water interface (**Fig. 9.10**). The two arrows in **Figure 9.10** show that oxygen moves in both directions. However, when the water temperature warms, more oxygen molecules gain the energy needed to escape the water and diffuse into the atmosphere. The result is a reduced amount of dissolved oxygen. Warmer waters also increase the activity of decomposer microbes in the water that consume oxygen. When these microbes actively decompose organic materials in the water, they can significantly deplete the oxygen in the water. The result is that during summer, salmon must adjust to life with less oxygen.

When salmon reach the ocean, they must adjust to a saltwater environment and a reduced level of dissolved oxygen. The graph in **Figure 9.11** shows that at all temperatures, salt water holds less oxygen than freshwater because the salt in the water physically takes up space that would be able to hold oxygen molecules.

After about three years of feeding in the cold ocean waters near Alaska, salmon return to reproduce in the same freshwater streams where they were hatched. Along the way they have to swim upstream against the current and jump waterfalls and other obstacles in fast-moving streams and rivers (**Fig. 9.12**). Running water has more oxygen than standing water in lakes, ponds, and swamps. As water moves over rocks and waterfalls, the churning, mixing, and splashing actions continuously bring more oxygen into the water from the atmosphere. For the same reason, the standing water in fish tanks and aquariums is often outfitted with aerators to maintain a high level of dissolved oxygen.

The Coho salmon illustrates how some aquatic animals adjust to the differences in dissolved oxygen that occur in different bodies of water at different times of the year. The most dissolved oxygen occurs in cold, running freshwater streams. How animals like salmon acquire oxygen under these changing conditions is presented in the next section.

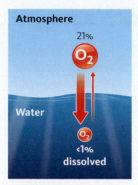

Figure 9.10 Movement of oxygen between water and the atmosphere. At the surface of a body of water, oxygen is constantly moving between the water and the atmosphere. Diffusion brings oxygen into the water (downward arrow) at a greater rate than it leaves the water (upward arrow). When the water warms, the amount of oxygen leaving the water (upward arrow) increases and therefore reduces the water's dissolved oxygen content.
© Cengage Learning 2014

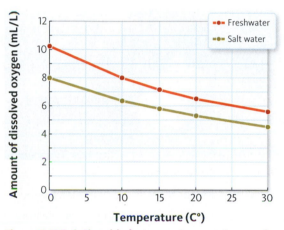

Figure 9.11 Relationship between temperature and dissolved oxygen. As the temperature increases, the amount of oxygen water can hold decreases. Freshwater holds more oxygen than salt water at all temperatures.
© Cengage Learning 2014

Figure 9.12 Moving water holds more oxygen than standing water. The turbulent water of a fast-moving stream is a challenge to swim against, but it holds much more oxygen than standing water.

- Photosynthetic organisms are the source of oxygen in the atmosphere, which, in turn, is the main source of oxygen dissolved in bodies of water. Aquatic photosynthetic organisms also contribute to the dissolved oxygen content of water.

- Gas molecules move passively down their concentration gradient by diffusion.

- Depending on physical factors, water contains about 20 to 40 times less oxygen by volume than air.

- Aquatic animals are adapted to different levels of dissolved oxygen.

- Inputs from sources and outputs by consuming organisms, as well as the temperature, amount of salts, and turbulence of the water, determine the level of dissolved oxygen.

1. Give two reasons why water contains so much less oxygen by volume than air.

2. What are the two sources of oxygen in a body of water?

3. Compare the percentage of oxygen and carbon dioxide in an animal's habitat at sea level and on a mountaintop. How do the atmospheric pressures compare?

4. Sea turtles leave the ocean, crawl above the high-tide line, dig a nest chamber, and bury a large number (50–100) of eggs (**Fig. 9.13**). The buried eggs will develop over the next 60 days, where they exchange gases with the air in the nest. (a) Which physical process moves gases into and out of the eggs? (b) How would the depth of the nest affect the rate of gas exchange with the aboveground atmosphere? (c) How will the percentage of oxygen and carbon dioxide in the nest change just after the eggs hatch and the turtles begin to actively dig their way to the surface? Explain your reasoning.

Figure 9.13 Sea turtle eggs.

5. In many parts of the country, the summer months are marked by higher temperatures and lower amounts of rainfall. How do these two environmental factors affect the flow of streams and the availability of oxygen for aquatic animals such as insects and other invertebrates?

9.3 Respiratory Systems Exchange Gases between the Environment and the Body

Humans have a large surface area over which to exchange gases with their environment. The respiratory system actively moves air in and out, or *ventilates*, the lungs and then exchanges gases with cells in the blood. Blood then transports gases to each body cell. The two processes work together to deliver oxygen at a rate fast enough to keep our tissues working. The most active tissues, such as nerves and muscles, have the greatest need for oxygen. The human brain, for example, demands seven times more oxygen than any other tissue. Because of this high demand, the brain is vulnerable to low levels of oxygen: a sudden or total lack of oxygen can cause unconsciousness within 5 to 10 seconds. It is also the reason that people get light-headed and feel dizzy at high altitude.

Respiratory system—the animal organ system that supplies body cells with oxygen and excretes carbon dioxide from the body; it involves actively moving air in and out of the lungs and the exchange of gases with cells in the blood.

a.

b.

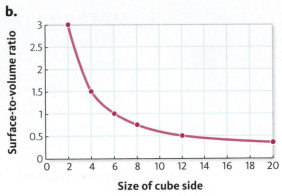

Figure 9.14 Flatworm surface-to-volume ratio.
(a) The flatworm's small body size and ribbonlike shape allow it to acquire oxygen directly from the water without specialized respiratory organs. (b) If you imagine an animal as a cube, the larger the animal, the lower the surface-to-volume ratio. A low surface-to-volume ratio limits the animal's ability to supply its cells with oxygen directly by diffusion.
© Cengage Learning 2014

Not all animals have a complex respiratory system. Simple animals like flatworms have a body shape that brings most of their cells in direct contact with the oxygen they need (**Fig. 9.14**). Flatworms can't grow too large, however, because large bodies have a much lower surface area to obtain oxygen *relative* to their size. Oxygen diffuses across the animal's surface and passes into the cells occupying its volume. The surface-to-volume ratio relates how much surface area is available to supply a given volume of cells with oxygen (and nutrients). If you imagine an animal as a simple cube having six sides, the graph in **Figure 9.14b** shows that the larger the animal, the lower its surface-to-volume ratio. In other words, larger animals have larger oxygen demands but a smaller surface area with which to meet those demands. In addition to gas exchange, other biological processes such as nutrient uptake and heat exchange are related to surfaces and surface-to-volume ratio.

More complex animals, like vertebrates, are too large for all of their cells to directly interact with the environment. Diffusion would be much too slow to meet each cell's needs. Animals have evolved a variety of specialized respiratory structures such as gills, lungs, skin, and air tubes to capture and deliver large amounts of oxygen to their cells. In fact, many animals have evolved more than one system. We've already seen that the adult tiger salamander exchanges gases through both lungs and skin, and some fish exchange gases using their lungs, skin, and gills. In the section ahead we will show how different respiratory systems and organs efficiently increase the ventilation and exchange of gases between the body and the environment.

Respiratory organs are adaptations for acquiring large amounts of oxygen

Humans require an efficient respiratory system that delivers large amounts of oxygen in a short period of time. The efficiency of a particular respiratory system is determined by how well it increases the movement of gases by diffusion and the size of its surface area. Gases in high concentration also exert high pressure in a mixture of molecules; thus gases move from an area of high pressure or concentration to areas of lower pressure or concentration. The steepness of the concentration gradient is one factor that determines the amount of gas that an animal exchanges with its environment. Five other factors influence the amount of gas exchange: (1) the density of the medium through which the gas diffuses—air, water, or a solid; (2) the

Surface-to-volume ratio—the ratio, or relationship, that determines how much surface area is available to supply a given volume of cells with oxygen and nutrients.

TABLE 9.2 Factors Influencing Gas Exchange

Factor		Gas Exchange	
		Increased	Decreased
Medium	Density	Gas = less dense	Liquid and solid = more dense
	Temperature	Warmer	Colder
Surface Area		Large	Small
Concentration or Pressure Gradient		Steep gradient = Large difference in concentration/pressure	Shallow gradient = Small difference in concentration/pressure
Distance Gases Travel		Short	Long
Contact Time		Long	Short

© Cengage Learning 2014

temperature of the gas molecules; (3) the surface area through which the gases pass; (4) the distance the gas molecules must travel; and (5) the time that gases move across the surface of an animal's respiratory structures into a moving fluid medium such as blood and body fluids. The longer these two moving fluids are in contact with one another, the greater the amount of gas exchange. Many respiratory systems are adapted to increase and extend the contact time by slowing down the flow of air or water when it meets blood.

Table 9.2 summarizes how these six factors influence gas exchange. If you read down the increased gas exchange column of the table, you'll see that the greatest amounts of exchange occur in *warm* air down *steep* pressure gradients across a *large* surface area for a short distance. As you will see, respiratory systems and organs increase their ability to acquire oxygen through modifying these six factors. Understanding this dynamic process is an important key to understanding how respiratory systems efficiently meet an animal's gas exchange requirements.

Insects and spiders exchange gases through a tubular network

Insects are a very diverse group of animals that have evolved a variety of ways to exchange gases with their environment. The majority of insects exchange gases directly with the atmosphere through a network of tubes called the *tracheal system.* The largest tubes, or trachae, connect directly to the atmosphere through valvelike openings called *spiracles* on the surface of their body. Tracheae branch into a series of smaller and smaller tubes that deliver oxygen within a very close distance to each body cell (**Fig. 9.15**). Many insects increase air flow through the tracheal system through body movements. The spiracles regulate air flow by opening and closing depending on the activity and the insect's oxygen requirements.

Insect tracheal systems take advantage of several factors that increase gas exchange. First, insects do not exchange gases with body fluids as vertebrates do; instead they use the atmospheric gas *medium,* where gases diffuse thousands of times faster than in fluid. Although insects have body fluids and a circulatory system, this system plays only a small role in gas exchange. Second, the tracheal system provides an enormous *surface area* that is permeable to oxygen and carbon dioxide. Finally, gases are delivered close to each body cell and thus have to diffuse only a short distance to their cellular destination. Together these factors allow insects to quickly

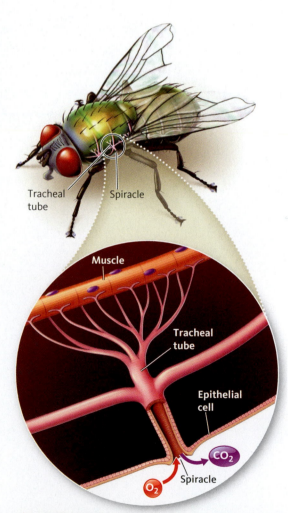

Tracheal tube
Spiracle

Muscle

Tracheal tube

Epithelial cell

CO_2

Spiracle

O_2

Figure 9.15 Tracheal system of insects. Insects obtain their oxygen directly from the atmosphere through a branched network of tubes called the tracheal system.
© Cengage Learning 2014

increase their oxygen delivery when they have a burst of activity to fly or run to escape predation. Blowflies, for example, can increase their oxygen delivery rate by 30 times, compared to vertebrates, which can only increase this rate by about 12 times.

Gills are highly efficient respiratory organs in water

Gills are structures that function to increase the surface area in contact between water and the body fluids that transport gases to and from body tissues. Many groups of animals that live in aquatic habitats, including marine worms, clams, lobsters, crabs, and fish, use gills to exchange gases. The tiger salamander larva uses gills that extend outward into the water. In addition to movement, the thousands of small tube feet of a starfish shown in **Figure 9.16** function as gills.

Let's look at the internal gills of a water-breathing fish to understand how these organs work. Always keep in mind that water, as a medium, contains low concentrations of oxygen and has a much slower diffusion rate than air. This is one reason that fish may spend as much as 20 percent of their energy extracting oxygen from water. Thus it is essential for survival that gills be very efficient to meet the oxygen demands of these active aquatic animals.

The basic structure of internal gills is remarkably similar in all fishes and sharks. Fish have four pairs of gills that open through five pairs of gill slits. A flap covers the gills of bony fishes, but sharks have no flap, making the slits visible (**Fig. 9.17**). Zoom in on an individual gill and you'll see that it has dozens of projections, called *gill filaments,* which give it a comblike appearance (**Fig. 9.18**). Blood flows through the filament where it absorbs oxygen from the water. A thin lining only a single cell thick covers each filament. The thinness of the filament means that gases travel only a *short distance* to enter or leave the fish's blood vessels. The filaments also increase the *surface area* of the gill many times. For example, a sea bass that weighs 44 pounds has a gill surface area of 100 square feet. Gill filaments from adjacent gills touch each other to form a fine mesh, which slows the

Figure 9.16 Breathing system of starfish. The starfish breathes through its tiny tube feet on the underside of its five arms. They function as gills, as well as enabling movement.

Figure 9.17 Shark gills are visible. Unlike other fishes, sharks and rays do not have a flap of tissue covering their gills. They ventilate their gills through constant motion with their mouth open.

Figure 9.18 Fish gill structure. (a) A fish's gills are a bright red color due to their thin covering and rich supply of blood vessels. (b) Each individual gill has a large surface area due to the ridges, or filaments, that give it a comblike appearance. Blood flowing through the filaments absorbs oxygen from the water.

Gills—respiratory structures used by many groups of aquatic animals that function to increase the surface area in contact between water and the body fluids for gas exchange.

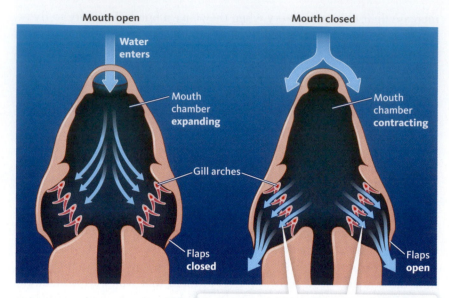

Figure 9.19 Fish gills are ventilated in two steps. Looking down on the head of a fish shows that it ventilates its gills in two steps. First, it draws water in through its mouth and closes its gill flaps. Second, it closes its mouth and opens its gill flaps. This action draws oxygenated water over its gills.
© Cengage Learning 2014

Mouth open · Water enters · Mouth chamber **expanding** · Gill arches · Flaps **closed**

Mouth closed · Mouth chamber **contracting** · Flaps **open**

As water moves over the gill, oxygen diffuses into the bloodstream. The result is a decreasing amount of oxygen in the water along its path over the gill.

Countercurrent exchange

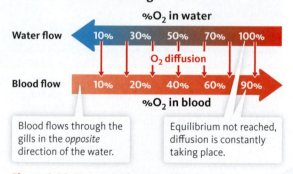

%O$_2$ in water

Water flow · 10% · 30% · 50% · 70% · 100%

O$_2$ diffusion

Blood flow · 10% · 20% · 40% · 60% · 90%

%O$_2$ in blood

Blood flows through the gills in the *opposite* direction of the water.

Equilibrium not reached, diffusion is constantly taking place.

Figure 9.20 Countercurrent exchange. Water moves over fish gills in one direction, and blood flows through the gills in the opposite direction. This results in a continual steep concentration gradient that moves oxygen from the water into the blood.
© Cengage Learning 2014

Lungs—inflatable internal respiratory structures used by vertebrates for gas exchange.

movement of water and increases the *contact time* for gas exchange. Refer back to **Table 9.2** and notice that the efficiency of gills is high because it increases the diffusion rate by altering three factors—surface area, distance, and contact time.

Gills also increase the *concentration gradient*. Water enters the mouth and passes over the gills and out through the gill slit (**Fig. 9.19**). As water moves over the gill, oxygen diffuses into the bloodstream. The result is a decreasing amount of oxygen in the water along its path over the gill. Fish have a remarkable adaptation called *countercurrent exchange,* in which blood flows through the gills in the *opposite* direction of the water. **Figure 9.20** shows how, with a countercurrent blood flow, at almost every point along the path, the higher percentage, or the greater concentration, of oxygen occurs in the water relative to the blood. This feature ensures that oxygen diffuses down its concentration gradient from the water into the blood along the length of the gill.

Fish use different methods for ventilation; for example, sharks and many fast-moving fish hold their mouths open as they swim, allowing large amounts of oxygen-containing water to move across their gills. However, many slow-moving or bottom-dwelling fish actively ventilate, or pump water across their gills, by gulping water. If you observe a fish in an aquarium, notice that it will continually alternate the opening and closing of its mouth and gill flaps. As the mouth is opened and the flaps are closed, water is drawn into the mouth and across the gills (see **Fig. 9.19a**). When the mouth is closed and the flaps are opened, water is forced across the gills and exits out the gill slits (see **Fig. 9.19b**). This two-step cycle draws a new supply of water over the gills even while the fish is stationary. Due to the gills' large surface area, short distance, steep concentration gradient, and long contact time, fish can extract up to 80 percent of the dissolved oxygen in water.

Lungs are internal organs for gas exchange in air

Most air-breathing vertebrates have inflatable internal respiratory structures, or **lungs**, for gas exchange. Lungs are remarkable adapta-

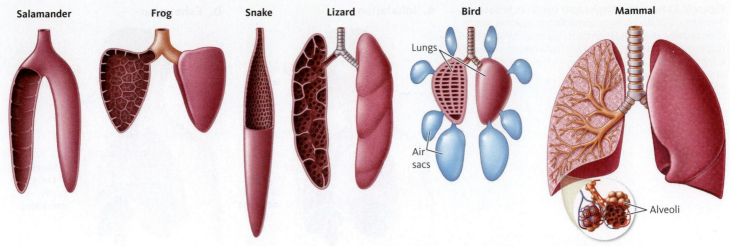

Salamander **Frog** **Snake** **Lizard** **Bird** **Mammal**

Lungs

Air
sacs

Alveoli

Figure 9.21 Vertebrate lungs. The lung structure of vertebrates varies from simple saclike structures in the salamander to extensive networks of tubes and sacs in mammals.
© Cengage Learning 2014

tions that expand an animal's range into drier habitats. The simplest lungs are those found in lungfishes that live in swamps and other aquatic habitats low on oxygen. In amphibians, lungs are more efficient structures for gas exchange than skin because they have a thinner epithelial lining, or a shorter distance to a rich supply of blood vessels, and a larger surface area.

Figure 9.21 shows that the structure of lungs varies greatly among vertebrates. Amphibian lungs are simple baglike structures, whereas frogs have spongelike lungs with a number of air pockets. Lizards have lungs that are further subdivided into sections, greatly increasing the surface area. Snakes have only a single lung to accommodate their slim body shape. Birds have a very complex respiratory system consisting of lungs complemented by an extensive system of small air sacs that even penetrate into bones.

Like birds, mammals have large oxygen requirements; their lungs are larger and divided into lobes with millions of tiny air sacs called *alveoli* (see **Fig. 9.21**). If you calculate the surface area of all of these air sacs in a pair of human lungs, it would be an area about 36 by 36 feet, or one side of a tennis court! When air enters the lungs of a bird or a mammal, with its higher body temperature, the air becomes *warmed* and *moistened,* increasing the diffusion rate of gases (see **Table 9.2**) and preventing the alveoli from drying out. However, there is a disadvantage to moistening the air—when an animal exhales it loses water. You can see this water loss on a cold day when your exhaled breath is visible.

The lungs of mammals and most other vertebrates are located in the chest area. The chest cavity is a closed subdivision of the trunk, sealed by membranes so that the only entrance and exit for gases is through the windpipe, or *trachea,* mouth, and nose. Mammals ventilate their lungs through the contraction and relaxation of a muscle called the **diaphragm** (**Fig. 9.22**). This muscle, along with muscles of the rib cage and abdomen, changes the *volume* of the chest cavity, which in turn changes the *pressure* on the lungs. When the volume increases, the pressure decreases, and vice versa.

Let's use the volume–pressure relationship we just introduced to explain how a mammal breathes. Take a deep breath. When you do this, the diaphragm muscle beneath your lungs contracts and moves downward,

Diaphragm—a large sheet of muscle used by mammals to ventilate their lungs.

Figure 9.22 How the diaphragm controls breathing. The muscular diaphragm controls the volume of the chest cavity and therefore controls breathing: (a) the diaphragm contracts and moves downward when air is inhaled, and (b) the diaphragm relaxes and moves upward when air is exhaled.
© Cengage Learning 2014

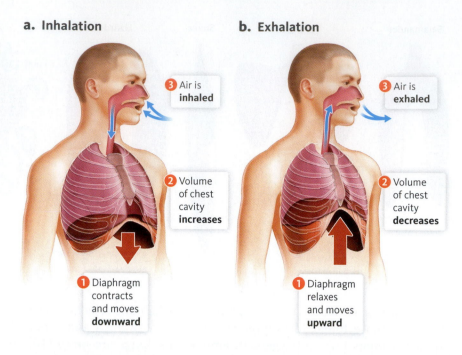

a. Inhalation

3 Air is **inhaled**

2 Volume of chest cavity **increases**

1 Diaphragm contracts and moves **downward**

b. Exhalation

3 Air is **exhaled**

2 Volume of chest cavity **decreases**

1 Diaphragm relaxes and moves **upward**

TechWizard/Shutterstock.com

Figure 9.23 Exhalation of a whale. A whale's blowhole is actually its nostrils, and when it exhales, it blows carbon dioxide and water vapor high into the air.

TABLE 9.3 Atmospheric Air vs. Expired Air

Gas	Atmospheric Air (%)	Expired Air (%)
O_2	20.95	16.4
N_2	78.09	79.5
CO_2	0.03	4.1

© Cengage Learning 2014

increasing the volume of the chest cavity (see **Fig. 9.22a**). This increased volume reduces the pressure on the lungs within the chest cavity, causing air to be inhaled and inflating the alveoli. When you exhale, the diaphragm muscle relaxes and moves upward, decreasing the volume of the chest cavity and increasing the pressure on the lungs as they push air out (see **Fig. 9.22b**). This two-step process ventilates the lungs by creating a gradient of atmospheric pressure between the inside and the outside of the body. The human lung is ventilated about 25,000 times a day with about 20,000 liters (5,280 gallons) of air!

Whales and other aquatic mammals have a strong rib cage and diaphragm muscle to be able to create large pressures in their enormous chest cavity against the water pressures surrounding their body. This pressure is seen when a whale surfaces and exhales, or blows water vapor, carbon dioxide, and other gases through its blowhole, often up to 25 feet into the air (**Fig. 9.23**).

Despite the large, warm, and moist surface area, there is a disadvantage to the way that mammalian respiratory systems are ventilated. For exchange to occur, gases have to be moved all the way to the ends of a branching network of tubes deep inside the lungs. This arrangement means that that after an exhalation, some air containing higher amounts of carbon dioxide is left in the lungs, which is then mixed with fresh inhaled air on the next breath. This mixing of "old air" with "fresh air" reduces the overall efficiency of the system. In fact, it takes about six normal breaths to totally replenish your lungs and airways with a fresh supply of air. The mixing of "old" and "fresh" air means that, contrary to popular belief, you actually exhale four times more *oxygen* than carbon dioxide (**Table 9.3**).

Birds are active animals that are adapted with a highly efficient respiratory system. They have a set of anterior air sacs toward the front of their bodies and another set of posterior sacs near the tail, with the lungs located between them (**Fig. 9.24a**). Some smaller air sacs are found in the bird's bones. All gas exchange occurs in the lungs, which have an intricate network of tubes and blood vessels. When a bird inhales, fresh air is drawn in through the mouth to inflate the lungs and the posterior air sacs (**Fig. 9.24b**). Gases are exchanged between the lungs and blood, and air saturat-

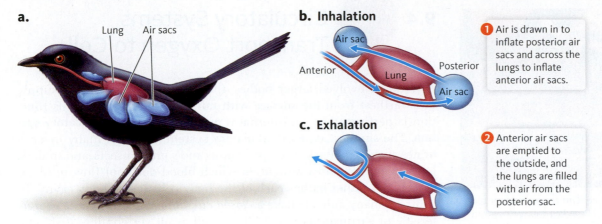

a.

Lung Air sacs

b. Inhalation

Air sac

Anterior Lung Posterior

Air sac

❶ Air is drawn in to inflate posterior air sacs and across the lungs to inflate anterior air sacs.

c. Exhalation

❷ Anterior air sacs are emptied to the outside, and the lungs are filled with air from the posterior sac.

ed with CO_2 then enters the anterior sacs. When the bird exhales, the anterior sac is emptied of its CO_2, the lungs are filled with fresh air from the posterior sacs, and the cycle begins again (**Fig. 9.24c**). Fresh, oxygen-rich air flows in a single direction through the lungs, which means that gases are exchanged during *both* inhalation and exhalation. Unlike the mammal's system, there is no mixing of "fresh air" and "old air"; the bird's system is much more efficient in delivering fresh, oxygen-rich air into the blood.

Animals have evolved a great diversity of highly efficient respiratory organs that increase the exchange of gases with their habitat. Lungs, gills, trachea, and the skin all function to meet the organism's oxygen requirements in a wide range of habitats. The next step is the transport of oxygen throughout the animal's body to reach every cell.

Figure 9.24 Birds have a highly efficient respiratory system. The respiratory system of a bird includes a pair of lungs and several air sacs that move air through the lungs. (a) A bird's respiratory system is a flow-through system that moves air across the lung surface when the bird inhales and exhales. (b) During breathing. Step 1—inhalation: air is drawn in to inflate posterior air sacs and across the lungs into the anterior air sacs. (c) Step 2—exhalation: the anterior air sacs are emptied to the outside, and the lungs are filled with air from the posterior sac. The cycle repeats itself.
© Cengage Learning 2014

9.3 check + apply YOUR UNDERSTANDING

- The surface-to-volume ratio of an animal relates how much surface is available to supply a given volume of cells with oxygen (and nutrients). The larger the animal, the lower its surface-to-volume ratio.

- Respiratory systems are adaptations to increase the amount and rate of gas diffusion to reach every cell in the body.

- Respiratory structures create steep pressure gradients, large surface areas, and short distances to blood or cells.

Figure 9.25 Slimy sculpin.

Nature's Images/Photo Researchers, Inc.

1. Why don't simple animals such as flatworms and sponges have respiratory systems?

2. Describe how each of the following factors influences the rate of diffusion: (1) temperature, (2) concentration gradient, and (3) the medium (air vs. water) in which gases travel.

3. On a whale watch expedition, you spot a humpback whale surfacing and blowing its gases out its blowhole. What gases compose the air that comes out of a whale's blowhole? Is its diaphragm contracted or relaxed? Use the gas pressure–volume relationship to explain how a whale blows.

4. Match the reptile with its lung structure:

   (1) A lizard that runs to catch its prey.

   (2) An iguana that lives in trees and feeds on leaves.

   (a) has a smaller baglike lung.

   (b) has large multichambered lungs.

   Which of the two lizards has a stronger diaphragm? Explain your choices.

5. The slimy sculpin is a fish that feeds and lives on the bottom of freshwater lakes (**Fig. 9.25**). Predict whether it has a small or a large gill surface area. Justify your prediction.

Figure 9.26 Earthworms breathe through their skin. When oxygen enters, their closed circulatory system, which consists of blood vessels and five small "hearts," keeps blood moving.

9.4 Circulatory Systems Transport Oxygen to Cells

As animals evolved larger bodies, they also needed a way to supply cells farthest from the surface with nutrients and oxygen. This function is performed by an internal transport system, or circulatory system. The simplest type of circulatory system, found in many invertebrates such as mollusks (except octopus and squid), insects, and spiders, is an *open circulatory system,* in which blood does not flow in blood vessels but rather bathes each of the body cells in the body cavity. In contrast, many animals have a *closed circulatory system* that is characteristic of earthworms (**Fig. 9.26**), as well as all vertebrates—fish, amphibians, reptiles, and mammals.

It may surprise you that the simple earthworm has a closed circulatory system with the same three features as circulatory systems found in complex animals such as humans. First, they have blood with cells that contain oxygen-binding proteins. In earthworms and humans, oxygen is quickly bound to an iron-containing protein called hemoglobin, a red-pigmented compound that gives the earthworm its pink color. Second, they have a distribution network of blood vessels that delivers oxygen to body cells. Third, both animals' circulatory systems have a pumping mechanism that keeps blood flowing and circulating. Unlike humans, earthworms do not have a distinct heart; instead they have five sets of muscular arteries that function as a heart to propel blood forward.

In the section ahead, we trace the path of oxygen through a human's closed circulatory system. We begin deep within the lung at the surface where air meets blood and then pump it forward in a single direction to transport and distribute oxygen throughout the body.

Oxygen moves into blood and binds to hemoglobin in red blood cells

Each minute, the lungs in a resting human move about 1.5 to 2.5 gallons of air from the terminal air sacs, or alveoli, into small, thin-walled blood vessels called capillaries (**Fig. 9.27a**). The layer of tissue between the blood capillaries and the alveolus, or the *blood–gas barrier* (**Fig. 9.27b**), must be extremely thin for efficient gas exchange. Capillaries are so narrow that blood flow slows to a crawl; the smallest capillaries are only the width of a single red blood cell. This slow movement of blood increases the contact time for gases to diffuse across the blood–gas barrier into red blood cells. Within a red blood cell, oxygen rapidly is attached to a molecule of hemoglobin (**Fig. 9.27c**). Red blood cells then deliver oxygen to all body cells.

Each human red blood cell contains about 200 to 300 million molecules of hemoglobin, which gives the cell—and blood—its bright red color. Each hemoglobin molecule has binding sites to carry up to four molecules of oxygen (see **Fig. 9.27c**). The mature red blood cells of most mammals are round disks that lack a nucleus and mitochondria. Without these organelles, the red blood cells of mammals can carry more oxygen than those of other vertebrates. In all other vertebrates—fish, amphibians, reptiles, and birds—red cells have nuclei and are larger and oval-shaped.

Circulatory system—the animal organ system that transports and distributes oxygen to body cells and carries carbon dioxide away from body cells. Closed systems consist of a pump and a network of blood vessels that carry blood; open systems do not have blood vessels but rather bathe each of the body cells in the body cavity.

Hemoglobin—a red-pigmented compound found in red blood cells that delivers oxygen to body cells.

Capillaries—narrow, thin-walled blood vessels that serve as the sites for exchange between the blood, tissue fluid, and body cells.

Red blood cells—blood cells that pick up and transport oxygen to body cells.

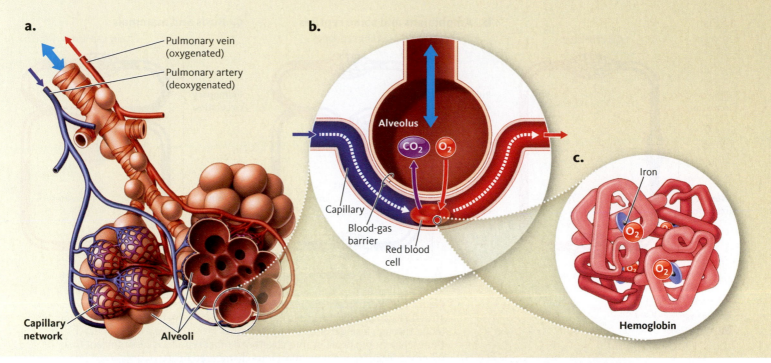

a.
- Pulmonary vein (oxygenated)
- Pulmonary artery (deoxygenated)

Capillary network

Alveoli

b.

Alveolus

CO_2 O_2

Capillary

Blood-gas barrier

Red blood cell

c.

Iron

O_2 O_2 O_2

Hemoglobin

Figure 9.27 Gas exchange at the blood-gas barrier.
(a) Oxygen and carbon dioxide are exchanged at the terminal ends of the respiratory system. (b) Gases are exchanged across a thin blood–gas barrier between the alveolus and the capillary blood vessel. (c) Oxygen then enters red blood cells where it loosely attaches to a molecule of hemoglobin.
© Cengage Learning 2014

Hemoglobin is the most common oxygen-transporting protein, used by vertebrates and some other groups of animals, as well as by some bacteria, fungi, and plants. The second most common oxygen-transporting protein, *hemocyanin,* is found in some mollusks such as the octopus and the squid, as well as lobsters, spiders, and snails. This oxygen-binding protein contains copper instead of iron, which causes the blood of these animals to appear blue in color.

The mammal circulatory system pumps blood through two circuits

The pattern of circulation and heart structure varies across vertebrate groups (**Fig. 9.28a**). Fish have a two-chambered heart that pumps blood that is low in oxygen, or *deoxygenated,* through the gills, where it picks up oxygen from the water and becomes *oxygenated.* From the gills, blood vessels then distribute oxygen throughout the body before returning to the heart in a single circuit. Amphibians and some reptiles have a three-chambered heart that pumps blood into two circuits—one to the lungs and one to the body. Notice in **Figure 9.28b** that this circulatory system mixes oxygenated and deoxygenated blood together, thus lowering the efficiency of the system to deliver large amounts of oxygen to tissues.

Birds and mammals have a very efficient system that uses a four-chambered heart, which completely separates oxygenated from deoxygenated blood in two circuits. The four-chambered heart consists of two receiving chambers called **atria** and two pumping chambers called **ventricles**. The heart functions as a double pump (**Fig. 9.28c**). The right side of the heart is a low-pressure pump that circulates blood to the

Atria (plural of *atrium*)—chambers of the heart that receive blood from veins.

Ventricles—chambers of the heart that pump blood into arteries.

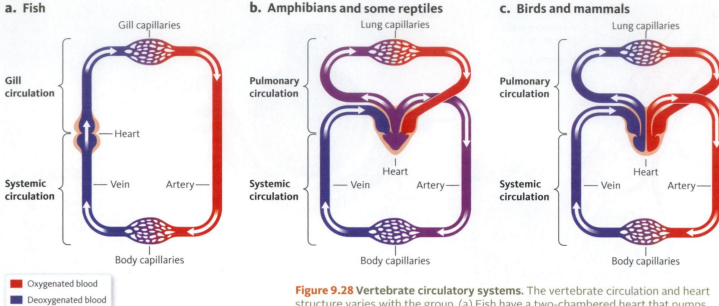

a. Fish

Gill capillaries

Gill circulation

Heart

Systemic circulation

Vein · Artery

Body capillaries

b. Amphibians and some reptiles

Lung capillaries

Pulmonary circulation

Heart

Systemic circulation

Vein · Artery

Body capillaries

c. Birds and mammals

Lung capillaries

Pulmonary circulation

Heart

Systemic circulation

Vein · Artery

Body capillaries

- Oxygenated blood
- Deoxygenated blood
- Mixed blood

Figure 9.28 Vertebrate circulatory systems. The vertebrate circulation and heart structure varies with the group. (a) Fish have a two-chambered heart that pumps blood through a single circuit, whereas other groups have a dual circuit. (b) The three-chambered heart in amphibians and most reptiles is less efficient than the (c) crocodilian, bird, and mammal four-chambered heart.
© Cengage Learning 2014

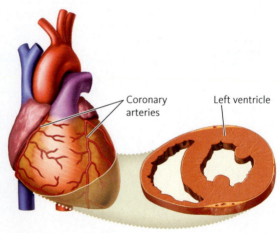

Coronary arteries · Left ventricle

Figure 9.29 The human heart. The human heart has a left ventricle with a much thicker wall of muscle to pump blood under high pressure throughout the body. This muscle is supplied with blood flowing through coronary arteries.
© Cengage Learning 2014

Arteries—strong, muscular blood vessels that carry blood away from the heart.

Tissue fluid—the fluid outside of blood vessels that surrounds the body cells and tissues. Tissue fluids exchange materials with both the cells and the blood.

Lymph—tissue fluid that flows through a system of vessels and collects in lymph nodes.

nearby lungs. The lower pressure is necessary to prevent the rupture of delicate lung capillaries. It consists of a receiving chamber, called the *right atrium,* and a pumping chamber, called the *right ventricle,* which pumps deoxygenated blood through the *pulmonary circuit.* The left side of the heart is a high-pressure pump that pushes blood to distant parts of the body through the *systemic circuit.* The left side consists of the left atrium and the left ventricle. If you look at a cross section of the two ventricles, you will see that the left side has a much thicker layer of cardiac muscle than the right side (**Fig. 9.29**). Each pump has a set of valves at each entry and exit point to prevent blood from flowing backward when the heart muscle contracts. For blood to keep flowing smoothly without a "traffic jam," both sides of the heart must pump *equal amounts* of blood into both circuits, but they must do this under *different* pressures.

Let's track a red blood cell containing its hemoglobin through the circulatory system starting at the left ventricle (see **Fig. 9.28c**). When the ventricle contracts, it pumps oxygenated blood into strong, muscular vessels called **arteries**, which carry blood away from the heart. Although most arteries are located deep in the body, when you feel your pulse, you are touching an artery near the surface. The pulse reflects the contraction and relaxation of your heartbeat. From the main arteries, blood flows through a branched network of arteries that gradually become narrower and narrower until they reach a network of capillaries (**Fig. 9.30**). These narrow arteries are wrapped in smooth muscle and act as a control point for the flow of blood into capillaries, diverting blood to places that need it most. The force of the flowing blood pushes water, nutrients, and oxygen out of the capillaries into the **tissue fluid** that surrounds the tissue cells. Most of the water and other substances flow back into the capillaries. The remaining fluid, called **lymph**, flows through a system of vessels and lymph nodes, and ultimately returns to the blood (see **Fig. 9.30**).

The carbon dioxide and other waste products made by cells move into the capillaries, and now the blood is deoxygenated. Blood then flows into more flexible blood vessels called **veins** for its return trip to the heart. The walls of veins are thinner than those of arteries, and they carry blood under low pressure. This is why they do not pulsate like arteries. Veins have small valves to prevent blood from pooling in the legs and feet. As you walk or stretch your legs, muscles in the legs contract and push blood upward to the heart. This is one reason that sitting for long periods of time can reduce blood flow through leg veins and cause blood clots. All the veins of the body carry blood back to the right atrium and then the right ventricle.

From the right ventricle, deoxygenated blood is pumped into an *artery* that brings blood to the lungs, where it becomes oxygenated once again. Unlike in the systemic circuit, arteries in the pulmonary circuit carry deoxygenated blood away from the heart toward the lungs. Next the oxygenated blood travels through a vein back to the left atrium and then into the left ventricle, which is where we started the cycle. Notice that in the systemic circuit, arteries carry oxygenated blood and veins carry deoxygenated blood, whereas in the pulmonary circuit the opposite is true.

Many people have the misconception that deoxygenated blood contains little to no oxygen. This is far from the case when the body is at rest; in fact, the hemoglobin molecule has unloaded only *one* of the four oxygen molecules. With three of the four oxygen molecules still bound to hemoglobin, deoxygenated blood is about 75 percent oxygenated when it enters the veins on its way back to the right side of the heart. Thus, when the body becomes physically active, the blood has a large reservoir of oxygen to meet the increased demand.

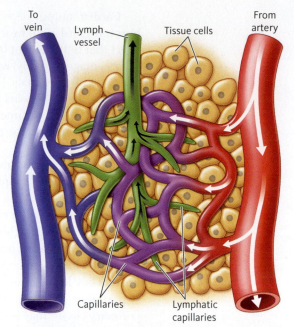

Figure 9.30 Capillary network. A capillary network receives oxygenated blood from an artery and returns deoxygenated blood into a vein.
© Cengage Learning 2014

The heart pumps blood to meet the body's demand for oxygen and energy

Active animals require a high rate of oxygen transport from their circulatory systems. The transport rate depends on the heart rate and the amount of blood the heart pumps, which in turn depends on the heart size. At rest, a human heart beats about 75 times a minute. Each heartbeat pumps about a quarter of a cup of blood, and thus it pumps a little over a gallon, or 5 quarts, of blood each minute. The amount of blood pumped each minute is called the *cardiac output*. Since you have 4 to 5 quarts of blood, your entire blood supply passes through each side of your heart every minute. This is quite impressive for an organ about the size of your fist and weighing less than a pound.

Veins—flexible blood vessels that carry blood back to the heart

Cardiac output depends on heart rate and the size of the heart. The hummingbird, with heart rates in the hundreds of beats per minute, has a heart that is 40 percent larger per body weight than those of mammals. As a percentage of body weight, bats have the largest and most muscular hearts of all mammals. Within seconds they can increase their heart rate from 500 to 1,100 beats per minute, but during hibernation, their hearts beat as little as four times per minute. In contrast, humans can more than double their heart rate from 70 to 180 beats per minute with strenuous exercise.

The heart itself is a muscular organ that contracts in constant rhythms, requiring a steady supply of blood and oxygen to its muscle and nerve cells. Amphibians easily meet the oxygen demands of their heart through an elaborate system of channels and spaces between the muscle fibers of the heart. However, in birds and mammals, the muscle layer is so thick and active that it has its own blood supply called the *coronary circuit* (see **Fig. 9.29**). The major arteries and veins can be seen sitting on the top of the heart. They lead to the capillary beds deep within the heart muscle itself. A partial or complete blockage of an artery that supplies oxygen in the coronary circuit may lead to the death of the muscle cells it feeds, resulting in a heart attack. Risk factors that lead to heart attack are presented in Chapter 21.

This section has tracked the route traveled by oxygen and carbon dioxide molecules between the atmosphere and a cell. Along the way, these molecules passed through the respiratory and circulatory systems. Oxygen has been the main focus of this chapter because it is required by all cells to function. The next section introduces you to factors that create demand for oxygen and energy.

9.4 check + apply YOUR UNDERSTANDING

- Circulatory systems transport and distribute oxygen throughout the body. They also transport carbon dioxide from cells to the lungs for excretion from the body.

- Red blood cells contain hemoglobin, which picks up oxygen and delivers it to body cells.

- Birds and mammals have a very efficient system that uses a four-chambered heart, which completely separates oxygenated from deoxygenated blood in two circuits.

- Gases are exchanged between blood and body cells in capillary beds close to each cell.

- Heart rate and heart size determine cardiac output, blood flow, and oxygen transport.

1. What are the three features of a closed circulatory system?
2. Describe how a capillary's structure is related to its function as a blood vessel where gases are exchanged with tissues.
3. Why is the human heart called a double pump?
4. Bats are the only mammals that fly. Predict whether the bat's blood–gas barrier is thinner or thicker than a nonflying mammal such as a mouse. Explain your reasoning.
5. Crocodile ice fish are found in the frigid waters around Antarctica (**Fig. 9.31**). They are the only known vertebrates that have no hemoglobin or red blood cells.
   a. Without hemoglobin, how do you think oxygen is transported from gills to cells?
   b. For each of the following choices, predict how the circulatory system of ice fish compensates for a lack of hemoglobin.

CONTINUED

Figure 9.31 Crocodile ice fish.

Uwe Kils

1. Smaller or larger heart size?

2. Narrower or wider blood vessels?

3. Lower or higher cardiac output?

c. Refer back to Section 9.2. Do you think these fish could live in warmer waters? Explain.

9.5 Oxygen Enters Cells to Meet an Animal's Energy Demands

Animals acquire chemical energy in the foods they eat and then "spend" it to maintain their internal organization and remain alive. This process of transferring energy from food nutrients to molecules that bring about cellular work is called metabolism. It's important to note that the body's metabolism follows the first law of thermodynamics, which states that energy is neither created nor destroyed but only changes form. When animals spend their energy, they transform the *chemical* energy of foods into other forms of energy such as *mechanical* energy (movement) and *electrical* energy (nerve impulses). Since the transfer of energy is not 100 percent complete, some energy is lost as thermal energy or *heat*. Some animals, like fireflies, transform chemical energy into *light* energy (**Fig.9.32**). This unique

Metabolism—the process of transferring energy from food nutrients to molecules that bring about cellular work.

Figure 9.32 Fireflies transform chemical energy into light energy. (a) Fireflies, also called lightning bugs, are known for their ability to emit light. During the summer night they flash in specific patterns to signal and attract a mate. (b) Cells in the rear end of the firefly use oxygen to transform the chemical energy in nutrients into light energy (orange arrows).
© Cengage Learning 2014

a.   **b.**

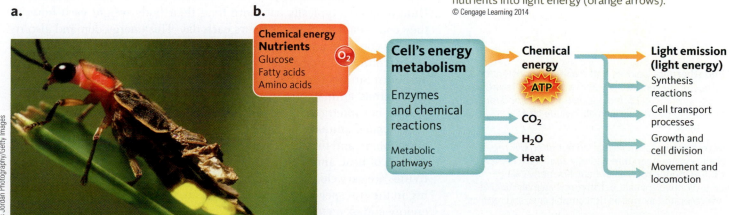

James Jordan Photography/Getty Images

reaction requires oxygen to produce the greenish yellow glow. What role does oxygen play in other energy transformations in animals?

In the section ahead, we link energy flow in organisms to the chemical steps needed to meet the energy demands of animal life. We begin by looking at how the hummingbird and the salamander meet their energy and oxygen demand, then delve into the cellular and chemical details of the process.

Energy demand depends on activity and body temperature regulation

Oxygen plays a central role in cell metabolism to transform chemical energy in food. Animals with a large energy requirement, or demand, also require a large amount of oxygen. It's a direct relationship—*increased energy demand requires increased oxygen supply.* Getting more oxygen to cells can be achieved through many different responses or adaptations such as a higher breathing frequency, deeper breathing, greater lung surface area, or faster heart rate and blood flow. You experience these respiratory and circulatory adjustments whenever you exercise or run to class.

At the organism level, hummingbirds have the highest rate of metabolism of all animals (about 30 times greater than humans), and amphibians such as salamanders have the lowest metabolism rate and lowest energy requirements of any terrestrial vertebrate. Two main factors determine the rate at which they burn fuel and demand energy from metabolism: activity and the way they regulate their body temperature.

Different activities demand different amounts of energy. Hummingbirds burn lots of fuel because they are very active. Flying is a very energy-demanding activity, but hummingbirds are the only birds that hover while they feed on flower nectar (**Fig. 9.33**). Hovering is an especially demanding type of flight because to remain in one place suspended in mid-air they need to beat their wings very fast, which means many muscle contractions and lots of energy expenditure.

Hummingbirds, like all birds and mammals, are endothermic animals; they generate heat internally as a by-product of metabolism. This internally generated heat typically elevates their body temperatures considerably above the temperature of the environment. Maintaining a constant body temperature over a wide range of conditions demands a lot of energy. In fact, about 60 percent of the energy they consume goes just to maintaining their body temperature. Endothermic animals must continually take in sugar and oxygen to "stoke the fires" of their metabolism. Hummingbirds typically eat more than their body weight each hour, as they continually feed for 12 hours each day. This energy demand also requires a constant supply of oxygen.

In contrast, salamanders have a much lower rate of metabolism and burn far less fuel than hummingbirds because they are *ectotherms.* Ectothermic animals produce less heat as a by-product of metabolism and do not *internally* regulate their body temperature. Formerly called *cold-blooded* animals, ectotherms include invertebrates, as well as fish, amphibians, and reptiles. They rely almost exclusively on environmental sources of heat, and their body temperature, rate of metabolism, and activities are very closely tied to the temperature of the environment. Basking in the sun speeds up their metabolism, which in turn increases their energy and oxygen demand.

Figure 9.33 Hummingbirds have a high energy demand. Hummingbirds hover to feed on flowers that provide concentrated nectar. To stay suspended in mid-air, a hummingbird's lift is provided only by the beating of its wings.

Dan Rodney/Shutterstock.com

Endothermic—animals that maintain a constant body temperature considerably above the temperature of the environment. Formerly called *warm-blooded* animals, endothermic animals include birds and mammals.

Ectothermic—animals that do not *internally* regulate their body temperature and rely almost exclusively on environmental sources of heat. Formerly called *cold-blooded* animals, ectothermic animals include invertebrates, as well as fish, amphibians, and reptiles.

Metabolism involves chemical pathways that make or use ATP

Energy demand in each body cell is met by the turning of the metabolic cycle. At the cellular level, metabolism consists of over 500 chemical reactions that occur as a series of steps in **metabolic pathways**. **Figure 9.34** shows that metabolism consists of energy-releasing pathways (shown on the right) that digest macromolecules and then break apart nutrients (glucose, fatty acids, amino acids) to *release* their energy, as well as pathways (shown on the left) that use energy and nutrients to build up, or create, new large macromolecules. For example, when animals eat sugars to fuel their energy requirements, the sugars enter pathways that break apart and release the energy in their bonds. On the other hand, when animals eat more sugars than they need, excess sugars enter pathways that use energy to build energy storage molecules such as fats and glycogen. Animals like hummingbirds with a high rate of metabolism rapidly break down nutrients to release energy to move their wings.

In **Figure 9.34**, note that energy-releasing pathways are coupled to energy-using pathways in a cycle of building and breaking down two molecules—ATP and ADP. Energy is captured and released through the addition and subtraction of phosphate groups. When energy is captured, a third phosphate group is added to **adenosine diphosphate (ADP)** to become **adenosine triphosphate (ATP)**. To release energy, a phosphate group is split from ATP to form ADP and P. The ADP–ATP cycle meets the energy demands of the cell and the body.

ATP is often described as the energy currency of the cell because the energy it stores in its bonds is used to build body structures or energy reserves. However, ATP is not just used to build or replace structures; some ATP is used for mechanical work such as contracting muscle or pulling apart chromosomes during cell division. ATP also performs electrical work to conduct nerve impulses or chemical work to move molecules against their concentration gradient (*active transport*).

You may have noticed a shift in how energy is measured. Recall from Chapter 8 that energy contained in food is measured in large units called *calories;* however, once food has been digested and enters metabolism on the molecular level, energy is now measured in the *number of ATP molecules* each nutrient molecule generates. In a grand sense, ATP is the molecule that drives life and is the molecular payoff for capturing and digesting food and breathing in oxygen.

Metabolism uses enzymes to break bonds and release chemical energy from nutrients

The energy contained in a molecule is locked into its chemical structure that is held together by its bonds. Thus the amount of energy obtained from a molecule depends on the amount of energy that went into building it and to how completely its structure is broken apart,

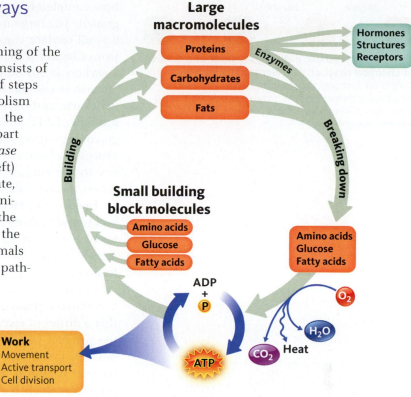

Figure 9.34 An overview of metabolism. Metabolism consists of energy-releasing pathways that transform chemical energy in nutrients into the chemical energy in ATP. The energy in ATP is then transformed into different kinds of cellular work such as active transport, muscle contraction, nerve impulses, and the energy-building pathways to make body structures. When energy demand is high, pathways shown on the right move rapidly.
© Cengage Learning 2014

Metabolic pathways—chemical reactions that occur as a series of steps in energy-releasing and energy-using pathways.

Adenosine diphosphate (ADP)—a large nitrogen-containing macromolecule with two phosphate groups. Captures energy from energy-releasing pathways by adding a third phosphate group to become ATP.

Adenosine triphosphate (ATP)—a large nitrogen-containing macromolecule with three phosphate groups that carries chemical energy to the sites where energy is used. Releases energy to energy-using pathways by splitting off a phosphate group to become ADP.

Figure 9.35 Metabolic pathways consist of a series of chemical reactions accelerated by enzymes. The simplified pathway shown starts with chemical A, which uses three different enzymes to produce two different intermediates and a product, chemical D.
© Cengage Learning 2014

Metabolic enzymes—enzymes that speed up chemical reactions in metabolic pathways. Operate inside cells (as opposed to digestive enzymes that work outside of cells). Function by breaking bonds or making bonds.

how completely its bonds are broken, or the degree to which it is degraded. The large macromolecules shown on the upper right of **Figure 9.34** all contain considerable amounts of energy in their bonds. Recall from Chapter 8 that the structure of a molecule of glucose consists of 6 carbon atoms bonded to 12 hydrogen atoms and 6 oxygen atoms. If glucose is completely degraded all the way to its constituent carbon, hydrogen, and oxygen atoms, it yields the most energy, or the largest number of ATP molecules. However, if fewer bonds are broken and the glucose is converted to a three-carbon molecule, for example, then this process has extracted less energy and yields fewer ATP molecules. The key to the complete breakdown of glucose lies in the molecule that enters metabolism in the right-hand side of **Figure 9.34**—oxygen.

Inside cells, **metabolic enzymes** accomplish the process of breaking bonds and making bonds. Recall from Chapter 8 that *enzymes* are proteins that speed up chemical reactions. In energy-releasing pathways, metabolic enzymes act on specific bonds to either break or build them. In doing so, enzymes convert substrate molecules into a product.

Enzymes are essential components of metabolic pathways. **Figure 9.35** shows a chain of chemical reactions in a metabolic pathway. Notice that a different enzyme catalyzes each reaction, and along the pathway the product of one chemical reaction becomes the substrate for the next chemical reaction. In this pathway, three enzymes combine to convert chemical A into chemical D. In this chain of chemical reactions, chemicals B and C are called *intermediates*. If a single enzyme in the series is not present or functional, then product D will not be produced. For instance, if enzyme 3 is not present, chemical C will accumulate and build up in the cell.

In the next section you will see a number of metabolic pathways that convert nutrients such as glucose and fatty acids into products. A very important enzyme channels the energy from a chemical reaction into bonding an additional phosphate group onto ADP to make the chemical energy storage molecule, ATP. The next section takes you through the three stages of metabolism and the metabolic reactions that use oxygen and produce carbon dioxide.

- Active animals with a high energy demand require a high rate of oxygen supply.

- Birds and mammals are endothermic animals that have high rates of metabolism and generate their body heat. Invertebrates, fish, amphibians, and reptiles are ectothermic animals that rely on environmental sources of heat.

- Metabolism includes energy-releasing and energy-using pathways that are coupled to a cycling of ADP and ATP.

- Metabolic enzymes speed up metabolic reactions that transfer energy from macronutrient molecules to ATP.

1. Which molecule captures the chemical energy from the breakdown of glucose?

2. What is the relationship between energy demand, oxygen supply, and rate of metabolism?

3. Compare digestive enzymes presented in Chapter 8 and metabolic enzymes presented in this section in terms of (a) where they operate, (b) what they do, and (c) their specificity.

4. Chorus frogs may be heard calling on warm nights in early spring for hours (**Fig. 9.36**). Watch and listen to the upland chorus frog at **http://www.musicofnature.org/home/upland_chorus_frog/**. Describe how four different kinds of energy are involved in calling for a mate. Do you think this is an energy-demanding activity?

5. Large birds like albatrosses have different activities that incur different energy demands. Place the following three activities in order from the highest to lowest (a) rate of metabolism and (b) heart rate. (1) Gliding–soaring in search of food. (2) Flapping wings to take off. (3) Incubating eggs on the nest.

Figure 9.36 Chorus frog.

9.6 Animal Cells Use Aerobic and Anaerobic Metabolism to Produce ATP

When it comes to oxygen, softshell clams lead a double life—they can live with or without it. Buried in the mud and sediments on the shore bottom, they generally live under very low oxygen conditions. During high tide, a clam is submerged under water. It opens the two halves of its shell and extends its two siphons to draw water across its gills (**Fig. 9.37a**). A clam's gills, unlike fish gills, filter microscopic food particles, as well as exchange gases with the water. With a fresh supply of dissolved oxygen, the clam's body cells use **aerobic metabolism** to make large numbers of ATP molecules to meet its relatively low energy demands.

> **Aerobic metabolism**—energy-releasing metabolic pathways that use oxygen to completely break down amino acids, fatty acids, and glucose to make large numbers of ATP molecules; also called *cellular respiration.*

a.

Water

Siphons

Shell

Foot

Sediment

b.

Figure 9.37 Clams use both aerobic and anaerobic metabolism. Clams live in the intertidal zone and are exposed to air during low tide and submerged under water during high tide. (a) During high tide, the clam is almost completely buried with only the tips of the siphons projecting above the sediment. The clam draws water in through one of its siphons to obtain oxygen and performs aerobic respiration. (b) During low tide, clams retract their siphons, close their shells, and use anaerobic metabolism.
© Cengage Learning 2014

Anaerobic metabolism—energy-releasing metabolic pathways that do not use oxygen; they incompletely break down glucose to make a small number of ATP molecules.

When the water recedes during low tide, clams become exposed to the dry air and predators (**Fig. 9.37b**). They respond by "clamming up," or tightly closing the two halves of their hinged shell together with strong cords of muscle. The closed shell essentially seals off their gills and soft bodies from their source of oxygen in the atmosphere. During this time they switch to **anaerobic metabolism** to produce ATP without oxygen. These metabolic pathways release only a fraction of the energy stored in a carbohydrate molecule. The low energy output is why many animals, including humans, use anaerobic metabolism only to supplement their energy needs for short bursts of activity lasting one to two minutes. However, clams are capable of keeping their muscles contracted without oxygen until the high tide returns about six or seven hours later. When they are closed, clams have a low rate of metabolism, low energy needs, and low oxygen requirements.

This daily switching between aerobic and anaerobic metabolism leads to some questions. How does oxygen help a cell produce more energy? What happens to the oxygen once it is inside the cell? Why is oxygen so essential to sustaining the life of the clam and other animals?

In the section ahead, you will find that the answers to these questions lie at the lowest levels of the hierarchy of organization—cell organelles, molecules, atoms, and even the subatomic particles, the protons and electrons. It is inside each cell where nutrient molecules and enzymes chemically interact in hundreds of metabolic pathways to release energy. All

Kuttelvaserova/Shutterstock.com

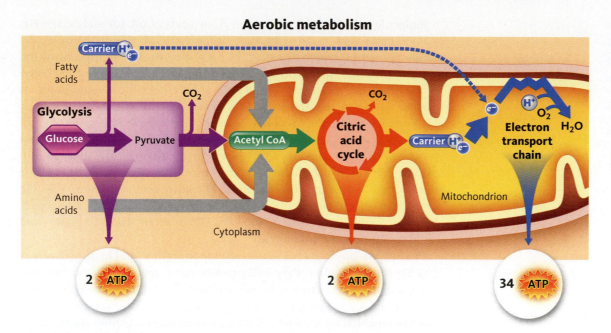

Aerobic metabolism

Glycolysis

Glucose → Pyruvate

CO_2

Fatty acids

Amino acids

Carrier H⁺ e⁻

Cytoplasm

2 ATP

Acetyl CoA

Citric acid cycle

CO_2

Carrier H⁺ e⁻

2 ATP

Electron transport chain

e⁻ H⁺ O_2 H_2O

Mitochondrion

34 ATP

of the feeding and breathing adaptations detailed in Chapter 8 and in this chapter pay off in energy. This section takes you to the animal's energy payday.

Aerobic metabolism unlocks large amounts of energy from nutrients

The pathway that operates most of the time in animal cells is aerobic metabolism. This process is also called *cellular respiration,* and it occurs in three stages: (1) glycolysis, (2) the citric acid cycle, and (3) the electron transport chain. As you work to understand this complex process, it's important to keep in mind the overall goal of the process—to transfer energy locked in the bonds of macronutrients to build ATP, which can then be used to drive life processes. All three major food nutrients—carbohydrates, fats, and proteins—can serve as fuel, but glucose is the preferred source of energy by cells and is the sole source of fuel used by the brain. Remember that glucose is the end product of carbohydrate digestion and is absorbed and transported into circulation. The pancreas secretes the hormone insulin, which regulates the entrance of glucose into cells from the blood.

In the cytoplasm, **glycolysis** converts glucose to pyruvate (a three-carbon compound) through a series of 10 different chemical reactions that occur as steps, one after the other (**Fig. 9.38**). A specific enzyme controls each step. During glycolysis, two important things happen. First, enzymes break apart glucose by removing its hydrogen atoms, which are then transported by special carrier molecules to the third stage of the process. Second, two ATP molecules are produced. The three products of glycolysis include a small amount of energy captured in ATP, some hydrogen atoms attached to their carrier molecules, and a chemical called *pyruvate.*

Pyruvate is a molecule that can be used to provide more energy because it then moves into the mitochondrion where it is converted to

Glycolysis—transfers energy in glucose to pyruvate (a three-carbon compound) through a series of 10 different chemical reactions. Some of the hydrogen atoms are stripped from glucose and attached to hydrogen-carrier molecules.

a molecule called acetyl coenzyme A, or acetyl CoA for short (see **Fig. 9.38**). Acetyl CoA is an extremely important intermediate molecule. From **Figure 9.38**, notice that acetyl CoA is a central branch point in the aerobic process—it can be produced from the breakdown of carbohydrates, fats, and proteins. Acetyl CoA then enters the second stage in aerobic metabolism—the citric acid cycle.

The **citric acid cycle**, also referred as the Krebs cycle, is a series of eight reactions that start and end with the same compound. The chemical reactions in this cycle convert *all* of the original carbon in the glucose into carbon dioxide, which diffuses from the cell into the blood where it is transported to the lungs for excretion. The citric acid cycle completes the breakdown of the *glucose* molecule and yields two more ATP molecules, hydrogen atoms, and carbon dioxide. What about the fatty acids and amino acids that enter the cycle?

Fatty acids are larger molecules with more carbon and hydrogen atoms (see **Fig. 9.38**); therefore they will produce more acetyl CoA and turn the citric acid cycle more times, generating more ATP than a molecule of glucose. Amino acids represent another potential source of energy. In animals like humans that eat a variety of foods, amino acids are often spared from being used for energy if the supply of carbohydrates and fats is adequate. However, in carnivorous animals that eat mostly protein, amino acids are abundant, so they are used as a source of energy. Amino acids are very diverse molecules and enter aerobic metabolism at many different points. To simplify matters, we only show them to be converted to acetyl CoA in **Figure 9.38**.

So far, glycolysis and the citric acid cycle have degraded one molecule of *glucose* with a net result of four ATP, hydrogen atoms attached to carrier molecules, and CO_2 molecules. Of all the atoms in a molecule of glucose that started the aerobic process, it is only the hydrogen atoms that make it to the third and final stage. It is in this final stage that the hydrogen atoms will interact with the oxygen from the atmosphere.

The **electron transport chain**, or third stage of aerobic metabolism, releases the energy in the *hydrogen atoms* from glycolysis and the citric acid cycle. Follow this step-by-step process as shown **Figure 9.39**. The process starts when carrier molecules release the hydrogen atoms at the inner mitochondrial membrane (Step 1). In this stage, hydrogen atoms are split into electrons and protons (H^+), which go their separate ways. The electrons, shown by the blue arrows in **Figure 9.39**, pass along a chain of large proteins (shown in purple), or the *electron transport chain,* located in the inner membrane of the mitochondria. As they are passed, they release small amounts of their energy. This energy is then used to move and concentrate the protons (H^+) on one side of the mitochondrial membrane (Step 2) and thus creates a steep concentration gradient. The concentrated protons then move down the gradient through a large enzyme that channels the energy to make a large number of ATP molecules (Step 3). If the cell is operating at maximum efficiency, 34 ATP molecules are produced for every molecule of glucose.

Oxygen now enters metabolism at the very last step in the process—it couples or links the movement of the electrons and protons together. **Figure 9.39** shows that in the final reaction, oxygen acts like an electron "magnet" and joins them with the protons to form *water* (Step 4). Without oxygen, the electrons back up like traffic on a highway, and the electron transport chain and the citric acid cycle come to a halt. Oxygen as the final electron acceptor in the electron transport chain makes the production of

Citric acid cycle—stage of aerobic metabolism that follows glycolysis; a series of eight reactions that yield two ATP molecules, hydrogen atoms attached to carriers, and carbon dioxide. Fatty acids and many amino acids enter metabolism in the citric acid cycle. Also referred as the *Krebs cycle.*

Electron transport chain—the third stage of aerobic metabolism; releases the energy in the hydrogen atoms from glycolysis and the citric acid cycle to yield a large number of ATP molecules.

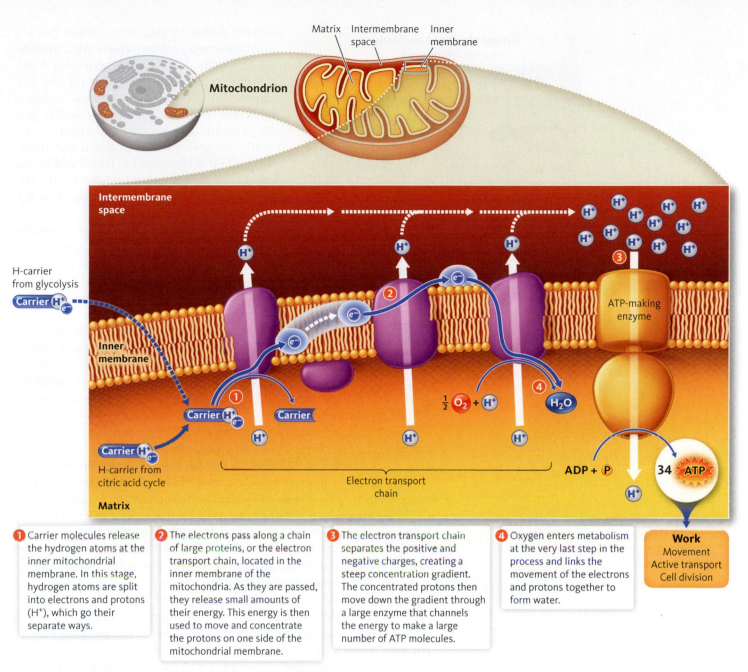

Mitochondrion

Matrix Intermembrane space Inner membrane

Intermembrane space

H-carrier from glycolysis

Carrier H⁺ e⁻

Inner membrane

Carrier H⁺ e⁻

Carrier

Carrier H⁺ e⁻

H-carrier from citric acid cycle

Matrix

② e⁻

① H⁺

Carrier H⁺ e⁻   Carrier

H⁺   H⁺

½ O₂ + H⁺

④ H₂O

Electron transport chain

③ ATP-making enzyme

ADP + P

34 ATP

H⁺

Work
Movement
Active transport
Cell division

❶ Carrier molecules release the hydrogen atoms at the inner mitochondrial membrane. In this stage, hydrogen atoms are split into electrons and protons (H⁺), which go their separate ways.

❷ The electrons pass along a chain of large proteins, or the electron transport chain, located in the inner membrane of the mitochondria. As they are passed, they release small amounts of their energy. This energy is then used to move and concentrate the protons on one side of the mitochondrial membrane.

❸ The electron transport chain separates the positive and negative charges, creating a steep concentration gradient. The concentrated protons then move down the gradient through a large enzyme that channels the energy to make a large number of ATP molecules.

❹ Oxygen enters metabolism at the very last step in the process and links the movement of the electrons and protons together to form water.

Figure 9.39 The electron transport chain. The electron transport chain occurs within mitochondria. Hydrogen atoms from carbohydrates, fats, and proteins are stripped of their electrons, which drive the production of ATP. Oxygen keeps the electron traffic flowing in the process. The end product is 34 ATP and water.
© Cengage Learning 2014

a large number (38) of molecules of ATP possible from each molecule of glucose. The electron transport chain generates virtually all (34 out of 38) of the ATP created during the three stages of aerobic metabolism. Recall that fatty acids enter metabolism in the citric acid cycle and thus require oxygen to release their large amount of energy—fats are burned by aerobic activities and aerobic metabolism. Glucose, on the other hand, can be broken down without oxygen—the subject we take up next.

Anaerobic metabolism generates small amounts of energy in the absence of oxygen

Animals evolved during a time when the atmosphere contained less oxygen than it has today and have adaptations to generate smaller amounts of ATP through anaerobic metabolism. There are several an-

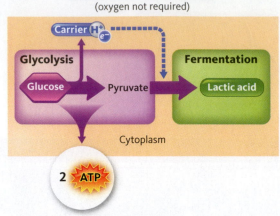

Anaerobic metabolism
(oxygen not required)

Figure 9.40 An overview of anaerobic metabolism.
In the absence of oxygen, animal cells have the ability to derive a small amount of energy through anaerobic metabolism, which includes glycolysis and fermentation pathways. Glucose is broken down incompletely to lactic acid, ATP, and hydrogen atoms attached to a carrier molecule.
© Cengage Learning 2014

Fermentation—second stage in anaerobic metabolism; a metabolic pathway that converts pyruvate to lactic acid or ethanol with the release of a small number of ATP molecules.

aerobic pathways that animals use, but the most common pathway starts with glycolysis and ends with fermentation (**Fig. 9.40**). Recall that glycolysis generates (1) two ATP molecules; (2) hydrogen atoms, stripped from the original glucose molecule; and (3) *pyruvate*. The accumulation of pyruvate in the cytoplasm will shut down glycolysis; therefore it is converted to lactic acid by the process of fermentation (see **Fig. 9.40**). Oxygen does not enter either glycolysis or fermentation pathways.

Human muscle can undergo anaerobic metabolism for short bursts of activity. For example, sprinting the 100-meter dash uses more oxygen than is immediately available, and thus muscle uses fermentation to generate ATP and the energy the sprint demands. The buildup of lactic acid will eventually inhibit enzyme function and muscle contraction, and lead to fatigue. After the sprint, the runner recovers by breathing deeply to bring in oxygen and release the buildup of carbon dioxide in the bloodstream and lungs.

Since fermentation *incompletely* breaks down glucose, its product, lactic acid, still contains a large amount of energy. Mammals recover the energy in lactic acid by recycling it into glucose in the liver. Although anaerobic metabolism does not produce large amounts of energy and is inefficient for long periods of time, it does allow muscles to contract in humans or clams even when oxygen is in limited supply. For microorganisms such as bacteria and yeasts with relatively low energy demands, the small amount of ATP that fermentation provides meets their needs. We will present another fermentation pathway in Chapter 15, showing you how yeasts convert glucose to ethanol in the beer and winemaking processes.

Throughout this section we have provided the background to answer the questions posed in the section opener. How does oxygen produce large amounts of energy for a cell? What happens to the oxygen once it is inside the cell? Why is oxygen so essential to sustaining the life of the clam and other animals? All of these questions are connected through the metabolic processes that generate and power the cell. The processes that power the work of living occur at the chemical level—chemical reactions that occur on membranes, mediated by enzymes, to ultimately transfer energy from ADP to make ATP. It all comes down to the chemical properties of oxygen atoms—oxygen is able to accept electrons from the foods we eat in the mitochondria of all cells. This is why you breathe.

Let's step back and pull together concepts presented in Chapter 8 and in this chapter. The process starts with the food you eat and the air you breathe. Nutrients such as glucose, fatty acids, amino acids, vitamins, and minerals are acquired as food by the digestive system, while oxygen is supplied by the respiratory system. Both nutrients and oxygen are then distributed to all cells by the circulatory system. These three body systems communicate and function together to meet the ever-changing energy demands of animals.

9.6 check + apply YOUR UNDERSTANDING

- Most animals use aerobic metabolism to produce ATP and the energy it contains.

- Aerobic metabolism involves three stages that break apart glucose into its carbon, hydrogen, and oxygen atoms. Hydrogen atoms are carried to an electron transport system that generates a large number of ATP molecules.

- Anaerobic pathways include glycolysis and fermentation that break apart glucose into lactic acid in the absence of oxygen. Nineteen times less ATP is made from anaerobic metabolism (2) than aerobic metabolism (38).

Figure 9.41 Sled dogs.

Marcel Jancovic/Shutterstock.com

1. List the reactants and products of glycolysis.

2. Animals breathe in O_2 and exhale CO_2. Where does the carbon in carbon dioxide come from?

3. Human red blood cells lack mitochondria when they reach maturity. How do they generate energy to live?

4. Long-distance sled dogs cover over 1,150 miles in about 10 to 17 days in the Iditarod Dog Race in Alaska (**Fig. 9.41**). They average between 5 and 15 miles per hour over the course of the race and may eat 10,000 calories a day. Which type of metabolism fuels the dog's muscles to sustain this activity? What kind of food would you feed the dog sled team— carbohydrates, fats, or proteins—to provide this large amount of energy? Explain.

5. Many species of plants produce compounds called cyanogens. When eaten, these compounds are absorbed into cells where they block the actions of proteins that operate in the electron transport chain. How will this action affect the animal eating it?

End of Chapter Review

Self-Quiz on Key Concepts

Oxygen and the Environment

KEY CONCEPTS: Over billions of years photosynthetic organisms produced oxygen that accumulated in Earth's atmosphere and in bodies of water. Most terrestrial habitats contain abundant sources of oxygen with the exception of soil and burrows. At increasing altitudes oxygen availability is constrained by low atmospheric pressure. Aquatic habitats show much more variation in oxygen availability. Depending on its temperature, salt content, and physical turbulence, water contains about 30 times less oxygen by volume than air.

1. Match each of the following **oxygen environments** with its description or example. Each term on the left may match more than one description on the right.

    Terrestrial habitat   a. Oxygen level is highly variable.
    Aquatic habitat   b. Amount of oxygen depends on temperature.
                            c. Oxygen availability changes with altitude.
                            d. Contains much lower amounts of oxygen.

2. Which of the following is *true* of high-altitude environments?
    a. Animals breathe air that has 10 percent oxygen rather than 21 percent at sea level.
    b. Animals breathe more oxygen in less volume than at sea level.
    c. The higher atmospheric pressure makes oxygen more available.
    d. Oxygen is less available due to a lower atmospheric pressure.

3. Which of the following aquatic habitats has the *greatest* amount of dissolved oxygen?
   a. a warm, stagnant swamp
   b. a cold, rushing stream
   c. a warm saltwater estuary
   d. about 500 feet down in the open ocean

Respiratory Systems—Ventilation and Gas Exchange

KEY CONCEPTS: Animals diffuse gases across the surface of their bodies or in specialized respiratory structures. Vertebrates meet a larger demand for oxygen with efficient respiratory systems that include skin, gills, and lungs. The respiratory system brings oxygen into the body and expels carbon dioxide through ventilation and gas exchange with blood and body fluids. Respiratory structures are adapted to increase the rate of diffusion by creating an enlarged surface area and a steep concentration gradient, decreasing the distance gases travel, and increasing the contact time between the gases and blood.

4. Match each of the following **respiratory organs** with its description or example. Each term on the left may match more than one description on the right.

Gills	a. Gases diffuse directly from the atmosphere into blood in insects.
Lungs	b. Gases diffuse into blood flowing in the opposite direction.
Skin	c. Restricts the animal to moist terrestrial environments.
Tracheal system	d. Moist internal respiratory structure that allow animals to inhabit drier habitats.

5. Which of the following conditions will *increase* the rate of diffusion?
   a. a short contact time between blood and gill
   b. a thick layer of skin to increase the distance between air and blood
   c. a steep concentration gradient due to a countercurrent flow
   d. diffusion in a dense medium like water

6. Which of the following is *false* regarding a bird's respiratory system?
   a. A bird's respiratory system is much more efficient than a mammal's.
   b. A bird's respiratory system has air sacs in its bones.
   c. A bird's respiratory system has fresh air in its lungs during exhalation.
   d. A bird's lungs mix old air with fresh air like a mammal's system.

Circulatory Systems and Gas Transport

KEY CONCEPTS: Animals transport oxygen and carbon dioxide in either an open or a closed circulatory system. Vertebrates have a closed system that includes four components: (1) blood cells that contain oxygen-binding proteins, (2) a distribution network of blood vessels, (3) thin capillaries that exchange materials with cells, and (4) a pumping mechanism that keeps blood flowing and circulating. Many invertebrates have open circulatory systems that do not include blood vessels.

7. Match each of the following **blood vessels** with its description or example. Each term on the left may match more than one description on the right.

Arteries	a. In the systemic circuit, they carry oxygenated blood.
Capillaries	b. In the systemic circuit, they carry deoxygenated blood.
Veins	c. They are thin vessels where oxygen enters circulation.
	d. In the pulmonary circuit, they carry oxygenated blood.
	e. Blood is pushed out of the heart into these vessels.
	f. Blood returns to the heart through these vessels.

8. Which of the following is *false* regarding the *right* and *left* sides of a mammal's heart?
   a. The right side pumps blood under higher pressure than the left.
   b. The right side has a thicker muscle layer than the left.
   c. The right side receives blood from a vein whereas the left side does not.
   d. The right side receives deoxygenated blood whereas the left side receives oxygenated blood.

9. The red-pigmented oxygen transporting protein found in mammals is ___, which contains the metal ___, and the blue-pigmented oxygen protein in spiders, snails, and lobsters is ___, which contains the metal ___.
   a. hemoglobin; iron; hemocyanin; copper
   b. hemocyanin; iron; hemoglobin; copper
   c. hemoglobin; copper; hemocyanin; iron
   d. hemocyanin; copper; hemoglobin; iron

Anaerobic and Aerobic Metabolism

KEY CONCEPTS: Animals have evolved aerobic and anaerobic pathways to generate energy under varying levels of oxygen. In the mitochondria, aerobic pathways completely break down glucose, fatty acids, and amino acids to carbon dioxide, water, and energy. Hydrogen atoms are carried from the first two stages—glycolysis and citric acid cycle—to the third stage—the electron transport chain. Oxygen is the final electron acceptor and is coupled to the production of a large amount of ATP and water. Anaerobic pathways incompletely break down glucose through glycolysis and fermentation with a small amount of energy and lactic acid as end products.

10. Match each of the following **stages of metabolism** with its description or example. Each term on the left may match more than one description on the right.

    Citric acid cycle
    Electron transport chain
    Fermentation
    Glycolysis

    a. stage where oxygen enters metabolism
    b. stage where fatty acids enter metabolism
    c. stage where glucose enters metabolism
    d. stage where pyruvate is converted to lactic acid
    e. stage where hydrogen carriers release hydrogen
    f. stage where water is produced by metabolism
    g. stage where the most ATP is produced

11. When broken down by metabolism, which nutrient produces the most energy?
    a. carbohydrate—glucose
    b. proteins—amino acids
    c. fats—fatty acids

12. Which atom in a glucose molecule loses its electrons to produce large amounts of ATP?
    a. carbon
    b. hydrogen
    c. oxygen

Applying the Concepts

13. The developing mosquito larva shown in **Figure 9.42** is an aquatic insect that pokes through the water surface with its spiracle to breathe air. Does it have a high or low energy and oxygen requirement? Explain how this adapts the mosquito to meet these requirements.

14. The hairy frog lives in fast-moving rivers and streams in western Africa. It has greatly reduced lungs. During the summer mating season, males develops long "hairy" strands on the skin on their hindquarters (**Fig. 9.43**). These hairs have a rich supply of blood vessels. Explain how these hairs meet the oxygen demands in the stage of life of this frog.

Figure 9.42 A mosquito larva.

Figure 9.43 Hairy frog.

Figure 9.44 The book lung of spiders.
© Cengage Learning 2014

Book lung

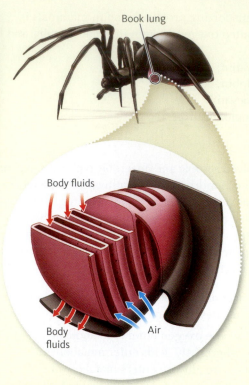

Body fluids

Body fluids

Air

15. Spiders have two narrow slits on their lower abdomen that each open to a respiratory structure called a *book lung*. Book lungs consist of a stack of air-filled hollow "pages," or thin narrow air pockets, that extend into a blood-filled chamber (**Fig. 9.44**). Within these air pockets, oxygen and carbon dioxide are exchanged with the spider's blood. Refer to factors in **Table 9.2** to explain how the book lung increases the rate of diffusion.

16. Without a sufficient supply of oxygen, fish cannot derive the energy they need to sustain life, so they suffocate, die, and float to the surface. Predict in which season of the year fish kills occur most often. Explain your reasoning.

Data Analysis

Metabolism by Lizards That Tolerate Freezing

The European common lizard lives in a wide range of habitats from sea level to high altitudes (**Fig. 9.45**). Its geographic range extends from northern Spain to north of the Arctic Circle. This ectothermic animal is adapted to the cold with an unusual ability to tolerate freezing temperatures below 0°C. In the winter it inhabits shallow burrows and encounters subzero temperatures. It has two mechanisms to tolerate freezing: (1) nearly 50 percent of its body fluids turn to ice, and blood circulation comes to a halt; and (2) it uses antifreeze compounds to keep some of its body fluids in a supercooled liquid state. The data below are from an investigation of lizards living above the Arctic Circle and their respiratory response to freezing. Lizards were gradually exposed to declining temperatures until they were in a frozen state at −2.5°C and then warmed to 3°C.

Figure 9.45 European common lizard.

Data Interpretation

In **Figure 9.46**, Graph a shows oxygen consumption as the lizards were slowly exposed to decreasing temperatures. Graph b shows oxygen consumption over time at three different states—supercooled, frozen, and thawed.

17. Describe how this reptile consumes oxygen as it progressively becomes cooler.

18. How long were the lizards able to live at subzero temperatures before shutting down their metabolism?

Critical Thinking

19. Would you expect to find differences in the body chemistry of lizards living in northern Spain?
20. What kind of metabolism and pathways is the lizard using when it is in a frozen state? What end product is accumulating in its tissues?
21. Do you think the lizards were respiring even though oxygen was not consumed at $-4.5°C$?
22. If you measured the lizard's CO_2 release, how do you think it would compare to O_2 consumption?
23. The brain of the lizard does not freeze. Refer back to Section 9.3 to explain why this might occur.

a.   **b.**

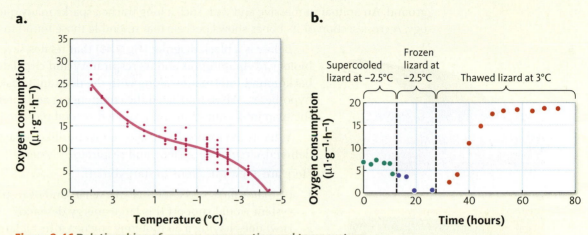

Figure 9.46 Relationships of oxygen consumption and temperature.
From Yann Voituron, Bruno Verdier, Claude Grenot, "The respiratory metabolism of a lizard (Lacerta vivipara) in supercooled and frozen states," Am J Physiol-Regulatory, Integrative and Comparative Physiology 283:1. Copyright © 2002 The American Physiological Society. Reprinted with permission.

Question Generator
The Respiratory and Circulatory Systems of Apatosaurus

Apatosaurus, which used to be called Brontosaurus, is a dinosaur that lived about 150 million years ago, during the Jurassic Period (**Fig. 9.47a**). It was one of the largest land animals that ever lived on Earth, with a length of 75 to 85 feet and weighing 20 to 30 tons. The animal had a neck about 30 feet long. Research has shown that *Apatosaurus* was most likely a grazing animal browsing low to the ground. An animal so massive and with such a long trachea sparks many questions about its physiology. A cross section of its bones shows pockets that resemble those found in bird bones (**Fig. 9.47b**).

a.

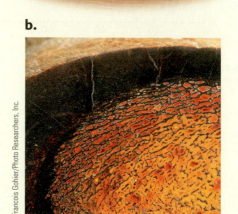

b.

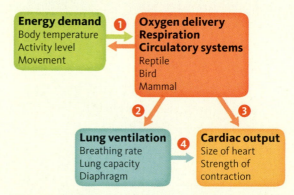

Figure 9.47 *Apatosaurus* and its bone structure.

Below is a block diagram (**Fig. 9.48**) that relates several aspects of the biology of *Apatosaurus* and its respiratory and circulatory systems. Use the background information along with the diagram to ask questions and generate hypotheses. We've translated the components and interaction arrows shown in the map into a question and provided several more to get you started.

What is the relationship between its birdlike bone structure and its energy demand, circulatory system, and respiratory system?

Here are a few questions to start you off.

1. If *Apatosaurus* was endothermic, what kind of respiratory and circulatory system would support this large energy demand?
2. How does having a 30-foot-long trachea affect a large animal's ventilation rate?
3. As a terrestrial animal that lived on savannahs similar to those inhabited by our modern-day elephant, would you expect it would meet some of its oxygen needs through its skin?
4. Would you expect its heart to be proportional in size to its body? Why or why not?

Use the diagram to generate your own research question and frame a hypothesis.

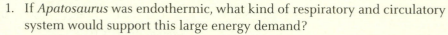

Figure 9.48 Factors relating to the respiratory and circulatory systems of *Apatosaurus.*
© Cengage Learning 2014

10 Animal Sensory Perception, Integration, and Movement

10.1 Battle of the nervous systems: bats versus moths in the New Mexico skies

10.2 Animal survival and reproduction depend on a flow of information

10.3 Sensory receptors detect information from the environment and send it to the brain

10.4 The brain and spinal cord interpret information

10.5 Cells of the nervous system are specialized to transfer signals

10.6 Muscles pull on rigid structures to produce movement

10.7 Animals are adapted to move in a variety of ways

10.1 Battle of the Nervous Systems: Bats versus Moths in the New Mexico Skies

Figure 10.1 Mexican free-tailed bats in flight. Each evening starting at dusk, thousands of bats emerge from caves. Using echolocation, they hunt for flying insects.

As twilight settles over the Chihuahuan Desert of southern New Mexico, one dark cave entrance comes alive with darting movements. An unknown silent signal is given, and in a swirling black cloud, hundreds of thousands of Mexican free-tailed bats *(Tadarida brasiliensis)* fly from their daytime roost into the evening sky in search of food (**Fig. 10.1**). The spectacular exit from Carlsbad Caverns can last for as long as three hours as wave after wave of bats emerges.

Bats are nocturnal animals, meaning they are most active at night—they spend their days in caves or sheltered areas grooming, sleeping, and resting (**Fig. 10.2**). What triggers this mass exit of the bats from the cave at sunset each night is still a bit of a mystery. Biologists believe part of the explanation is that the bats roosting near the cave entrance perceive the setting of the sun as a primary signal. As those bats begin to fly away from the cave, the other bats see the departure as their signal to leave.

Hunting at night can be a real challenge. Imagine standing in a pitch-black room filled with furniture and other people, with everyone trying to find an object as small as a coin. Complicate the problem by tossing the coin through the air and trying to catch it. This is the difficulty facing insect-eating bats each night. The problem becomes more complex when you consider that bats locate and capture their airborne meals, often several hundred feet in the air, while flying fast and avoiding obstacles such as trees and other bats.

How do bats accomplish this seemingly impossible task? They are one of several species of mammals, including dolphins and some whales, that have evolved a very precise echolocation sensory system for navigating, hunting, and communicating. Echolocation is a type of biological sonar that uses high-frequency sound waves. Bats generate sound waves using their larynx and vocal cords at frequencies well beyond the range of human hearing. They produce these high-frequency sounds in short, loud bursts and listen for the return echo. The wavelength of the sound produced is perfectly adapted to bounce off moth-sized insects, creating the return echo. Smaller insects, other bats, and trees reflect the high-frequency waves differently, allowing the bats to listen with their ears for the direction of the echo from their preferred food source.

Flying burns a lot of energy, but bats are excellent fliers, and their specialized senses help them locate high-energy food sources. Free-tails, like 70 percent of all bats, eat insects, dining mostly on moths and other flying insects. Free-tailed bats are not large; adults weigh only about 1/2 oz—the equivalent of three nickels—and have a wingspan of about 11 inches. Although small, each adult typically eats the equivalent of half its body weight in insects each night (nursing females will eat their full body weight in insects). The tens of thousands of bats combine to eat more than three tons of insects each night. Because these bats consume so many insects, their impact on local insect populations is significant. Their feeding has more than an ecological impact. About half of their diet consists of agricultural pests that damage cotton and alfalfa crops, making bats an economically important partner for farmers.

Once the bat detects a moth using echolocation, it responds by quickly zeroing in on the moth's exact location. Among mammals, only bats have

Figure 10.2 Bats roosting. Like many species of bats, Mexican free-tailed bats find shelter that will provide protection from the environment and predators.

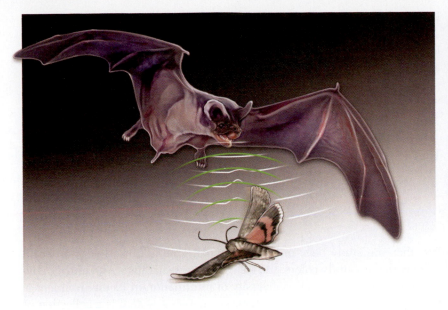

Figure 10.3 Bats use echolocation to detect prey. Insects evolved about 250 million years before bats. When bats appeared around 50 million years ago, many insects learned to hear and avoid them.
© Cengage Learning 2014

evolved the ability to fly long distances, and the free-tails are the speed-sters of the bat world. They have long, narrow wings covered in a thin membrane of skin that allows them to sustain speeds of 25 mph and ac-celerate to a top speed of 47 mph. Flight requires a specialized wing shape, strong pairs of muscle groups, a refined ability to sense a rapidly changing environment, and the ability to make continuous small corrections while moving in three dimensions. Bats need lightweight bones, a light body weight, and the ability to sustain high levels of aerobic activity.

The predator–prey interactions in the nighttime skies over New Mexico represent the outcome of millions of years of evolutionary advances in nervous systems and sensory capabilities. The insects that the bats feed on are not helpless prey; in fact, they have evolved their own specialized senses and signals that protect them from their predators (**Fig. 10.3**). Owlet moths emit their own ultrasound pulses, a type of high-frequency clicking produced by their wings. When bats hear this sound, they abruptly turn and stop their pursuit. Originally, the clicking sounds were thought to jam the bats' echolocation radar, but biologists now know that it really is a signal to the bat that the moth is poisonous and to be avoided. Other moth species have evolved the ability to mimic this clicking sound as a survival mechanism. Still other groups of moths along with a number of other fly-ing insects have developed very specialized ears whose main function is to listen for bat echolocation frequencies. When they hear the bats' echoloca-tion call they quickly escape being eaten by rapidly dodging the oncoming bat. Their battle continues to another day.

In this chapter, we'll explore the relationship between the structure and function of the nervous system, the skeletal muscles, and the bones they interact with. Nerves stimulate skeletal muscles to contract, and shortening of the muscles results in movement. The nervous system serves as an information-collection, signal-transport, regulation, and evaluation system. Although most animals have a nervous system, its complexity and arrangement differ from one group to the next. Sensory nerves provide

Echolocation—a form of biological sonar that uses high-frequency sound waves to locate food and to navigate.

Mikhail Melnikov/Shutterstock.com

Figure 10.4 Protective adaptations of the owlet moth. Owlet moths emit ultrasound pulses as a means of protection from bats. They also detect bats with the fine, hairlike structures on their antennae. These structures are filled with hundreds of thousands of specialized sensory cells that help owlet moths obtain information about the surrounding environment.

Nervous system—a collection of tissues and specialized cells that provides information about the environment, signaling pathways throughout the body, and the ability to integrate information input and generate responses.

Neurons—structural and functional units of the nervous system, specialized for carrying information signals from one location to another.

Nerves—bundles of neurons that provide a signal pathway to and from an animal's brain. Each nerve is a cablelike structure that contains the signal-input regions from many adjacent neurons.

Central nervous system (CNS)—the nervous system in vertebrate animals comprising the brain and spinal cord, which receive and integrate information delivered from various types of sensory neurons.

Brain—in animals, a large concentration of nerve cells that act as the information-processing center, controlling the other organ systems of the body either by activating muscles or by causing secretions of chemicals such as hormones.

information about both the external and internal environments and may be sensitive to stimulation from different conditions in the habitat. As we've learned from the evolutionary battle between the Mexican free-tailed bat and owlet moths, integrated biological systems combined with adaptations provide animals with the tools necessary to find food and defend themselves in order to survive and succeed in the habitats where they live.

10.2 Animal Survival and Reproduction Depend on a Flow of Information

Whether you study bats, moths, or elephants, all animals live in a world filled with the sounds, colors, and tastes of diverse habitats. An animal's ability to sense and react to this diversity of information in its environment is made possible by its **nervous system**. The nervous system is a collection of tissues and specialized cells that provides information about the environment, signaling pathways throughout the body, and the ability to integrate information input and generate responses. Specialized cells of the nervous system called chemical receptors allow male owlet moths to find and recognize a potential mate, even when that mate is far away (**Fig. 10.4**). Their chemical receptors are very sensitive to *pheromones,* chemicals released by female moths of the same family that signal males that they are ready to mate.

The ability to detect and respond to chemicals in the air is one example of how animals rely on their nervous systems. In moth antennae, each of the thousands of small hairlike structures contains cells with chemical receptors specialized to collect specific chemical information, such as the presence of pheromones. The male moth can angle its antennae to zero in on the location of the female. The ability to find prey using echolocation or to detect chemicals in the air, combined with the ability to process these signals and respond, are examples of adaptations that improve an animal's ability to survive and reproduce. In the following section, we'll examine how the nervous system is structured, the three basic types of nerve cells, and how information flows through these specialized cells in the body.

Information flows through sensory and motor pathways to and from the brain

Bats are exceptional aerial acrobats with superb sensory perception, able to dodge trees and other bats while flying in the darkest of nights. Their ability to accomplish these feats arises from the specialization of their nervous system and the proper functioning of nerve cells, called **neurons**. Neurons are the structural and functional units of the nervous system, specialized for carrying information signals from one location to another. When linked together in a cablelike arrangement, neurons become an information pathway through all the tissues of the body. Let's begin by looking at how signals are routed through information pathways.

When an echo returns to the bat's ear, the sound stimulates specialized sensory neurons, generating a signal that must be passed along to other neurons in the information pathway. These nerve signals continue along the nervous system pathway to the brain, where they are processed, and new signals are sent to stimulate muscle groups that control flight. Bundles of neurons, called **nerves**, provide the signal pathway to and from an animal's brain. Each nerve is a cablelike structure that contains the axons from many neurons. These collections of axons are often referred to as *nerve fibers,* which are dedicated to carrying information in only one direction, either toward or away from the brain (see **Fig. 10.5**). The axons are surrounded by a layer of connective tissue for protection. Based on a wide variety of sensory inputs processed by the bat, at just the right moment motor neurons signal the bat's jaw muscles to contract.

The nervous system in vertebrate animals is divided into two primary parts: the **central nervous system (CNS)** and the peripheral collection of *sensory* and *motor neurons* found throughout the rest of the body (**Fig. 10.6**). The brain and spinal cord make up the central nervous system, which receives and integrates information delivered from various types of sensory neurons. These structures are composed of large concentrations of specialized *interneurons,* cells that are linked together, integrate signals, and produce appropriate responses.

The **brain** represents the information-processing center. It controls the other organ systems of the body either by activating muscles or by causing secretions of chemicals such as hormones. Different regions of the brain specialize in the evaluation of information about the body and its environment. In addition, the brain stores information such as memories and instinctive behavioral responses to environmental stimuli. For example, when a bat detects a moth in flight, information is evaluated in the brain and compared to memory patterns of similar encounters.

Signals, such as the perception of pressure as the bat grabs onto a rock with its feet, originate with sensory neurons. These signals are passed to larger bundles of nerves located in the spine called the

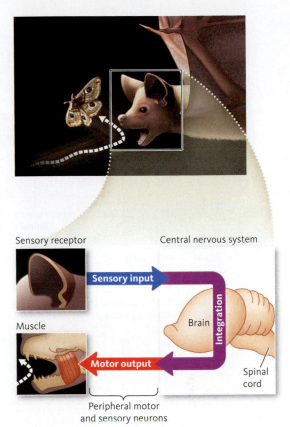

Figure 10.5 Example of a nerve impulse pathway. In this example, sound energy stimulates sensory nerves in the ear of the bat. The nerve signal is routed to the brain, where the signal is processed and new signals are generated. These new signals from the brain are routed to motor neurons in the jaw muscles, causing them to snap shut, capturing the moth in mid-flight.
© Cengage Learning 2014

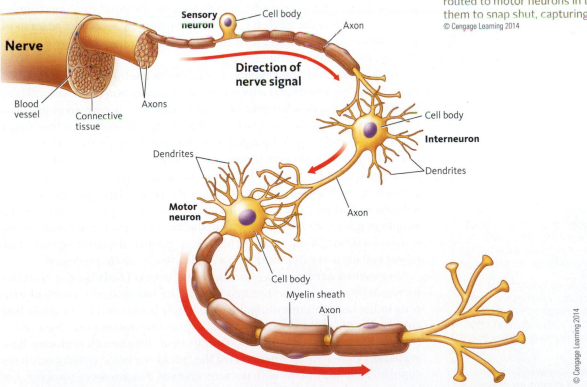

Figure 10.6 Types of neurons. Three types of neurons make up the vertebrate nervous system: sensory, interneuron, and motor. Note that the sensory and motor neurons have a myelin sheath covering, which greatly increases the speed of the nerve impulse.

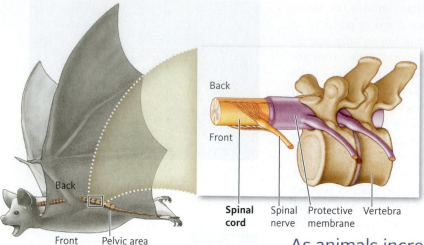

Figure 10.7 The spinal column protects the spinal cord. The spinal cord runs from the pelvic area to the base of the brain and is surrounded by a protective membrane and the vertebrae bones.
© Cengage Learning 2014

spinal cord (Fig. 10.7). Functionally, the spinal cord, protected by the flexible bony vertebrae, serves as the primary signal pathway between the brain and the rest of the body as well as a location where many reflex reactions are controlled. At many points along the spinal cord, pairs of nerves branch off to the left and right sides of the body, where they connect with sensory or motor neurons, whereas others connect with glands and internal organs. Membranes made from tough connective tissue surround the spinal cord, protecting it against damage and infection.

As animals increase in complexity so do their nervous systems

So far we've examined the complex nervous system of vertebrate animals such as bats. Now let's look at how nerve tissue and nervous systems evolved in simple animals (Fig. 10.8). Sponges are the simplest of all animals; they lack a networked nervous system. Instead, sponges rely on collections of sensory cells located in the epithelial tissues that detect and respond to their environment without the ability to process those signals. The simplest nervous systems evolved in radial animals such as jellyfish. Their nervous systems consist of nerve nets, a mesh of nerves spread throughout their bodies that detects stimuli and controls cells that contract, allowing the animal to change shape or direction. Certain neurons in their epithelial tissues are specialized to detect mechanical stimulation. When touched, they signal certain tissues to contract and others to relax—changing the shape of the animal. Although nerve nets are less complex than vertebrate nervous systems, the rapid, rhythmic contractions controlled by the jellyfish's nerve nets are a highly successful adaptation. Jellyfishes are able to detect light from the sun, and they can move quickly through the ocean to capture small prey.

At the next level of evolutionary complexity are the bilateral animals—those that have a front and a back end, as well as up and down sides. These animals show the development of a head, along with a corresponding concentration of nerve bundles (see Fig. 10.8). In the absence of a true brain, clusters of neurons, called ganglia *(singular, ganglion)*, perform basic functions of integrating sensory inputs and controlling limb movement. For example, ganglia in the heads of flatworms integrate input from light-sensitive eyespots as well as nerves located along the body. Flatworms prefer dark habitats so that they can escape predators, so the perception of light by the simple ganglia in their heads results in the movement of the animals toward a dark location. Humans have clusters of ganglia along the spinal cord, where certain nerve impulses such as reflex reactions are processed.

Segmented earthworms (Annelids) and insects (Arthropods) illustrate increasingly complex nervous systems that include concentrations of neurons in the head, representing the first simple brains and nerve cords that run the length of the back. Although segmented worms don't have a distinctive head, a concentration of neurons near their mouth performs integrated nervous system functions for the rest of the body. Arthropods have sophisticated body plans with nervous systems that are more complex. For instance, segmented body sections perform specific tasks, which include collecting environmental input from intricate sensory organs such as compound eyes and specialized antennae. Octopuses and squid are two kinds

Spinal cord—functionally, bundles of nerves located in the spine that serve as the primary signal pathway between the brain and the rest of the body, as well as a location where many reflex reactions are controlled.

Nerve nets—the simplest nervous systems that evolved in radial animals such as jellyfishes and hydras, consisting of a mesh of nerves spread throughout their bodies that detect stimuli and control cells that contract, allowing the animals to change shape or direction.

Ganglia *(singular, ganglion)*—clusters of neurons that perform basic functions of integrating sensory inputs and controlling limb movement in the absence of a true brain.

a. Porifera (sponge)

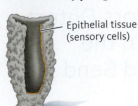

Epithelial tissue (sensory cells)

b. Cnidarian (jellyfish)

Nerve net

c. Planarian (flatworm)

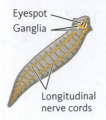

Eyespot
Ganglia

Longitudinal nerve cords

d. Annelida (earthworm)

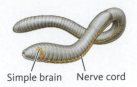

Simple brain   Nerve cord

e. Arthropod (grasshopper)

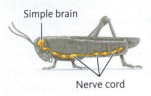

Simple brain

Nerve cord

f. Mollusk (octopus)

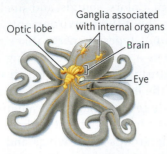

Optic lobe

Ganglia associated with internal organs

Brain

Eye

Figure 10.8 Evolution of nervous systems. Nervous systems have evolved from simple sensory cells in the epithelial tissues of sponges to nerve nets that connect these specialized cells. In more complex animals, concentrations of nerve cells, called ganglia, act as local integration centers. Worms and insects were the first animals to evolve true brains with concentrated regions of neurons where most information processing takes place.
© Cengage Learning 2014

of mollusks that have evolved sophisticated nervous systems and relatively larger brains, which give them an amazing ability to solve problems (see **Fig. 10.8**). In captivity, octopuses have learned to unscrew bottles to get at food inside and to open the lids on their aquariums to escape. Sensory perception is highly evolved in this group; in fact, the giant squid has the largest eyes of any animal—up to one foot in diameter.

The complexity of the nervous system is a function of the specialization of nerve cells and the development of sensory organs. As we move forward in this chapter, we'll investigate in more detail how individual nerve cells act to receive and process information and how they relay information along the communication pathway of the central nervous system. In addition, we will learn how nerve cells integrate with muscle groups and bones to produce coordinated movement in animals.

10.2 check + apply YOUR UNDERSTANDING

- The nervous system is a group of specialized tissues and cells that collects and transmits information about the environment and an organism's body.

- Increasingly complex organisms have a corresponding increase in the complexity of their nervous system—becoming more centralized and specialized.

- The nervous system in vertebrate animals is divided into two primary parts: the central nervous system (CNS) and the peripheral collection of *sensory* and *motor neurons* found throughout the rest of the body.

1. Describe the two primary systems that make up the overall nervous system in most vertebrates.

2. What is the primary role of a neuron?

3. How do neurons relate to nerves?

4. Describe the changes that occurred in the transition from simple animal nervous systems to more complex vertebrate systems.

5. Explain why it is important that there are links between each of the three types of nerve cells in a body.

10.3 Sensory Receptors Detect Information from the Environment and Send It to the Brain

Red-tailed hawks flying over the prairie must constantly evaluate their surroundings. Colors, sounds, and smells provide a constant stream of information about their environment. Sights and sounds are examples of *sensory input,* types of energy that stimulate sensory neurons. When the hawk sees a garter snake moving through the grass, its eyes are actually capturing light energy, which stimulates thousands of sensory neurons in the eyes (**Fig. 10.9**). The signal then travels along nerves to the brain where the image is evaluated. Recognizing the snake as a good meal, the hawk's brain stimulates thousands of motor neurons to move muscle groups in its wings and adjust its flight path to swoop down quickly to grab the snake with its sharp claws. A range of other senses, such as balance, the ability to feel pressure on its feet as it holds the snake, and its sense of smell, all contribute to the hawk's ability to find and capture its next meal.

In the previous section, we looked at the structure of the nervous system, nerve cells, and the general role of sensory and motor neurons. As we learned, sensory neurons detect information about the environment (internal or external) and communicate that information to the brain for integration and response. In the section ahead, we'll present five different types of sensory receptors and show how they adapt animals to hunt, find a mate, or escape predators in their environment. **Table 10.1** lists the five categories and the types of input that stimulate them.

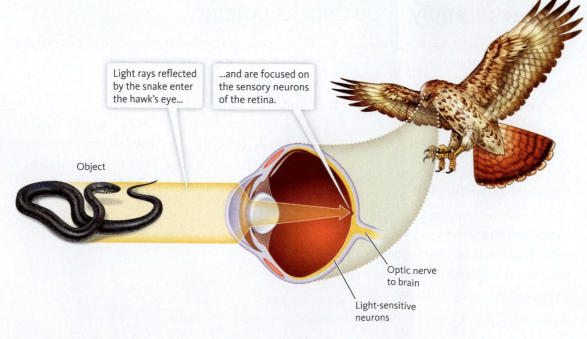

Figure 10.9 The pathway of light to the retina of a hawk. The red-tailed hawk relies heavily on its senses of vision and smell to catch its next meal. Notice the pathway the reflected light energy takes to the sensory nervous tissue located in the retina on the inside surface of the eye.
© Cengage Learning 2014

Object

Light rays reflected by the snake enter the hawk's eye...

...and are focused on the sensory neurons of the retina.

Optic nerve to brain

Light-sensitive neurons

TABLE 10.1 Five Categories of Sensory Receptors

Type of Sensory Receptor	Function
Photoreceptors	Concentrated in the eyes to detect specific wavelengths of light energy
Mechanoreceptors	Found in skin, ears, and internal organs; sensors that detect pressure, acceleration, and spatial information about limb and body position; in the ear they detect sound waves and help with balance
Pain receptors	Concentrated in the skin and some internal organs; these sensors detect tissue damage or noxious chemicals; The brain registers their activity as pain
Chemoreceptors	Detect specific types of chemicals dissolved in fluids or in the air; taste buds are an example
Thermoreceptors	Located in the skin; these sensors detect heat and cold and changes in temperature

© Cengage Learning 2014

Nerve signals have three important characteristics

Three general characteristics of a sensory neuron signal are important to understand as you learn how these specialized cells function. The first characteristic is the *type of energy* that the sensory neurons perceive, such as light or sound. The second is *signal intensity,* which is a product of the frequency of the signal being sent and the number of sensory nerve cells stimulated by the same event. As you grip your pencil harder, more pressure-sensitive nerve cells are stimulated, increasing the intensity of the signal sent to the brain. The third characteristic is *signal duration,* or how long the signal is sent. Sensory neurons can send signals only when they're stimulated, and they stop transmitting when the stimulus is removed. For example, as you release the pencil, the sensation of pressure stops. Continuous stimulation, like pressure from a sock on your foot, actually can cause sensory nerves to stop sending signals after a short period. You initially feel the pressure of the sock, but shortly after the sensation diminishes or is lost.

Animals detect light energy with photoreceptors

Our ability to see the world around us results from the types of light-sensitive cells located in our eye, the structure of the eye, and the development of the visual processing area in our brain. Photoreceptors are sensory cells stimulated by various wavelengths of light energy. For example, our eyes perceive light energy in wavelengths between 400 and 700 nm (1×10^{-9} meters, a very small unit of measure). This range of wavelengths includes the entire array of colors along with black and white (**Fig. 10.10**). Individual wavelengths represent particular colors, while white is the blending of the entire visible spectrum. Objects appear black because of combinations of dark pigments which absorb most wavelengths and reflect relatively little light. For comparison, bees see shorter wavelengths, called ultraviolet, down to 300 nm, an adaptation to colors and patterns reflected by the flowers they feed on (**Fig. 10.11**). When photoreceptors are stimulated, the absorbed light energy generates sig-

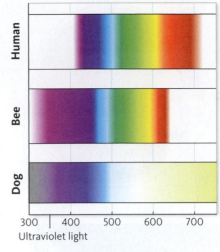

Wavelength (nm)

Figure 10.10 Comparison of bee, human, and dog vision. Bees and humans can see many of the same colors. Bees also have an adaptive advantage and can see wavelengths in the ultraviolet range, which is important information they use to find their next meal. Not all animals have the same color vision as humans. Contrary to myth, dogs are not color-blind; they just have two kinds of cone cells, which are sensitive to different wavelengths.
© Cengage Learning 2014

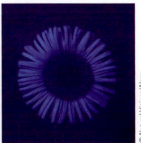

Figure 10.11 Comparison of bee and human perception of a flower. (Left) Flower as seen by humans. (Right) Same flower as it appears to a bee. The subtle patterns revealed by ultraviolet wavelengths help the bee recognize potential food sources.

Photoreceptors—specialized cells that are sensitive to light energy.

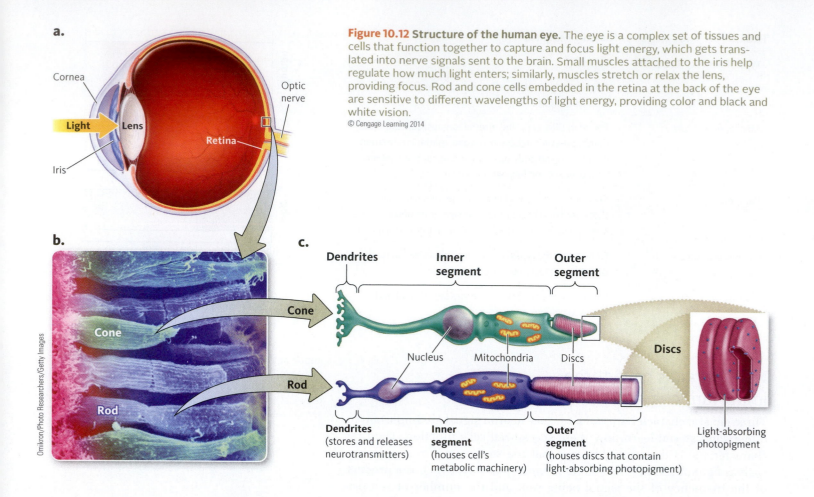

a.

Cornea

Light

Lens

Iris

Optic nerve

Retina

Figure 10.12 Structure of the human eye. The eye is a complex set of tissues and cells that function together to capture and focus light energy, which gets translated into nerve signals sent to the brain. Small muscles attached to the iris help regulate how much light enters; similarly, muscles stretch or relax the lens, providing focus. Rod and cone cells embedded in the retina at the back of the eye are sensitive to different wavelengths of light energy, providing color and black and white vision.
© Cengage Learning 2014

b.

Omikron/Photo Researchers/Getty Images

Cone

Rod

c.

Dendrites   **Inner segment**   **Outer segment**

Cone

Nucleus   Mitochondria   Discs

Rod

Dendrites
(stores and releases neurotransmitters)

Inner segment
(houses cell's metabolic machinery)

Outer segment
(houses discs that contain light-absorbing photopigment)

Discs

Light-absorbing photopigment

nals in different neurons leading to the brain, which are then interpreted as objects and colors.

Vertebrates have evolved a complex eyeball with a single lens. Light enters through the front (called the *cornea*) and passes through a clear lens, which allows the image to be focused on the back of the eye. The eyeball is enclosed and filled with fluid that helps the eyeball keep its shape and protects the tissues inside. The lens is located at the front on the eyeball (**Fig. 10.12**). The lens is flexible, changing shape as small muscles pull on it, allowing light to be focused on photoreceptors at the back of the eye. The amount of light let into the eye is controlled by small muscles attached to the colored portion, called the *iris*. The iris opens or closes in response to light conditions, an important characteristic because more light stimulates greater numbers of photoreceptors.

Photoreceptors called *rods* and *cones,* named for their shape, are embedded in the *retina,* a thin layer of neuron-rich tissue that coats the inside of the eyeball (**Fig. 10.12b, c**). Cone cells provide us with color vision and work only during the day. Humans have six million cone cells that give us sensitivity to green, red, and blue light. Our color vision is very sharp because of the high density of cones near the back of the eye, allowing us to distinguish thousands of slight changes in wavelengths that give colors their various shades.

Rod cells, which provide us with our grayish–black and white vision, have a different shape and only work in low light levels or at night. Although rods are more numerous than cones, numbering about 150 million in each eye, they are more dispersed around the retina and provide vision that is less clear than the color vision from cone cells.

Not all vertebrates see the world like we do because of different combinations of cones and rods. Animals such as cats, which are adapted to hunt at night, have greater numbers of rods than cones, whereas dogs have evolved two kinds of light-sensitive cones, which give them a different view of the world.

Hawks see prey while they are flying high in the sky, so they have evolved eyes designed for high visual acuity—the ability to detect small objects within the field of vision. Improved visual acuity results from adaptations related to the size and shape of the eye, as well as the numbers and types of cones and rods. Hawks are able to distinguish their prey two or three times farther away than humans can. How are hawk eyes different from our own? First, their eyes are large relative to their head size (**Fig. 10.13**). Large eyes allow for a larger visual field as well as more light to be gathered, providing greater stimulation of the photoreceptors and more information to be evaluated by the brain.

Like most predators, hawks have binocular vision, a necessary trait for hunters that must precisely judge ever-changing distances to moving prey. Just like your eyes, hawk eyes are rotated toward the front of the face so that the visual signals from the two eyes overlap (**Fig. 10.14**). This overlap sends slightly offset images to the brain, which is able to interpret them as a three-dimensional image providing *depth perception,* the ability to accurately judge distance. Smaller birds that are prey for hawks tend to have their eyes set on the sides of the head, allowing them to watch for danger in a wider field of view.

Insects and spiders have compound eyes made from many individual lenses fused together into one larger structure (**Fig. 10.15a**). Compound eyes can be quite complex. For example, dragonflies have eyes with as many as 30,000 individual lenses (**Fig. 10.15b**). Light enters through individual corneas, passes through a fixed crystalline cone which acts as a lens, stimulating the light-sensitive photoreceptor cells (**Fig. 10.15c**). Nerve impulses pass along the axon and arrive at a nerve ganglion that interprets the multiple images the compound eye provides. Although insects lack the ability to focus their eyes, each lens sends an independent signal to the brain, providing a tremendous amount of visual input from many directions at the same time. Although there is wide diversity in the structure and capabilities of animal eyes, each animal relies on the selective sensitivity of photoreceptors to light energy to provide it information about their environment.

Figure 10.13 Hawks have excellent binocular vision. The structure and placement of their eyes on the face give them the ability to see small objects with binocular vision while they are flying high in the sky.

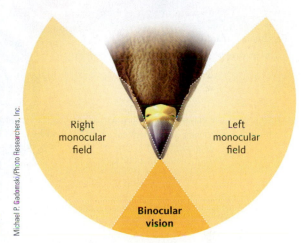

Figure 10.14 Hawk eye placement. Placement of the eyes on the front of a hawk's face provides about 70° of overlap of the signals from each eye (dark orange area), giving them excellent binocular three-dimensional vision necessary for depth perception.
© Cengage Learning 2014

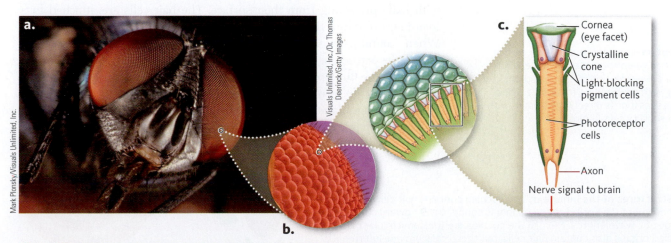

Figure 10.15 Insects have compound eyes. The eye is made up of thousands of individual lenses fused together. Each lens provides a slightly different visual signal for the brain to interpret.

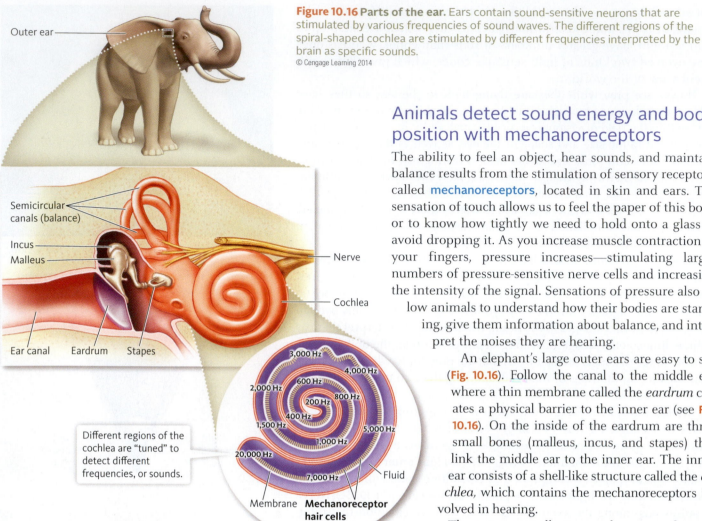

Figure 10.16 Parts of the ear. Ears contain sound-sensitive neurons that are stimulated by various frequencies of sound waves. The different regions of the spiral-shaped cochlea are stimulated by different frequencies interpreted by the brain as specific sounds.
© Cengage Learning 2014

Outer ear

Semicircular canals (balance)

Incus
Malleus

Nerve

Cochlea

Ear canal   Eardrum   Stapes

3,000 Hz
4,000 Hz
2,000 Hz   600 Hz
200 Hz   800 Hz
400 Hz
1,500 Hz   5,000 Hz
1,000 Hz
20,000 Hz
7,000 Hz   Fluid

Different regions of the cochlea are "tuned" to detect different frequencies, or sounds.

Membrane   **Mechanoreceptor hair cells**

Animals detect sound energy and body position with mechanoreceptors

The ability to feel an object, hear sounds, and maintain balance results from the stimulation of sensory receptors, called **mechanoreceptors**, located in skin and ears. The sensation of touch allows us to feel the paper of this book or to know how tightly we need to hold onto a glass to avoid dropping it. As you increase muscle contraction of your fingers, pressure increases—stimulating larger numbers of pressure-sensitive nerve cells and increasing the intensity of the signal. Sensations of pressure also allow animals to understand how their bodies are standing, give them information about balance, and interpret the noises they are hearing.

An elephant's large outer ears are easy to see (**Fig. 10.16**). Follow the canal to the middle ear where a thin membrane called the *eardrum* creates a physical barrier to the inner ear (see **Fig. 10.16**). On the inside of the eardrum are three small bones (malleus, incus, and stapes) that link the middle ear to the inner ear. The inner ear consists of a shell-like structure called the *cochlea,* which contains the mechanoreceptors involved in hearing.

The outer ear collects sound waves, a form of air pressure, and channels them through the canal to the middle ear. Sound waves vibrate the eardrum, which passes the vibrations along to the three small bones linked to the cochlea (**Fig. 10.17**). Here, the vibrations stimulate mechanoreceptors and become nerve impulses sent to the brain, which interprets the impulses as sound.

The mechanoreceptors inside the cochlea are hair cells, sensory neurons with hairs projecting from their surface. The hair cells are attached to a membrane that runs through the middle of the fluid-filled spiraling cochlea. When sound waves enter the structure, they cause the fluid to move, which in turn causes the membrane attached to hair cells

Mechanoreceptors—sensory receptors located in skin and ears that, when stimulated by the environment, allow for the perception of sound and physical touching along with contributing to balance. The strength of the sensation is directly associated with the number of sensory cells stimulated.

Side to side   Semicircular canals   Vestibular nerve   Cochlear nerve

Up/down

Left/right   Cochlea

Figure 10.17 Structures of the inner ear. The cochlea contains hair cells that are sensitive to movement. The spiral shape channels sound waves toward cells sensitive to specific wavelengths. Next to the cochlea are the semicircular canals. These fluid-filled organs also contain hair cells, but these cells provide information on orientation and balance.
© Cengage Learning 2014

to vibrate. The movement displaces fluid which bends the hair cells, causing them to send signals to the brain. Loud sounds exert greater pressure than do softer sounds and produce greater intensity signals to the mechanoreceptors.

Animals are able to detect different frequencies because of the arrangement of the cochlea and the density of mechanoreceptors. Frequencies differ from sound intensity and refer to the oscillation rate of pressure waves. We perceive the frequency of sound waves as pitch, like a musical note. The spiral section of the cochlea contains regions sensitive to different frequencies. Different regions of the cochlea are "tuned" to detect different frequencies. **Figure 10.16** shows where along the curved cochlea different frequencies are perceived.

In vertebrates, hearing and balance are senses located in adjacent structures in the inner ear. In a similar manner to hearing, balance relies on the stimulation of mechanoreceptors. Three *semicircular canals* located next to the cochlea contain receptor cells stimulated by movement. The canals are situated in three different axes so that movement in one direction stimulates only one set of mechanoreceptors (see **Fig. 10.17**). The stimulation of hair cells in this organ provides a sense of balance and signals our brain that we are upright. When we move, the hairs bend in a different direction sending different signals. The brain integrates input from each ear, the eyes, and receptors throughout the body, allowing us to keep objects in focus and maintain awareness of our bodies' orientation as we move. The spinning motion of a carnival ride may make us dizzy or cause motion sickness because of conflicting sensory input between our eyes and the spinning motions detected in our semicircular canals.

Fish rely on a different arrangement of mechanoreceptor cells, called *lateral lines,* embedded in their skin along the length of their bodies. The sensitivity and density of these sensory neurons provide critical information about their environment (**Fig. 10.18**). The lateral line consists of hair cells recessed in small openings along the body. Collections of hair cells linked to nerves are protected by a gelatinous covering and provide the fish with a range of information like its orientation in the water, allowing it to swim upright, and the ability to detect movement or vibrations. The sensory cells of the lateral line help fish find food, avoid predators, and move in a seemingly effortless way in large groups of fish.

Animals detect chemicals, pain, and temperature with a diverse set of receptors

The perception of taste and smell in vertebrates results from stimulation of sensory receptors by specific chemicals (**Fig. 10.19**). Smells and tastes are detected when chemicals bind to membrane proteins found on the surface of **chemoreceptors** located in the sinuses or mouth. The binding action of the chemical stimulates the chemoreceptor cells. Intensity of a smell or taste is related to the number of chemoreceptors stimulated at the same time. As more chemicals bind, more signals are sent to the brain and interpreted as a stronger sensation.

In vertebrates, the chemoreceptors for smell and taste are concentrated in the sinuses and the mouth. Humans have approximately 50 million chemoreceptor cells located in hairlike structures high up in the nasal passages, whereas a dog may have 300 million (see **Fig. 10.19**). The various combinations of chemicals detected stimulate different sets of sensory

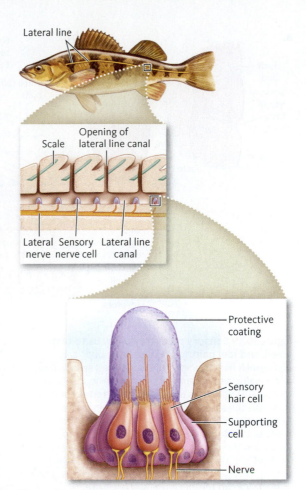

Figure 10.18 Hair cells on their skin provide sensory information for fish. The lateral line embedded in tissues along the length of the fish's body contains hair cells that are sensitive to pressure changes in the water surrounding the fish and also provide information on orientation and balance.
© Cengage Learning 2014

Chemoreceptors—sensory receptors that, when stimulated by chemicals in the environment, allow for the perception of taste and smell in vertebrates.

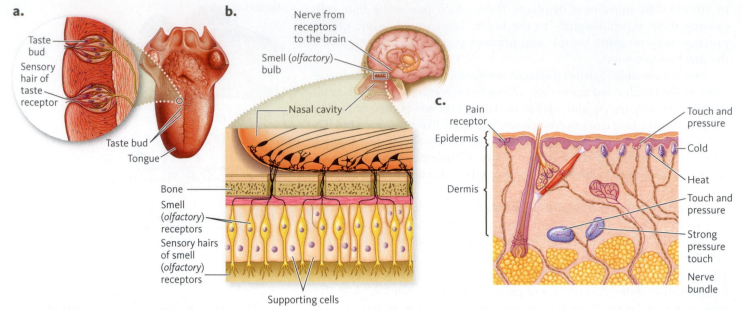

a.
Taste bud
Sensory hair of taste receptor
Taste bud
Tongue

b.
Nerve from receptors to the brain
Smell (*olfactory*) bulb
Nasal cavity
Bone
Smell (*olfactory*) receptors
Sensory hairs of smell (*olfactory*) receptors
Supporting cells

c.
Pain receptor
Epidermis
Dermis
Touch and pressure
Cold
Heat
Touch and pressure
Strong pressure touch
Nerve bundle

Figure 10.19 Sensory receptors allow us to taste, smell, and feel pain. (a) Human taste buds can distinguish five basic chemical qualities: *sweet, sour, bitter, salty,* and *umami* (amino acids/monosodium glutamate). All other *tastes* are detected in combination with the smell (olfactory) receptors in the sinuses. In this way, (b) our sense of smell makes a significant contribution to our sense of taste. (c) Pain receptors located in the skin and organs are sensitive to different kinds of stimulation, such as hard bumps, swelling, tearing, and pinching. Other sensory receptors provide information on temperature and pressure.
© Cengage Learning 2014

Pain receptors—sensory receptors that, when stimulated, allow for the perception of pain caused by tissue damage such as cuts or sharp blows to the skin.

Thermoreceptors—sensory receptors located in the skin that, when stimulated, allow for the perception of hot and cold.

cells, which in turn are interpreted as unique smells. Taste is similar to smell, but the chemoreceptors are located in taste buds, pits located on and around the tongue (see **Fig. 10.19a, b**). Although located in a different spot, the chemoreceptors work the very same way as those found in the sinuses. However, they are far fewer in number, and they are sensitive to fewer chemical combinations.

In insects, chemoreceptors for taste are generally located near the mouth, although bees, ants, and wasps have taste receptors located on their antennae. The fine, hairlike structures along the length of the antennae of the gypsy moth contain chemoreceptors sensitive to thousands of chemicals. Butterflies and moths have more than a million chemoreceptors on their antennae, which can be angled in many directions, serving as a type of chemical radar used to detect pheromones up to one mile away.

The perception of pain results from stimulation of specialized sensory receptors caused by tissue damage, such as cuts or sharp blows to the skin (**Fig. 10.19c**). Several different kinds of **pain receptors** are located in a variety of locations, such as the skin, muscles, joints, bladder, and digestive tract. As more receptors are stimulated, larger numbers of signals are sent and interpreted as more intense pain. The tips of most animal paws or fingers have high densities of these receptors, which is why even a small paper cut can seem intense.

Pain is an uncomfortable feeling that tells you something may be wrong with a part of your body. Pain comes in many different forms. Sometimes it's just a nuisance, like a mild headache; at other times, it can be debilitating. *Somatic pain* is pain familiar to all of us, felt on the skin, muscle, joints, bones, and ligaments. Somatic pain is generally sharp and localized. If you touch or move the area that hurts, the pain will increase. The receptors for this type of pain are very sensitive to temperature (hot or cold), vibration, and stretch (in the muscles). They also are sensitive to swelling that occurs when you cut yourself, sprain a muscle, or burn your skin. Muscle cramps are a type of somatic pain caused by a lack of oxygen in the muscle that hurts.

The ability of animals to sense heat and cold is based on the stimulation of sensory neurons, called **thermoreceptors**, located in the skin (see

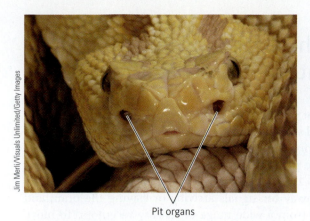

Pit organs

Figure 10.20 Pit organs help some snakes sense the world around them. Rattlesnakes are able to sense heat energy using specialized thermoreceptors, called pit organs, located on the front of their face. These organs provide a distinct advantage for hunting at night or in dark animal burrows underground. The thermal image of the woman on the right gives us an idea of the heat energy collected by the snake's pit sensors.

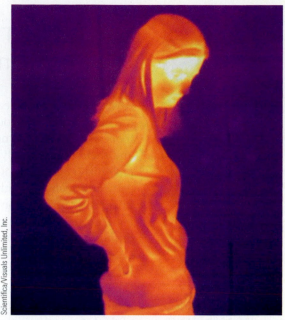

Fig. 10.19c). The distribution of these sensors is not even. Your fingertips and the inside of your mouth have higher densities of these neurons that help protect you from burning or freezing these tissues. Thermoreceptors sense temperatures and can be sensitive to variations as little as one degree.

Rattlesnakes have a unique kind of thermoreceptor arrangement. The eastern diamondback, the largest of all rattlesnakes, has thermoreceptors recessed in a pit organ on the front of its face to detect the warmth of potential prey or threats (**Fig. 10.20**). The pit organ acts like night vision goggles for the snake. This allows the snake to paint a visual picture of heat sources in its environment, a very successful adaptation for animals that hunt warm-blooded or endothermic prey such as rodents at night.

10.3 check + apply YOUR UNDERSTANDING

- Information about an animal's external and internal environments is collected by sensory receptors linked to nerves.

- Sensory receptors provide information when stimulated by specific types of energy.

- There are five basic types of receptors common to most animals.

1. In your own words, describe the role of sensory receptors in the animal world.

2. List the five types of sensory receptors and what types of signals stimulate them.

3. Compare the role of the sensory neurons inside our ears that have hairs on their surface with the hairs on the insect's body.

4. Wolves are active hunters during the day and at night. They use their sensitive sense of smell and keen vision to locate prey at great distances. Speculate about the types of photoreceptors found in their eyes and the density of chemoreceptors in their noses.

5. If you lost hair cells at the very beginning of the spiral of the cochlea, what kind of hearing loss would you experience?

Figure 10.21 The brain and spinal cord of elephants interpret signals and relay responses. Information about their environment helps elephants keep track of one another and find food, watering holes, and locations for dust baths.

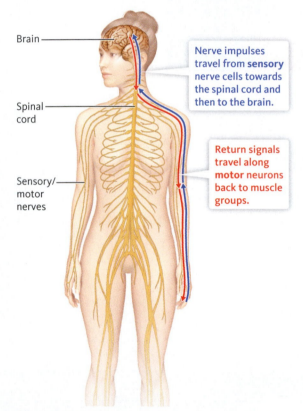

Brain

Spinal cord

Sensory/ motor nerves

Nerve impulses travel from **sensory** nerve cells towards the spinal cord and then to the brain.

Return signals travel along **motor** neurons back to muscle groups.

Figure 10.22 Nerve impulses travel both to and from the brain. Nerve impulses travel from sensory nerve cells toward the spinal cord and then to the brain. Return signals travel along motor neurons back to muscle groups.
© Cengage Learning 2014

Reflex arcs—localized, involuntary, spontaneous nervous system loops located in certain portions of the animal body. Reflex arcs are actions that occur without signal processing in the brain and are almost instantaneous interactions between nerves and muscle groups initiated by a stimulus.

Brain stem—a protected area located underneath the brain that is dedicated to vital housekeeping functions in the body, such as blood pressure, respiratory rate, heart rate, coughing, and reflex reactions such as vomiting.

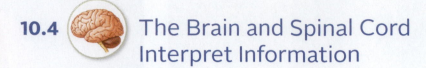

10.4 The Brain and Spinal Cord Interpret Information

African elephants can eat up to 400 pounds of food each day. On the savannas of Africa, where food sources can be widely dispersed, they must rely on their senses, their memory about the location and types of food available, and their ability to solve problems such as selecting a route and avoiding potential dangers (**Fig. 10.21**). The elephant's brain and spinal cord work together, providing a pathway for signals arriving or leaving the brain and integrating information from a wide range of sensory receptors. To move between feeding sites, the elephant must coordinate movements of many different muscle groups that control leg movement, the head, and the trunk. Coordinated movements result from activities in the brain that continually integrate sensory inputs, such as pressure, balance, and visual cues. These capabilities are associated with highly developed sections of the vertebrate brain dedicated to thought. In this section, we'll explore the functional role of the brain and spinal cord in interpreting signals and relaying responses.

The spinal cord is a signal pathway and controls some reflex movements

The sensory and motor nerves connect to nerve communication pathways to the brain at the spinal cord. In animals, the spinal cord is divided into many segments, with right and left pairs of nerves branching off and leading to various portions of our trunk or limbs. The sensory and motor nerves end at these segments or junction points along the spinal cord, where they branch off (**Fig. 10.22**). Each junction represents a link between the central nervous system and the wide variety of sensory neurons and muscle groups.

In addition to acting as a communication pathway, the spinal cord works in controlling reflex reactions. **Reflex arcs** are involuntary, spontaneous actions that occur without signal processing in the brain. Reflex arcs are almost instantaneous interactions between nerves and muscle groups, generated by a stimulus. When the doctor taps a rubber hammer on your kneecap, she is testing one of your body's reflex arcs. The tapping stimulates stretch-sensitive mechanoreceptors in the knee that send signals to the spinal cord. The sensory neurons end directly on motor neurons that stimulate muscles in the thigh that make your knee jerk (**Fig. 10.23**). Quickly releasing a hot pan is an example of a reflex arc. All animals have similar kinds of reflex arcs. The direct link between sensory and motor neurons occurs within the spinal column and bypasses the signal integration, which occurs at the brain. This represents an effective evolutionary adaptation designed to protect the body from harm. This adaptation is important because the reflex arc response is faster than one involving integration and signal generation from the brain.

The brain stem regulates many bodily functions and the cerebellum has a role in motor function and coordination

The vertebrate brain contains many specialized regions and subdivisions where specific activities take place. Tucked underneath the brain in a highly protected location at the top of the neck is the **brain stem**,

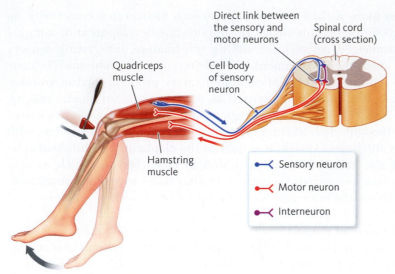

Figure 10.23 Reflex arcs link sensory and motor neurons at the spinal cord.
© Cengage Learning 2014

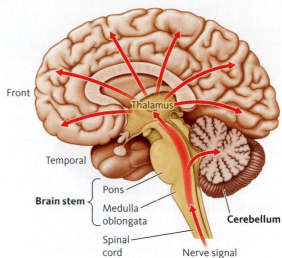

Figure 10.24 Body-regulating housekeeping functions are located in a protected area on the underside of the brain. The pons and the medulla oblongata, located on the underside of the brain, play vital roles in controlling activities such as respiration and blood pressure. The cerebellum controls fine motor activities.
© Cengage Learning 2014

an area of the brain dedicated to vital housekeeping functions in the body. The thickened region at the bottom of the brain shown in **Figure 10.24** is the brain stem. This area contains the *medulla oblongata* and *pons,* two regions that coordinate involuntary body regulating functions such as blood pressure, respiration rate, heart rate, coughing, and reflex reactions such as vomiting. Damage to the brain stem can cause critical health problems or death.

Bundles of interconnected neurons run the full length of the brain stem connecting to the spinal cord at the bottom and a signal relay center called the thalamus at the top. Important activities such as control over the sleep–wake cycle are located in this nerve bundle. A certain amount of nerve-signal filtering also occurs here, a necessary function because the brain is unable to process every one of the signals from millions of sensory neurons arriving simultaneously. For instance, while you are sleeping, background noises are filtered, whereas barking dogs or alarm clocks are processed.

The **cerebellum** is a large, bell-shaped outgrowth of the pons that has extensive neuron connections with other regions of the brain (see **Fig. 10.24**). Because of its location, it is relatively well protected from damage compared to the front and temporal regions and brain stem. This region plays an important role in muscle control, some mental functions such as attention and language, and some emotional functions such as fear and pleasure responses. A chief function is control of fine motor skills, required in activities such as writing, coordinated movement, focusing the eyes, or buttoning a shirt. Because balance is an important part of movement, the cerebellum works very closely with the part of the brain that controls balance, receiving input from the semicircular canals near the eardrum.

The cerebrum coordinates sensory and motor functions

One of the most notable features of the vertebrate brain is the wavy, infolded outer layer called the **cerebrum**, a characteristic that strongly distinguishes mammals from other vertebrates. The folding of tissue

Cerebellum—a large, bell-shaped outgrowth of the pons that plays an important role in muscle control, some mental functions such as attention and language, some emotional functions such as fear and pleasure responses, and fine motor skill control of muscles.

Cerebrum—the large, infolded region at the top/front of the brain. This region functions in behavior, movement, sensory processing, communication, problem solving, and memory.

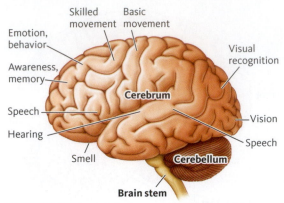

Figure 10.25 The cerebrum. In animals that exhibit complex behaviors, the folded area of the cerebrum contains regions devoted to processing a range of sensory input as well as areas dedicated to higher-order processes such as problem solving.
© Cengage Learning 2014

creates more surface area and allows each neuron to connect with up to 10,000 others. This structure supports the development of a highly interconnected network for storing information, integrating sensory input about the environment, and solving complex problems. The cerebrum contains regions dedicated to memory, problem-solving, communication, and movement, as well as sensory areas for smell, hearing, and vision (**Fig. 10.25**). In this region higher-order processes associated with thought and learning take place, capabilities associated with highly intelligent vertebrates such as the elephant. The infolded cerebrum also contains areas dedicated to characteristics such as self-awareness, considered by many scientists to be a unique human trait. In addition, the brain stores critical information such as memories and the instinctive behavioral responses.

The areas of the cerebrum where sensory information from locations such as skin and internal organs are interpreted are shown in blue and red in **Figure 10.26**. The left side of the brain controls the right side of the body and vice versa. Notice that the face and mouth, with their rich supply of nerves, encompass a large portion of the sensory area. The ability to blink, whistle, and show many emotions with facial expressions results from having such a large area of the cerebrum devoted to this part of the body along with the larger number of receptors and small muscles in these areas. In contrast, only a small region of this area receives sensory input from the legs. The motor areas (shown in red in **Figure 10.26**) initiate motor

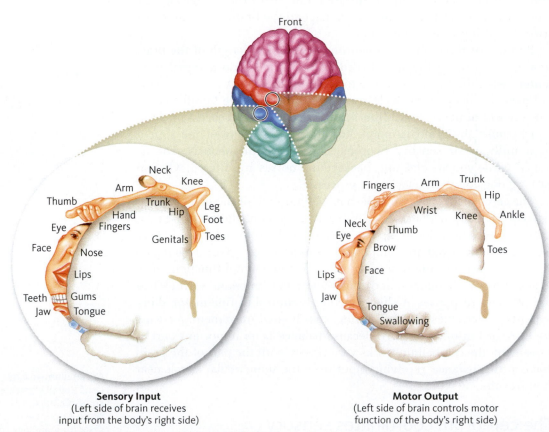

Sensory Input
(Left side of brain receives input from the body's right side)

Motor Output
(Left side of brain controls motor function of the body's right side)

Figure 10.26 The specialized hemispheres of the human brain. The left and right halves of the brain control motor (red region) functions on opposite sides of the body. The relative size of the feature in the drawing indicates how much of that region of the brain is devoted to sensing (darker blue region) and controlling that area. For example, the face has a large area of control compared to the tongue, which provides us with excellent fine motor control or dexterity of this fleshy tool.
© Cengage Learning 2014

Human	Elephant	Mouse	Fish

Figure 10.27 Comparison of brain development in mammals and fish. Large animals like elephants and humans have large, well-developed cerebrums and the ability to solve complex problems and retain memories. The brains of mice and fish lack this tissue and the abilities it provides.
© Cengage Learning 2014

signals to skeletal muscles in these specific areas. Damage to a specific area of the sensory or motor region of the brain affects those particular sensations or movements.

Not all vertebrate brains have the same level of development (**Fig. 10.27**). Animals thought of as highly intelligent, such as humans and dolphins, possess complex cerebral development in the large, infolded regions. Both dolphins and elephants have the ability to solve problems and process memories associated with a well-developed cerebrum. Not all animals have well-developed infolding of the cerebrum; this area is nearly absent in small mammals such as mice. Although mice can seem crafty and able to perform tasks like learning mazes, they lack complex behaviors, such as communication, that are related to development of the cerebrum.

10.4 **check + apply** YOUR UNDERSTANDING

- The brain and spinal cord are composed of specialized neurons that receive and integrate information and select responses.

- The brain is a large collection of several trillion integrating neurons linked together in an extensive array operating in highly specialized regions.

1. In your own words, define what a reflex arc is.
2. What is the primary role of the cerebellum in the brain?
3. Examine the brains of the elephant, mouse, and fish shown in **Figure 10.27**. Would you expect a fish to have well-developed problem-solving abilities? Would you expect the mouse to feel emotions to the same degree as elephants? Defend your prediction.
4. A patient has feeling in his mouth and tongue but is having trouble speaking, swallowing, and smiling on the right side of his face. Where might you expect to find a tumor that would cause this problem?
5. A dog that has been injured has damage to its medulla oblongata and pons. What types of difficulties would the animal be experiencing?

10.5 Cells of the Nervous System Are Specialized to Transfer Signals

Figure 10.28 Mexican free-tailed bats.

Mexican free-tailed bats rely heavily on their highly developed echolocation to navigate the night skies (**Fig. 10.28**). Their ability to accomplish this feat arises from the complexity of their nervous system and adaptations derived from millions of years of natural selection. As we've learned, neurons play a vital role in the ability of animals to sense and respond to their environments, regulate homeostasis, process information, and move. Each of these critical processes depends on the proper function of nerve cells. In this section, we'll examine the nervous system at the cellular level, exploring the structure of neurons and the process of how nerve signals are generated and sent along nerve pathways. The functioning of the neurons not only allows animals to sense the environments and control their actions but also is at the base of how we store memories, learn, and feel.

Three categories of nerve cells share roles and characteristics

Nerve cells play a common role in the animal world. From zebras to zebra fish, nerve cells share the same function of communicating information. They also share characteristics such as external shape and internal arrangement. Recall that there are three broad classes of nerve cells, the basic unit of the nervous system, and all are dedicated to communicating information (**Fig. 10.29**). *Sensory neurons* convey information from the tissues and organs of the body toward the central nervous system. *Motor neurons* convey information from the central nervous system out to muscle, gland, or other nerve cells. Motor neuron cell bodies are located in the spinal cord with cell membranes that can extend all the way out to sensory receptors located in the skin. *Interneurons* are found in the central nervous system. They function by relaying information between sensory and motor neurons, as well as within the brain and spinal cord.

The branchlike extensions found on most neurons look odd compared to other body cells, but they have many of the same characteristics. Like epithelial or bone cells, neurons have a cell body that contains a nucleus with DNA, along with familiar organelles including mitochondria and endoplasmic reticulum (**Fig. 10.30**). Inside, you'll find cytoplasm and metabolic activities occurring just as in a muscle cell in the heart or an epithelial cell of your skin. Although structurally similar to other cells, functionally a nerve cell's only role is to transmit signals.

A typical sensory or motor neuron has a distinctive, elongated shape, with the cell body located near one end (see **Fig. 10.30**). Interneurons are much more compact, with branching patterns radiating out from

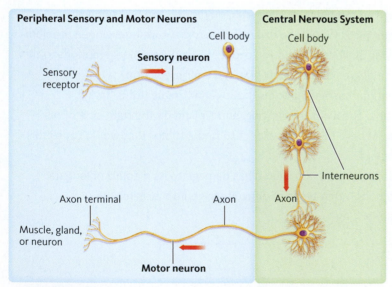

Figure 10.29 Three classes of neurons. Sensory neurons are linked to interneurons of the spinal cord and brain, whereas motor neurons carry signals away from the brain and spinal cord.
© Cengage Learning 2014

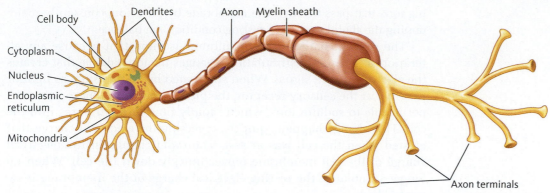

Figure 10.30 **Structure of a nerve cell.** Neurons have distinctive shapes, but they all engage in one activity: sending signals from one location to another. While they may have strange shapes, they are cells that contain all the necessary internal organelles and structures found in other cells.
© Cengage Learning 2014

the entire cell body. The branchlike structures near the cell body are dendrites, where neurons receive information from adjacent neurons. When a neuron is stimulated, the signal enters at the dendrites, moves toward the cell body, and then proceeds along the long, ropelike axis of the neuron called the *axon*. Sensory and motor neurons have a myelin sheath, an outer covering that increases the speed of nerve signals. The axon is an extension of the cell membrane involved in sending nerve signals to neighboring cells. The length of the axon varies by location in the body and by the animal. At the end of the axon are the axon terminals, peglike structures where signals are passed to adjacent neurons or muscles.

Nerve signals develop from action potentials

Nerve signals begin when one or more neurons respond to a stimulus and generate an electrical impulse—a change in the electrical charge across the cellular membrane of all cells. There are variations on how nerve signals develop, so we'll focus on a common example. At the cellular membrane, the change in electrical charge from the stimulation is called an **action potential**. The electrical charge isn't very strong, only about 100 millivolts, or 1/1,000 of a volt. For comparison, the voltage of the electricity in the wires of your home is about 120 volts. For each neuron, the action potential is an all-or-nothing event, similar to turning on a light switch. A stimulation either is sufficient to generate an action potential at an individual neuron or it is not.

To understand an action potential, we first have to set the stage for the cellular membrane conditions when the neuron is at rest, before it is stimulated. At rest, neurons have a slightly negative internal electrical charge. The charge results from a greater concentration of positively charged sodium ions located outside the cell and fewer positively charged potassium ions inside the cell (**Fig. 10.31**). The difference in ion concentrations results from an active transport sodium–potassium pump, a collection of membrane proteins that move sodium ions out of the cell and potassium ions into the cell, using ATP to fuel the process. Even at rest when no action potential is being sent, the sodium–potassium pump is active, transport-

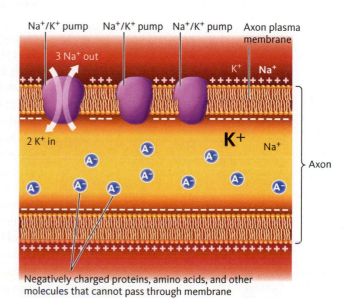

Figure 10.31 **Action potential.** This cutaway of an axon shows events at the plasma membrane of the neuron. Large proteins (purple) use energy to move potassium (K^+) and sodium (Na^+) ions rapidly across the membrane against their concentration gradients. This helps maintain a charge difference between the inside and outside of the neuron, which sets the stage for the rapid creation of an electrical signal, or action potential.
© Cengage Learning 2014

Action potential—nerve signals begin when one or more neurons respond to a stimulus and generate an action potential—a rapid and temporarily irreversible change in the charged ion concentration across the cellular membrane of all cells. The electric charge created by the movement of charged ions isn't very strong.

ing ions that pass through the membrane by diffusion, constantly maintaining the slightly negative resting-condition ion balance.

The resting charge of the membrane is important because any alteration, as a result of stimulation, changes the ion balance that creates the electrical nerve signal. When stimulated by factors such as light or pressure at the sensory receptor, the plasma membrane becomes more permeable to sodium ions, which rapidly flow through pores into the cell. This change happens quickly, reversing the electrical charge that existed when the cell was at rest. The rapid change in the electrical charge of the cell membrane moves quickly down the cell. When no longer stimulated, the resting electrical charge of the membrane is re-established.

Once generated, a nerve signal travels in only one direction, from the receiving dendrites to the cell body and down the axon. Think of nerves as one-way streets, where information signals are restricted to a single lane and all travel in one direction toward their destination. The movement of sodium ions across the membrane creates a short period of time when the neuron is returning to its resting potential, guaranteeing that the signal can move in only one direction. Keep in mind that the action potential may or may not be sufficient to stimulate the next neuron along the pathway and continue as a nerve signal, either toward or away from the CNS, or the ganglion in invertebrates. Strong inputs that stimulate more sensory neurons, or sustained signal inputs, produce greater sensory perception.

Not all action potentials travel at the same speed; in fact, structural differences along the axon of some neurons actually speed up nerve signals. The axon has a series of thick and thin regions (see **Fig. 10.30**). The thick regions are support cells wrapped around the axon like a jellyroll. They provide metabolic assistance as well as protection to neurons. This cellular covering, called a *myelin sheath,* allows the action potential to move quickly from the axon of the neuron over the surface of the myelin sheath from one node to the next. This greatly speeds up the rate at which a nerve signal moves by avoiding the slower ion-generated electrical signal. Nerve signals in myelin-covered axons travel at speeds of 450 mph, whereas neurons without the axons have signals that move at little more than 1 mph. The speed difference is an important adaptation for survival and protection; for instance, the withdraw reflex reaction when you touch a hot pan reduces the potential for a bad burn.

Signals between neurons involve neurotransmitters

As the bat moves through the night sky, its sensory neurons provide a continual stream of information for its brain to interpret and respond to. These nerve signals must be passed from one neuron to another in a continuous chain to and from the brain. Let's consider an example neuron involved in sending signals to muscles in the bat's wing in response to sensory input related to balance. When the action potential reaches the end of an individual neuron, several things can happen. If the signal is too weak to stimulate the next nerve, the action potential stops. However, if the frequency of stimulation is high enough or the number of cells stimulated is large enough, the action potential can stimulate the next neuron in the signal pathway. Most nerve cells interact with many others; thus, signals result from the combined input of several neurons experiencing the same stimulus.

Nerve cells don't actually touch one another; there is a very small fluid-filled gap between adjacent neurons, called a *synapse* (**Fig. 10.32**). Although

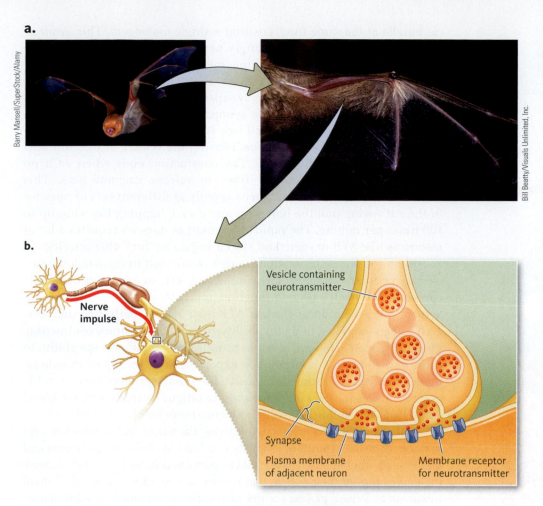

a.

b.

Nerve
impulse

Vesicle containing
neurotransmitter

Synapse

Plasma membrane
of adjacent neuron

Membrane receptor
for neurotransmitter

Barry Mansell/SuperStock/Alamy

Bill Beatty/Visuals Unlimited, Inc.

Figure 10.32 The path of a nerve impulse. Nerve impulses travel along nerves from the brain to muscles in the wing. At junctions between nerve cells *(synapses)*, different neurotransmitters carry the electrical impulse of an action potential from one cell to the next.
© Cengage Learning 2014

the gap between neurons is tiny, the nerve signal must be transmitted across this space in order for it to continue along its pathway. To flap its wings and remain on the right flight path to catch the moth, a bat must generate nerve impulses in the brain that travel across a series of synapses along the pathway to the muscles. At each synapse, the upstream neuron releases a chemical signal called a **neurotransmitter**, which crosses the gap and stimulates the downstream neuron cell membrane, changing its membrane potential (**Fig. 10.32b**). In effect, the neurotransmitter "carries" the electrical signal across the synapse, allowing the communication from the bat's brain to the muscle to continue on its way.

Motor neurons stimulate muscle cells at neuromuscular junctions

The skeletal muscles located in the bat's wing have a different kind of neuron junction where the end of the motor neuron axon comes in close contact with muscle cells (**Fig. 10.33**). The synapses between motor neurons and muscle fibers are called *neuromuscular junctions,* where neurotransmitters stimulate muscle fibers to contract. Neurologists have identified more than 30 different types of neurotransmitters, and the most common is a compound called acetylcholine (ACh), which is found in neurons that stimulate skeletal muscles. Like all neurons, motor neurons have many axon terminals that branch out from the cell membrane. These terminals synapse with many different muscle cells within a mus-

Neurotransmitter—a chemical signal that crosses the gap between adjacent neurons and stimulates the downstream cell. In effect, the neurotransmitter "carries" the nerve signal across the synapse, allowing the communication to continue on its way.

Motor unit

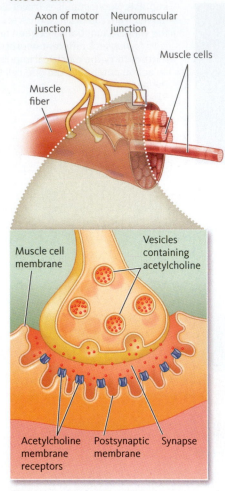

Axon of motor junction

Neuromuscular junction

Muscle cells

Muscle fiber

Muscle cell membrane

Vesicles containing acetylcholine

Acetylcholine membrane receptors

Postsynaptic membrane

Synapse

Neuromuscular junction

Figure 10.33 Motor nerves. Motor nerves end at the neuromuscular junction, where they come in close contact with bundles of muscle cells, called motor units.
© Cengage Learning 2014

Dirk Ercken/Shutterstock.com

Figure 10.34 Poison dart frog. The poison dart frog is colorful but deadly. Small amounts of secretions that collect on its skin can kill large animals in a very short period of time.

cle bundle at the same time, creating a single *motor unit*. This arrangement is important because a single nerve signal has the potential to stimulate the contraction of multiple muscle cells at the same time.

When an action potential arrives at the terminal end of a motor neuron, ACh is released into the synapse. ACh moves down its concentration gradient and crosses the synapse gap, coming in contact with the membrane of the muscle cell (see **Fig. 10.33**). ACh then binds to receptors on the plasma membrane, causing the release of calcium ions and the contraction of muscle. The contraction ends when ACh no longer stimulates the muscle fiber to release calcium ions. This contract-and-relax process happens rapidly as different sets of muscles in the bat's wing pull the bones up and down, flapping the wing up to 100 times per minute. The rapid movement of muscles requires a lot of resources like ATP, oxygen, and stored sugars to "fuel" this activity.

When continual stimulation keeps a motor unit in contraction, muscle fatigue follows because low supplies of ATP, oxygen, and metabolic compounds force the cells to function anaerobically. Recall that the anaerobic process produces far less ATP than aerobic pathways, and lowered energy supplies limit the ability of cells to continue contracting. Physically, muscle fatigue is felt as muscles get "tired" and the ability to exert muscle tension declines. You experience this feeling of muscle fatigue when you hold something heavy for a long period of time. Aerobic exercise makes muscles more resistant to fatigue by increasing the blood supply and the number of ATP-producing mitochondria in each cell.

There are adaptive advantages to varying the size of motor units. We tend to think of the bat in terms of its ability to echolocate or fly, but its fingers and toes also are nimble, and bats are able to perform delicate tasks, such as feeding and caring for their young, that require precise muscle control of small motor units. Where precise control of muscle movement is needed, it's an adaptive advantage to have a single neuron stimulate a small motor unit that contains a few muscle cells and has an increased control over which fibers contract. In contrast, the large flight muscles located in the bat's breast require less precision in their movement, and thus each motor neuron stimulates larger numbers of cells in each of these larger motor units.

Many chemicals can interfere with nerve signals

Many chemical compounds found in nature can interfere with the development of action potentials and the neurotransmitter signaling between neurons. What would happen if a chemical interfered with the movement of ions across the cell membrane of neurons? The Columbian poison dart frog is a small frog native to parts of Central and South America (**Fig. 10.34**). These brightly colored amphibians are easy to spot against the green vegetation. In fact, their bright color is a warning signal to would-be predators, identifying them as a dangerous meal. If these warning signs are ignored, animals that prey upon these frogs die. The skin of these frogs secretes a chemical that, when consumed, permanently alters the shape of the sodium ion pumps in the cell membrane of neurons, allowing ions to pass through freely. When absorbed into the bloodstream of an animal, the toxin short-circuits the development of action potentials in motor neurons, such as those necessary for breathing. With their respiratory muscles disabled, victims stop breathing and die. The toxin is so deadly that just a few grams will kill an adult human in less than half an hour. The toxin doesn't affect the frogs because they have developed sodium-ion protein channels with a different shape, and, as a result, the toxin cannot bind to their modified ion pump.

Humans have discovered a number of natural chemical extracts from plants as well as artificially created compounds that speed up or slow down the nervous system. Recreational drugs, alcohol, and caffeine all have an effect on the nervous system, typically in a negative way by interfering with the development of action potentials or the transmission of signals at nerve synapses. Cocaine, methamphetamine, and ecstasy are examples of drugs that are classified as stimulants, causing control centers in the brain to speed up respiration rate, heart rate, and blood pressure. Caffeine, a compound found in coffee and energy drinks, is also a stimulant, although its effects on the body typically are limited to warding off drowsiness in regular users. For people who are sensitive to the effects of caffeine, it can cause a feeling of nervousness, restlessness, and even dehydration. Alcohol is an example of a compound classified as a depressant, a compound that slows activities in the nervous system. For example, alcohol slows the heart rate, lowers blood pressure, and inhibits the release of ACh in motor neurons—slowing muscle reaction times. In the brain, depressants such as heroin or morphine inhibit communication between interneurons, slowing down the ability to process information and think.

Damage to nerve cells and tissues can be permanent

Sports injuries, automobile accidents, or even bad falls can permanently damage neurons of the brain and spinal cord resulting in paralysis or loss of certain brain functions. However, unlike other cells, at maturity most neurons in adults cannot divide. When they are damaged or die from an injury, they cannot be replaced through mitosis. Not all neuron damage is fatal to the cell; if the cell body is undamaged, the cell can regenerate the dendrites or axons. This is why a finger that is cut off can be successfully reattached. However, when neurons located in the spinal cord are severely damaged, all the muscles supplied by those nerves below the injury become paralyzed and are unable to contract voluntarily.

10.5 check + apply YOUR UNDERSTANDING

- Cells of the nervous system are specialized to transfer signals.

- There are three broad categories of neurons, which play similar roles and have similar characteristics.

- Neurotransmitters act as chemical links between adjacent neurons.

- Motor neurons stimulate groups of muscle cells called motor units to contract at once. The junction between the motor neuron and motor unit is called the neuromuscular junction.

- Certain types of damage to nerve cells can be permanent.

1. What are the three categories of neurons?
2. Describe the parts of a neuron and how they function.
3. Myasthenia gravis is a disorder that blocks acetylcholine (ACh) receptors at the neuromuscular junction. Would this cause a constant contraction or muscle weakness and relaxation? Explain.
4. When the dendrites of a sensory neuron are stimulated, what are the steps that follow that result in the development of an action potential?
5. What is the advantage of having one motor neuron stimulate dozens of individual muscle fibers at the same time? What would be the disadvantage?

10.6 Muscles Pull on Rigid Structures to Produce Movement

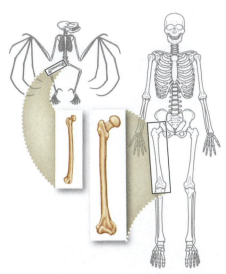

Figure 10.35 Comparison of thigh bones. Similar groups of bones and muscles can be found in most vertebrates. Notice that the thigh bones may look different, but they work the same way in both bats and humans.
© Cengage Learning 2014

Most vertebrates, including Mexican free-tailed bats and humans, share the same general skeletal muscle groups and bones. In many cases, these bone–muscle groups produce similar types of movements, such as the bending of legs and arms. For example, the bones in the leg of the bat are much smaller but have direct counterparts in your legs (**Fig. 10.35**). In fact, similar groups of muscles are attached to those corresponding bones and do the same job.

Muscles alone do not move the body; movement happens when muscles contract and pull a rigid structure such as a bone. When a bat needs to flap its wings, move its jaw muscles to chew, or grab onto a rock to roost, specific groups of muscles contract and pull against bones to produce the desired action. Although invertebrates lack an internal skeleton, they also rely on the contractions of muscles to produce movement. Soft-bodied animals such as worms have evolved a *hydrostatic skeleton,* which consists of closed or fluid-filled chambers located throughout the animal's body (**Fig. 10.36a**). Alternating contractions compress these chambers, changing their shape and producing movement. Insects such as crickets have evolved an *exoskeleton,* a rigid outer body covering that provides structure and support (**Fig. 10.36b**). Muscles are attached to the inner surface of the exoskeleton, pulling against it to produce movement. In the following section we'll look at the structure of vertebrate animal muscles, how muscle tissue contracts, and how muscles work with skeletons to produce movement.

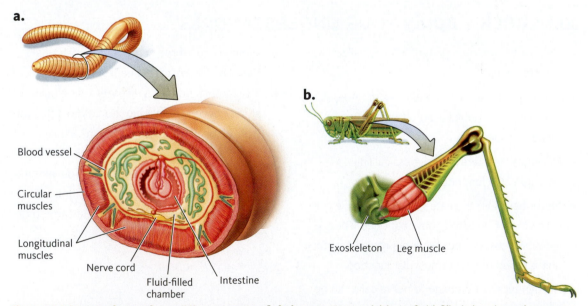

a.

Blood vessel

Circular muscles

Longitudinal muscles

Nerve cord

Fluid-filled chamber

Intestine

b.

Exoskeleton   Leg muscle

Figure 10.36 Invertebrates have different types of skeletons. Worms (a) have fluid-filled chambers that act as a hydrostatic skeleton. Insects (b) have a hard outer exoskeleton.
© Cengage Learning 2014

Skeletal muscles work with bones and joints to produce movement in vertebrates

Bone tissue fulfills many roles in the body, such as mineral storage, blood cell production, support for the body, and *a site for muscles to attach*. Muscles don't attach directly to bones; instead, straplike connective tissues called tendons link them together (**Fig. 10.37**). As muscle tissues contract, they get shorter and pull against tendons attached to bones, producing movement. Muscles only contract and shorten; thus they can pull bones in only one direction. Pairs of opposing muscles, such as the biceps and triceps of your upper arm, give limbs the ability to move in opposite directions.

Skeletal muscles and their associated bones in limbs typically interact as levers. Levers are rigid objects, such as the bones in your arm, that work with the elbow, which can then increase the muscle force applied to lifting an object. The bone, as a rod, represents the rigid structure of the lever rotating near a pivot point called a joint. The bones of the lower and upper arm represent the rods; your elbow is the joint where the two adjacent bones come together (see **Fig. 10.37**). Muscle contraction acts on both rods of the lever to produce the bending or twisting movement of a limb. Most limb movements are very complex, and for the purposes of illustration, we've simplified this process.

Bones move and rub continuously against each other, generating friction. Over time, friction can damage the ends of bones that carry a lot of weight, such as the hip and leg bones. To protect against such damage, the ends of bones at joints are covered in a thick layer of collagen tissue called *cartilage*, which acts as a shock absorber. The joint is filled with fluid, which helps lubricate and cushion bone-to-bone connections. These bone junctions include straplike connective tissues called ligaments, which hold bones in place and stabilize joints. Most joints, including your knee, have multiple ligaments. The ligaments run parallel to the bones and wrap around the knee joint, providing tremendous stability given the amount of use these joints undergo (**Fig. 10.38**). Each of these components—nerves, muscle, bone, tendon, and ligament—must work together to produce the desired movement.

Muscles are built from bundles of muscle fibers and bundles of contracting proteins

Of the three types of muscle tissue, skeletal muscle is the most abundant, accounting for 36 to 42 percent of an adult human's body weight. The human body has more than 600 skeletal muscles that vary tremendously in size. For instance, when you walk, the hamstring, one of three large muscles on the back of your thigh, pulls the thigh backward and raises your knee. At the other extreme, muscles less than half an inch long attach to your upper eyelid. Large or small, each skeletal muscle performs an important function and works in the same way.

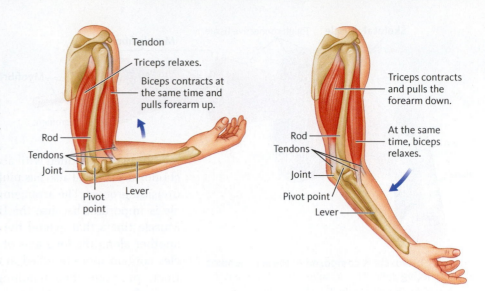

Figure 10.37 Muscle and bone work together to provide movement. The muscle and bone arrangement provides a mechanical advantage. In a rod and lever arrangement, the fixed bone is the rod and the moving bone represents the lever. Muscles pull on bones, rotating limbs at pivot points called joints. Opposing pairs of muscles pull limbs in opposite directions.
© Cengage Learning 2014

Knee joint ligaments

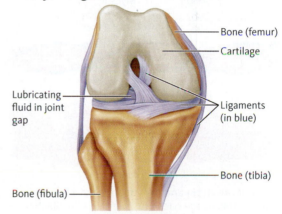

Figure 10.38 The role of ligaments. Bone-to-bone connections are stabilized by ligaments (shown in blue). Fluid in the joint and protective cartilage provide protection against wear and tear for the ends of the bones.
© Cengage Learning 2014

Tendons—straplike connective tissues that link muscles to bones. As muscle tissues contract, they shorten and pull against tendons attached to bones, producing movement.

Joint—the location where two bones involved in movement come together. The joint is filled with fluid, which helps lubricate and cushion bone-to-bone connections.

Ligaments—straplike connective tissues that hold bones in place and stabilize joints. Most joints, including those in the knee, have multiple ligaments.

Skeletal muscle—one of three types of muscle tissue in the animal body that contract and result in movement of limbs as well as structures such as the lips and eyes.

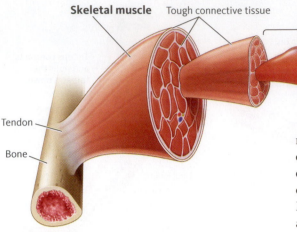

Figure 10.39 Muscle is composed of parallel bundles of contracting cells. The arrangement of the muscle fibers in rows of parallel bundles helps orient the direction of the muscle contraction in only one plane of motion.
© Cengage Learning 2014

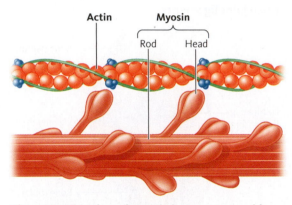

Figure 10.40 Actin and myosin interact to provide contraction. Inside each muscle fiber are collections of actin and myosin proteins. When stimulated by nerve impulses, calcium and ATP help bind the two long fibers together, ratcheting them along very rapidly. This action shortens the muscle fibers and, in large groups, the entire muscle.
© Cengage Learning 2014

Muscle fibers—long, cylindrical cells found in skeletal muscles composed of bundles of specialized contracting protein fibers wrapped in a covering of tough connective tissue. Larger, more powerful muscles contain more bundled muscle fibers.

Skeletal muscles are constructed of individual cells known as **muscle fibers**, which are long cylindrical cells composed of bundles of specialized contracting protein fibers wrapped in a covering of tough connective tissue (**Fig. 10.39**). Each muscle cell has a plasma membrane and is filled with abundant endoplasmic reticulum and mitochondria. Each cell also has multiple nuclei, a result of an incomplete cell division process. The arrangement of muscle fibers within a larger muscle is important because the fibers contract in only a single direction. Muscle fibers that extend the length of the muscle lie parallel to one another along the long axis of the muscle. Larger, more powerful muscles contain more bundled muscle fibers. The cylindrical shape of the fibers promotes their bundling together in a ropelike fashion, adding strength and protection to the muscle.

Muscle fibers in turn are filled with bundles of *myofibrils,* collections of contracting proteins that run parallel along the length of the fiber and are an important part of the muscle fiber's cytoskeleton. In essence, each muscle cell is a nested bundle of bundles (see **Fig. 10.39**). The repeating bundles of filaments form units of contraction called *sarcomeres,* a feature that gives skeletal muscle a unique striped appearance. Each sarcomere contains a thick filament made from the protein *myosin,* which has clublike heads projecting from it. The thin filament is made from globular *actin* molecules arranged in strands (**Fig. 10.40**). The thick and thin strands lie next to one another, and during contraction the myosin heads bind to the actin molecules. This construction is important to the ability of the strands to interact at the molecular level.

Skeletal muscles shorten when the thin actin filaments slide over the thicker myosin filaments in a process called the *sliding filament model* of muscle contraction. The club-shaped heads on the myosin filament bind with actin molecules and ATP, causing the binding site to change shape and rotate forward. This interaction happens rapidly, quickly shortening the sarcomere. When motor neurons stimulate large bundles of muscle fibers at the same time, the entire muscle contracts.

Motor neurons signal muscle fibers to contract and move bones

In order to flap its wings in rhythm, the bat must be able to precisely signal the contraction of individual muscle fibers. This process relies on the ability of motor neurons to communicate the action potential to muscle cells. The link between motor neurons and muscles occurs at the *neuromuscular junction* where the axon terminals come into close contact with the surface of one or more muscle fibers (**Fig. 10.41**). The collection of muscle fibers stimulated by a single motor neuron is a *motor unit,* which contracts at the same time. Motor units may be large, like those in your leg muscles, where a single motor neuron may control thousands of individual muscle fibers, or they may be small, like those in the eye that pull on the lens and provide precise control in focusing light. Signals from motor neurons play an important role in muscle contraction by initiating the stimulus that results in contraction. Keep in mind that skeletal muscle control is voluntary and depends upon conscious thought to promote the contraction of specific muscles.

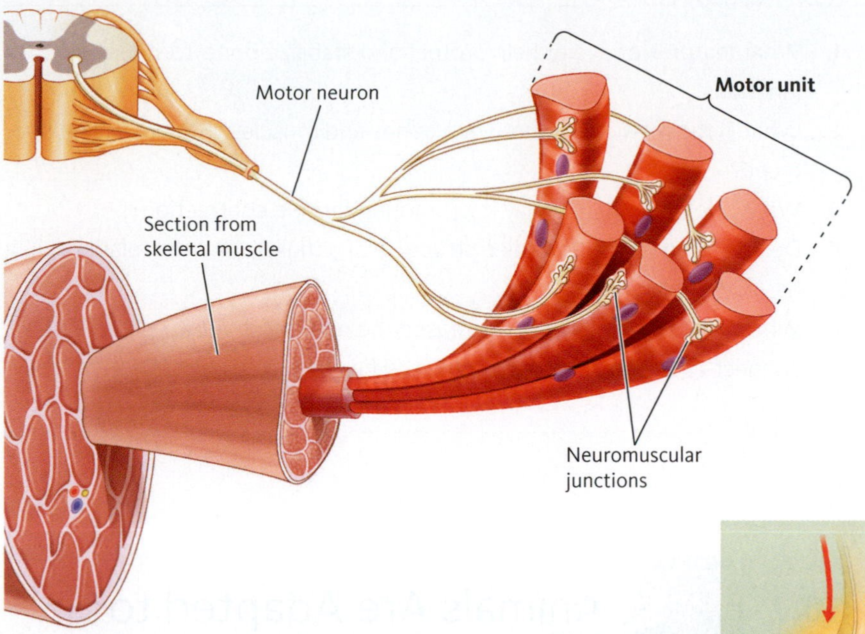

from spinal cord

Motor neuron

Section from
skeletal muscle

Motor unit

Neuromuscular
junctions

Figure 10.41 Motor unit. A motor unit consists of a motor neuron and the muscle fibers it stimulates. Large motor units generate large, powerful muscle contractions. Small motor units stimulate the contraction of smaller collections of muscle fibers.
© Cengage Learning 2014

You may think it's natural to turn a page in this book without thinking about it, but your brain is actively involved in generating the nerve signals sent to your arms and hands that make this happen. When stimulated by a motor neuron, individual skeletal muscles contract. As more power is needed to accomplish a task, more motor neurons send signals, recruiting more muscle bundles to contract.

Recall that when the action potential reaches the axon terminal at the neuromuscular junction ❶, the neurotransmitter ACh is released into the synapse (**Fig. 10.42**). The binding of ACh to the outer membrane ❷ of the muscle cell causes a change in its electrical charge, resulting in an action potential in the muscles' cells, which causes protein channels to open and the release of calcium ions from the endoplasmic reticulum inside the cell ❸. Calcium ions bind to proteins ❹ that in turn bind to additional proteins, which expose binding sites that lead to the shortening of bundles of muscle fibers and the muscle. When the stimulation from a motor neuron ceases, energy is used to pump calcium back inside the endoplasmic reticulum of the muscle fiber ❺, myosin binding sites on actin are re-covered, and the contraction stops and the fibers return to their resting state.

Muscle contraction relies on the presence of ATP to fuel the muscle cells. As we learned in Chapter 9, ATP is produced during aerobic respiration at the mitochondria, a process that requires the availability of oxygen. Using the energy of ATP, the heads on myosin fibers bend slightly and expose a new binding site for the actin fiber. The binding to actin triggers the release of a molecular springlike action in the myosin head, which snaps back toward the tail, producing the pulling force that powers the thin actin strand over the thicker myosin fibers.

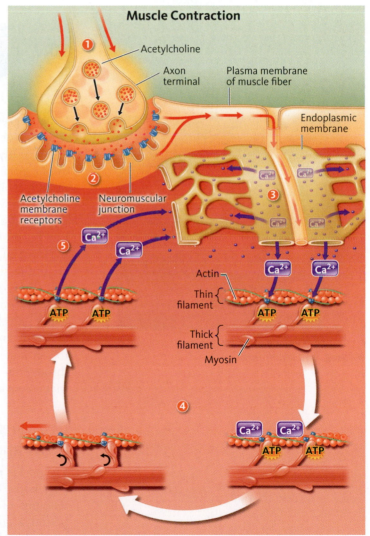

Muscle Contraction

Acetylcholine

Axon terminal

Plasma membrane of muscle fiber

Endoplasmic membrane

Acetylcholine membrane receptors

Neuromuscular junction

Actin

Thin filament

Thick filament

Myosin

Figure 10.42 How muscles work. Muscle contraction occurs when messages sent along motor neurons arrive at neuromuscular junctions, where the signal is transferred to the muscle cells by the neurotransmitter acetylcholine (ACh). The stimulation of the muscle fiber results in the release of calcium ions from the endoplasmic reticulum, and they assist the actin and myosin to bind and ultimately contract the muscle.
© Cengage Learning 2014

- Skeletal muscles contract, pulling on bones to produce the movement of limbs.

- Muscle tissue is composed of bundles of contracting proteins, which shorten when stimulated by motor neurons.

1. What features of a joint help protect and stabilize bone-to-bone junctions?

2. What is the relationship between bones and muscles that allows limbs to move?

3. What role does oxygen play in promoting muscle contraction?

4. Describe the layered, bundled structure of a muscle and the relationship it has with motor neurons.

5. What would happen if you decreased the number of motor neurons that connect to the muscles controlling the fingers in your hands?

10.7 Animals Are Adapted to Move in a Variety of Ways

Canada geese have a complex nervous, muscular, and skeletal system adapted for walking, swimming, and flying (**Fig. 10.43**). In spite of their ability to fly, they actually spend most of their time walking and feeding on the ground. To fly they have evolved large breast muscles that contract, pulling their wings up and down, which allows them to cover long distances. To swim, their strong leg muscles and webbed feet pull their bodies through the water. To walk, they need an integrated set of sensory receptors linked to specialized areas of the brain to coordinate their ability to walk or run.

In this section, we'll explore different ways animals are adapted to specific kinds of movements. Complex motor coordination relies on the ability of the nervous system to initiate muscle contractions in multiple limbs while perceiving and balancing weight distribution, the force of gravity, and resistance from air, water, or soil. We'll take you from flying high in the air and climbing the tops of trees all the way to digging and crawling through an underground tunnel. Let's begin by looking at those animals that crawl and dig their way through soil.

Crawling and digging animals have strong arms and shoulders

The pocket gopher of the Great Plains is a champion digger. Its small, compact body, large claws, and strong muscles help it swiftly move large amounts of soil (**Fig. 10.44**). Only 12 inches long and weighing barely a pound, a pocket gopher can move up to 8,000 pounds of soil each year. Like most diggers, the pocket gopher spends nearly all of its life below ground creating burrows. A single den's tunnels and chambers can reach out 600 feet.

Pocket gophers have a well-adapted set of features to accomplish their digging feats (see **Fig. 10.44**). They use long, curved claws to scratch and loosen even the most compacted dirt. To scratch and dig, they have large, strong arm and shoulder muscles attached to thick, sturdy bones.

Figure 10.43 Geese are well adapted to flying, walking, and swimming.

John S. Sfondilias/Shutterstock.com

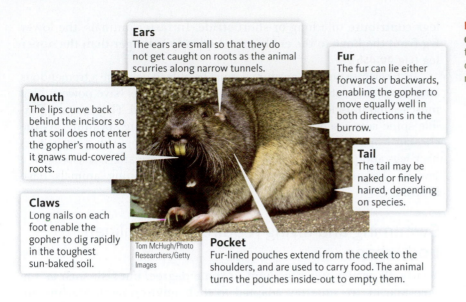

Ears
The ears are small so that they do not get caught on roots as the animal scurries along narrow tunnels.

Mouth
The lips curve back behind the incisors so that soil does not enter the gopher's mouth as it gnaws mud-covered roots.

Fur
The fur can lie either forwards or backwards, enabling the gopher to move equally well in both directions in the burrow.

Tail
The tail may be naked or finely haired, depending on species.

Claws
Long nails on each foot enable the gopher to dig rapidly in the toughest sun-baked soil.

Tom McHugh/Photo Researchers/Getty Images

Pocket
Fur-lined pouches extend from the cheek to the shoulders, and are used to carry food. The animal turns the pouches inside-out to empty them.

Figure 10.44 Adaptations of the pocket gopher for digging in the soil. The telltale mounds of dirt signal the presence of the pocket gopher, a champion digger common to gardens, lawns, and farms across much of the U.S.

In contrast, earthworms don't have large muscles or clawed appendages, yet they are also amazingly successful at moving through the soil. Using the combination of radial and longitudinal muscles in the body wall, they squeeze and extend, twist and turn their long, thin bodies through remarkably tight holes. Each of the worm's sections has tiny, clawlike bristles that help anchor each section for leverage. Using their bristles in combination with waves of muscular contractions and the slimy mucus that coats the body, they are able to maneuver themselves through the soil.

Running requires hip, leg, and foot strength

Running is an important adaptation that allows animals to escape predators or catch slower prey. The cheetah rarely runs farther than one third of a mile, yet it covers that distance more swiftly than any land animal on Earth, at 70 mph (**Fig. 10.45a**). For comparison, a pronghorn antelope of the western prairies, considered a very fleet-footed animal, reaches speeds of 45 mph, a galloping horse 40 mph, and the swiftest human a paltry 22 mph.

The cheetah's body is built for speed. Cheetahs have a lean build; long, thin legs; a long tail to counterbalance their weight while running; and a small, aerodynamic skull. The length and proportion of

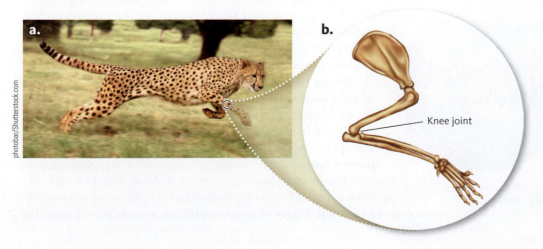

photobar/Shutterstock.com

Knee joint

Figure 10.45 Cheetah. Cheetahs are well adapted for their high-speed sprints. One unique adaptation common among fast running animals is that their knee to ankle bones (b) are longer than the upper leg bones. The powerful muscles attached to these leg bones, along with a lean body and weight balancing tail, help move them along at up to 70 mph for short distances.

legs contribute to a long or short stride. In faster animals, the lower leg from the knee to the ankle joint typically is longer than the upper leg (**Fig. 10.45b**).

To accelerate quickly, an animal must be able to generate a tremendous amount of force at the ankle, knee, and hip. Cheetahs have powerful leg muscles and muscle/bone attachment that favor speed. Their curved, flexible spine acts essentially like a spring when they are sprinting. As the cheetah leaps forward, its arched spine straightens, pushing against its back legs and propelling the front of the animal forward. These are just a few of the many anatomical adaptations fast-running animals have evolved.

Swimming and diving require a streamlined body shape

Nearly all vertebrates can swim to some degree, but fish such as the blue marlin are superbly adapted to their environment. Marlins are among the fastest fish in the ocean; they reach speeds of 37 mph over short distances. Like most fish, marlins are highly streamlined with small scales that reduce friction. Adding to their streamlining is their lack of a functional neck; instead their head is molded smoothly into their body. In addition, they have no external ears or reproductive parts to cause drag. Powerful muscle groups attached to their tails help move them through the water. The marlin's skeleton has many bones attached to its spine; each is a strong, flexible attachment point for powerful skeletal muscles (**Fig. 10.46**). Its muscles are thick and high in fat content, which is necessary to sustain the energy needed for long-distance swimming. Like all sprinting animals, marlins have special types of muscle fibers in their skeletal muscles. Their skeletal muscles have a high proportion of *fast-twitch muscles* which are well adapted for anaerobic performance such as sprint swimming in fish. This adaptation is also found in the muscles of marine mammals that dive for prolonged periods, in birds, and in mammals that sprint and hop.

To dive deep, animals must be adapted to sense and respond to tremendous changes in water pressure, oxygen availability, cold temperatures, and low light conditions. Water pressure doubles at only 30 feet of depth; the pressure is 15 times greater at a depth of 500 feet than at the surface. Sperm whales, the largest toothed whales, dive to depths of 9,000 feet, where pressures exceed 4,000 pounds of pressure per square inch, enough to crush a car. Deep-diving animals have bodies and bones strong enough to withstand the pressure and flexible enough to be pressed in. The sperm whale has several adaptations that help it cope, for example, a flexible rib cage, which allows its lungs to collapse inward as pressure rises (**Fig. 10.47**).

Diving requires a specialized respiratory system to supply the muscles with needed oxygen. Diving mammals must be able to hold their breath for extended periods; the sperm whale can hold its breath for 90 minutes. During deep dives, the sperm whale's metabolism slows to conserve oxygen. Sperm whales have an abundant amount of *myoglobin*, a protein molecule that stores oxygen in skeletal muscle tissue. On long dives, myoglobin slowly releases oxygen, allowing aerobic respiration to continue longer in muscle cells. As in most diving animals,

Figure 10.46 Marlin. Marlins are fast, powerful fish whose streamlined body and strong skeleton and muscle groups allow them to move easily through the water at high speed.

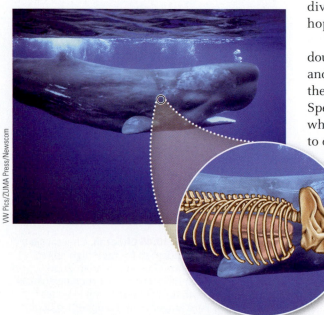

Figure 10.47 Sperm whale. The sperm whale dives to depths of 9,000 feet, where its collapsible ribcage helps it adapt to the crushing pressure.

the sperm whale's blood has a very high number of red blood cells, which contain critical oxygen-carrying hemoglobin. During dives, oxygenated blood is directed toward the brain and other essential organs as oxygen levels drop in the blood.

Flying and gliding use the forces of lift and control

Three existing animal groups have the ability for active flight: birds, bats, and insects. Flying is an active process involving the flapping of wings, whereas gliding is more passive, relying on adaptations that catch the air and slow an animal's descent to the ground. As we learned with the Mexican free-tailed bat, flyers require many adaptations that include lightweight bodies and strong but light bones. Birds, bats, and even the pterosaurs of 200 million years ago share a thin, lightweight bone structure in their wings (**Fig. 10.48**).

The mechanics of bird flight are remarkably similar to those of an aircraft. In fact early aircraft wings were copied from birds' wings. Lift is produced when air flowing over the curved top of the animal's wing lowers the air pressure above the wing, while higher-pressure air flows underneath the flat wing bottom. Feathers are lightweight, provide protection, and contribute to thermoregulation by insulating the skin (**Fig. 10.49**). Their aerodynamic shape and network of interlacing hook-like fibers are important features that help displace air during flight while the wings are flapping and aid in controlling the direction of flight.

To sustain flight, birds need enough muscle power to lift their bodies and provide propulsion through the air. The necessary power comes from two main flight muscles attaching to each wing (**Fig. 10.50**). The main flight muscle is the large breast muscle that powers the downstroke. This muscle is so large that it may make up to 35 percent of the bird's body weight. The wishbone, located in the chest, strengthens the skeleton of the chest to help the bird withstand the rigors of flight. A far smaller muscle, which has a tendon that curves around to attach to the top of the shoulder bone, or humerus, powers the upstroke. Ducks have wings that contain all the same bones and similar muscle groups found in human arms, but in

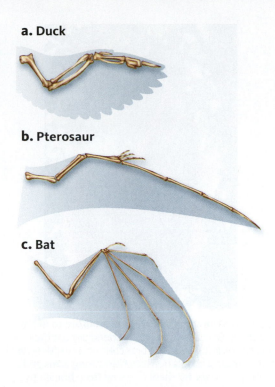

a. Duck

b. Pterosaur

c. Bat

Figure 10.48 Comparison of wing bones of a duck, a pterosaur, and a bat. The long, thin bones in the wings of a duck, a pterosaur, and a bat are very similar. Each animal has roughly the same number of strong, lightweight bones in a similar arrangement. The similarity between them is related to their common adaptation for flight.
© Cengage Learning 2014

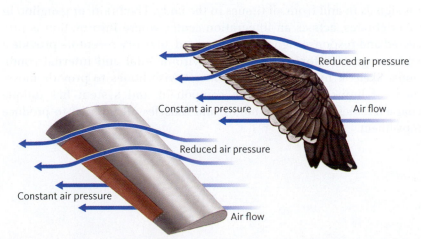

Reduced air pressure

Constant air pressure

Air flow

Reduced air pressure

Constant air pressure

Air flow

Figure 10.49 Feathers allow flight in birds. Feathers evolved from scales and have become delicate aerodynamic adaptations that allow birds to fly. Note how similar the form of the bird wing (right) is to an airplane wing (left).
© Cengage Learning 2014

Flight muscles

Wishbone

Figure 10.50 Flight muscles of a duck. The flight muscles of a duck attach to both the upper and lower surfaces of its upper arm bones. The opposing groups of muscles allow the duck to flap its wings up and down, providing lift.
© Cengage Learning 2014

Figure 10.51 Orangutans are well adapted to life in the trees. Because orangutans spend almost their entire lives living up in the branches of trees, they have many physical adaptations such as strong arms and hands necessary for them to swing from branch to branch.

Figure 10.52 Hand bones of the orangutan. Notice the large, long, curved fingers of the orangutan's hand bones. These help the orangutan grab and firmly hold branches as its moves through the trees.

addition to the two large flight muscles, they have 48 additional wing and shoulder muscles that help produce the wide range of movements of the wing and its feathers.

Climbing and swinging require depth perception and long arms

The ability to climb has evolved many times in dozens of different animal groups. Climbing gives animals access to food and shelter and protection from predators. Climbers must have a strong, sure grip and the muscle power to pull their bodies upward.

Orangutans are superb climbers and swingers, which is important because they spend 95 percent of their lives in trees (**Fig. 10.51**). Adult males may be five feet tall and weigh 180 pounds, a size that requires powerful muscles to lift them swiftly up a tree. It's estimated that an orangutan's arm muscles are seven times stronger than a human's, a fact easily seen in the orangutan's effortless ability to dangle from branches by a single arm. Orangutans have a long reach and are able to swing themselves through the trees by reaching and pulling from one supporting branch to another. The arms of orangutans are twice as long as their legs, and much of the arm's length results from the length of the lower arm bones rather than the upper. The long fingers and toes, combined with the opposable thumb on their feet, allow them to grip and climb with all four limbs (**Fig. 10.52**). In addition, their fingers and toes are permanently curved, another adaptation that allows them to strongly grip branches.

Animal adaptations for movement are very diverse. We've examined just a few of these adaptations, and there is a common evolutionary thread. Natural selection has produced body shapes, sensory capabilities, and muscle and bone modifications suited to an animal's lifestyle and environment. Your understanding of the general patterns of movement provides you with a greater ability to understand the anatomical and biological adaptations needed for any one type of movement.

In this chapter we've explored the structure and function of various nervous systems and muscles and their functional adaptations in the animal world. The nervous system serves as a communication pathway for signals to and from all tissues in the body. The brain, or ganglion in invertebrates, acts as an integration center where information is processed and response signals are generated. Sensory receptors provide a wide range of information about environmental and internal conditions. Skeletal muscles work in concert with bones to provide movement of limbs. Although each component and system has unique characteristics, the nerves, bones, and muscles all interact to produce movement.

- Natural selection has played a significant role in the development of adaptations for movement in the animal world.

- Characteristics such as body shape, limb structure, muscle size and placement, and the evolution of sensory input such as vision are dominant factors in determining successful adaptations for movement.

1. What types of adaptations have birds evolved that make them successful flyers?

2. What types of adaptations have evolved for animals such as fish that swim?

3. What is the relationship between the length of an animal's legs, its stride, and its ability to run fast?

4. Predict whether bats that feed exclusively on fruit would have evolved an echolocation system (**Fig. 10.53**).

5. Ostriches are large, flightless birds native to Africa. They stand as tall as six to nine feet, weigh as much as 350 pounds, and prefer open savannas as their habitat (**Fig. 10.54**). Because ostriches cannot fly, they have evolved other adaptations such as a powerful set of legs, a long neck, and large eyes with excellent vision. Speculate what role these adaptations have in helping them survive.

Merlin Tuttle/BCI/PhotoResearchers/Getty Images

Figure 10.53 Fruit bat.

iStockphoto.com/Peter Malsbury

Figure 10.54 Ostrich.

End of Chapter Review

Self-Quiz on Key Concepts

The Nervous System Coordinates Activities Throughout the Body

KEY CONCEPTS: In vertebrate animals, the central nervous system consists of a brain and spinal cord that work together to integrate sensory input and coordinate all the activity in the body. The bundles of nerve cells form communication pathways between the central nervous system and sensory and motor neurons throughout the rest of the body.

1. Match each of the following **information collection** or **transfer** terms with its description.

Nerves	a. cells that stimulate muscle cells to contract
Neurons	b. an interconnected network of specialized cells
Nervous system	c. cells dedicated to collecting information such as pressure and
Motor neuron	temperature
Sensory neuron	d. cells that sense input, transmit information, and process signals
	e. bundles of neurons that function in transmitting signals

2. Match the following **nervous system** term with the best description.

Central nervous system	a. nerve cell clusters in invertebrates that integrate information
Brain	b. consists of the brain and spinal cord
Ganglia	c. large collections of neurons concentrated in the head

3. The portion of the brain that participates in higher-order thought processes such as problem solving is called the _____.
   a. pons
   b. cerebellum
   c. cerebrum
   d. medulla oblongata

4. Vertebrates have a tubular bundle of nerves running along their backbone called a _____ that serves as a pathway between the brain and the _____.
   a. brain; central nervous system
   b. ganglia; peripheral nervous system
   c. spinal cord; central nervous system
   d. spinal cord; peripheral nervous system

Nerve Cells Sense the Environment, Communicate, and Signal Muscles to Contract

KEY CONCEPTS: Neurons are specialized cells that interconnect to form communication pathways to transmit chemical–electrical signals throughout the animal body. Most animals have a broad array of sophisticated sense organs that allow them to perceive information about their external and internal environments. Animal senses include sight, hearing, touch, taste, smell, balance, temperature, and pain. Different types of sensory cells respond to physical or chemical stimuli and send signals to the central nervous system where they are received and interpreted.

5. Match each of the following **nervous system** terms with the best description.

Sensory receptors	a. nerve cells involved in the sense of taste and smell
Pain receptors	b. nerve cells adapted to receive specialized kinds of input
Chemoreceptors	c. nerve cells stimulated by pressure
Mechanoreceptors	d. nerve cells stimulated when you bruise your arm

6. The ropelike extended portion of a nerve cell called the _____ is often wrapped in a protective layer of cells.
   a. dendrite
   b. neural membrane
   c. ganglia
   d. axon

7. A nerve impulse called a(n) _____ is caused by the movement of _____ ions from outside the nerve cell membrane to inside, creating an electrical gradient wave that sweeps down the cell (refer to **Fig. 10.31**).
   a. neurotransmitter; hydrogen
   b. myosin agent; helium
   c. action potential; sodium
   d. contraction; zinc

8. World-class figure skaters often include dramatic spins as a part of their routines. Skaters must train continuously to overcome the loss of balance associated with confusing sensory input when they are spinning rapidly. Which organ located in the inner ear works in perceiving balance?
   a. cochlea
   b. auditory ganglia
   c chemoreceptors
   d. semicircular canals
   e. retina

9. Vultures are common scavengers that feed on the remains of dead animals. They often soar over a thousand feet in the air looking for a meal. What type of sensory neuron is likely responsible for their ability to perceive their next meal while soaring at these heights?
    a. lateral line
    b. mechanoreceptor
    c. chemoreceptor
    d. photoreceptor
    e. thermoreceptor

Muscle Tissue Contracts and Works with Bones to Move Limbs

KEY CONCEPTS: Muscle cells consist of specialized contracting proteins organized into bundles of motor units that contract in response to nerve signals (voluntary and involuntary control). Contraction forces created by muscle groups pull against bones of the skeleton to produce movement. Skeletal muscles are attached to bones by cords of connective tissues called tendons, whereas ligaments provide connections between adjacent bones. Simple animals lack internal skeletons, instead relying on internal hydrostatic skeletons or hard exoskeletons to support movement.

10. Match each of the following **component of movement** terms with the best description.

Joint	a. a thin, contracting protein fiber
Muscle fiber	b. a place where two bones come together
Sarcomere	c. a collection of contracting proteins
Actin	d. a muscle cell that contains multiple nuclei

11. A male white rhinoceros may weigh as much as 7,500 pounds, and the bones of its legs are under immense twisting forces as it runs and turns. What type of tissue helps connect and stabilize the tremendous forces at the bone-to-bone connections?
    a. tendon
    b. ligament
    c. cartilage
    d. smooth muscle tissue

12. Nerve signals are passed between adjacent cells by the release of a chemical called a(n) _____ into the gap called the _____.
    a. neurotransmitter; synapse
    b. calcium surge; dendrite
    c. axon; axon terminals
    d. sodium pump; synapse

Animals Have Evolved Many Forms of Movement

KEY CONCEPTS: The coordinated actions of muscles, bones, and the nervous system produce a range of movement in animals. Movement generally requires a sense of coordination and the ability to move through the habitat in which the animal lives. Animals have adapted a wide range of physical features adapted for swimming, flying, walking/running, crawling/sliding, or jumping.

13. Match the **type of movement** with the adaptation.

Swimming and diving	a. strong feet and lower leg muscles
Running and jumping	b. strong arms with precurved fingers
Digging and crawling	c. a flexible rib cage that is able to compress
Climbing and swinging	d. a compact body with strong front arms

Figure 10.55 Flying squirrel.

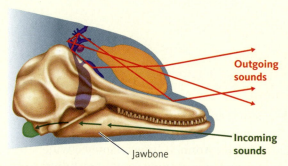

Figure 10.56 American alligator.

14. Many members of the squirrel family are excellent gliders (**Fig. 10.55**). They use this method of movement to glide from one tree to another. This method requires them to have which of the following adaptations?
    a. strong flight muscles
    b. lightweight body and bones
    c. a heavy, strong set of muscles and bones
    d. the ability to withstand strong changes in pressure

15. The American alligator is a large reptile that is most at home in the water but suns itself on riverbanks. What characteristic of the alligator (**Fig. 10.56**) indicates that it lacks the adaptation seen in sprinting animals?
    a. its streamlined body structure
    b. its short legs, especially the short lower leg structure
    c. the long tapered snout
    d. the leathery scale-covered skin

16. Sea lions dive as deep as 600 feet into the ocean to get their catch of fish and squid. Which of the following adaptations do sea lions likely possess to help them dive for extended periods?
    a. a fluid-filled hydrostatic skeleton, combined with a large heart
    b. a compressible spine, combined with large eyes
    c. an ability to slow their metabolic rate and high myoglobin levels in muscles
    d. large flipper muscles and fur for warmth in the cold ocean

Applying the Concepts

17. To navigate and find food, dolphins use a type of echolocation similar to that used by bats. They emit a focused, wide-angle beam of high-frequency clicks and listen for the echo. To make the sounds, dolphins move air through bony nasal structures and a specialized structure within their forehead (**Fig. 10.57**). The bones of the skull and various structures all play important roles in forming the beam of sound, while the lower jaw acts as the primary echo reception pathway to the internal ear. Based on this background information, predict what category of sensory receptors helps the dolphin detect its prey.

Outgoing sounds

Incoming sounds

Jawbone

Figure 10.57 Pathway of echolocation in dolphins.
© Cengage Learning 2014

18. The moon-walking inchworm is the caterpillar life stage of the geometer moth (**Fig. 10.58**). Inchworms have feet at both ends of their body. To move forward, their front feet grab a leaf and pull the back end up. The back legs then grab the leaf and allow the front legs to reach out, creating the impression that the inchworm is measuring its journey inch by inch. Based on this information, describe the type of skeleton the inchworm might possess. What kind of rigid structure do its muscles pull against as the inchworm moves across the edge of leaves?

Figure 10.58 The inchworm is the caterpillar life stage of the geometer moth.

19. Gorillas are close relatives of orangutans. They have large, powerful muscles, but they spend their lives living on the ground rather than in the trees. How might the muscles in the arms and hands of gorillas be different from those of orangutans?

20. Zebra fish are common aquarium pets; they also are often used in research on vision. In the wild, zebra fish inhabit shallow, sunny portions of coral reefs in tropical oceans. A biologist at Florida State University discovered a gene important for development in most animals that also affects the ratio of rods to cones in the eyes of this fish. The retina of a normal zebra fish has both rods and cones, whereas the mutant fish has almost all rods in its retina. What types of visual differences would you expect to find if you tested the two types of fish? How well adapted would the mutant fish be to the shallow, sunny waters of the coral reef habitat?

Data Analysis

Bite Force and Muscles in a Great White Shark

Australian researchers have measured a great white shark's bite and found it to be about three times as powerful as a lion's and just over half as strong as the bite of a *Tyrannosaurus rex* (**Fig. 10.59**). These sharks can reach lengths of 20 feet and weigh as much as 4,600 pounds. Why is bite force important for such a large animal? As a top predator, sharks require a large bite force to kill and dismember large prey such as seals, sea turtles, and tuna.

Bite force is a complex movement to estimate because it depends on body size, head shape, tooth shape, structure of jaw bones, and the size and attachment of jaw muscles—all important adaptations for the shark to obtain its food. It also varies from the front of the mouth to the back. As in all sharks, the jaws of the great white are made from more flexible cartilage rather than stiffer bone, an advantage when attacking large animals that may squirm to get away. Similar estimates have been made for extinct animals like the megalodon shark, a giant ancestor of the great white (**Fig. 10.60**). These many variables account for differences among researchers and the bite force values

Figure 10.59 Jaw muscles of a great white shark.
© Cengage Learning 2014

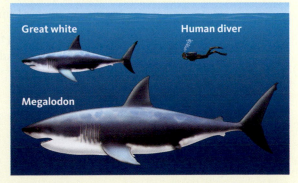

Figure 10.60 Size comparison of great white shark, megalodon, and human.
© Cengage Learning 2014

TABLE 10.2 How Hard Can an Animal Bite?

Animal	Animal Age/Size	Body Weight (lb)	Front of Mouth Bite Force (lb)	Rear of Mouth Bite Force (lb)
Great white shark	Immature	530	1,602	3,131
	Sexual maturity	930	2,341	4,577
	Largest ever measured	7,300	9,320	18,216
Megalodon	Small (estimate)	105,140	55,522	108,514
	Large (estimate)	227,510	93,127	182,201
Tyrannosaurus rex	Large (estimate)	14,900	1,450	3,000

© Cengage Learning 2014

(Wroe S, Huber DR, Lowry M, McHenry C, Moreno K, Clausen P, Ferrara TL, Cunningham E, Dean MN, Summers AP. 2008. Three-dimensional computer analysis of white shark jaw mechanics: how hard can a great white bite? J Zool 276, 336–342; and Rayfield EJ. 2004. Cranial mechanics and feeding in Tyrannosaurus rex. Proc Biol Soc 271, 1451–1459.)

shown in **Table 8.1** and here in **Table 10.2**. Use this background information and the data in **Table 10.2** to answer the questions below.

Data Interpretation

21. What is the relationship between the body weight of a shark and its bite force?

22. Which area of the mouth generates the greatest bite force? What is the relationship in bite force between the front and back of the mouth? Is this true for sharks and *T. rex*?

23. Based on the data in **Table 10.2**, which end of the mouth has the greatest force? Look at the photos of the shark jaw and speculate why that portion of the jaw has the greatest bite force.

Critical Thinking

24. Examine the computer-generated model of the muscles in the mouth of the great white shark (**Fig. 10.59**). What is the relationship between muscle mass and the location of the greatest bite force?

25. *Tyrannosaurus rex,* a top predator that lived 65 million years ago, had a bony skeleton and jaws similar to those of humans. Is there a bite force advantage to having a bony skeleton?

26. Bite force is related to the size of the muscles pulling on the jaws. How much more muscle mass might the smallest megalodon have had compared to the largest great white shark ever caught?

Question Generator
How the Sand-Diving Lizard Keeps Cool

The shovel-nosed, or sand-diving, lizard lives in the sandy Namib desert in Africa. Living on the exposed, hot sand can pose challenges to remaining cool, but this lizard has developed an interesting set of adaptations to solve this problem. It performs a "thermal dance" when the sand becomes too hot for its feet. It props itself up on its tail, lifts a front foot and the opposing back foot, holds them up for a while, then sets them down and holds up the other pair (**Fig. 10.61**). The lizard repeats the dance until the sand cools down a bit, or, when all else fails, the lizard dives into the loose sand and "swims" down to a cooler level.

Michael & Patricia Fogden/Minden Pictures

Figure 10.61 Sand-diving lizard. The sand-diving lizard lives in one of the hottest climates on Earth. To cope with the heat, the lizard does a thermal hot-footed dance to keep its feet from burning on the hot sand.

The physical adaptations of the sand-diving lizard include a streamlined snout that eases its entry into the sand and minimizes resistance as the lizard swims through the sand; nostrils on top of the snout to keep sand out; and splayed legs and long, thin toes. When the lizard dives beneath the sand, air trapped between the sand grains allows the lizard to remain buried up to two feet deep for 24 hours or longer. If the lizard finds itself on the surface when rain falls, it's in trouble because the sand becomes too firm for it to penetrate. Stranded above ground, it becomes easy prey for predators.

Use the block diagram (**Fig. 10.62**) that illustrates several aspects of the interaction between the nervous and muscular systems that are key to the survival of this lizard. Use the background information along with the block diagram to ask questions and generate hypotheses. Let's start off by translating the block diagram elements and their links into some example questions.

What is the relationship between the lizard's ability to sense its environment, process that information, and generate responses that include movement to cooler and safer areas under the sand?

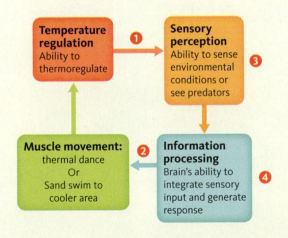

Figure 10.62 Factors relating to the muscular and nervous systems of the sand-diving lizard.
© Cengage Learning 2014

Here are a few questions to start you off.

1. What types of sensory neurons would you expect to find if you examined the lizard's feet? (Arrow #1)

2. What is the relationship between these sensory neurons and the lizard's thermal dance? (Arrow #2)

3. Describe the placement of the lizard's eyes and its ability to spot predators. (Block #3)

4. Does the brain of this tiny lizard have a large, infolded cerebrum? (Block #4)

Use the block diagram to generate your own research questions and frame a hypothesis.

11 Animal Mating and Reproduction

11.1 The Coho salmon's drive to reproduce

11.2 Reproduction is an important mechanism of evolution

11.3 Different reproductive strategies enhance evolutionary fitness in animals

11.4 Reproductive systems produce and unite gametes

11.5 Cellular mechanisms of sexual reproduction include meiosis and gametogenesis

11.1 The Coho Salmon's Drive to Reproduce

The Coho salmon *(Oncorhynchus kisutch)* has a remarkable life journey that is guided by one purpose—to survive in order to reproduce (**Fig. 11.1**). Depending on where the fish is born, some salmon swim more than 1,000 miles, battling predators, waterfalls, dams, and upstream currents to reach their freshwater breeding grounds. If the salmon are some of the 3 percent that survive their amazing voyage to their breeding grounds, within a week they build several nests, select their mates, breed, and die. The salmon's journey is an extraordinary one in which each fish follows its instinct for reproduction, contributing to the survival of the species.

The life cycle of the salmon includes several stages and lasts about three years (see **Fig. 11.1**). It begins in the spring, when fertilized eggs develop into hatchlings (called fry), which start their life in freshwater streams feeding on terrestrial and aquatic insects along with small crustaceans. The eggs hatch when the rivers are low, and as the rivers rise the fry slowly swim downstream, feeding in the freshwater habitat for a

Figure 11.1 The life cycle of the Coho salmon. As the fish mature in the freshwater stream (top right) they make their way toward the ocean. In the ocean they focus on growing by feeding on other fish. When they reach sexual maturity at three years of age, they swim back to the freshwater stream where they were born to spawn and die.
© Cengage Learning 2014

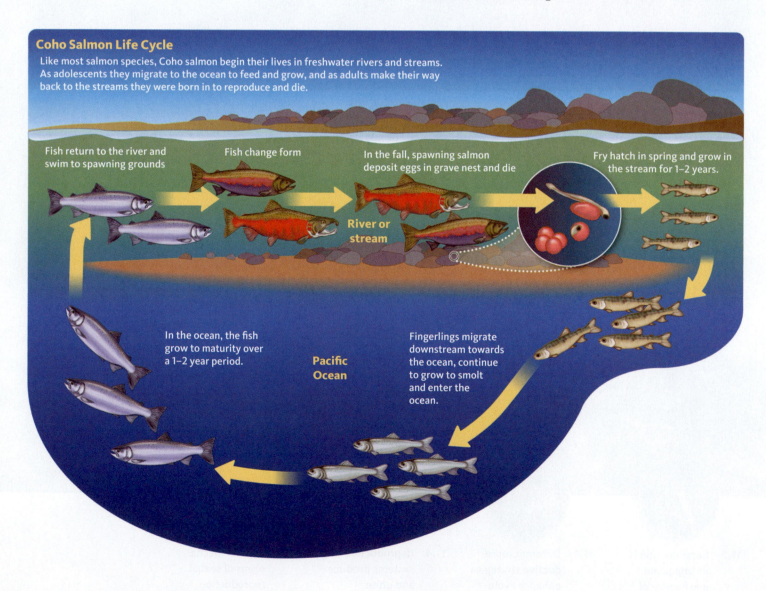

Coho Salmon Life Cycle

Like most salmon species, Coho salmon begin their lives in freshwater rivers and streams. As adolescents they migrate to the ocean to feed and grow, and as adults make their way back to the streams they were born in to reproduce and die.

Fish return to the river and swim to spawning grounds

Fish change form

In the fall, spawning salmon deposit eggs in grave nest and die

River or stream

Fry hatch in spring and grow in the stream for 1–2 years.

In the ocean, the fish grow to maturity over a 1–2 year period.

Pacific Ocean

Fingerlings migrate downstream towards the ocean, continue to grow to smolt and enter the ocean.

year or two while slowly migrating toward the ocean. At this point in their migration, when they are at about 25 percent of their adult size and are referred to as *smolt,* they develop a new set of physiological characteristics that prepare them for the saltwater environment to which they are migrating. Without these changes, the fish would not be able to maintain proper water and salt balance when they begin their ocean voyage.

About 10 percent of the original hatch survives to reach the sea; most are eaten by predators. At sea they primarily feed on fish to complete their growth and development cycle, reaching two feet in length and weighing about 10 pounds. The salmon migrate long distances to feeding grounds along the North Pacific coast until they reach maturity (**Fig. 11.2**). During these stages of development, the fish become sexually mature, change to a reddish color, and develop a hooked and elongated snout, features more pronounced in males than in females (see **Fig. 11.2**). At this point, hormones signal the fish that it is time to reproduce. Using an internal sun compass and their sense of smell, they begin the long journey back to the stream where they were born. Salmon live a short, difficult life. Fewer than 100 fish of the original 3,000 eggs complete the journey, and the adults die just days after they reproduce.

Once the fish have reached their native grounds, they spawn by depositing and fertilizing the eggs. The female uses her tail to build a nest in the sandy river bottom. The sizes of the nest vary greatly, but in general they are about nine inches deep and a few feet in length. Meanwhile the males battle each other to claim their territory. The larger males outcompete the smaller males and choose a female mate that has selected an optimal nesting site in terms of water flow and nest protection.

The mating process begins with a brief courtship ritual. To select a female, the male swims around and crosses over the female a few times. This behavior signals that the two will spawn together. The two fish tap tails, quiver, and rapidly release their eggs or sperm together. Within a minute, the female covers the nest and moves forward to dig another. In total, she will release three to four thousand eggs. Males and females mate with several partners, so the male either stays close to fertilize another batch of the same female's eggs or goes off to select another mate. The fish spawn for 30 to 40 continuous hours until they are exhausted. The females actually cut their tails to the bone in the process of digging so many nests. After spawning, the females guard their territory to prevent the destruction of their eggs by females that arrive a little later in the season.

Because the salmon do not feed during their upstream migration, once they deplete their body fat reserves, they drift downstream and die from starvation and other natural causes. Scientists have discovered a cue that triggers the spawning and death. They have observed rapid aging in the blood, tissues, and organs of salmon due to a change in various glands of the fish. This change begins shortly after the fish reach matu-

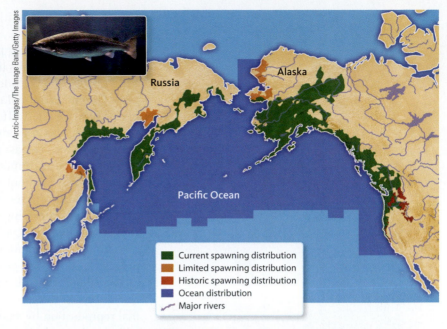

Arctic-Images/The Image Bank/Getty Images

Figure 11.2 Distribution and feeding grounds of Coho salmon. These fish are found in the Pacific Ocean from the northwest coast of North America to the Northeast coast of Russia (dark blue water). The green areas show the current breeding and spawning grounds of the salmon. The salmon are born in these spawning grounds, migrate out to the ocean to feed and return to their birthplace to reproduce before they die.
© Cengage Learning 2014

rity, perhaps also initiating their migration back to the spawning grounds.

Another interesting facet of salmon reproduction is the strategy to reproduce just once in its lifetime. As with all aspects of life, organisms have a life history that includes different strategies to survive and reproduce. A general trend in animal reproduction is that animals that live shorter lives invest more of their efforts in reproduction. For instance, salmon, mayflies, and cicadas reproduce just once in their lifetime. Because they have only one chance to pass on their genes, they channel their energy and resources into large numbers of offspring to ensure that sufficient numbers reach maturity without any parental care. The offspring of animals that reproduce many times throughout their lifetime usually have a higher probability of surviving to maturity because of more parental care and a slower development cycle.

Throughout evolution, each species has developed survival strategies programmed into the genes that dictate its reproductive cycle. The genetics of the salmon controls this organism's drive to reproduce but also creates limitations for the species. In the chapter, we will explore the process of animal reproduction by examining both asexual and sexual reproduction. We will start by exploring mating strategies and behaviors of animals as they are tied to evolutionary fitness. We also will look at the various systems, structures, and processes that produce and unite gametes in the reproductive process. We will conclude the chapter by looking at the cellular mechanisms of sexual reproduction known as meiosis and gametogenesis.

11.2  Reproduction Is an Important Mechanism of Evolution

Throughout the chapters in this unit, we have examined day-to-day challenges that animals face to obtain food, exchange gases, grow, and survive. Like all organisms, animals face the challenge of **reproduction**, which is the process that creates new individuals of the same kind from previously existing individuals. Reproduction has allowed life to continue as an unbroken chain from the first cells for billions of years. It is so important that almost everything an organism does helps it survive so that it can reproduce.

In the early chapters, we emphasized the diversity of life as a result of evolution. Recall from Chapter 3 that genetic variation and natural selection shape evolutionary change, adaptations, and fitness. A successful species is made up of individuals that survive the current environment and produce offspring with similar characteristics. These offspring survive and eventually pass the successful genes and characteristics to the next generation. Natural selection and reproduction are mechanisms by which evolution occurs. It is through reproduction that organisms pass on their genetic characteristics to ensure the evolutionary fitness of the species.

Animals have many adaptations and diverse ways of reproducing— from the hydra that copies itself without mating, to the salmon that lays thousands of eggs that are covered in sperm, to the blue whale that houses the fertilized egg inside her body for almost a year before giving birth to an infant. This section will examine the diverse modes of animal reproduction that occur asexually or sexually.

Reproduction—the process that creates new individuals of the same kind from previously existing individuals.

Asexual reproduction—the process that creates offspring without the fertilization of the egg. In most cases, the offspring are genetically identical to the parent and develop through the process of mitosis.

Asexual reproduction creates genetically identical offspring

Sea stars, sponges, hydra, and flatworms are animals that can reproduce asexually without the uniting of sperm or egg cells. Through **asexual reproduction**, an animal creates offspring that are genetically identical to itself through the process of mitosis. Three types of asexual reproduction result in identical offspring. Hydra, corals, and sponges sometimes reproduce by *budding*. The bud, which is identical to the parent, can detach and reattach to a new location, or it can remain attached to the parent to form a new colony (**Fig. 11.3**). Flatworms and segmented worms often reproduce through *fragmentation,* where the body simply splits into segments and a new organism grows into a fully mature individual from each fragment of the parent (**Fig. 11.3c**).

The third type of asexual reproduction is more complex than the previous examples. Some insects, reptiles, amphibians, and fish can develop from an unfertilized egg in a process called *parthenogenesis.* Parthenogenesis is considered asexual reproduction because an egg and sperm do not unite. However, because this mode does involve an egg, the offspring is not identical to the parent. Recall from Chapter 3, the production of gametes involves the generation of genetic variation through meiosis. Honeybees are social insects that live in complex colonies with a caste system consisting of the queen, smaller female workers, and male drones (**Fig. 11.4a**). The queen sexually reproduces once in a lifetime and holds the sperm inside her body for her lifespan of three years or so. The workers build a honeycomb nest where the queen deposits her eggs, but she only selectively deposits the sperm into some of the cavities (**Fig. 11.4b**). The cavities with eggs and sperm represent sexual reproduction and develop into females. Honeybees also reproduce asexually—those cavities with only eggs undergo asexual parthenogenesis and develop into male drones.

Why would the honeybee reproduce in such a fashion? There are several advantages to asexual reproduction. In the case of honeybees, the colony of more than 20,000 bees works together to gather nectar, make honey, feed larvae and physically maintain the nest. The female workers live about a month, and the male drones live for two months. Given the large number of bees and the shortness of their lifespan, it is advantageous for the queen to reproduce in a quick and energy-efficient manner. When the queen reproduces asexually, she does not need to devote time or energy to finding a mate and mating. The honeybees illustrate an important advantage to asexual reproduction: it is a rapid means of reproduction. In other animals, asexual reproduction also serves as a rapid means of reproduction. Nonmotile animals such as coral and sponges use budding, especially when they are isolated from other sexually reproducing individuals.

There are also disadvantages to asexual reproduction. The production of offspring that are identical copies of the parent limits the genetic variation of the group. The greater the number of genetically similar individuals in a population, the more difficult it is for the population to adapt to a changing environment or to pathogens or parasites. As selection pressures in the environment change, if all the individuals of the population have the same genetics and cannot adapt, they will die. As with everything an animal does, there are tradeoffs, meaning that some adaptations are compromises. In terms of reproduction, the adaptation of asexual reproduction uses less energy and time to produce offspring, but the offspring are genetically similar, and the population is less likely to adapt to changes in the environment.

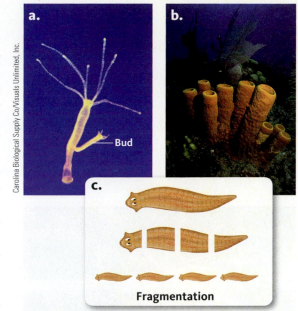

Fragmentation

Figure 11.3 Asexual reproduction. Many animals reproduce asexually through budding or fragmentation. The buds on both this hydra (a) and sponge colony (b) can break off to establish themselves in a new location. (c) This flatworm illustrates segmentation in which each of these pieces can give rise to new individuals.
© Cengage Learning 2014

a.

Female honeybee (worker)   Queen   Male honeybee (drone)

b. Queen bee laying eggs   Eggs

Figure 11.4 Honeybees reproduce sexually to create females and asexually to create males. (a) Female honeybee (referred to as a worker), queen honeybee, and male honeybee (called a drone). (b) Queen laying eggs.
© Cengage Learning 2014

Sexual reproduction generates diversity among offspring

Most animals reproduce sexually. **Sexual reproduction** is the creation of genetically different offspring from the fusion of male and female gametes. **Gametes** are reproductive cells, such as egg cells in females and sperm cells in males, which fuse during *fertilization* to form a zygote. There is a competitive advantage in producing offspring with genetic contributions from two individuals rather than one. The process that creates the gametes and the random fertilization of eggs by sperm allow new gene combinations to come together in the next generation. Organisms with new gene combinations may have a competitive advantage, such as being resistant to a particular pathogen. Sexually reproducing populations are able to change and adapt through time to their environment. Compared to asexual reproduction, a disadvantage to sexual reproduction is that it requires an investment of energy and time to find a mate and produce gametes. In addition, in the process of attracting a mate, animals often draw attention to themselves, making individuals more vulnerable to predation.

As we have seen, honeybees, like many animals, are able to reproduce both asexually and sexually. Let's return to their hive to compare the costs and benefits of both forms of reproduction. Asexual processes produce male drones, and sexual processes create two types of females—virgin queens and female workers. Shortly after birth, virgin queen bees leave the hive to reproduce with male drones from another colony. Each virgin queen will mate with a dozen or so different males, collecting and storing their sperm. To ensure genetic variation within the honeybee species, mating occurs outside of the hive with bees from other colonies. If the males were to mate with their own queen, they would be reproducing with their mother, perpetuating similar genetic material. To further increase the genetic variation of the hive, a queen bee will mate with multiple mates in a one- to two-day period, storing the sperm in a sac until she is ready to fertilize eggs.

Recall from Chapter 3 that the most fit or adapted organisms withstand a variety of selective pressures at each stage of their life cycle to live to reproductive age and pass their genetic information to their offspring. The amount of genetic variation in a population increases the evolutionary fitness of a species. An example of the consequences of a lack of genetic variation can be seen in the recent epidemic of honeybee colony collapses. One of the culprits is a parasitic mite, which feeds off fluids of the bee (**Fig. 11.5**). Some bee colonies are better adapted to identify these invaders and aggressively attack them. Researchers have analyzed the DNA of bees from various honeybee hives across the United States to help determine the hardiness of the nests. The colonies with greater genetic variation show more resistance to these mites. In other words, there is a direct relationship between greater genetic variation and a higher survival rate of the colony. Later in this chapter, we will explain the mechanisms by which sexual reproduction introduces genetic variation.

a.

Varroa mite

Maryann Frazier/Photo Researchers/Getty Images

b.

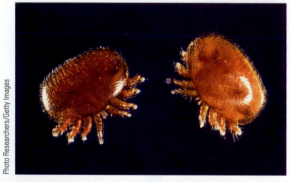

Photo Researchers/Getty Images

Figure 11.5 Varroa mite, a honeybee parasite. Bee colonies have been collapsing. (a) One of the culprits is this parasitic mite that feeds on honeybees. (b) Close up of the Varroa mite.

Sexual reproduction—the creation of genetically different offspring from the fusion of male and female gametes.

Gametes—reproductive cells, such as egg cells in females and sperm cells in males, that fuse during fertilization to form a zygote.

- Asexual reproduction is a fast mode of reproduction that creates genetically identical offspring through mitosis (budding and fragmentation). Parthenogenesis is considered asexual reproduction because eggs and sperm do not reunite.

- Sexual reproduction is energy-intensive but generates diversity among offspring and is an important mechanism of evolution.

- Sexual reproduction creates genetic variation by combining the genetic material of two individuals through the fusion of sperm and egg cells (gametes).

- There is a competitive advantage to producing offspring with genetic contributions from two individuals rather than one.

1. How do asexual and sexual reproduction differ from one another?
2. Describe two advantages and two disadvantages of sexual reproduction.
3. In your words, describe how evolution and sexual reproduction are dependent on each other.
4. List three reproductive strategies that honeybees utilize.
5. Komodo dragons sometimes lay unfertilized eggs that develop into offspring. Which type of reproduction is this lizard using?

11.3 Different Reproductive Strategies Enhance Evolutionary Fitness in Animals

a.

b.

Brown garden snails, one of several land snails, have several interesting adaptations for reproduction that aid in their survival as a species. First, these snails use their antennae and perhaps their sense of smell to find a mate. Once they find a mate, they perform a courtship ritual that lasts for several hours and includes crawling over and touching each other (**Fig. 11.6a**). One of the pair then shoots a sharp dart into its partner, releasing a chemical that increases the survival of the sperm and leads to more successful offspring production (**Fig. 11.6b**). As hermaphrodites, each snail carries both male and female reproductive structures. A mating pair exchanges sperm through a genital pore located on the right side of the head, each snail mating with several different partners and collecting sperm in an internal sac. Because each snail is competing with several other mates to fertilize the eggs, the "love dart" provides a reproductive advantage if the dart hits the partner in the proper location. After the eggs are fertilized, they pass through the snail, exiting with a nutrient-rich jelly coating that will nourish and protect each developing embryo. The eggs are deposited in a damp area where they will stay moist and complete the development cycle.

Ultimately the various *reproductive strategies* of animals determine which genes are passed to the next generation. For instance, some of the reproductive strategies used by the garden snails include mating with multiple partners, using the "love dart," and fertilizing their eggs

Figure 11.6 Snails have both male and female reproductive structures. (a) Two garden snails performing a courtship ritual that includes touching each other and (b) shooting a "love dart" to improve reproductive success.

internally. To reproduce successfully, animals have evolved an arsenal of adaptations related to behavioral, structural, social, and energy aspects of life. Animals use these various adaptations to attract and choose a mate, engage in courtship behavior, make gametes, communicate, store energy, and compete for a mate.

Sexual reproductive strategies synchronize the release of gametes

For sexual reproduction to be successful, viable eggs and sperm must be brought together at the same time. Many animals use water as a vehicle to carry the sperm to the eggs. Sponges, corals, sea urchins, lobsters, fish, and frogs are among the many animals that release large quantities of eggs and sperm into the water (**Fig. 11.7**). Because all of these animals are reproducing in the water where currents can wash the gametes away, how do they ensure the eggs are fertilized by the sperm?

The aquatic environment presents a challenge for microscopic gametes to find each other. Another challenge is that gametes are short-lived, with only a small window of time in which fertilization can occur. One strategy to maximize the number of offspring is to produce and release more gametes. Sponges release millions of sperm, but only a few will fertilize the eggs of a nearby sponge (see **Fig. 11.7**). Sea urchins gather in large groups to release both sperm and eggs at the exact same time, increasing the chances of fertilization. Male and female salmon simultaneously release a few hundred gametes with each mating.

Salmon, sponges, sea urchins, and other animals use **external fertilization**; that is, they fertilize the eggs outside the body. This type of fertilization depends on several strategies for success. First, the release of the male and female gametes is closely timed to ensure the eggs are fertilized. Because external fertilization occurs in the water (sperm rely on the water to swim to the eggs), some aquatic animals time the release of their gametes with the phase of the moon. Others use the tide as an important indicator. Lobsters and frogs rely on water temperature, which is determined by the season. In all of these examples, the animals receive a cue from the environment that triggers an internal chemical response that signals the time for reproduction.

Many land animals depend on a breeding season and have evolved to reproduce at a certain time of the year to achieve the best reproductive success. Animals often migrate at the same time to common breeding grounds where they select their mates. Most mammals and birds have a breeding season characterized by abundant supplies of food to support the developing offspring. In the arctic, birds lay eggs in May and June so that the young can take advantage of the early July hatch of mosquitoes, butterflies, and other insects. Triggered by day length and the phases of the moon, deer go into "heat" and mate in November. Their offspring are born in May and June when the conditions are less harsh, food supply is abundant, and the newborn has enough time to grow and develop in order to survive its first winter.

Internal fertilization unites gametes in a moist, protected environment

Most land animals have strategies that allow them to overcome the challenge of getting swimming sperm to the eggs and then keeping them in a moist environment to grow and develop. Like snails, most reproduce by

Fred Bavendam/Minden Pictures

Figure 11.7 Reproductive strategy of sponges. These sponges use external fertilization to sexually reproduce. They release millions of sperm into the water to fertilize eggs from other nearby sponges.

External fertilization—union of sperm and eggs outside the body, usually in water or a moist environment.

a.

b.

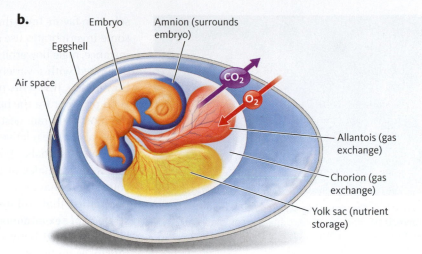

Embryo Amnion (surrounds embryo)

Eggshell

Air space

CO₂

O₂

Allantois (gas exchange)

Chorion (gas exchange)

Yolk sac (nutrient storage)

Brandon Blinkenberg/Shutterstock.com

Figure 11.8 The amniotic egg is an important evolutionary adaptation. (a) Birds lay amniotic eggs protected by a hard, porous shell. (b) The inside of the egg and its membranes that aid in nutrient storage (yolk), gas exchange (allantois and chorion), and suspension and cushioning of the embryo (amnion). Note the arrows indicating gas exchange through diffusion. Oxygen enters the egg for respiration, and carbon dioxide diffuses out as a waste product.
© Cengage Learning 2014

internal fertilization: the sperm is transferred inside the female's body, where the protected egg is fertilized. This strategy consumes more energy and results in fewer offspring, but it enhances the organism's reproductive success by protecting the gametes and maintaining a moist environment for the developing young. Organisms such as reptiles, birds, and mammals use internal fertilization, which entails a period of internal embryonic development or laying eggs protected by shells.

These animals also have an important evolutionary adaptation called the **amniotic egg** (**Fig. 11.8**). The amniotic egg is often called a "self-contained pond" because it has a food source and protective membranes to protect and aid the developing embryo. You can view life as a set of chemical reactions that take place in water, and the amniotic egg allows the embryo to be surrounded by water. The egg's three membranes keep water close to the developing embryo, facilitate gas exchange, and allow nutrient and waste processing. Amniotic eggs permit larger and more developed young at birth and permit animals to completely move away from water. By contrast, frog eggs are a few millimeters in size, and frog larvae are less than a centimeter in length.

Competition for mates results in sexual selection

Animals use different strategies for mating to enhance their evolutionary fitness. Because some individuals are better than others at competing for a mate, mate selection plays an important role in which genes are passed on to the next generation. Some animals mate with one partner for life, others are loyal to one mate each season, and others, like the honeybee and land snail, mate with multiple partners in a short period of time.

Animals that fertilize their eggs internally spend much more time attracting and selecting a mate than do animals that disperse their gametes into the environment. Because the purpose of reproduction is to pass on genes, a large amount of energy is spent attracting a mate that shows the most evolutionary fitness. Sexual selection favors traits that increase reproductive success. Fireflies use visual cues in their blinking lights, some birds use song calls, and other birds such as peacocks use an elaborate display of feathers to get the attention of a partner.

Recall from Chapter 5 the elaborate behaviors associated with mating rituals and mate selection to bring the male and female into close contact for copulation, or sexual intercourse. Also, remember from Chapter 3 that sexual

Internal fertilization—union of the sperm and egg inside the female's body. This strategy consumes more energy and results in fewer offspring, but it enhances the organism's reproductive success by protecting the gametes and maintaining a moist environment for the developing young.

Amniotic egg—an egg found in reptiles, birds, and mammals that has a food source and protective membranes to protect and aid the developing embryo; often called a "self-contained pond" of fluid that supports the growth and development of the embryo.

Figure 11.9 The courting strategy of a male bowerbird. Male bowerbird building an elaborate nest of twigs and bright blue objects to attract a mate.

selection favors traits that increase reproductive success. Most vertebrates and some invertebrates use social mechanisms to identify and attract a mate. For instance, male bowerbirds entice females by building intricate twig structures decorated with a variety of objects such as feathers, snail shells, leaves, and flowers (**Fig. 11.9**). This nest (or bower) is an indicator that the bird will be a good provider for the hatchlings; therefore, the female selects a mate that has the most elaborate nest. The construction of bowers evolved to persuade females about the relative quality of the males in the population.

Many animals show different physical characteristics between the males and females of the species (**Fig. 11.10**). This dynamic plays a role in courtship behavior and sexual selection, ultimately promoting traits in a population that will increase an organism's mating success. This common strategy of sexual dimorphism can include differences in size, coloration, or body structure between the sexes. Coho salmon display sexual dimorphism, as do lions and seals (see **Fig. 11.10**).

Sexual dimorphism is a result of the strategies that animals use for mate selection. In animal groups apart from mammals and birds, the larger females tend to produce more eggs and therefore contribute more offspring (and alleles) to the next generation. Males select the largest females because size is a sign of reproductive fitness. In honeybees, the female mates with several males and stores the eggs. In this case the female is larger than the male to support the production and storage of eggs.

In some animal groups, the male puts on an elaborate display to attract females, often defending himself against other interested males. This type of sexual selection favors males that are larger in size and have competitive features such as horns and antlers that make them more competitive. Elephant seals mate only once every year, gathering together on the beach. Males must compete for females, and the largest males win the battle and mate with a harem of several females. In this example, sexual selection drives physical traits that stay in the population over successive generations. Adaptations for sexual selection increase an organism's fitness and the probability that adaptive genes are passed to the next generation.

Sexual dimorphism—a difference in physical characteristics between the males and females of the same species; examples are differences in size, coloration, or body structure.

a.

b.

Figure 11.10 Examples of sexual dimorphism. Many animals show sexual dimorphism including differences in size, coloration, or body structure between the sexes to increase mating success. (a) The showy mane of the male lion is used for protection in fights, as well as to look big and bold. (b) Reproductive effort in elephant seals is limited by nutrient reserves available at the onset of breeding. Males start the season weighing 2 to 8 times as much as females. The male also has a trunk that amplifies its mating call and condenses water to conserve it. During the three-month breeding season, males fast from food and water, losing over 30% of their body weight.

Energy expenditure for reproduction differs among animals

An important aspect of an animal's reproductive strategy is the amount of energy it invests in reproduction. Different species exhibit different life expectancies and reproductive rates. For instance, female mosquitoes live about a month and lay hundreds of eggs before dying, whereas gorillas live up to 40 years, bearing two to three offspring in their lifetime. An animal's life cycle is influenced by characteristics such as growth rates, developmental patterns, timing and rate of reproduction, and life span. These lifetime patterns, called the life history of an organism, show that two strategies of energy expenditure to ensure reproductive success have evolved in the animal kingdom. Some animals, like mosquitoes, salmon, and frogs, invest all of their energy in producing massive amounts of eggs as a way of having a small number of surviving offspring, whereas others like gorillas and humans invest their energy in making fewer eggs but providing more parental care (Fig. 11.11).

Let's look more closely at how different species use their energy for reproduction. In humans, energy typically is devoted to producing one or two children who are highly developed at birth (Fig. 11.12). Like chimpanzees, human babies are highly developed because of the maternal resources dedicated to a single offspring's development over a long period of time. After birth, human mothers nurse babies for several months, and both parents care for them for many years. In contrast to humans, the Coho salmon and other fish spawn and deposit up to 3,000 eggs in one season. About 30 percent of these eggs produce a viable offspring that is less than an inch long when it hatches after only four weeks of development (Fig. 11.12c). No parental care occurs because the adults die soon after spawning, and the young salmon must fend for themselves. Many insects, fish, and small mammals such as rodents share these life history traits. For every

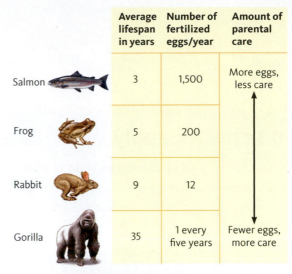

	Average lifespan in years	Number of fertilized eggs/year	Amount of parental care
Salmon	3	1,500	More eggs, less care
Frog	5	200	
Rabbit	9	12	
Gorilla	35	1 every five years	Fewer eggs, more care

Figure 11.11 Relationship between amount of parental care and number of eggs. Animals like frogs and salmon, that produce larger numbers of eggs, are not able to provide the parental care that animals producing fewer eggs can provide.
© Cengage Learning 2014

Life history—series of changes an organism undergoes throughout its lifetime patterns.

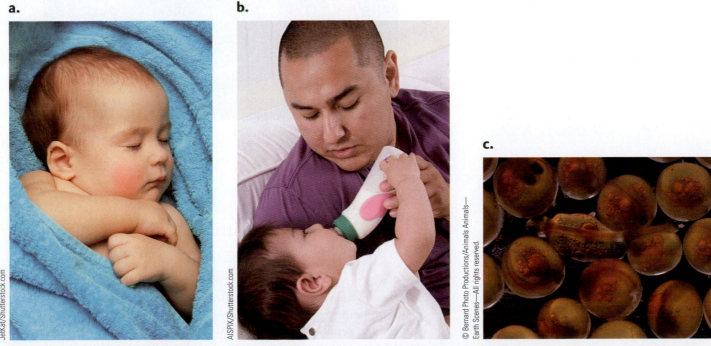

a.

b.

c.

Figure 11.12 Humans and other mammals expend energy on one or two highly developed offspring; salmon and other animals produce numerous, less developed offspring. (a) Human babies are born highly nourished and developed at birth. (b) Baby receiving care from parent. (c) Salmon hatchlings receive no parental care. They feed on egg sacs for the first couple of months of life.

animal, the energy expenditure spent on achieving reproductive success represents different adaptations resulting from millions of years of natural selection processes. Regardless of the strategy, in general the number of individuals in a population generally stays the same whether an animal produces thousands of eggs or just a few.

11.3 check + apply YOUR UNDERSTANDING

- Animals have many adaptations and strategies to ensure that their genes are passed on to the next generation, enhancing evolutionary fitness.

- Some animals use external fertilization, whereas others use internal fertilization to synchronize the release of gametes.

- The amount of energy invested in gamete production, mate selection, number of offspring, and parental care differ among animals and provide examples of strategies for reproductive success.

- Competition for mates results in sexual selection.

1. What is meant by external fertilization?
2. Why is the amniotic egg considered an important adaptation for reproduction?
3. What is the advantage for an animal to use internal fertilization?
4. Animals use several reproductive adaptations and strategies to ensure evolutionary fitness. Summarize these strategies using the land snail as an example.
5. **Figure 11.13** shows two birds of paradise in their mating ritual, which includes an elaborate dance. List three ways that the male and female differ from each other. Identify which bird is the male and which is the female. Explain how this sexual dimorphism enhances reproductive success.

Phil Savoie/Minden Pictures

Figure 11.13 Birds of paradise.

11.4 Reproductive Systems Produce and Unite Gametes

No matter what strategies an animal uses for reproduction, the journey of the sperm to the egg is a battle against long odds. Human females start life with close to 2 million eggs, yet only 400 of those eggs survive to be released one at a time each month over a period of about 40 years. Human males will produce upward of a trillion sperm in their lifetime. In a typical ejaculation,

200 million sperm wind their way through the ducts of the male, exit the penis, and enter the female reproductive tracts on the way to the egg. About 200 sperm complete the journey, and maybe one of those sperm will be accepted and taken in by the egg. As a result of this improbable but successful journey, humans, like other animals, have evolved ways to increase the likelihood of passing their genes on to the next generation.

Although the reproductive strategies, patterns, and behaviors of animals are diverse, the similarity between insect, amphibian, reptile, bird, and mammal reproductive systems may surprise you (**Fig. 11.14**). In all animal groups, the reproductive system functions to produce and transport gametes and to produce hormones that maintain the structure and proper functioning of reproductive systems. In mammals, the female reproductive system also serves an important role in nurturing developing offspring.

The functions of the reproductive system are carried out by the gonads, ducts, glands, and other accessory organs. The gonads, consisting of the *ovaries* and *testes,* are the primary reproductive organs and are responsible for producing eggs, sperm, and hormones. These hormones function in the maturation of the reproductive system, the development of sexual characteristics, and the regulation of the physiology of the reproductive system. Other organs, ducts, and glands serve secondary roles—transporting and sustaining the gametes—and, in mammals, nourishing the developing offspring. In this section, we will explore the structures and functions of the familiar human reproductive system and compare it to that of other animals.

Male reproductive structures produce and deliver sperm

The male reproductive system in mammals serves three functions: (1) to produce and transport sperm in a protective fluid, (2) to discharge sperm within the female reproductive tract, and (3) to produce and secrete the male sex hormones responsible for maintaining the male reproductive system. Unlike the female, whose sex organs are located entirely within the pelvis, the male has reproductive organs that are both inside and outside the pelvis (**Fig. 11.15**). The male genitals include a pair of testes, the duct system made up of the epididymis and the vas deferens, the secretory glands including the seminal vesicles and prostate gland, and the penis. All of these organs work together to produce, nourish, and send the sperm on its journey to the egg.

Let's look at the details of each of the human male reproductive organs (see **Fig. 11.15**). After reaching sexual maturity or *puberty,* each male testis makes close to 100 million sperm every day. These immature cells are stored in the epididymis, where they continue to develop into mobile sperm cells. During sexual arousal, blood flow is altered in the penis, causing an erection.

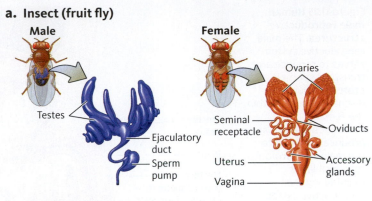

a. Insect (fruit fly)

Male — Testes, Ejaculatory duct, Sperm pump

Female — Ovaries, Seminal receptacle, Oviducts, Uterus, Accessory glands, Vagina

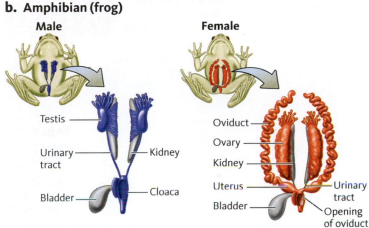

b. Amphibian (frog)

Male — Testis, Urinary tract, Kidney, Bladder, Cloaca

Female — Oviduct, Ovary, Kidney, Uterus, Bladder, Urinary tract, Opening of oviduct

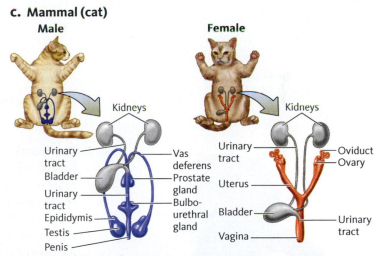

c. Mammal (cat)

Male — Kidneys, Urinary tract, Bladder, Urinary tract, Epididymis, Testis, Penis, Vas deferens, Prostate gland, Bulbo-urethral gland

Female — Kidneys, Urinary tract, Uterus, Bladder, Vagina, Oviduct, Ovary, Urinary tract

Figure 11.14 Comparison of reproductive structures. Although the reproductive strategies, patterns, and behaviors of animals are diverse, the structure of their reproductive systems is similar. Notice in each of these diagrams the series of paired gonads (testes and ovaries), ducts, and glands that are used to produce and release gametes. (a) Insect. (b) Amphibian. (c) Mammal.
© Cengage Learning 2014

Gonads—primary reproductive organs (ovaries and testes) that are responsible for producing eggs, sperm, and hormones.

Figure 11.15 Human male reproductive structures. The male reproductive system serves to produce and transport sperm in a protective fluid, discharge sperm within the female reproductive tract, and produce and secrete the male sex hormones responsible for maintaining the reproductive system.

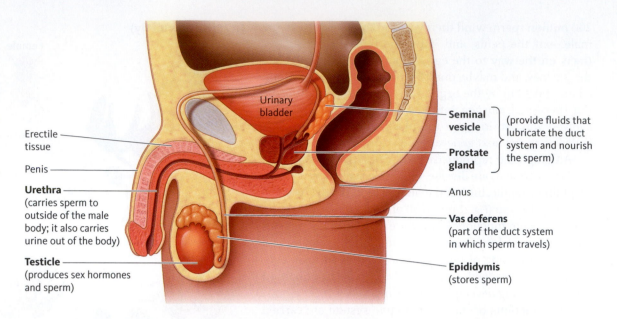

Urinary bladder

Erectile tissue

Penis

Urethra (carries sperm to outside of the male body; it also carries urine out of the body)

Testicle (produces sex hormones and sperm)

Seminal vesicle

Prostate gland

(provide fluids that lubricate the duct system and nourish the sperm)

Anus

Vas deferens (part of the duct system in which sperm travels)

Epididymis (stores sperm)

TABLE 11.1 Sperm Production and Delivery of a Few Mammals

Animal	Average Sperm/Ejaculate	Travel Time to the Egg
Mouse	50,000,000	15 minutes
Guinea pig	80,000,000	15 minutes
Human	280,000,000	Up to 1 hour
Sheep	1,000,000,000	Up to 5 hours
Cow	3,000,000,000	3 minutes
Pig	8,000,000,000	15 minutes

© Cengage Learning 2014

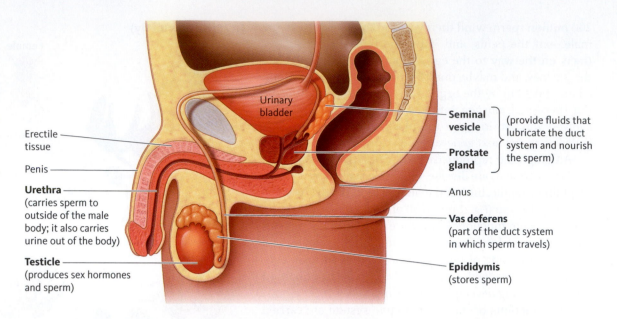

Figure 11.16 Male copulatory organs of a seed beetle. The promiscuous seed beetle uses the bristles on its penis to remove sperm deposited by previous males.

Semen—fluid that protects, nourishes, and helps to deliver the sperm.

The function of the erection is to enable the penis to physically enter the female reproductive tract and release sperm. During ejaculation, smooth muscles in the reproductive system contract and move the sperm from the epididymis through the duct system and out through the urethra of the penis. Along the way, the sperm mixes with fluids from secretory glands forming semen, which protects, nourishes, and helps to deliver the sperm. Once the sperm reach the female reproductive tract, components of the semen help to combat the more acidic environment inside the female and move the sperm toward the egg.

Because one of the functions of the male reproductive system is sperm production, it is interesting to examine characteristics of different animals (examples in Table 11.1). Some sperm reach the egg very quickly, whereas others take several hours. In humans, the sperm needs to travel about four inches to reach the egg, but in pigs the sperm travels over three feet. One way the pig increases its chances of reproductive success is through the massive production of sperm as well as the use of uterine contractions to propel the sperm along the long female reproductive tract.

You may also find it intriguing to look at different adaptations for delivering sperm and its genome to the next generation. Some animals that mate with multiple partners in a season sometimes have structures, like a bottlebrush, that displace the previous partner's sperm (**Fig. 11.16**). Once male gorillas establish their territory there is little need for competition, so they have a proportionally small penis (2 inches) compared to their large body size (400 pounds). In honeybees, the sole function of male drones is to mate; they get only one chance in their short lifetime. In fact, the act of sexually reproducing is a dramatic and fatal one for them. During the final stages of mating, the male's reproductive organs along with some abdominal tissues are ripped from his body and left inside the female. From the colony point of view, the male has now fulfilled his function, and there is no reason for additional resources to be devoted to these bees.

A final function of the male reproductive system is to produce and secrete the male sex hormones responsible for maintaining the male reproductive system. **Testosterone** is an important developmental and reproductive hormone responsible for many activities in the male reproductive system. The formation of sperm is controlled through the interactions of several hormones, including testosterone. This key hormone stimulates the growth of the male reproductive organs during development and sexual maturation. It also is responsible for secondary sexual characteristics such as facial hair, increased muscle size, and voice changes associated with puberty. Remember that these sexual characteristics are important factors in many animal species with regard to mate selection. Other animal sexual characteristics driven by increased testosterone levels include antlers, bright coloration, and manes.

Testosterone levels are influenced by a complex chain of events that can be altered by physiological responses, lifestyle habits, and environmental exposures. Because testosterone production is the result of the interaction of several hormones, occasionally something goes wrong, and the result is infertility or low sex drive. The process that regulates the amount and timing of testosterone release starts in the brain as part of the endocrine system (**Fig. 11.17**). Both luteinizing hormone and follicle-stimulating hormone have critical roles in the reproductive system. In males, **luteinizing hormone (LH)** stimulates cells in the testes to produce testosterone. **Follicle-stimulating hormone (FSH)** stimulates the growth of follicles in the ovary and induces the formation of sperm in the testis. LH and FSH work together to regulate the development, growth, pubertal maturation, and reproductive processes of the body.

Female reproductive structures produce eggs

In mammals, the female reproductive system consists of organs located inside the body in the pelvic region (**Fig. 11.18**) and mammary glands located in the breasts. The female system has evolved to produce and transport eggs to the site of fertilization, carry the developing fetus, and give birth. In humans, the female reproductive system contains three main parts: (1) the vagina, which acts as the receptacle for the male's sperm and also the birth canal, (2) the uterus, which holds and nourishes the developing fetus, and (3) a pair of ovaries, which produce the eggs. The breasts are also an important mammalian reproductive organ during the parenting stage of reproduction.

Eggs, some of the largest cells of the body, are formed and released by the ovaries. The released egg travels from the ovary down the ovi-

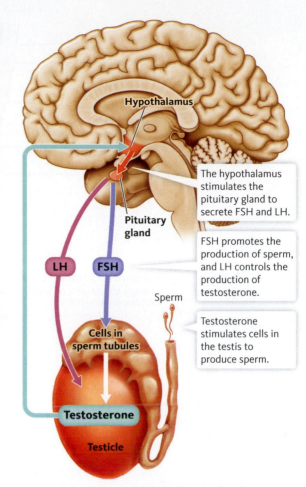

Figure 11.17 Hormonal links between the pituitary and testis. The pituitary gland in the brain receives signals to release two hormones: luteinizing hormone (LH) and follicle-stimulating hormone (FSH). These hormones act on specialized cells, which respond by releasing testosterone and triggering production of sperm.
© Cengage Learning 2014

The hypothalamus stimulates the pituitary gland to secrete FSH and LH.

FSH promotes the production of sperm, and LH controls the production of testosterone.

Testosterone stimulates cells in the testis to produce sperm.

Testosterone—important developmental and reproductive hormone responsible for many activities in the male reproductive system. Stimulates development of male secondary sexual characteristics, such as increased bone and muscle mass, and growth of body hair.

Luteinizing hormone (LH)—a reproductive hormone present in both males and females. In males, LH stimulates cells in the testes to produce testosterone. In females, LH controls the ovarian and uterine cycles, and it acts on cells in the ovaries to produce estrogen and progesterone.

Follicle-stimulating hormone (FSH)—a reproductive hormone present in both males and females that stimulates the growth of follicles in the ovary and induces the formation of sperm in the testis.

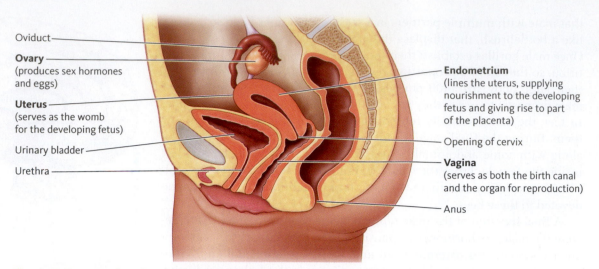

Oviduct

Ovary
(produces sex hormones
and eggs)

Uterus
(serves as the womb
for the developing fetus)

Urinary bladder

Urethra

Endometrium
(lines the uterus, supplying
nourishment to the developing
fetus and giving rise to part
of the placenta)

Opening of cervix

Vagina
(serves as both the birth canal
and the organ for reproduction)

Anus

Figure 11.18 Human female reproductive structures. The female reproductive system produces and transports eggs to the site of fertilization, carries the developing fetus, and gives birth.
© Cengage Learning 2014

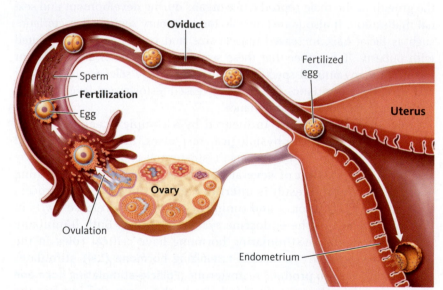

Oviduct

Sperm

Fertilization

Egg

Ovulation

Ovary

Fertilized
egg

Uterus

Endometrium

Figure 11.19 Release of an egg. The egg is released and travels down the oviduct, where it is fertilized by the sperm. At pregnancy, the fertilized egg implants in the uterus.
© Cengage Learning 2014

duct (also referred to as the fallopian tube). In animals that use internal fertilization, fertilization occurs in the oviduct (**Fig. 11.19**). In animals such as birds that do not have a uterus, the egg continues down the oviduct and is deposited outside of the body. In animals such as mammals that have a uterus, the fertilized egg travels to the uterus, where it implants in the uterine wall (see **Fig. 11.19**). Because the egg is a stationary cell, it relies on the smooth muscles of the reproductive tract and the ciliated epithelial cells of the oviduct to move it down the mucus-coated tract of the oviduct.

The egg has either of two fates: it is unfertilized, or it is fertilized and may develop into an embryo. In humans, if the egg is fertilized, development starts right away as the egg continues its weeklong journey to the uterus. In most mammals, if the egg is fertilized, it implants itself in the lining of uterus. Sometimes an ectopic pregnancy occurs in which the egg implants in the oviduct, but the fetus is not able to survive in this situa-

Figure 11.20 The ovarian cycle. The ovarian cycle begins when follicle-stimulating hormone (FSH) from the pituitary gland stimulates the development of the egg in the ovary (1). The egg is surrounded by follicle cells that secrete estrogen, which stimulates the lining of the uterus to thicken (steps 2 and 3). As estrogen levels increase and FSH output decreases, another hormone (luteinizing hormone [LH]) is secreted by the pituitary to complete the development and release the egg from the ovary (ovulation) (steps 4 and 5). If fertilization and implantation occur, additional hormone signals are sent that prevent the next ovarian cycle from starting. As in this diagram, if fertilization does not occur, the lining, mucus, and blood are shed from the body during menstruation and the cycle starts again at Day 1.
© Cengage Learning 2014

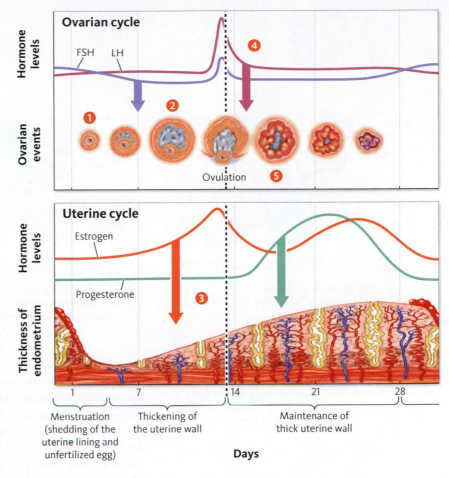

tion. If the egg remains unfertilized, it continues down the tract through the cervix and out the vagina. If implantation occurs, membranes from the egg merge with the epithelial lining of the uterus, forming the placenta, which provides nutrient and waste exchange to the developing fetus.

The female reproductive system includes several external organs collectively referred to as the vulva. These organs protect the reproductive tract from trauma and infection and include erectile tissue (such as the penis) with nerve endings that have thousands of mechanoreceptors for erotic sensations. In many animals, the muscular contractions from sexual arousal are needed to force the sperm and egg together. In some animals, such as camels, rabbits, and cats, release of the egg is dependent on a hormone response due to direct stimulation from the male rather than cyclical hormone levels.

As in the male reproductive system, hormones regulate many aspects of the female reproductive system. Maturation of the reproductive system is driven by increased levels of two hormones at puberty. As levels of LH and FSH increase, they trigger the release of estrogen. Estrogen acts on the various tissues and organs in the female reproductive tract, causing the breasts, ovaries, uterus, and vagina to mature.

Once sexual maturity is reached, additional hormones regulate the maturation and release of eggs, and, in mammals, hormones signal the female's body to receive the fertilized egg. In mammals, these events happen on a cycle. The formation and release of eggs that occur in the ovaries is called the ovarian cycle and is tightly coordinated with the uterine cycle, which prepares the uterus to receive the egg.

In mammals, the production of eggs is not continuous. In humans this cycle of releasing a mature egg occurs consistently throughout the year approximately every 28 days; in rodents the cycle lasts only about six days per year. Most mammals ovulate only during a specific breeding season when they are "in heat," which means they are receptive to mating. Bears, foxes, and wolves have only one breeding season a year to allow their offspring to grow enough to survive the following winter. Rabbits do not go into heat and are receptive to breeding throughout the year.

The cycles may differ among species, and the release of an egg, or ovulation, is the result of complex hormonal signals (**Fig. 11.20**). The pitu-

Placenta—organ in the uterus of mammals that provides nutrient and waste exchange to the developing fetus.

Estrogen—hormone that acts on the various tissues and organs in the female reproductive tract, causing the breasts, ovaries, uterus, and vagina to mature.

Ovarian cycle—series of changes in mammals that result in formation and release of eggs; occurs in the ovaries.

Uterine cycle—series of changes in mammals that prepare the uterus to receive the egg.

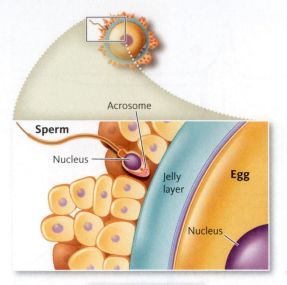

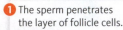

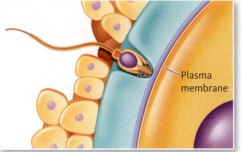

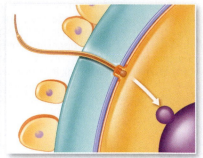

① The sperm penetrates the layer of follicle cells.

② Enzymes in the acrosome are released, creating a path to the plasma membrane of the egg cell.

③ The sperm nucleus enters the egg cytoplasm and fuses with the egg's nucleus.

Figure 11.21 Fertilization in mammals. The fertilization process brings about conception in three steps: (1) contact, (2) penetration, and (3) fusion of the egg and sperm nuclei. Pregnancy (not shown here) begins when the fertilized egg is implanted in the uterine wall. In other animals, the sperm directly contacts the egg cell without a jelly layer or follicle cells attached.
© Cengage Learning 2014

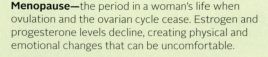

Menopause—the period in a woman's life when ovulation and the ovarian cycle cease. Estrogen and progesterone levels decline, creating physical and emotional changes that can be uncomfortable.

Conception—the stage of reproduction that is marked by fertilization and the formation of a zygote.

itary releases two hormones (LH and FSH) that promote ovulation and the production of the sex hormones (estrogen and progesterone) by the ovaries. These sex hormones stimulate the breasts and uterus to prepare for possible fertilization and pregnancy. If pregnancy does not occur, hormone interactions signal the lining of the uterine wall to break down. Humans and other primates shed the uterine lining through menstrual bleeding, whereas most animals do not have a menstrual cycle and absorb the lining instead.

Many women take synthetic versions of estrogen and progesterone as contraceptive methods to prevent pregnancy. These hormones are delivered as pills, skin patches, shots, or implants that function to prevent ovulation and thus prevent pregnancy. By changing the natural balances of estrogen and progesterone, the synthetic hormones deceive the body into thinking it is pregnant. A high level of progesterone is maintained, which inhibits secretions of FSH and LH from the brain. The result is that no new eggs develop and ovulation does not occur, so an egg cannot be fertilized and pregnancy is prevented.

Mammals are born with a finite number of eggs, and eventually ovulation and the ovarian cycle cease. In humans, this change, referred to as **menopause**, occurs at around 50 years of age. Because the ovarian cycle is controlled by hormones, menopause also triggers hormone interactions. Estrogen and progesterone levels decline, creating physical and emotional changes that can be uncomfortable. Women undergoing menopause are susceptible to osteoporosis because estrogen also plays a role in bone strength, and bones get weaker as levels decline. Also, as estrogen levels decline, cholesterol levels rise, increasing the risk of cardiovascular disease.

Mammals nurture their young through pregnancy

One characteristic of mammals is the amount of nurturing and parental care that they invest in their offspring. **Conception** occurs when the egg and sperm unite and fuse together by the fertilization process (**Fig. 11.21**), forming a zygote. During fertilization, the sperm penetrates the layer of

TABLE 11.2 Gestation Periods of Various Mammals

Mammal	Number of Gestation Days	Litter Size	Average Adult Female Weight (pounds)
Rat	21	10	.5
Rabbit	31	6	2.5
Pig	114	10	150
Human	266	1	135
Blue whale	314	1	5,000
African elephant	640	1	265

© Cengage Learning 2014

follicle cells. The sperm acrosome releases enzymes that create a path to the egg's plasma membrane. The sperm nucleus next enters the egg's cytoplasm and fuses with the egg's nucleus, marking conception. The fertilized egg travels down the oviduct toward the uterus. **Pregnancy** begins when the zygote implants into the uterine wall (see **Fig. 11.19**). During pregnancy, or *gestation,* the developing fetus grows within the uterus of the mother, which provides protection and nourishment as the fetus develops. Animals have different gestation periods. In general, the larger the animal, the longer the gestation period (**Table 11.2**) and the more well developed the offspring is at birth. In many animals, more developed newborns can become mobile sooner and therefore have an increased likelihood for survival. The size of the litter also affects the length of the gestation period. In general, animals with larger litters have a shorter gestation period.

There are three types of mammalian pregnancy. Mammals such as the platypus lay eggs, marsupials such as the kangaroo give birth to relatively undeveloped young that complete development in a pouch connected to the mother's abdomen, and mammals such as rabbits and humans retain their young through a placental connection in the uterus for longer periods. All of the animals listed in **Table 11.2** use a placenta to nourish their young.

The placenta is an organ that connects the developing fetus to the mother (**Fig. 11.22**). The placenta is rich in maternal blood vessels and provides a large surface area for the exchange of gases, nutrients, and wastes between the mother and fetus. The umbilical cord connects arteries from the fetus to the placenta, ensuring that the maternal and fetal blood supplies do not mix. The placenta also synthesizes hormones that prevent ovulation throughout pregnancy.

At the end of the gestation period, hormones released by the pituitary gland initiate labor and uterine contractions. Early in labor, a woman's "water breaks" when the amniotic sac ruptures, releasing the half-liter or so of fluid it contained. As labor progresses the cervix opens, and the contractions become more frequent and more powerful until the fetus is born through the vagina. Shortly after birth, the placenta detaches and sheds from the uterus. In humans, caregivers cut the umbilical cord; however, in the wild, animals sever the cord with their teeth. Throughout pregnancy, the same hormones that act on the ovaries and uterus during pregnancy prepare the mammary glands for lactation, or the production of milk. At birth, the mammary glands start to release milk so that the mother can nourish the newborn.

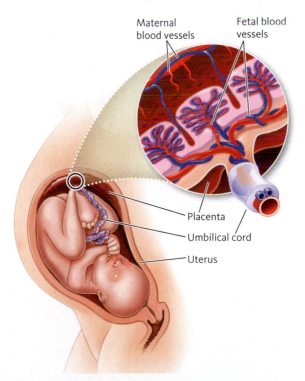

Maternal blood vessels

Fetal blood vessels

Placenta

Umbilical cord

Uterus

Figure 11.22 The placenta. The placenta connects the developing fetus to the mother and provides for the exchange of gases, nutrients, and wastes. At birth, the placenta weighs about one pound to support a seven-pound baby.
© Cengage Learning 2014

Pregnancy—condition that begins when the zygote implants into the uterine wall.

- The animal reproductive system consists of gonads, ducts, glands, and accessory organs that work together to produce, deliver, and unite gametes (sperm and eggs).

- Male reproductive structures produce and deliver sperm.

- Female reproductive structures produce eggs.

- In mammals these systems also nourish and protect the developing embryo from fertilization through birth.

- Reproductive systems also release hormones that control sexual maturation, mating behavior, gamete reproduction, pregnancy, and birth.

1. List the four basic components of animal reproductive systems.
2. What are the two primary roles of the gonads in both the male and female reproductive systems?
3. How does the placenta play a key role in the success of mammals?
4. What is the pathway that the sperm takes from the testes through implantation in the female at pregnancy?
5. Based on the following information, predict whether the following animals have a shorter or longer gestation period compared to each other. Ground hogs live about 6 years and give birth to one litter of about five young each year. Wolves live about 10 years and give birth to litters of four to seven pups each year. Hippos live about 45 years and give birth to one offspring every couple of years. Support your prediction.

11.5 Cellular Mechanisms of Sexual Reproduction Include Meiosis and Gametogenesis

Recall that *gametes* are reproductive cells, such as egg cells in females and sperm cells in males, which fuse during fertilization to form a *zygote.* With an understanding of the sexual reproduction systems and organs that produce, foster, and unite the gametes, you are now ready to learn about the cellular and genetic mechanisms involved in producing genetically unique sperm and egg cells. Reproduction is essential for a species to sustain its population. In order to do so, sexually reproducing animals receive one set of chromosomes from each parent (**Fig. 11.23**). Humans have a total of 46 chromosomes, 23 from the mother and 23 from the father. The 46 chromosomes constitute 23 pairs, which is a full set of genetic material. This state is referred to as diploid. To maintain this normal chromosome count, when the egg and the sperm unite, each must contain only one set of 23 chromosomes; otherwise the total number of chromosomes would double with each generation. Gametes with one set of 23 chromosomes are haploid. After a sperm cell fertilizes the egg, their genomes combine, resulting in a zygote that is diploid (see **Fig. 11.23**).

If each of your cells has 46 chromosomes, how are the chromosomes reduced to 23 in gamete cells? Meiosis is a special type of cell division in which a diploid cell divides to produce four haploid gametes. An important aspect of meiosis is that it "mixes" genes to create new gene combinations on the gametes' chromosomes. (This will be discussed more in the following paragraphs.) After fertilization, the zygote has genes for all traits from both parents and the mixed genes

Diploid—the number of chromosomes in most cells in the body, representing a full genetic set of chromosomes with contributions from both parents.

Haploid—the number of chromosomes in a gamete, representing half the number of chromosomes after fertilization.

Meiosis—a special type of cell division in which a diploid cell divides to produce four genetically unique haploid gametes.

from two parents, thus providing genetic variation that is important in sexual reproduction.

As with all of the processes of life, the intricacies of the cellular processes of reproduction are amazing. However, despite the accuracies of these processes, errors can occur in meiosis as well as in the process that forms the gametes. It is estimated that one in five fertilized eggs in humans contains chromosomal abnormalities. These abnormalities often result in miscarriages or stillborns. In fact, 35 percent of miscarriages may be caused from these errors, and 5 to 10 percent of stillborns have chromosomal abnormalities. These abnormalities usually occur during the formation of eggs, and the number of errors increases as women get older. The complex process of cell division that creates the gametes is responsible for these mistakes.

Meiosis creates genetic variation

Specialized cells in the male testes and female ovaries produce gametes through the cellular-division process of meiosis. Only those cells destined to become gametes undergo meiosis. This process is important not only for creating haploid sperm and egg cells but also for creating genetic variation. Think of your family: although you may have traits similar to those of your parents and siblings, each of you is genetically unique, except for identical twins. Meiosis is the basis for these differences.

Let's begin the process by looking at a cell destined to become a sperm cell. Diploid cells contain matched pairs of chromosomes from each parent called **homologous chromosomes**. In animals, all of the chromosomes of a pair are homologous to each other, carrying genes for the same traits except for the pair of sex chromosomes. For instance, of the 23 pairs of chromosomes in humans, chromosomes 1 through 22 are homologous, with one copy from each parent. The last chromosome pair is the X and Y chromosomes, which determine sex. These chromosomes are not homologous because the X chromosome does not carry the same genes as the Y chromosome. Through the cell-division process of meiosis, the homologous chromosomes (pairs 1–22) are separated, dividing the diploid cell into haploid cells.

So, how does meiosis reduce the number of chromosomes? **Figure 11.24** illustrates the process of meiosis, showing a cell with only one pair of homologous chromosomes for a total of two chromosomes. The red chromosome is from one parent and the blue is from the other. The first step of meiosis is replication of the homologous pair using the DNA replication process you learned about in Chapter 7. After the DNA is replicated, the chromosomes then go through two divisions, referred to as *meiosis I* and *meiosis II,* which result in four haploid cells. In meiosis I, the homologous chromosomes separate into two daughter cells. At this point, each cell only has one blue or red chromosome: it is haploid. These haploid cells then must undergo meiosis II, which separates the replicated sister chromatids. The end result is four haploid cells.

Meiosis not only preserves the genome size from one generation to the next, it also introduces genetic variation. To understand these mecha-

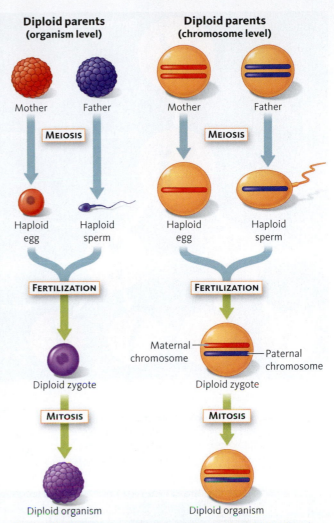

Figure 11.23 Cellular mechanisms of sexual reproduction in animals. The three processes of meiosis, fertilization, and mitosis work together in the life cycle of an organism (left column). Meiosis produces haploid gametes that unite in fertilization. The fertilized egg is a diploid zygote containing genetic contributions from each parent. The right column shows meiosis at the chromosome level. For simplification, the parents are shown with only one chromosome.
© Cengage Learning 2014

Homologous chromosomes—the matched pairs of chromosomes (one from each parent) found in diploid cells.

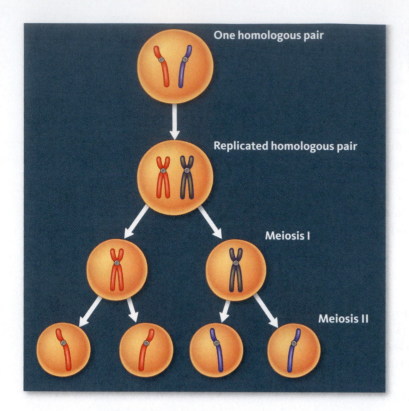

One homologous pair

Replicated homologous pair

Meiosis I

Meiosis II

Figure 11.24 A simplified look at meiosis. In meiosis, the replicated homologous chromosomes in a diploid cell go through two cellular divisions to produce four haploid cells.
© Cengage Learning 2014

Figure 11.25 A detailed look at meiosis. Meiosis consists of eight stages, shown below. In animals, it starts with a diploid cell that has undergone DNA replication and ends with four haploid gametes.
© Cengage Learning 2014

nisms, we need to look more closely at the process of meiosis, which consists of four phases: prophase, metaphase, anaphase, and telophase (**Fig. 11.25**). It is important to understand that the events that create genetic variation occur in all sperm and egg cells before they come together at fertilization. The first source of variation occurs when the homologous chromosomes align and exchange segments of their chromosomes (**Fig. 11.26a**). This shuffling process, which occurs during prophase I, is called **crossing over**. Crossing over introduces new genetic combinations that are unique from both of the parents.

A second source of genetic variation occurs in meiosis I when the maternal and paternal homologous pairs randomly align in the center of the cell. The chromosomes then separate and move to opposite poles, and the cell splits into two. As a result, each daughter cell gets a random assortment of chromosomes from each parent. When three homologous chromosomes line up in metaphase I, four different outcomes result (see **Fig. 11.26b**). After meiosis II, these four combinations yield potentially eight differ-

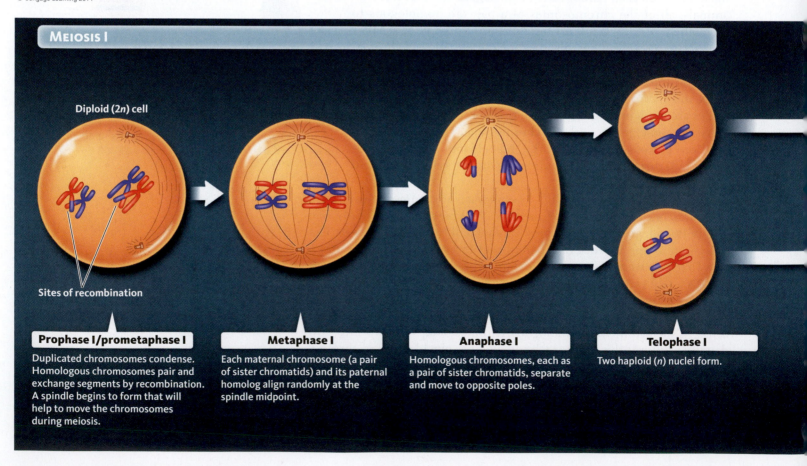

MEIOSIS I

Diploid (2*n*) cell

Sites of recombination

Prophase I/prometaphase I

Duplicated chromosomes condense. Homologous chromosomes pair and exchange segments by recombination. A spindle begins to form that will help to move the chromosomes during meiosis.

Metaphase I

Each maternal chromosome (a pair of sister chromatids) and its paternal homolog align randomly at the spindle midpoint.

Anaphase I

Homologous chromosomes, each as a pair of sister chromatids, separate and move to opposite poles.

Telophase I

Two haploid (*n*) nuclei form.

ent outcomes (see **Fig. 11.26b**). This random assortment of maternal and paternal chromosomes generates an impressive amount of genetic variation that evolution can then act upon. In fact, in humans with 23 different chromosomes, about 8.4 million combinations can occur. Given these huge numbers of potential outcomes, it is easy to see how you and your siblings are different.

A final source of genetic variation in sexual reproduction occurs at the fertilization stage. In humans, typically only a few sperm complete the journey to the egg. Exactly which sperm fertilizes each egg is a random event. Continuing our exploration of genetic variation in humans, if each sperm and each egg has one of 8 million possible genetic combinations, then theoretically there are 64 trillion genetic combinations of offspring! This huge number does not include genetic variation produced by crossing over. It is safe to conclude that of the trillions of sperm and hundreds of eggs in a man and a woman, no two gametes will have the same genetic information.

As you can imagine, the steps of meiosis are tightly regulated, but errors do occur. If improper chromosome separation occurs in meiosis I or meiosis II, then the gamete either has an extra chromosome or is missing one (**Fig. 11.27**). This is known as **nondisjunction**. If this gamete combines with another gamete, the resulting zygote will have an incorrect number of chromosomes. Very few of these cases result in a successful pregnancy. In cases where the chromosome is small or one of the sex chromosomes is involved, the baby may come to term but most likely will exhibit developmental abnormalities.

Down syndrome is an example of a nondisjunction that has three copies of chromosome 21. Nondisjunction occurs more frequently in

Crossing over—source of genetic variation that occurs in meiosis during prophase I, when the homologous chromosomes align and exchange segments of their chromosomes.

Nondisjunction—improper chromosome separation that occurs in meiosis I or meiosis II, resulting in a gamete with either an extra or a missing chromosome.

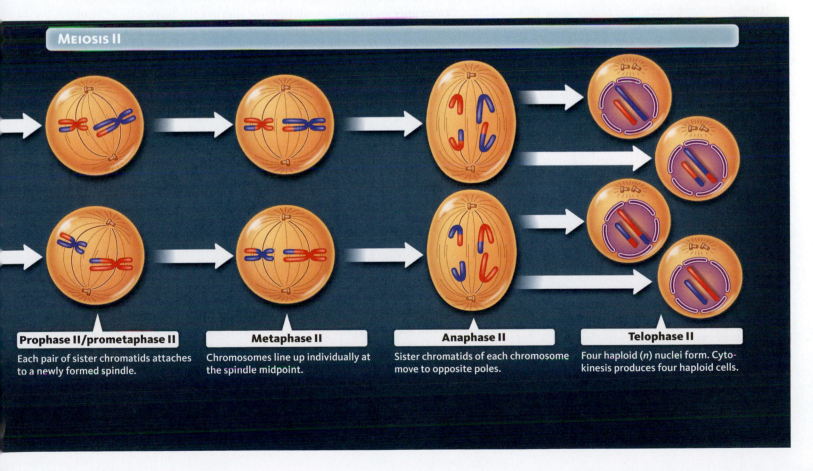

MEIOSIS II

Prophase II/prometaphase II
Each pair of sister chromatids attaches to a newly formed spindle.

Metaphase II
Chromosomes line up individually at the spindle midpoint.

Anaphase II
Sister chromatids of each chromosome move to opposite poles.

Telophase II
Four haploid (*n*) nuclei form. Cytokinesis produces four haploid cells.

a.

Homologous chromosomes pair. | Chromatids fuse, break and repair. | After crossing over, a segment of DNA has been exchanged between chromosomes.

b.

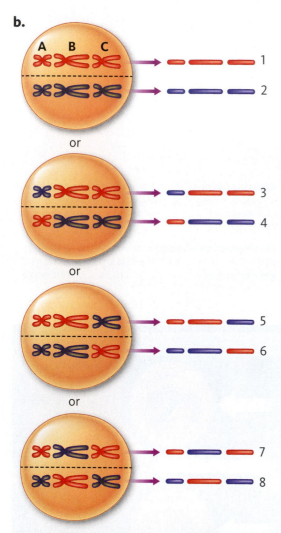

A B C

1
2

or

3
4

or

5
6

or

7
8

Figure 11.26 Genetic variation and meiosis. (a) Crossing over. (b) Four potential genetic outcomes from random assortment during meiosis I that result in eight potential genetically unique gametes after meiosis II.
© Cengage Learning 2014

Gametogenesis—process for forming and developing haploid gametes from diploid germ cells in the gonads; involves the coordinated events of meiosis, mitosis, and cell differentiation.

Spermatogenesis—process occurring in the testes that forms sperm.

Oogenesis—process occurring in the ovaries, producing ova (or eggs).

older individuals. For instance, the risk of a 22-year-old mother giving birth to a child with Down syndrome is about one in 2,000, compared to a one in 30 risk if the woman is 45 years old (**Fig. 11.28**).

Gametogenesis produces eggs and sperm

Males and females have different processes for forming and developing gametes. Both processes start with diploid germ cells in the gonads that undergo mitosis, creating specialized cells that then undergo meiosis and cell differentiation, forming the haploid gametes (**Fig. 11.29**, see page 368). This process of **gametogenesis** involves the coordinated events of meiosis, mitosis, and cell differentiation.

In males, sperm production begins at puberty and continues throughout a man's adult life. The formation of sperm, called **spermatogenesis**, takes about 74 days in humans. Prior to puberty, male germ cells found in the testes undergo mitosis to provide a continual source of cells that give rise to sperm. Through mitosis the male germ cell gives rise to spermatocytes that then undergo the two cellular divisions of meiosis, creating the sperm (see **Fig. 11.29**). The end result is four haploid immature sperm cells that undergo cell differentiation and mature into sperm cells.

The immature sperm cells produced in the testes are undeveloped and cannot swim or penetrate the egg. They enter the epididymis (see **Fig. 11.15**), where maturation and cell differentiation take place. Although the shape and size of sperm cells vary from species to species, all mature sperm cells have a head, midpiece, and tail (**Fig. 11.30**). The head region contains the nucleus (containing the haploid genome) and the acrosome, a modified lysosome (cellular organelle that digests macromolecules) that covers the tip and contains enzymes that break down the outer surface of the egg. The midpiece is wrapped with several mitochondria that produce ATP to power the sperm cell. The tail is a flagellum that allows the sperm to propel forward. After meiosis, changes in hormone levels alter gene expression and trigger cell differentiation, which includes formation of the acrosome, loss of excess cytoplasm, and formation of a flagellum (see **Fig. 11.30**).

In females, egg production begins during early *fetal* development. The process producing ova (or eggs), called **oogenesis**, occurs in the ovaries. Similar to the formation of sperm cells, female germ cells mature via mitosis, yielding a diploid oocyte (see **Fig. 11.29**). In some bony fishes and amphibians that produce large quantities of eggs each year, mitosis provides a pool of oocytes. In mammals, females are born with all of the oocytes that they will ever produce. The oocytes remain in meiosis I until hormones at sexual maturation signal the completion of meiosis and release of the mature egg cell in subsequent ovarian cycles. In humans, each month about 1,000 oocytes mature, but most die before ovulation. Mature eggs are released with the ovarian cycle, and meiosis in the egg cell is completed only after the egg is fertilized by a sperm.

Eggs are large, storing nutrients to nourish the zygote. They become large cells because the cellular divisions are unequal in oogenesis, producing

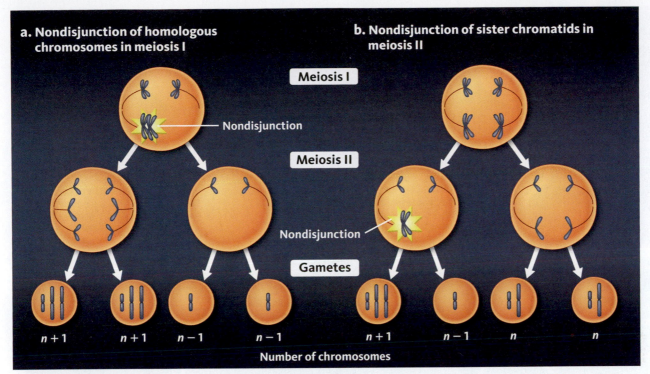

a. Nondisjunction of homologous chromosomes in meiosis I

b. Nondisjunction of sister chromatids in meiosis II

Meiosis I

Nondisjunction

Meiosis II

Nondisjunction

Gametes

$n+1$ $n+1$ $n-1$ $n-1$ $n+1$ $n-1$ n n

Number of chromosomes

Figure 11.27 Errors in meiosis. Nondisjunction occurs when the chromosomes do not separate properly in either meiosis I or meiosis II. When this error occurs, cells will have either too many or too few chromosomes.
© Cengage Learning 2014

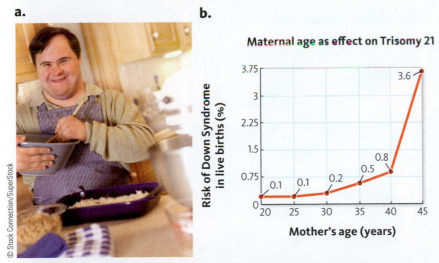

a.

b.

Maternal age as effect on Trisomy 21

Risk of Down Syndrome in live births (%)

3.75

3.6

3

2.25

1.5

0.8

0.75

0.5

0.1 0.1 0.2

0

20 25 30 35 40 45

Mother's age (years)

© Stock Connection/SuperStock

Figure 11.28 Down syndrome, trisomy of Chromosome 21. (a) A person with Down syndrome, which is a result of nondisjunction. (b) The risk of Down syndrome in a child greatly increases with the mother's age.
© Cengage Learning 2014

one oocyte and three nonfunctional cells called polar bodies. The oocyte is bigger than the polar bodies because the oocyte gets almost all the cellular organelles and cytoplasm as a food supply (see **Fig. 11.29**). Eventually the polar bodies degenerate and disappear, leaving one haploid functional gamete (the oocyte) at the end of meiosis in females.

Eggs come in various shapes and sizes but typically are the largest of all animal cells (**Fig. 11.31**). The largest component of the egg cell is cytoplasm. The cytoplasm stores nutrients for the developing embryo. Most

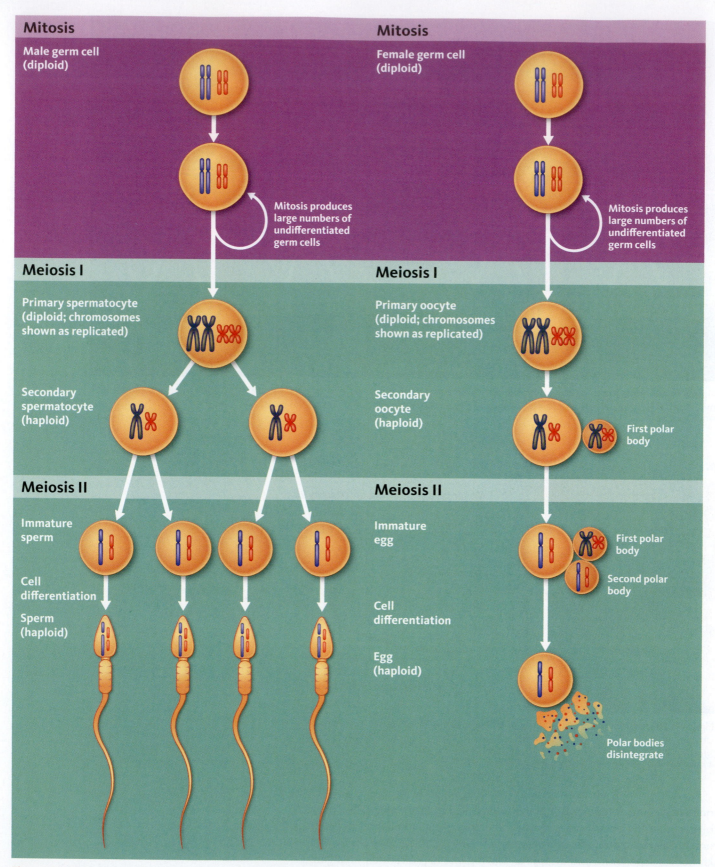

Figure 11.29 Gametogenesis. Spermatocytes and oocytes undergo meiosis and cell differentiation to form the sperm and egg cells. Note that these cells are not drawn to scale. The egg cell is the largest animal cell and the sperm cell is the smallest.
© Cengage Learning 2014

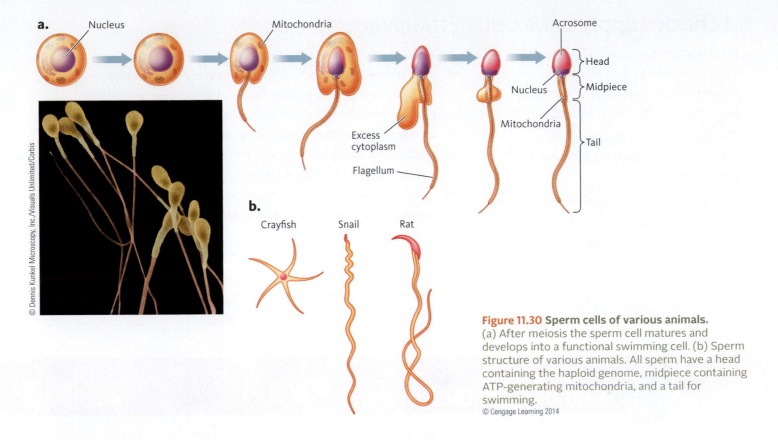

Figure 11.30 Sperm cells of various animals.
(a) After meiosis the sperm cell matures and develops into a functional swimming cell. (b) Sperm structure of various animals. All sperm have a head containing the haploid genome, midpiece containing ATP-generating mitochondria, and a tail for swimming.
© Cengage Learning 2014

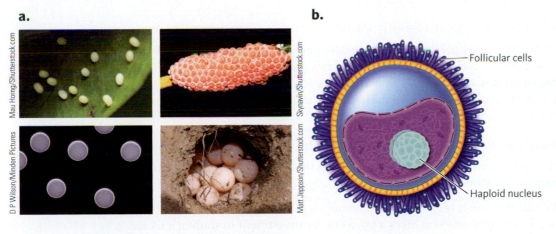

Figure 11.31 Egg cells. (a) Collage of various egg cells. Shown clockwise are eggs from various insects, snail, turtle, and sea urchin. (b) Human egg surrounded by follicle cells that protect the egg and release hormones.
© Cengage Learning 2014

eggs also have an outer coating of jelly or a shell that provides protection and prevents desiccation. In mammals, the egg is surrounded by follicle cells that help to nourish and protect the egg cell and release hormones that control the reproductive cycle. If the egg is fertilized, the sperm penetrates this outer layer to gain access to the cell membrane in order to fertilize the egg to produce the next offspring.

It is fitting to complete Unit 2 on Adaptations of Animals with the many aspects of reproduction in animals. This unit started with an overview of structure and function of various systems, tissues, and cells. Each chapter further explored the inner workings of animals at these different levels of hierarchy. Animal reproduction is a culmination of interactions between animals in their habitat and their various organ systems, cell types, and cellular mechanisms. Reproduction includes development and growth, interactions of nutrient and gas exchange in the developing fetus, and important sensory response in selecting a partner and mating.

- Meiosis is a cellular division process that reduces the diploid number of chromosomes in half to produce gametes, which have a haploid number of chromosomes.

- This reduction stabilizes the number of chromosomes through the following generations.

- Eggs and sperm are formed through gameto-genesis, a process that involves meiosis, mitosis, and cell differentiation.

1. What are the three source of genetic variation from sexual reproduction?

2. To reach the egg, a sperm swims an inch in eight minutes, beating its tail 16,000 times! Once it reaches the egg, the sperm enters the egg cell to fuse its chromosomes with the chromosomes of the egg. Describe the cellular structures that support these functions for the sperm.

3. Meiosis involves the separation of homologous chromosomes. In your own words, describe what homologous chromosomes are.

4. Briefly compare and contrast oogenesis and spermatogenesis (for each process include the function, where and when it starts, and the products of the process).

5. Speculate why women in their 40s are more likely to give birth to a child with Down syndrome. Why are most chromosomal abnormalities from oogenesis and not spermatogenesis?

End of Chapter Review

Self-Quiz on Key Concepts

Sexual Reproduction, Genetic Variation, and Evolutionary Fitness

KEY CONCEPTS: Sexual reproduction creates genetic variation by combining the genetic material of two individuals. Each sperm and egg cell has a genetically unique combination of genes as a result of meiosis and is an important source of genetic variation in the sexual reproductive process. New gene combinations can improve the evolutionary fitness of an organism by increasing the raw material upon which natural selection acts.

1. Match the following **mode of reproduction** with the best definition or example.

   Asexual reproduction   a. creates offspring that are genetically identical
   Sexual reproduction   b. creates genetically different offspring via the fusion of gametes

2. Which statement regarding reproduction is *false*?
   a. Sexual reproduction takes a large energy investment in finding a mate.
   b. Sexual reproduction creates genetically diverse offspring.
   c. Gametogenesis and the random fertilization of eggs by sperm allow new gene combinations to come together in the next generation.
   d. Animals either use asexual or sexual reproduction but never both.

3. From an evolutionary point of view, which animal is most fit?
   a. a bear that lives 30 years and has a total of 10 offspring
   b. a bear that lives 20 years and has a total of 12 offspring
   c. a strong, large bear that has dominant control over a very large territory
   d. a bear that rears two young and gives good parental care so that the young will survive into adulthood

Strategies for Reproduction

KEY CONCEPTS: Animals have many adaptations and strategies to ensure that their genes are passed on to the next generation. Some animals utilize asexual reproduction, a quick means of reproducing that lacks genetic variability. Others invest more resources in sexual reproduction. The amount of energy invested in gamete production, mate selection, number of offspring, and parental care are strategies for reproductive success.

4.   Match each of the following **adaptations** with its definition, characteristic, or example.

Internal fertilization   a. mammalian structure that nourishes the developing fetus
External fertilization   b. marked by different characteristics between males and females
Sexual dimorphism   c. utilized by most amphibians and fishes
Amniotic egg   d. egg is fertilized inside the female's body
Placenta   e. aids reproduction on land, protecting and aiding the developing embryos in a pool of water

5.   Humans, like honeybees and Coho salmon, use different reproductive strategies. Which of these does not apply to humans?
a.   sexual dimorphism
b.   placenta to nourish the fetus
c.   lots of parental care
d.   external fertilization

6.   Which of the following is considered an advantage of asexual reproduction?
a.   It creates genetic variation.
b.   Less energy is expended.
c.   The offspring are genetically identical to one another.
d.   It allows the offspring to quickly adapt to changing environmental conditions.

Reproductive Systems

KEY CONCEPTS: The animal reproductive system consists of gonads, ducts, glands, and accessory organs that work together to produce, deliver, and unite gametes (sperm and eggs). In mammals, these systems also nourish and protect the developing embryo from fertilization through birth. Reproductive systems also release hormones that control sexual maturation, mating behavior, gamete reproduction, pregnancy, and birth.

7.   Match each of the following **reproductive organs** with its definition, characteristic, or example.

Ovary   a. organ that produces testosterone and sperm
Testes   b. gamete-forming organs in males and females
Gonads   c. organ that produces estrogen and eggs

8.   Match each of the following **reproductive events** with its definition, characteristic, or example.

Uterine cycle   a. implantation of the fertilized egg into the uterus
Ovarian cycle   b. when the sperm and egg unite (fertilization)
Conception   c. includes egg development and ovulation
Pregnancy   d. prepares the uterine wall for implantation

9.   After sperm cells are produced, they are mainly stored in the:
a.   vas deferens.
b.   epididymis.
c.   penis.
d.   prostate.

10. Fertilization of the egg by the sperm normally occurs in the:
a.   ovary.
b.   oviduct.
c.   uterus.
d.   vagina.

Meiosis and Gamete Production

KEY CONCEPTS: Meiosis is a cellular division process that reduces the diploid number of chromosomes in half to produce gametes, which have a haploid number of chromosomes. This reduction stabilizes the number of chromosomes through the following generations. The eggs and sperm are formed through gametogenesis, a process that involves meiosis, mitosis, and cell differentiation.

11. Match each of the following **processes** with its definition, characteristic, or example.

Meiosis   a. source of genetic variation in meiosis
Gametogenesis   b. cellular division process that creates haploid gamete cells
Crossing over   c. improper separation of chromosomes during meiosis
Nondisjunction   d. cellular process involving mitosis, meiosis, and cell differentiation

12. A human cell has 46 total or 23 pairs of chromosomes. How many chromosomes does it have after meiosis I and meiosis II?
    a. 46 after meiosis I and 23 after meiosis II
    b. 46 after meiosis I and 46 after meiosis II
    c. 23 after meiosis I and 23 after meiosis II
    d. 46 after meiosis I and 46 after meiosis II

13. Spermatogenesis and oogenesis differ in that:
    a. oogenesis produces one functional ovum, whereas spermatogenesis produces four functional spermatozoa.
    b. spermatogenesis begins before birth.
    c. oogenesis produces four haploid cells, whereas spermatogenesis produces only one functional spermatozoon.
    d. oogenesis begins at the onset of sexual maturity.

Applying the Concepts

14. Birds, reptiles, and amphibians have a cloaca that is the opening for the intestinal, reproductive, and urinary tracts (see **Fig. 11.14** for comparison of systems). Male turtles have a penis housed inside the cloaca. When turtles mate, the male deposits sperm into the female's cloaca. In terms of reproductive organs, in females the cloaca would be similar to what mammalian organ?

15. The green sea turtle mates every couple of years. Once she starts to lay eggs, she does not mate again until the next season. She stores the sperm in her oviducts for up to four years until she is ready to lay her eggs. During the nesting season, every 13 days she drags her 300- to 400-pound body out of the water to lay 100 to 200 eggs per nest. Genetic studies show that sometimes the entire nest is fathered by one individual and sometimes the nest has multiple fathers. What might be a reproductive advantage of storing sperm? What might be a disadvantage?

16. Anabolic steroids are synthetic hormones that mimic testosterone in the body. These steroids stop the release of LH and FSH. How would this affect both the male and female reproductive cycles?

Data Analysis

Rates and Factors of Infertility

Infertility is the inability of a couple to get pregnant after one year of trying. Risk factors for infertility are age, obesity, being underweight, smoking, exercise (too much or too little), alcohol and drug use, sexually transmitted diseases, and poor nutrition. According to the 2005 National Survey of Family Growth, 12 percent of American couples have impaired *fecundity*, meaning they experience difficulties conceiving or bringing the pregnancy to term.

Data Interpretation

17. According to **Table 11.3** and **Figure 11.32**, what age group of women has the most difficult time conceiving and giving birth?

18. Use the data in **Table 11.3** to calculate the percent change from 1982 to 2002 in each of the age groups. Which group had the largest change?

Critical Thinking:

19. Why do you think the 35- to 44-year-old group represents the largest distribution of impaired reproductive function?

20. Let's look at the data in a different manner. Calculate the percent change of each age group relevant to 1982. For instance, the 35- to 44-year-old group increased by 2.9 percent from 1982 to 2002, which is a 29 percent increase over 1982 (2.9/10 × 100%). Interpret these calculations and draw conclusions.

21. About half of cases that have problems conceiving are due to male infertility. Sperm quantity and quality are affected by several factors. What effect would low testosterone levels have on male infertility?

22. Obesity is an increasing trend in America. Studies tie obesity in both men and women with infertility. In women, obesity increases the levels of estrogen. How might these higher levels of estrogen affect the female reproductive system?

TABLE 11.3 2005 National Survey of Family Growth Data

Percentage of Women Aged 15–44 Years with Impaired Fecundity				
National Survey of Family Growth, 1982–2002				
	Survey Year			
Age Range	1982	1988	1995	2002
15–24 years	4.3%	4.8%	6.1%	6.9%
25–34 years	10.0%	9.6%	11.2%	12.9%
35–44 years	12.1%	10.6%	12.8%	15.1%

© Cengage Learning 2014

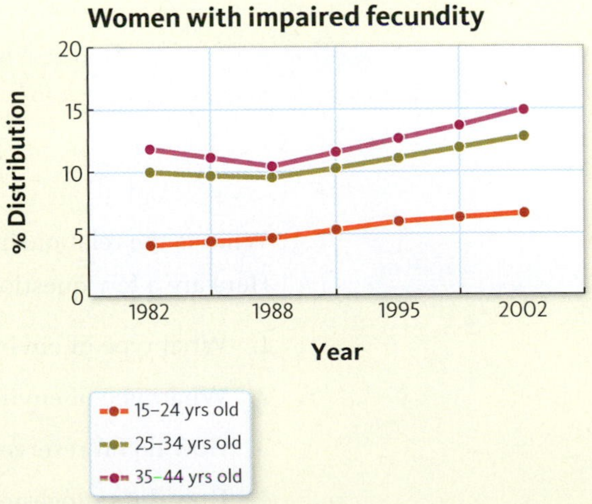

Women with impaired fecundity

- 15–24 yrs old
- 25–34 yrs old
- 35–44 yrs old

Figure 11.32 Impaired fecundity in women, 1982–2002
© Cengage Learning 2014

Question Generator
Delayed Implantation in Black Bears

American black bears are the most common bear in North America (**Fig. 11.33**). Black bears use a reproductive strategy in which they mate in early summer but delay the implantation of the fertilized eggs until November, when they enter their winter den. This delay allows the female to concentrate on building fat reserves throughout the summer and fall, doubling her weight in preparation for a long period of dormancy. If her body cannot support the litter, the fertilized eggs will not implant. In November, if conditions are favorable, progesterone levels increase, preparing the uterine wall for implantation.

Figure 11.33 Black bears are pregnant and give birth during their winter dormancy period.

The fertilized eggs implant and the embryos begin to grow and develop. This development stage is brief at 63 to 70 days compared to the overall gestation period of 220 days. The cubs are born during this dormant period. When born, they are 6 to 8 inches in length, blind, hairless, and helpless. The mother nurses the cubs in the den until March, when the cubs emerge in search of food. The cubs remain with their mother for about a year and a half, at which point they go their separate ways.

Below is a block diagram (**Fig. 11.34**) that relates several aspects of the biology of American black bears and their reproduction. Use the background information along with the diagram to ask questions and generate hypotheses. We've translated the components of the block diagram into several questions to get you started.

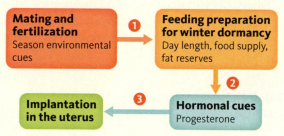

Figure 11.34 Factors relating to the reproductive cycles of American black bears.
© Cengage Learning 2014

What is the relationship between mating, fertilization, hormone cues, and implantation? Here are a few questions to start you off.

1. What type of environmental cues might trigger mating? (Arrow #1)

2. What type of environmental cues might trigger winter dormancy? (Arrow #1)

3. How do fat reserves influence progesterone levels? (Arrow #2)

4. How does progesterone cause implantation into the uterus? (Arrow #3)

Use the block diagram to generate your own research question and frame a hypothesis.

12 Plant Growth, Structure, and Function

12.1 The plant behind your café latte

12.2 Plants display a variety of forms, life histories, and activities

12.3 Plant cells are the structural and functional units of the plant body

12.4 Plant growth and differentiation produce new cells and tissues

12.5 Roots, stems, and leaves extend the surface area of the plant body

12.6 Plants defend themselves against attack from animals and microbes

12.1 The Plant Behind Your Café Latte

If you are enjoying a fresh cup of aromatic, flavorful coffee right now, you are not alone. More than 500 billion cups are consumed annually, making coffee one of the world's most popular beverages. In the United States, each person drinks an average of 22 gallons of coffee each year in the form of a café latte, cappuccino, espresso, iced coffee, or regular cup of brewed coffee. Americans consume one-fifth of the world's coffee, making them the number one coffee consumers in the world. This level of consumption drives a thriving coffee industry.

The story behind your cup of coffee begins with a coffee plant *(Coffea arabica)* growing on a large plantation. The highlands of Brazil produce about one-third of the world's coffee, and this single species accounts for about three-quarters of the world's coffee crop. Arabica coffee, as it is called, grows best in tropical and subtropical areas of the world (**Fig. 12.1**). In particular, coffee plants are adapted to warm climates with year-round temperatures about 70°F; at least 60 inches of rainfall a year; and rich, porous soil. The best coffee-growing regions are marked by a nine-month rainy season followed by a three-month dry season, when the coffee beans are harvested and dried.

The remaining coffee comes from a closely related species, *Coffea canephora,* which is also called robusta coffee. Robusta is more tolerant of warm temperatures and is cultivated in poorer soils closer to the equator. Robusta coffee tends to be bitter, less flavorful, and higher in caffeine, containing about 50 percent more caffeine than arabica coffee. It is a good choice for your double espresso.

A cultivated coffee plant is a small tree or a shrub 6 to 12 feet tall. It consists of the three recognizable vegetative organs—stems, leaves, and roots—common to most plants. It has a main trunk and horizontal branches that support its large, elliptical-shaped leaves, which are dark green and shiny due to a waxy coating (**Fig. 12.2**). In woody plants such as the coffee tree, the leaves are the sites for photosynthesis and sugar production. Water used in photosynthesis is absorbed from the soil by the root system. Root systems are surprisingly large to "mine" the soil of small amounts of minerals. In fact, most minerals are present in the soil in concentrations of "parts per million," so roots grow to impressive lengths to obtain these small quantities. In the case of the coffee plant, if you laid all of the roots of a single plant end to end, they would extend over 15 miles. A coffee tree's roots extend six feet outward from the trunk, and almost all of them are close to the surface, where most water and minerals are found. Because the roots constantly remove nutrients from the soil where they grow, coffee plantations are heavily fertilized with additional supplies of nitrogen, calcium, and magnesium to support plant growth.

Strong and healthy vegetative organs provide the sugars and water necessary to make flowers, fruits, and, ultimately, beans. Coffee beans that are ground to a fine powder for brewing your cup of coffee are actually the seeds of the plant. Coffee seeds, like all seeds of flowering plants, grow and develop within fruits, which in turn develop from the female part of the flower. Let's track the development of the coffee beans from the flower.

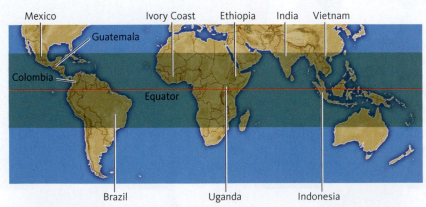

Figure 12.1 Map of the Coffee Belt. Arabica coffee is grown in tropical and subtropical regions of the world. The coffee-growing countries form the coffee belt, where warmer temperatures and large amounts of rainfall favor the growth of coffee plants. The top 10 coffee-producing countries are labeled.
© Cengage Learning 2014

Leaf

Flowers

Fruit cluster

Ripe fruit

Harvested

Seed

Processed
Roasted

Roasted beans

About three or four years after planting, several flowers develop within each bud along the stems. During the flowering period, the coffee plantation is in bloom (see **Fig. 12.2**). Once the flowers open, their fragrant odor and sugary nectar attract bees to begin the sexual reproductive process.

About 15 weeks after the plant flowers, the coffee fruit rapidly absorbs water and sugar and enlarges into a round, berrylike fruit. Each fruit forms an outer skin and a fleshy, sweet pulp that encloses its two seeds. When it is immature, the fruit is green; as it ripens, it turns yellow and then crimson (see **Fig. 12.2**). Finally, after seven to nine months, during the dry season, the coffee fruits are bright red, glossy, and firm and are ready for harvest and processing into coffee beans for market.

Your cup of coffee is a complex mixture of hundreds of chemicals, such as sugars, oils, acids, and caffeine that combine to produce the coffee's color, flavor, aroma, body, and acidity. Many people love their morning cup of coffee because it increases their alertness. As the most widely used stimulant in the world, the caffeine in coffee increases heart rate, blood pressure, and breathing. Depending on the method of processing and the blend of coffee, an eight-ounce cup of coffee contains about 150 mg of caffeine, compared to instant coffee with about 50 mg and decaffeinated coffee with just 2 mg. For comparison, a 12-ounce cola soft drink contains roughly 50 mg of caffeine, and many energy drinks contain as much as 250 to 300 mg of caffeine.

Figure 12.2 Coffee: from flower to bean. The coffee plant is a shrub or small tree with a main trunk that supports a canopy of leaves. Below ground it produces an extensive root system for the absorption of water and minerals. Coffee flowers bloom, are pollinated, and begin to form fruits and seeds, which are harvested when they are red and ripe (called "cherries"). After being sorted, dried, and removed from the fruit, the coffee seed is roasted. Roasted beans are ground and brewed to make a cup of coffee.
© Cengage Learning 2014

From the plant's perspective, the production of caffeine repels or deters animals, such as insects and vertebrates, from eating the seeds. Coffee leaves also make caffeine, the bitter taste of which causes animals to avoid eating these plant parts too. Caffeine also inhibits the growth of some bacteria, fungi, and viruses. In fact, caffeine is one of the many chemical weapon adaptations that allowed original coffee plants to survive in nature before they were domesticated and farmed as a crop.

In this chapter we will focus on the plants and their products that are all around you. We will explore how the various structures function in the life of flowering plants, such as the coffee plant, at several levels of the biological hierarchy—cells, tissues, and organs. We will begin with the basics of plant structure at the organism level and zoom in to explore how the lower levels are integrated into the growth and survival of the whole plant.

12.2 Plants Display a Variety of Forms, Life Histories, and Activities

Far from the highlands of Brazil, the pink lady slipper orchid lives on the forest floor of a Vermont woodland. The lady slipper is a member of one of the most recently evolved plant families—the orchids. This wildflower consists of two large leaves, a very short stem, and a coarse fibrous root system. It lives for many years, and each spring it produces a single pink flower with a complex slipperlike shape (**Fig. 12.3**). One of its three petals forms an inflated pouch that attracts and guides bees through the pollination process. Although the coffee plant and the lady slipper orchid both produce flowers and live long lives, their bodies have different shapes and tissues.

Plants such as the coffee and the lady slipper can be placed into two broad categories based on the production of woody tissues. The coffee plant is a woody plant, which adds layers of woody tissues to its stems and roots each year to increase in diameter. You know woody plants as trees and shrubs such as oaks, cottonwoods, and willows. Woody plants also include cactuses, which produce woody tissues to support their heavy water-filled stems. The lady slipper orchid does not actively grow in width by adding wood; it is an herbaceous plant. This group includes many of the crop plants that feed us, as well as the ferns on the forest floor and other wildflowers. Although some herbaceous plants live for decades, in general, they grow faster and live shorter lives than woody plants.

In the section ahead we begin our exploration of plants by placing familiar plants into categories based on their overall shape or form and their patterns of growth and reproduction through time. We then present a brief overview of how plants such as the coffee plant and lady slipper grow, survive, and reproduce in their habitats.

Plants exhibit many different woody and herbaceous growth forms

When botanists analyze the plant communities in an area, they begin by grouping plants by their shape, structure, and appearance, or growth form. In forest habitats, trees are the most common growth form. Trees are woody plants that consist of a large main vertical, woody trunk that sup-

iStockphoto.com/Terry Wilson

Figure 12.3 Lady slipper orchid. The lady slipper orchid is an herbaceous plant that produces a single flower each spring.

Woody plant—plant that adds layers of woody tissues to its stems and roots each year so that it increases in diameter; it grows in length through apical meristems and in diameter through lateral meristems (vascular and cork cambiums).

Herbaceous plant—a plant that does not actively grow in diameter by adding wood; it grows only through apical meristems.

Growth form—a plant's shape, structure, and appearance.

Trees—woody plants that consist of a large main vertical, woody trunk that supports numerous smaller branches and a canopy of leaves; generally tall plants.

ports numerous smaller branches and a canopy of leaves. Broad-leaved, or deciduous, trees, such as maples and coffee trees, produce thin, flat leaves. Many deciduous trees drop and rebuild their entire set of leaves each year, whereas evergreen trees such as conifers have needlelike leaves and gradually replace them over several years. Shrubs, on the other hand, are woody plants that do not have a main trunk; rather, they have many similar-sized branches that grow near the base of the plant. Shrubs are shorter in height and often live in the shade of trees in the understory of the forest. Dry, arid habitats have succulent plants such as cactuses that store large amounts of water in their roots, stems, and leaves. Water storage gives succulents a swollen or bulging appearance (**Fig. 12.4**).

Plants that show a grasslike growth form are herbaceous plants with slender stems that bear narrow leaves. The 3,500 species of grasses are the dominant vegetation in many habitats such as prairies and home lawns. Wheat, rice, and Kentucky bluegrass all show this growth form. Grasses are flowering plants, and their flowers are generally wind-pollinated, so they are not colorful or showy to attract insects.

Vines such as pumpkins, ivy, and morning glory use the support of the soil to creep along the ground, or they use woody plants or other sturdy objects to grow and climb upward toward sunlight. This saves them from investing heavily in building lots of strong support tissues, which enables them to rapidly grow and extend their stems along the forest floor into tree and shrub canopies. Vines use a variety of structures to attach and climb on their supports. Some vines coil, or twine, around supporting plants, whereas other species, such as Boston ivy, use special sticky pads (**Fig. 12.5**)

A plant's life history reflects its pattern of growth and reproduction

An individual flowering plant begins its life as a seed, grows for a period of time, and then goes through a stage of reproduction. Ultimately it enters a period of decline before its life ends. The time a plant spends in each of these stages of growth and reproduction determines its *life history*. Each plant species has a particular life history, which includes how long it lives. Plants that grow, flower, and produce seeds within a year are called annuals. Annuals are herbaceous plants that hold nothing back, packing all of their life into a single growing season. With only one opportunity to reproduce, they grow rapidly and produce large numbers of seeds that are small and easily dispersed. Many annuals are adapted to disturbed habitats by rapidly capturing resources and building roots, shoots, and flowers. Their ability to grow and reproduce quickly gives them a competitive edge in gardens, fields, and along roadsides.

In many areas of the country, mullein is an example of a biennial plant (**Fig. 12.6**). Like annuals, biennials are herbaceous plants with a single period of reproduction. However, biennials flower in the

Figure 12.4 Saguaro cactus, a succulent. Succulent plants such as this saguaro cactus store enormous amounts of water in their stems. Cactuses produce woody tissues to support their large weight.

Figure 12.5 Boston ivy. Vines, like this Boston ivy, use adhesive pads to attach and quickly cover the sides of buildings.

Shrubs—woody plants that do not have a main trunk but have many similar-sized branches that grow near the base of the plant; shorter in height than trees.

Succulent—plants such as cactuses that store large amounts of water in their roots, stems, and leaves, which gives them a swollen or bulging appearance; often found in dry, arid habitats.

Grasslike—herbaceous plant growth form marked by slender, green stems that bear several long narrow leaves.

Vines—weak-stemmed plants that use the support of the ground, woody plants, or other sturdy objects to grow toward the sunlight.

Annuals—herbaceous plants that grow, flower, and produce seeds within a single growing-season.

Biennials—herbaceous plants that have a single period of reproduction in the second and final year of their life; lifespan spread out over two years.

Figure 12.6 The biennial life cycle of mullein.

During the first year, mullein seeds germinate and grow a circle of velvety leaves. During the second year, the mullein plant produces a tall flower stalk and completes its life cycle. Seeds are dispersed at the end of the second growing season.
© Cengage Learning 2014

1st Growing season

2nd Growing season

Photo by Robert Noyd

Jerome A. Krueger

| Spring | Summer | Fall | Winter | Spring | Summer | Fall |

Seed germination — **Vegetative growth** Roots and shoots — Dormancy or Overwinter — **Reproductive growth** Flowers, fruits, and seeds — **Seed dispersal**

second, and final, year of their life. The first year is marked by growth of its leaves and roots. Mullein produces a circle of velvety leaves on a short stem close to the ground and has an extensive root system (see **Fig. 12.6**). Together these structures produce and store energy gained in the first year. By being close to the ground, its leaves can draw on the warmer soil temperatures and keep producing sugar throughout most of the winter. In the spring of the second year, mullein uses its stored energy to rapidly grow a tall flower stalk (see **Fig. 12.6**). Depending on the environment and competition, a typical mullein plant will produce as many as 100,000 to 200,000 seeds. Cabbage is a crop plant that is also a biennial.

Perennial plants live for more than two years and have repeated episodes of reproduction. Coffee trees produce a new crop of flowers, fruits, and beans each year for many years. The lady slipper is also a perennial herbaceous plant. Perennials add new stems and roots onto what they have already established in years past. Trees, shrubs, and all woody plants are perennial plants. However, not all perennials are woody; many grasses and wildflowers are perennials that live long lives (**Fig. 12.7**). The faster growing

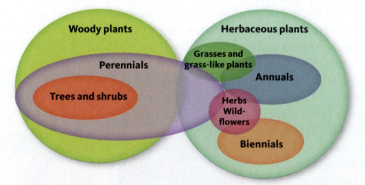

Woody plants — Herbaceous plants — Perennials — Grasses and grass-like plants — Annuals — Trees and shrubs — Herbs Wild-flowers — Biennials

Figure 12.7 Relationships among growth forms and life histories of plants. This Venn diagram shows how woody and herbaceous plants have evolved in numerous growth forms and life histories. Each circle or oval represents a group of plants based on life history, growth form, or whether or not they form woody tissues. Overlapping circles show how they relate. For example, the ovals for annuals and biennials are enclosed within the circle for herbaceous plants and do not overlap at all with the woody plants circle. This means that these plants are herbaceous and do not form woody tissues. Note that these are generalizations and exceptions often can be found.
© Cengage Learning 2014

Perennials—plants that live for more than two years and have repeated episodes of reproduction; may be herbaceous or woody.

herbaceous perennials in favorable habitats often begin flowering right away in the first season. However, small, hardy alpine perennials may not flower for 10 or 15 years, and some woody plants may not reproduce for decades.

Plants perform life activities for survival, growth, and reproduction

Plants, regardless of whether they are cactus, mullein, or a coffee tree, perform the same basic processes to drive their life activities (Table 12.1). At the top of the list is photosynthesis, which is the process by which plants use light energy to transform a gas (CO_2) and a liquid (H_2O) into sugars. In an economic sense, sugars represent a plant's "income."

Absorption and transport of the raw materials for photosynthesis dominate a plant's activities. Leaves contain green-colored pigment molecules that absorb light. But first they must be supported into a position to capture light, and this is the function of stems. In addition to light, photosynthesis requires carbon dioxide (CO_2) and the absorption of water and minerals. Terrestrial plants exchange gases with the atmosphere; carbon dioxide diffuses into a leaf down its concentration gradient, and oxygen diffuses out of the leaf. Roots require oxygen for the production of ATP, which is then used to absorb dissolved minerals from the soil. Submerged aquatic plants acquire all their substances directly from the water. Once inside the plant, water and minerals are transported from roots to leaves, through a specialized "plumbing" system of tubelike cells. The power of the sun pulls a column of water from the ground all the way to the leaves and out into the atmosphere. In redwood trees, water is pulled a distance of nearly 400 feet. Along the way water may be placed in storage for future use. Leaves use the water during photosynthesis to make sugars, which are transported in the opposite direction to roots and other sites for storage. Sugars are also used to fuel growth to expand the surface area in which to absorb water and light for photosynthesis.

Not all plants are green and perform photosynthesis to make their own sugar; some steal it from other plants. In fact, more than 4,000 species of flowering plants are parasites that tap into the water, mineral, or sugar supplies from other plants. Some parasites, such as the orange dodder plant (Fig. 12.8), have little to no chlorophyll and are completely dependent on their host for sugars, water, and minerals. Other parasitic plants contain green chlorophyll pigments for photosynthesis and obtain only water and minerals from their host plants.

Plants adjust their growth by perceiving and responding to information from the environment. Information such as temperature or the amount of light and water stimulate specialized receptors that, in turn, signal changes in growth or ways to cope with stress. For example, vines such as the dodder respond to contact with a nearby plant and wrap around it. Continued survival of the plant depends on producing signaling molecules to attract cooperative bacteria and fungi in the soil or defensive chemical weapons to thwart herbivores, hungry insects, parasitic plants, or infectious pathogens.

TABLE 12.1 Plant Life Activities

Activity	Structures
1. Photosynthesis: production of sugars	Leaves, stems
2. Aeration and exchange of gases with environment (O_2, H_2O, CO_2)	Roots, stems, leaves
3. Support and positioning of shoots	Stem
4. Absorption of water and minerals	Roots
5. Transporting water to leaves/sugars from leaves	Roots, stems, leaves
6. Storage of energy and water	Root, stems
7. Growth and expansion of surface area	Roots, stems
8. Self-defense against herbivores, parasites, and pathogens	Roots, stems, leaves
9. Response to environmental information	Roots, stems, leaves
10. Reproduction and dispersion of seeds or spores	Flowers, fruits, seeds

© Cengage Learning 2014

iStockphoto.com/Nancy Nehring

Figure 12.8 Orange dodder. This salt marsh dodder forms a tangled orange mass as it tightly wraps its stems around wetland vegetation. It sinks its roots into its green host plant to obtain sugars, water, and minerals.

Photosynthesis—process by which plants use light energy to transform a gas (CO_2) and a liquid (H_2O) into sugars (chemical energy).

Many plants are able to reproduce asexually through growth and mitosis and sexually through meiosis to create genetically diverse offspring in the form of spores and seeds. This genetic diversity brought about rapid evolutionary change and speciation. Plant evolution has spanned the past 450 million years from their ancestors, the green algae, to the wide diversity of forms, structures, and adaptations that cover our landscape today.

Finally, plants that are well adapted to their habitat perform their life activities efficiently and are able to respond to environmental changes. This topic is discussed in Chapter 13. Like animals, plants function at all levels of the biological hierarchy of scale. The cellular level of the plant and how cells acquire energy, store substances, and support and defend themselves are the topics of the next section.

12.2 check + apply YOUR UNDERSTANDING

- Plants are grouped into woody and herbaceous categories depending on whether or not they produce woody tissues.

- Plants appear in a variety of growth forms, such as trees, shrubs, succulents, grasslike plants, and vines.

- Plants live for different periods of time. Annuals complete their life cycle in a single year, biennials live for two years, and perennials grow and reproduce over many years.

- Plant life activities are centered on photosynthesis. Growth, transport, support, absorption, and storage activities all contribute to increasing surface area for the capture and transformation of light energy.

iStockphoto.com/serahcus

Figure 12.9 Mistletoe.

1. You see a woody plant with dozens of similar-sized branches arising from near the base. Which growth form is it?

2. Which letter on the table below represents a group of plants with a short life, fast growth rate, and high seed production? Refer to the Venn diagram in **Figure 12.7** to determine which letter on the table represents a group of plants that does not exist in nature.

	Herbaceous	Woody
Annual	A	B
Perennial	C	D

3. What do annuals and biennials have in common?

4. Which of the 10 life activities listed in **Table 12.1** are related to a plant's ability to conduct photosynthesis?

5. Mistletoes are plant parasites that attach to trees and shrubs (**Fig. 12.9**). Based on the photo, what are mistletoes obtaining from their host plant: (a) sugars; (b) water and minerals; (c) sugars, water, and minerals? Explain.

12.3 Plant Cells Are the Structural and Functional Units of the Plant Body

The dark, thick, green leaves of the coffee plant are composed of millions of cells that perform photosynthesis. Round-shaped photosynthetic cells in the lower layers of the leaf are loosely arranged around large air spaces (**Fig. 12.10**). Let's zoom in on one of these "spongy" cells to explore the structure and organization of plant cells in general (see **Fig. 12.10**).

The first thing you may notice is the rigid, boxlike structure of these coffee leaf cells. These cells are cemented together and are unable to move or change their position—they are locked in. The rigidity of the cells comes from the **cell wall** that encloses the contents of the cell and determines its size and shape. Inside the cell wall are the plasma membrane, the cytoplasm, and a single nucleus. The eukaryotic coffee nucleus is bounded by a membrane and contains almost as many chromosomes (44) as a human cell (46). The cytoplasm contains a variety of organelles and is dominated by a large central vacuole. The **vacuole** is the storage site for water, pigments, and toxins. You may notice other familiar organelles, such as mitochondria and components of the endomembrane system, namely, Golgi complexes, ribosomes, and long channels of endoplasmic reticulum. Finally, this cell is specialized for photosynthesis and contains dozens of green **chloroplasts** that produce sugars to meet the plant's energy needs.

Plant cells contain many of the same structures that we saw in animal cells presented in Chapter 6 (far right column in **Table 12.2**). Organelles perform the same functions in both kinds of cells. However, plant cells do not have lysosomes to digest and recycle worn-out cell parts.

> **Cell wall**—provides rigidity to plant cells; encloses the contents of the cell and determines its size and shape.
>
> **Vacuole**—plant cell organelle that stores water, pigments, and toxins; a large central vacuole that occupies the majority of the cell volume.
>
> **Chloroplasts**—an organelle in a plant cell that is specialized for photosynthesis; a type of plant plastid that contains the green pigment chlorophyll.

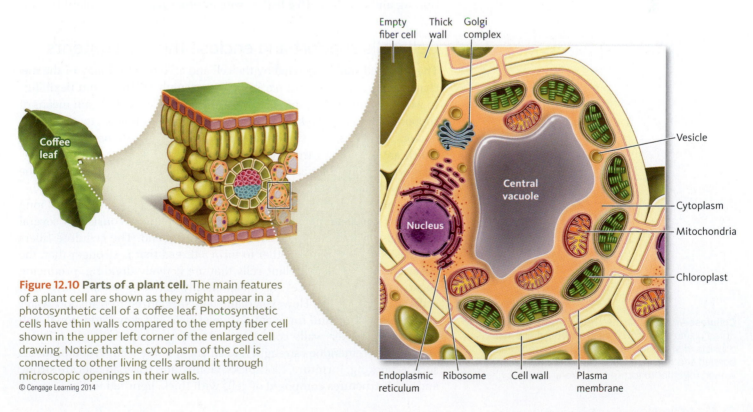

Figure 12.10 Parts of a plant cell. The main features of a plant cell are shown as they might appear in a photosynthetic cell of a coffee leaf. Photosynthetic cells have thin walls compared to the empty fiber cell shown in the upper left corner of the enlarged cell drawing. Notice that the cytoplasm of the cell is connected to other living cells around it through microscopic openings in their walls.
© Cengage Learning 2014

Empty fiber cell · Thick wall · Golgi complex · Vesicle · Central vacuole · Cytoplasm · Nucleus · Mitochondria · Chloroplast · Endoplasmic reticulum · Ribosome · Cell wall · Plasma membrane · Coffee leaf

TABLE 12.2 Major Components of Plant Cells

Cell Structure		Function(s) in Plant Cells	Animal Cell
Cell wall		Protects, supports	
Plasma membrane		Controls entry and exit of molecules into and out of cell	X
Mitochondria		Transforms energy in nutrients (sugars, fatty acids, amino acids) into energy (ATP)	X
Endomembrane system	Endoplasmic reticulum	Builds lipids and proteins; ribosomes build proteins	X
	Golgi complex	Packages substances for export; builds cell wall	X
	Vesicles	Transports substances throughout cell and to plasma membrane; assists cell wall building	X
Central vacuole		Provides support; water and mineral storage; recycling of worn-out organelles; defense (toxins); communication (pigments)	
Plastids	Chloroplasts	Performs photosynthesis	
	Chromoplasts	Communicates	
	Leucoplasts	Stores energy	
Cytoskeleton	Microtubules	Moves chromosomes during cell division; intracellular transport	X
Nucleus	Chromatin	Stores genetic information	X
	Nucleolus	Part of chromatin that codes for ribosomes	X
Ribosomes		Build proteins	X

© Cengage Learning 2014

This function is performed inside their vacuole. In the section ahead, we will focus on three structures that make plant cells different from animal cells: cell walls, a central vacuole, and multipurpose organelles called *plastids*. Together, the millions of plant cells are the structural and functioning units of the coffee leaf as well as other parts of the plant body.

Cell walls support and enclose the cell contents

The plant cell wall is secreted by the cell and accounts for many of the mechanical properties of plant parts, such as strength, stiffness, and flexibility. In woody plants such as the coffee tree, the cell walls are the main means of support that holds leaves out to capture sunlight and flowers to attract pollinators. Cell walls, for example, are responsible for the stems of palm trees, which are incredibly strong yet are flexible to bend and resist hurricane-force winds. Cell walls vary in thickness and composition, depending on the role the cell plays in the structural support of the plant part.

The plant cell wall is a framework of fibrous threads composed of long strands of **cellulose** molecules. Each strand of cellulose consists of several thousands of glucose molecules bonded end to end. The cellulose fibers are twisted and wound together to form a thread that is stronger than the same thickness of steel. Plant cells that are actively dividing, producing sugars, storing substances, or metabolically active generally have thin walls that are flexible. However, plant cells specialized to transport water and support the plant, for example, wood cells, lay down additional wall layers. The thicker walls contain a cementing compound called *lignin*, which adds tremendous strength and stiffness to the wall. Wood products such as flooring, furniture, toothpicks, violins, and baseball bats are strong and rigid structures composed of cells with thick, lignified walls.

Cellulose—macromolecule that forms the framework of the plant cell wall; a complex carbohydrate that consists of several thousands of glucose molecules bonded end to end as long fibers that are twisted and wound together to form a very strong material; fiber.

The central vacuole stores water, toxins, and pigments

When you admire a beautiful purple flower in bloom or enjoy the crisp lettuce leaves in your salad, you are appreciating the role played by the plant cell's central vacuole. In these instances, the vacuole is the large, fluid-filled organelle that often occupies most of the plant cell (see **Fig. 12.10**). The vacuole in the cells of the flower petal is filled with water, which supports and expands the petal. Therefore, keeping your cut flowers blooming is mostly a matter of maintaining water flow to the flower petals to keep their vacuoles filled.

As with all cell structures, the contents of a vacuole are strongly related to its function. Vacuoles generally don't make the substances they contain; instead they receive them from other parts of the cell and store them. The vacuoles in a lemon or other citrus fruits store organic acids that give lemon meringue pie its characteristic sour taste. In seeds, vacuoles store proteins and minerals for the growing embryo. Storage in plant cells is an active function where substances are constantly moving into and out of vacuoles. Their membranes are areas of intense trafficking in molecules.

The vacuole serves as the storage site for toxic and defensive compounds such as the oils in cells of poison ivy. Caffeine in coffee beans and nicotine in tobacco leaves are toxic to herbivores that eat them. Here the vacuole serves to defend the plant. The pharmacy is filled with compounds stored in plant vacuoles, from the morphine in pain relievers to the codeine in cough syrups and the atropine in eye drops.

Finally, plant pigments are molecules that give color to a plant part. Not all pigments are stored in vacuoles, but red, white, and blue pigments often are stored there. The white color of coffee flowers is a pigment found in the vacuoles of its petals. The blue color of blueberries is from a blue pigment placed in vacuoles. Here the pigments function to communicate that the flower is open for pollination and the fruit is ripe and filled with sugar. The vacuole also stores red pigments. The red coloration of many carnivorous plants lures insects into their traps. Red and blue pigments combine to produce purple pigments, which absorb ultraviolet light and act as a protective sunscreen.

Plastids are organelles for photosynthesis, communication, and energy storage

Depending on the role they play, plant cells also contain three specialized organelles called **plastids**—chloroplasts, chromoplasts, and leucoplasts. Plastids are found in all vegetative and reproductive plant structures and are classified based on the types of pigments they contain. Plastids, like mitochondria, contain a small circular piece of DNA and can divide and multiply within a cell.

Chloroplasts are green plastids that have a complex system of internal membranes that are important in the process of photosynthesis (see **Fig. 12.10**). It has been estimated that a single chloroplast contains about 600 million molecules of chlorophyll, giving these disc-shaped organelles a strong green color. When you consider that an individual cell contains between 50 and 100 chloroplasts, the light-absorbing power of a leaf becomes staggering. Photosynthesis and chloroplast structure are presented in more detail in Chapter 13. The vibrant green landscape around you, as well as green plant foods such as cucumbers, green peppers, and celery, all contain cells with an abundance of chloroplasts (**Fig. 12.11**).

Plastids—specialized plant organelles that include chloroplasts, chromoplasts, and leucoplasts; found in all vegetative and reproductive plant structures; classified based on the chemical compounds (pigments, starch) they contain; contain a small circular piece of DNA and can divide and multiply within a cell.

Leucoplasts

Elena Elisseeva/
Shutterstock.com

Chloroplasts

P&R Fotos/age fotostock

Martin Kreutz/Panther Media/
age fotostock

Figure 12.11 Three types of plastids. A typical party platter contains plant cells with all three kinds of plastids. Broccoli, cucumbers, and celery are rich in chloroplasts, the color of tomatoes and carrots is due to the chromoplasts in their cells. Colorless leucoplasts are found in the cauliflower, and white portions of the cucumber and celery stalks.

Stefan Sollfors/Science Faction/Getty Images

Figure 12.12 Stages of banana ripening. As bananas ripen, their green chloroplasts are transformed into yellow chromoplasts. Also the material that binds cell walls together breaks down, which causes the bananas to soften.

Chromoplasts are cell organelles filled with carotenoid pigments that impart yellow and orange colors to plant parts. They are abundant in yellow fruits, flowers, and roots such as yams and carrots. In flowers and fruits, chromoplasts attract insect pollinators and fruit-eating animals. Thus, chromoplasts play an important role in plant reproduction. In humans, carotenoids in yellow and orange vegetables and fruits are excellent sources of vitamin A in the diet (see **Fig. 12.11**). Vitamin A is required for the production of light-absorbing pigments in your eye and the ability to see at night, so eating carrots does lead to you being able to see better in the dark!

Leucoplasts are plastids that lack pigments and the internal system of membranes characteristic of chloroplasts. They are organelles that store energy in the form of long molecules of glucose called *starch.* An abundant amount of starch is stored in leucoplasts found in stems, roots, and seeds. The bulk of the coffee bean is composed of cells filled with starch. This is also true of corn seeds, so when you are enjoying a bowl of popcorn while viewing a movie, you are feasting on leucoplasts used to store starch in the corn kernel.

Plastids are striking for their ability to change from one type to another. This change occurs during the ripening of fruits such as tomatoes and peppers. Green, unripe bananas ripen into yellow bananas (**Fig. 12.12**). Chloroplasts are transformed into chromoplasts over the course of a few days. This change in color is a signal to fruit-dispersing animals that the fruit is ripe, sweet, and ready to eat.

12.3 check + apply YOUR UNDERSTANDING

- Plant cells are eukaryotic cells enclosed by a cell wall, with cytoplasm containing many organelles, structures, and a membrane-bound nucleus.

- The cell wall is composed of cellulose and varies in thickness in different types of cells. Walls support the contents of the cell and the plant body against gravity.

- The central vacuole stores water and other substances such as colorful pigments, minerals, and toxins. Through water storage, it presses on cell walls and contributes to support.

- Plant cells may contain one or more kinds of plastids: photosynthetic chloroplasts, yellow–orange chromoplasts, and starch-storing leucoplasts. One kind of plastid can be changed into another.

1. Name three structures found in plant cells that are not in animal cells.

2. How do the walls of storage cells differ from cells specialized to transport water?

3. French fries are made from potatoes, an underground organ that stores starch. Which plastid dominates the fries that you eat at the local fast-food restaurant? Are the cell walls thick or thin?

4. *Eucalyptus* leaves have cells that produce sugars through photosynthesis and also have toxins to deter herbivores (koalas are able to eat and resist their leaf toxins; see **Figure 8.3**). Which two organelles in *Eucalyptus* leaf cells perform these two functions?

5. Spiderwort is a plant that produces purple flower petals and yellow pollen sacs (**Fig. 12.13**). For each of these two flower parts, name the pigment and the cell structure responsible for its color.

J S Sira/age fotostock

Figure 12.13 The flower of a spiderwort.

12.4 Plant Growth and Differentiation Produce New Cells and Tissues

One of the most important events for the coffee grower is the development and opening of flower buds. A bud is an undeveloped part of a plant that grows into a new branch (stem), leaf, or flower. The buds that develop into flowers start the series of events that leads to the coffee bean. The branch produces two kinds of buds: a *terminal* bud at its tip and *lateral* buds at the base of each leaf. A cluster of five lateral buds bursts open and forms flowers, which can potentially form a total of 20 fruits (**Fig. 12.14**). Coffee plants are known to be finicky. If the flower buds are subjected to temperatures only a few degrees too warm or too cold, or they lack sufficient water, they die. This means that the plant produces no flowers, no fruits, and no coffee beans.

In the section ahead, we will explore the events and processes occurring inside buds as well as other growing points of the plant. We will start with the very first growing points that emerge from a seed and then zoom in to look at growth on the cellular level. Here cells are actively building new cells complete with walls, cytoplasm, organelles, and DNA. If the environment is favorable with lots of light and water, growth is extensive. In this way, the plant's current environment is imprinted onto its structure.

Plant growth occurs at specific locations in the plant body

About 50 to 60 days after they are sown, coffee seeds germinate and initiate the growth of a new plant. Germination means that the growing point, or meristem, emerges from the seed (**Fig. 12.15**). The first structure to emerge is the root, which grows and elongates by adding new layers of cells to its tip, or root apical meristem (see **Fig. 12.15**). Cells of the root meristem respond to gravity and grow downward. On the op-

Bud—an undeveloped part of a plant that grows into a new stem, leaf, or flower.

Meristem—a growing point of a plant; grows by increasing the number of cells (cell division) and the size of cells (cell expansion).

Apical meristem—growing points at the root and shoot tips and in buds; in roots it is the part that perceives and responds to gravity and grows downward; shoot apical meristem grows upward, increasing the height of the plant.

Figure 12.14 Lateral buds of a coffee tree. Coffee growers protect lateral tree buds because these structures develop into flowers and fruits for harvest. (a) In coffee trees, five lateral buds are located at the base of each leaf. (b) These buds then develop into a cluster of flowers. (c) After pollination, the flowers produce many fruits and coffee beans for market.
© Cengage Learning 2014

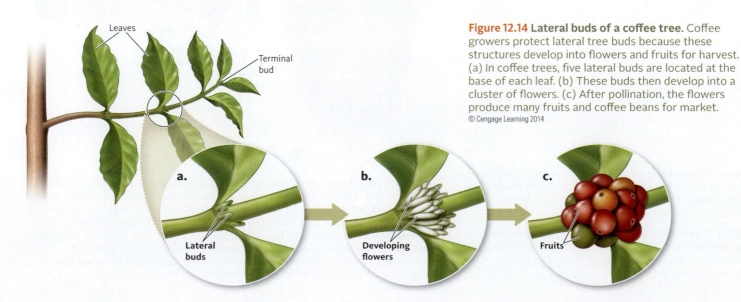

Leaves

Terminal bud

a. Lateral buds

b. Developing flowers

c. Fruits

Shoot apical meristem

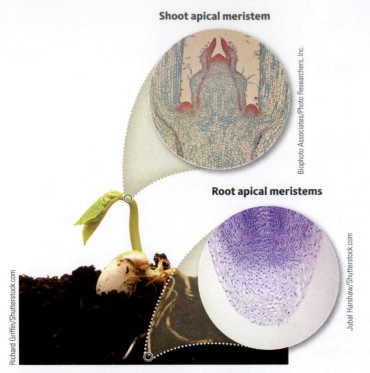

Biophoto Associates/Photo Researchers, Inc.

Root apical meristems

Jubal Harshaw/Shutterstock.com

Richard Griffin/Shutterstock.com

Figure 12.15 Seed germination. The first sign of growth from a germinating seed is its root apical meristem. The root tip adds new layers of cells and begins to establish the plant's root system. The shoot apical meristem extends the plant upward and forms buds and leaves.

posite side of the seed a shoot apical meristem grows upward, increasing the height of the plant in the same way. The terminal bud is one of several shoot apical meristems of the woody coffee plant (**Fig. 12.16**). As the shoot apical meristem grows, it forms branches and leaves. Differences in the rate and duration of growth in different areas of the plant produce its particular growth form. Growth continues at each root tip below ground and at each shoot tip above ground as long as conditions are favorable (see **Fig. 12.16**).

During the first year of life, woody plants begin to produce two types of lateral meristems. Each forms a ring or cylinder of dividing cells from the tip to the base of the plant. As in apical meristems, the processes of cell division, expansion, and differentiation occur in lateral meristems. Trees and shrubs in seasonal climates form the annual growth rings that you can count to determine how long the plant has lived. This wood-producing meristem is called the **vascular cambium** (see **Fig. 12.16**). On the outside of the woody plant, another lateral meristem called the **cork cambium** continually divides to form the protective outer layer of the

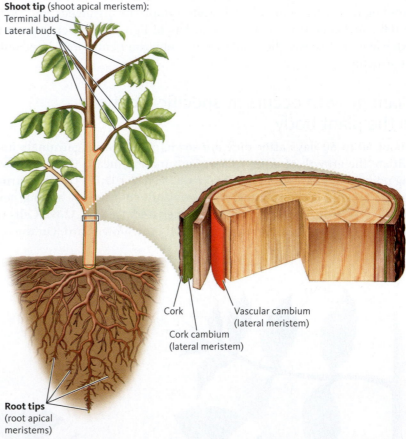

Shoot tip (shoot apical meristem):
Terminal bud
Lateral buds

Cork

Vascular cambium (lateral meristem)

Cork cambium (lateral meristem)

Root tips (root apical meristems)

Figure 12.16 Apical and lateral meristems of a woody plant. A woody plant grows at its shoot and root tips by way of apical meristems and grows in thickness by way of lateral meristems. The shoot apical meristems occur within buds at the tips and along the branches. The cork cambium continually divides to form the outer protective layer in contact with the environment. The vascular cambium produces new layers of wood, which increase the transport and storage of water and provide a strong base of support. This increases the capacity of the plant for stronger support and increased water transport.
© Cengage Learning 2014

Vascular cambium—a lateral plant meristem found in woody plants; produces xylem (wood) to the inside and phloem to the outside of roots and stems of woody plant.

Cork cambium—a lateral meristem found in woody plants; divides to form a protective outer layer of dead cells (cork) and an inner layer of living storage cells (parenchyma).

plant. While the dead cork forms the very outer layer of the plant, the familiar term **bark** actually refers to all of the tree's tissues outside of the vascular cambium. Peeling the bark off a tree often removes the thin vascular cambium and exposes the underlying wood. Together, these lateral meristems produce new cells in a circular pattern, producing a tree that grows larger in girth or diameter.

Herbaceous plants only add new growth through apical meristems; they lack the ability to produce wood through lateral meristems. Woody plants, on the other hand, grow at their tips by apical meristems and become thicker with the formation of the two cambiums. These meristems remain active throughout the life of the plant and produce new layers of cells each growing season. This pattern of growth through its life, or *indeterminate* growth, is different than the *determinate* growth pattern shown by animals. Recall from Chapter 7 that animals grow to a particular size and then stop growing when they reach maturity.

Plant growth is caused by cell division and expansion

At the cellular level, growth at meristems is accomplished by increasing the number and size of cells and creating different types of cells. Three distinct cellular processes, which overlap in their location and timing, bring about the orderly process of growth. Let's focus on the root apical meristem as it grows and produces a massive surface area for absorption of water and minerals.

The cell cycle in plants is similar to that presented in Chapter 7 for animals (**Fig. 12.17**). The first process, *cell division,* increases the number of cells at the tip. The two daughter cells produced during division have identical genetic complements and occur in the same four steps, or phases, as animal cell division (see **Fig. 12.17**). During interphase, DNA is replicated and chromosomes are duplicated. During the mitosis phase of the cell cycle, chromosomes are condensed during prophase, attached to microtubules and aligned during metaphase, and then moved to opposite poles during anaphase. During telophase, the nuclear membrane and cell wall form to separate the two daughter cells. The development of the plant cell wall is an important difference between animal and plant cell division. Daughter cells, in turn, become dividing cells and add a second layer to the tip. Through successive rounds of mitotic cell division, roots continue to add new layers of cells to the growing tip. Ultimately, the shape of the plant part, whether it becomes a round root or a flat leaf, is determined by the direction in which new cells are produced. However, cell division is only part of the story.

New cells increase in size through a process of *cell expansion,* when the cells swell and stretch. Meristematic cells have thin walls and are rich in cytoplasm. As the cell matures, the vacuole fills with water to

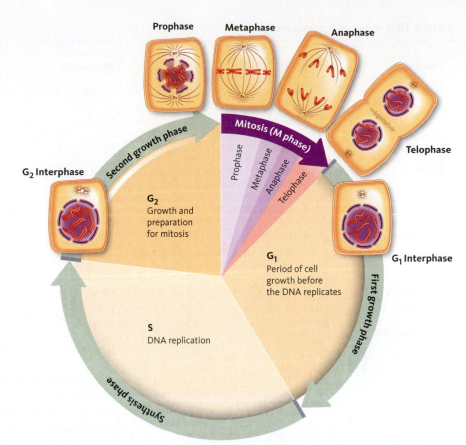

Figure 12.17 The cell cycle produces growth at meristems. Plant growth increases the number of cells through cell division. At meristems, cells go through the cell cycle to produce new daughter cells, which then become new dividing cells. This pattern results in the continual addition of new layers of cells. The phases of mitosis are shown at the top right.
© Cengage Learning 2014

Bark—consists of all tissues outside of the vascular cambium.

TABLE 12.3 Principal Tissues in Plants

Tissues	Function	Cell Status at Maturity
Parenchyma	Photosynthesis (shoots) Storage (roots and shoots) Flexible support (shoots)	Living
Support (fibers)	Rigid support	Dead
Xylem	Water and mineral transport Rigid support	Most are dead
Phloem	Sugar transport	Living
Epidermis	Protection against water loss (shoots) Absorption of water and minerals (roots) Gas exchange (stomata)	Living
Cork (woody roots and shoots)	Protection against water loss Gas exchange	Dead

© Cengage Learning 2014

Turgor pressure—the outward push exerted by water against the cell wall; plays a key role in cell expansion during growth and in flexible support of herbaceous plants and their parts.

Epidermis—plant tissue on the outside edge of the plant body that is in direct contact with the environment; generally a single layer of cells; its functions depend on its location; remains throughout the life of herbaceous plants; is replaced in roots and stems in the first year in woody plants.

Cork—protective plant tissue on the outside of roots and stems of woody plants; produced by cork cambium; functional and dead at maturity.

Fibers—extremely long, narrow plant cells with thick, lignified walls; provide structural support to plant body; generally dead at maturity; abundant underneath the epidermis of stems and help the plant bear bending stresses. In broad-leaved woody plants, fibers lend rigid support to strengthen wood.

Parenchyma—plant tissue consisting of large thin-walled cells; versatile tissue that stores water and starch in leucoplasts in roots and stems; performs photosynthesis in leaves.

create turgor pressure, which is an outward push exerted by water against the thin cell wall to slowly stretch or expand the volume of the cell. This process increases the cell to many times its original size and accounts for the largest portion of plant growth. In fact, the expansion of cells results in the root growing and pushing its way through the soil with enough force to crack a rock. Thus, plants grow by the absorption of water, which is one reason why crops are irrigated and habitats with the most rainfall have the greatest amount of plant growth and vegetation.

Differentiation produces specialized cells and tissues

Farther away from the tip, a fully expanded cell now matures by the process of *differentiation*, when it changes from a simple meristem cell to a complex, specialized cell. A cell's structure reflects a close relationship to its function: a support cell thickens its walls and a storage cell forms vacuoles and leucoplasts. Ultimately, the cell's development and final structure depend on its location within a particular plant organ—roots, stems, or leaves. Within these organs you'll find differentiated, or mature, cells that are specialized into different kinds of cells, or cell types, that work together within a tissue to perform a specific function such as transport or support (Table 12.3). To perform their function, some plant cell types are genetically programmed to die as they mature. Cells must die before they function to transport water and provide rigid support for the plant. Different tissues are located in different areas within the plant body. We will look briefly at the different areas, starting on the outside and moving toward the middle of a herbaceous plant organ.

The outside edge of the plant is composed of cells that are in direct contact with the environment and thus function to acquire water and minerals, exchange gases, and protect the plant from water loss and against pathogens. A single layer of cells called the epidermis accomplishes this first line of defense and remains throughout the life of herbaceous plants (Fig. 12.18a). The epidermal tissue of woody plant roots and stems is replaced in the first year by the actions of the cork cambium, which produces an inner layer of living storage cells and an outer layer of dead cells called cork. Cork is another example of how dead cells play important functional roles in plants.

Inside the epidermis or cork in woody plants are a variety of cell types and tissues. Support tissue consists of long, narrow, thick-walled cells called fibers (Fig. 12.18b). Fibers are abundant underneath the epidermis of stems and help bear the stresses of bending as the plant sways in the wind. In broad-leaved, woody plants, fibers lend rigid support to strengthen wood. Some fibers are soft and are used in clothing fabrics such as cotton, linen, and rayon. Along with fibers are large thin-walled cells that form parenchyma tissue (Fig. 12.18c). This versatile tissue stores

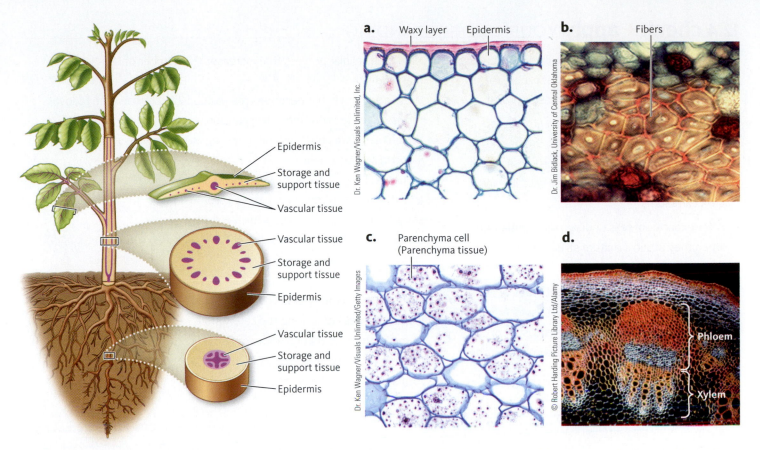

a. Waxy layer Epidermis
Dr. Ken Wagner/Visuals Unlimited, Inc.

b. Fibers
Dr. Jim Bidlack, University of Central Oklahoma

c. Parenchyma cell (Parenchyma tissue)
Dr. Ken Wagner/Visuals Unlimited/Getty Images

d. Phloem Xylem
© Robert Harding Picture Library Ltd/Alamy

Epidermis
Storage and support tissue
Vascular tissue
Vascular tissue
Storage and support tissue
Epidermis
Vascular tissue
Storage and support tissue
Epidermis

Figure 12.18 The tissues composing a typical herbaceous plant. (a) The outer surface is composed of the closely packed cells of its epidermis that secrete a waxy layer on all aboveground parts of the plant. (b) In many plants, thick-walled fibers (shown in light brown) support the plant. (c) Filling in the middle of the plant organ is a site for storage known as parenchyma tissue. The purple specks are starch grains located inside leucoplasts. (d) Transport tissues occur together in discrete bundles of vascular tissue. Within the vascular bundles, xylem develops to the outside and phloem tissue develops to the inside of the stem or root. In leaves xylem occurs on the top of vascular bundles called veins.
© Cengage Learning 2014

water and starch in roots and stems; it also performs photosynthesis in leaves. The turgor pressure from water storage also plays an important role in supporting the herbaceous plant. A wilted plant is incapable of supporting leaves and flowers.

The central portion of a plant organ is composed of special conducting cells that transport water or sugars. This is the plant's **vascular system**, which runs throughout the plant and connects roots, stems, and leaves. In the leaf you see the vascular system as a network of veins. Each vein is composed of two tissues: xylem and phloem (**Fig. 12.18d**). **Xylem** tissue is composed of hollow, dead cells with a thick rigid wall; it is an ideal structure for a "pipe" that carries water. Xylem cells in the root are part of a pipe that runs all the way from near the root tip to the leaves. Wood is composed of xylem, so a tree's transport tissue also provides a strong, rigid support for the plant. **Phloem** tissue is a separate column of long, narrow cells that carries sugars from sites where sugars are made or stored to the rest of the plant. Both phloem and xylem run all the way from the leaves to near the root tips. Let's now apply the functional cell types and tissues to build the next level of the hierarchy: the organs.

Vascular system—plant tissue system that is composed of conducting tissues that transport water and minerals (xylem) and sugars (phloem); runs throughout the plant and connects roots, stems, and leaves.

Xylem—plant tissue that transports water and minerals throughout the plant; starts near the root tip and flows toward the leaves; mostly composed of hollow, dead cells with rigid walls; dominates the structure of wood; occurs and functions with phloem in vascular bundles; produced to the inside of woody plant by the vascular cambium.

Phloem—plant tissue that carries sugars throughout the plant; flows from sites where sugars are made or stored to places where it is used or stored; occurs and functions with xylem in vascular bundles; produced to the outside of woody plant by the vascular cambium.

- Plants grow at specific places called meristems. At meristems, cells divide to increase cell number and expand to increase cell size.

- All plants grow in length through growth at root and shoot apical meristems. Stems and roots of woody plants grow in thickness through lateral meristems called vascular and cork cambiums.

- After expansion, cells undergo differentiation, when they attain a mature structure and specialized function.

- Plants are composed of different types of cells that compose different tissues. Tissues are groups of cells that protect, defend, support, store, and transport substances throughout the plant.

1. Name the cellular process that increases the number of cells in meristems.

2. Write a sentence that describes the relationship between the following terms: water, growth, and cell expansion.

3. Which meristem produced the following: (a) cork in a wine bottle, (b) the wood in a violin, (c) the head of cabbage (bud)?

4. A chemical used to control weed growth acts to disrupt the formation and attachment of microtubules to chromosomes in meristematic cells. Which cellular process does this herbicide inhibit in the control of growth?

5. A stalk of celery supports its leaves. Refer to **Figure 12.19** and the celery cross section to identify tissues composing the celery stalk. Name two tissues that form the red-stained circles (#1). Name the tender tissue that forms the bulk of the white portion of the stalk (#2).

Figure 12.19 Cross section of a celery stalk.

12.5 Roots, Stems, and Leaves Extend the Surface Area of the Plant Body

If you are a coffee enthusiast interested in growing your own coffee plant for the freshest cup imaginable, then you can expect to work for at least three years before reaping your first crop of beans. This is how long it takes for a coffee plant to grow and develop the extensive root and shoot system (**Fig. 12.20**) needed to produce flowers, fruits and coffee beans. Your hard work and patience may pay off because a well-managed coffee tree can yield coffee beans for up to 80 years or more.

The coffee tree consists of a main trunk that grows upward to support many horizontal branches that may branch again and again. This growth results in structure that positions leaves to capture sunlight and avoid shading other leaves. The total area of its leaves has been measured to be 16 feet by 16 feet (264 square feet). This same branching pattern occurs below ground to build a massive root system with an even larger surface area: 72 feet long by 72 feet wide (5,000 square feet)! Plants such as coffee trees invest their sugar "income" to build roots, stems, and leaves to maximize their surface area to capture even more light, water, and minerals. This is how plants compete with each other.

Growing a successful and profitable crop of coffee beans requires an understanding of how plant organs are organized and built to carry out their specialized activities (see **Table 12.1**). In the section ahead, we will explore how roots, stems, and leaves contribute to the growth and survival of the plant in its habitat.

Plants build extensive root systems to anchor themselves and absorb water and minerals

The root apical meristem begins to establish a branched network of roots called its **root system**. In woody plants, the first root to emerge develops and thickens to become a **taproot** that ultimately anchors a massive system of thick woody and thin threadlike roots. Notice that the coffee tree's root system grows only a few feet deep (see **Fig. 12.20**) yet extends outward far beyond the trunk. A plant's root system functions primarily in the rich topsoil near the surface, where it absorbs water and mineral nutrients from the large pool of decaying vegetation. In prairie grasses and herbs, the fertile topsoil is deeper and therefore the root systems grow to greater depths; in some cases roots grow 10 feet deep. Grasses do not have a single large taproot like trees; instead they form a deep **fibrous root system** consisting of many similar-sized roots that all grow from the base of the stem.

Let's zoom in and look at a magnified root tip (see **Fig. 12.20**). The root apical meristem continually divides and expands to produce a **root cap** of epidermal cells to protect it as it grows through the soil. Root cap cells secrete a slimy substance that allows the growing tip to navigate around soil particles in search of new water sources. Farther back from the tip, mature epidermal cells extend outward, forming numerous **root hairs** that directly contact soil particles and absorb the water and minerals they contain by osmosis. Root hairs are delicate and live for only a few days before they are replaced by the growing root tip. Building new root hairs is critical because they absorb 90 percent of the plant's water supply. The remaining 10 percent of water is absorbed directly into the root through the epidermis.

The path of water passes from the root hair into the **cortex**, a region of the root that specializes in storing starch (**Fig. 12.21**). When it reaches the innermost layer of cells of the cortex, a layer of cells, or **endodermis**, controls the passage of minerals and other molecules that enter the xylem. Water moves across this layer by osmosis. However, many minerals cross the endodermis by active transport. Recall that active transport requires energy, which is fueled by aerobic metabolism in mitochondria. This is the reason why endodermal cells require oxygen and sugars to produce enough ATP for this transport process. In fact, roots receive almost half of all sugars produced by the leaves.

Once water and minerals are transported across the endodermis, they enter xylem tissue. The strong xylem "pipes" transport the water and dissolved minerals upward through its stems using turgor pressure, to the leaves for photosynthesis and the meristems for cell expansion. In turn, roots receive and use large supplies of sugars delivered through phloem to fuel the active transport of minerals and the storage of starch. In this way the root and shoot systems depend on each other to function.

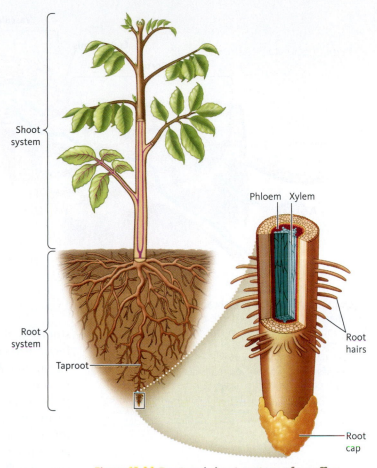

Figure 12.20 Root and shoot system of a coffee plant. Out of view belowground, the coffee plant builds a taproot system that anchors it to its location, stores starch, and transports the water it absorbs upward through the xylem. Phloem delivers sugar to the growing root tip, which fuels cell division and the constant rebuilding of root hairs and root cap cells that were shed into the soil.
© Cengage Learning 2014

Root system—a branched network of roots of a plant.

Taproot—a thick primary root specialized for storage.

Fibrous root system—a network of many similar-sized roots that all grow from the base of the stem.

Root cap—dome of epidermal cells that protects the growing root as it grows through the soil; secretes a slimy substance that allows the growing tip to navigate around soil particles in search of new water sources.

Root hairs—long, narrow extensions of epidermal cells that directly contact soil particles and absorb the water and minerals they contain; absorb 90 percent of the plant's water supply; delicate and live for only a few days before being replaced by the growing root tip.

Cortex—the outer region of the root that specializes in storing starch.

Endodermis—the innermost layer of cells of the cortex; controls the passage of minerals and other molecules into the xylem.

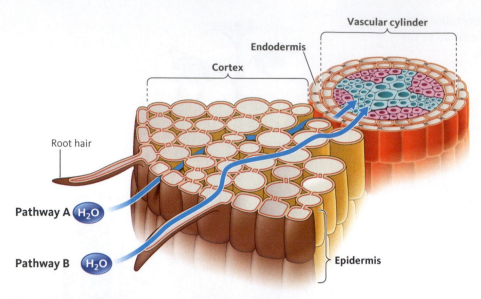

Figure 12.21 **Path of water through the root.** Water and minerals are absorbed into the root through root hairs and epidermal cells. Water moves either rapidly *between* cells (pathway A) or more slowly *through* cells (pathway B). When it reaches the endodermis, water, along with dissolved minerals, is forced to go through cells where the root regulates which substances enter the xylem and which do not. Water passively follows its concentration gradient "downhill" by osmosis.
© Cengage Learning 2014

Figure 12.22 **Strawberry plant runners.** The strawberry plant builds horizontal stems called runners to extend its body outward rather than upward to gain access to light, water, and minerals.

Annual growth ring—a ring of cells produced by woody plants each growing season; consists of a light-colored ring of larger, wider xylem cells (springwood) followed by a dark ring of narrower, thicker-walled cells (summerwood).

Stems provide the plant with access to resources

A plant's stem is important to its ability to compete for resources. Tall stems provide access to light by placing the leaves of one plant above the leaves of competitors, as well as access to wind currents to disperse pollen or seeds. Placing organs into position requires strong support tissues that include xylem cells and fibers with thick cell walls. Some stems require little support tissue because they grow horizontally along the surface or underground to provide access to more space and resources (**Fig. 12.22**).

Stems connect the roots at one end of the plant body to leaves on the other end and serve as the "middleman" in the plant's storage and distribution system. By receiving hormonal signals from different parts of the plant, stems control the transport of water to leaves and sugars to other plant organs. Plants that live in habitats where the next rainfall is unpredictable store water in their stems. Although all living plant tissues store small amounts of water, many plant stems such as the cactus are specialized to store large amounts of water when it is plentiful for later use when it is scarce. The sharp spines on many varieties of cactus act to protect the plant's water stores. In addition to storing water, stems store starch. Specialized underground stems called *tubers* are packed with starch in parenchyma tissue. The potato chips you eat for a snack are sliced-up stems loaded with stored starch.

The stems of most herbaceous plants are flexible and green. When you look at its internal structure, the bulk of the stem contains parenchyma tissue, which functions in flexible support, energy storage, and photosynthesis (**Fig. 12.23a, b**). Vascular tissues occur in discrete bundles for transport. Support within the stem is accomplished through turgor pressure on cell walls and with stiff fibers. The stem's outer epidermis secretes a layer of wax and has openings for the diffusion of carbon dioxide.

The stems of woody plants have a structure dominated by tissues specialized for support and water transport, the thick-walled xylem tissue (see **Fig. 12.23c**). In temperate climates, each spring the vascular cambium adds layers of new xylem cells toward the inside and new phloem cells to the outside of the tree. The wet spring season produces larger, wider xylem cells, whereas the drier summer months tend to produce narrower, thicker-walled wood. The result is an annual growth ring. Each year the tree produces a light-colored ring of springwood followed by a dark ring of dry summerwood (**Fig. 12.23c**). Years with abundant amounts of rainfall are indicated by wider rings, whereas dry years are shown by narrower growth rings. This pattern allows foresters to look back in time and determine the years that the tree experienced favorable and unfavorable growth conditions. Phloem occurs outside of the vascular cambium in the inner bark region of the woody stem.

The outer bark of the woody stem is produced by the cork cambium, which divides and expands into several layers of protective cork cells. A ring of phloem transports sugars just outside the wood and vascular cambium and beneath the cork (see **Fig. 12.23c**). The cork layer has openings that allow oxygen to diffuse into the living phloem and vascular cambium for aerobic metabolism and ATP to fuel growth and sugar transport.

Leaves are the major photosynthetic organs of the plant

A plant invests considerable amounts of energy and materials into building its leaves. Much larger than a small coffee tree, a large sugar maple tree has about 100,000 leaves, with a total surface area of about 9,000 square feet for absorbing sunlight and carbon dioxide in the process of photosynthesis. Through absorbing carbon dioxide in photosynthesis, the plant absorbs the carbon it needs to build all of its organic molecules, such as starch, cellulose, proteins, and lipids.

Coffee and broad-leafed maple leaves are thin, flattened organs organized like a sandwich (**Fig. 12.24**). The upper and lower layers are the epidermis. In leaves as in stems, the outer cell walls of epidermal cells secrete layers of wax to reduce water loss, protect against ultraviolet radiation, and block entry of pathogens. The epidermis also contains openings called **stomata**, which allow carbon dioxide to diffuse into the leaf from the atmosphere for photosynthesis (see **Fig. 12.24**). When stomata are open they also allow water vapor to diffuse out of the leaf. Each opening is controlled by a pair of *guard cells* that respond to environmental cues such as light, wind, and humidity. For example, when roots come into contact with dry soil, they send hormone signals that cause guard cells to close and reduce water loss before any adverse effects of water shortage arise.

The middle portion, or **mesophyll** region, of the leaf is composed of parenchyma, support, and vascular tissues (see **Fig. 12.24**). The mesophyll is dominated by large numbers of parenchyma cells that carry out photosynthesis. To allow for exchange of gases between the atmosphere and photosynthetic cells, about a third of the volume of a leaf is composed of air spaces. Xylem tissue delivers water to the photosynthetic parenchyma cells, and phloem tissue exports sugars from the leaves to other parts of the plant. The continuous movement of water saturates the air spaces of the leaf. When stomata are open, water moves out of the leaf into the atmosphere, and the leaf becomes cooler to keep the temperature within the limits for photosynthesis.

Leaf structure reflects the environment in which the plant develops. Leaves on the outside of a tree or in an open field are subjected to high light intensities. These sun leaves have a thicker mesophyll layer and a higher rate of photosynthesis than do shade leaves. Sun leaves require greater amounts of carbon dioxide and thus will have a higher concentra-

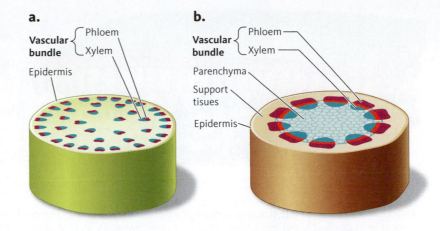

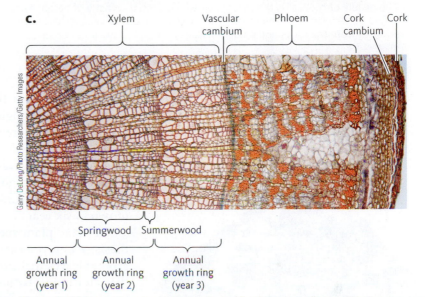

Figure 12.23 Stem structure of woody and herbaceous plants. The tissue organization of herbaceous and woody stems is shown here. (a, b) Herbaceous stems have large areas for storage and photosynthesis. (c) In woody stems, the epidermis is replaced by cork, and the vascular cambium produces xylem (wood) and phloem tissues. The large amount of wood provides strong support and increased capacity for water transport, which allows woody plants to grow tall and support a large canopy of leaves. Large numbers of leaves produce greater amounts of sugars and energy for the tree. This photograph shows three annual growth rings.
© Cengage Learning 2014

Stomata (singular form, **stoma)—**openings in the epidermis of leaves and herbaceous stems for gas exchange; allow carbon dioxide to diffuse into the leaf from the atmosphere for photosynthesis; opening is controlled by a pair of guard cells that respond to environmental cues such as light, wind, and humidity.

Mesophyll—the middle portion, or region, of a leaf; dominated by large numbers of parenchyma cells that carry out photosynthesis; also is composed of support and vascular tissues.

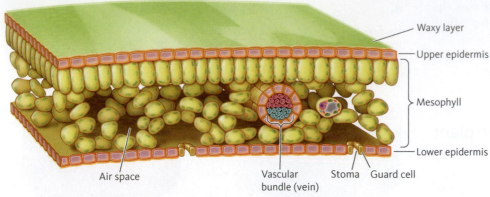

Figure 12.24 **Anatomy of a leaf.** Cross section of a deciduous leaf shows the organization of its tissues. The upper and lower epidermis secrete a protective waxy layer that prevents water loss. The middle of the leaf is the mesophyll, which consists mostly of parenchyma cells that are specialized to perform photosynthesis. The vascular bundle, or vein, includes xylem tissues that deliver water from the roots and phloem tissues that export sugars to other parts of the plant.
© Cengage Learning 2014

Figure 12.25 **Cross section of a conifer needle.** The round shape of a conifer leaf adapts it to retain more water than a thin, broad leaf. Conifer leaves also have thicker layers of wax on their surfaces.

tion of stomata. Shade leaves in the understory of a tropical rainforest, for example, compensate for lower light intensities with thinner, wider leaves with a greater surface area.

Conifer needles have a more compact shape than the broad-leaved trees (**Fig. 12.25**). This shape exposes less of the leaf surface to light but helps retains more water. As an analogy, think of a wet towel and the evaporation of water. The towel dries very slowly when it is rolled up compared to a towel that is laid out flat and exposed. Deciduous leaves are flat, with a much greater exposure to light and air, and therefore have much higher rates of photosynthesis than conifer leaves; however, they also lose more water. Thus, deciduous, broad-leaved trees have higher water requirements than conifers. Conifer needles have much more support tissue and are stiffer than broad deciduous leaves and therefore are more durable and "expensive" to build. The "cheaper" broad leaves photosynthesize more rapidly, pay back their investment quickly, and are shed at the end of the growing season, whereas the higher-cost conifer needles are retained for a longer period of time of two to six years.

In many plant species, leaves serve functions other than photosynthesis. When you eat onion rings, you are eating modified fleshy underground leaves that store starch (**Fig. 12.26**). Tendrils are specialized leaves that attach and twine around a structure to support a climbing plant. Carnivorous plants have leaves adapted for the capture and digestion of insects and other small arthropods. The leaves of the Venus flytrap are modified into a snap-trap mechanism that actively captures insects that land on its surface. Leaves of another carnivorous plant—the pitcher plant—curl, swell, and fuse into a pitcherlike tube. This remarkable plant is featured in Chapter 13.

Figure 12.26 **Leaves compose the edible part of an onion.** The fleshy leaves of an onion bulb store starch.

- Root systems anchor the plant in place, absorb water and minerals, and transport them upward to the growing meristems and leaves.

- Stems support and position leaves and flowers to gain resources. They are positioned between roots and leaves and serve transport and storage functions for the plant.

- Leaves are the main organs that make sugars for the plant, and their structure reflects the habitat in which they live.

Figure 12.27 Yucca.

1. Which root structures absorb the vast majority of the water for the plant? In which part of the root system are they located?

2. Name two functions of roots that are also performed by herbaceous stems.

3. Which tissue comprises the greatest proportion of the structure and function of a woody stem? If you stand next to a tree, which part of the trunk has more annual growth rings: the part next to your head or the part next to your knees?

4. Shade-grown coffee is grown under a canopy of diverse species of shade trees that support a greater biodiversity of animals and plants. Sun-grown coffee is grown in direct sunlight in open rows like a cornfield. Predict whether shade-grown or sun-grown coffee (a) uses more water, (b) has a higher concentration of stomata, and (c) produces more coffee beans.

5. Yucca plants, like many members of the agave family, are long-lived perennial plants found in dry areas of the southern United States (**Fig. 12.27**). Predict whether (a) the waxy layer is thick or thin, (b) there is a large or a small number of stomata on the upper epidermis, (c) leaves have a thick, densely packed or thin mesophyll layer, (d) the plant has large or small amounts of support tissues (fibers), (e) leaves that are cheap or costly to build and therefore short- or long-lived. Defend your predictions based on its habitat or the figure.

12.6  Plants Defend Themselves against Attack from Animals and Microbes

At the base of the food web, plants produce a large variety of organic molecules such as starch, sugars, oils, and proteins, which provide the energy to other organisms that can gain access to them. Plants are constantly under attack from herbivorous animals, plant parasites, and pathogens such as fungi, bacteria, and viruses. For example, more than 850 insect pests and dozens of pathogens attack the coffee tree and cause disease. A *diseased* plant is one whose normal functioning and activities are inhibited by pathogens or pests. Coffee leaf rust, a fungal pathogen, causes major destruction to the coffee plant's capacity to make sugar by photosynthesis (**Fig. 12.28**). The reduction in sugar causes reduced growth and reproduction and ultimately reduces the amount of coffee beans produced by the tree.

An outbreak of coffee leaf rust disease occurs because three factors interact: a capable pathogen, a susceptible host plant, and a suitable environment. Growers and plant pathologists work closely together to reduce disease by increasing the genetic resistance of the crop plant, reducing

Figure 12.28 Coffee leaf rust. This yellow-orange rust is a fungus that infects coffee leaves and reduces photosynthesis. Water and warm temperatures favor colonization and the release of hundreds of thousands of spores that are dispersed into the air.

Fruit blotch
disease
(fungus)

Leaf spot
disease
(fungus)

Root knot
disease
(roundworm)

Scot Nelson, University of Hawaii at Manoa

Scot Nelson, University of Hawaii at Manoa

Scot Nelson, University of Hawaii at Manoa

Figure 12.29 Coffee plants can be attacked by numerous pathogens and parasites. All parts of a plant are subject to attack by many different parasites and pathogens. All of the diseases shown would not occur on the same tree.
© Cengage Learning 2014

sources of the pathogen, or changing the environment in which the pathogen thrives. For example, proper pruning of coffee trees opens up the branching pattern and reduces the humidity around leaves. This practice reduces the moisture available for the fungus to grow and spread.

In nature, the battle between plants and their pathogens has been going on for millions of years. Natural selection is very intense in this aspect of a plant's life; well-defended plants tend to leave many more offspring than do poorly defended plants. Although there is an enormous pool of potential pathogens in nature, most plants are resistant to attack by the majority of microbes they encounter. However, each plant species, like the coffee plant, is vulnerable to attack by a select group of pathogens that have evolved the ability to infect different organs at different stages in their life cycle (**Fig. 12.29**). In the section ahead, we will explore how animals and pathogens attack and infect plants and how plants defend themselves.

Plants are constantly under attack from herbivores and pathogens

Many herbivorous animals have teeth and mouthparts that crush the plant's cell walls, allowing them access to the living contents of cells and tissues. The cytoplasm of the living plant cell contains proteins, carbohydrates, fats, minerals, and water. Recall from Chapter 8 that cellulose- and lignin-comprising plant cell walls are indigestible to animals and are processed in different ways in ruminant (cattle, deer) and nonruminant animals (horses).

Pathogens that attack plants belong to the same groups of organisms that cause disease in humans: bacteria, viruses, fungi, protozoa, and nematodes. In some cases an individual plant may be attacked by thousands of individuals of a single pathogen at the same time. Plants are also attacked by many species of parasitic plants, such as dodder or mistletoe (see **Figs. 12.8** and **12.9**, respectively). Each group has a different set of characteristics and capabilities to cause disease.

The disease process begins with an organism breaking through plant surfaces to gain access to the plant's tissues. Parasitic plants, fungi, and nematodes most commonly gain access by penetrating directly through the outer cell walls of the epidermis. Stomata and other natural plant openings serve as other entry points for bacterial and fungal pathogens. Under favorable moisture conditions, coffee leaf rust spores germinate and then enter leaves through stomata. Finally, all bacteria, most fungi, and some viruses can enter plants through wounds made by broken branches, sand blasting, hail damage, animal feeding, and fire. Some viruses are injected into plant hosts when insects pierce the epidermis and suck sugars from the phloem tissue.

After entering and gaining access to plant tissues, pathogens absorb plant nutrients, grow, and multiply. They begin to spread to surrounding tissues. Some infections are limited to a small, local area or a particular organ; others spread throughout the entire plant. Bacteria that get into xylem tissues in the stem may spread to other areas and eventually cause the plant to wilt. Some fungi and bacteria produce toxins that injure host cells and produce some of the disease symptoms. Just as in animals, viruses that invade plants require a living cell to reproduce and spread. Successful infections often result in visible symptoms, such as discolored or malformed leaves or stems or dead areas or spots on leaves (see **Fig. 12.28**). These symptoms indicate that the battle has just begun.

Plants are well defended by structural barriers and chemical weapons

Plants have defenses in place to prevent attack as well as defenses to respond to an attack. The first lines of defense are the structural barriers that plants have on their surface. These structures include the amount and type of waxes that compose the outer layer. Waxes on leaves and fruit surfaces prevent formation of a film of water in which pathogens may be able to swim or multiply. A thick mat of hairs on some shoots serves the same purpose. Just below the waxy layer are the tough outer cell walls of the epidermis. Although some fungal pathogens can "drill" through the wall, most do not have the enzymes that would allow them to force their way in. This makes the cell wall an impenetrable and highly effective barrier to most pathogens.

The second line of defense includes chemical weapons that either deter or poison attackers. Plants, herbivores, and pathogens have interacted with each other in a sort of "biological arms race," where plants produce a toxic compound and then herbivores and pathogens eventually evolve the ability to resist the toxin. Recall from Chapter 8 that koalas evolved resistance to the toxins made by the *Eucalyptus* trees and in Chapter 3 that passion-vine butterflies were selected for by their ability to resist toxins and eat the leaves of passion-vine plants. Through random mutations and subsequent natural selection, plants in turn evolve the ability to produce new toxins in a continual back-and-forth arms race. This coevolution over millions of years has led to more than 100,000 different plant compounds. On the other side of the coin, many of these same groups of chemicals are used not to deter animals but to attract them. Humans use many of these compounds as drugs, pesticides, flavoring agents, and perfumes.

Alkaloids are a diverse group of chemicals that repel herbivores by their bitter taste and stimulate their nervous systems. The caffeine found in coffee, tea, and energy drinks is an example of an alkaloid. This group includes familiar chemicals such as nicotine in tobacco, morphine from opium poppies (**Fig. 12.30**), cocaine from the coca plant, and mescaline from the peyote cactus. In an interesting twist, some animals such as the poison dart frog and the passion-vine butterfly not only resist the effects of alkaloids from the plants they feed on but actually make use of them as protection against potential predators that learn to avoid animals with bad tastes.

Terpenoids are a large group of oily compounds that deter feeding by herbivores. To humans, however, many terpenoids are found in our spice cabinet as nutmeg, ginger, and oregano to flavor foods. The oil in cedar wood repels termites and moths and often is used to line cedar chests and closets to protect clothing. In conifers, terpenoids such as turpentine are formed to seal wounds and prevent infection. These highly flammable

Kodda/Shutterstock.com

Figure 12.30 Opium poppy. The alkaloid morphine and terpenoids occur in latex or a milky fluid that oozes from wounded opium plants. This sticky fluid heals wounds and deters animals from feeding. Terpenoids are defensive compounds that are common in plants such as opium, spurges, and milkweeds.

plant compounds are one reason why forest fires in pine stands are extremely hot and difficult to control. Finally, the milky fluid found in milkweeds and spurges is a rubbery, latex-type of terpenoid that repels insects and heals wounds.

The final group of chemical weapons is the cyanogens. Many members of the rose, bean, grass, and sunflower families produce these highly toxic compounds. Cyanogen, a compound produced in small amounts by many plants, is confined to the vacuole. Once the seed or leaf tissue is crushed by an herbivore, a chemical reaction releases gaseous hydrogen cyanide. In animals, this poison blocks the electron transport chain in the mitochondria, halting aerobic metabolism and ATP production. The toxin affects the herbivore's nervous and circulatory systems and can result in convulsions or death. Interestingly, plants are unaffected by their own cyanogens because they have slightly different proteins in their electron transport system.

This chapter has introduced you to the way plants grow; develop roots, stems, and leaves; adapt to their environment; respond and communicate; and defend against attack. Whether it is the coffee and pancakes you have for breakfast, a hike you take in the woods, the changing colors of autumn leaves, or the rose you receive, this chapter has tried to deepen your appreciation of the colorful plant world around you. Chapter 13 will build on your understanding of structure by delving deeper into how a plant functions in its habitat.

12.6 check + apply YOUR UNDERSTANDING

- Disease occurs when three factors interact with each other: a capable pathogen, a susceptible host, and a favorable environment (for example, moisture, temperature).

- Most plants are resistant or immune to attack by the majority of microbes with which they come into contact. Plants are prone to attack by many different pathogens on different organs and at different stages in their life cycle. Each group of pathogens has a different set of characteristics and capabilities to cause disease.

- Plants defend themselves with structural and chemical barriers that deter or poison the attacker.

1. What are the three interacting factors that cause plant disease?
2. What are the two lines of defense that plants have to prevent and respond to an attack?
3. Name three ways in which pathogens gain access to plant tissues.
4. How does each of the three groups of chemicals defend the plant against attack?
5. Conifers are susceptible to root rot caused by *Armillaria*, a fungal pathogen that is present in the soil and causes disease only when roots are weakened or damaged by some type of stress, such as drought. From this description, which of three interacting factors (pathogen, host, environment) tips the balance in favor of disease? Why would a stressed plant be more susceptible to attack?

End of Chapter Review

Self-Quiz on Key Concepts

Plant Growth, Growth Forms, and Life histories

KEY CONCEPTS: Plants grow into a variety of herbaceous and woody forms from the activities of meristems. Apical meristems produce roots, stems, leaves, and reproductive structures. Woody plants also have lateral meristems that grow in diameter through the production of wood and bark. Plants with short life histories allocate a greater share of their resources to rapid growth, acquisition of resources, and reproduction than do longer-lived perennials.

1. Match each of the following **plant life histories** with its description or example. Each term on the left may match more than one description on the right.

   Annual plant   a. plant that flowers in the second growing season
   Biennial plant   b. plant that lives it life cycle in a 12-month time frame
   Perennial plant   c. plant that lives for more than two years
                             d. trees and woody plants that tend to grow more slowly
                             e. plant that overwinters solely as seeds

2. Match each of the following **meristems** with its description or example. Each term on the left may match more than one description on the right.

   Cork cambium   a. forms the outer bark of a woody plant
   Root apical meristem   b. forms the leaves of a plant
   Shoot apical meristem   c. forms the bulk of a woody stem
   Vascular cambium   d. forms the hairs that absorb water from soil particles
                             e. has a cap to protect it from friction as it grows

3. Which life process(es) is accomplished only by woody plant stems and not by other organs?
   a. photosynthesis
   b. storage
   c. transport
   d. support and access

4. Which of the following growth forms does *not* form woody tissues?
   a. vines
   b. grasses
   c. succulents
   d. shrubs

Plant Cells and Tissues

KEY CONCEPTS: The plant body is composed of about only a dozen types of cells, which combine to form three tissues—dermal, ground, and vascular systems—that occur in all organs. Cells have three unique structures: a cell wall composed of cellulose, a central vacuole, and specialized plastids.

5. Match each of the following **tissues** with its description or example. Each term on the left may match more than one description on the right.

   Epidermis   a. stores starch, water, and toxins
   Parenchyma   b. protects against water loss
   Support tissue   c. absorbs water from the soil
   Phloem   d. transports sugars
   Xylem   e. consists of fibers
                      f. transports water and minerals
                      g. performs photosynthesis

6. Match each of the following **plant cell structures** with its description or example. Each term on the left may match more than one description on the right.

Cell wall
Chloroplast
Chromoplast
Leucoplast
Vacuole

a. contains starch for energy storage
b. composed of fibers of cellulose
c. provides turgor pressure for cell expansion in meristems
d. is abundant in cells that form the bulk of corn kernels
e. contains purple pigments and toxins
f. contains the green pigment that absorbs light in photosynthesis
g. contains carotenoids that are yellow-orange in color

7. Which of the following compounds in cell walls gives them rigidity?
   a. lignin
   b. cellulose
   c. starch
   d. hemicellulose

8. Which of the following cellular processes increases the number of cells in meristems?
   a. cell differentiation
   b. cell expansion
   c. cell maturation
   d. cell division

Plant Organs: Roots, Stems, and Leaves

KEY CONCEPTS: A plant's organs are adapted to function in its environment. The cells and tissues of the plant organs collectively account for the production, transport, and storage of sugars made by photosynthesis. Stems and leaves maximize surface area for the absorption of light and carbon dioxide. Roots form a vast underground network of threads that absorb water and mineral nutrients for the plant.

9. Match each of the following **plant organs** with its description or example. Each term on the left may match more than one description on the right.

Leaf
Root
Stem

a. contains fewer fibers than other plant organs
b. provides access to water and mineral nutrients
c. provides access to light and carbon dioxide
d. onion rings are an example of this organ
e. French fries are an example of this organ
f. contains parenchyma cells that perform photosynthesis

10. Which of the following is *false* regarding the sun and shade leaves?
    a. Sun leaves have a thicker mesophyll than shade leaves.
    b. Sun leaves are generally narrower than shade leaves.
    c. Sun leaves produce more sugars than shade leaves.
    d. Sun leaves have fewer stomata per square area than shade leaves.

11. Fibrous root systems:
    a. are generally deeper than taproot systems.
    b. are found only in large, woody plants.
    c. are the first roots to arise from the seed.
    d. have a large, fleshy root for storage.

Plant Defenses against Herbivores and Pathogens

KEY CONCEPTS: Plants have defenses in place to prevent an attack as well as defenses to respond to an attack. In all cases, plants defend themselves and deter herbivores and pathogens by forming structural barriers and producing toxins to poison their enemies and inhibit their growth.

12. Match each of the following **chemical defenses** with its description or example. Each term on the left may match more than one description on the right.

Alkaloids
Cyanogens
Terpenoids

a. nicotine, caffeine, and morphine are examples
b. compounds that are made by pine trees to seal wounds
c. compounds that block respiration in herbivores
d. compounds that affect the nervous system of herbivores
e. spices and flavoring agents such as essential oils

13. Which components interact in the development of plant disease?
   a. pathogen—environment—habitat
   b. plant—pathogen—fungus
   c. plant—pathogen—environment

14. How do viruses gain entry into plant cells?
   a. through stomata
   b. through wounds
   c. directly through the epidermis
   d. through openings in the cork in woody stems

Applying the Concepts

15. Plant and animal cells have structures and organelles that reflect their function. Consider three cells: a leaf mesophyll cell (see **Fig. 12.10**), a root parenchyma cell, and an animal stomach (epithelial) cell (see **Fig. 6.8**). Which of these cells has (a) mitochondria for ATP production respiration, (b) chloroplasts for photosynthesis, (c) leucoplasts to store starch, (d) ribosomes to make proteins, (e) Golgi complexes to secrete substances outside of the plasma membrane, (f) a nucleus containing chromosomes?

16. The aspen leaf miner is an insect pest of quaking aspen trees. In the spring, eggs hatch into small, wormlike larvae that bore into the leaf and start to feed on, or mine, the mesophyll tissue. The leaf miner snakes through the leaf (**Fig. 12.31**) and causes leaves to prematurely dry up and drop from the tree. Explain how plant growth will be directly affected by the leaf miner. How will leaf miner damage alter the size of wood cells appearing in the tree's spring- and summer-wood? Will the annual growth ring be wide or narrow?

Figure 12.31 Aspen leaf miner.

17. A girdled tree is one in which a ring of bark is removed entirely around the trunk of a tree. All that remains is the xylem (**Fig. 12.32**). How will this affect the tree's sugar transport? How will the roots receive sugars from the leaves? Eventually, how will water transport be affected?

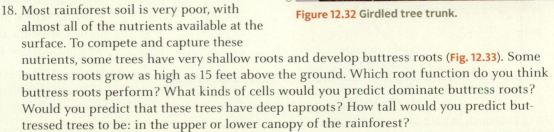

Figure 12.32 Girdled tree trunk.

18. Most rainforest soil is very poor, with almost all of the nutrients available at the surface. To compete and capture these nutrients, some trees have very shallow roots and develop buttress roots (**Fig. 12.33**). Some buttress roots grow as high as 15 feet above the ground. Which root function do you think buttress roots perform? What kinds of cells would you predict dominate buttress roots? Would you predict that these trees have deep taproots? How tall would you predict buttressed trees to be: in the upper or lower canopy of the rainforest?

Figure 12.33 Buttress roots.

Data Analysis

The Growth of Giant Pumpkins

The pumpkin plant is a sprawling vine that produces large leaves. Its flowers, which are pollinated by honeybees, result in the familiar orange-colored fruits. Pumpkin fruits normally grow to be about 9 to 18 pounds for market, but enthusiastic growers breed and grow giant pumpkins that can weigh more than a thousand pounds (**Fig. 12.34**). The current world record is a 1,725-pound Atlantic giant pumpkin, which won the Ohio Valley Giant Pumpkin Growers' annual weigh-off in October 2009.

Figure 12.34 Giant pumpkin.

Data Interpretation

Figure 12.35 shows a graph of the growth of the weight of a single pumpkin over a 90-day period from July 8 to October 5 in the northeastern United States. Use this background information and the data presented in **Figure 12.35** to answer the questions below.

19. The growth of the giant pumpkin over the July to October period occurs in three different stages: early, mid, and late. Describe these stages in terms of when they start and end and how fast the pumpkin grows.

20. Does a pumpkin grow at a constant rate, or does it roughly add the same amount of weight each day?

21. Giant pumpkins are the fastest growing of all fruits. In which time period did it grow the fastest? Approximately how much weight did the pumpkin grow each day from day 40 to day 50? What was its daily weight gain from day 70 to day 80?

Critical Thinking:

22. Some of the largest cells in plants are found in the flesh of ripe fruits such as the pumpkin. Turgor pressure is the process that brings about large plant cells. Which cell organelle is most responsible for the expansion of pumpkin flesh cells?

23. Growth occurs by both cell division and cell expansion. Which of these processes dominates the early phase of the fruit growth? Which of these processes do you think caused the increase in pumpkin weight from day 45 to day 55?

24. Do you think that giant pumpkins show a faster rate of cell division than normal-sized pumpkins?

25. Name at least two environmental factors that strongly influenced the growth rate of this giant pumpkin.

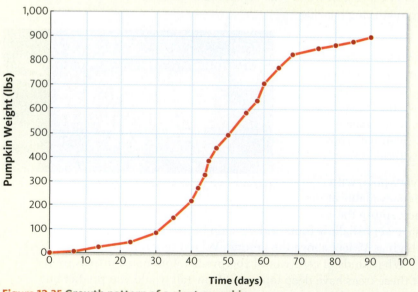

Figure 12.35 Growth pattern of a giant pumpkin.
Data interpolated from graph in Hu DL, Richards P, Alexeev A. 2011. The growth of giant pumpkins: how extreme weight influences shape. Int J Non-Linear Mech, 46, 637.

26. Do pumpkins show an indeterminate or determinate pattern of growth? How does the pumpkin's growth curve compare to that of animal growth?

27. Giant pumpkin growers remove all of the other flowers, fruits, and trailing parts of the vine to increase the size of the pumpkin. Explain why you think this works.

28. Do you think the growth curve of other fruits such as tomatoes, strawberries, and bananas would show the same shape? Why or why not?

Question Generator

Early Springtime on the Forest Floor

The spring mermaid is an annual plant that lives on the floor of temperate deciduous forests across eastern North America (**Fig. 12.36**). Its life history pattern adapts it to live for a short period of time between snowmelt in the spring and canopy closure a month later. During this brief period of time, temperatures generally are cold, with little competition for light, water, and minerals. Once the towering trees develop their leaves, the light reaching the forest floor drops considerably. Mermaid seeds germinate in the winter, but the first leaf doesn't develop until after the ground has thawed in April. It develops a few short roots that are confined to the top few inches of the soil. Flowering begins a few weeks later, and the plant dies in early June.

Below is a block diagram (**Fig. 12.37**) that relates several aspects of the biology of the spring mermaid living on the forest floor in early spring. Use the background information along with the diagram to ask questions and generate hypotheses. We've translated the components of the block diagram into several questions to get you started.

Figure 12.36 Spring mermaid.
(From McKenna MF, Houle G. 2000. Why are annual plants rarely spring ephemerals? New Phytologist 148, 295–302.)

© Carol Gracie

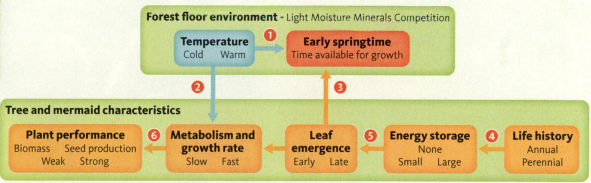

Figure 12.37 Factors relating to the growth of the spring mermaid.
© Cengage Learning 2014

Here are a few questions to start you off.

1. How will a warm spring influence the forest trees and the time available for mermaid growth (Arrow #1)?

2. How do cold temperatures affect a plant's metabolism and growth rate (Arrow #2)?

3. How does the leaf emergence of the trees affect the time available for mermaid growth (Arrow #3)?

4. How much energy does an annual and a perennial store over the winter (Arrow #4)?

5. How is the amount of energy stored related to timing of leaf emergence (Arrow #5)?

6. Do faster-growing plants produce more seeds than slower-growing plants (Arrow #6)?

Use the block diagram to generate your own research question and frame a hypothesis.

© Cengage Learning 2014

13 Plants Functioning in Their Habitat

13.1 Fly soup for the purple pitcher plant

13.2 Plants are adapted to respond to environmental stress

13.3 Plants produce sugars by photosynthesis

13.4 Plants use sugars for energy and as building blocks for growth

13.5 Soils provide plants with water and minerals

13.1 Fly Soup for the Purple Pitcher Plant

The sun has been up for an hour over the New Jersey Pine Barrens as a black fly emerges from the trees and begins buzzing near the edge of the bog in search of its next meal (**Fig. 13.1**). It picks up a promising scent with the sensitive chemoreceptors on its antennae and immediately changes direction to a curious looking set of plants. Zeroing in on the source of the enticing odor, the fly lands on the leaf of a purple pitcher plant *(Sarracenia purpurea)* (**Fig. 13.2a**). The fly has made a deadly mistake.

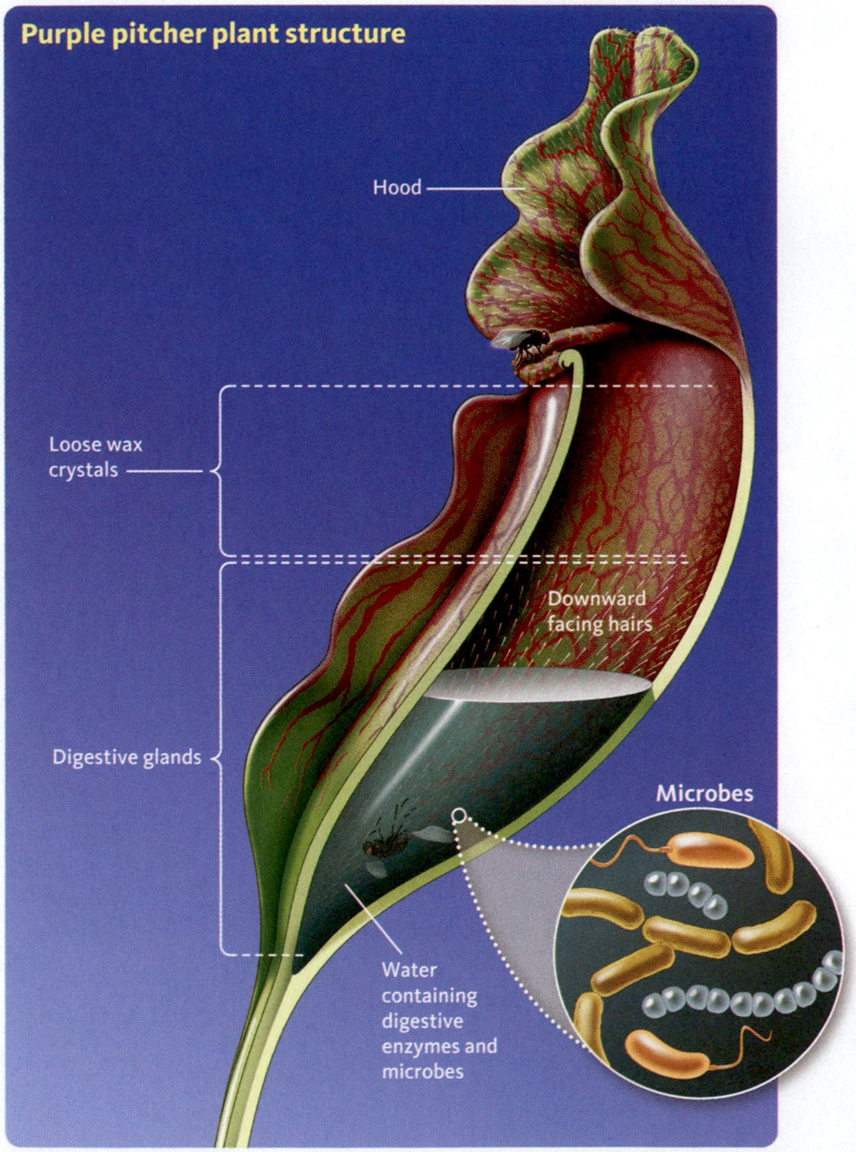

The lip of the pitcher plant leaf is wet with drops of sweet-smelling nectar that the fly ingests using its strawlike mouthparts. The fly has no way of knowing that the nectar contains plant extracts called alkaloids that disorient the fly. Within moments, the fly topples backward, landing in the water reservoir inside the plant (see **Fig. 13.2a**). It struggles to escape, but its path out of the pitcher is blocked by downward-pointing hairs just long enough to prevent the fly from climbing around them. The sides of the pitcher also have a covering of soft wax that easily breaks off and gums up the footpads of the fly, preventing it from acquiring a foothold. This struggle continues for an hour, but eventually the fly tires, falls back, and drowns in the pool of water.

This scenario plays out each hour of the growing season in wetlands across northeast North America, where pitcher plants attract various flies, beetles, midges, and mosquitoes to feed on the spiked nectar. The purple pitcher plant is an evolutionary oddity—one of 630 species of carnivorous plants that obtain some of their nutrients for growth, development, and reproduction by trapping and consuming insects or protozoans.

The adaptations of the purple pitcher plant and other carnivorous plants evolved as a selective advantage for plants growing in nitrogen-poor soils, enabling them to grow in places like wetlands. Nitrogen is critical for building many biomolecules such as proteins, DNA, and the chlorophyll molecules used in photosynthesis. The pitcher plant has the same dietary needs as other plants for minerals such as phosphorus and potassium that it uses to construct cells and aid in metabolic activities. Most plants rely exclusively on their roots to obtain these elements directly from the soil, but in nutrient-poor soils, there are simply not enough of these important elements to support the growth of large plants.

Purple pitcher plant structure

Hood

Loose wax crystals

Downward facing hairs

Digestive glands

Microbes

Water containing digestive enzymes and microbes

Figure 13.1 The structure of the purple pitcher plant. The purple pitcher plant is named for the unique color and shape of its leaf—the defining feature of the pitcher plant, in both form and function. What seems to be a stalk is actually a single, large modified leaf that folds in on itself, flaring out at the top. The plant produces nectar from glands along the rim that attracts insects. Curious insects that venture further into the plant are trapped by loose, waxy cells and downward-facing hairs. Ultimately, digestive enzymes and microbial communities in the fluid combine to reduce prey into smaller molecules—some of which are absorbed by the plant to provide critical nutrients.
© Cengage Learning 2014

a.

Bill Beatty/Visuals Unlimited, Inc.

b.

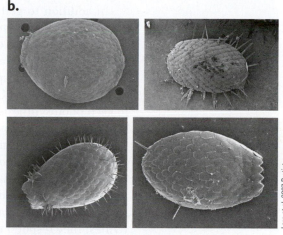

Lara et al. 2007 Protist

Figure 13.2 The purple pitcher plant supports a community of microbes. (a) When mature, the leaf of the pitcher plant forms a vase that catches rainwater, creating a pool of water at its base. (b) A small collection of the kinds of microbes that form one part of a complex food chain living in the water at the base of the pitcher plant leaf.

This environmental pressure has led to unique adaptations to trap and use the nutrients found in insects.

Nutrient shortages cause problems, and plants typically react to these situations by limiting growth or altering the timing of activities such as reproduction. Carnivorous pitcher plants have evolved the ability to obtain elements from the decomposing bodies of insects, an advantageous way to avoid stresses that might otherwise limit their success. Pitcher plants still have roots that grow through the soil and obtain other minerals and water to help satisfy the plant's dietary requirements.

The highly modified leaves of the pitcher plant are very distinctive. The vaselike shape occurs during the early development of the leaf, which folds over and fuses together as it begins growing from the plant's base. As the leaf enlarges, the epidermal cells on the inside sprout downward-facing hairs and secrete the soft wax that prevents insects from escaping. As the leaf matures, the vaselike shape allows the plant to begin collecting water, and the lip of the leaf begins to secrete the tasty spiked nectar.

The pitcher plant doesn't eat the insects that fall into its trap; instead, the small pool of water at the base of the pitcher creates a perfect micro-habitat for a community of organisms that exist in a mutualistic relationship with the pitcher plant. The plant relies on this diverse community of bacteria, protozoans, and mosquito and midge larvae to help provide it with nutrients (**Fig. 13.2b**). Members of this miniature food web feed on one another and on the bodies of decaying insects that drown in the pool of water. Fly larvae are the top predators, feeding on the dead insects and breaking them down into smaller molecules and elements, some of which are absorbed by the plant to fuel growth and development. Feces and other waste products from the organisms also wash into the pool, providing the plant with more usable forms of nitrogen and phosphorus.

Similar to the insects it traps or the microorganisms it supports in its fluid, the pitcher plant is a living organism that needs sugars. The trapped fly was trying to obtain sugar-rich nectars to help fuel its metabolic pathways to produce ATP. To generate ATP, pitcher plants use their mitochondria to process sugar molecules into energy during aerobic metabolism.

In addition to getting some nutrients from dead insects, the purple pitcher plant also captures light energy during photosynthesis. Photosynthesis is an important energy-transformation pathway for most plants. Using photosynthesis, plants construct energy-rich molecules like sugar compounds that are transported through vascular tissue to cells throughout the body of the plant. Growing in warm sunny places allows pitcher plants to keep photosynthesis running, supplying abundant sugars.

Pitcher plants provide an example of the variety of ways organisms adapt to less than ideal conditions. Since plants are rooted to a single location, they have evolved unique structures and adaptations to survive everchanging environmental conditions such as light, temperature, and water supplies. Environmental changes pose problems that interfere with the ability of plants to continue growth, development, and reproductive activities; they respond to change by alternating their life processes, like allocating more resources to help reach sunlight. Since soils, made from different types of parent rock, are highly variable from one location to the next, plants grow roots, extending them out into the soil to acquire the mineral elements and water needed for the plant's survival.

In the chapter ahead, we will build on the plant structure from the previous chapter to help us take a closer look at how plants function in diverse habitats. We'll focus on how plants deal with environmental stresses created from events like floods or fires. We'll show you how plants build energy-rich carbohydrates during photosynthesis and link those molecules to ATP production through aerobic metabolism. The final section will help you understand the interaction between plants and the soils they grow in. Let's begin by taking a closer look at how plants have become adapted and respond to stresses posed by the environments they live in.

Figure 13.3 Poison ivy vine. Poison ivy makes use of its ability to cling to the trunks of trees as an avenue of growth to reach sunlight from the shade of the overhead trees.

Susan Leavines Harris/Photo Researchers, Inc.

Environmental stress—an environmental factor that reduces a plant's ability to carry out metabolic activities at optimal levels.

13.2 Plants Are Adapted to Respond to Environmental Stress

We've seen how the pitcher plant has successfully adapted to nutrient-poor soils. In very similar ways, other plants, including poison ivy—have adapted to a variety of environmental challenges, including low light conditions, to thrive in less than ideal habitats. Like many plants, poison ivy grows best where light is abundant and shaded areas are limited. However, in the pine forests of New Jersey, the dense tree canopy intercepts up to 90 percent of the sunshine before it reaches the forest floor. For "light-hungry" plants, this cool, moist habitat of intense shade is less favorable for growth (**Fig. 13.3**).

A lack of light is one example of an **environmental stress**—an environmental factor that reduces a plant's ability to carry out metabolic activities at the most favorable levels. As with animals, plants must respond to environmental conditions by altering their growth patterns and metabolism. In *response* to the stress of deep shade, the poison ivy vine directs more energy to shoot growth in an effort to reach the available light. In the section ahead, we'll explore ways that plants cope with environmental stresses in their habitat. This will help you take a "plant's-eye" perspective on the many hazards they face every day.

Environmental stresses affect plant growth and survival

Plants spend their lives growing in one location, and when a resource is in short supply, it creates stress; to survive they must respond by altering their patterns of growth, development, or reproduction. Although each species is adapted to live within its *ecological niche,* or range of environmental conditions and resource availability, no single niche supplies these resources in unlimited amounts. Environmental stress may involve factors such as light, temperature, water, humidity, or minerals. Recall from Chapter 4 (**Table 4.2**) that plants, like animals, require a wide variety of resources such as light, water, gases such as CO_2 and oxygen, and a diverse range of minerals. Since resources are unevenly distributed across the landscape, their availability in any one location helps to explain the diversity, abundance, and distribution of plants in an ecosystem.

A wide range of evolutionary adaptations allow plants to survive and respond to environmental stress (for example, the ability of poison ivy to redirect resources to stem growth in low light conditions). As long as the stress does not reach lethal levels for long periods of time, the plant usually recovers. Plants respond to small changes in temperature, moisture, or light availability by adjusting which tissues receive resources. For example, during times of low water availability, plants will redirect resources to root growth to seek out moisture in the soil. What are lethal stress levels for plants? It depends on the developmental stage of the plant, the season, the type and extent of the environmental stress, and the genetics of the plant. For instance, because very young plants lack a well-developed root system, they are extremely sensitive to water or mineral nutrient shortages. Stress caused by such limitations during this phase of development will often kill young plants. Mature plants with tissues that are heartier and contain stored nutrients, such as the stems of woody trees, are less likely to be damaged by short-term limitations of resources. By altering their growth patterns and shifting resources to root growth, mature plants may survive repeated occurrences of high-stress events like water shortages.

Environmental stress can also occur from an oversupply of resources, as well as from shortages. In late spring in northern climates, low-lying land often experiences temporary flooding from winter snow melt and spring rains. Excess water is unable to quickly drain downriver, and the ground rapidly becomes saturated, meaning that the air spaces in the soil are now occupied by water, causing the diffusion of oxygen to be slowed. Recall from Chapter 9 that a lack of oxygen causes animals and plants to switch to anaerobic pathways, which produce far less energy. While aquatic plants are able to withstand long periods using anaerobic metabolism, some fast-growing plants like corn that have high energy needs can survive for only a few days under saturated soil conditions (**Fig. 13.4**). Willows and other plants that grow in river bottoms are well adapted to low soil-oxygen conditions and can survive for months with their root zone flooded.

Figure 13.4 Flooded soils lack oxygen for plant metabolism. Corn is poorly adapted to survive even short periods of flooding; unless the water level recedes in a few days, this field of corn will die.

Plants respond to many different stresses

Extended periods of water shortages, or *droughts,* are a common occurrence in many ecosystems. Drought threatens the survival of most plants by causing metabolic reactions to slow down or stop. Some species of mosses and spike mosses can tolerate extremely low levels of moisture; however, if you allow your houseplant to go without water for a long time, it wilts and eventually dies (**Fig. 13.5**). Without water, metabolic pathways like photosynthesis shut down, and the plant cannot produce the high-energy sugar molecules necessary to fuel ATP production and construct new cells for growth or reproduction. In severe cases, plants will begin to use water stored in their vacuoles, causing them to wilt as the cells lose structural support.

During the growing season, plants are highly susceptible to cell damage from freezing temperatures. Rapid drops in temperatures sometimes occur early or late in the growing season and quickly cause considerable damage or even death to healthy plants. Like animals, perennial plants *acclimate,* or gradually respond to changing temperature patterns by altering their cellular chemistry. As autumn temperatures steadily begin to drop over several months' time, plants adapt by packing sugars into their cells, which actually lowers their freezing point. This adaptation increases the ability of their cells to survive low temperatures. A sudden drop in temperature that creates ice or frost doesn't allow plants to acclimate, and the water-filled cells in tender structures like leaves and flowers are typically killed.

Fires create an ecological *disturbance* by partially, or in some cases completely, removing plants within a habitat (**Fig. 13.6**). You may recall from Chapter 4 that some plants are well adapted to survive the intense heat of fire. Most species of pine trees have evolved in habitats where fires are relatively frequent. For example, the pitch pine has thick, insulating bark that allows it to easily survive the small fires that break out in the Pine Barrens of New Jersey. Such fire-adapted trees respond quickly to the altered postfire environment, which includes more light reaching the ground, less competition for water, and exposed mineral soils. Undamaged seeds that have lain dormant respond by sprouting, while the surviving roots of existing plants like the pitcher plant quickly produce new shoots. In a few short weeks, new green plants cover the formerly scorched earth.

The environment of a plant's habitat has a tremendous impact on plants and their communities. Long-term environmental patterns of factors like rainfall, temperature, and light availability shape plant structure and require plants to respond to a wide range of changing conditions. Natural selection has led to the development of unique adaptations that have allowed plants to thrive in the entire range of ecological niches. It is important to consider that adaptations don't insulate plants from stress; events like long-term flooding or fire help shape the survival and distribution of plants. In the next section, we'll explore photosynthesis as a means of energy transformation.

© Nigel Cattlin/Alamy

Figure 13.5 Plants wilt when water is scarce. Without water, herbaceous plants lose the ability to remain upright and wilt.

Konstantin Mironov/Shutterstock.com

Figure 13.6 Fires create an ecological disturbance. Grass fires typically burn at much lower temperatures than forest fires because there is far less fuel available to create the heat of a large wildfire. Although the aboveground biomass of grasses might be removed by the fire, they can quickly recover. Pine trees with thick bark are not threatened by these fires.

- Environmental conditions are not fixed, and plants must adapt and respond to these changes.

- Changes that occur slowly allow plants to acclimate to the new conditions.

- Fire is a major force of disturbance, creating new environments for plants to exploit.

1. How is a response different from an adaptation?
2. Why does a houseplant die when left unwatered for an extended period?
3. Explain how plants respond to resource limitations.
4. Describe why a late spring frost might be deadly to a plant.
5. The resurrection plant is a type of spike moss that is descended from the earliest plants, which lived in highly unpredictable habitats with dramatic swings in moisture availability. During a drought, the plant loses moisture and shrivels up into a compact ball (**Fig. 13.7**). When the rains return, the plant tissues rehydrate, and the plant resumes normal activities. Describe the reasons this represents either an adaptation or a response activity or both.

© Science Photo Library/Alamy

Figure 13.7 The resurrection plant. The resurrection plant is a type of club moss that lives in desert habitats. This plant is known for its ability to survive almost complete drying out; during dry weather in its native habitat, its stems curl into a tight ball and uncurl again when exposed to rains that wet the soil.

13.3 Plants Produce Sugars by Photosynthesis

As you hike through the woods, you notice a rich understory of plant life on the forest floor, including small trees, vines, ferns, and herbs such as wild sarsaparilla. Sarsaparilla is a perennial herb that is adapted to living in the shade produced by the tall trees above it (**Fig. 13.8**). With a leafy roof over their heads blocking direct sunlight, sarsaparilla plants must efficiently capture the small amount of light that makes its way to them. Oriented like an umbrella, their large leaves are able to capture light at different angles as the sun moves across the sky. Although they are adapted to live on a low light "budget," they must have a small amount of light to produce the sugars needed to fuel their growth, reproduction, and survival. How do plants transform light energy into the chemical energy of sugars?

In the section ahead, you will find that, just as with aerobic metabolism, these transformations occur at the lowest levels of the hierarchy of organization—cells; molecules; atoms; and even the subatomic particles, the protons and electrons. Inside the leaf cell specialized green pigment molecules capture light and efficiently transfer the energy to other chemi-

Photo by Robert Noyd

Figure 13.8 Wild sarsaparilla thrives in a shady environment. Wild sarsaparilla is at home in the shade of an overhead hardwood canopy. It is adapted to use the fleeting amounts of sunlight that directly strike its leaves as the sun moves across the sky. During photosynthesis O_2 is released from the leaf, while CO_2 moves into leaf tissues. Recall that during aerobic respiration O_2 is needed in the leaf for use in metabolism, while CO_2 is released. In mature plants there is a net release of O_2 and a net uptake of CO_2.
© Cengage Learning 2014

Figure 13.9 The two stages of photosynthesis.
Photosynthesis consists of two major stages linked to one another by their inputs, locations, and need for resources. The key is the availability of sunlight, which provides the energy to drive the production of sugars. Intermediate molecules produced during photosynthesis carry energy in their chemical bonds. This energy is used to rearrange other molecules in producing sugars.
© Cengage Learning 2014

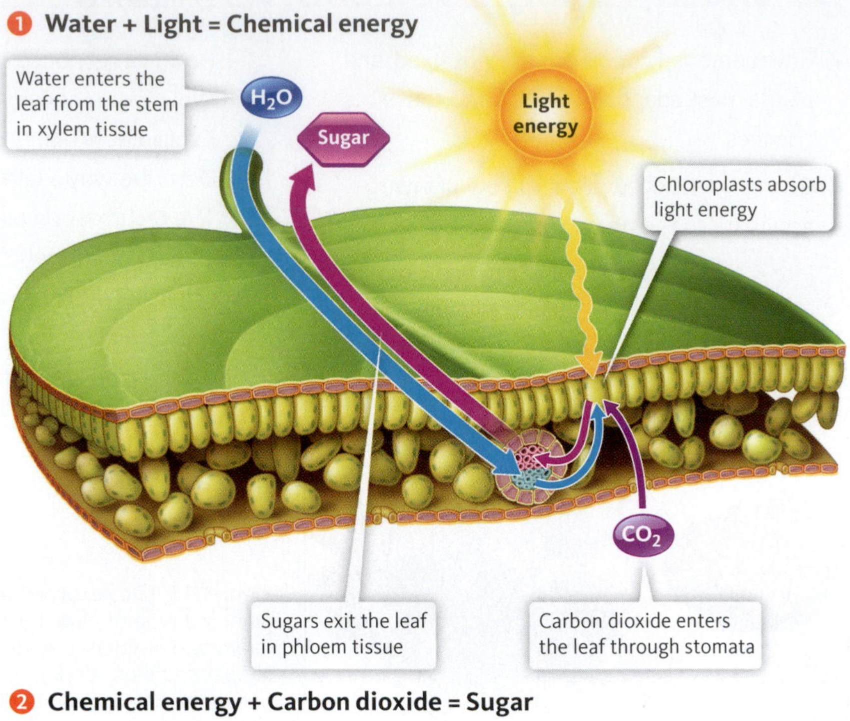

1 Water + Light = Chemical energy

Water enters the leaf from the stem in xylem tissue

H_2O

Sugar

Light energy

Chloroplasts absorb light energy

Sugars exit the leaf in phloem tissue

Carbon dioxide enters the leaf through stomata

CO_2

2 Chemical energy + Carbon dioxide = Sugar

cals and ultimately to the bonds in a sugar molecule. This chemical process is called photosynthesis.

To perform photosynthesis and build energy-rich sugar molecules, land plants must also obtain two other essential ingredients—water from the soil and carbon dioxide from the atmosphere (**Fig. 13.9**). The photosynthesis process includes dozens of sequential chemical steps, each accomplished with the aid of a specific enzyme. Our goal in this section is not to present all the complex chemical reactions involved but rather to present the important steps of the process that allow you to connect how plants make a living in the habitats where you find them—even on the forest floor.

The photosynthesis "machine" is the chloroplast

All green parts of a plant—green flower pieces; unripened fruits; herbaceous stems; and leaves, the main photosynthetic organs—perform photosynthesis. All of these parts contribute to the overall sugar "income" of plants.

Let's zoom in on a sarsaprilla leaf to find the photosynthetic "machinery" where the chemical reactions of photosynthesis take place. Underneath the thin, waxy epidermis of the leaf are layers of cells that contain organelles called chloroplasts (**Fig. 13.10a**). A typical leaf cell has between 20 and 100 chloroplasts, and one square inch of leaf surface has an estimated 300 million chloroplasts. You can imagine the immense power available to make sugar when a large maple tree has as many as 100,000 leaves. During the day, chloroplasts are continually moving within the cell to maximize the collection of light energy from the sun. The wide leaves and umbrella shape of sarsaparilla expose the leaf surface to capture as much light as it can. This adaptation is a characteristic of many plants living in the understory.

Inside a chloroplast are hundreds of interconnected membranes stacked like coins (**Fig. 13.10b**). Zooming in further, we see large clusters of

Photosynthesis—a metabolic process common to plants and certain other organisms whereby specialized green pigment molecules capture light and efficiently transfer the energy to other chemicals and ultimately to the bonds in a sugar molecule.

Chloroplasts—the organelles where the chemical reactions of photosynthesis take place.

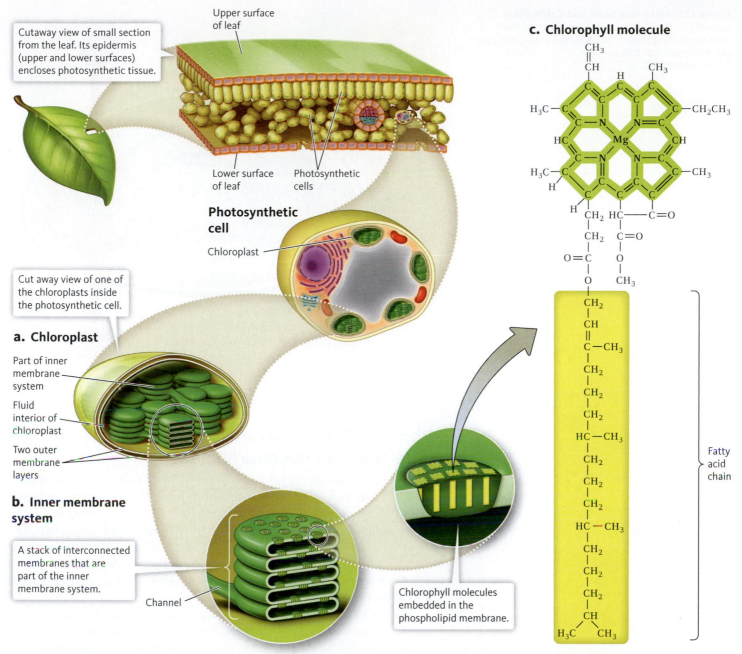

Upper surface of leaf

Cutaway view of small section from the leaf. Its epidermis (upper and lower surfaces) encloses photosynthetic tissue.

Lower surface of leaf

Photosynthetic cells

Photosynthetic cell

Chloroplast

Cut away view of one of the chloroplasts inside the photosynthetic cell.

a. Chloroplast

Part of inner membrane system

Fluid interior of chloroplast

Two outer membrane layers

b. Inner membrane system

A stack of interconnected membranes that are part of the inner membrane system.

Channel

Chlorophyll molecules embedded in the phospholipid membrane.

c. Chlorophyll molecule

Fatty acid chain

Figure 13.10 Chloroplasts are the organelles of photosynthesis. Inside the mesophyll tissue of the leaves are cells that contain chloroplasts, the organelles where photosynthesis takes place.
© Cengage Learning 2014

green pigments called chlorophyll (**Fig. 13.10c**). A chlorophyll molecule is shaped like a tennis racket with its fatty acid chain. The head of the racket contains four nitrogen atoms surrounding a central atom of magnesium. To build the billions of chlorophyll molecules in its leaves, a plant must acquire nitrogen and magnesium from the soil.

The chlorophyll molecules perform two functions: (1) they absorb light, and (2) they transfer that energy to other compounds. Chlorophyll pigments are adapted to absorb a specific range of light wavelengths similar to the ones we see with our eyes. Chlorophyll absorbs the violet-blue and orange-red light from the sun and reflects more green light, a characteristic that gives leaves their green color. Shade-adapted plants like wild sarsaparilla have slightly larger and fewer chloroplasts, but they contain higher concentrations of chlorophyll than leaves growing in full sunlight.

Chlorophyll—green pigments embedded in the surface of chloroplasts. A chlorophyll molecule is shaped like a tennis racket with its fatty acid chain. The chlorophyll molecules perform two functions: (1) they absorb light, and (2) they transfer that energy to other compounds.

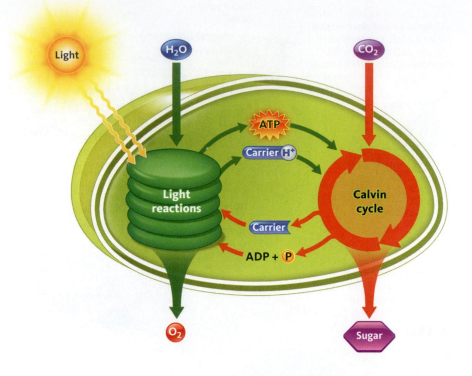

Photosynthesis requires raw materials obtained from the environment

Now that we have identified the photosynthetic structures in the chloroplast, we can look at how plants acquire the necessary energy and raw materials from their environment to build sugars. Sugars are organic macromolecules called carbohydrates, which are constructed from carbon, hydrogen, and oxygen atoms. *Carbon* and *oxygen* are acquired from carbon dioxide (CO_2) in the atmosphere and the *hydrogen* from water (H_2O) in the soil. Energy and a variety of enzymes are needed to assemble the atoms into larger molecules.

The chemical reactions that build sugars occur in two stages (**Fig. 13.11**). The first stage is called the light reactions because the absorption of light by chlorophyll drives the process. The chemical products from this stage, in turn, drive the second stage, called the Calvin cycle. Like an assembly line, the Calvin cycle is dependent on the light reactions for a continual supply of its parts.

In most plants (90 percent), both stages of photosynthesis take place during the day, when light is available, and shut down at night. This cycle presents a unique challenge for the plant, because to acquire carbon dioxide from the air, the leaf opens its pores, or stomata. As a result, water vapor evaporates rapidly from the leaf into the atmosphere. To keep photosynthesis running, a continuous supply of water must be delivered to the leaves. The water comes from the uptake of water from the soil through the roots. Thus the ability of roots to absorb water plays a major role in photosynthesis. The stage is now set with both the "machinery" (chloroplasts), the energy (light), and the raw materials (CO_2 and H_2O) to come together. We now begin with the first step—the light reactions.

Light reactions—the first of two stages in photosynthesis characterized by a complex set of reactions in which the absorption of light energy is used to transform water and other resources into ATP, and a hydrogen carrier molecule is used during the second stage.

Calvin cycle—the second stage of photosynthesis, a set of linked chemical reactions that do not directly require sunlight but instead rely on the products of the light reaction.

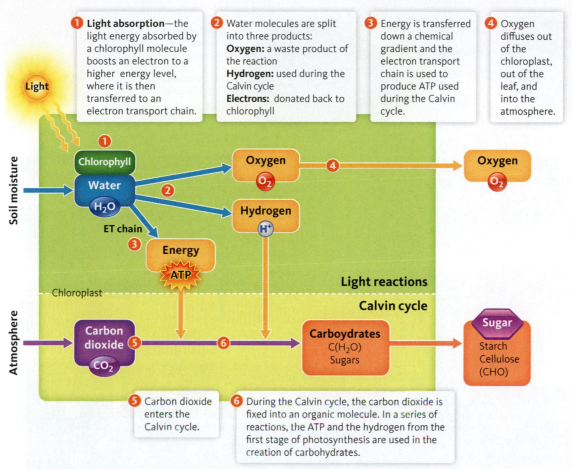

① **Light absorption**—the light energy absorbed by a chlorophyll molecule boosts an electron to a higher energy level, where it is then transferred to an electron transport chain.

② Water molecules are split into three products:
Oxygen: a waste product of the reaction
Hydrogen: used during the Calvin cycle
Electrons: donated back to chlorophyll

③ Energy is transferred down a chemical gradient and the electron transport chain is used to produce ATP used during the Calvin cycle.

④ Oxygen diffuses out of the chloroplast, out of the leaf, and into the atmosphere.

⑤ Carbon dioxide enters the Calvin cycle.

⑥ During the Calvin cycle, the carbon dioxide is fixed into an organic molecule. In a series of reactions, the ATP and the hydrogen from the first stage of photosynthesis are used in the creation of carbohydrates.

Figure 13.12 The steps of photosynthesis. This block diagram can help you follow the complex set of processes involved in photosynthesis. You can see the links between the two stages and the required inputs and outputs.
© Cengage Learning 2014

The light reactions absorb light, split water, and produce hydrogen and ATP

Figure 13.12 shows the molecules involved in the light reactions stage. Three major events mark this stage: (1) the harvesting of light energy by chlorophyll, (2) the splitting of water, and (3) the transfer of electrons through an electron transport chain to produce ATP and a hydrogen carrier.

Light harvesting is the process of capturing light energy and converting it to chemical energy so that it can be stored in biological molecules (**Fig. 13.13**). Clusters of chlorophyll molecules absorb light energy. The light energy is so strong that it causes a chlorophyll molecule to lose an electron, which is then passed through a chain of reactions. This lost electron must be replaced, and this is where water comes into the story.

Water enters a large enzyme complex on the inner chloroplast membrane, where its chemical bonds are broken, splitting the molecule into hydrogen and oxygen atoms. The hydrogen atoms are then stripped of their electrons, which are used to replace the chlorophyll's lost electron to keep the process going. The oxygen from the water is released and moves out of the leaf into the atmosphere where it becomes the air that we breathe. What happens to the electron that passed from water to chlorophyll to the electron transport chain? What happens to the hydrogen atoms split from water?

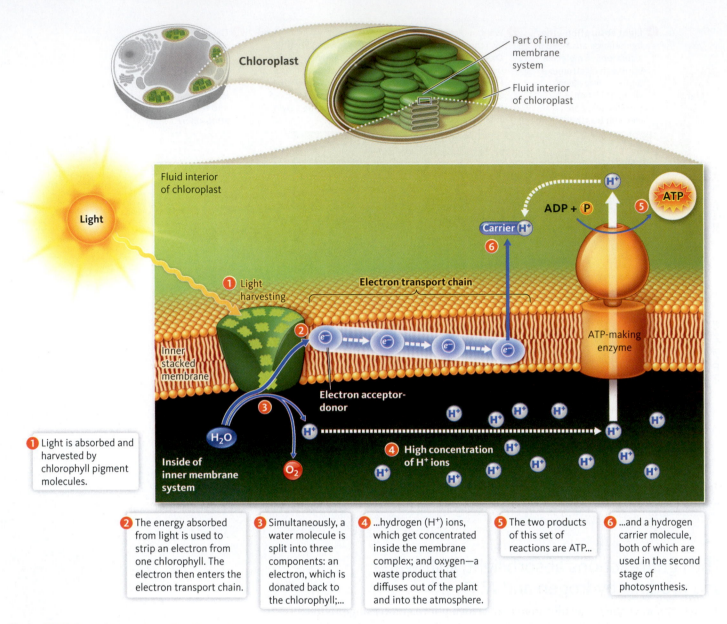

Part of inner membrane system

Chloroplast

Fluid interior of chloroplast

Fluid interior of chloroplast

Light

1 Light harvesting

Electron transport chain

ADP + P

Carrier H⁺ **6**

5 ATP

2 Electron acceptor-donor

ATP-making enzyme

Inner stacked membrane

3

H₂O

H⁺

Inside of inner membrane system

O₂

4 High concentration of H⁺ ions

H⁺ H⁺ H⁺ H⁺ H⁺ H⁺ H⁺ H⁺ H⁺ H⁺ H⁺ H⁺

1 Light is absorbed and harvested by chlorophyll pigment molecules.

2 The energy absorbed from light is used to strip an electron from one chlorophyll. The electron then enters the electron transport chain.

3 Simultaneously, a water molecule is split into three components: an electron, which is donated back to the chlorophyll;...

4 ...hydrogen (H⁺) ions, which get concentrated inside the membrane complex; and oxygen—a waste product that diffuses out of the plant and into the atmosphere.

5 The two products of this set of reactions are ATP...

6 ...and a hydrogen carrier molecule, both of which are used in the second stage of photosynthesis.

Figure 13.13 The light reaction. This figure shows a cross section from the inner membrane complex, where the set of reactions involved in the first stage of photosynthesis takes place.
© Cengage Learning 2014

In a process similar to aerobic metabolism in the mitochondria, the chloroplast has a collection of large proteins called the *electron transport chain* (see Section 9.6). As the electron stripped from the chlorophyll is passed along the chain, small amounts of its energy are used to create a high concentration of hydrogen atoms on one side of the inner membrane. The hydrogen ions now flow "downhill" through an enzyme complex that captures the energy to build the energy carrier molecule, ATP. A hydrogen carrier molecule that cycles between the two stages of photosynthesis then picks up the hydrogen atoms and delivers them to the Calvin cycle.

Figure 13.12 serves as a map to follow the complex events of photosynthesis. The light reactions are represented in Steps 1 through 4, shown in red circles. As a preview of the Calvin cycle, notice that carbon dioxide (CO_2) makes an appearance where it is combined with hydrogen (H) to make a carbohydrate (CHO). Here's how plants use the excess carbon dioxide humans place into the atmosphere.

The Calvin cycle reactions convert carbon dioxide to sugars

The second stage of photosynthesis is the *Calvin cycle,* a set of linked chemical reactions that do not directly require sunlight but instead rely on the products of the light reaction (see **Fig. 13.12**). In the chloroplast, this stage occurs in its fluid-filled regions surrounding the membrane stacks.

Let's begin the Calvin cycle by introducing one of the most important enzymes on Earth. *Rubisco* is the abbreviated name of an enzyme that captures atmospheric carbon dioxide and incorporates it into a five-carbon molecule needed to drive a chemical reaction known as *carbon fixation* (Step 5 in **Fig. 13.12**). Because this stage uses the carbon dioxide from the air to create biological molecules, this reaction is arguably the most important chemical reaction on Earth. From an ecological perspective, rubisco single-handedly moves carbon from the atmosphere into the food chain. In fact, every molecule in your body that contains carbon passed through this enzyme and reaction! This reaction is so successful that photosynthetic organisms remove roughly 10 percent of all atmospheric carbon per year. Although this is a tremendous amount of carbon, in the last 50 years plants have been unable to keep up with the increased levels of carbon emitted into the atmosphere through the burning of fossil fuels.

After rubisco has captured carbon dioxide from the atmosphere, the following steps in the Calvin cycle use the energy in ATP and the hydrogen from the light reaction to rearrange intermediate molecules into a carbohydrate, which then becomes the building blocks of sugars, cellulose for cell walls, and starch to store energy (**Fig. 13.14**). It is important to recognize that for autotrophs like plants, the sugars can be combined with other substances, such as minerals from the soil, to build amino acids, nucleic acids, and in fact every organic molecule used in living things.

These carbohydrates are critical for the growth of the plant. As illustrated in **Figure 13.14**, some by-products remain after these reactions cycle back to the inner membrane for reuse in the light reaction, reducing the plant's need to continually replace materials necessary for the chemical reactions.

The overall reaction for photosynthesis is $H_2O + CO_2 \rightarrow$ carbohydrate $+ O_2$. Keep in mind that the energy from the sun is required to drive the reaction. To assess whether you understand the flow of molecules from the environment through the photosynthetic process, return to **Figure 13.13** and account for the fate of each of the chemical substances that enter and leave the process. For example, be sure you can follow an electron from the hydrogen in a water molecule all the way through the light reactions and the Calvin cycle to a sugar, starch, or cellulose molecule. This will help you reach the goal of this section of connecting the moisture in a plant's habitat to how it is used in photosynthesis.

Adaptations allow plants to live in warm, dry habitats

Photosynthesis provides an illustration of biochemical as well as structural adaptations. As you've learned, acquiring enough resources for growth and survival is a difficult task for plants. Water loss during gas exchange in support of photosynthesis is a daily predicament experienced by plants growing in habitats where water is especially limited. Roughly 15 percent of plants have evolved alternative metabolic pathways that increase their

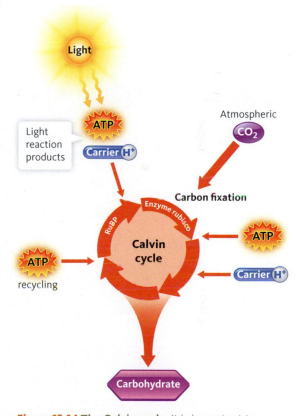

Figure 13.14 The Calvin cycle. It is important to remember the direct link between the two stages of photosynthesis. The ATP and hydrogen carrier are used during the Calvin cycle. The Calvin cycle produces sugars used to support plant life activities and some materials that are recycled back to the light reaction and the beginning of the cycle.
© Cengage Learning 2014

Figure 13.15 Two variations on the photosynthetic metabolic pathways allow plants to thrive in warm, dry climates. The adaptations change either the location or timing for the Calvin cycle. The light reactions occur in just the same manner as in other plants; however, the Calvin cycle may occur in other tissue types or at different times. Plants that use these pathways are well adapted to warm, dry climates. Minimizing their water loss provides them an advantage over other plants in these habitats.

© Cengage Learning 2014

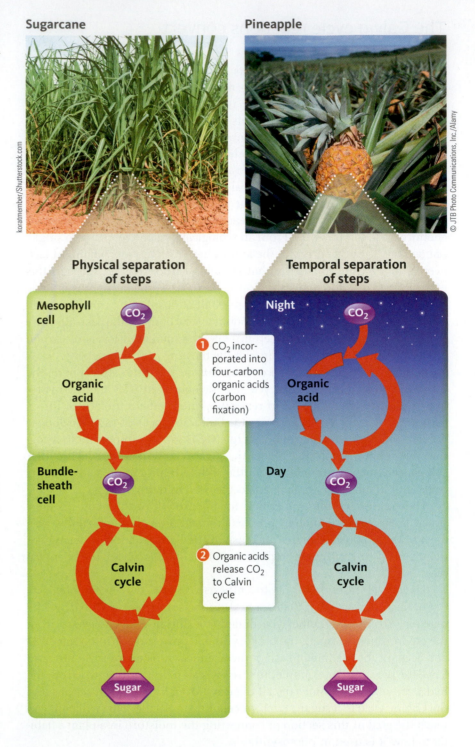

Sugarcane

Pineapple

koratmember/Shutterstock.com

© JTB Photo Communications, Inc./Alamy

Physical separation of steps

Mesophyll cell

CO_2

Organic acid

Bundle-sheath cell

CO_2

Calvin cycle

Sugar

Temporal separation of steps

Night

CO_2

Organic acid

Day

CO_2

Calvin cycle

Sugar

❶ CO_2 incorporated into four-carbon organic acids (carbon fixation)

❷ Organic acids release CO_2 to Calvin cycle

survival and growth rates and dramatically reduce water losses (**Fig. 13.15**) in warm, dry climates. The ability to photosynthesize more efficiently under warm, dry conditions gives these plants a significant competitive advantage over other plants.

To limit water loss in some plants, the Calvin cycle is separated either *physically* from the light reaction, or the two stages are separated by *time*. For instance, in many common garden weeds, corn, and sugar cane, the Calvin cycle is physically separated from the light reaction, taking place outside the chloroplast in adjacent cells called *bundle-sheath cells* (see **Fig. 13.15**). The wide distribution of plants that use this type of photosynthesis reflects the competitive advantage of this adaptation in certain habitats; for example, 50 percent of grasses, which tend to grow in warm, dry climates, use this alternative photosynthetic pathway.

Another adaptive photosynthetic pathway occurs in about 10 percent of all vascular plants, including cacti, pineapple, and plants growing in desert climates. In this pathway, the light reactions and the Calvin cycle occur within the same cells. However, the light reaction is separated from the Calvin cycle in *time*. The light reactions continue during daylight, whereas the carbon fixation during the Calvin cycle only begins at night when these plants open their stomata to take in carbon dioxide. This separation of the light reaction from the Calvin cycle allows stomata to remain shut during the day, providing a tremendous benefit in conserving scarce water resources when temperatures are highest and humidity levels the lowest.

13.3 check + apply YOUR UNDERSTANDING

- Plants transform light energy into chemical energy in the bonds of carbohydrate molecules by the process of photosynthesis.

- Photosynthesis is a two-stage process, with the stages interdependent on one another.

- About 15 percent of all plants have evolved an adaptive photosynthetic pathway that minimizes water loss and makes them better able to survive in warm, dry habitats.

1. What are the two stages of photosynthesis, and how do they interact?

2. Explain how a plant absorbs light energy from the sun. To what type of light are plant photosynthesis pigments sensitive?

3. What are the three products produced when a water molecule is split during the light reaction stage of photosynthesis? Where do they end up?

4. You are able to observe a plant growing in the hot, dry climate of central Arizona. Using a gas detection meter, you determine it is taking in carbon dioxide during the night and releasing oxygen during the day. Speculate about the photosynthetic pathway employed by this plant.

5. Foresters are conducting experiments using elevated carbon dioxide levels in various forested regions of the country. These experiments mimic the potential projected increases in carbon dioxide from humans burning fossil fuels. Speculate about the impact on tree growth in these experiments.

13.4 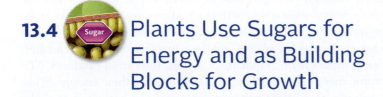 Plants Use Sugars for Energy and as Building Blocks for Growth

The edge of a northern Iowa wetland in springtime is an interesting place to experience ecology on the prairie. Migratory birds have started to return, and the sight of new cattail leaves rustling in the wind is a sure sign that the prairie wetland is alive with new growth. Although most plants can't survive for long in wet soils, the common cattail is well adapted to life in the oxygen-poor wetland soils (**Fig. 13.16**). In fact, cattails thrive in

Figure 13.16 Cattails. Cattails are a common wetland plant. They have adapted in ways that allow them to outcompete other plants in wet soils.

Josiah J. Garber/Shutterstock

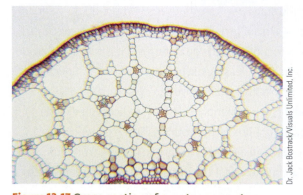

Dr. Jack Bostrack/Visuals Unlimited, Inc.

Figure 13.17 Cross section of a water arum stem. Cross section of a water arum or sweet flag stem showing the air chambers (aerenchyma tissue) typical of aquatic plants. Enlarged 100 times normal size.

water up to 2 feet deep, growing nearly 10 feet tall, a characteristic that makes them highly competitive in wetland habitats.

Plants require sugar and oxygen to fuel their activities. In fact, between a third and half of all sugars made in the leaves are transported to the roots, which are growing and absorbing minerals. Getting oxygen into actively growing root cells is especially challenging for aquatic plants that live in wet and flooded soils. Oxygen dissolved in water moves 10,000 times slower than in air, making it difficult for plant roots to get enough of this critical resource. To avoid this problem, cattails and other wetland plants are adapted with air chambers in their stems and roots, which serve as pathways for oxygen from the atmosphere to reach tissues below the waterline (**Fig. 13.17**).

As springtime temperatures rise and days lengthen, cattails begin rapid cellular division producing the new season's growth. The broad, stiff leaves of the cattail are a bright green color and provide a large surface area for photosynthesis to take place. The sugar molecules produced during photosynthesis move through phloem tissue to meristems in the roots and the leaves. The energy the cattail acquires through photosynthesis is spent to do the work of metabolism. Plants use the chemical energy in the chemical bonds of sugar molecules for growth, maintenance, reproduction, and storage. To support these energy-requiring activities, the cattail produces ATP through aerobic metabolism. Although plants acquire their energy differently than animals, they share very similar metabolic pathways to make ATP. In this section, we will follow the path of sugar from the chloroplast, where it is made, to the mitochondria in the cells, where it is used.

Sugars are transported throughout the plant body

Sugar molecules produced during photosynthesis are relatively abundant in the mature, green leaves of the cattail. The mature leaves represent a *source* of energy-containing sugars available for parts of the plant that have high energy demands or the ability to store sugars for future use (**Fig. 13.18**). In contrast, young expanding leaves, buds, flowers, and fruits are energy *sinks*, or parts of the plant that use sugar (**Fig. 13.19**).

To transport the sugars from sources to sinks, plants use their network of phloem tissue. Recall from Chapter 12 that phloem cells are living tissues with thin primary cell walls that transport sugars in solution, called *sap*. Water makes up a significant portion of the sugar-rich plant sap, which moves into the phloem tissues via osmosis from higher to lower concentration gradients and transports the products of photosynthesis around the plant. Without enough water, plants are limited in their ability to produce sap and move sugars.

The movement of sugars begins when carbohydrates created during photosynthesis are moved out of the chloroplast and into the cytoplasm of the cell. Here they are converted into sucrose, one form of sugar common in plants. The second step involves moving the sugar molecules a very short distance, from the mesophyll cells where they were created into the phloem tissues where they will move to other parts of the plant. At the source point, usually the leaf where photosynthesis occurs, the loading of sugars into phloem occurs against a concentration gradient—since loading requires the input of energy it occurs through an active transport process. The third step involves phloem transport from source to sink tissues such as roots or new tissues.

Once in the phloem, sap transport between the leaf and other parts of the plant is an energy-passive process called *pressure flow*. The difference in sugar concentration between sources and sinks creates a pressure imbalance, causing the sugar-rich solution to flow away from the leaves through the phloem without any additional energy input. The greater the concentration differences between source and sink, the quicker the sugar solution will move to those tissues. Fast-growing plants can transport their sap at rates of 20 to 40 inches per hour, quickly moving their high-energy resources to high-demand tissues. When the sugar-rich sap arrives at sink tissues, active transport processes unload the energy-rich solution from the phloem for local use or storage. At the sink site, sugars are used to make ATP through aerobic metabolism, stored as starches, or converted into other organic compounds needed throughout the plant to support its metabolic activities.

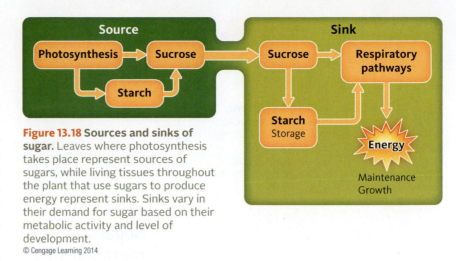

Figure 13.18 Sources and sinks of sugar. Leaves where photosynthesis takes place represent sources of sugars, while living tissues throughout the plant that use sugars to produce energy represent sinks. Sinks vary in their demand for sugar based on their metabolic activity and level of development.
© Cengage Learning 2014

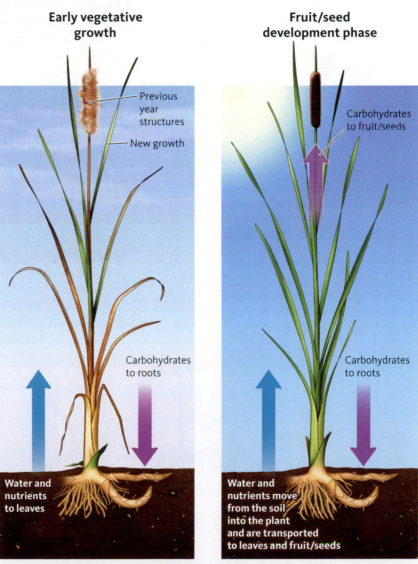

Figure 13.19 Nutrient needs change as plants develop. The need for nutrients changes during the development of a plant. During early development sugars move to the roots, while aboveground tissues need mineral nutrients and water from the roots. At maturity, the needs are different with carbohydrates, mineral nutrients, and water moving to developing fruits/seeds.
© Cengage Learning 2014

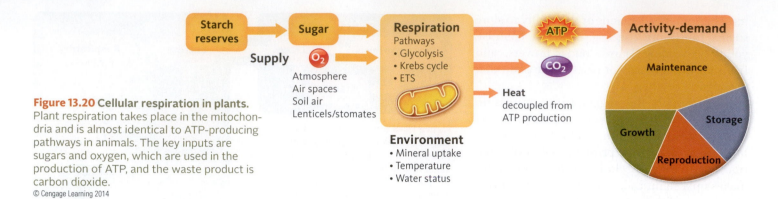

Figure 13.20 Cellular respiration in plants. Plant respiration takes place in the mitochondria and is almost identical to ATP-producing pathways in animals. The key inputs are sugars and oxygen, which are used in the production of ATP, and the waste product is carbon dioxide.
© Cengage Learning 2014

Sugars are used to produce ATP by aerobic metabolism and anaerobic metabolism

Aerobic metabolism is the primary pathway that occurs in mitochondria found in all the living tissues of the plant. In this regard, plants and animals share a common pathway for transferring energy locked in the bonds of macronutrients like sugars to build the ATP used to drive life processes. Recall that aerobic and anaerobic pathways in animals were explored in Chapter 9.

Unlike photosynthesis, which only occurs during daylight hours, aerobic metabolism operates continuously in living cells as long as resources are available. Using an analogy from economics, aerobic metabolism represents the plant's expenditures, or "cost of living." Instead of money, a plant spends the sugar it makes or saves as part of its overall energy budget (**Fig. 13.20**). In temperate habitats, plants like cattails and poison ivy may spend as much as 60 percent of their daily gain in photosynthetic carbon through aerobic metabolism, a maintenance function. In tropical regions that figure may rise to as high as 80 percent, leaving little for long-term storage or use in other metabolic pathways such as reproduction and growth.

In order for aerobic metabolism to produce energy-rich ATP, a number of resources must be available. Sugar molecules such as glucose are the preferred fuel for aerobic metabolism, but fats and proteins can also be used (**Fig. 13.21**). Oxygen gas is also required for this metabolic pathway to

Aerobic metabolism

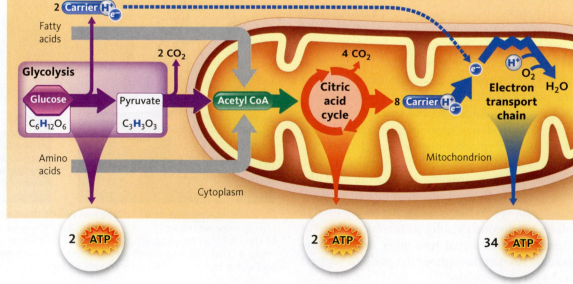

Figure 13.21 Aerobic metabolism. Aerobic respiration relies on the presence of sugars and oxygen to fully utilize the energy in glucose in the production of ATP. The three stages—glycolysis, citric acid cycle, and electron transport—all take place within the mitochondria.
© Cengage Learning 2014

fully utilize the energy stored in sugars. These resources, oxygen and sugar molecules, come together in the mitochondria of cells.

Limits on the availability of sugar or oxygen decrease the aerobic metabolism rate and amount of ATP produced. This is especially important for wetland plants with roots immersed in water-saturated soils or standing water where oxygen availability in the root zone acts as a limiting resource. Whereas plants like the common cattail have specialized structures to move oxygen to underwater tissues, other terrestrial plants like corn shut down aerobic metabolism when resources are limited due to water-saturated soils. This is why corn is vulnerable to flooding stress.

The chemical summary illustration for aerobic metabolism shown in **Figure 13.21** is very similar to the one used to describe aerobic metabolism in animals in Chapter 9. It shows the input of glucose and oxygen, yielding ATP molecules along with carbon dioxide and water. The arrows reveal where the molecules and atoms involved in the process are used during the metabolic reactions. For instance, the oxygen gas that diffused into the leaf is used in the final phase. Sugar molecules enter the same three stages as in animal cells—(1) glycolysis, (2) the citric acid cycle, and (3) the electron transport chain (see **Fig. 13.21**). When plants use other molecules, such as fats or amino acids from proteins, glycolysis is bypassed completely. Plant structures like seeds, for instance, contain fats and oils that are used for energy production.

The rate of aerobic metabolism depends heavily on environmental conditions such as temperature, oxygen availability, tissue type, and life stage of the plant. Each of these factors exerts a controlling influence by either speeding up or slowing down overall ATP production. For example, in general, the rate of aerobic metabolism doubles for every 10°F of temperature increase. The increase results from a speed-up in chemical reactions caused by the addition of heat energy. In addition, different tissues engage in aerobic metabolism at different rates; in meristems, where cell division is rapid, rates will be relatively high to support the energy needs of these metabolically active tissues. On the other hand, the mature tissues of parenchyma cells dedicated to starch storage in the stem will have low metabolic rates because their energy needs are much lower.

Plants switch to anaerobic metabolic pathways to produce smaller amounts of ATP when oxygen is limited or absent, as in the flooded cornfield example (**Fig. 13.22**). Anaerobic pathways do not require oxygen, take place in the cytoplasm of the cell, and, as in aerobic metabolism, begin with glucose molecules. This alternate pathway is inefficient when compared to aerobic metabolism pathways and does not supply enough energy to meet the basic demands, providing just 5 percent of the ATP energy yield from each molecule of sugar. The reason is that anaerobic metabolism incompletely breaks down sugars, and its product still contains large amounts of energy that the plant is unable to use (see **Fig. 13.22**). Plants exposed to poor environmental conditions that force them to rely on anaerobic metabolism will eventually die; for example, flooding stress is a strong agent of natural selection, and only plants well adapted to this condition will survive.

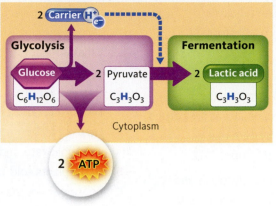

Figure 13.22 Anaerobic metabolism. Anaerobic respiration differs from aerobic respiration by taking place in the absence of oxygen. Without oxygen, the energy in glucose is not fully utilized, and far less ATP is produced. Like animals, plants have a difficult time surviving on anaerobic respiration for any length of time.
© Cengage Learning 2014

- Most plants use aerobic metabolism to produce ATP as a readily usable form of energy.

- Aerobic metabolism involves three stages that break apart glucose into its carbon, hydrogen, and oxygen atoms. Hydrogen atoms are carried to an electron transport system that generates a large number of ATP molecules.

- Anaerobic pathways include glycolysis and fermentation that break apart glucose into lactic acid in the absence of oxygen. A small number of ATP molecules are made relative to aerobic metabolism.

1. Which two molecules must be present at the mitochondria in order for aerobic metabolism to take place?

2. What is meant by loading and unloading in the process of sugar transport?

3. Why is the presence of both water and oxygen in the soil important to supporting rapid plant growth in the spring?

4. How is aerobic metabolism related to environmental temperature, metabolic rate, and activity level of the tissues?

5. Houseplants sold in stores are usually grown inside commercial greenhouses where the soils and environmental conditions are tightly controlled to maximize plant growth and health. What might happen to these greenhouse plants if you were to use soil dug from your neighborhood yards and grow the plants outside, where environmental conditions are not controlled?

13.5  Soils Provide Plants with Water and Minerals

Figure 13.23 Houseplants receive water and nutrients from humans. Houseplants are subject to the physical and biological environments we create for them. The soil, sunlight, water, and lack of competition all combine to influence the health of the plant.

J. Krueger

The potted philodendron lives a unique existence apart from its wild relatives like the purple pitcher plant. Set on a shelf near the window, this houseplant is entirely dependent on the environmental conditions you create for it (**Fig. 13.23**). The temperature of the room, the amount of sunlight streaming through the window, regular watering, the potting soil it's planted in, and fertilizers are critical factors in its ability to grow and survive.

Like its wild relatives, its roots grow and "mine" the soil in search of mineral nutrients used to build molecules important in metabolism and building the plant body. In plants, minerals are elements that are chemically bound to soil particles, which in solution allow them to move across plant membranes into the plant. Nitrogen, phosphorus, and potassium are examples of mineral nutrients the philodendron needs in relatively large amounts because of their role in so many processes. For example, phosphorus is a component of both DNA and the membranes of all cells. Still other minerals like iron are required in far smaller amounts, although they may play just as critical a role in metabolic reactions. To meet the plant's nutritional needs and maintain growth and metabolic reactions, homeowners periodically rejuvenate the limited quantity of minerals in the pot with commercial fertilizers, or plant food.

Water is equally critical to the philodendron's survival. On warm days near the window, the temperature rises, increasing the plant's metabolic rate and water loss through its open stomata that provide a pathway for gas exchange. Regular watering wets the soil, allowing the uptake of mineral nutrients in solution, as well as providing this important resource used for structural support, sugar transport in fluid sap that moves through vascular tissue, and photosynthesis.

In the section ahead, we will examine the relationship plants like the philodendron have with the soils they grow in. We'll explore the various properties of soils, which influence the plants growing in them; how plants move their valuable carbohydrates from one location to another; and water loss through evaporation.

Soils vary in physical, chemical, and biological properties

We spend our lives walking on top of soils, but few of us have considered how important they are in the life of plants. There are many thousands of different soil types, but they all share three basic characteristics: (1) they are derived from rocks and minerals that were broken down into smaller and smaller particles over time; (2) they contain the decayed remains of plants, animals, fungi, and microbes, or organic matter; and (3) they are a habitat for microorganisms, fungi, invertebrates, and larger animals that modify soil characteristics over time.

Although we might think of soil as a solid mass, depending on the type, as much as 50 percent of a soil's volume is actually occupied by air that fills the spaces between the soil particles. A physical property of soil that is important to all plants is its *texture,* a description of the relative mix of different sized particles in a soil. Soil is a mixture of three different sized particles—sand, silt, and clay (**Fig. 13.24**). Why is texture important? A soil's texture determines the amount of water it can hold, as well as the amount of oxygen it can supply to roots. For instance, the mostly sandy soils in the Pine Barrens, where the purple pitcher plant grows, have less ability to hold water, but they have the largest pore size, or distance between the particles, providing a lot of space for oxygen (see **Fig. 13.24**). In contrast, clay soils are typically highly compacted, leaving little room between the particles for oxygen necessary to support aerobic metabolism in roots. Wet clay soils are difficult for roots to penetrate in order to absorb minerals that are tightly bound to soil particles.

Another important chemical property of a soil is its pH, a measurement of the number of free hydrogen ions in the water solution of the soil. The pH influences the ability of plants to absorb minerals because the acidity changes their chemical structure, making it easier or more difficult for the plant roots to absorb water and minerals. The pH of a soil is influenced by the type of material the soil is derived from, as well as atmospheric deposits in dust and rainfall.

Soil also contains organic matter, the decayed remains of plants, animals, and microbes. On average, decayed matter makes up about 5 percent of soil volume, giving the soil a dark color (**Fig. 13.25**). *Compost* sold at a garden center is made from decayed organic matter. Many species of fungi, microbes, and

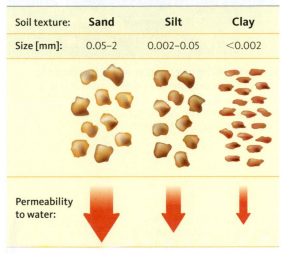

Soil texture:	Sand	Silt	Clay
Size [mm]:	0.05–2	0.002–0.05	<0.002

Permeability to water:

Figure 13.24 Soils have different textures. Soils are classified based on their texture, a measurement of the relative amounts of the three sizes of particles present in the soil.
© Cengage Learning 2014

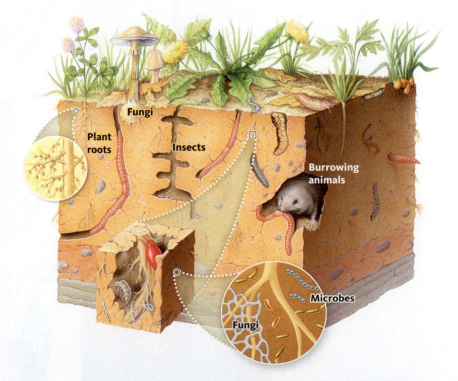

Figure 13.25 Many types of organisms participate in recycling soil nutrients. Beneath the soil surface are entire communities of microbes, fungi, and animals large and small. Each plays a role in modifying the characteristics of soils, like pore spaces, as well as in recycling nutrients.
© Cengage Learning 2014

Figure 13.26 Organic matter in soil improves it. Organic matter in the soil helps absorb and hold water, making it available for plants over an extended period of time. The dark color in this soil is mainly due to decaying plant matter.

invertebrates like worms feed on decaying materials, speeding up this recycling process (**Fig. 13.26**). Organic matter has a significant impact on plant growth because its spongelike quality increases the soil's ability to hold water.

Water transport is powered by transpiration from leaves

During the growing season, when plants are growing quickly and producing flowers and fruit, they have a high demand for water. For instance, corn requires 600,000 gallons per acre each growing season, the equivalent of over 20 inches of rain, while a large oak tree moves over 150 gallons of water up its stem each day during the hot summer months. Each drop of water comes from the soil, where it is absorbed by the roots and moved upward to all other plant tissues. Most of the water absorbed by the plants eventually is lost to the atmosphere through evaporation from the leaves.

Why does it take so much water to grow corn? To answer this question, let's start at the surface of the leaves on a sunny summer day. To obtain the CO_2 required to support photosynthesis, a plant must open its stomata (**Fig. 13.27**). When open, the moisture-rich tissues inside the corn leaf are

Figure 13.27 Water movement through plants. Water movement through plants is driven by the water concentration gradient between tissues.
© Cengage Learning 2014

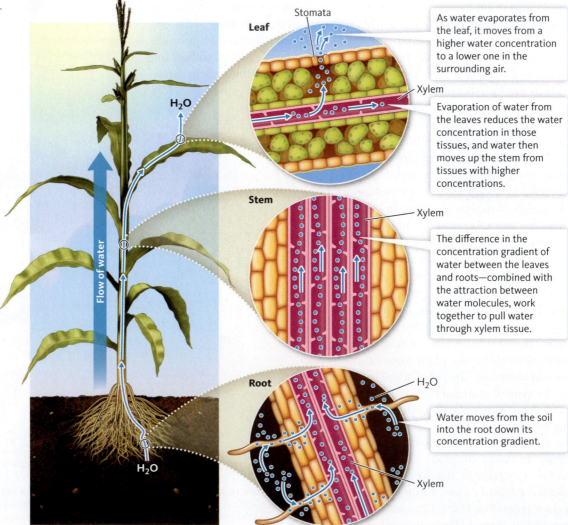

Flow of water

H₂O

Leaf

Stomata

As water evaporates from the leaf, it moves from a higher water concentration to a lower one in the surrounding air.

Xylem

Evaporation of water from the leaves reduces the water concentration in those tissues, and water then moves up the stem from tissues with higher concentrations.

Stem

Xylem

The difference in the concentration gradient of water between the leaves and roots—combined with the attraction between water molecules, work together to pull water through xylem tissue.

Root

H₂O

Water moves from the soil into the root down its concentration gradient.

H₂O

Xylem

exposed, and water evaporates. Although water loss in dry habitats can lead to death, it is also a necessary process because it pulls water up from the soil and through the plant to the leaf.

Let's follow the pathway of water from the soil and through the plant shown in **Figure 13.27**. Water in the soil pore spaces comes into contact with the growing root, moving across its membrane into the root down the water concentration gradient. Once inside the xylem root tissue, water movement is a passive process requiring no input of ATP. Recall from Chapter 1 that water molecules stick to one another. This property, called *cohesion*, is what makes raindrops form and creates surface tension that permits lightweight insects like water striders to stand on the surface of a pond rather than sink to the bottom (**Fig. 13.28**). Although single hydrogen bonds are weak, the collective strength among billions of molecules in a single drop of water is substantial.

In order for water to move from the roots to the leaves of the corn plant, it relies on the tension between water molecules. Tension provides a continuous link between the water in the leaves all the way down the strawlike xylem tissue to the roots (see **Fig. 13.27**). As water evaporates from open stomata, a process called **transpiration** pulls the thin column of water up from the soil into the roots, up the stem, and to the leaves. As much as 90 percent of the water absorbed in the root zone is lost through stomata via transpiration. In habitats like grasslands where water is available in limited amounts, water stress created during gas exchange places a high metabolic cost on plants and greatly limits their ability to grow. Consequently, corn crops grown in grassland ecosystems like central Nebraska require substantial amounts of irrigation. You can see why plants devote significant energy for root growth to seek out water sources.

Plants require mineral resources to grow

The many types of plant tissues in the purple pitcher plant are constructed using a wide variety of biomolecules, each of which is made from unique combinations of elements. The various amounts of elements needed represent its dietary requirement. In the spring, the fast-growing pitcher plant has a high need for lots of nutrients, whereas far fewer are required during the winter when the plant isn't growing. As we've learned, habitats rarely provide resources in unlimited quantities, limiting the ability of plants to make the entire range of products dependent on that resource and ultimately affecting some life activity.

How do plants come by these various elements? We've learned that during gas exchange, plants acquire carbon from the carbon dioxide in the atmosphere, while the roots of plants seek out mineral nutrients and water. In other cases, like that of the carnivorous pitcher plant, some elements and molecules are obtained from the digested remains of other organisms.

The purple pitcher plant's roots are continually growing through the wetland soils in search of nutrients. Like all plants, it requires nitrogen to make the green pigment chlorophyll, amino acids in proteins, and enzymes critical to cellular metabolism and growth (**Table 13.1**). To obtain nitrogen, plants must continually access new sources of this critical nutrient in the soil, a process requiring significant resources and energy to grow new root tissues. As the roots push forward through the soil, they come in contact with soil particles and water that contain small amounts of minerals in solution. The water and minerals cross cell membranes and enter the plant passively or actively with the spending of energy. Once inside the

optimarc/Shutterstock.com

Figure 13.28 Cohesion between water molecules creates surface tension. This property allows lightweight insects like this water strider to walk on the surface of the water without sinking.

Transpiration—the evaporation of water out of plant leaves into the atmosphere through open stomata. As water evaporates from the leaf, a thin column of water is pulled all the way through the plant from the roots up the stem and to the leaves.

TABLE 13.1 Mineral Functions in Plants

Mineral Element	Function
Nitrogen (N)	Structural components: Amino acids—proteins Nucleic acids—DNA, RNA ATP/ADP Chlorophyll Defensive chemicals
Potassium (K)	Opening/closing of stomata
Calcium (Ca)	Provides strength to cell walls
Magnesium (Mg)	Key element in chlorophyll
Phosphorus (P)	ATP/ADP Nucleic acids—DNA, RNA Sugar-phosphates Phospholipids in cell membranes
Sulfur (S)	Part of proteins in electron transport chain of light reactions in photosynthesis
Iron (Fe)	Part of proteins in electron transport chains in (1) light reactions of photosynthesis and (2) aerobic metabolism

Micronutrients: chlorine (Cl), copper (Cu), manganese (Mn), zinc (Zn), molybdenum (Mo), and boron (B)

© Cengage Learning 2014

plant's xylem tissue, the minerals are transported in solution and distributed throughout the plant.

Each species of plant has a unique need for each type of mineral, and a single plant doesn't need each of the many minerals in its diet in the same amounts. Minerals needed in large amounts, like nitrogen, are called *macronutrients*. Macronutrients like nitrogen, phosphorus, and potassium are used in so many different biomolecules in every cell of the plant that there is a high demand for them (see **Table 13.1**). Plants quickly experience metabolic problems if macronutrients are limited, interfering with their ability to carry out metabolism. In contrast, *micronutrients* are needed in relatively small amounts. For example, chlorine is an essential part of an enzyme that splits water during the light reactions of photosynthesis. However, the amount plants use is so small that it's measured in parts per million. Even so the unavailability of micronutrients in a habitat can limit plant metabolism and ultimately survival.

In this chapter, we've explored how a plant acquires light, water, and minerals from its environment and how it uses these resources to support growth and survival. The availability of resources in the plant's habitat has a significant impact on its life activities and plays a significant role in the long-term natural selection processes that lead to adaptation. To create high-energy sugar molecules, plants transform the energy in sunlight using a metabolic process called photosynthesis. Using the aerobic metabolism pathway, plants produce ATP molecules to fuel maintenance, growth, development, and reproduction. Anaerobic metabolism takes place under low levels of oxygen, producing far less ATP. In the next chapter, we'll see how plants use their sugars to build reproductive structures—flowers, fruits, and seeds.

- Soils have different physical and biological characteristics that influence the plants growing there.

- Water enters and is transported through the plant passively.

- Most water acquired by the plant is lost to transpiration.

- Soils provide mineral resources that are taken up in solution by roots.

1. In your own words, describe why organic matter in soils might be important to plant growth.

2. What is the relationship between soil pore space, soil oxygen, and aerobic metabolism in roots?

3. Why is the amount of available phosphorus in the soil important to plants?

4. Specialty gardeners have perfected the art of growing miniature versions of plants called *bonsai*. They regularly trim away parts of the root system to assist in dwarfing the plant (**Fig. 13.29**). Can you suggest why trimming the roots causes a plant to dwarf rather than reach full size?

5. **Figure 13.30** shows a plant with a nutrient deficiency that causes the older mature leaves to change from their characteristic green color to a much paler green. As the deficiency progresses these older leaves become uniformly yellow. Using your knowledge of plant nutrient requirements and **Table 13.1**, propose at least two minerals that may be lacking in the soil of this plant. Explain your reasoning.

Figure 13.29 A bonsai tree.

Figure 13.30 Evidence of a nutrient deficiency in leaves.

End of Chapter Review

Self-Quiz on Key Concepts

Plants Are Adapted to Environmental Conditions and Resource Availability

KEY CONCEPTS: Each plant species is adapted to survive within a range of environmental conditions such as light, temperature, moisture, and nutrient availability. These important environmental conditions are continually changing in daily, seasonal, and yearly patterns. Too little or too much of any resource creates stress to which the plant must be able to respond or it will die. Plants are consumers of resources, and each plant has a specific diet—a complex set of nutritional requirements, including oxygen, carbon dioxide, water, minerals, and sunlight to meet its growth, development, and reproductive needs.

1. Match each of the following **ecological terms** with its description or example.

   ecological niche

   stress

   disturbance

   a. an event such as a fire that alters the plant community
   b. the range of environmental conditions and resource availability to which a plant is adapted
   c. an environmental factor that reduces a plant's ability to carry out metabolic activities

2. A plant's ability to tolerate a change in environmental conditions obtained over many generations of selection would be considered which of the following?

   a. response
   b. adaptation

3. Flooding of a crop field early in the growing season creates problems for which metabolic function?
   a. photosynthesis
   b. aerobic metabolism in roots
   c. Both are correct.

4. Plants respond to short-term environmental stress by:
   a. redirecting resources.
   b. accelerating photosynthesis.
   c. migrating.
   d. adapting.

Plants Use Photosynthesis to Produce Sugars

KEY CONCEPTS: Life on Earth is solar powered. Using the sun's energy, along with water, plants convert carbon dioxide from the atmosphere to sugar molecules. The process of photosynthesis is divided into two broad stages: the light reactions and the Calvin cycle. The light reactions transform the sun's energy into chemical energy. In the process, oxygen is released into the atmosphere, and intermediate energy molecules are made. In the Calvin cycle, these intermediate molecules are used to convert carbon dioxide to sugar molecules. Plants that live in warm, dry environments have photosynthetic adaptations to minimize water loss.

5. Match each of the following **photosynthesis terms** with its description or example.

   chloroplast   a. the primary pigment that collects energy from sunlight
   chlorophyll   b. the first stage of photosynthesis
   light reaction   c. the organelle where photosynthesis takes place
   Calvin cycle   d. includes carbon fixation

6. Which of the following is an important characteristic of the light reaction?
   a. produces sugar
   b. involves the enzyme rubisco
   c. takes place within the inner membrane of the chloroplast
   d. uses ATP and energy carrier molecules during the reactions

7. Alternative photosynthetic pathways have evolved to help plants survive in:
   a. Arctic climates.
   b. warm, dry habitats.
   c. wetlands.
   d. high altitude.

Plants Use Sugars for Energy and as Building Blocks for Growth

KEY CONCEPTS: When oxygen is present, plants engage in aerobic metabolism to completely break down simple sugars into carbon dioxide and water. When oxygen is absent, plants use anaerobic metabolism and convert simple sugars into alcohol in a process called *fermentation*. While aerobic metabolism produces greater quantities of ATP, anaerobic metabolism can produce sufficient amounts to allow plants to survive for a limited amount of time.

8. Which mechanism accounts for the passive movement of sap from sources to sink tissues in a plant?
   a. diffusion
   b. osmosis
   c. cohesion
   d. bulk flow

9. Which of the following parts of a fast-growing plant has the *highest* aerobic metabolism rate?
   a. root apical meristem
   b. parenchyma storage tissue
   c. dormant seeds
   d. stem xylem tissue

10. Anaerobic metabolism produces comparatively less ATP than aerobic metabolism because:
    a. it does not use sugar as an input.
    b. sugars are not fully broken down.
    c. the electron transport chain runs slower.
    d. the oxygen is not fully broken down.

Plants Rely on Minerals, Oxygen, and Water in Soils to Support Growth

KEY CONCEPTS: Plants are rooted to a single location, growing in soils that have very specific characteristics, including the mix of soil particle sizes, available mineral elements, decaying organic matter, and a range of microbes and animals that live in the soils. Using their extensive root systems, plants absorb water, minerals, and oxygen from the pores between soil particles. If soil resources needed by the plant are not available in the right amounts, plants limit their cellular metabolism, decreasing their growth, development, and reproduction.

11. Match each of the following **nutrient terms** with its description or example.

    soil texture   a. a mineral nutrient needed in relatively large amounts by plants
    mineral availability   b. a mineral nutrient needed in relatively small amounts by plants
    compost   c. influenced by the pH of the soil; directly affects absorption of minerals
    macronutrient   d. a mixture of soil particles and decaying organic matter
    micronutrient   e. evaporation of water, primarily through open stomata in a leaf
    transpiration   f. a description of the relative amounts of sand, silt, and clay in a soil

12. Which of the following nutrients is considered a macronutrient in plants?
    a. chlorine
    b. nitrogen
    c. iron
    d. helium

13. Which of the following soils would have the largest percentage of pore spaces to best support aerobic metabolism in plant roots?
    a. sandy soils
    b. silt soils
    c. clay soils

14. Which of the following similarities do chloroplasts share with mitochondria?
    a. Both engage in photosynthesis.
    b. Both produce sugar molecules for use in other parts of the plant.
    c. Both support a complex electron transport chain set of reactions.
    d. Both are found in all living cells of the plant.

15. The movement of water through a plant:
    a. is passive.
    b. requires active transport.
    c. is driven by mineral nutrition.
    d. is a function of habitat type.

Applying the Concepts

16. Based on your knowledge of plant structure, what impact would low phosphorus availability have on the growth of your lawn?

Figure 13.31 Soybean plants can partner with nitrogen-providing bacteria.

Courtesy of C.R. Crozier

17. Rhizobium is a beneficial species of bacteria that forms a mutualistic relationship with plant roots, helping the plant to acquire nitrogen. Which of these soybean plants (**Fig. 13.31**) is successfully "infected" with rhizobium and why?

18. Coal-fired power plants are a common sight in many communities. Over the past century, decades of pollution contributed to a phenomenon called *acid rain.* Acid rain is caused by emissions of sulfur dioxide and nitrogen oxides, which react with the water molecules in the atmosphere to produce acids. Eventually, the acids fall as rain, altering the pH of soils. How would the nutrient status of downwind plant communities be affected?

Figure 13.32 Rose bush.

Sandra Voogt/Shutterstock.com

19. Imagine you are standing next to the leaves of a rose bush (**Fig. 13.32**). In your hands is a digital gas analyzer used to measure changes in oxygen and carbon dioxide levels at the surface of the leaf, right at the stomata. Describe what the meter might tell you about gases moving into/out of the leaf related to photosynthesis during the day. Predict what the meter would read at nighttime.

20. The century plant, or agave, is an evergreen plant commonly used as an ornamental in the dry, hot regions of the southwestern United States (**Fig. 13.33**). Its leaves have a thick, waxy covering, which gives them a bluish-gray color. Agaves lose very little water through transpiration, and if you were to look at their stomata with a magnifying glass, you would see they were closed during daylight hours. How does this provide a competitive advantage to the agave? How does it limit water loss?

Figure 13.33 Thick, waxy leaves of the Agave, or century plant.

Chris Curtis/Shutterstock.com

Data Analysis
Carbon Gain in Yucca

Yucca glauca is an evergreen plant common in prairies dominated by grasses that use the warm, dry climate–adapted photosynthetic pathways (**Fig. 13.34**). The stiff, angled leaves of yucca all grow from a center, circular rosette and emerge at a steep angle. The evergreen and steep angled leaves, combined with its ability to photosynthesize during cool weather and with low sun angles during the spring and fall, are critical to its ability to compete for sunlight with grasses in the prairie habitat (**Fig. 13.35**). An experiment was run to determine if yucca was able to gain a significant portion of its carbon during the spring and fall when its competitors were dormant. The area of the experiment in northeast Kansas has warm, dry summers that favor the growth of grasses over the yucca.

Examine the graphs in **Figure 13.36** that show the relationships between growth (bottom) and three environmental factors—light levels, air temperature, and rainfall. For each graph, the x-axis represents the month of the year. The top graph records changes in light intensity over time for a flat surface (dashed line) and the angled leaf surface (solid line); the second graph records changes in temperatures; the third, rainfall. The bottom graph illustrates how the yucca grew in response to all of these environmental factors combined together.

Figure 13.34 Spike shaped, angled leaves of *Yucca glauca*.

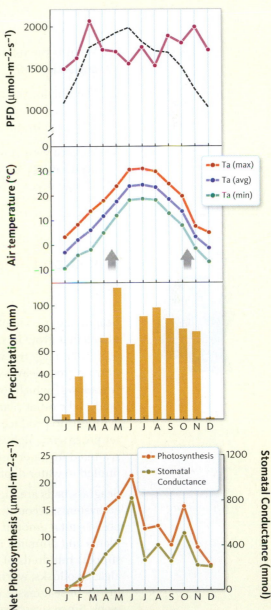

Figure 13.36 Relationships between growth and three environmental factors.

Adapted with permission from Maragni, L., Knapp, A., and McAllister, C., "Patterns and Determinants of Potential Carbon Gain in the C3 Evergreen Yucca glauca (Liliaceae) in a C4 Grassland," 87(2):230–236. Copyright © 2000 by The Botanical Society of America, Inc.

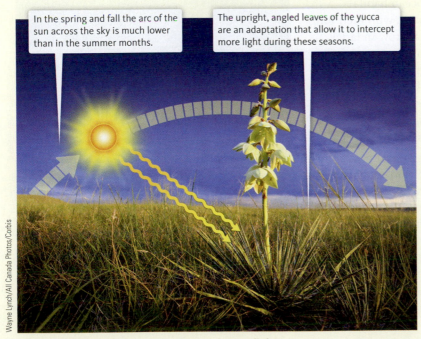

In the spring and fall the arc of the sun across the sky is much lower than in the summer months.

The upright, angled leaves of the yucca are an adaptation that allow it to intercept more light during these seasons.

Figure 13.35 How the yucca competes for sunlight.

Data Interpretation

21. In which month of the year did *Yucca glauca* have the highest photosynthetic rate?

22. What were the temperature and moisture conditions when this high growth rate occurred? How do these numbers compare to those in the rest of the growing season?

23. Were the angled surfaces of the yucca leaves receiving more light than the flat surfaces in the spring and fall?

24. In general terms, did a significant amount of growth occur between November and April (outside of the frost-free period of the year, indicated by the arrows in the air temperature plot)? Is the yucca's adaptation to photosynthesize during these periods an important characteristic?

Critical Thinking

25. As a cool weather–adapted plant, is the yucca well adapted to avoid problems associated with high temperatures and transpiration?

26. What is the relationship between the higher growth periods and the light intercepted by the angled leaves?

27. Plant biologists working on climate change in prairie ecosystems are concerned about how increased temperatures may influence plant diversity. How might increased temperatures affect the yucca in this scenario?

28. If yucca were a warm, dry photosynthesis-metabolism–adapted plant, would there be the same need for its cool-weather growth adaptation? How would you expect the growth patterns to change if yucca used this adaptive pathway?

Question Generator

Seagrasses Live in Underwater Meadows

Seagrasses are one of a few species of flowering plants that grow in shallow ocean habitats (**Fig. 13.37**). They resemble grasses with their long, narrow green leaves and often grow in large "meadows," which look like underwater grass-lands. Like other plants, seagrasses use photo-synthesis to produce sugars, but living underwater poses some light stresses. Light is quickly scattered in water, and seagrasses are generally limited to water depths of less than 20 feet.

Seagrass beds are highly diverse and productive ecosystems, accounting for 15 percent of the ocean's total amount of carbon stored in the plants and animals. These meadow habitats support hundreds of species that include fish, algae, mollusks, worms, and nematodes. Herbivory of seagrasses is an important link in food webs. Since seagrasses grow in shallow waters close to shore, these habitats are vulnerable to damage from human activities and pollution.

Figure 13.37 Seagrasses live in shallow water.

© Michael Patrick O'Neill/Alamy

Below is a block diagram (Fig. 13.38) that relates several aspects of the biology of seagrasses to their photosynthetic rate and growth. Use the background information along with the block diagram to ask questions and generate hypotheses. We've translated the components and interaction arrows shown in the map into a question and provided several more to get you started.

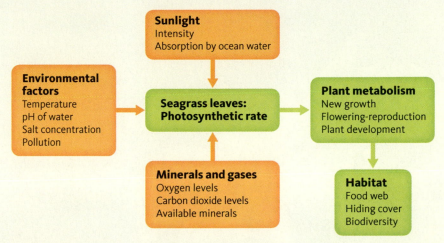

Figure 13.38 Factors relating to the photosynthetic rate and growth of seagrasses.
© Cengage Learning 2014

Here are a few general questions to start you off.

1. Would you expect seagrasses to have well-developed root systems for collecting water?
2. What effect does the depth of the water have on the availability of light?
3. How might seasonal variations in water temperature and hours of sunlight per day influence the ability of seagrasses to produce new growth?
4. How might pollution from human activity along the shoreline affect the biodiversity found in seagrass habitats?
5. Dense beds of seagrasses grow in the bottoms of shallow bays. Would you expect these soils to be nitrogen-rich?

Use the block diagram to generate your own research questions and frame a hypothesis.

14 Plant Reproduction

14.1 The sex life of the saguaro cactus, a symbol of the American Southwest

14.2 Plants use both asexual and sexual methods of reproduction

14.3 Flowers are the organs for sexual reproduction

14.4 Pollination and fertilization produce seeds

14.5 Seeds are dispersed and grow in new locations

14.1 The Sex Life of the Saguaro Cactus, a Symbol of the American Southwest

The setting sun in the western sky creates a silhouette of a plant recognizable to many—the saguaro cactus *(Carnegiea gigantea)*. This plant has come to symbolize the desert southwest because it lives nowhere else. Its geographic range is restricted to a small part of the Sonoran desert where there is scant rainfall, the summers are hot, and freezing temperatures in winter rarely occur (**Fig. 14.1**). This state flower of Arizona can grow more than 50 feet tall, weigh more than six tons, and live for more than 150 years. During its long life, it grows slowly and survives through its adaptations for water storage, but how does this giant cactus find a mate and reproduce in this harsh desert habitat? This is where our story begins.

The sex life of the saguaro begins after it grows upward for 35 to 50 years. When it finally reaches adulthood, it forms its first flower with the arrival of warm spring temperatures and longer days in April and May. Supported by energy stored in its succulent stems, the cactus produces dozens of large, funnel-shaped white flowers (**Fig. 14.2**). Each saguaro flower contains both male and female sexual organs (see **Fig. 14.2**). The male organ makes sperm-producing **pollen grains**, and the female organ makes **egg cells** within numerous ovules contained in its enlarged ovary. Flowers are built for sexual reproduction, which means they bring about the fertilization of an egg by a sperm cell to produce a genetically unique embryo within a seed. How do saguaro flowers accomplish this?

About two hours after sunset, the flowers, which are held high off the ground, open and expand. They remain open until the afternoon of the following day. To signal all potential pollinators to their location, the flowers release a strong, musty scent into the air. Their open flowers offer large amounts of sugary nectar and pollen as a reward to any nocturnal animal that visits and hopefully cross-pollinates, or transfers, some of the pollen to the flower of another cactus.

The odor and nectar are irresistible to two species of bats: the lesser long-nosed bat and the Mexican long-tongued bat. These bats have a long snout and brush-tipped tongue, which make them well adapted to reach deep into the flower to obtain nectar. In the process their faces are covered with large amounts of yellow pollen that is then deposited into the next flower they visit (**Fig. 14.3**). Like many mammals, bats require large supplies of energy-rich sugars and proteins in their diet. The saguaro is an excellent source of food, and bats travel long distances to obtain it. In one study, bats flew 18 miles from their roosts to an area that typically contains thousands of saguaro flowers. Here they foraged in small groups, visiting about 80 to 100 cactus flowers each night, often returning to the same flower several times. In this way, the saguaro has provided a vital source of food for the bat, and in turn it has promoted the movement of pollen to other saguaro flowers.

It is now early June, and a month or so has passed since the saguaro flowers were pollinated by bats, bees, and birds. Pollen grains deposited on the stigma have produced sperm cells that then travel down the style to fertilize egg cells contained in ovules deep within the ovary of the flower. This event has caused the once cream-colored sepals, petals, and stamens to wither and fall from the cactus. The flower's ovary enlarges into a green-colored fruit that weighs about the same as a golf ball (about 1.6 ounces).

Figure 14.1 Geographic range of the saguaro cactus. Saguaros inhabit a small geographic range in the southwestern United States because they are tolerant of dry climates and intolerant of cold temperatures.
© Cengage Learning 2014

Pollen grain—structure that carries male genome to the female sex organs of a flower and produces sperm cells; produced in the male sexual organs of the flower.

Egg cell—haploid female reproductive cell (gamete); after fertilization it becomes a diploid zygote.Egg cell—haploid female reproductive cell (gamete); after fertilization it becomes a diploid zygote.

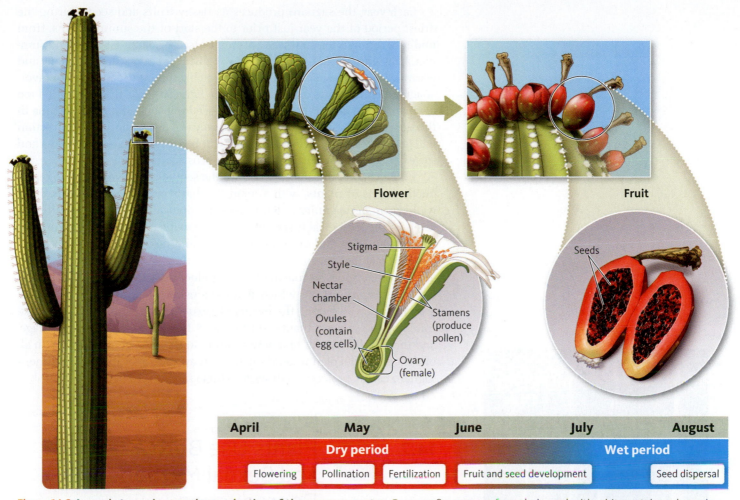

Flower

Stigma

Style

Nectar chamber

Ovules (contain egg cells)

Stamens (produce pollen)

Ovary (female)

Fruit

Seeds

April	May	June	July	August
Dry period			Wet period	
Flowering	Pollination	Fertilization	Fruit and seed development	Seed dispersal

Figure 14.2 Annual stages in sexual reproduction of the saguaro cactus. Saguaro flowers are funnel-shaped with white petals and sepals that secrete large amounts of nectar. About 10 P.M. each night in May, the flowers open. Pollen is released at the top of thousands of stamens and received by the stigma of the female part of the flower. Pollination occurs in late spring during the dry period of the year. Fruits mature during July and ripen into bright red, fleshy fruits that split open to disperse up to 2,000 tiny, black seeds. Dispersal is timed during the wet period when moisture is available for germination and seedling growth.
© Cengage Learning 2014

Eventually, the green fruit ripens into a red-colored, fleshy fruit that splits open to reveal a sweet center brimming with more than 2,000 tiny, black seeds (see **Fig. 14.2**). Over its long reproductive life (about 100 years), a single healthy saguaro will produce approximately 40 million seeds.

Saguaro seeds are dispersed in many ways. Most fall to the ground while they are still within the ripe fruit. There at the base of the parent plant, the brightly colored fruit is a source of water and sugar for a number of animals, such as insects, rodents, and larger mammals. The open fruits left intact on the cactus provide food for birds and bats. The lesser long-nosed bat, which pollinated the flowers a few months ago, now returns to feast on the water-filled, fleshy fruit pulp. Seeds are eaten along with the nutrient-rich pulp, and they remain intact as they pass through the animal's digestive tract before they are deposited at a new location. Birds such as the white-winged dove often disperse seeds away from the parent plant to areas underneath other plants, where the seeds are sheltered from predators and the heat of the desert. These lucky seeds have an increased chance of survival.

Figure 14.3 Bat pollination of saguaro cactus. Bats are one of several animals that are adapted to pollinate saguaros. Bats are attracted by the plentiful amounts of sugar-rich nectar and protein-rich pollen.
SearchNet Media/Flickr/Getty Images

Each year, the saguaro produces its fleshy fruits and seeds during the driest period of the year just prior to the start of the summer rains, from mid-July through the first week of August (see **Fig. 14.2**). The rain and summer heat ensure that seeds will have sufficient moisture available and warmth for germination and seedling establishment. In addition to favorable temperatures, saguaro seeds require light, so seeds that are buried too deeply will not germinate. This seed-to-seedling stage is a perilous time in the life of a new saguaro. Of the thousands of seeds, only a tiny fraction will germinate into a seedling, which will be vulnerable to herbivores and the extreme temperatures of the desert ecosystem.

In the chapter ahead, we will trace the sequence of events in the reproductive cycle of plants, with a main emphasis on flowering plants. Here we have an opportunity to apply several important concepts on different levels of the biological hierarchy, such as genetic diversity and natural selection, signaling and recognition, mitosis and meiosis, growth and development, metabolism, and ecological interactions. Animal sensory systems also play an important role in plant reproduction. We will begin with asexual plant reproduction, which is an extension of the plant's vegetative growth, before we trace the events of sexual reproduction that are illustrated by the saguaro plant—starting with flowers to the fruit seed stage, and then to the growth of a new seedling in the next generation. With all of the energy devoted to reproduction, on average over its long lifetime, a saguaro cactus replaces itself with a single seed.

Figure 14.4 **Aspen clone.** A large stand of aspen trees consists of trees connected by a shared root system.

Asexual reproduction—the process that creates offspring that are genetically identical to the parent through the process of mitosis; forms clones.

Clone—a group of organisms that are genetically identical copies of a single parent; offspring from asexual reproduction; a stand of aspen trees that are genetically identical.

Sexual reproduction—the creation of genetically different offspring from the fertilization of male and female gametes.

14.2 Plants Use Both Asexual and Sexual Methods of Reproduction

Quaking aspen trees are champions when it comes to creating genetically identical copies of themselves through asexual reproduction. Aspens grow throughout the Rocky Mountains, and in the fall they turn a brilliant gold color. All of the aspen trees shown in **Figure 14.4** grew from the roots of a single tree and are considered a genetically identical stand of trees, or a clone. Foresters estimate that a giant aspen clone in Utah contains 47,000 trees and covers an area about the size of 84 football fields. Although they have the exact same DNA, each individual tree in a clone is an independent physiological individual that grows and develops in response to the particular environmental factors in its habitat.

Aspen trees, like saguaro cactuses, also generate genetic variation through sexual reproduction. Each aspen clone is a stand of trees consisting of a single sex; all the trees produce either male flowers or female flowers (**Fig. 14.5**). Male flowers produce large amounts of pollen, which contains cells that produce sperm cells. Female flowers produce egg cells. Unlike saguaros, which depend on bats to carry their pollen, aspens use the wind and create "pollen clouds" that drift to female flowers in an adjacent clone. A successful pollination and fertilization between sperm and egg results in the production of a genetically unique seed. By using both asexual and sexual means of reproduction, the aspen combines the advantages of both—the ability to rapidly grow into open spaces and the genetic diversity to preadapt the aspen to new habitats it may encounter in the future. In the section ahead, we will describe how other plant species re-a.

b.

© Bob Gibbons/Alamy

Ned Therrien/Visuals Unlimited/Getty Images

Figure 14.5 Male and female aspen catkins (flowers). Aspens produce long, spikelike clusters that consist of either male or female flowers. (a) Male flowers make pollen. (b) Female flowers catch the wind-borne pollen.

produce both asexually and sexually and how this fits into its overall reproductive, or life cycle.

Asexual reproduction is an extension of vegetative growth

Asexual reproduction is also called *vegetative reproduction* because a plant reproduces itself through the growth of its vegetative organs—roots, stems, or leaves. For example, ferns and other herbaceous plants reproduce through horizontal underground stems called *rhizomes* (**Fig. 14.6**). Willows are woody trees that form dense thickets along rivers, creeks, and ponds through rhizomes.

Horticulturists use vegetative reproduction to propagate and multiply the number of plants for market. Cutting plant stems propagates many house and landscape plants. Growers can apply specific plant hormones to promote cell division and root development. In both cases, cell division and differentiation produce a new plant that is genetically identical to its parent.

Many aquatic plants can be physically cut into pieces and reproduce through fragmentation. Eurasian milfoil is a floating plant found in freshwater lakes that reproduces through small stem fragments. Each stem fragment rapidly grows into a new plant. Eventually milfoil forms dense populations that shade out native plants and disrupt the aquatic food chain. In many northern states, boats and propellers are inspected to prevent the spread of milfoil fragments from one lake to another.

The advantage of rapid vegetative reproduction is offset by a distinct disadvantage—the reduction of genetic variability of a population. All members of a plant clone, for example, have the same genetic resistance or susceptibility to insect pests and pathogens. A change in the ability of a pathogen to infect one plant may destroy the entire population. Recall from Chapter 3 that natural selection acts on the genetic diversity of a population, so from an evolutionary perspective, asexual reproduction reduces natural selection and the ability of the population to adapt to new habitats and conditions. This is one reason why most perennial plants use both methods of reproduction.

Sexual reproduction involves a series of well-timed events

In conifers and flowering plants, the sexual union of sperm and egg cells produces seeds, which are then dispersed to new locations. This reproductive process is accomplished in five basic steps:

1. A plant builds sexual organs (cones or flowers) that produce pollen and female sex cells.
2. The process of pollination carries the male genome to the female part of the flower.
3. The pollen makes sperm cells, which then travel to fertilize the egg cell.
4. The embryo grows and is nourished by storage tissues within the seed.
5. The seed is dispersed into a favorable habitat, where it will germinate and grow into a seedling as part of the next generation.

Each step has a number of events that must be well timed to occur. Natural selection and chance act on each step of the process. Plants that successfully pass their particular alleles to the next generation lead to the plant's increased fitness or adaptation to the habitat.

Plants invest a considerable amount of energy, raw materials, and time in sexual reproduction. The five steps and the many events that occur must be timed with the environment to give seeds a chance to survive and grow. For example, building flowers during freezing cold temperatures would bring the reproductive process to a halt. Such timing would be especially disastrous for annual and biennial plants, which have only a single period of reproduction in their lifetime and do not reproduce asexually. Thus the sexual method of reproduction involves greater risk of failure than reproduction through building new roots or shoots.

A plant's sexual reproductive success relies heavily on the environment and the functioning of the plant's roots, stems, and leaves. It's easy to ignore the contribution of roots to the reproductive success of plants; however, water and cell expansion keep the flower parts physically supported during the period of blooming. Water also plays an important role in the swelling of plump, juicy, fleshy fruits. Larger, more colorful fruits may lead to greater dispersal and fitness. A year with plenty of rainfall and warm sunny days leads to an abundance of wildflowers, to every hiker's delight.

Plants have a unique reproductive cycle involving two generations

The reproductive or sexual cycle of plants is different from that of animals. In animals, specialized cells in the reproductive organs undergo meiosis to produce gametes (sperm and eggs). These two cells then unite through fertilization to form a zygote, which grows into an embryo that ultimately develops into an adult animal. Reproduction in animals results from a single generation of adults. Plants, on the other hand, have a slightly more complex cycle that involves *two* generations of adult plants—one

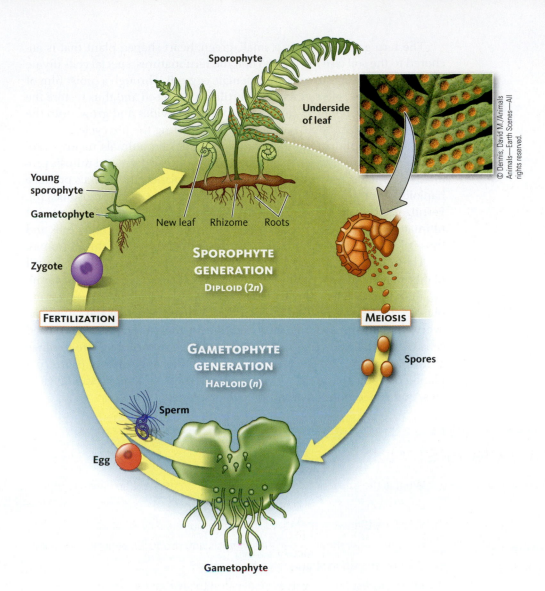

Sporophyte

Underside
of leaf

Young
sporophyte

Gametophyte

New leaf   Rhizome   Roots

Zygote

**SPOROPHYTE
GENERATION**
DIPLOID (2*n*)

FERTILIZATION

MEIOSIS

**GAMETOPHYTE
GENERATION**
HAPLOID (*n*)

Spores

Sperm

Egg

Gametophyte

Figure 14.6 Alternation of generations in the life cycle of a fern.
The fern life cycle consists of two generations: the sporophyte and the gametophyte. This overview shows where meiosis and fertilization occur in the plant reproductive cycle. Ferns grow new leaves (fiddleheads) and reproduce asexually from horizontal stems called rhizomes. Sporophytes produce spores on the underside of their leaves by meiosis. Spores germinate and grow by mitosis into a small heart-shaped structure called a gametophyte. Male and female sex organs develop and produce sperm and eggs. The multiflagellate sperm swims to and fertilizes the egg, which grows by mitosis into a young sporophyte, a fiddlehead, and then a mature fern.
© Cengage Learning 2014

generation produces eggs and sperms (gametes), and the other generation produces specialized reproductive cells called spores. To meet the challenge of understanding plant reproduction, let's track through the life cycle of a fern, a plant familiar to many people (see **Fig. 14.6**). All plants progress through the same stages as the fern.

The plant life cycle is called *alternation of generations* because it alternates between a spore-bearing generation, or *sporophyte,* and a gamete-bearing generation, or the *gametophyte.* The plant that is recognizable as a fern is the sporophyte generation. Its roots, rhizome, and leaves are composed of cells that have two sets of chromosomes and are diploid (2*n*) (see **Fig. 14.6**, top). As a fern matures, special cells on its leaves undergo meiosis to form a large number of spores. The microscopic spores are visible as large clusters on the underside of leaves of many ferns (see **Fig. 14.6**). Recall that meiosis shuffles the genetic material and reduces the number of chromosomes in half. So each spore, which is a single cell, is genetically unique and contains one set of chromosomes—it is haploid (*n*). Fern spores are dispersed, and those that land in a favorable location will start to divide and grow into another plant or gametophyte.

Spore—in plants, a reproductive cell that is genetically unique; it is dispersed into new habitats and grows into a new individual.

The fern gametophyte is a small, green, heart-shaped plant that is anchored to the soil (see **Fig. 14.6**). When the fern matures, special cells divide to produce male sperm cells and a female egg cell. Through a moist film of water, the sperm cells swim to and fertilize an egg cell and thus restore the diploid state of the life cycle. The zygote now divides and grows into the familiar fern, or sporophyte plant, thus completing the life cycle.

To briefly compare animal and plant cycles, in animals meiosis produces *gametes*, whereas in plants meiosis produces *spores* and mitosis produces eggs and sperm cells. The cycles of both plants and animals produce haploid egg and sperm cells that contain a single set of chromosomes (n). Fertilization produces a diploid ($2n$) zygote that grows into an embryo and ultimately into the fern roots, stems, and leaves. Thus, your body cells and those of the leafy fern are similar by containing two sets of chromosomes.

The fundamental concepts of plant reproduction are the same for all plants, and we have chosen to trace the path of events for the most familiar, numerous, and dominant plants in most ecosystems—the flowering plants. The trees, grasses, and wildflowers that you see in the landscape every day are the sporophyte generation in their life cycle. Specialized cells within their flowers produce spores, which in turn divide to produce eggs and sperm. As we progress through the angiosperm (flowering plants) reproductive cycle, we will point out these structures (spores, gametes) and generations (sporophyte, gametophyte).

14.2 check + apply YOUR UNDERSTANDING

- Many plants can reproduce both asexually and sexually.

- Asexual reproduction occurs when a plant reproduces itself through the growth of its vegetative organs—roots, stems, or leaves. Offspring are genetically identical to the parent.

- Sexual reproduction involves a series of steps that include the fertilization of an egg by a sperm cell. The union of egg and sperm generates genetic diversity among offspring.

- Plants have a life cycle that alternates between a gamete-producing gametophyte generation and a spore-producing sporophyte generation. Meiosis produces genetically unique spores.

1. What is the impact of sexual reproduction on the genetic diversity of a population?

2. What are the two generations in the reproductive cycle of plants? Which of these generations undergoes meiosis and produces genetic variation? Which generation results from fertilization?

3. After a forest fire, how does clonal growth give aspens a competitive advantage over conifers that only reproduce by sexual reproduction?

4. Several factors influence the success of plant reproduction. Give an example of how a plant's shoot system influences the success of its reproductive system.

5. **Figure 14.7** shows aspen trees scattered on the side of a mountain in Colorado. It was taken in the autumn when the golden and orange aspen leaves changed color. Do you think that this distribution was the result of sexual or asexual method of reproduction? Explain your thinking.

Figure 14.7 **Aspen trees in autumn.**

14.3 Flowers Are the Organs for Sexual Reproduction

During the spring, many people visit Washington, D.C., for the blooming of the cherry blossoms (**Fig. 14.8**). Each spring, the famous cherry trees explode into a sea of pale pink and white blossoms that begin the two-week-long Cherry Blossom Festival. The festival brings in about a million visitors per year and $184 million for Washington-area businesses. The 3,000 ornamental flowering cherry trees were gifts from Japan in 1912 and are bred for their beautiful flowers. The cherry trees are so significant to tourism that the National Park Service has set up a special website and webcam to allow viewers to track the progress of flower buds and the development of the cherry blossoms. The peak bloom varies from late March to early April. The precise timing of flowering causes botanists to ask many questions. Which environmental factors signal the cherry trees to bloom? How do the trees perceive these signals? How do cherry trees keep track of the seasons so that they all flower at the same time? How do flowers function in the life of cherry trees?

In the section ahead, we describe the sequence of events that marks the start of the sexual reproductive cycle of flowering plants—the development and emergence of flowers and the sex cells they produce. We answer the questions regarding how plants accurately keep track of time and make the dramatic transition from the vegetative to the reproductive stage of their life.

Figure 14.8 The cherry blossoms in Washington, D.C., bloom each spring. The beautiful cherry blossoms attract millions of people to Washington, D.C., each spring. Internal rhythms (biological clock) and environmental cues trigger their development.

Plants time their flowering from internal and environmental cues

Plants grow and reach sexual maturity when they achieve the ability to flower. The saguaro first flowers when it is seven feet tall and 35 to 50 years old. The transition from vegetative to reproductive structures involves changes in the expression of specific genes in cells of the meristem. Genes that produce leaves are silenced, while genes that produce flowers are activated. These genetic switches are caused by an internal biological clock as well as a variety of cues from the environment.

One of the important environmental cues that triggers the switch to flowering is the amount of daylight. In Chapter 13, you examined light as a source of *energy* that drives photosynthesis; however, light is also a source of *information.* Plants perceive the duration, quality, and quantity of light falling on their surfaces to inform them about their surroundings, the time of day, and the season of the year.

Many plant species have a characteristic photoperiod in which they flower in response to the specific lengths of day and night. Light receptors in leaves actually measure the length of the darkness, but plants were originally named based on the length of the daylight hours from sunrise to sunset. Plants that typically flower when the day is longer (the night period is shorter) than a critical day length are classified as *long-day plants.* The critical day length varies from one species to another. These plants typically flower in the spring in temperate habitats and include daffodils, irises, and grasses. In contrast, *short-day plants* flower when the day length is shorter than its critical day length. Short-day plants such as asters and goldenrods flower in August and September in northern climates. Short days and long nights trigger their flowering. For people with seasonal al-

Photoperiod—the daily cycle of light and darkness that influences flowering in many plants.

Figure 14.9 **Common ragweed.** Ragweed is a short-day plant that causes allergic symptoms in the late part of the summer when the plant disperses large amounts of pollen into the air.

Sepals—the outermost parts of the flower that enclose the developing flower bud; often green and in the same number as petals; colorful in some plants to attract the attention of pollinators.

lergies, their symptoms of sneezing, runny nose, and watery eyes are caused by an abundance of pollen dispersed into the air at particular times of the year. So if you have allergy symptoms in June, it probably is due to grasses, wind-pollinated trees, or other long-day plants rather than short-day plants such as ragweed or sagebrush, which release their pollen in August and September (**Fig. 14.9**).

Another major environmental cue to flowering is temperature. The saguaro only flowers when the temperature is warm. Cooler spring temperatures delay the flowering of the cherry blossoms in Washington, D.C. Other plants flower only after their buds have been exposed to a period of cold temperatures followed by warm temperatures. This shift in temperature allows the plants to synchronize their flowering time with favorable moisture and growth conditions in the spring and avoid flowering in the fall or winter. Along with their internal rhythms, plants perceive and integrate light, temperature, and moisture conditions to determine the spring, summer, and autumn seasons.

A complete flower consists of four whorls of floral organs

Flowers are the sexual reproductive structures of flowering plants, or angiosperms, and are one of the distinguishing features of this group. Flowers are attached to the stem by a stalk that ends with a swollen base called the *receptacle.* The receptacle anchors the attachment of four whorls of modified leaves that develop within the floral meristem—the outermost sepals, the petals, the male stamens, and the central female carpel. These four floral organs compose the parts of a *complete* flower (**Fig. 14.10**).

The sterile parts of the flower do not produce sex cells and consist of the sepals and petals. **Sepals** are the outermost parts of the flower and en-

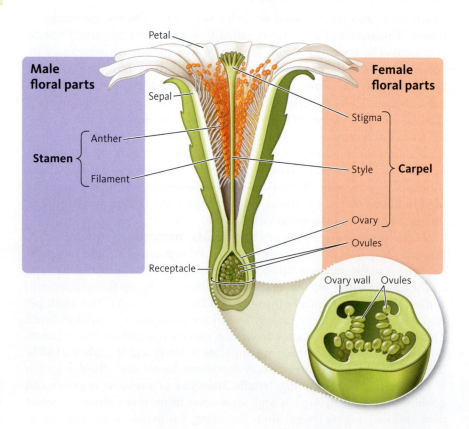

Figure 14.10 **The parts of a complete saguaro flower.** A complete flower contains four whorls of floral organs: sepals, petals, stamens, and carpels. A cross section through an ovary reveals numerous ovules.
© Cengage Learning 2014

close the developing flower bud. In many plants the sepals are green, but in other flowers, such as the orchids, the sepals are colorful to attract the attention of pollinators. **Petals** form the whorl just inside the sepals and often are colorful, with a sweet fragrance to attract pollinators. In many flowers like the saguaro, special glands at the base of the petals secrete nectar, a rich, sugary liquid. In the saguaro, the sepals and petals are the same shape and color and are not distinctly different from one another (see **Fig. 14.10**).

The fertile parts of the flower are the male stamens and the female carpel. **Stamens** are male floral organs that consist of a filament that supports a saclike structure called the **anther**. The anther produces pollen grains, which carry the male genome to the female part of the flower. The female **carpel** is located in the central portion of the flower and develops last. The carpel consists of three regions: the top is the *stigma,* a surface that receives pollen; the middle is the *style,* which elevates the stigma; and the bottom is an enlarged **ovary**. Within the ovary are compartments that support ovules, which will mature into seeds. A flower may consist of a single carpel (often called a *pistil*) or many carpels fused together.

Flowers from the 250,000 different angiosperms vary in the presence or absence of floral organs as well as their number, shape, color, and arrangement. For instance, quaking aspen flowers lack sepals and petals and have only one kind of sex organ, either male stamens or female carpels (see **Fig. 14.5**). Most flowers are bisexual because they contain both male and female sex organs.

The production of gametes involves meiosis and mitosis

Although flowers bring color to our lives, their sole function in the life of an angiosperm is to create genetic diversity through the production, release, and fertilization of sex cells. At the cellular level, genetic recombination is the result of meiosis and fertilization, a set of complex processes that are common to all sexually reproducing organisms. Let's recap the main features of meiosis and mitosis before we put them into play in the plant's reproductive cycle. Recall from Chapter 11 that meiosis produces three outcomes: (1) it reduces the number of chromosomes by half; (2) it introduces genetic variation through crossing over of sections of DNA and the random assortment of chromosomes; and (3) it produces four genetically unique daughter cells. This is in contrast to the process of mitosis, which faithfully duplicates the genetic material and produces two identical daughter cells, each with the same number of chromosomes and the same alleles as the parent cell. Recall that mitosis drives the process of growth and asexual reproduction, and meiosis drives genetic variation and reproduction. As we step through the production of sex cells in flowering plants, notice that both processes play a role.

The Production of Pollen   Now let's zoom into an anther, located on top of the stamens in a saguaro flower (**Fig. 14.11**). The anther is composed of four pollen sacs, or *cavities,* where groups of diploid (2*n*) cells develop. Recall that a diploid cell contains two sets of chromosomes, one set from each parent. These diploid cells undergo meiosis to form four haploid cells called *spores.* This is the reason why the familiar flowering plant is the

Petals—modified leaves of a flower that form the whorl just inside the sepals, protecting the reproductive organs; often colorful, with a sweet fragrance to attract pollinators.

Stamens—fertile male flower parts that produce pollen; consist of a filament that supports the pollen-producing anther.

Anther—the top part of the stamen (male) that produces pollen grains.

Carpel—female reproductive organ of a flower located in the central portion of the flower. The carpel consists of three regions: the top is the *stigma,* a surface that receives pollen; the middle is the *style,* which elevates the stigma; and the bottom is an enlarged ovary.

Ovary—structure at the base of a carpel that contains ovules; after pollination and fertilization, it will swell to become a fruit.

sporophyte generation of the two-generation life cycle; its cells are diploid and produce haploid spores in its sex organs, or flowers. Remember that because they are products of meiosis, each spore contains a different and unique set of genetic information.

Next, each spore nucleus divides by mitosis to produce two cells and develops a resistant outer coat, forming the pollen grain. Thus each pollen grain carries two cells. Eventually one of the cells inside of the pollen grain divides by mitosis to generate the male gametes, or two sperm cells. This is the reason why the pollen grain is referred to as the gametophyte generation in the life cycle. When the development process of pollen grains is complete, the walls of the anther pollen sacs break down, and the sacs split open to release the pollen. This process is summarized in **Figure 14.11**.

Ovules and Egg Cells   To start to trace the steps in the formation of an egg cell, we need to look inside the female ovary at an ovule before the saguaro flower is ready to bloom. The two processes here are similar to those that produce pollen. Follow the steps shown in **Figure 14.11** as they proceed from top to bottom. A large diploid cell is surrounded by layers of protective tissues that form an opening at one end of the ovule.

In the first step, this diploid cell undergoes meiosis to form four haploid spores. In most plants, only one of the spores continues its development. This spore divides to form several cells, one of which is the female gamete, or *egg cell.* The other important cell, called the *central cell* because of its location in the middle of the ovule, plays a key role in the development of the seed's food-storage tissue.

The process described has produced tens of thousands of genetically unique pollen grains and ovaries filled with ovules. The stage is now set for the blooming of the flower and the release and transfer of pollen to a receptive stigma.

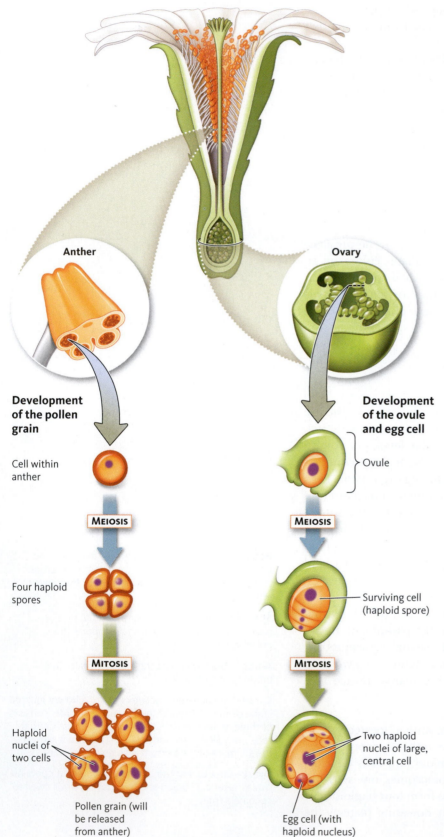

Figure 14.11 Pollen and egg development. The development of pollen grains occurs in the anther, and the development of egg cells occurs within the ovules of the flower. Notice that each pollen grain contains cells produced by meiosis and mitosis and is therefore genetically different from all other pollen grains released from the anther. In a similar process, cells in the ovule undergo meiosis and mitosis to produce a genetically unique set of cells, one of which will form the egg. Like pollen, every egg has a unique set of genes and capabilities.
© Cengage Learning 2014

Anther

Ovary

Development of the pollen grain

Cell within anther

MEIOSIS

Four haploid spores

MITOSIS

Haploid nuclei of two cells

Pollen grain (will be released from anther)

Development of the ovule and egg cell

Ovule

MEIOSIS

Surviving cell (haploid spore)

MITOSIS

Two haploid nuclei of large, central cell

Egg cell (with haploid nucleus)

- Plants flower in response to an internal clock and external cues such as light, temperature, or growth and development.

- A flower consists of four whorls of floral organs: two sets of sterile organs (sepals and petals) and two sets of fertile floral organs (male stamens and female carpels).

- Pollen grains develop within anthers, where meiosis occurs to produce male spores.

- Egg cells develop within ovules within the ovary of the female carpel.

1. In which months do long-day plants flower in the northern United States?
2. List the four floral organs that compose a complete flower in the order of their development.
3. Place the following cells and processes in their correct sequence: egg cell, spores, meiosis, mitosis.
4. Consider two pollen grains that are released from a flower's anther. Are the two nuclei inside a single pollen grain genetically identical or different? Are the two pollen grains carrying identical or different sets of genetic information? Why is the pollen grain called the gametophyte generation of the plant's life cycle? Explain.
5. There are about 100 carpels clustered in the center of a cinquefoil flower (**Fig 14.12**). Each carpel contains a single ovary and a single ovule (see **Fig 14.12**). How many egg cells are contained in each ovule and in each flower? How many times did meiosis occur in each flower? How many haploid spores were initially produced? How many proceeded through mitosis?

Figure 14.12 Cinquefoil.

14.4 Pollination and Fertilization Produce Seeds

When we think of floral scents, we usually associate them with sweet, pleasant perfumes. However, some flowers, such as the dead-horse arum, attract pollinators by deceiving them into thinking that they are dead animals (**Fig. 14.13**). When their flowers open, they release a foul-smelling stench that attracts blowflies, beetles, and flies from miles away. Remarkably, this species also generates heat that vaporizes the odor and further attracts the insects into its flower to contact its pollen. These insects come to the flower to lay eggs and, in doing so, bring about pollination. **Pollination** is the transfer of pollen from the anthers of the male to the female stigma. To be successful and make lots of seeds, each flower must attract and receive many pollen grains. In general, plants such as the saguaro and the dead-horse arum attract animals such as bats and blowflies, or they make use of the wind.

Flowers are built and function for sexual reproductive success, that is, for the production of a large number of vigorous seeds. To produce a single seed, an egg cell located in the ovule must be pollinated by a single pollen grain in a 1:1 relationship. In the section ahead, we will trace the events of pollination and fertilization that unite the male and female gametes to produce a package containing the starting point for a new plant—a seed (**Fig. 14.14**).

Figure 14.13 Dead-horse arum flower. The unusual flower of the dead-horse arum has a leaf-like structure mimicking the rear end of a horse. The flower's brown "tail" in the center consists of numerous flowers that release a foul odor to attract blowflies, which are deceived into thinking that the plant is a rotting carcass in which to lay their eggs.

Pollination—the transfer of pollen from male anthers to a female stigma.

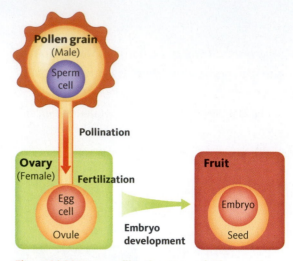

Figure 14.14 From pollination to seed. Three major reproductive events occur in sequence to produce a seed within a fruit: pollination, fertilization, and embryo development. Chemical signals from the growing embryo trigger the transformation of the ovary into the fruit and the ovule into the seed. Without pollination and fertilization, the seed and fruit will not develop.

© Cengage Learning 2014

Animal-pollinated flowers attract pollinators with a floral display

Animal pollination uses insects, birds, and bats to carry pollen from flower to flower. This type of pollination occurred in the earliest flowering plants about 125 million years ago. Even today, animals pollinate approximately 90 percent of all angiosperms, including many fruit crops such as apples, blueberries, and oranges. Reproductive success or failure often hinges on the ability of a flower to attract a pollinator through its floral display, which includes its size, shape, color, markings, scent, and the rewards it offers.

Think of a floral display as a roadside billboard that attracts your attention and sends you a message. Large billboards with attractive colors attract notice far more than small, drab-colored signs. In the same way, large, colorful flowers stand out in the landscape and generally attract more pollinators and produce more seeds than smaller flowers. However, large flowers are very costly to produce and maintain.

Flowers are shaped into tubes, bells, or bowls. Bowl-shaped flowers such as the buttercup provide easy access to pollen and nectar and are pollinated by a wide diversity of insects, such as bees, beetles, and flies. Some flower shapes have evolved to conform to the size of the specific pollinator so that they brush against anthers or deposit pollen on a stigma. For example, the shape of the saguaro flower remarkably matches the head size and tongue length of the bat species that pollinate it.

Many colors are associated with particular groups of pollinators. Hummingbirds are attracted to red flowers, and bats and moths are attracted to white flowers. The reason lies in how different animal groups perceive their world. Recall from Chapter 10 that animals have different lenses and visual receptor systems and thus do not see the same colors that humans do. Bees, for example, distinguish only four different colors in the visible spectrum: yellow, blue–green, blue, and ultraviolet light. This is why bees are attracted to blue, purple, and yellow flowers (**Fig. 14.15**).

Finally, a floral display is connected to a reward, which attracts and strengthens the relationship. Animals visit flowers to eat, to meet with members of the opposite sex and mate, and to lay their eggs. The most common foods provided to animal pollinators are pollen and nectar. Bats and birds visit saguaro flowers for their nutritious nectar. In addition to

Figure 14.15 How humans and bees view the same flower. Bees have a visual system that sees in the ultraviolet range of the spectrum. The yellow flowers in the photo appear to bees as purple and dark purple, shown at right.

Sylvie Bouchard/Shutterstock.com

© Natural Visions/Alamy

sugars, nectar contains amino acids, lipids, and minerals as well as small amounts of toxins and defensive chemicals to repel herbivores, pathogens, and nectar "thieves" that steal nectar without providing pollination "services."

Wind-pollinated flowers release and capture large amounts of pollen

About 10 percent of flowering plants are *wind-pollinated,* that is, they depend on wind currents to transfer pollen between plants and flowers. Many trees and grasses that live in open habitats with regular wind currents are wind-pollinated. Unlike animal-pollinated flowers, the flowers of wind-pollinated grasses, aspens, and oaks are green, brown, or drab colored. They do not rely on large, showy floral displays or inviting scents, or offer nectar as a reward. The structure of the flower and how these flowers are arranged on a stem, or its inflorescence, reflect its function—to liberate into the air large amounts of pollen that will be captured by female stigmas (**Fig. 14.16**). Petals would interfere with this function and are either small or absent on these flowers. Anthers are exposed to air currents by long filaments and styles that have a large, feathery surface area.

Wind-pollinated plants are responsible for our hay fever and other seasonal allergies marked by sinus inflammation, congestion, sneezing, runny nose, and watery eyes. Among North American plants, weeds are the most prolific producers of allergenic pollen. Ragweed is the major culprit, but other species that cause allergies are sagebrush, pigweed, lamb's quarters, and Russian thistle.

Pollination starts the race to fertilize the egg

Whether carried by the wind or on the body of an animal, many pollen grains do not survive the journey. In most plants, pollen lives and remains viable for only a short time. For example, pollen may live for only 30 minutes in some wind-pollinated plants or about a day in some insect-pollinated flowers.

When the pollen comes into contact with the stigma, a complex series of chemical signals occurs (**Fig. 14.17**, step 1). Proteins pass from the pollen to the cells of the stigma and indicate whether the pollen is compatible with it. Some plant species such as the saguaro are self-incompatible, where the stigma rejects pollen that carries its own genetic information. However, most species are self-compatible and accept their own pollen along with pollen of other plants of its species. The acceptance or rejection by the female has a major influence on the genetic composition of the seeds produced and the genetic diversity of the

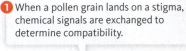

Figure 14.16 Many woody plants are wind pollinated. The male inflorescence of common hazel consists of numerous flowers that only produce stamens. These hanging inflorescences produce clouds of lightweight pollen that catch air currents and travel to female flowers. Relative to animal-pollinated flowers, wind-pollinated flowers are smaller in size, have longer filaments, have stigmas with a larger surface area, and do not produce nectar.

© blickwinkel/Alamy

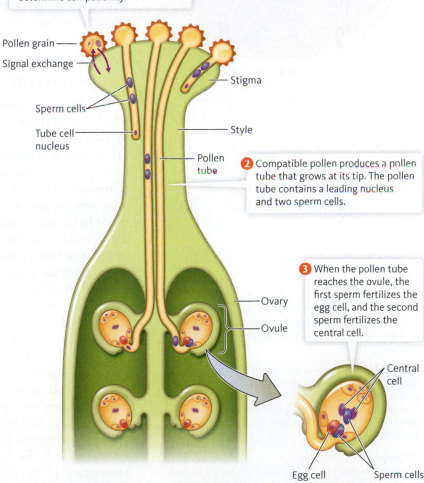

1 When a pollen grain lands on a stigma, chemical signals are exchanged to determine compatibility.

Pollen grain

Signal exchange

Sperm cells

Tube cell nucleus

Stigma

Style

Pollen tube

2 Compatible pollen produces a pollen tube that grows at its tip. The pollen tube contains a leading nucleus and two sperm cells.

3 When the pollen tube reaches the ovule, the first sperm fertilizes the egg cell, and the second sperm fertilizes the central cell.

Ovary

Ovule

Central cell

Egg cell

Sperm cells

Figure 14.17 Pollen-stigma interactions. Pollination leads to fertilization. The female stigma recognizes and controls the interaction. Compatible pollen germinates and produces one of the fastest growing cells in nature—the pollen tube. The tube cell's nucleus is guided to the entrance to the ovule where it delivers two sperm cells for double fertilization.

© Cengage Learning 2014

Inflorescence—a group of flowers in a definite arrangement on a plant stem.

Figure 14.18 Coconut "eyes." The three "eyes" on the base of a coconut show where the pollen tubes entered the ovule before the coconut seed developed.

population. Recall from Chapter 3 that genetic variation is the raw material on which natural selection acts.

A compatible pollen grain absorbs water, oxygen, and carbohydrates. Absorbing the nutrients activates a set of genes that produce enzymes to start making the ATP needed for growth. One of the two cells in the pollen grain swells and emerges through a pore in the pollen grain as a slender single-celled tube, or **pollen tube** (see **Fig. 14.17**, step 2). The other cell in the pollen grain divides by mitosis to form two sperm cells. Flowering plant sperm cells do not have flagella and, like all plant cells, contain a nucleus along with organelles such as ribosomes, Golgi complexes, and mitochondria.

The pollen tube is a single cell that grows rapidly through the style. In fact, the pollen tube cell grows at rates close to a centimeter per hour, among the fastest of any cell in nature. The cytoplasm constantly moves materials to the tip to maintain strong water pressure and build the cell wall, plasma membrane, and cytoskeleton Once the pollen tube reaches the opening of an ovule inside the ovary of the flower, it stops growing and then discharges its two sperm cells into the ovule, which contains two principal reproductive cells: the egg cell and the central cell (see **Fig. 14.17**, step 3). The three openings where pollen tubes entered the ovules are visible on a coconut seed as dark circles (**Fig. 14.18**).

Fertilization produces an embryo contained within a seed

One of the distinguishing characteristics of angiosperms is the way in which its two sperm cells perform **double fertilization** within an ovule. The first sperm cell entering the ovule fertilizes the egg cell to form a zygote. The second sperm cell fertilizes the central cell to form an energy-storage tissue called the **endosperm**. This seemingly simple act requires chemical communication and integration among all cells in the ovule, the pollen tube, and the sperm it carries.

At fertilization, fusion of the nuclei from the sperm and egg cells restores the diploid ($2n$) number of chromosomes in the zygote. The male sperm cell contributes half of the genetic information of the zygote; the other half comes from the female egg cell. Each ovule is fertilized by a different set of sperm from different pollen grains, making each zygote a unique genetic individual—the generation of variation by sexual reproduction.

The diploid zygote begins a series of cell divisions that will produce a multicellular **embryo**, which is a young plant growing inside of an ovule (**Fig. 14.19**). From the very beginning of its development as an embryo, the plant is organized with one end (a meristem) growing into the shoot and the other end growing and developing into the root. The first leaves to develop inside the seed are the *cotyledons*. The developing embryo requires a great deal of water and nutrients to support rapid cell division. The endosperm is a tissue that stores energy in the form of starch and oils for the seed. It functions to supply the growing embryo with fuel. The embryo and endosperm are enclosed by the walls of the ovule, which eventually form a protective seed coat to become the **seed**.

The plant has invested a large amount of energy and resources to get to this point. It has built flowers, produced sperm-producing pollen grains, transferred pollen to the stigma, exchanged signals, and grown a fast-growing pollen tube that delivered two sperm to the ovule. Within the ovule, fertilization has resulted in the growth of a new plant, the embryo. It is provisioned with nutrients and a protective coat in its subsequent journey to a new habitat.

Pollen tube—a slender single-celled tube that develops from a pollen grain; grows from the stigma and down the style to penetrate the ovule and deliver sperm (male gamete) to unite with an egg cell (female gamete).

Double fertilization—a distinguishing process in angiosperms where the two sperm nuclei from the pollen grain unite with two cells in the ovule. The first sperm cell fertilizes the egg cell to form a zygote, and the second sperm cell fertilizes the central cell to form the endosperm.

Endosperm—food-reserve tissue of the seeds of flowering plants that surrounds the embryo; stored energy is absorbed as the embryo develops.

Embryo—a young plant that grows from a zygote (fertilized egg); in seed plants it grows inside an ovule and then a seed.

Seed—a multicellular structure that consists of an embryo and food reserves (endosperm) and is enclosed by a protective seed coat; originates as an ovule that contains a fertilized egg.

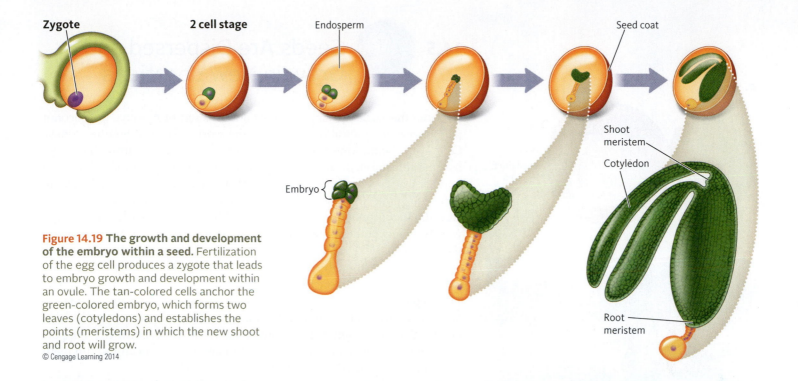

Zygote   **2 cell stage**   Endosperm   Seed coat

Shoot meristem

Cotyledon

Embryo {

Root meristem

Figure 14.19 The growth and development of the embryo within a seed. Fertilization of the egg cell produces a zygote that leads to embryo growth and development within an ovule. The tan-colored cells anchor the green-colored embryo, which forms two leaves (cotyledons) and establishes the points (meristems) in which the new shoot and root will grow.
© Cengage Learning 2014

14.4 check + apply YOUR UNDERSTANDING

- A flower's display includes color, shape, scent, size, and shape. Specific pollinators are attracted to different colors, shapes, and markings.

- Animal-pollinated flowers have a showy display of color, shape, and size, with sticky or spiked pollen.

- Wind-pollinated flowers are drab-colored, produce large amounts of pollen, and do not offer nectar as a reward.

- Pollination and recognition of compatibility lead to the production of sperm, pollen tube growth, and fertilization of the egg cell and central cell. A zygote and nutritive food storage tissue are formed.

1. Place the following six steps in the correct order: pollination, fertilization, sperm formation, pollen germination, embryo development, pollen tube growth.

2. List three reasons why animals visit flowers. How do flowers attract the animal to return with pollen from another flower?

3. A pumpkin produces 500 seeds. How many pollen grains were required to produce them? How many pollen tubes and sperm cells were required?

4. Two groups of flies are pollinators. One group has short tongues, the other long tongues. Match the two descriptions of the following flowers to their group of fly pollinators: (a) a flower with nectar at the end of a long floral tube, and (b) a flower with an open bowl shape with readily accessible pollen and nectar. Explain your thinking.

5. Many species in the violet family produce complete flowers that never open (**Fig. 14.20**). These closed flowers successfully produce fruits and disperse seeds. Make the following three predictions: (a) Is this violet species self-compatible or self-incompatible? (b) Are seeds from these flowers more likely to be genetically similar or different than the parent plant? (c) Will all of the seeds be genetically identical to one another? Explain.

© Bob Klips

Figure 14.20 Closed violet flower.

a.

Husk

Skin

Shell (fruit wall and seed coat)

Meat

© Sunsetman/Dreamstime.com

b.

Kevin Schafer/Danita Delimont Photography/Newscom

Figure 14.21 Coconuts use water to disperse fruits. (a) The coconut fruit is a fibrous husk that encloses a large seed inside a hard shell. The white coconut meat and milk are its food-storage tissue. (b) The coconut palm grows along coastlines of the tropics and disperses its fruit in the water, where the coconut floats with the ocean currents.

Fruit—a ripened ovary that develops from flowers; contains seeds and other parts of the flower.

Vegetable—a root, stem, or leaf of a plant that is eaten by humans.

14.5 Seeds Are Dispersed and Grow in New Locations

Far from the saguaro cactus in the Sonoran desert of Arizona, the coconut palm grows in tropical areas around the world—Hawaii, Southeast Asia, Central and South America, and the coast of Africa. Flowers from these trees become transformed into very large fruits that can weigh more than five pounds. The coconuts sold in supermarkets are actually large seeds that, in nature, are protected by a thick husk and a hard, brown fruit wall (**Fig. 14.21a**). How did the coconut tree migrate and establish itself on beaches all over the world? One hypothesis states that coconuts were dispersed by floating on the surface of the ocean where they drifted for miles across the open ocean (see **Fig. 14.21b**). Although some coconuts germinate in the water, many will wash up onto shore where they germinate and establish a new coconut tree. This hypothesis rests on the assumptions that a floating coconut could drift for thousands of miles and that the embryo in the seed would remain viable and be able to germinate and grow upon arrival. Do experiments support this hypothesis, or is it more likely that humans carried coconuts with them as they migrated across the globe?

A variety of observations, experiments, and computer simulations tried to answer this question. Wind and water currents could push a coconut on average about 25 miles per day. At this speed it would take the coconut seven months to cross the Pacific Ocean, and it would not stay afloat that long. In fact, laboratory experiments have shown that coconuts can float and then germinate after a maximum of 110 days in seawater. Biologists concluded that it is extremely unlikely that coconuts could have floated on wind and water current across the Pacific Ocean, and they must have been helped in some way by humans.

In the section ahead, the reproductive cycle of a flowering plant will draw to a conclusion. Along the way, flowers have developed and opened, spores have divided into gametes, and pollination and fertilization have occurred. The stage is now set to describe the final stages that take us from seed and fruit development and their movement, or *dispersal*, into new and favorable habitats.

Fruits grow and develop from flowers

Familiar fruits such as apples, cherries, peaches, and peanuts have one thing in common—they all develop from flowers and contain seeds. In the grocery store or the kitchen, the term *fruit* usually means a plant part that is sweet tasting. Although many fruits are sweet, this definition excludes fruits such as nuts, beans, and tomatoes. From a botanical perspective, a **fruit** is a ripened ovary together with other floral parts that may be attached to it. It's important to distinguish the botanical and common usage for the terms *fruit* and *vegetable*. In common usage a fruit is a part of the plant that is sweet tasting. This is why peas and beans, for instance, are commonly referred to as vegetables, when botanically speaking they are ripened ovaries, or fruits. **Vegetables** such as lettuce, carrots, and potatoes come from a plant's vegetative organs—roots, stems, or leaves.

The growth and development of a fruit are complex processes that begin with ovary development within the flower (**Fig. 14.22**). Successful pollination and fertilization cause sepals and petals to collapse and fall away

a.
Fertilized flower

© Daniel Borzynski/Alamy

b.
Developing fruit

Robert Pickett/Visuals Unlimited, Inc.

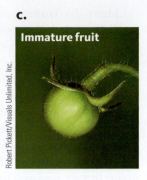
c.
Immature fruit

Robert Pickett/Visuals Unlimited, Inc.

d.
Mature fruit

Gabrielle Hovey/Shutterstock.com

Figure 14.22 Tomato fruit development. This series of four photos shows the transformation of the development of a fruit. (a) It begins with a pollinated tomato flower; (b) the small ovary sheds its flower parts and enlarges; (c) the fruit grows and expands; (d) the ripened ovary absorbs water, enlarges, and changes color and flavor to become a large, red, fleshy tomato fruit enclosing numerous seeds.

and the stamens to wither and deteriorate. The fruit then goes through a series of rapid cell divisions that increase the number of cells. The size and shape of a banana or a pear depends on the number and direction of these cell divisions.

Cell division then ceases and is followed by cell enlargement. Water and food reserves accumulate, and the fruit reaches its mature size and shape (see **Fig. 14.22**). During this time, the ovary wall changes into the fruit wall, which becomes soft in fleshy fruits (such as tomatoes) and dry in dry fruits (such as nuts).

The final stage of development involves the ripening of the fruit (see **Fig. 14.22**). Ripening occurs after the seeds have completed their development. Ripening changes the fruit's color, texture, and flavor, which make the fruit attractive and tasty for animals. On the cellular level, the change in color in fleshy fruits is the result of green chloroplasts changing into red chromoplasts. The softening of the fruit is accomplished by enzymes that break down cell walls and the "glue" between them. Another set of enzymes converts starch to sugars, giving the fruit a sweet flavor.

Fruit development is regulated by several different hormones. A group of fleshy fruits, such as apples, bananas, and tomatoes, releases **ethylene**, a gaseous hormone that activates genes to turn on the ripening process. Fruits produce ethylene even after they have been harvested. The apples, bananas, and grapes in a fruit bowl are releasing ethylene gas to speed up the ripening process. This type of ripening has led to the saying that "one rotten apple ruins the whole bunch." Seeds produce hormones that stimulate the fruit to enlarge and mature. However, cultivated fruits such as bananas, pineapples, and grapes are exceptions to this rule. These fruits have been selected and cultivated for their ability to develop into fruits that don't contain seeds.

Fruits are adaptations for seed dispersal

Colorful, fleshy fruits are adapted for dispersal by animals. Seeds often are spat out or travel within animals and pass through their digestive system, where they subsequently are regurgitated or defecated in conditions favorable for germination. In these cases, natural selection favors seeds with a tough, resistant seed coat that allows them to survive passage through the physical and chemical stresses of an animal's digestive system. Many types of dry fruits also travel on the outside of animals with which they make contact, using seed adaptations that include hooks, spines, burrs, or substances that stick to hair or feathers. In this type of dispersal, selection favors smaller seeds that are less noticeable to the animal, which will eventually pick off the seeds and discard them.

Ethylene—a gaseous hormone that causes some fruits to ripen.

Many dry fruits are also adapted for wind dispersal. Wind-dispersed seeds, like wind-dispersed pollen, have modifications that reduce their size and weight and increase their surface-to-volume ratio to slow their descent to the ground, resulting in dispersal farther from the parent plant. Wind-dispersed fruits are characteristic of dry, open habitats such as fields or along the sides of roads. Other dry fruits such as nuts and grains are eaten and dispersed by animals.

How a seed is dispersed influences how far it travels. Observations show that birds and ocean currents can carry seeds great distances; however, the vast majority of seeds move only short distances from the parent plant. In fact, research has shown that 99 percent of tree seeds were found beneath the crown of the parent tree in an area called the *seed shadow.* Seeds that fall within the seed shadow compete among themselves and with the parent plant for light, water, and minerals. Seed predators, or animals that eat seeds, generally search for seeds near the parent plant and ignore seeds outside of the seed shadow. The few seeds that are dispersed farther away may fall into habitats where they can germinate, survive, and eventually outcompete resident plants.

From an evolutionary perspective, sexual reproduction and seed dispersal are essential processes for adapting a plant population to its environment. Dispersal moves a unique set of genes (alleles) to new locations and into new populations. Recall from Chapter 3 that this migration, or gene (allele) flow, is a major contributor to the genetic diversity of a plant population and thus provides the raw material on which natural selection operates.

Dormancy prevents germination when the chance of survival is low

The completion of the reproductive cycle and one of the most critical periods in the life of a plant is the transition from dispersed seed to seedling (**Fig. 14.23**). For saguaros this is the stage with the highest mortality. To continue into the next generation, a seed must resume its growth, or **germinate**, in the right place at the right time under the right conditions. Sometimes seeds are dispersed and ready to grow as soon as environmental conditions are favorable. However, most seeds are dispersed in a state of **dormancy**, or arrested growth, and will not germinate even if the conditions are right. Dormancy is an adaptation with great survival value for a

Germination—growth of an embryo within a seed that breaks through the seed coat; often occurs after a period of dormancy.

Dormancy—a state of arrested growth; an adaptation with great survival value for plants, fungi, some protists, and bacteria.

Figure 14.23 Seed germination and seedling growth. The first sign of germination is the emergence of the root through the seed coat. The expanding and growing shoot pulls the cotyledons and first leaves above the soil into the sun and begins to establish the seedling as an independent plant.

Bogdan Wankowicz/Shutterstock.com

plant. For example, if warm temperatures triggered seed germination, then many seeds might begin growth during a spell of warm autumn weather and then be killed by the cold winter frost. Plants that live in seasonal habitats often require a specific period of cold temperature before germination. This is how seeds perceive that the season is spring and conditions are safe for them to germinate and grow.

Each plant species has a unique set of germination requirements, which include temperature, oxygen, and water. Some seeds, such as the saguaro, require light, whereas others are inhibited by light. For many small seeds with few food reserves, light has great survival value to prevent germination when the seeds are buried deeply in the soil.

Seeds germinate and become established as seedlings

When conditions are favorable, the seed absorbs water and swells, and germination begins. Internally, nutrients stored in the endosperm or cotyledons are broken down and transported to the embryo. The rate of metabolism increases dramatically as each cell uses the nutrients and oxygen for energy production to support cell division and expansion. Cells of the shoot and root apical meristems move rapidly through the cell cycle with DNA replication and mitosis. Once germination has begun, there is no turning back; the embryo is committed to growth or death. Externally, germination is marked by the appearance of the root breaking through the seed coat on its way to establishing a root system. **Figure 14.23** shows the stages in the germination and establishment of a bean plant.

The seedling has established itself when it no longer relies on the nutrient reserves in its seed because it now has leaves and can sustain itself through photosynthesis. At this early stage of development, the seedling is especially vulnerable to drought, freezing, competition, and herbivores. Many seedlings are washed away by rain or are buried by earthworms. The successful establishment of the young seedling concludes the reproductive cycle for the flowering plant. During its lifetime, on average, only a single seed will be dispersed, break dormancy, germinate, survive, grow, compete, and reproduce. In other words, a plant has replaced itself.

14.5 check + apply YOUR UNDERSTANDING

- Flower parts become fruit parts. A fruit is a ripened ovary together with other floral parts that may be attached to it.

- Seeds consist of an embryo and nutritive food-storage tissue, enclosed in a seed coat.

- Most seeds are dispersed close to the parent plant. Animals commonly disperse fleshy fruits, whereas wind and animals disperse dry fruits.

- Seed dormancy improves the chances that a seed will germinate under conditions that favor survival and reproduction.

1. What is the relationship between a fruit and a flower?

2. What is the advantage for the seed to be dispersed beyond the seed shadow?

3. How does requiring a period of freezing temperature adapt a seed for survival?

4. A shipment of bananas contains a packet of a powdered chemical that absorbs ethylene gas. How will this affect the fruits?

5. Refer back to the fern spores discussed at the end of Section 14.2. How are seeds and spores similar in their functions? How are they different in their structures?

End of Chapter Review

Self-Quiz on Key Concepts

Plant Reproductive Success

KEY CONCEPTS: Plants can multiply through asexual or sexual reproduction. Asexual reproduction occurs through the rapid growth of vegetative organs and produces no genetic variation in the population. Sexual reproduction unites sperm and eggs through fertilization and produces high levels of genetic variation. The process of plant sexual reproduction involves producing specialized structures that are timed and coordinated with other plants, animals, and the environment.

1. Match each of the following **methods of plant reproduction** with its description or example. Each term on the left may match more than one description on the right.

   Asexual   a. an aspen clone
   Sexual   b. shoots of an aquatic plant breaking apart and colonizing a new pond
                  c. method of reproduction used by annuals
                  d. marked by fertilization
                  e. generally true when a plant has flowers, fruits, and seeds
                  f. results in large amount of genetic diversity

2. Match each of the following **reproductive cells** with its description or example. Each term on the left may match more than one description on the right.

   Egg cell   a. the male sex cell
   Sperm cell   b. formed in ovules inside the ovary of a flower
   Spore   c. a single reproductive cell that will divide and grow into another plant
                  d. travels down the pollen tube
                  e. produced by the sporophyte generation in the plant life cycle
                  f. formed by meiosis in the life cycle of plants

3. Dandelions produce flowers and seeds without meiosis and fertilization. This is an example of _____ reproduction, and all of the seeds will be genetically _____ .
   a. asexual: identical
   b. asexual: different
   c. sexual: identical
   d. sexual: different

4. The aspen tree, flowering cherry tree, and saguaro cactus represent which generation in their life cycle?
   a. diploid gametophyte
   b. diploid sporophyte
   c. haploid gametophyte
   d. haploid sporophyte

Flowers and Pollination

KEY CONCEPTS: Environmental and internal cues cause flowers to develop from shoot meristems. Flowers are the sexual organs, with four sets of parts that function for the transfer of pollen (pollination) from male to female parts and for fertilization. Animal pollinators are attracted to characteristics such as a flower's size, height, shape, color, markings, and scent and to the rewards the flower offers. Wind-pollinated flowers are adapted to disperse and capture large amounts of pollen through the air.

5. Match each of the following **flower parts** with its description or example. Each term on the left may match more than one description on the right.

Petal   a. sterile part of flower that usually is colorful
Carpel   b. outermost parts of the flower that enclose the developing flower bud
Sepal   c. consists of a stigma, style, ovary, and ovules
Stamen   d. produces pollen grains
               e. consists of an anther and filament
               f. will be transformed into a fruit after fertilization
               g. fertile part of flower where seeds develop

6. Match each of the following **pollination systems** with its description or example. Each term on the left may match more than one description on the right.

Animal   a. Flowers are drab colored
Wind   b. Flowers have a foul odor
               c. Flowers living in open habitats
               d. Large amounts of pollen are produced
               e. Stigma has a large surface area for pollen capture
               f. Flowers have large petals
               g. Pollen is released from a great height

7. Match each of the following **compatibility systems** with its description or example. Each term on the left may match more than one description on the right.

Self-compatible   a. Rejects pollen from own flower or flowers on same plant
Self-incompatible   b. Accepts pollen from own flower or flowers on same plant
               c. Creates high levels of genetic diversity
               d. Creates low levels of genetic diversity

8. Which part of the pollen grain functions in protection, transfer, and recognition by the stigma?
   a. generative cell
   b. tube cell nucleus
   c. outer pollen coat
   d. cytoplasm of the tube cell

9. Which of the following cellular processes produces an egg cell and sperm cells?
   a. cell differentiation
   b. meiosis
   c. cell expansion
   d. mitosis

Fertilization and Seed Development

KEY CONCEPTS: Compatible pollen grains germinate, produce sperm, and grow pollen tubes to fertilize an egg cell. Fertilization is followed by seed development, which includes growth of the embryo, filling of the endosperm with nutrients, and development of a seed coat. Seeds produce hormones that trigger fruit development.

10. Match each of the following **seed parts** with its description or example. Each term on the left may match more than one description on the right.

Embryo   a. Consists of the shoot and root meristems
Endosperm   b. Encloses the entire seed
Seed coat   c. Is the storage tissue for the seed
               d. Is the result of the first fertilization event
               e. Is the result of the second fertilization event
               f. Cotyledons are parts of this seed part

11. Which of the following is the correct sequence in *fertilization*?
    a. pollen tube growth, pollen germination, fertilization of egg cell, sperm cell formation
    b. sperm cell formation, pollen tube growth, pollen germination, fertilization of egg cell
    c. pollen germination, pollen tube growth, sperm cell formation, fertilization of egg cell
    d. pollen germination, sperm cell formation, pollen tube growth, fertilization of egg cell

12. Which of the following is *false* regarding a *pollen tube*?
    a. consists of a chain of cells that divide in a single direction
    b. grows by adding cell wall material at its tip
    c. arises from the germination of a pollen grain
    d. grows rapidly down the style into an ovule

Fruit and Seeds: Dispersal and Dormancy

KEY CONCEPTS: The ovary of the flower develops into the fruit, which functions to disperse the seeds it contains into favorable sites for germination. Fleshy fruits attract animals for seed dispersal, and seeds of dry fruits catch wind currents or become attached to animal fur. Seeds usually are dispersed in a dormant condition that prevents germination when the chance of survival is low. Seeds break dormancy when environmental conditions are favorable and germinate into seedlings. Seedlings establish their independence from their nutrient reserves contained in the seed when they build roots, stems, and leaves.

13. Match each of the following **dispersal mechanisms** with its description or example. Each term on the left may match more than one description on the right.
    Animal   a. fruits with hooks, spines, and burrs
    Water   b. fruits with a sticky surface
    Wind   c. fleshy, sweet-tasting fruits
                       d. fruits with a large air pocket
                       e. lightweight with appendages

14. How far do most seeds disperse from the parent plant?
    a. not far at all—only a few feet and yards away
    b. a moderate distance—hundreds of feet and yards away
    c. long distances—miles away

15. Which of the following seed-germination cues prevents seeds from germinating too deeply in the soil?
    a. temperature
    b. light
    c. moisture
    d. oxygen

16. Ethylene is a gaseous hormone that activates genes for which process?
    a. ovary development
    b. fruit ripening
    c. fruit dispersal
    d. fruit enlargement

Applying the Concepts

17. Poinsettias are beautiful houseplants that are commonly sold between Thanksgiving and Christmas (**Fig. 14.24**). The poinsettia initiates flowering when it has sustained six weeks of nights that were longer than 11.5 hours (critical day length of 12.5 hours). Is it a short-day or a long-day plant? In the United States, do poinsettias enter their photoperiod around the vernal equinox (March 20–21) or the autumnal equinox (September 22–23)? Name an animal that is attracted to the poinsettia's red "leaves."

Figure 14.24 Poinsettia flowers.

18. Farmers plow their fields in preparation for planting crops (**Fig. 14.25**). Plowing brings buried seeds to the surface. Which environmental factors surrounding a weed seed does plowing alter? Seeds with which germination requirements will be favored? How can this practice control weeds?

19. Burdock is an herbaceous biennial that lives in fields and pastures. It was the inspiration for the hook-and-loop fabric fastener (Velcro®). In nature, its adaptive hooks catch and tightly stick to a passing animal's fur. The animal carries the burr some distance before it stops to remove it. Look at **Figure 14.26** and determine whether this adaptation is one of the burdock's receptacle, fruit, or seed. How would the height of the stem affect the kinds of animals that disperse the burr? Would a large- or small-sized burr be favored to move the burr out from the seed shadow? Explain.

Figure 14.25 Tractor plowing field.

Figure 14.26 Burdock fruits.

20. Pine trees are wind-pollinated plants. The distance that a pollen grain travels is determined by three factors that are observed in nature: (1) the height at which it is released, (2) the air flow or wind currents that it catches, and (3) how long the pollen grain can stay aloft. A pine pollen grain has two winglike projections that contain gas (**Fig. 14.27**). Which of the three factors is maximized by this structure?

Figure 14.27 Pine pollen grain.

Data Analysis

Butterfly Feeding and Nectar Concentration

Most day-flying butterflies have relatively low energy requirements because they have low rates of metabolism and are slow fliers. The European skipper feeds on nectar from thistles, clover, alfalfa, and vetch as its source of food energy (**Fig. 14.28**).

In the field, skippers select flowers that provide them with a rich source of sugar. Flower nectar varies in the concentration of sugars that it provides to its pollinators. Ecologists who study butterfly–flower interactions are interested in understanding what concentration of nectar skippers prefer. Do male and female skippers differ in their food preferences? To answer these questions, plant ecologists measured the duration of feeding and volumes of uptake of artificial nectar at different sugar concentrations for both male and female butterflies.

From Pivnick KA, McNeil J. 1985. Effects of nectar concentration on butterfly feeding rates for Thymelicus lineola and a general feeding model for Lepidoptera. Oecologia 66, 226–237.

Data Interpretation

Figure 14.29 shows (a) the duration of feeding and (b) volume of nectar ingested at increasing sugar concentrations for male and female skippers. The experiment was conducted at 25°C (77°F).

21. Describe the relationship between the length of time a skipper fed and the nectar concentration. At which nectar concentration do females begin to spend much more time ingesting nectar than do males?

22. What is the relationship between the sugar concentration and the volume that the skipper ingests?

Critical Thinking

23. Why do think females consume more nectar than males?

24. It was stated that generally butterflies only ingest dilute sugars. Do these data support or refute this conclusion?

25. Why do you think that the greater the concentration, the lower the volume ingested by the skipper? Do you think there's a limit to the ability of a butterfly to suck up nectar through its long proboscis? Explain your reasoning.

26. Which sucrose concentrations are most costly for plants to produce? How is the butterfly placing selective pressure on the flower's nectar concentration?

27. How do a butterfly's energy requirements compare to those of a hummingbird? Predict whether the concentration of sucrose that these fast-flying birds prefer is greater or lesser than that of butterflies.

Figure 14.28 Skipper drinking nectar.

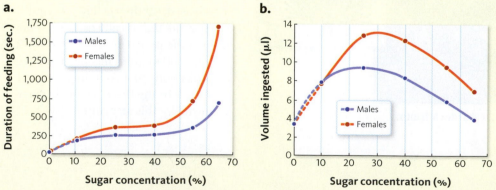

Figure 14.29 Graphs that show relationships between butterfly feeding and nectar concentration.
© Cengage Learning 2014

Question Generator
Strangler Fig Fruit Dispersal and Bats

The strangler fig is an important keystone species in many tropical ecosystems because it is an important food source for a wide range of animals, which disperse its seeds. Different species of strangler figs produce fruits of different sizes, colors, ripening patterns, and aromas. One species produces green, aromatic fruits that ripen over a 5- to 8-day period. Another species produces red odorless fruits that ripen over a 2- to 3-week period (**Fig. 14.30**). Each fruit contains dozens of seeds with a thick seed coat that pass through an animal's digestive tract without being harmed.

Fig fruits and their seeds are eaten and distributed by a number of bat and bird species. Bats feed at night and primarily use scent to detect and find ripe, green fig fruits (see **Fig. 14.30**). They take small chunks of fruit and fly away to digest the juices and pass them and the seeds efficiently within a half hour. Birds feed on fruit during the day and use visual cues to find red fig fruits. They stay and eat the fruit for longer times, and they scatter most of the seeds through their feces near the parent plant.

Below is a block diagram (**Fig. 14.31**) that relates several aspects of the biology of strangler fig seed dispersal by fruit bats and birds. Use the background information along with the block diagram to ask questions and generate hypotheses. We've translated the components of the block diagram into several questions to get you started.

Figure 14.30 Red strangler fig fruits are eaten by bats.

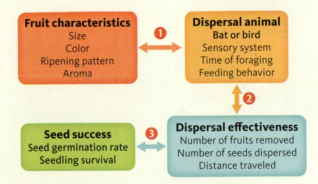

Figure 14.31 Factors relating to strangler fig seed dispersal.
© Cengage Learning 2014

Here are a few general questions to start you off. The number of the question refers to the number on the block diagram above.

1. Which fruit characteristics are most important to the dispersal animal? (Arrow #1)

2. Which fruit-dispersal animal is most effective at moving large numbers of seeds long distances? (Arrow #2)

3. What is the relationship between the number of fruits removed and the number of seeds that germinate? (Arrow #3)

4. How does the fig's fruit-ripening pattern affect feeding behavior? (Arrow #1)

5. How does the fig's aroma affect the number of fruits removed by bats? (Arrows #1 and #2)

Use the block diagram and the characteristics in each box to generate your own research question and frame a hypothesis.

From Korine C, Kalko KV, Herre EA. 2000. Fruit characteristics and factors affecting fruit removal in a Panamanian community of strangler figs. Oecologia 123, 560–568.

15 The Biology of Fungi

15.1 Explosive spore ejection of the hat-thrower fungus

15.1 Explosive Spore Ejection of the Hat-Thrower Fungus

The hat-thrower fungus *(Pilobolus kleinii)* is one of many species of fungi adapted to live and grow on the feces, or dung, of herbivores such as cows, horses, moose, and bison. As disgusting as it sounds, dung has everything a fungus needs to live: water, carbon, nitrogen, vitamins, and minerals. Structurally, the dung pile is held together by bits of undigested straw embedded in a matrix of mucus, which is an ideal habitat for fungi such as *Pilobolus*. These dung-loving fungi are decomposers that break down the remains of the herbivore's meal and transform its chemical energy into the energy they need to live.

Pilobolus has evolved the ability to survive in two very different environments: the dung in the open pasture and the dark, warm, fluid-filled digestive system of a cow (**Fig. 15.1**). During its time inside the cow, the fungus is a **spore**, the reproductive and dispersal stage of its life cycle. However, once the fungus is eliminated from the body with the dung, specific chemical cues contained in the cow dung trigger spore germination and growth of cellular filaments called **hyphae** (singular, *hypha*) (**Fig. 15.2**). In contrast to animals, fungi such as *Pilobolus* secrete enzymes that digest large food molecules *outside* their bodies and then absorb the small building block molecules back into their hyphae. This external type of digestion is an essential feature of all fungi.

The fungus grows into an interconnected network of hyphae, or **mycelium**, that branches outward throughout the dung to find and absorb nutrients (see **Fig. 15.2**). Water plays a critical role in the metabolism and transport of substances to the growing hyphal tips. *Pilobolus* also requires a specific chemical compound, found only in dung, that helps it bind and absorb iron, an essential mineral necessary for aerobic metabolism. This growth requirement restricts *Pilobolus* to grow only on dung; the compound is rarely found anywhere else. *Pilobolus* spends its life migrating back and forth between the cow and the dung it eliminates in the pasture. The question is how *Pilobolus* gets from the dung back into the cow. The answer lies in the way the fungus explosively ejects its reproductive spores.

Fungi such as *Pilobolus* have evolved a variety of ways to disperse their spores to new habitats. Recall from Chapter 2 that a fungal spore is a reproductive cell that is dispersed, germinates, divides, and grows into a new fungus. Hundreds of spores are contained within a black *spore sac* located on top of a stalk, like a hat. On the dung, the black hat sits atop a bubble on the end of a stalk less than an inch tall (see **Fig. 15.2**). Amazingly, the tiny bubble acts like a lens and focuses the sun's rays onto photoreceptors that cause the stalk to bend toward the sun.

Spore—a reproductive cell that is dispersed, germinates, divides, and grows into a new fungus; may arise by mitosis (asexual spore) or meiosis (sexual spore).

Hyphae—thin, microscopic cellular filaments that make up the main body of a multicellular fungus; grow into a highly branched network (mycelium).

Mycelium—an interconnected network of hyphae that branch throughout a substrate to find and absorb nutrients; vegetative part of the fungus.

Digestive system of a cow

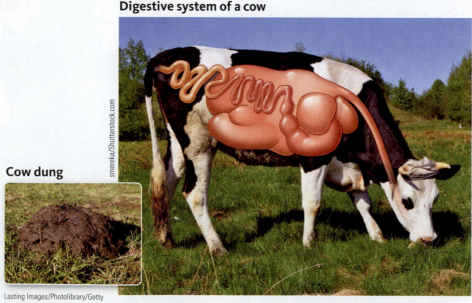

smereka/Shutterstock.com

Cow dung

Lasting Images/Photolibrary/Getty Images

Figure 15.1 Dung fungi have adapted to living in two very different environments. The life of a dung fungus, such as the hat-thrower, includes time in the open pasture and time within the cow's digestive system. Cows do not graze near their own dung, so to get back into the cow, the fungus must throw its spores far from the pile into the surrounding pasture.
© Cengage Learning 2014

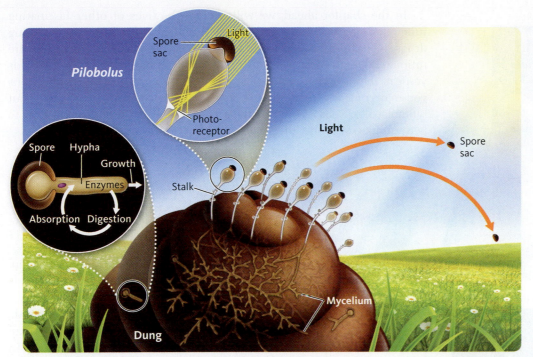

Figure 15.2 Growth and spore dispersal of the hat-thrower fungus. The hat-thrower fungus is one of dozens of species that colonize and live on dung. Its spores germinate into a growing hypha that secretes enzymes that digest and absorb the dung's nutrient molecules. Its reproductive hyphae form spores in a black spore sac (hat) that bends toward the light and shoots its spores far away from the dung pile.
© Cengage Learning 2014

The bubble also fills with water and builds up pressure. Meanwhile, as the spores mature, the lower wall of the spore sac weakens and breaks away from the bubble with a ring of sticky mucus. This action relieves the pressure in the bubble, causing a surge of water upward as a forceful burst that carries the black spore sac with it. The ejection speed has been clocked at 46 feet per second to a distance of about 6 feet. The ring of mucus from the spore sac allows the ejected sac to stick to a blade of grass when it lands. This remarkable adaptation ejects the hat (spore sac) filled with spores into a sunny, open space of the pasture, and hence the name hat-thrower fungus.

Cows typically avoid grazing near their own dung, which means they select grasses in the areas where spore sacs have attached. Here the distance traveled by the spore sac pays off. The cows eat the dispersed spore sacs along with the blades of grass; thus the hat-thrower fungus makes its way back into the cow. Within the cow, the spore sac has a tough, waxy wall that resists breakdown by digestive juices. The spore sac then passes out as part of the dung, where the spores will grow and start the process all over again.

Fungi constitute their own kingdom of organisms that obtain their nutrition from other living things and have eukaryotic cells enclosed by a cell wall. Like plants, they reproduce by spores and grow from a fixed location. These superficial similarities to plants are why they have been traditionally studied in plant biology courses. In a similar way, we have grouped them together in this unit. However, we now know through genetic and biochemical research that fungi are more closely related to animals than plants. For example, both animals and fungi obtain nutrients and store energy in similar ways using similar macromolecules.

a.

b.

Figure 15.3 *Candida* **occurs in two different growth forms in humans.** (a) The unicellular yeast form is a normal inhabitant of the skin, mouth, digestive and urinary tracts, and vagina of healthy persons. (b) The pathogenic filamentous form causes a yeast infection.

Yeast—a general term for a single-cell fungus; most reproduce by budding and are capable of metabolizing sugar by alcoholic fermentation; a single-celled type of growth form.

Filamentous—a multicellular growth form of fungi composed of a network of hyphae; in contrast to the single-celled yeast growth form.

Fruiting body—the structure of a fungus that produces spores; a mushroom is an example of a fruiting body.

Cell wall—composed of chitin, proteins, and carbohydrates; provides mechanical strength and support along with flexibility; works with the plasma membrane to control the exchange of substances into and out of the cell; also prevents the cell from bursting if a large amount of water enters by osmosis.

Chitin—a complex carbohydrate that includes nitrogen atoms in its structure. It is a strong compound that forms the cell walls of fungi.

In the chapter ahead, we will explore the lives of other fascinating fungi that cause disease and allergies, and also produce mind-altering drugs, foods, and lifesaving antibiotics. These remarkable organisms are intimately involved in your life as a college student, whether in the foods you eat, the beverages you drink, or the athlete's foot or yeast infections you may contract. Get ready to apply the knowledge you've gained about biomolecules, cells, growth, nutrition, metabolism, and reproduction to better understand the fungal lifestyle.

15.2 Fungi Grow as Filaments and Single Cells

At some point during their lifetime, most women will experience the intense burning, itching, and pain of a vaginal yeast infection. The culprit is a single-celled fungus in the genus *Candida*. Surprisingly, *Candida* is a normal resident on our skin and in the mucous membranes of the mouth, digestive tract, and vagina. Here, populations of *Candida* remain small because of the intense competition from bacteria for limited nutrients. In its normal **yeast** growth form (**Fig. 15.3a**), *Candida* is a harmless single-celled fungus. However, when triggered by a change in the temperature, pH, or hormonal environment of the vagina, *Candida* produces a multicellular, tubelike filament, or hypha (**Fig. 15.3b**). With a new set of genes switched on, the **filamentous**, or hyphal, growth form produces new proteins that can stick to and penetrate the epithelial cells lining the vagina. This invasion provokes an inflammatory response of the epithelial tissues, which leads to the irritating and painful symptoms. Women treat their infection with antifungal medicines, which are able to block the fungi from making an essential component of their cell membranes and inhibit the transformation of the fungus from the harmless yeast to the pathogenic hyphal growth form.

Although several human fungal pathogens are capable of switching from the yeast to the hyphal form, most fungi are harmless and grow as *either* hyphae or single-celled yeasts. Only about 1 percent of known fungi are classified as yeasts. The other 99 percent are filamentous with tubular cells. Each of these growth forms is adapted for a particular lifestyle and the materials it lives on. The section ahead introduces you to the structure of the fungal body, its growth form, and the cells that compose it. The success of fungi is partly attributed to their highly flexible patterns of growth.

Filamentous fungi consist of a highly branched network of hyphae

When most people think of fungi, they usually don't think of yeast infections. Instead, they remember mushrooms on their pizza or maybe squeezing a puffball fungus on the hiking trail. A mushroom is a **fruiting body**, or a structure that produces *spores* (**Fig. 15.4**). The mushroom, or fruiting body, represents a small fraction of the entire fungus; the majority of the fungus lives under the surface of the soil or within the material it grows on, such as wood. If you take a closer look underneath the fruiting body, you will find that it is connected to a highly branched network of microscopic threads of hyphae, or its mycelium. This fungal body is a highly ordered network of cells that grows and spreads throughout the soil in search of nutrients and water (see **Fig. 15.4**). In a similar manner to plant roots, the tubular hyphae create a large surface area for the breakdown and absorption of building block molecules. In the chapter opener, you saw that the hat-thrower fungus builds an extensive mycelium that runs through, breaks down, and absorbs nutrients in cow dung.

The fungal mycelium continues to grow and reproduce as long as it has a source of nutrients and environmental conditions are favorable. For one individual fungus living in a forest in eastern Oregon, the conditions have been favorable for thousands of years! During this time, hyphae of the honey mushroom fungus have infected and killed many trees as it has grown to cover an area of about 2,200 acres. It is considered one of the largest organisms on Earth. The single fungus is estimated to weigh over 150 tons and continues to grow to this day. The specialized fruiting bodies will disperse their spores, degenerate, and their cells will die. However, the underground mycelium lives on. How much longer will it live? It is difficult to say because fungi, unlike animals, don't show a well-defined life span; their growth and life are open-ended.

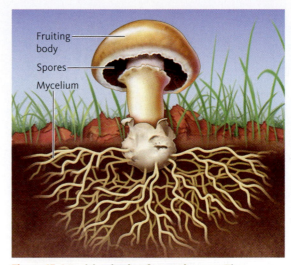

Figure 15.4 Fruiting body of a mushroom. The mushroom fruiting body is only a small portion of the fungal body. Growing throughout the soil is an extensive network of hyphae that forms the mycelium.
© Cengage Learning 2014

Fungal cells have walls made of chitin

The filamentous fungal body is composed of microscopic cellular threads, or hyphae (**Fig. 15.5a**). If we zoom in, we see that a hypha is composed of eukaryotic cells that have virtually all of the same organelles and structures as animal cells (**Fig. 15.5b**). Overall, the fungal cell has a nucleus that is enclosed by a nuclear envelope. The cytoplasm contains mitochondria, ribosomes, vacuoles, and a network of endoplasmic reticulum. The Golgi complex packages cell wall components, and vesicles deliver them to the growing tip. Energy is stored in lipid bodies and glycogen organelles. A cytoskeletal network supports and maintains the cell's shape and plays an important role in cell division, synthesis of the cell wall, and moving different structures throughout the cell. Although plant cells also have many of these structures, fungal cells do not have chloroplasts and therefore must obtain their nutrients from other organisms.

The most distinguishing feature of the fungal cell is its **cell wall** (see **Fig. 15.5b**). The wall provides mechanical strength and support along with flexibility. Cell walls of fungi are not composed of cellulose as in plant walls; rather, they are composed of proteins and carbohydrates, including chitin. **Chitin** is a complex carbohydrate that includes nitrogen atoms in its structure. It is a strong compound that forms the outer shells of lobsters and crabs. The cell wall works with the plasma membrane to control the exchange of substances into and out of the cell, and it prevents the cell from bursting if a large amount of water enters by osmosis.

Fungal hyphae grow at the tips by the process of mitosis. Each cell at the leading edge of a hy-

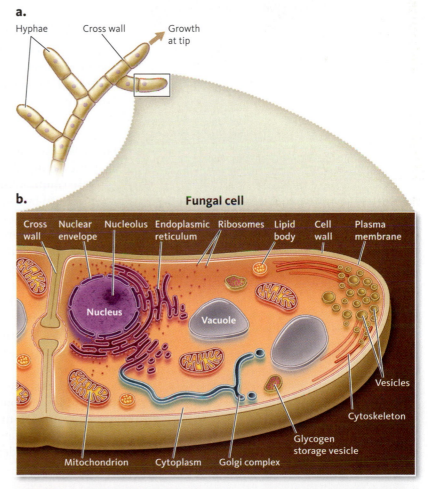

Figure 15.5 Fungal hyphal cell structure. The mycelium is composed of hyphae that are divided into cells. A magnified view of the cell shows its eukaryotic structures and organelles. The cell at the tip of the hypha is actively building new cell walls and cytoplasm as it grows. It also has numerous ribosomes, Golgi complexes, vesicles, and mitochondria. Lipids and glycogen are stored in specialized organelles. Transport of nutrients and communication between cells occurs through an opening in the hypha's cross wall.
© Cengage Learning 2014

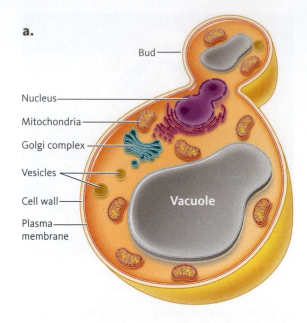

a.

Bud

Nucleus

Mitochondria

Golgi complex

Vesicles

Cell wall

Plasma membrane

Vacuole

b.

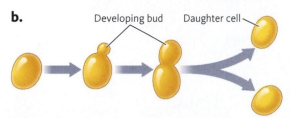

Developing bud   Daughter cell

Figure 15.6 Yeast cell structure. Yeasts are fungi that consist of a single cell. (a) The eukaryotic yeast cell is enclosed by a cell wall and contains the same organelles and structures as the fungal cell shown in **Figure 15.5**. (b) Most yeasts reproduce by budding, a process where the nucleus divides by mitosis and a new cell buds off and grows into a daughter cell. Developing buds can also be seen in **Figure 15.3a**.
© Cengage Learning 2014

Budding—form of asexual reproduction that produces a daughter cell that is genetically identical to parent organism; common in yeasts.

pha is in interphase, where it continually builds new organelles, structures, and cytoplasm and extends its walls. When the cell reaches a critical length, it triggers DNA synthesis, doubling its chromosomes, and then enters the mitosis phase of the cell cycle. During mitosis, the nucleus splits in two, with each nuclei moving apart. An inner cross wall is laid down between the two nuclei to form two new daughter cells, each with a single nucleus. In some fungi, the cross wall has an opening to allow the flow of substances to the growing tip. The new tip cell grows in length and repeats this process. This continual process of cell division extends the fungal body (mycelium) into its source of nutrients and energy.

The fungal mycelium is a highly efficient and flexible nutrient-capturing system. However, growing and building a large, interconnected network of hyphae is expensive; it requires a continual supply of raw materials such as chitin, sugars, amino acids, and phospholipids. These supplies need to be transported to the tip of the growing hypha, where they are assembled into new cell walls, plasma membranes, and cellular structures. Growth requires a continuous supply of energy in the form of ATP.

Yeasts are single-celled fungi

There are more than 1,200 species of yeasts, which are fungi that consist of a single cell instead of a string of cells attached end to end. In contrast to the hyphal growth form, yeast has a smaller surface area for absorbing nutrients and is adapted to living in fluids on plant and animal surfaces (**Fig. 15.6a**).

Yeasts do not make enzymes that are capable of breaking down large, complex molecules such as starch, cellulose, or proteins. Instead, they digest and absorb simple sugars as their energy source and therefore are common in nature where sugars are found. Large populations of yeasts can be found on the outer surfaces of leaves and stems, which secrete small amounts of sugars. Yeasts grow in nectar found in flowers, where they are swallowed and live in the gut of honeybees and nectar-feeding butterflies, bats, and birds. The skin of sweet, juicy fruits such as grapes has small cracks that ooze sugars onto the fruit's surface and also support large yeast populations. In fact, the naturally occurring yeasts that live on the surface of grapes are used in the wine-making process.

Most yeasts grow and undergo asexual reproduction by a process called **budding** (see **Fig. 15.6b**). During interphase the cell produces a small bud on its surface and replicates its DNA and chromosomes. Once it enters mitosis, the nucleus splits. One of the two nuclei enters the bud, and the bud becomes a daughter cell. In this case, mitosis produces two different-sized but genetically identical daughter cells. The cell wall is built, and eventually the bud grows and breaks free to become an independent cell.

15.2 check + apply YOUR UNDERSTANDING

- Fungi exhibit two growth forms: a multicellular, filamentous body composed of threadlike hyphae and single-celled yeasts.

- Filamentous fungi are most common. The fungal body consists of an interconnected network of mycelium.

- A threadlike hypha is composed of eukaryotic cells that are surrounded by a cell wall and contain many of the same organelles and structures as animal cells.

- Fungi grow at their tips and continue to grow as long as they have a source of nutrients and the environmental conditions are favorable.

1. Name the growth form that (a) is shown by most fungi, (b) is single-celled, (c) forms hyphae, and (d) consists of a branching network that generates a large surface area.

2. In addition to carbon, hydrogen, and oxygen, which chemical element is required to make chitin and cell walls?

3. From the fungi's point of view, is building a large, interconnected mycelium a small or large expense of energy? Justify your answer.

4. A friend tells you that beer is brewed when special yeasts break down the starch from grains such as barley and wheat. Do you believe him? Explain why or why not.

5. More than 40 species of fungi are bioluminescent, or give off light energy from the metabolic breakdown of nutrients. Which part of the fungus shown growing out of wood in **Figure 15.7** is generating light: the mycelium or the fruiting body? Predict if the light dims or glows more brightly with increased moisture. How would the light respond to increased oxygen?

Figure 15.7 Bioluminescent fungi.

© Ben Nottidge/Alamy

15.3  Fungi Obtain Nutrients from Living and Nonliving Sources

When you think of fungi, you generally think of them as decomposers such as the hat-thrower fungi. However, it may surprise you that many species of fungi are *predators* that capture, kill, and digest small animals. One species creates a sticky network of hyphae that act like a spider web to capture prey; the hyphae of another species form specialized trapping rings (**Fig. 15.8**). When a roundworm, or nematode, passes through a ring and touches the edge of the hypha, the cells rapidly swell and constrict the loop, like a cowboy's lasso. This rapid action ensnares and holds the nematode before these fungal predators penetrate and digest their prey. In addition to nematodes, fungi catch other soil organisms such as springtails, amoebae, and rotifers. Nematode-trapping fungi, as they are called, can be found in soils throughout the world from the Equator to the poles.

Fungi, like animals, are heterotrophs that depend on other organisms as their source of nutrients and energy. Biologically, this source may be a *living* organism, such as a nematode, the cells of the female reproductive tract, or the photosynthetic cells of a leaf. Living sources are referred to as hosts. The source may also consist of *nonliving* material, such as cow dung,

© Carolina Biological/Visuals Unlimited/Corbis

Figure 15.8 Nematode trapped by predatory soil fungi. Predatory soil fungi create specialized loops of hyphae to capture small roundworms called nematodes. When the nematode slides into the loop and touches the hypha, the loop closes and tightens around the worm. This photo shows a nematode that has been trapped by several loops. Once trapped, the fungus will then penetrate and digest the nematode.

> **Host**—an organism that is infected by or fed upon by another living organism; a living source of nutrients and energy.

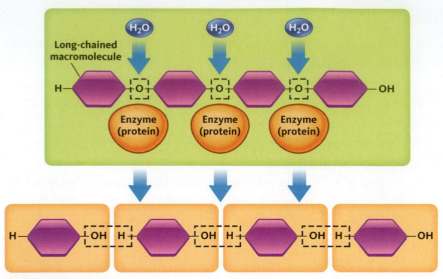

Figure 15.9 Hydrolysis. Fungi require water and enzymes to break down long-chained macromolecules into smaller building block molecules. Note that water is added and split by enzymes to add an H and an OH to the ends of the products of the chemical reaction (building-block molecules).
© Cengage Learning 2014

Building block molecules

Figure 15.10 Fungi attack other fungi. The lobster mushroom has an orange-colored fungus colonizing and covering its fruiting body. The fungal parasite causes the mushroom to resemble a cooked lobster and gives it a seafood-like flavor. Be sure to always consult an expert before eating any mushroom found in the wild.

Substrate—nonliving source of nutrients and energy for fungi; examples include cow dung, hair, wood, and fallen leaves.

Hydrolysis—the splitting of long-chained macromolecules into their building block molecules through the addition of water and enzymes.

hair, or fallen leaves. These nonliving sources are called substrates. Fungi are similar to other organisms; they compete for space, nutrients, and water in their habitat.

Fungi cannot obtain nutrients without water. Recall from Chapter 8 on animal digestion that all of the major nutrient macromolecules (carbohydrates, lipids, and proteins) have bonds that can be broken by enzymes, but there's a catch: water is needed to break long-chained macromolecules into their building block molecules by the process of hydrolysis (Fig. 15.9). In addition to enabling digestion, water is needed to dissolve and absorb building block molecules back into the hypha and then move them to the growing tip. Thus water is an essential part of the fungal lifestyle to grow and digest host cells and substrate molecules.

The section ahead discusses the various roles that fungi perform in ecosystems such as predators, parasites, mutualists, and decomposers. In their ecological roles, fungi live in close contact with their hosts, substrates, and competitors. Each fungal species is adapted to a specific set of environmental factors in its microhabitat (temperature, oxygen, moisture, pH) that make up its niche.

Fungal predators, parasites, and pathogens obtain nutrients from living hosts

Most fungi are not predators like the nematode trappers described earlier. However, many species are *parasites* that acquire their energy and nutrients from living hosts such as animals, plants, or other fungi. An example of a fungus parasitizing another fungus is the lobster mushroom. The lobster mushroom is not a specific species of mushroom; rather, it is a mushroom covered by a fungal parasite. The parasitic fungus forms an orange powder on the surface of many mushroom hosts and causes some of them to have a lobsterlike appearance, flavor, and texture. In this case, a fungal parasite converts the inedible mushroom into an edible delicacy (Fig. 15.10).

When a parasite causes disease, it is called a *pathogen.* Disease is marked by specific symptoms, such as the burning and soreness of a yeast infection or the itching and redness caused by the athlete's foot fungus. The coffee leaf rust fungus that causes leaf spots and reduced growth in coffee trees is an example of a plant pathogen. Even mushroom growers have to maintain sanitary conditions to prevent various fungal pathogens from invading and reducing their crop. In freshwater ecosystems, fungal pathogens infect frogs through their skin and reduce their hosts' breathing and energy production. These pathogens have been implicated as a major factor in the decline and extinction of frog populations worldwide. To successfully attack a living host such as a frog or human, a parasite must

Diverse groups of fungi exchange nutrients with photosynthetic partners

Many groups of fungi live and interact with photosynthetic organisms such as plants, green algae, and cyanobacteria. In this mutualistic relationship, fungi receive sugars from their partner in exchange for water, minerals, or protection. Both partners benefit from the relationship.

Fungi commonly form partnerships with plants. In fact, fungi have been found living within plant leaves as well as in the soil connecting to plant roots. This particular root–fungus partnership is called a mycorrhiza. The fungal mycelium extends out from the root into the surrounding soil, where the hyphae expand the plant's surface area for water and mineral absorption (Fig. 15.11). In some types of mycorrhizas, the fungus connects this nutrient and water "pipeline" directly inside root cells where water and minerals, especially phosphorus, move from the fungus into the root, and sugars move from the plant into the fungus. The fungus uses the sugar to fuel growth of its mycelial network, and the plant uses the minerals and water to grow more vigorously and better tolerate environmental stresses such as drought.

Fungi also partner with populations of green algae or cyanobacteria as organisms called lichens. Lichens are commonly found on trees, fallen logs, or soil, or encrusted on rocks and gravestones. Fungi that form lichens are not different from those that are free-living: they grow as filamentous fungi. The partnership begins when a fungus overgrows or encapsulates the algae or cyanobacteria as it grows on its substrate. Of the estimated 15,000 lichens, the vast majority (85 percent) involve a fungus and green algae. However, a small number of lichens consist of fungi living with both green algae and cyanobacteria. In these lichens, there is a partnership among microorganisms from three different kingdoms.

The bulk of the lichen body is formed by the fungal mycelium (Fig. 15.12). The algal cells are enmeshed near the top of the lichen, where they receive light needed to convert carbon dioxide into sugars. Sugars are

Dana Richter/Visuals Unlimited, Inc.

Figure 15.11 Mycorrhizal fungi on the roots of a pine seedling. Very thin mycorrhizal hyphae extend the absorptive surface of plant roots by penetrating into smaller spaces of the soil particles. Hyphae promote plant growth by absorbing water and nutrients from soil.

Mycorrhiza—a symbiotic (mutually beneficial) partnership between fungi and the roots of plants; fungus receives sugars and plant receives water and minerals.

Lichens—a mutualistic partnership between fungi and populations of green algae or cyanobacteria; commonly found on trees or fallen logs, in soil, or encrusted on rocks and gravestones.

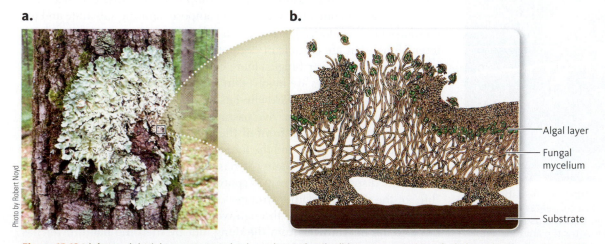

a.

Photo by Robert Noyd

b.

Algal layer

Fungal mycelium

Substrate

Figure 15.12 Lichens. (a) Lichens grow on bark, rocks, and soils. (b) A cross section of a lichen shows a green algal cell layer near the surface of the lichen, which captures light and makes sugars. The algae are enmeshed in a network of fungal hyphae that protects them from dehydration and anchors the lichen to the substrate. Lichens disperse to new locations by releasing algae wrapped in a fungal hyphae from the lichens' upper surfaces.
© Cengage Learning 2014

Figure 15.13 Fruiting bodies of a wood-decay fungus. (a) Fungi that decompose and form fruiting bodies on wood have the genetic information to break down cellulose. Wood, like all substrates, is composed of a unique set of macromolecules. **(b)** To decompose a particular substrate, a fungus must have a set of genes that makes the necessary enzymes to break specific bonds.
© Cengage Learning 2014

Photo by Robert Noyd

b.

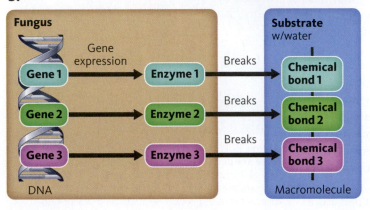

transferred from the photosynthetic partner to the fungus; in return the fungus absorbs water when available.

Living on a rock or on the bark of a tree is a challenging habitat in which to obtain water and other resources, and lichens have evolved the ability to withstand long periods of drought. When rain falls, the fungal mycelium rapidly absorbs water and dissolved nutrients, allowing photosynthesis by its algal partner to begin. Lichen growth depends heavily on the supply of water, and lichens are especially abundant in areas with large amounts of rainfall, fog, or dew, such as the Pacific Northwest and the southeastern states.

Decomposers obtain nutrients by breaking down nonliving substrates

The nonliving part of the natural world is a rich source of nutrients for fungi. In all ecosystems, both terrestrial and aquatic, fungi decompose and recycle materials in cow dung, fallen leaves, and tree trunks (**Fig. 15.13a**), insect skeletons, dead animals, and cast-off antlers. As they transform the chemical energy trapped in these materials, the fungi take their cut, so to speak, using the energy from the nutrients to build their bodies and fuel their activities.

Chemically, each substrate has a unique molecular profile and favors fungi with a particular set of capabilities or adaptations. Dead animal tissues such as feathers, antlers, and skin are composed of the structural protein keratin, a substrate rich in amino acids but low in carbohydrates. Decomposing these tissues favors fungi that have the genes to make enzymes to break the bonds of keratin (see **Fig. 15.13b**). On the other hand, wood is rich in carbohydrates and low in amino acids and proteins. Wood-decay fungi have been selected on this substrate because they have enzymes to break the bonds of cellulose and not proteins such as keratin. This is why you'll find a different community of fungi decomposing antlers than you will on wood, leaves, or fruits. You will even find different species and communities of fungi on different types of animal dung. Each substrate in nature is unique; therefore, it exerts strong selection pressure on the decomposer fungi that colonize it.

Decomposers compete with other decomposers for the substrate and for the nutrients it contains. For any given substrate, there may be several different fungi that are capable of decomposing it. What makes one fungus, such as the hat-thrower fungus, more competitive than other dung fungi? Although many factors are involved in the competitive ability of a fungus, the hat-thrower is able to quickly find its resource, to grow, and to reproduce. The hat-thrower's spores survive passage through the cow's digestive system, which places it at the front of the line, so to speak. On the dung, its spores quickly germinate, and the fungus rapidly builds a mycelium to capture nutrients. With the energy it has acquired, the hat-thrower fungus makes and disperses a new crop of spores. This strategy favors it early in the "life" of the dung pile. Other stronger competitors may take longer to find the dung pile in the pasture through wind-dispersed spores and eventually displace the hat-thrower from the dung, but by then the hat-thrower has moved on. Another way that some fungal decomposers get out in front of their competition is by first starting out as a pathogen, killing their host, and then dismantling its remains (**Fig. 15.14**). Fungi also compete with each other by engaging in chemical warfare (see Section 15.5).

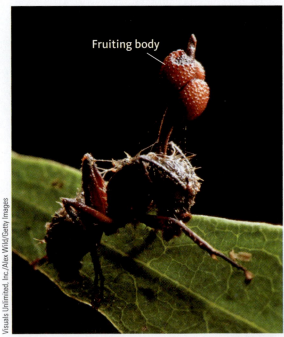

Visuals Unlimited, Inc./Alex Wild/Getty Images

Fruiting body

Figure 15.14 The zombie-ant fungus. The pathogenic fungus *Ophiocoryceps* infects a carpenter ant and grows within its brain. The infection causes the ant to become disoriented and die. After the ant dies, the fungus becomes a decomposer and digests the insides of the ant. Within a short time, the fungus produces a stalked fruiting body that grows out of the ant's head and body to disperse spores onto the forest floor, where other ants can be infected.

- Fungi act as predators, parasites, and pathogens that attack plants, animals, fungi, and bacteria. The living host provides carbohydrates, lipids, proteins, and water.

- Fungi make and secrete digestive enzymes and require water to digest and grow on host cells and substrates.

- Some fungi live in partnerships with photosynthetic organisms: on plant roots as mycorrhizas or with green algae and cyanobacteria as lichens.

- Decomposers break down and recycle nonliving materials in terrestrial and aquatic ecosystems.

1. Name two types of relationships where fungi harm their living hosts and two mutualistic relationships where a fungus receives sugars from a photosynthetic organism.

2. What is the function of digestive enzymes and water in the life of a fungus?

3. After several days, sweet, juicy strawberries are attacked by *Rhizopus,* a common fungus that contaminates fruits (**Fig. 15.15**). For each of the pairs of adaptations listed, select the one that most likely is characteristic of *Rhizopus:* (a) produces lots of spores OR fewer spores, (b) has genes that produce enzymes that break down proteins OR genes that make enzymes that break bonds of sucrose, (c) has a fast growth rate OR a slow growth rate.

4. Fungi are abundant in freshwater ecosystems where they decompose submerged substrates such as fallen plant parts and fragments of dead organisms. (1) In streams bordered by deciduous trees, predict in which season you'd find the highest numbers of fungi and spores in the water. Think or refer back to oxygen and aerobic organisms in aquatic environments discussed in Section 9.2. (2) Predict whether populations of aquatic fungi would be greater in a fast-moving stream or in a pond. Explain the connection between population size and oxygen.

5. The human toenail fungus invades the living nail bed underneath the nonliving toenail. The fungus causes symptoms that include a thickened, yellow, or cloudy nail. Is the toenail fungus acting as a parasite, pathogen, or decomposer? Predict whether you can kill the nail fungus by only treating the dead tissue of the toenail.

Figure 15.15 Fungi attack strawberries.

15.4  Fungi Reproduce by Making and Dispersing Spores

If you are one of the estimated 6 percent of people in the United States who are sensitive to mold spores, then you may pay attention to the allergy report on the morning weather forecast. All across the country, scientists collect air samples and then count and map the number of spores to alert people of the potential for experiencing the misery of seasonal allergies (**Fig. 15.16a**). Warmer temperatures and periods of moisture stimulate molds to make and release large numbers of spores; thus mold counts tend to be higher in the southeastern parts of the country, especially in summer.

The term *mold* generally refers to simple fungi that produce incredible numbers of spores from the tips of their hyphae (see **Fig. 15.16b**). In general,

a.

b.

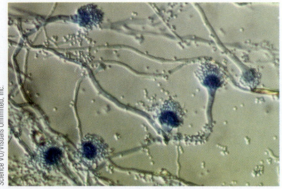

Figure 15.16 Mold spore map. (a) The weather forecast in many areas of the country includes a map that shows the number of mold spores in a given volume of air. The higher the mold spore counts, the greater the potential for experiencing irritating allergy symptoms. (b) Allergies are caused by a number of fungi, such as *Aspergillus*, that release tremendous numbers of tiny, round spores into the air.

molds don't make large, complex fruiting bodies. One of the most common mold spores floating in the air is *Aspergillus,* a genus of fungi that causes allergies and sinus infections and can trigger asthma attacks. Of the thousands of molds that people encounter each day in their basements, on their food, or in the air they breathe, only a small percent causes allergy symptoms.

The hat-thrower fungus is a marksman compared to *Aspergillus,* which releases incredible numbers of spores into the air. It and other mold spores are raining down on the book you are reading and the surfaces around you—they literally are falling from the sky. With each breath you take, you inhale fungal spores. A cubic meter of air can contain thousands of spores. In the section ahead, we will examine the two basic functions that spores perform in the life of fungi: reproduction and dispersal to new habitats and substrates. Spores are adaptations that have allowed fungi to spread out over Earth's surface for the past 600 million years.

Spores are specialized structures for asexual or sexual reproduction

Many molds such as *Aspergillus* produce huge numbers of asexual spores to find new substrates. The spores fall onto a suitable substrate and then grow and accumulate the necessary energy and materials for reproduction. The mold's hyphae switch to the reproductive phase; they grow upward and their tips swell and bear many spore-producing cells. Through rapid cell division, each of these cells produces a chain of spores one after the other (**Fig. 15.17**). The hundreds of asexual spores are produced by mitosis and are genetically identical.

The reproductive cycle of fungi often includes fruiting bodies that produce *sexual* spores. The sex life of a fungus such as the poisonous death cap mushroom (**Fig. 15.18**) starts when two genetically different spores, each with a haploid, or single, set of chromosomes germinate and their hyphae grow through the soil (see **Fig. 15.18**). When they encounter each other, chemical signals exchanged between them determine if they are *compatible*. A compatible interaction causes the two hyphae to fuse together and mate (see **Fig. 15.18**). The cytoplasm of the two individuals merges together, but the two nuclei remain separate and the hyphae contain cells with two different sets of genetic information. This new binucleate mycelium grows and divides throughout the soil and develops into a *fruiting body* (see **Fig. 15.18**). The fruiting body of the death cap mushroom is the fungi's structure of sexual reproduction.

The fruiting body of different species of fungi produces a fertile surface where special spore-producing cells are borne. In mushrooms such as the death cap, the fertile surfaces are the gills; other fungi produce sexual

Figure 15.17 *Aspergillus* spore formation. Aspergillus makes large numbers of round, asexual spores by mitosis at the tips of special reproductive hyphae. This photo sequence shows the "spore-making machinery" of *Aspergillus.*
C.W. Mims: Ultrastructural analysis of conidiophore development in the fungus Aspergillus nidulans using freeze-substitution. Protoplasma, vol. 144/2. Springer-Verlag, Wien, 1988

DAVID M PHILLIPS/Photo Researchers/ Getty Images

© The Weather Channel Companies, LLC.

Science VU/Visuals Unlimited, Inc.

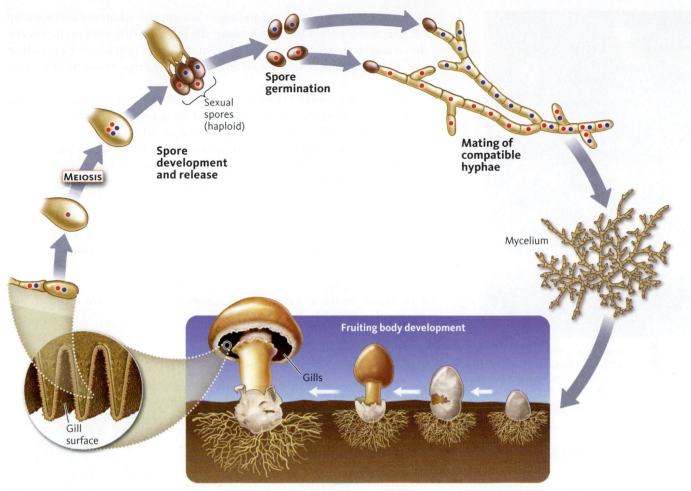

Figure 15.18 Mushroom life cycle. The sexual phase of a mushroom begins with the mating of two genetically different hyphae in the soil as shown by the red and blue nuclei at the top. Moving clockwise, the merged hyphae form a mycelium that develops into a fruiting body, or mushroom. On the edge of each gill of the mushroom, meiosis occurs to produce four genetically unique haploid spores. These spores are called sexual spores because they are a product of meiosis. Compare this process with the asexual spores produced by mitosis shown in **Figure 15.17**.
© Cengage Learning 2014

spores on the surface of teeth, pores, or cups. The prized edible morel mushroom produces it spores on the surface of its cap, whereas the puffball produces its spores within a ball-like sac. Along this surface, each binucleate spore-producing cell goes through meiosis to form four genetically unique *sexual* spores (see **Fig. 15.18**). Through the production of sexual spores, the large fruiting bodies you encounter in nature function to generate considerable genetic diversity on which natural selection acts.

Fungi use wind, rain, and insects to disperse their spores

Fungi have evolved many remarkable methods of dispersing their spores to carry their genes to new locations and find new substrates. The hat-thrower fungus *Pilobolus* responds to light and uses water pressure to forcibly eject its spores onto grasses a few yards away.

In damp forests, fruiting bodies that resemble a bird's nest grow on the surface of leaf litter, wood chips, and mulches. The hollow, cuplike structure contains several egg-shaped bodies that contain thousands of spores (**Fig. 15.19**). This fungus uses the energy contained in a raindrop to splash

Figure 15.19 Bird's nest fungus. The bird's nest fungus uses the energy of raindrops to eject its "eggs," which are sacs containing spores. The eggs of some species have a long strand that catches and sticks to nearby objects.

iStockphoto.com/Ruud de Man

Figure 15.20 Stinkhorn fungus. Stinkhorns attract flies with their foul odor. The flies then disperse the stinkhorn spores to new locations.

the "eggs" out of the cup, and the eggs then stick to whatever surface they hit. Dispersed only a few yards away, the egg covering eventually breaks down and the sexual spores are released. As you might predict, fungi that use splash cups are found in habitats that receive large amounts of rainfall each year.

A stinkhorn is easy to identify by its phallic-shaped fruiting body with its long, spongy stalk and enlarged head for a cap (**Fig. 15.20**). The spore-bearing surface is located at the top of its head. When mature, a stinkhorn releases enzymes that cause the outer surface to break down into a foul-smelling slime that contains millions of sexual spores. Flies find the odor irresistible and visit to feed on this slimy mixture. Spores that stick to these flies are carried off to new locations in the forest.

These are just a few examples of how natural selection has favored fungi that produce extraordinary numbers of spores and have developed specialized means to disperse them to local and distant habitats. However, only a tiny fraction ever reaches a site that fits its ecological niche and germinates to produce a new mycelium. In cases where they fail to land in a favorable location, many fungal spores remain dormant within their thick wall, waiting for an opportunity.

15.4 check + apply YOUR UNDERSTANDING

- Fungi produce spores to carry out two basic functions in their life cycle: reproduction and dispersal to new habitats/substrates.

- Asexual spores are made by mitosis and function to disperse the fungus to new locations via air currents.

- Sexual spores are made by the union of hyphae from two genetically different individuals. Meiosis occurs within large fruiting bodies and genetically unique spores are dispersed.

1. What are the two basic functions that spores perform in the life of a fungus?

2. Name three ways a fungal *spore* is dispersed to new hosts or substrates.

3. Mitosis and meiosis are processes that occur in the life of a *stinkhorn*. Which of the two genetic processes—mitosis or meiosis—occurs in the following events: (a) spore germination and the growing hyphae in the soil, (b) the process within the spore-producing cell on its cap, (c) the development of the fruiting body, (d) the growth of the mycelium through the soil?

4. Predict whether the following fungal cells are haploid or diploid: (a) a spore made inside a bird's nest fungus fruiting body or "egg," (b) cells in the stalk of the stinkhorn, (c) cells at the growing tip from a germinated spore.

5. The distance that spores travel is affected by the shape of the fruiting body (**Fig. 15.21**). Predict which of the two fruiting bodies disperses its spores the greater distance: the mushroom on the right or the puffball on the left. Explain.

Kenneth H Thomas/Photo Researchers/Getty Images

Ari N/Shutterstock.com

Figure 15.21 Puffball and bolete mushrooms.

15.5 Fungi Produce Potent Toxins, Antibiotics, and Ethanol

Eating a single death cap mushroom may be enough to kill you (**Fig. 15.22**). The death cap accounts for greater than 90 percent of the fatalities caused by mushroom poisoning in the United States. It is often mistaken for a similar-looking harmless mushroom or for an edible puffball in an early stage of development. The death cap grows in the leaf litter under deciduous trees, where it decomposes plant litter in the soil. In the fall the mycelium develops fruiting bodies that make two of the most potent toxins known.

If you eat a death cap mushroom, in the days following your meal the toxins prevent protein synthesis in your liver and kidneys, causing them to fail. Destruction of these two vital organs often leads to heart failure, convulsions, and coma. At this point, you have only a 10 percent chance of survival because it is too late for effective medical treatment.

Why do mushrooms make toxins? What evolutionary advantage do toxins provide the fungus? The answers are not completely clear, but the fruiting body functions to disperse spores, and the toxins likely deter animals or insects from eating the mushrooms and preventing reproduction. Many animals such as snails eat mushrooms, and insects lay their eggs on fruiting bodies where they hatch into hungry larvae.

The section ahead discusses fungi as master chemists in the living world. Whether it is the toxins that deter humans from eating them, the antibiotics to inhibit other microbes, or the enzymes to produce ethanol for beer, wine, and spirits, the metabolic pathways of fungi produce a vast array of powerful chemical substances.

Fungal toxins damage cells in the animal body

Many fungi produce harmful chemicals, or toxins, as part of their normal life activities. These poisonous substances damage another organism's structure or disturb its functioning. However, not all "poisonous mushrooms" cause the level of destruction of the death cap mushroom. *Psilocybe* mushrooms produce a toxin that affects neurotransmitters in the central nervous system. This toxin attaches to receptors in the synapse that cause psychoactive effects, visions, and hallucinations. These "magic" mushrooms have been part of shamanic or religious rites in cultures in the desert southwest and Mexico for centuries.

Ergot is a fungus that produces toxic alkaloids similar to the defensive compounds produced by plants (see Section 12.6). This fungus survives extreme conditions, not as a spore but as a structure consisting of a large mass of hyphae along with nutrient reserves. This dark, elongated structure, called an *ergot,* produces a potent toxin. The fungus infects flowers such as wheat, barley, and rye, in which it forms black ergots in the grain kernels (**Fig. 15.23**). During the Middle Ages, toxic ergot was harvested along with the rye grains and ground into flour and baked into bread. The ergot toxins constrict blood vessels, which leads to muscle spasms and seizures. The burning sensation in a person's limbs was called the Holy Fire and was thought to be punishment for sins. In 1692, ergot disease is thought to have led people to accuse others of being possessed by the devil, which led

Martin Fowler/Shutterstock.com

Figure 15.22 Death cap mushroom. The death cap mushroom is one of the world's most deadly mushrooms. It has white gills and a band of tissue around its white stalk and grows out from a cup. This mushroom is aptly named because its toxins destroy the liver and kidneys of humans, leading to organ failure and death.

Infected grains

David O. Cavagnaro/Peter Arnold/Getty Images

Figure 15.23 Ergot fungus. The ergot fungus infects the seeds of cereal crops and turns them black. The fungus produces a potent toxin that can lead to hallucinations and constriction of blood flow to the limbs.

a.

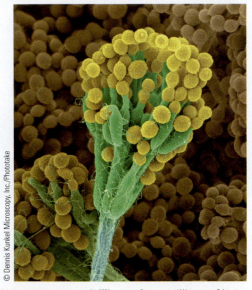

b.

Figure 15.24 A *Penicillium* colony. Millions of human lives have been saved by the chemical substance produced by *Penicilium*. (a) *Penicillium* is a common blue-green mold that grows on a variety of decaying materials in nature and is easily grown in culture. (b) Like *Aspergillus*, *Penicillium* produces chains of round spores on the tips of reproductive hyphae. Its spores are universally present in air.

Antibiotics—chemicals that interfere with the metabolism of bacteria and other fungi and thus inhibit their growth or kill them.

to the Salem Witch Trials in the Massachusetts colonies. Today, the medical benefit of ergot is the class of drugs called *ergotamines,* which are used to treat millions of people with migraine headaches.

The human food supply is vulnerable to toxins made by *Aspergillus,* a mold that grows on corn kernels, grains, coffee beans, and nuts. Perhaps you are aware of peanut butter or dog food recalled due to the presence of this powerful toxin. As this fungus grows, it releases toxins called *aflatoxins.* Aflatoxins pose a threat during the processing, storage, or transport of food. When humans, horses, poultry, or livestock eat the contaminated products, the toxin is processed in the liver, where it disrupts the cell's mitochondria, lysosomes, and endoplasmic reticulum. In the mitochondria, aflatoxins interfere with the electron transport system and therefore inhibit cellular energy production. These fungal toxins have been classified as one of the most potent carcinogens known and are constantly monitored by food producers and inspectors.

Fungi make antibiotics that inhibit their microbial competitors

A fallen log or leaf on the forest floor is an intense battle zone where fungi compete for its resources. Some species are very aggressive and destroy their competition simply by overgrowing them. Other fungi bombard their competitors with toxins, enzymes, or acids. One chemical compound made by fungi is of special interest to humans—antibiotics. **Antibiotics** are chemicals that interfere with the metabolism of bacteria and other fungi, and thus inhibit their growth or kill them outright. Hundreds of different fungi make antibiotics, but asexual molds such as *Aspergillus* and *Penicillium* are prolific producers of these compounds. The blue–green mold *Penicillium chrysogenum,* the original source of penicillin, was accidentally discovered in 1928 (**Fig. 15.24**).

Unless you are one of the 10 percent of the population allergic to penicillin, you probably have taken it at some time in your life to treat a *bacterial* infection. Antibiotics do not work against *viral* infections such as the common cold because they target structures found in bacterial cells that are not found in viruses. Penicillin disrupts the building of the bacterial cell wall. Without this ability, bacteria cannot reproduce and are vulnerable to counterattack by the body's immune system. Other antibiotics inhibit bacterial ribosomes and their gene expression, or they block a step in the synthesis of critical enzymes. These compounds inhibit or kill bacteria that compete with fungi for the substrate molecules they require.

Humans have taken advantage of this discovery to inhibit or kill bacterial pathogens. Today most penicillin is synthesized in the laboratory to increase production, broaden the number of bacteria that it can treat, and reduce the chances of antibiotic resistance. The fungi that produce antibiotics have saved countless human lives since their discovery and production.

Yeasts ferment sugars into ethanol and carbon dioxide without oxygen

Many adults enjoy watching sports on a big-screen television with friends while eating a hot pizza and drinking a cold beer (**Fig. 15.25**). These food products, in part, can be attributed to the metabolic reactions inside yeasts.

Through the process of **alcoholic fermentation**, yeasts derive energy from sugars and produce ethanol, or grain alcohol, and carbon dioxide (**Fig. 15.26**). Notice that the process is similar to *lactic acid fermentation* discussed in Chapter 9, except the final products are different. Both types of fermentation take place without oxygen, or are anaerobic.

The pizza-baking and beer-brewing processes highlight several aspects of the biology of yeasts. Pizza dough is made by adding baker's yeasts to sugar in warm water. Because the yeasts can only metabolize simple sugars, they absorb the sugars and begin aerobic metabolism. In another bowl, flour (starch) and salt are mixed together to form the dough. The activated yeasts now metabolize the sugars without oxygen and produce carbon dioxide, which causes the dough to expand and rise as the dough is kneaded. When the pizza dough is placed in the oven, the heat causes carbon dioxide gas bubbles to be trapped in the crust (see **Fig. 15.25**). In addition to gas, the yeasts produce ethanol, which evaporates during baking.

Beer is the third most popular drink in the world after water and tea. It is made from four common ingredients: water, starch, hops, and brewer's yeasts. Brewer's yeast is the same species as baker's yeast; however, different strains produce different characteristics. For example, baker's yeasts are selected to grow fast and to rapidly produce carbon dioxide gas, whereas brewer's yeasts metabolize and reproduce more slowly. The starting material, or substrate, for brewing beer is starch found in barley or wheat grain. Because yeasts cannot metabolize starch, the barley must first be "malted" or converted to simple sugars. The malted barley is boiled with hops that give a bitter flavor to the beer before it is placed into fermentation tanks. Fermentation occurs without oxygen, and the yeasts produce ethanol (alcohol) and bubbles of carbon dioxide. In some cases, more carbon dioxide is added before bottling.

Nearly all fungi require *oxygen* and are aerobic organisms. Aquatic species that live underwater in ponds and streams use the dissolved oxygen in the water for their metabolism. The guts of herbivores such as cows, sheep, and elephants contain anaerobic fungi that colonize and break down ingested plant materials in the absence of oxygen. However, yeasts and a few other fungi can switch from aerobic to anaerobic metabolism as conditions change. Recall that the anaerobic fermentation process *incompletely* breaks down glucose to ethanol, which still contains a large amount of remaining energy. A 12-ounce bottle of beer still contains about 150 calories of energy primarily from ethanol. Although anaerobic metabolism *does not produce large amounts of energy*, microscopic yeast cells or rumen fungi do not have large energy requirements.

Throughout this chapter, we have discussed many different fungi that are part of our lives and show remarkable adaptations to fit their ecological niche. Not all organisms called molds are fungi. For example, slime molds are fungal-like protists that live in similar habitats and produce spores within fruiting bodies like fungi (**Fig. 15.27**). However, slime molds do not secrete digestive enzymes to decompose substrates, or form cell walls made of chitin like fungi. Water molds are set apart from true fungi because they produce spores with flagella and have cell walls with complex carbohydrates rather than chitin. Like fungi, slime molds and water molds have extraordinary lifestyles, body structures, nutritional habits, and reproductive strategies.

Figure 15.25 Pizza and beer are made using yeast. A meal consisting of pizza and beer involves two types of yeasts that ferment sugars to produce ethanol and carbon dioxide. Baker's yeast in the pizza dough creates the bubbles that make the crust rise, and brewer's yeast makes ethanol and bubbles in the beer.

Anaerobic metabolism
(oxygen not required)

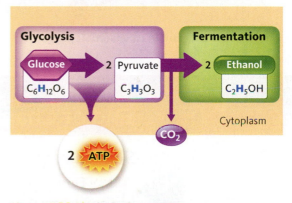

Figure 15.26 Alcoholic fermentation is an anaerobic pathway. Through glycolysis and fermentation, yeasts transform the energy in sugars into ATP and produce carbon dioxide and ethanol as by-products.
© Cengage Learning 2014

Figure 15.27 Slime mold fruiting bodies. This slime mold resembles a fungus because it lives on wood and produces bright orange fruiting bodies that produce spores. Unlike fungi, when a spore germinates, it produces cells with flagella that merge to form a slime-like structure that feeds on bacteria. Their role in the environment is not as decomposers or pathogens. They are classified as protists.

Alcoholic fermentation—metabolic pathway used by yeast to derive energy from sugars without oxygen; produces ethanol (alcohol) and carbon dioxide.

- Fungi produce a large variety of chemicals for survival and reproduction. Humans have exploited many chemicals from fungi for medicine, recreation, and industry.

- Fungi produce a variety of toxins and antibiotics to inhibit animals and other microorganisms from eating them as well as to defend themselves against attack.

- Yeasts perform fermentation where they break down sugars into ethanol and carbon dioxide.

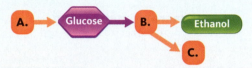

Figure 15.28 Diagram of an anaerobic pathway.
© Cengage Learning 2014

1. Fungal toxins affect different animal body organs and systems. How do the toxins produced by the death cap differ from *Psilocybe* and ergot with regard to their effects on the body?

2. How do antibiotics target bacteria to inhibit their growth? Why are antibiotics ineffective against viruses?

3. Some fungi break down cellulose by fermentation. Look at the anaerobic pathway in **Figure 15.28**. In which block—A, B, or C—would you place (1) cellulose, (2) carbon dioxide, (3) pyruvate?

4. Hikers and campers often pack dehydrated or dried foods, such as bananas and apricots, as well as food placed in airtight packages. Explain how these packaged foods are prevented from being attacked by fast-growing molds.

5. *Neocallmastix* is an anaerobic fungus that lives in the intestines of herbivores. Predict whether this fungus (a) produces enzymes to break down cellulose to glucose; (b) has cells with mitochondria; (c) produces carbon dioxide, ethanol, or lactic acid; (d) has cell walls containing chitin.

End of Chapter Review

Self-Quiz on Key Concepts
Fungal Body Structure, Growth Forms, and Growth

KEY CONCEPTS: Fungi show two growth forms: a multicellular filamentous body and a single-celled yeast. Filamentous fungi are composed of threadlike cells that form a branched, interconnected network of hyphae that is adapted for finding and absorbing food. In some groups, the mycelium forms complex tissues and large fruiting bodies that make and disperse spores. Fungi grow at their tips and continue to grow as long as they have a source of nutrients and the environmental conditions are favorable.

1. Match each of the following **fungal body structures** with its description or example. Each term on the left may match more than one description on the right.

    Fruiting body
    Hypha
    Mycelium
    Yeast

    a. an individual thread or filament composed of a chain of cells
    b. part of the fungal body that bears spores
    c. a network of interconnected filaments
    d. a single-celled fungus or growth form
    e. usually found in flower nectar and other sugary substrates

2. Which fungal cell structure supports and maintains the cell's shape and plays an important role in cell division and synthesis of the cell wall?
    a. Golgi apparatus
    b. mitochondria
    c. vesicles
    d. cytoskeleton

3. Which term or phrase describes the growth of filamentous fungi?
    a. determinate and closed
    b. a fixed life span like that of many animals
    c. infinite
    d. open-ended under favorable conditions

4. Which fungal growth form lives in sugary tree sap or immersed in the nectar of flowers?
   a. filamentous fungi
   b. yeasts
   c. fruiting bodies
   d. mycelium

Fungal Nutrition and Ecological Relationships

KEY CONCEPTS: Fungi live in close contact with their living and nonliving sources of nutrition. Fungi act on living hosts as predators, parasites, and pathogens that attack plants, animals, fungi, and bacteria. The living host provides carbohydrates, lipids, proteins, and water. Decomposers make and secrete digestive enzymes to begin the process of extracellular digestion.

5. Match each of the following **nutritional relationships** with its description or example. Each term on the left may match more than one description on the right.

   Decomposer   a. a fungus that derives nutrients from a living host
   Lichen   b. a partnership between a soil fungus and a plant root
   Mycorrhiza   c. a fungus that causes disease and presents symptoms
   Parasite   d. a fungus that breaks down the proteins in leather
   Pathogen   e. a fungus that breaks down nonliving materials in a substrate
   Predator   f. a fungus that captures and kills small animals
                     g a fungus that provides water and minerals to a plant in exchange
                        for sugar
                     h. a fungus that partners with green algae or cyanobacteria to form
                        an independent organism

6. Yeasts use which of the following biological molecules as a source of energy?
   a. lignin
   b. cellulose
   c. glucose
   d. starch

7. Which of the following is necessary for a fungus to break down a particular biomolecule in a substrate?
   a. a specific gene
   b. a specific enzyme
   c. water
   d. all of the above

Fungal Reproduction

KEY CONCEPTS: Fungi reproduce by making spores. Asexual spores are made by mitosis and function to disperse the fungus to new locations via air currents or insects. Sexual spores are made by meiosis within large fruiting bodies, like the familiar mushroom, and also allow the fungus to survive unfavorable conditions. Like sexual reproduction in animals and plants, sexually produced spores function to increase genetic diversity and increase evolutionary fitness.

8. Match each of the following **kinds of spores** with its description or example. Each term on the left may match more than one description on the right.
   Asexual spores   a. formed by meiosis
   Sexual spores   b. formed by mitosis
                     c. formed and released on gills, pores, or surfaces
                     d. produces genetic diversity
                     e. spores that are all genetically identical
                     f. spores that generally cause allergies in humans

9. Which of the following is *false* regarding large fruiting bodies such as mushrooms?
   a. Meiosis occurs in cells on the gills of the mushroom cap.
   b. The mycelium differentiates and organizes into distinct tissues.
   c. Hyphae that form the mushroom have two separate nuclei in each cell.
   d. They produce both asexual and sexual spores.

10. Which of the following characteristics determines whether two hyphae are compatible for merging together?
    a. They are of the same species.
    b. They are genetically *different*.
    c. They are haploid.
    d. All of the above.

Fungal Metabolism and Products

KEY CONCEPTS: Fungi produce a wide variety of chemicals for survival and reproduction. Toxins and antibiotics are made to inhibit animals and other microorganisms from eating them as well as to defend themselves against attack. Yeasts produce ethanol by the anaerobic process of fermentation. Humans have exploited many chemicals from fungi for medicine, recreation, and industry.

11. Match each of the following **fungal products** with its description or example. Each term on the left may match more than one description on the right.
    Antibiotic   a. a product of yeast fermentation
    Ethanol   b. a chemical that inhibits the growth of competing microbes
    Toxin   c. a chemical that destroys cells and organs of animals and plants
                    d. penicillin, a product made by the fungus *Penicillium*
                    e. a compound that is produced by the death cap mushroom

12. Some fungi produce a toxin that causes hallucinations in humans. Which of the following organs/organ systems is affected by such toxins?
    a. liver
    b. brain
    c. kidneys
    d. stomach

13. Antibiotics inhibit or kill bacteria. How do antibiotics work?
    a. They destroy the nucleus and DNA.
    b. They prevent the bacteria from making a cell wall.
    c. They prevent the bacteria from making toxins.
    d. All of the above.

Applying the Concepts

14. When the fruiting body of the inky cap mushroom begins to mature, digestive enzymes are released and the mushroom digests itself into a sticky, black liquid (**Fig. 15.29**). This self-digestion process causes the closely packed gills to pull apart and release millions of sexual spores into the air. However, the inky liquid traps many spores. If they germinate and grow, which spores are more likely to form a new fruiting body: those in the inky fluid or those released into the air? Why?

Figure 15.29 Inky cap mushrooms.

iStockphoto.com/Matauw

15. White-nose syndrome, a disease that has killed millions of bats in Europe and the northeastern United States, is caused by a fungal pathogen (**Fig. 15.30**). The fungus grows as a white, powdery substance on the noses, wings, and ears of bats. (1) Predict what environmental factors favor the growth of the fungus. (2) Do you think this fungus also infects bats that roost in trees instead of caves? (3) How do you think the fungus is spread? (4) At the end of an outing, people who explore caves disinfect their gear and clothing. Why do they do this?

Figure 15.30 Bat infected with white-nose syndrome.

16. Two jelly fungi look very similar but live in different substrates. *Dacrymes* is found on the wood of dead conifer trees (**Fig. 15.31**). *Tremella* is a parasite on other fungi that live on living hardwood trees. Which of the two jelly fungi produces the following enzymes: (a) enzymes to degrade chitin, (b) enzymes to degrade cellulose, (c) enzymes to degrade proteins in living cytoplasm?

a. **b.** **c.**

Figure 15.31 Jelly fungi fruiting bodies. Jelly fungi are known for the jelly-like texture of their fruiting bodies. Sexual spores of these three species of jelly fungi are produced on their bright yellow surfaces (a) *Dacrymyces palmatus* (b) *Calocera cornea* (c) *Dacrymyces stillatus*.

Data Analysis

Jelly Fungi and Wood Decay

There are about 80 species of jelly fungi, all of which grow on wood (see **Fig. 15.31**). Biologists performed a laboratory experiment to determine the extent and type of decay that three of these jelly fungi cause in various kinds of wood. Researchers used a soil-block test. Small blocks of hardwoods and conifer wood were saturated with water and then inoculated with equal amounts of fungi and placed into a sterile coarse-textured sandy soil. After 12 weeks, wood blocks were removed, weighed, and chemically analyzed. The results are shown in **Table 15.1**.

TABLE 15.1 Chemical Analysis of Wood Decay by Jelly Fungi

Species	Wood Substrate	Percent Moisture	Percent Weight Loss	Percent Change in Lignin	Percent Change Cellulose
Dacrymces capitatus	Hardwood	207	36	−31	−52
Dacrymces capitatus	Conifer	126	15	−12	−29
Calocera cornea	Hardwood	104	4	−0.4	−1
Calocera cornea	Conifer	120	19	−0.6	−34
Dacrymces stillatus	Hardwood	150	20	−17	−37
Dacrymces stillatus	Conifer	145	5	0	0

From Seifert, K.A. 1983. Decay of wood by the Dacrymycetales. Mycologia 75(6): 1011–1018.

Wood-decay fungi require oxygen and produce two kinds of rot depending on the component of the plant cell wall they degrade. Brown rot fungi selectively degrade and metabolize cellulose but do not have the enzymes to break down the cementing compound called *lignin*. White rot fungi degrade both cellulose and lignin and develop a white color.

Data Interpretation

17. Which jelly fungus on which wood caused the greatest decay? What is your evidence?

18. Which jelly fungi cause brown rot? Which cause white rot?

19. As a general rule, most brown rot fungi affect conifers, whereas white rot fungi occur more frequently on hardwoods. Do the data support this generalization?

Critical Thinking

20. Investigators observed that the decay was concentrated in the springwood rather than the summerwood (**Fig. 15.32**). Why do you think this was the case?

21. The percent moisture data show that after 12 weeks the wood gained water. The wood is a source of nutrition and energy for the fungus. Thinking of the metabolic process that the jelly fungi use to convert the wood into energy, where did the water come from? In this same metabolic process, where do the carbon atoms in the cellulose go?

22. Why do you think the investigators used a loose, coarse soil rather than a packed clay soil? What impact would this have on the metabolism of the wood-decay fungi?

23. You are buying a home and notice that the window frames have dried, jellylike fruiting bodies on their surface. You know that window frames are generally constructed with conifer wood. Are you alarmed? Do these fungi indicate any other problems with the window?

Springwood

Summerwood

Biodisc/Visuals Unlimited, Inc.

Figure 15.32 Conifer tree rings.

Question Generator
Animals and Underground Truffles

The deer truffle is a fungus that is mycorrhizal with conifers and less commonly with hardwoods. It grows about 2 to 3 inches below the soil surface, just below the layer of decaying organic matter. It forms firm, round, tan-colored fruiting bodies (**Fig. 15.33**). Dark, blackish-brown spores fill the entire cavity of the fruiting body, which has a thick, rindlike outer covering.

Animals, especially deer, squirrels, and voles, eat the truffles as part of their diet. The animal digests the spore-bearing tissue but not the spores, which are defecated some distance from the parent. Dung beetles or earthworm beetles may carry spores down into the soil.

Below is a block diagram (**Fig. 15.34**) that relates several aspects of the biology of the deer truffle that fruits below the forest floor. Use the background information along with the block diagram to ask questions and generate hypotheses. We've translated the components of the block diagram into several questions to get you started.

Figure 15.33 Deer-truffle fruiting bodies.

© Josef Hlasek, www.hlasek.com

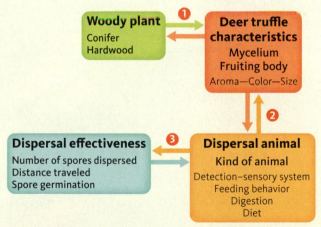

Figure 15.34 Factors relating to the growth and spore-dispersal of deer truffles.
© Cengage Learning 2014

Here are a few questions to start you off.

1. What nutrients are exchanged between the woody plant and the fungus? (Arrow #1)

2. Are some kinds of animals more likely to find and eat truffles than others? (Arrow #2)

3. Which kinds of animals disperse deer truffles spores the farthest? (Arrow #3)

4. How far do animals disperse deer-truffle spores? (Arrow #3)

Use the diagram to generate your own research question and frame a hypothesis.

16 Physical and Chemical Cycles and the Biosphere

16.1 A changing climate and life for the elkhorn coral

16.1 A Changing Climate and Life for the Elkhorn Coral

Figure 16.1 The Dry Tortugas. The crystal-clear waters of the Dry Tortugas are located in the Gulf of Mexico, 75 miles west of Key West. The warm waters, abundant nutrients, and array of ecological niches support a diverse community of organisms. This national park preserves rare coral species and provides a safe haven for birds, turtles, and dozens of species of fish.

Imagine that you have an opportunity to take a scuba-diving trip in the Gulf of Mexico at Dry Tortugas National Park, a location famous for its bird and marine life as well as pirates and sunken ships (**Fig. 16.1**). Hidden beneath the tropical waters of the park, brightly colored fish nibble on a wide range of aquatic invertebrates scurrying across the rocky reef. Slipping underwater from your dive boat for a closer look, you realize that the rocks are actually a community of living animals called corals. The entire reef is made from billions of mostly dead individual corals bound together by their hard calcium carbonate exoskeletons, giving them their rocklike appearance. Most coral reefs, including those in the Dry Tortugas, are found in warm waters near the equator, where light and water depth, temperature, and nutrients favor growth. Coral reefs consist of small outcrops ranging from a few square feet to the vast Great Barrier Reef in Australia, which encompasses over a thousand square miles.

Rising from the bottom along the reef's edge are several branched yellow–orange corals called elkhorn coral *(Acropora palmata).* Stretching up to five feet across and resembling a set of elk antlers, these corals dwarf the dozens of other colorful species that inhabit the reef. Corals live in extensive colonies of genetically identical organisms produced by asexual reproduction. Each tiny individual, called a *polyp,* is only a few millimeters in diameter (**Fig. 16.2**). However, over time and millions of generations, they combine to form large reefs in tropical waters around the world. Corals, relatives of soft-bodied jellyfish, are simple animals consisting of a slender body dominated by their tentacles, stomach, hard exoskeleton, and a partner—photosynthetic algae (see **Fig. 16.2**). The hard exoskeleton is produced and secreted by the polyp and provides structure and protection. Although corals are able to catch small fish or plankton using their tentacles equipped with harpoonlike stinging cells (see **Fig. 16.2**), most corals obtain up to 95% of their energy and nutrients through a symbiotic relationship with photosynthetic algae. The hard exoskeleton of the coral body provides protection to the algae, and the algae produce most of the carbohydrates used by the corals. Their dependence on symbiotic algae makes corals very sensitive to any environmental change that threatens the life or function of the algae.

Corals are an ecological keystone species, supporting tremendous biodiversity by creating a rich blend of habitats for invertebrates and fish. Elkhorn coral were once one of the most abundant species in the Caribbean and the Florida Keys. However, in the last 30 years, nearly 95% have died. As a result, fish diversity in some areas has been reduced by half. There are many threats to the survival of corals, such as disease, extreme hot or cold temperature events, predation, climate change, storm damage, and changes in the chemistry of ocean water. Each of these factors has

Varina and Jay Patel/Shutterstock.com

contributed to the significantly diminishing reproductive success and survival rate of these corals. Because coral reefs grow slowly, natural recovery takes decades, or may never occur due to permanent changes to their habitat. In 2006, the elkhorn coral was one of two coral species added to the endangered species list.

As you continue your dive, moving slowly along the reef, you notice startling color changes. Large sections of the elkhorn coral have lost their vibrant colors and are white; these sections are dead (**Fig. 16.3**). Marine biologists who study corals have determined that this change, called bleaching, is often tied to environmental changes such as increases in water temperature or acidity. Small but sustained increases in water temperature cause corals to expel their symbiotic algae. Unable to cope with the loss of nutrients provided by the photosynthetic algae, the polyps soon die off. What remains are the bleached exoskeletons of the coral. In these dead zones, fish and other organisms that make coral reefs the most productive ecosystems on Earth are nearly absent. In 1998, an El Niño event, part of a naturally occurring multiyear global weather pattern, resulted in a widespread increase in tropical ocean temperatures that sparked the worst coral-bleaching event ever observed. More than 16% of the world's reefs died that year; greater than 90% were killed in certain areas of the Indian Ocean.

Many reefs lie close to populated coastal areas, making them vulnerable to chemicals used in agriculture and suburban runoff. Nitrogen and phosphorus, two of the main nutrients contained in fertilizers, are water-soluble and rapidly move out of soils and into rivers and groundwater. Drainage patterns carry these nutrients into the ocean and create areas with significantly altered water chemistry. An increase in the availability of nitrogen and phosphorous changes the food web by altering the populations of photosynthetic organisms that can live there. For example, with increased nutrients, different algae species grow over the corals, blocking sunlight necessary for the photosynthetic partner to produce nutrients. These new algae in turn attract and support a different mix of animal life, thus changing the biodiversity of the habitat. Such human activities affect the natural nutrient cycles and ultimately the plant and animal life living there, because even small changes in nutrient availability have large impacts on coral reefs.

Like other living organisms, corals rely on access to carbon to create a wide range of macromolecules. Consider the movement of carbon atoms from the environment, through a living organism, and back to the environment. Elkhorn corals acquire carbon for their exoskeletons and other organic molecules from products of photosynthesis via their symbiotic algae. The chemical reactions in photosynthesis are part of the global movement of carbon atoms between living and nonliving reservoirs, locations where the elements are held. The carbon atoms reside in these reservoirs for various lengths of time. For instance, the hard exoskeleton of the

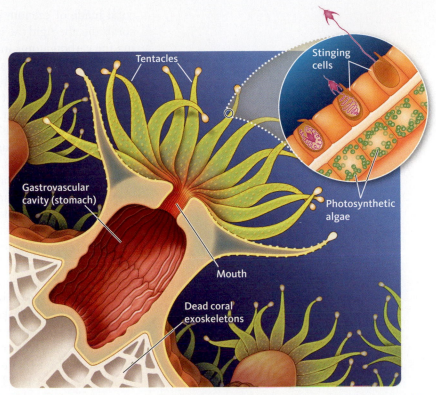

Figure 16.2 **Anatomy of a coral.** This cross section of a typical hard coral shows its simple body plan, which consists of tentacles it uses to acquire small food, an internal gastrovascular cavity that acts as a stomach for digesting food, and the symbiotic algae. Many corals are able to use specialized barbs that they rapidly eject to spear small organisms they can feed on.
© Cengage Learning 2014

© Stephen Frink Collection/Alamy

Figure 16.3 **Dying coral.** The bright white color of this elkhorn coral is a clear indication that it is dying or dead. The absence of the green pigment chlorophyll creates the bleaching, indicating the absence of the symbiotic algae as a food-producing partner.

coral made of calcium carbonate ($CaCO_3$) includes carbon atoms. When corals die, the rigid calcium carbonate exoskeleton can remain intact, holding the carbon atoms in place for centuries.

Coral reefs and ocean waters are part of a marine ecosystem that represents the largest carbon sink, a location where minerals are stored or retained for a period of time on Earth, absorbing one third of all the carbon emitted by human activities each year. The ability to fix carbon and hold it for long periods has a positive influence on global temperatures, because atmospheric carbon dioxide contributes to the greenhouse effect. After millions of years, powerful geologic forces, such as volcanic eruptions, recycle carbon back into the atmosphere, just one example of how the carbon cycle is completed.

The elkhorn coral, along with all life on Earth, relies on the movement of water and a wide range of critical substances between living and nonliving locations. In this chapter, we'll connect the role of chemical cycling to maintaining the stability of organisms such as corals, their ecosystems, and the relationship to biosphere-level movement of elements. We will show how water and chemical cycling connect to the balance of ecosystems and individual habitats. We'll take a closer look at elements such as carbon, nitrogen, and phosphorus, which are important parts of macronutrients and are key components in a wide range of biomolecules such as proteins and nucleic acids. We'll also explore the role human activities, such as the burning of fossil fuels, have played in altering the heat-trapping characteristic of our atmosphere, influencing global temperature patterns and climate. Let's begin our exploration by studying the importance of the water cycle.

16.2 The Water Cycle Is Critical to Life at All Levels

Because elkhorn corals are surrounded by ocean water, it's difficult to understand how the global cycling of water is important to this stationary species. Water, like all elements in our environment, is not static; in fact, all elements exist within a set of globally linked cycles (**Fig. 16.4**). Thus, the flow of resources between *reservoirs,* or temporary locations where the substances reside, links together habitats, ecosystems, and our entire biosphere.

Water may reside in *biological* reservoirs such as the bodies of plants and animals, or in *nonliving* reservoirs such as the atmosphere, oceans, soil, or rocks. Corals also rely on water motion caused by wind and currents to bring them drifting food and to carry away wastes. Most marine animals are very sensitive to the salt concentration in ocean water, and changes caused by evaporation and rainfall can have a significant impact on cellular metabolism in their bodies.

The cycling of water takes many forms, including summer rainstorms, glacial ice, and evaporation from plant leaves. Why are we

Figure 16.4 Two phases of the water cycle. This dramatic photo of a building thunderstorm shows two phases of the water cycle, the liquid ocean, which evaporates to produce water vapor that forms clouds.

Peter Wollinga/Shutterstock.com

concerned with the movement of water in our ecosystems? Like corals, we depend on water as our most important resource; all life on Earth requires some access to water to survive. The availability of water and its quality determine how many and what kinds of organisms can be supported in a habitat. In a similar way, we depend upon the natural movement of elements between reservoirs to maintain the natural balance of the global environment. This movement includes a large range of natural and human-altered processes that move molecules such as water between reservoirs and make them available or unavailable to the living world.

At the organism level, water is the most abundant component by weight. Your own body is 65% water, and a tomato is greater than 90% water (**Fig. 16.5**). Recall from Chapter 12 that plants absorb water through their roots, moves it up the stem and into the leaves and fruit, where it reenters the atmosphere as it evaporates through the openings in the leaf surface called stomata. Movement of water through the plant happens quickly. Greater than 90% of a plant's water is lost to the atmosphere within hours or days, eventually returning to Earth in the form of precipitation. In most organisms, water is critical for cellular metabolism, and the water balance of body fluids is tightly controlled. In spite of this, your body loses two to three quarts (about six pounds) of water each day because of normal body functions. To survive, you must replace this fluid.

Understanding the availability and movement, or *cycling*, of resources between reservoirs provides us with clues about how ecosystems operate, how the living world responds to changes, and how human activities influence our world. In the following section, we'll look at the physical processes that affect the timing and movement of water between reservoirs as well as the availability of freshwater to support life.

The sun powers the movement of water between reservoirs

In developed countries, water is conveniently available with the turn of a tap. We rarely consider how the water got to our home or where it comes from. The wells that draw water out of local aquifers and the pipes that deliver the water to our houses are a small part of the global water cycle (**Fig. 16.6**). Water molecules are always in movement, whether they move rapidly, as in a roaring river, or slowly, like the advancing sheets of thick glacial ice in Antarctica that travel only a few inches each year. Although this movement between reservoirs, or **water cycle**, moves water in different states, such as liquid or vapor, the balance of water on Earth is constant over time; it simply shifts from one reservoir to another. As water changes states, it moves from the ocean, into the atmosphere, to the land, through freshwater ecosystems, and back to the ocean (see **Fig. 16.6**). Studying the water cycle allows us to understand how the movement of water among reservoirs, and its change of state between liquid, vapor, and solid, influences the living world. Although water may change its physical state as it cycles, it is unique in that it's the only resource that doesn't change chemically as it cycles.

The movement of water between reservoirs relies primarily on energy from the sun. Heat from sunlight warms the surface of the ocean and evaporates the water from a liquid form to water vapor. Solar heat drives water evaporation from the soil through plant transpiration; water also cycles into the atmosphere when ice and snow evaporate directly into wa-

Figure 16.5 The tomato is part of the water cycle. The water cycle is not only out there somewhere in the biosphere, but right in front of you in your salad. Water absorbed from the soil fills the cells of the tomato—accounting for 90% of its weight. You are part of the water cycle, with 65% of your weight coming from the water in your tissues.

Water cycle—the continual movement of water on, above, and below the surface of Earth, including water contained in living organisms. Energy from the sun powers much of the movement of water. Water typically changes states between liquid, vapor, and ice at various places in the water cycle.

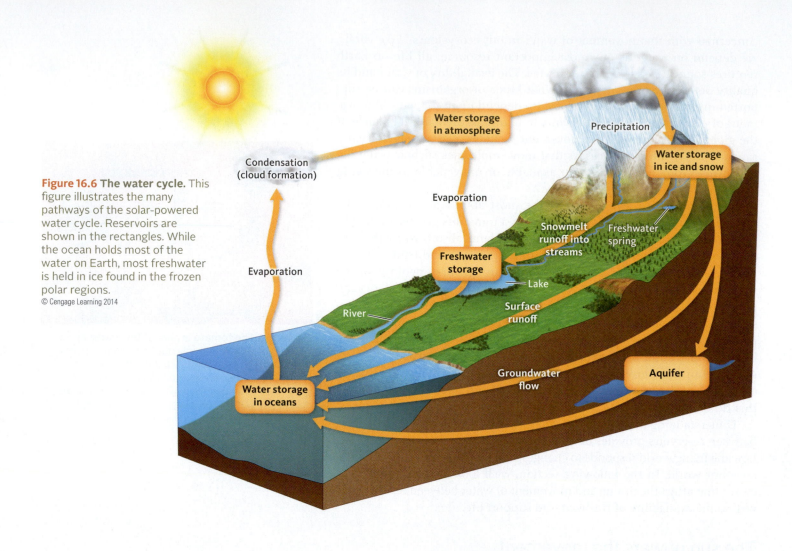

Figure 16.6 The water cycle. This figure illustrates the many pathways of the solar-powered water cycle. Reservoirs are shown in the rectangles. While the ocean holds most of the water on Earth, most freshwater is held in ice found in the frozen polar regions.
© Cengage Learning 2014

Water storage in atmosphere

Condensation (cloud formation)

Precipitation

Water storage in ice and snow

Evaporation

Snowmelt runoff into streams

Freshwater spring

Evaporation

Freshwater storage

Lake

River

Surface runoff

Water storage in oceans

Groundwater flow

Aquifer

Figure 16.7 Physical states of water. Water has three different physical states—liquid water; ice, which is the solid form of water; and water vapor, a gas common in our atmosphere that we know as humidity. The conversion from one state to another requires energy input or loss such as heat from the sun.
© Cengage Learning 2014

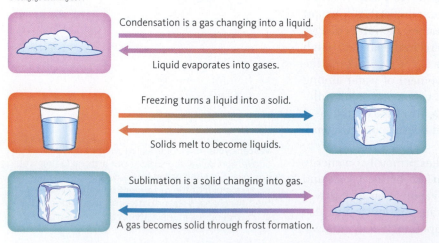

Condensation is a gas changing into a liquid.

Liquid evaporates into gases.

Freezing turns a liquid into a solid.

Solids melt to become liquids.

Sublimation is a solid changing into gas.

A gas becomes solid through frost formation.

ter vapor (**Fig. 16.7**). This water vapor is the humidity you feel on a warm afternoon after a rain shower.

Rising air currents turn water vapor into liquid droplets or ice crystals light enough to float, forming clouds. In some cases, small water droplets collide and form larger water droplets that become too heavy for the air to support them, and they fall as precipitation. Precipitation can be rain, snow, or hail. In colder regions, snow can accumulate, forming ice caps and glaciers, which can store frozen water for thousands of years. Snowpack in temperate climates thaws and melts when spring arrives. This constant recirculation provides fresh water to terrestrial ecosystems, supporting life on land.

Precipitation falls back into oceans or on land, where gravity causes the water to flow downhill over the ground as surface runoff. A portion of this runoff enters rivers, eventually moving water toward the oceans. The rest soaks into the soil, where a portion infiltrates deep into the ground to fill aquifers (saturated subsurface rock). Groundwater can also stay close to the surface and seep back into lakes, rivers, or oceans or find openings in the land surface and emerge as freshwater springs. Over time, water completes the cycle between these reservoirs and begins the process once more.

Water often defines the boundaries of ecosystems

On land, the location and movement of water is often used to define the way we describe ecosystems. *Watersheds* are geographic regions where all precipitation drains into specific rivers or lakes. Watersheds can be as small as a mountain valley feeding a single stream or as large as the Missouri River watershed, covering all or part of 10 states and extending into Canada, approximately 1,245,000 square miles (**Fig. 16.8**). Large watersheds such as the Missouri drain immense amounts of water and have large variations in the types of habitats they support. Much of the Missouri River watershed is prairie grasslands; however, along its western edges it supports ponderosa pine forests in the Black Hills of South Dakota as well as in Montana.

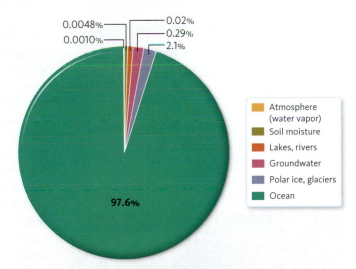

Figure 16.8 The Missouri River watershed. The Missouri River watershed (dark brown) is a geographic basin that drains surface water from a large section of the U.S. An example of a watershed with a common landscape, the region is dominated by grasslands and includes much of North America's Great Plains (green).
© Cengage Learning 2014

Most water is not salt-free, clean, or available

Ninety-seven percent of all water on Earth, 300 million cubic miles of it, is found in oceans (**Fig. 16.9**). The Pacific Ocean is the largest body of water, covering half of the planet and holding half of all ocean water. Ocean water is salty, or saline, with a typical quart holding around 2.5 teaspoons of salt. Marine plants and animals have evolved a wide range of physical and metabolic adaptations that allow them to thrive in salt water. For example, seagrasses chemically alter the salt content of their cells, allowing them to take in water, whereas sea turtles have special glands they use to excrete excess salt. All animals require small amounts of salt to maintain ion balance at the cellular level, but the amount of salt in seawater exceeds the amount most terrestrial animals are able to metabolize and excrete. This makes seawater unavailable for drinking or use at the cellular level. Whereas drinking saltwater is safe after the salt is removed, drinking untreated seawater would create a metabolic crisis in your body by quickly dehydrating you at the cellular level. As you learned in Chapter 6 (Section 6.4, **Fig. 6.17**), saltwater in the blood and tissue fluids actually draws water out of each cell, a function of osmosis, causing rapid water loss and potentially leading to death.

The remaining 3% of water on Earth is composed of **freshwater**. Polar ice caps and mountain glaciers are the largest reservoirs, composing 68% of this resource. Nearly 90% of ice is found in Antarctica, where ice sheets at the South Pole are two miles thick (**Fig. 16.10**). This large quantity of freshwater is unavailable for most organisms to use. You would think an enormous amount of water is present in our lakes and rivers, but as you

0.0048%
0.0010%
0.02%
0.29%
2.1%

Atmosphere (water vapor)
Soil moisture
Lakes, rivers
Groundwater
Polar ice, glaciers
Ocean

97.6%

Figure 16.9 A comparison of various water reservoirs. This figure illustrates the dramatic difference in the percentage of total water stored in ocean water and freshwater reservoirs.
© Cengage Learning 2014

Freshwater—naturally occurring water stored in ice, lakes, rivers, and underground as groundwater in aquifers. Freshwater excludes saltwater found in oceans and is generally low in concentrations of dissolved salts.

Figure 16.10 **Antarctica.** The continent of Antarctica is mostly covered in thick sheets of ice that have resided there for hundreds of thousands of years. Ice sheets near the South Pole can be two miles thick.

TABLE 16.1 Average Residence Time for Water

Antarctica snowpack	20,000 years
Oceans	3,200 years
Glaciers	20–100 years
Seasonal snow cover	2–6 months
Soil moisture	1–2 months
Groundwater (shallow)	100–200 years
Groundwater (deep)	10,000 years
Lakes	50–100 years
Rivers	2–6 months
Atmosphere	9 days

© Cengage Learning 2014

Figure 16.12 **Polluted drinking water is common in some parts of the world.** Polluted sources of drinking water are a major health hazard for a large portion of the world's population. Limited sources of freshwater polluted by human activities create special challenges in developing countries where disease and waterborne parasites are still common.

Figure 16.11 **A center-pivot irrigation system.** Center-pivot irrigation systems are common tools that farmers on the Great Plains use to irrigate their crops. However, the ecological price of this activity is the tapping of aquifers for increasingly more water, in some cases lowering groundwater levels by hundreds of feet.

see in **Figure 16.9**, they hold just 0.02% of all the water on the planet. Most terrestrial life depends on this small fraction of available freshwater for survival.

In the U.S., about two-thirds of the available freshwater is used to support the irrigation of agricultural crops (**Fig. 16.11**). In many places, water for irrigation is drawn from deep groundwater reservoirs, allowing farmers to grow plants where normally not enough precipitation falls. This agricultural process has been successful in the central and western United States, but as a result, more groundwater is removed than replenished, a process that rapidly depletes many deep reservoirs. In coastline ecosystems, excessive groundwater withdrawals allow saltwater to seep into and contaminate drinking water supplies.

How long water, or any other resource, stays in one form or location is called its *residence time.* Understanding a resource's residence time provides information about the stability or mobility of substances, their potential influence on environments, and their availability to organisms. The longer the duration of residence time in any one location, the less available that resource becomes to support organisms in a different habitat. A shorter residence time indicates that a substance changes location more frequently. For example, freshwater locked in the ice and snowpack of Antarctica has a lengthy residence time of 20,000 years, which is a very long time compared to that of water in the atmosphere, which has a relatively short residence time of nine days (**Table 16.1**). Water vapor in the atmosphere resides for only a few days but can fall as snow over cold mountain ranges or in the Arctic, where it then stays for decades or even thousands of years.

Contaminants from human activities make many sources of water unfit for human consumption. The lack of clean water is one of the major challenges facing a large percentage of people in developing countries (**Fig. 16.12**). Common sources of contamination include microbial pollution in sewage and animal waste, agricultural pesticides and fertilizers, and storm water runoff from urban areas. Contamination alters the quality of surface water and groundwater, often introducing toxins that disrupt the communities of organisms living in these areas. If pollution levels are dangerously high and widespread, species may be completely displaced or killed off in local areas.

- Water and other elements cycle through the living and nonliving worlds.

- Water resides in sinks for various amounts of time—from minutes to millions of years.

- Water drainage patterns are often used to define ecosystem boundaries.

- Freshwater is a rare commodity; most water is found in the ocean or in ice.

1. Create a diagram that traces a drop of water through an example water cycle.
2. Describe in your own words what a *watershed* is.
3. What is meant by residence time when discussing the water cycle?
4. What is the role of gravity in the water cycle?
5. Over millions of years, the amount of energy received on Earth from the sun cycles between low and high. Recognizing that the sun plays a major role in the water cycle, how would you expect a period of high solar energy to impact the amount of liquid water on Earth?

16.3 Carbon, Nitrogen, and Phosphorus Cycle through the Macromolecules of Living Organisms

The rainforest along the foothills of western Costa Rica is home to thousands of species of plants and animals, and the dense canopy of trees seems to stretch on forever (**Fig. 16.13**). As we have learned, the carbon atoms in each organism's biomolecules were initially captured during photosynthesis. Some of those carbon atoms remain in the wood of the canopy trees for dozens of years, even after the trees die. The carbon in leaves that fall from branches is quickly broken down and recycled by insects, fungi, or bacteria. These and similar atoms such as nitrogen and phosphorus are used in the construction of biomolecules in the structures of new organisms, which ultimately die and continue the nutrient cycle.

Carbon is critical because it is the backbone of familiar biomolecules such as carbohydrates, lipids, and proteins. Your body is about 9.5% carbon, and each carbon atom, either in your body or in a rainforest plant, arrived there through the same process: the **carbon cycle**. As you learned in Chapter 13, the process of carbon fixation during photosynthesis cycles carbon between the atmospheric reservoir into living organisms like plants and trees.

Carbon does not exist just as a few atoms in biomolecules or as carbon dioxide gas in the atmosphere. Carbon also dissolves in water and forms a component of sediment on the ocean floor and in bogs and eventually in rocks. Understanding how and why carbon moves between reservoirs provides us with information on how ecosystems operate and carbon's potential influences on our climate. In the following section, we'll explore the global carbon cycle, its major reservoirs, and how humans have affected this process. Let's begin with identifying the reservoirs and examining how photosynthesis influences annual carbon dioxide (CO_2) levels in the atmosphere.

gary yim/Shutterstock.com

Figure 16.13 A Costa Rican rainforest. The rainforests of Costa Rica contain an immense amount of biodiversity. Each of the species relies on nutrient cycling to gain access to key nutrients. In turn, each plant has a unique lifespan and ability to hold those nutrients over time. Some die and decay quickly, cycling elements such as carbon rapidly. In contrast, some trees live for decades and hold onto their resources for long periods of time. Given the rapid growth rates of many rainforest plants, the soils are comparatively nutrient poor, with most minerals residing in the tissues of living plants.

> **Carbon cycle**—the global biogeochemical cycle in which carbon is exchanged among the living and nonliving portions of Earth. It is one of the most important cycles of Earth and allows for carbon to be recycled and reused throughout Earth and all of its organisms.

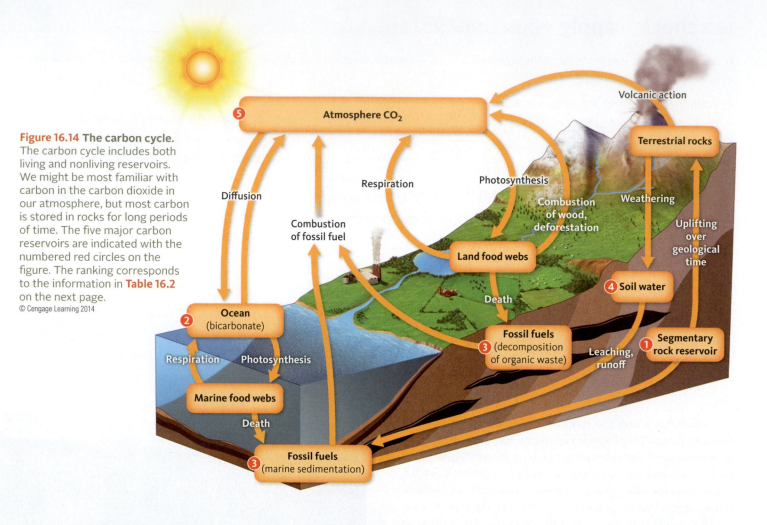

Figure 16.14 The carbon cycle. The carbon cycle includes both living and nonliving reservoirs. We might be most familiar with carbon in the carbon dioxide in our atmosphere, but most carbon is stored in rocks for long periods of time. The five major carbon reservoirs are indicated with the numbered red circles on the figure. The ranking corresponds to the information in **Table 16.2** on the next page.
© Cengage Learning 2014

Figure 16.15 Carbon-rich sedimentary rocks. In many cases the carbon appears as dark bands, where millions of years ago large amounts of vegetation were compressed and formed layers in the sediment. Similar carbon-rich sediments are the source of the petroleum and oil we burn in our cars.

Most carbon is stored in inorganic reservoirs

Earth has five major reservoirs for carbon, each interconnected by a series of pathways that govern the timing and movement of carbon between them (**Fig. 16.14**). The largest reservoir consists of sedimentary rocks (**Fig. 16.15**). The world's oceans contain the next largest amount of carbon, mostly dissolved in water as bicarbonate (HCO_3-) (**Table 16.2**). The atmosphere is a relatively small reservoir but has more carbon than the total carbon in the living biomass, the carbon held in macromolecules in our bodies as well as in the bodies of other organisms. Various chemical, physical, geologic, and biological processes influence the annual movements of carbon between these reservoirs. By studying carbon's movement, we can better understand its potential impacts on the living world.

The level of carbon dioxide gas in our atmosphere is approximately 0.04% of the volume of the gases we breathe. This may seem to be an insignificant amount, but even such small levels of carbon have a dramatic impact on the living world—through photosynthesis. Sources of atmospheric carbon include CO_2, the end product of cellular respiration from living organisms, carbon inside Earth brought up by volcanoes, the burning of organic matter such as forest fires, and the burning of fossil fuels.

As with water, the residence time or the length of time carbon stays within a reservoir influences its availability to organisms. Carbon held in rocks remains there for long periods of time, in some cases in excess of 150 million years (Table 16.3). Volcanoes and geologic uplifting move carbon out of the crust and return it to the surface or atmosphere. Carbon dioxide typically resides in the atmosphere for three to five years until it is taken up by plants or other photosynthetic organisms. In contrast, carbon moves rapidly through the living world. For example, as we learned in Chapter 12, annual plants such as most commercially grown sunflowers live just a few months before their bodies die and begin to decompose, returning the acquired carbon to the environment. In contrast, the carbon in the cell walls of wood in trees may take centuries to return to the soil through decomposition.

Most of the annual cycling of carbon takes place between the atmosphere and plants or other photosynthesizing organisms. Each year, 120 gigatons of carbon dioxide is removed from the atmosphere. Since 1958, climatologists have been measuring the amount of carbon dioxide in the atmosphere near the top of Mauna Loa in Hawaii. Even though Mauna Loa is an active volcano, the atmospheric research center is located on its north slope, where prevailing winds and clear skies eliminate the influence of volcanic emissions. The measurements reveal an important trend that atmospheric carbon dioxide is increasing (Fig. 16.16). The burning of fossil fuels to power our cars and produce electricity accounts for a significant portion of the line's upward slope.

Human activities impact natural patterns in the carbon cycle

Since the beginning of the Industrial Revolution in the mid-1800s, humans have altered the natural patterns of the carbon cycle. The carbon-rich fossil fuels we burn originated from plants that grew 300 million years ago. Vast amounts of carbon from plants accumulated under anaerobic conditions and over millions of years created pockets of liquid oil or rocklike coal. Each year we extract this carbon from underneath Earth's surface and burn it in engines and power plants, releasing nearly 30 billion tons of CO_2 into the atmosphere annually. As carbon emissions from human activities have increased over the past 15 years, the atmospheric concentration also has increased (Fig. 16.17). This means we are adding carbon to the atmosphere faster than natural processes such as photosynthesis can remove it. This results in the yearly increase in atmospheric carbon that we can measure at the Mauna Loa research station.

Increasing carbon dioxide levels in the atmosphere have a direct negative effect on oceans. Although photosynthetic algae and plants are able to absorb huge quantities of carbon from the atmosphere, these increases in carbon dioxide levels actually alter the chemistry of ocean

TABLE 16.2 Major Carbon Reservoirs

Reservoir	Mass in Billions of Tons
Earth's crust (sedimentary rock)	100 million
Oceans (bicarbonate)	39,700
Oil and natural gas	4,000
Soil (microbes and decaying organic matter)	1,500
Atmosphere (carbon dioxide)	750

© Cengage Learning 2014

TABLE 16.3 Residence Time of Carbon in Various Reservoirs

Land plants	5 years
Atmosphere	3–5 years
Soils	25 years
Fossil fuels	650 years
Oceans	350 years
Carbonates in rock	150 million years

© Cengage Learning 2014

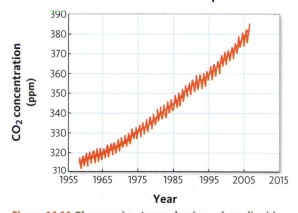

Carbon dioxide in the atmosphere

Figure 16.16 Changes in atmospheric carbon dioxide levels over time. Since recordings began in 1958, the data from the Mauna Loa atmospheric research site have documented the steady increase in atmospheric carbon dioxide levels. Data indicate that this rise is tied to increasing human emissions from the burning of fossil fuels.

© Cengage Learning 2014

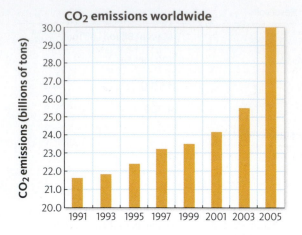

CO₂ emissions worldwide

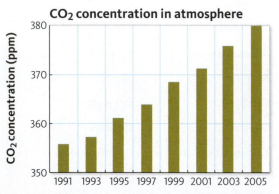

CO₂ concentration in atmosphere

Figure 16.17 Relationship between CO₂ emissions and CO₂ concentrations in the atmosphere. This graph links the burning of fossil fuels and the steady increase in atmospheric carbon dioxide levels. Climate scientists believe that atmospheric levels of carbon dioxide may double in the next century if we continue to increase our demands for petroleum products.

© Cengage Learning 2014

Nitrogen cycle—the global movement of nitrogen, where nitrogen is converted between its various chemical forms. The movement between reservoirs can be carried out via both biological and nonbiological processes. Microbes are a critical living resource in the cycling of this element.

water. Seawater absorbs most of the carbon dioxide produced by human activities and transforms it into carbonic acid, which in turn increases the acidity of the water. Since the advent of the Industrial Revolution, pH levels of the ocean have decreased from 8.2 to 8.1. Although this seems to be a small change, it represents a tenfold increase in acidity. Marine animals such as corals and mollusks that form calcium-based shells are very sensitive to the pH of ocean water. Increasing acidity weakens their protective shells, making them more vulnerable to disease, changes in their environment, or predators.

Nitrogen cycling relies on a diversity of microbes

Nitrogen, like carbon, is an important element necessary to support life. As you learned in earlier chapters, all living cells require nitrogen to form critical macromolecules such as nucleic acids and enzymes in order to carry out life activities. Although nitrogen is abundant in the atmosphere, accounting for about 78% of the total gases, most organisms cannot tap into this vast reservoir. Plants and animals can't use nitrogen gas directly for their metabolism. In addition, the nitrogen found in soils is highly soluble and is rapidly washed down and away from roots. These two conditions help make nitrogen availability a limiting factor in plant growth. This uptake of nitrogen, or *assimilation*, becomes a problem for animals because a high percentage of the nitrogen in their bodies originates in plants.

Numerous groups of soil-dwelling bacteria and archaea transform atmospheric nitrogen into forms that plants can absorb. This process of *nitrogen fixation* is the primary pathway between nitrogen reservoirs in the nonliving and living worlds (**Fig. 16.18**). Many of the nitrogen-fixing microbes live freely in the soil; in fact, each teaspoon of soil contains several hundred thousand of these microbes. Certain species of plants, such as members of the pea and bean family, form a symbiotic relationship with these microbes, growing small protective nodules on their roots that house and provide nutrients to the microbes in exchange for the nitrogen they fix (**Fig. 16.19**). Similar processes occur in aquatic environments.

Decomposer bacteria and fungi chemically convert organic forms of nitrogen found in macromolecules, such as nucleic acids and proteins, to simpler inorganic forms such as ammonium, another form of nitrogen available to plants. However, other groups of microbes transform the inorganic nitrogen in the soil back into nitrogen gas. The gas rises up through pores in the soil and is released into the atmosphere, where it can again be fixed by microbes for plants. Microbes such as bacteria derive energy and nitrogen for themselves in the process. Each group of nitrifying microbes produces key enzymes needed to complete a chemical conversion. The end product of their metabolic pathways is excreted into the soil and serves as a source of energy and nitrogen for the next group of microbes in the cycle. The combination of these and other transformations is called the **nitrogen cycle**.

As with carbon, the nitrogen cycle has been affected by human activities. Since the 1950s, the widespread use of agricultural and lawn fertilizers along with increases in livestock ranching, discharge from wastewater treatment plants and septic tanks, and the burning of fossil fuels have combined to alter the natural patterns of nitrogen cycling. The large-scale production of nitrogen fertilizers requires large amounts

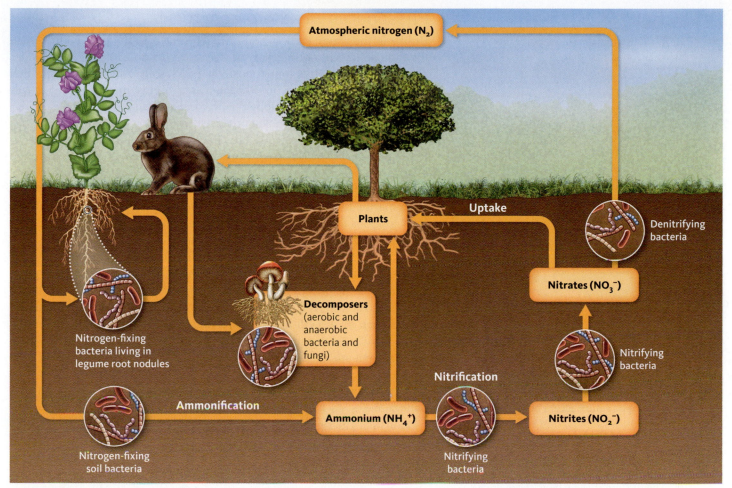

Figure 16.18 The nitrogen cycle. The microbial world links the nonliving reservoir of nitrogen in the atmosphere to the living world. Nitrogen is used by plants to make chlorophyll (see **Fig. 13.10**) and enzymes that run photosynthesis and cellular respiration. Bacteria and archaea live in the soil converting nitrogen between several forms, some of which become available to plants. Some species form symbiotic relationships with many plants in the bean family.
© Cengage Learning 2014

Bean root nodules

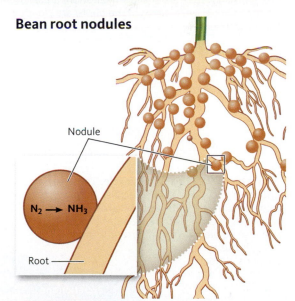

Figure 16.19 Bean plant root nodules. Symbiotic microbes in the bean family live in root nodules, which are small tissue growths that bulge from the roots. Inside the nodules, the microbes convert nitrogen gas to a form usable by the plant. In return, the plant provides a safe habitat for the microbes and sugars, necessary to support nitrogen fixation by the bacteria, an aerobic respiration–intense process.
© Cengage Learning 2014

of energy to create a usable form of nitrogen for plants. Why are these activities considered such a problem? Left unchecked, they add nitrogen to surface water and groundwater systems well beyond normal levels, altering the natural system by supporting plants that otherwise might not survive nitrogen limitations. Thus, nutrient enrichment in the water is greatly increased, which becomes a challenging issue in freshwater and coastal aquatic habitat management. For example, nitrogen added to agricultural fields often washes into lakes and streams, resulting in large algal blooms. As the algae die and are decomposed by bacteria, available oxygen is depleted from the water, creating extremely low oxygen conditions that result in fish kills (**Fig. 16.20**).

Mississippi River

Figure 16.20 Nitrogen runoff alters ocean water chemistry. This satellite photo shows a portion of the Gulf of Mexico and the Mississippi River. Each year, large amounts of nitrogen dissolved in surface water are carried downriver and deposited in the shallow waters off the coast of Louisiana. The light green ocean waters are shallow near-shore areas, while the bright white areas in the photo are clouds in the atmosphere.

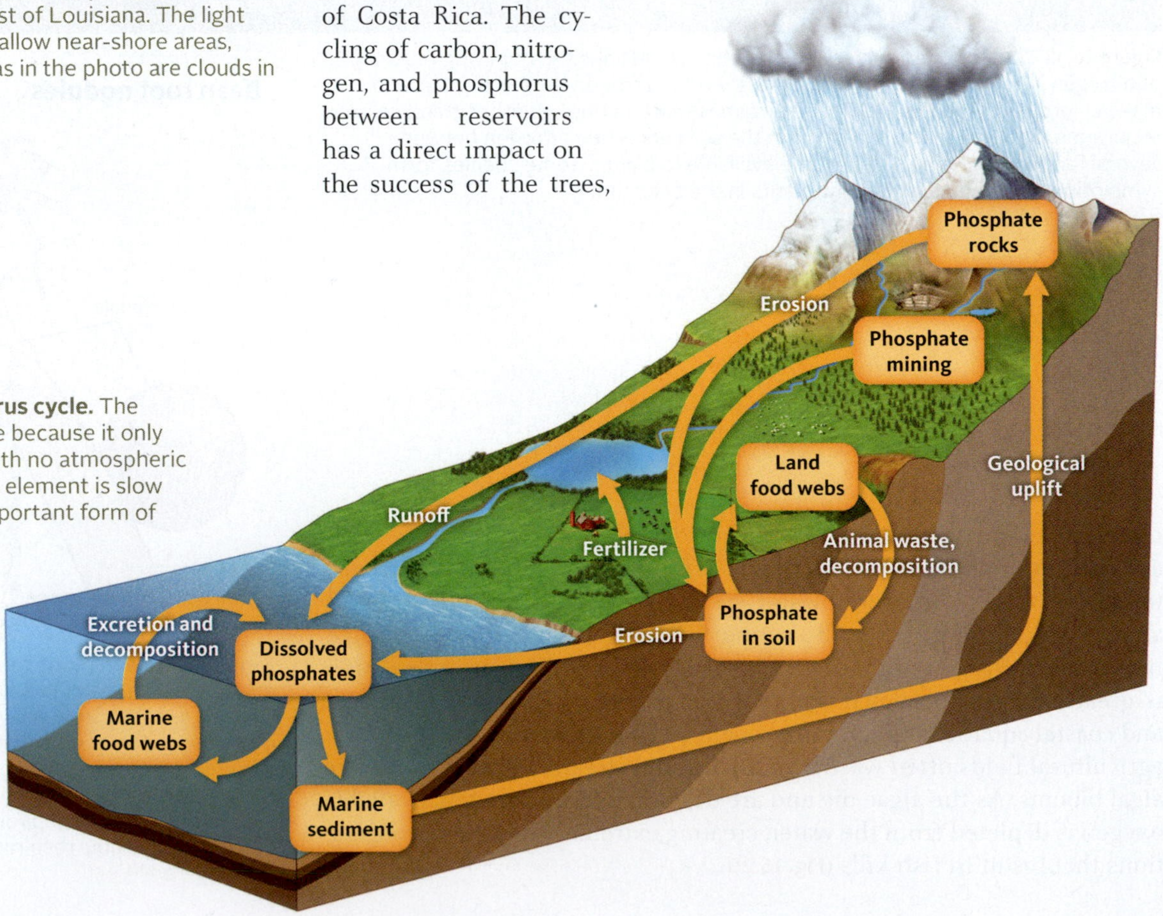

Phosphorus weathers from rocks

Unlike in the water, carbon, and nitrogen cycles, in the phosphorus cycle the atmosphere is not a reservoir because phosphorus is a solid under normal conditions. It is found mostly in rocks, soils, and sediments, which comprise its major reservoir. Phosphorus is highly insoluble and moves quickly only when it is combined with oxygen in the form of phosphate.

As rocks containing phosphorus weather into smaller particles, phosphorus is deposited into soils, streams, and rivers. In terrestrial ecosystems, phosphorus weathered from rock enters the food web through plants that absorb this essential macronutrient through their roots. Eventually, some phosphorus is carried by rivers into the oceans, where it may enter marine food webs or be deposited into sediments on the sea bottom. Over millions of years, portions of the sea floor are uplifted and exposed to erosion, beginning the phosphorus cycle once again (**Fig. 16.21**).

In many ecosystems, phosphorus is a limiting factor, and for centuries people have added fertilizers rich in this element to stimulate plant growth. Lack of phosphorus limits an organism's ability to produce biomolecules such as ATP and phospholipids, which ultimately limits life activities such as growth and reproduction. Today, modern agriculture depends heavily on the addition of phosphorus to fields in order to promote plant growth and generate the high crop yields that help feed a large population. Just as with nitrogen, excess phosphorus runs off from fields or through the soil, entering surface and groundwater systems where it is a pollutant.

Consider the significance of nutrient cycling to the rainforest trees growing on the mountains of Costa Rica. The cycling of carbon, nitrogen, and phosphorus between reservoirs has a direct impact on the success of the trees,

Figure 16.21 The phosphorus cycle. The phosphorus cycle is unique because it only includes land and water with no atmospheric reservoir. Recycling of this element is slow because no biologically important form of phosphorus is gaseous. When phosphorus is deposited in oceans, it eventually becomes part of marine sediments. Once there, it may take millions of years to form rock and uplift as mountains and finally erode again to become available to living things.

© Cengage Learning 2014

which themselves are a part of the cycle. To create a wide range of carbon-rich biomolecules, rainforest trees rely on the power of the sun to fuel photosynthesis. This is the primary pathway for carbon to enter the living world from nonliving reservoirs. This is vastly different from the biological relationship between trees and the soil microbes active in the nitrogen cycle that convert nitrogen into usable forms for plants. The largest reservoir of phosphorus is in sedimentary rocks, where gradual breakdown over time releases small amounts of phosphorus into the environment. Humans have altered these chemical cycles in many ways, adding large quantities of phosphorus to agricultural lands as fertilizer. In the next section, we'll examine how human activities have altered the carbon cycle and contributed to changes in our climate.

16.3 check + apply YOUR UNDERSTANDING

- Most carbon is stored in rock; however, the carbon in the atmosphere has a direct impact on global temperatures through the greenhouse effect.

- Burning of fossil fuels by humans has added carbon to the atmosphere at rates greater than it can be removed by natural processes.

- Most nitrogen is stored in reservoirs in forms that are not usable by the biological world. Metabolic processes in microbes convert nitrogen into forms usable by plants.

- The largest reservoir of phosphorus is in rocks. As the rocks weather, phosphorus becomes available to organisms.

1. What group of organisms plays a crucial role in the nitrogen cycle?
2. What characteristics do the carbon and nitrogen cycles share?
3. What would be the impact on the carbon cycle of a worldwide increase in the number of active volcanoes?
4. Refer to **Figure 16.18** to predict the impact of the following human activities on populations of nitrifying bacteria in the soil: (1) Addition of wastewater or sewage, (2) Addition of large amounts of nitrate fertilizers to the soil. Would the populations increase or decrease. Defend your prediction.
5. Soil contamination from toxic industrial chemicals has been a major problem since the onset of the Industrial Revolution. On one site of a former chemical plant the contamination drastically reduced microbial populations in the soil. Which of the following chemical cycles, *nitrogen* or *phosphorus*, would be most affected in an attempt to revegetate native plants on the site?

16.4  Climate Change Causes Global Ecological Concerns

Ingrid Visser/age fotostock

Figure 16.22 Melting polar ice is threatening the polar bear's habitat. Polar bears are threatened by climate change that is causing ice to melt across the Arctic. As sea ice diminishes, polar bears must swim longer distances to find food and reach denning sites. This causes stress that can have a negative impact on reproduction capacity and the overall health of the animals.

Although most polar bears are born on land, they spend much of their life hunting seals along the opening and edges of the polar ice pack (**Fig. 16.22**). They rely on the seal's skin and fat reserves, the portions of the seal's body that offer the most nutrition needed for bears to survive the cold climate. For the bears, these long winter months are good times; their food sources are plentiful, and the animals are able to gain the weight they need to survive the lean times ahead.

In spite of their adaptation to the Arctic environment, several populations of polar bears are declining in numbers. In recognition, the U.S. Fish and Wildlife Service and several international agencies have listed polar bears as a threatened species, thus providing the bears with additional protection. Climate scientists who study the Arctic have observed evidence of a several-decades-long retreat of Arctic ice likely tied to increasing temperatures. The International Union for Conservation of Nature (IUCN) has identified this change in the Arctic and the related melting sea ice as the most significant threats to the species.

The key danger posed by temperature increases to polar bears is malnutrition or starvation due to habitat loss and reduced hunting success. The presence of summer ice is an important physical characteristic in the polar bear's habitat. Because these bears hunt from sea ice, environmental changes that alter ice patterns have a significant impact on the ability of polar bears to survive. Rising temperatures melt pack ice earlier in the year, forcing the bears to migrate to shore before they have built sufficient fat reserves. Insufficient food, in addition to poorer body condition in bears of all ages, leads to lower reproductive rates in females as well as lower survival rates in cubs and juvenile bears.

A global effort is underway to determine the cause of the increasing temperatures in the Arctic. In the early 1990s, the United Nations established a panel of experts to investigate changes in global climate patterns. A 2007 U.N. report attributes human impacts that alter the natural cycling of elements such as carbon as the likely cause. Abundant scientific evidence reveals that our climate is undergoing a rapid change, a condition that we are only now beginning to address. As we discussed in Section 16.3, humans are adding carbon to the atmosphere at rates higher than nature's ability to remove it under natural conditions, creating an imbalance that likely contributes to increasing temperatures.

In the following section, we'll investigate the way climatologists have determined how climates such as the polar bear's habitat in the Arctic have changed in the last century as well as over very long periods of Earth's history, and the effect that certain atmospheric gases such as carbon dioxide have on surface temperatures. Finally, we'll explore how a changing climate affects ecosystems and the distribution of species.

Climate change occurs over long periods of time

It's important to understand the plight of the polar bear in light of what scientists know about climate change and habitats. Climate generally describes weather patterns that occur in a location or region over decades or longer. This differs from *weather,* which refers to short-term environmental conditions including temperature, moisture, and wind. The passing of the seasons results from a complex interaction of physical patterns of shifting air and ocean temperatures, precipitation patterns, ocean currents, and ice and snow accumulations. Observation reveals that none of these factors remains constant from one day or season to the next, and environmental factors can change dramatically over long periods. Understanding climate is important because of its significant impact on the living world, which influences the evolution and adaptation of organisms to habitats and thus the organisms we find in any given location.

Climate is not constant, and many factors influence shifting patterns. These factors include physical changes such as the movement of conti-

Climate—generally describes weather patterns that occur in a location or region over decades or longer.

nents over millions of years, the energy output of our sun, and the orientation and rotation of Earth. Consider the energy output of our sun, which provides light for photosynthesis and energy to keep our atmosphere warm enough to support life. Lower energy outputs from our sun, which occur in predictable cycles, decrease the amount of heat on Earth. Climate scientists believe periods of lower solar energy have resulted in extended ice ages in our planet's past.

Global climate change is a term you've likely heard. The term refers to global changes in the statistical patterns of weather over periods that range from decades to millions of years. They include measurable changes compared to past temperatures that the Arctic and many other regions are now experiencing. Climate change is not uniform in its impact. Certain effects such as warming or precipitation changes may be limited to a specific region. In recent years, especially in the context of governmental environmental policies, the term *climate change* describes variations in modern climate over the last 150 years since the onset of the Industrial Revolution, influenced by higher levels of atmospheric carbon dioxide. Changing environmental conditions alarm biologists because of the direct impact of these conditions on habitats. Ecosystems such as the Arctic that support polar bear populations are especially vulnerable to climate change because they depend upon cold temperatures and cannot tolerate a broad change in their environment.

The evidence of a changing climate is recorded in many places

The evidence of long-term climate change is preserved in various physical and biological places. Although we might concentrate on warming temperatures in certain locations, consider that accurate thermometers to document temperature patterns have only been widely available since the mid-1800s. To understand how climate has varied over long periods, climatologists study other characteristics that leave a physical record, called *proxies,* which provide reliable information that can be calibrated and compared against other evidence. In this way, proxies provide us a way to indirectly measure climate change.

The information collected from many proxies such as ice cores or tree rings tells an important, consistent story of climate change that has significantly influenced the evolution and distribution of organisms (**Fig. 16.23**). Over the last 160 years, proxy data records reveal a warming climate that mirrors that recorded by thermometers. This convincing relationship fosters the worldwide distress over human impacts to the carbon cycle and increasing atmospheric CO_2 to higher levels compared to the geologic record.

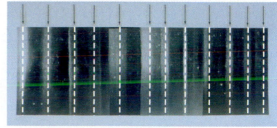

Anthony Gow, United States Army Corps of Engineers, Cold Regions Research and Engineering Laboratory/NOAA/NCDC

Figure 16.23 Ice cores preserve seasonal information. One section of an ice core showing layers from the mid-1800s. This ice core shows layers of ice accumulated since the mid-1800s (layers are read left to right). Just like tree rings, ice cores preserve seasonal information. This section contains 11 annual layers indicated by the arrows. Summer layers are less dense and appear lighter in color than winter layers. Notice the year-to-year variation in winter/summer snowpack. Over thousands of years, this evidence provides clues about changes in precipitation trends and climate.

Greenhouse gases influence temperatures

Stand inside a greenhouse in the summer and you'll quickly appreciate the ability of glass to trap heat. In a similar way, certain gases in the atmosphere have the ability to capture heat and affect global temperatures (**Fig. 16.24**). How does this process work? Incoming solar radiation enters as short waves (1), strikes Earth's surface, and bounces off as long-wave radiation or *heat* (2). Water vapor, carbon dioxide, and other greenhouse gases absorb some incoming solar radiation; they also trap heat in the atmosphere (3), retransmitting some back toward the surface (4). This

Global climate change—refers to global changes in the statistical patterns of weather over periods that range from decades to millions of years; in modern times it refers to those human activities that have had a direct impact on global factors, such as temperature.

Greenhouse gases—atmospheric gases such as water vapor and carbon dioxide, which absorb incoming solar radiation and trap heat in the atmosphere.

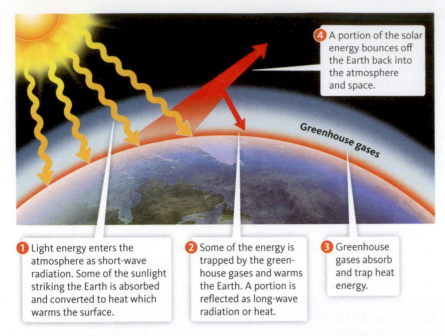

④ A portion of the solar energy bounces off the Earth back into the atmosphere and space.

Greenhouse gases

① Light energy enters the atmosphere as short-wave radiation. Some of the sunlight striking the Earth is absorbed and converted to heat which warms the surface.

② Some of the energy is trapped by the greenhouse gases and warms the Earth. A portion is reflected as long-wave radiation or heat.

③ Greenhouse gases absorb and trap heat energy.

Figure 16.24 The greenhouse effect. The physical process of trapping heat energy in the Earth's atmosphere is called the Greenhouse Effect. Water vapor, CO_2, and other gases act as greenhouse gases, contributing to our very survival. Without this process, Earth would be a cold, harsh place that would not support the life around us. Excess greenhouse gases have contributed to the global climate change seen in the last 150 years.
© Cengage Learning 2014

Global warming—refers to a long-term increase in temperatures near Earth's surface; in typical use it refers to climate changes associated with humans.

physical process, called the *greenhouse gas effect,* has contributed to our very survival; without it, Earth would be a cold, harsh place that could not support life.

Each greenhouse gas has unique properties (**Table 16.4**). One factor that makes them special is their contribution to the greenhouse effect by trapping heat and its residence time in the atmosphere. Because each gas has distinct environmental effects, scientists must account for the individual characteristics of each gas when attempting to predict the long-term impact of the pollutants humans have added to the atmosphere.

Since the onset of the Industrial Revolution, greenhouse gases added to the atmosphere by humans have been substantial. Activities such as driving cars and even turning on lights can contribute to carbon emissions. Carbon dioxide is a well-documented pollutant produced during the burning of fossil fuels such as coal and petroleum for energy. Human activities producing carbon dioxide have soared in the last 150 years (**Fig. 16.25**). The result is an atmosphere containing heat-trapping gasses at what may be the highest levels in the last 400,000 years.

Following two decades of rigorous scientific analysis, climatologists have determined that the human-caused increase in greenhouse gases is the primary reason for documented warming temperatures. The term global warming, a long-term increase in temperatures near Earth's surface, refers to this climatic shift. Data collected from satellites, weather stations, and a wide range of research sites reveal a global increase in temperatures during the last century (see **Fig. 16.25**). Like climate, the effects of global climate change vary in different regions of the world. For example, warming is more noticeable in the northern hemisphere than near the equator.

Predicting future climate conditions is an incredible challenge. They're based on limited information and rely on computer modeling that must

TABLE 16.4 Greenhouse Gases in the Atmosphere

Greenhouse Gas	Preindustrial Level	Current Level	Residence Time	Contribution to Greenhouse Effect
Carbon dioxide	280 ppm	390 ppm	Variable, 6 years+	9%–26%
Methane	700 ppb	1,745 ppb	12 years	4%–9%
Nitrous oxide	270 ppb	314 ppb	114 years	Unknown
Ozone	Unknown	10–40 ppb	22 days	3%–7%
Water vapor	Unknown	Variable	9 days	36%–72%
Chlorofluorocarbons (CFC 12)	0	553 ppt	100 years	Unknown

Note: Contribution to the greenhouse effect is derived from the amount of greenhouse gas, its residence time in the atmosphere, and its ability to absorb heat energy.
ppb = parts per billion; ppm = parts per million; ppt = parts per trillion
© Cengage Learning 2014

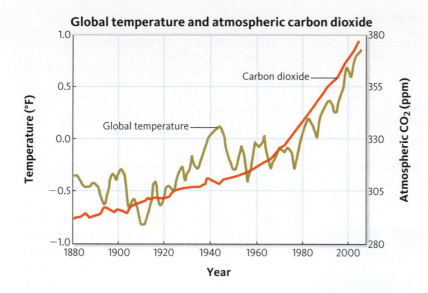

Global temperature and atmospheric carbon dioxide

(Graph showing Temperature (°F) on left axis ranging from -1.0 to 1.0, and Atmospheric CO₂ (ppm) on right axis ranging from 280 to 380, plotted against Year from 1880 to 2000. Two curves are labeled "Carbon dioxide" and "Global temperature.")

Figure 16.25 Relationship between global temperature and carbon dioxide levels in the atmosphere. This graph shows the relationship between increased average temperatures and levels of carbon dioxide. As CO_2 levels rise, so do the temperatures. This relationship is causing increasing concern about the future of many habitats, such as those of the polar bear that depend upon cooler conditions. A warming planet means changes for all ecosystems, and organisms will have to tolerate these changes or migrate away from them.
© Cengage Learning 2014

simulate complex global climate patterns. Important scientific questions remain about how much warming will occur, how fast it will occur, and how the warming will affect the rest of the climate system, including precipitation patterns and storms.

The difficulty in making predictions about future climatic conditions has led many people to dismiss the notion of modern climate change. In response, the United Nations, governments, and science organizations have partnered to educate the public. Overwhelming evidence from scientists working in the biological and geologic disciplines has established how temperatures, atmospheric gases, ecosystem changes, and human development interact to alter our planet.

Climate change affects ecosystems and species distribution

Climate change presents special challenges for the living world. Organisms must be able to survive under these varying conditions, adapt, or relocate, or ultimately they will die. Depending upon the rate of change, an organism's genetic ability to survive or adapt may be limited. Consider the rate of Arctic ice melt and the impact on polar bear populations. In addition to creating nutritional stress by making it harder for the bears to hunt seals, melting ice is expected to impact the ability of pregnant females to find and build maternity dens. As the distance increases between the ice and the coast, females must swim longer distances to reach den sites. This physical stress may lead to lower reproductive success, further reducing polar bear populations.

Melting ice from glaciers on land contributes to *sea level rise,* which has the potential to flood coastline ecosystems. Evidence from past interglacial periods reveals that sea levels have been as much as 300 feet higher than today (**Fig. 16.26a**). Climate scientists estimate that a six-foot rise in sea level by 2100 will move coastlines dozens of miles inland (see **Fig. 16.26b**). The impact in south Florida will be dramatic: the Everglades shown at the southern tip will be submerged under 6 ft. of water. Botanists are concerned about the ability of coastal plant communities such as mangroves to respond and migrate into the newly created coastlines.

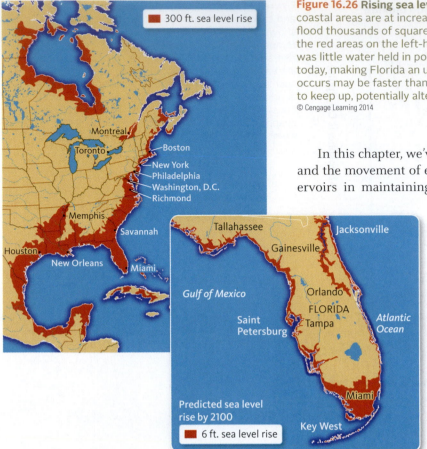

Figure 16.26 Rising sea levels could flood low-lying coastal areas. Low-lying coastal areas are at increased risk from a rise in sea level. A six-foot increase would flood thousands of square miles in the southeast United States. You can see from the red areas on the left-hand map that during prehistoric periods where there was little water held in polar ice, the sea level was as much as 300 feet higher than today, making Florida an underwater state! The speed at which sea level rise occurs may be faster than the ability of slowly migrating organisms, such as plants, to keep up, potentially altering the future diversity of these ecosystems.
© Cengage Learning 2014

In this chapter, we've examined the critical role of the cycling of water and the movement of elements between various living and nonliving reservoirs in maintaining ecological balance and supporting life. A wide range of natural and physical processes is involved in the typically slow movement of resources from one reservoir to the next. For instance, the sun provides much of the energy to power the movement of water, but ocean water may have a residence time of several thousand years. We also learned that climate change is a natural phenomenon, and evidence of its effect is recorded in locations such as ice cores in Antarctica. Humans have altered the natural patterns of element cycling, adding greenhouse gases to the atmosphere and contributing to climate change in the form of increased temperatures. Climate change alters the environmental conditions in ecosystems, which act as strong forces of natural selection on populations.

16.4 check + apply YOUR UNDERSTANDING

- Over long periods of time, changes in climate are normal.

- Quantifiable records of climate change are recorded in many places.

- Greenhouse gases are necessary to support life on Earth and have different characteristics.

- Additions of greenhouse gases by humans to the atmosphere are likely the cause of current global temperature increases.

- Climate change impacts the survival and distribution of organisms.

1. Why are proxies so valuable to the study of climate change?
2. Distinguish between climate and weather.
3. Organism A has a narrow ecological niche, and Organisms B has a wide ecological niche. Which organism is more susceptible to climate change? Explain.
4. What is the relationship between a rise in sea level and global warming?
5. If humans were able to curb their greenhouse gas emissions today (with the exception of water vapor), would you expect to see an immediate drop in the rate of global warming? Why or why not? Use the information in **Table 16.4** to help you make your case.

End of Chapter Review

Self-Quiz on Key Concepts

Water and Minerals Cycle Between Living and Nonliving Reservoirs

KEY CONCEPTS: The cycling of water, carbon, nitrogen, phosphorus, and other substances influences the distribution and abundance of life on Earth. Water and most minerals move slowly in global cycles, from environmental reservoirs, into food webs, and then back to reservoirs. Human activities such as farming and burning of fossil fuels have an impact on these cycles, often speeding up the cycling process between reservoirs.

1. Match each of the following **biogeochemical cycling** terms with its description.

    Reservoir
    Watershed
    Carbon budget
    Nitrogen fixation

    a. balance of the exchange of carbon between carbon reservoirs
    b. primary pathway between nitrogen reservoirs in the living and nonliving worlds
    c. geographic region where all precipitation drains into specific rivers or lakes
    d. temporary location where the elements or molecules reside

2. What is the largest reservoir of phosphorus?
    a. living organisms
    b. accumulations in polar ice
    c. sediments and rocks
    d. the atmosphere

3. Which group of organisms brings nitrogen into the living part of the global nitrogen cycle?
    a. microbes
    b. marine/freshwater photosynthetic organisms
    c. terrestrial plant communities
    d. mammals

4. Most of Earth's freshwater is:
    a. found in the oceans.
    b. located in rivers and lakes.
    c. locked in polar and glacial ice.
    d. in the bodies of living organisms.

5. Carbon is released into the atmosphere by:
    a. burning fossil fuels.
    b. photosynthesis.
    c. the acidification of ocean water.
    d. nuclear power production.

Climates are in Constant Flux; Human Activities Influence the Rate of Change

KEY CONCEPTS: Geologic and physical evidence indicates that global climate change is a normal process that stretches back to the earliest ecosystems. Climate change is a complex phenomenon involving many physical factors as diverse as the energy output of our sun to the burning of fossil fuels. These factors change the climates of our planet over decades, centuries, or longer periods of time. Atmospheric gases such as carbon dioxide, water vapor, methane, and others are known as greenhouse gases, which trap heat in our atmosphere and contribute to global warming. In the last century, the burning of fossil fuels by humans, along with other activities, has had a dramatic impact on the increase of certain greenhouse gases and a likely impact on higher global temperatures.

6. Match each of the following **global climate change**-related terms with its best description.

Weather   a. the ability of chemicals in the atmosphere to absorb heat energy

Global warming   b. long-term statistical patterns in weather conditions

Climate   c. long-term increase in temperatures near Earth's surface

Greenhouse gas effect   d. short-term environmental conditions such as temperature and precipitation

7. The link between global warming and species distribution results in:
   a. habitat relocation and species migration.
   b. growth of the polar ice cap and extinction.
   c. growth of the polar ice cap and species migration.
   d. lowering of most sea levels and species migration.

8. Which of the following is considered a greenhouse gas (see **Table 16.4**)?
   a. water vapor
   b. methane
   c. nitrous oxide
   d. all of the above

9. Which of the following is considered an alternative source of information for documenting climate change?
   a. thermometer readings from the past century
   b. rainfall records for the past year
   c. tree ring growth records
   d. soil phosphorus records for a county

10. Which of the following factors is linked to long-term climate change?
    a. movement of continents
    b. orbit of Earth
    c. energy output of the sun
    d. all of the above

Applying the Concepts

11. An extended global cold period called the Younger Dryas occurred about 12,800 years ago, a time when humans were becoming established in North America. One hypothesis links the onset of the Younger Dryas to humans hunting to extinction 114 large species of herbivorous animals such as mammoths, which produced large quantities of methane (**Fig. 16.27**). Explain why this hypothesis has a scientific basis and could be true.

12. Global warming has many environmental effects, including increased rates of evaporation from oceans and lakes as well as transpiration from plants. What influence might this have on climate and why?

Figure 16.27 Mammoths.

13. Large-scale volcanic activity can influence climate patterns for years due to volcanic emission of huge quantities of gases and ash many miles into the atmosphere. Whereas greenhouse gases absorb solar energy, ash and certain volcanic gases reflect solar energy back into space. In 1815 Mount Tambora in Indonesia erupted, the largest eruption in the last 10,000 years. The climatic effects were rapid, causing an extremely cold spring and summer in 1816, known as "the year without a summer." New England and Europe were hit exceptionally hard with snow and frost in June, July, and August, which resulted in food shortages. Based on the relationship between greenhouse gases and climate, describe in your own words the influence of the gases and particles emitted during this eruption.

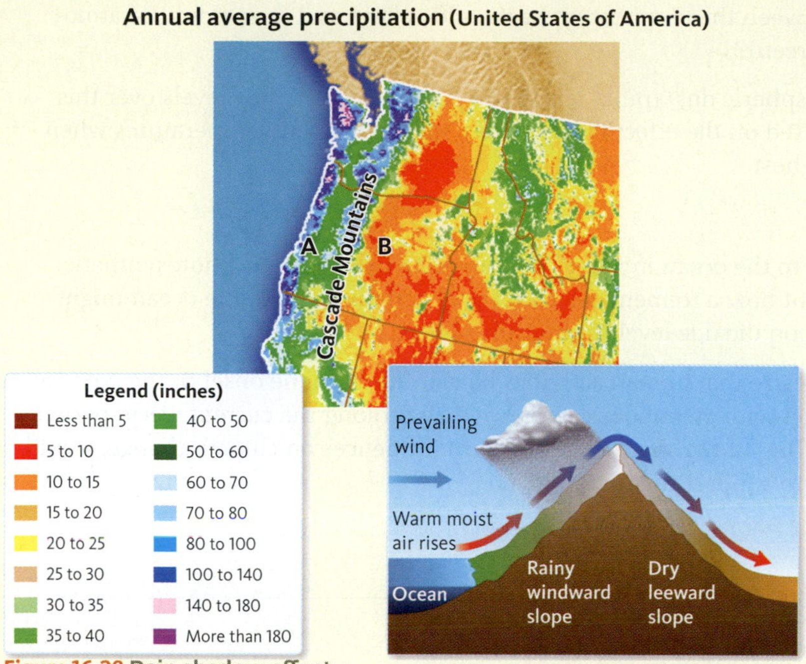

Annual average precipitation (United States of America)

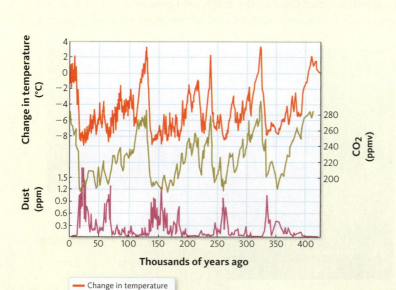

Legend (inches)

■ Less than 5	■ 40 to 50
■ 5 to 10	■ 50 to 60
■ 10 to 15	■ 60 to 70
■ 15 to 20	■ 70 to 80
■ 20 to 25	■ 80 to 100
■ 25 to 30	■ 100 to 140
■ 30 to 35	■ 140 to 180
■ 35 to 40	■ More than 180

Figure 16.28 Rain shadow effect.
© Cengage Learning 2014

Prevailing wind

Warm moist air rises

Ocean | Rainy windward slope | Dry leeward slope

14. Geography can have a dramatic impact on the water cycle. Air currents blowing up against mountains cause air to rise and rapidly cool, causing moisture to fall as precipitation. As a result, one side of the mountain range receives more moisture and the other receives less rain, a condition called a *rain shadow effect*. Based on your understanding of plant communities from Unit 3 and the water cycle, speculate about this influence on the water cycle at the regional level. Using the map in **Figure 16.28** as your guide, explain which area, A or B, experiences the rain shadow effect from the Cascade mountains and how that influences the plant communities.

15. Land clearing for short-term agricultural use is responsible for most of the deforestation in the Amazon rainforests of South America. Farmers cut and burn large patches of the forest in order to plant crops. Predict the effect this action has on the carbon and water cycles. Justify your prediction.

Data Analysis

Using Proxy Data to Determine the History of Climate Change

Climate scientists have spent decades researching the history of climate change on Earth. To do so, they rely on proxy data derived from the measurements of substances such as ice that are captured in place in order to help reconstruct the long periods of climate that occurred before humans recorded this information. The Vostok ice core from Antarctica provides data for the longest in length of time, reaching back 420,000 years. Examination has revealed a number of past glacial cycles (**Fig. 16.29**).

The graph contains information on the variation in climate over a period of 420,000 years. It contains three elements: the first is the presence of dust (pink); the second is carbon dioxide (green); and both are used to reconstruct the third element, temperature (orange). The development of this graph is based on the relationship between atmospheric temperatures and the quantities of greenhouse gases as well as the ability of atmospheric dust to reflect solar radiation.

Data Interpretation

16. Based on the temperature graph (orange), how many temperature peaks and lows have occurred in the past 420,000 years?

Figure 16.29 Relationships among global temperatures, carbon dioxide, and atmospheric dust.
© Cengage Learning 2014

Change in temperature
CO_2
Dust

Thousands of years ago

17. Is there a relationship between the temperature peaks and valleys and the amount of atmospheric carbon dioxide (green)?

18. Does the amount of atmospheric dust (pink) tend to be at high levels or low levels over this extended time period? Based on the effect that dust has, speculate about temperatures when dust levels are at their highest.

Critical Thinking

19. The addition of nutrients to the ocean increases the productivity of aquatic photosynthetic organisms. Speculate about how a tremendous increase in carbon fixers in the ocean might influence the current carbon dioxide levels in the atmosphere.

20. Climate scientists hypothesize that human activities have influenced the onset of the next ice age. By warming the planet with greenhouse gases, we may prolong the current warm period. Based on your understanding of the short- and long-term influences on climate change, would you say there is evidence to support this claim?

Question Generator
Ethanol Production and the Movement of Nitrogen between Reservoirs

Ethanol is an alcohol fuel product produced through a yeast fermentation process using corn, a renewable crop grown across much of North America. In an effort to decrease our reliance on fossil fuels created from oil, ethanol production has increased rapidly. The biofuel has increased in popularity because this homegrown product can be renewed each year as a crop. Annually more than four billion gallons are produced in factories located in areas of high corn production, primarily the mid-western states (**Fig. 16.30**).

To increase production, each season farmers add nitrogen fertilizer to provide this critical nutrient to fuel plant growth and fruit (corn kernel) production. The nitrogen fertilizers are produced in chemical plants using a high-heat and high-pressure process that burns large quantities of another fossil fuel: natural gas (CH_4). Greater than 96% of all corn crops are given nitrogen fertilizer, adding up to some 12,500 tons in the United States.

Nitrogen added to cornfields is water-soluble and can rapidly leach into groundwater or nearby lakes and streams. In the Mississippi and Ohio Rivers, watersheds where most ethanol is produced, this leaching results in large amounts of nitrogen making its way downriver all the way to the Gulf of Mexico. At each stage of transport, the increased levels of nitrogen alter aquatic habitats, creating ecological impacts that promote the invasion by exotic plants and animals while choking off and killing native species.

Below is a block diagram (**Fig. 16.31**) that relates several aspects of the biology of nitrogen transport resulting from ethanol production.

Figure 16.30 Ethanol is a biofuel made from corn.

Use the background information along with the block diagram to ask questions and generate hypotheses. We've translated the components of the block diagram into several questions to get you started.

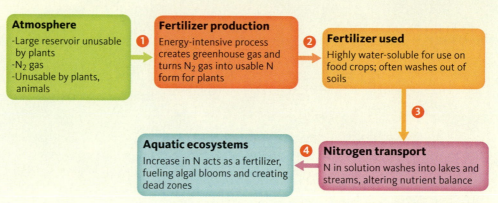

Figure 16.31 Factors relating to effects of ethanol production on nitrogen levels in the atmosphere.
© Cengage Learning 2014

Here are a few general questions to start you off.

1. How are long-term temperature shifts in your community linked to the production of nitrogen fertilizer? (Arrow #1)

2. How are corn kernels linked to the supply of nitrogen in the atmosphere? (Arrow #2)

3. How are seasonal changes in precipitation and plant growth linked to N transport? (Arrow #3)

4. Why are ecosystems in the Gulf of Mexico affected by the farming practices in faraway places like Minnesota? (Arrow #4)

Use the block diagram to generate your own research questions and frame a hypothesis.

17 Preserving Biodiversity through Conservation Biology

17.1 Grizzly bears and wolves in the Greater Yellowstone Ecosystem

17.1 Grizzly Bears and Wolves in the Greater Yellowstone Ecosystem

Crashing through the heavy brush, the mother grizzly bear with her two young cubs in tow quickly moves to the protective edge of the tree line. Close behind, a pack of seven gray wolves is intent on separating her from one of the cubs and making it their next meal.

This dramatic predator–prey scene was far more common when grizzlies and wolves roamed a larger portion of North America than they do today (**Fig. 17.1**). As settlers pushed west, animals such as bears and wolves were considered dangerous predators and were killed on sight. By 1926, no wolves were left in Yellowstone National Park, and grizzly bears were confined to remote mountain regions. Their loss eliminated a critical set of predators in the natural food chain and their natural functions within the ecosystem.

Many conservation initiatives date back to the 1800s; however, by the 1960s, a widespread ecological awakening occurred in the United States as more people began to realize that not only does every species have an intrinsic value worthy of preservation but that each is also

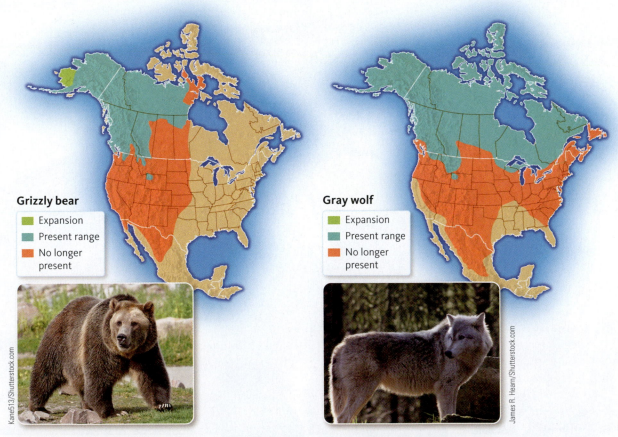

Grizzly bear

- Expansion
- Present range
- No longer present

Gray wolf

- Expansion
- Present range
- No longer present

Figure 17.1 Grizzly bears and wolves once roamed much of the U.S. Before settlers moved west, grizzly bears and gray wolves roamed the entire colored areas shown in the two maps. Now they remain only in the blue areas, with modest expansion during the last 20 years in the green areas. Human hunting and settlement pressures that fragmented the landscape and isolated remaining populations have greatly reduced their range today.
© Cengage Learning 2014

important to the functioning and health of their entire ecosystem. The field of **conservation biology**, which deals with the protection of biological diversity, grew out of this movement. In Yellowstone National Park, a dedicated group of biologists worked for over a decade to return the gray wolf to this ecosystem. Between 1995 and 1997, 31 wolves from wild packs in Alaska and Canada were reintroduced. This restoration project has been highly successful, and by the end of 2011, nearly 100 descendants of those first wolves have dispersed into various packs throughout the park (**Fig. 17.2**).

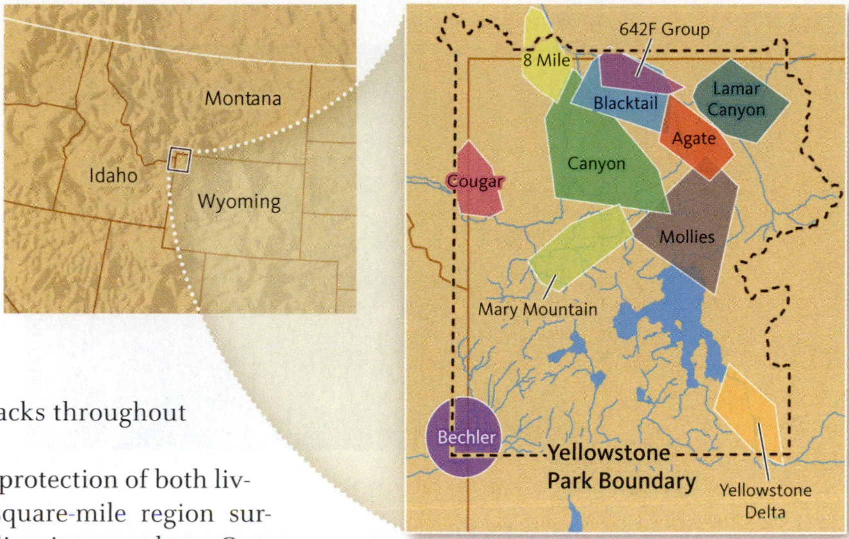

The preservation of biodiversity involves the protection of both living organisms and their habitat. The 30,000-square-mile region surrounding Yellowstone National Park and extending into northern Canada is known as the Greater Yellowstone Ecosystem (GYE), a mixture of federal, state, and private lands that provide different levels of protection for the diverse array of organisms found there. Wolves within the boundaries of the park are protected from hunting, but outside they may be killed if they're caught stalking livestock. In Montana and Idaho, limited hunting is allowed to help control the numbers of wolves in certain areas. This ecosystem is home to coyotes, elk, aspen trees, and many species that require specialized protection in order to survive. It is one of the few refuges from humans where we can observe wolf–grizzly bear interactions today (see **Fig. 17.2**). The conservation efforts aimed at preserving the GYE has produced one of many large, protected ecosystems where animals are able to live their lives free from human disturbance and predation.

The nearly 3,500 square miles of Yellowstone National Park were set aside in 1872 to preserve the natural grandeur of its geysers and scenic vistas, but in recent years the park has become an important refuge for many organisms such as bears and wolves. Its area, although more expansive than the state of Delaware or Rhode Island, is still too small in many ways. Large, highly mobile animals such as wolves and grizzlies need big, resource-rich areas in which to hunt as well as protected corridors for following food sources during seasonal migrations. Part of the GYE conservation strategy is a proposal to create a protected corridor of land along a 1,900-mile line between Yellowstone National Park and the Canadian Province of Yukon in northwestern Canada. A component of the international grizzly bear conservation strategy, the Yellowstone-to-Yukon corridor project, seeks to ensure that grizzly bears can move safely between core habitats. Although this effort has yet to be implemented, groups of scientists and other environmental advocates in government agencies and private organizations are actively engaged in working to see this project fulfilled.

Threats to biodiversity include the invasion of exotic or nonnative species. Invasive species often overrun habitats, causing tremendous damage to food webs and disrupting important biological interactions. For example, in autumn the mother grizzly devotes a lot of time looking for high-fat food sources to help her family survive the long winter. Seeds of the whitebark pine represent one of the preferred food sources during the fall. In years of poor cone production, bears will fight to the death when compet-

Figure 17.2 Wolf packs in Yellowstone. The Greater Yellowstone Ecosystem covers portions of three states and the broader biogeographic range for animals such as wolves, grizzly bears, and bison that are highly mobile. Protecting entire ecosystems provides habitat for the target animals as well as all the other organisms living there. In just a few years, the original reintroduction of 31 wolves has grown and dispersed throughout Yellowstone National Park. Wildlife biologists have given each pack a unique name, and each wolf has been given its own identifier to help track its life cycle and movement.
© Cengage Learning 2014

Conservation biology—the field of biology that deals with the protection of biological diversity.

a.

Christine Hayot/Dorena Genetic Resource Center/USDA Forest Service

b.

© Cliff Keeler/Alamy

c.

Courtesy Kate Kendall/USGS

Figure 17.3 Blister rust is a threat. (a) Blister rust is an invasive disease that infects and eventually kills the whitebark pine, whose seeds are an important food source for grizzly bears. The disease is rapidly reducing the availability of this high-fat food source in the Yellowstone ecosystem. (b, c) Healthy whitebark pine can produce lots of seeds each year, thus providing food for many forest animals including the grizzly.

ing for the seeds. Unfortunately, whitebark pine is threatened by blister rust, an invasive fungus accidentally introduced into North America around 1900 that reduces the health of the trees and limits their seed production. Eventually, the disease kills trees and thereby alters the plant community, creating an ecological shift that can significantly affect the long-term health of bears (**Fig. 17.3**). Much of the area infected by blister rust overlaps the grizzly bear's range. This poses special challenges for wildlife biologists seeking to implement policies that support bear survival and highlights the fact that changes in one seemingly unrelated organism can drastically influence another.

Gaining a better understanding of how humans develop and implement conservation activities will help you consider how your actions can directly affect the conservation of biodiversity in your area. In the chapter ahead, we'll examine strategies used to preserve, protect, and restore biodiversity on Earth. We'll build on your understanding of the hierarchy of links between a species and its ecosystem, investigating the important interactions that often are degraded when species are in peril. We'll also investigate the special properties of locations that have a high concentration of biodiversity. Let's begin by exploring how we can respond with specific actions to tackle tough policy choices concerning which organisms to protect and which to ignore, perhaps at our own peril.

17.2 The Goal of Conservation Biology Is Preserving Biodiversity and Healthy Ecosystems

One of nature's most inspiring sights is a large bird of prey swooping low and, without missing a wing beat, effortlessly plucking a trout from the water (**Fig. 17.4**). The American bald eagle is a majestic bird, used as the symbol of the freedom and strength of our country since 1787. The wingspans of these predators can reach over seven feet, which allows them to soar thousands of feet in the air. Adult eagles, with their easily recognizable white heads, may live as long as 30 years in the wild. As top-order predators in their food web, they play an important ecological role in maintaining species balance within a habitat.

Like grizzly bears and wolves, bald eagles once were actively hunted. This hunting was so successful that bald eagle populations were reduced to dangerously low levels, and by the late 1800s, they were nearly extinct in the lower 48 states. In addition to the toll from hunting, since the 1940s pollution and other toxins made their way through the bald eagles' food chain and accumulated to unhealthy levels in their bodies. The collection of toxins in blood and other cells, called *bioaccumulation,* interfered with reproduction and resulted in early death, hampering recovery of the newly protected species.

Over decades, bald eagle populations in the lower 48 states grew smaller and more isolated from one another. By 1960 there were just 450 breeding pairs, down from an estimated 25,000 pairs only 200 years earlier. Small, isolated populations pose special problems for species; they typically have low genetic diversity, which reduces their overall fitness or ability to withstand new threats to their survival. Because of this, they are highly vulnerable to disease or other hazards that could cause their extinction.

Successful eagle conservation began in the 1940s with laws prohibiting hunting. Slowly, the dangers from habitat destruction, pollutants, and toxins were recognized, along with the positive value of conservation actions at the habitat and ecosystem levels. In addition to the benefits of a hunting ban, chemicals harmful to eagle reproduction, such as the synthetic pesticide DDT (dichlorodiphenyltrichloroethane), were removed from use. Although habitat loss remains the greatest threat, the eagle population in the lower 48 states has climbed to more than 11,000 nesting pairs (**Fig. 17.5**). This is one of many examples of the successful impact of combined conservation efforts on a single species.

iStockphoto.com/KenCanning

Figure 17.4 American bald eagle. Bald eagles are skillful hunters, able to swoop low over the water and seize a fish as large as four pounds with their sharp claws. Strong flyers with excellent eyesight, bald eagles can cover large territories in order to find food.

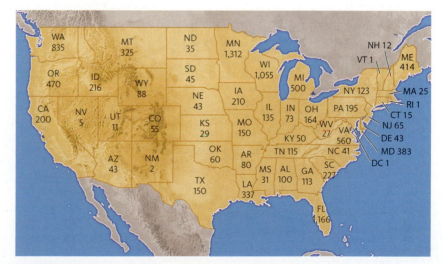

Figure 17.5 Eagle population recovery. The numbers of bald eagles have been steadily increasing since the 1940s. Today, you have a chance to see this magnificent flyer in all 48 continental states as well as Alaska.
© Cengage Learning 2014

Preserving biodiversity conserves healthy ecosystems

Biodiversity—refers to the richness of living systems, which includes the diversity of all organisms, the diversity within and among species and populations, and the diversity of ecosystems.

Ecosystem services—the economic valuation of the collection of physical resources found in a functioning ecosystem combined with the contribution of each species found there. Ecosystem services are grouped into broad categories, such as nutrient and water cycling, carbon sequestration from photosynthesis, food production, and recreational benefits.

Native populations in any habitat evolved through a long process of natural selection, creating a complex set of biological relationships such as food webs. As species are eliminated, food webs begin to collapse. The severity of the ecological impact from a loss of species diversity depends on the functional role and the overall contribution each species plays in that unique food web.

The richness of living systems, called biodiversity, has several components, including the diversity of all organisms, the diversity within and among species and populations, and the diversity of ecosystems. Consider that each species is important to its habitat, and the presence of a variety of species helps to increase the capability of an ecosystem to be resilient in the face of a changing environment. The loss of certain organisms, such as top-order predators, in a food chain may have more ecological impact than the loss of one rare species because of these predators' critical ecological role and their impact on a food web. Preserving broad-scale biodiversity is important to maintaining biological balance and thus natural, healthy ecosystems. For example, bald eagles are both predators and scavengers. Eliminating them causes the loss of these two important ecological roles, which may not be filled by other animals in that ecosystem.

The collection of physical resources found in a functioning ecosystem combined with the contribution each species makes is referred to as ecosystem services. Ecosystem services are grouped into broad categories, such as providing nutrient and water cycling services, carbon uptake through photosynthesis, food production, and recreational benefits. Ecosystems also provide cultural services, offering benefits that are nonmaterial, such as a sense of place and spiritual well-being. Table 17.1 lists examples of the types of services a healthy ecosystem provides to benefit both biodiversity and human welfare.

Placing values on ecosystem services provides a familiar method for comparing costs and benefits, allowing governments to justify spending money on managing and improving habitats. Because each species has a unique functional role and traits, its contribution to ecosystem services is unique. Consider that all species are not necessarily ecological equivalents; each has a slightly different niche and contributes to the ecosystems services in a slightly different manner.

The rapid rate of global climate change threatens biodiversity by shifting habitat conditions. Climate changes alter the biodiversity by forcing organisms to adapt or to migrate to hab-

TABLE 17.1 Ecosystem Goods and Services

Healthy ecosystems carry out a diverse array of processes that provide both goods and services to humanity. Here, goods refer to items given monetary value in the marketplace, whereas the services from ecosystems are valued but are rarely bought or sold.

Ecosystem processes include:

Maintenance of energy fluctuation, dissipation, climate modulation

Maintenance of hydrologic fluctuation, water cycle, water quality

Biological productivity, plant pollination

Maintenance of biogeochemical cycling, storage, mineral–gaseous cycles, water–air quality

Decomposition, weathering, soil development stability, soil quality

Maintenance of biological diversity—food webs and genetic diversity

Absorbing, buffering, diluting, detoxifying pollutants

Ecosystem "goods" include:

Food

Construction materials

Medicinal plants

Wild genes for domestic plants and animals

Tourism and recreation

Ecosystem "services" include:

Maintaining hydrological cycles

Regulating climate–carbon balance

Cleansing water and air

Maintaining the gaseous composition of the atmosphere

Pollinating crops and other important plants

Generating and maintaining soils

Storing and cycling essential nutrients/elements

Absorbing and detoxifying pollutants

Providing beauty, inspiration, spiritual well-being, recreation

Ehrlich, P. R., and A. H. Ehrlich, 1991. *Healing the Planet*. Addison Wesley, New York.

Lubchenco, J., S. A. Navarrete, B. N. Tissot, and J. C. Castilla, 1993: Possible ecological responses to global climate change: near shore benthic biota of Northeastern Pacific coastal ecosystems. In: Earth System Responses to Global Change: Contrasts Between North and South America [Mooney, H.H., B. Kronberg, and E.R. Fuentes (eds.)]. Academic Press, San Diego, CA, USA, pp. 147–166.

Richardson, C. J. 1994. Ecological functions and human values in wetlands: a framework for assessing forestry impacts. Wetlands 14:1–9.

itats that are more suitable. For instance, the decline of whitebark pine in the GYE forces grizzly bears to seek out other food sources. This is one reason why conservation of the GYE and its range of habitats and wildlife corridors can be so important to the preservation of a species (**Fig. 17.6**)

Invasive species represent another impact on ecosystems that can have a damaging influence on native biodiversity, displacing native species or causing their health to decline. Invasive species lack normal population checks, such as competitors from their native habitat, natural enemies, or diseases. Food chains, community structure, and ecological interactions can be permanently altered, eventually jeopardizing the health and ecosystem services all organisms depend upon. Zebra mussels were accidently introduced to the Great Lakes in the late 1980s (**Fig. 17.7**). In spite of educational programs and expensive control measures, these relatively small mussels have spread throughout the five Great Lakes and many locations across the country, displacing most native mussel species.

Endangered and threatened species require special protection

The Endangered Species Act was signed into law in 1973 in recognition of the special challenges faced by species that are nearing extinction. The goal is to prevent the extinction of plants and animals and to recover and maintain those populations by removing or lessening threats to their survival. The act establishes two categories of protection based on the species' ability to survive in the near term. An **endangered species** is one that is near extinction throughout all or a significant portion of its biogeographic range, whereas a **threatened species** is likely to become endangered in the near future. These designations are more than just labels; the law specifies levels of protection for elected species.

Figure 17.6 Importance of wildlife corridors. The Yellowstone-to-Yukon conservation initiative aims to preserve enough land along a 1,900-mile-long corridor to allow grizzly bears and other large animals to migrate along historically important pathways between core habitats.
© Cengage Learning 2014

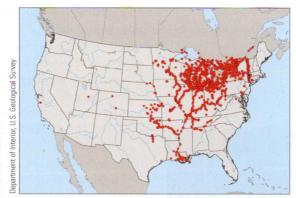

Figure 17.7 Spread of zebra mussels. To make boaters aware of their potential role in spreading the invasive zebra mussel, state and local governments have developed extensive public education campaigns that show the visible damage that can be done when a nonnative species such as zebra mussel spreads through an ecosystem. The purpose is to slow the spread of zebra mussels (red dots on map) to waters that are not yet infested in order to preserve native species.

Endangered species—an organism that is near extinction throughout all or a significant portion of its biogeographic range.

Threatened species—an organism likely to become endangered in the near future.

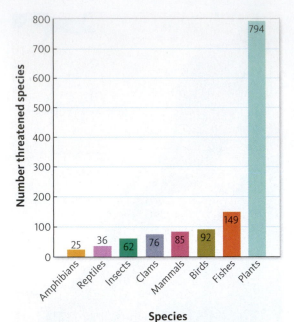

Figure 17.8 **Endangered and threatened species (U.S., 2012).** Endangered and threatened species include many different types of animals as well as plants. This graph shows the numbers of species currently listed, a small fraction of those in danger from human activities. The U.S. Fish and Wildlife Service is responsible for developing management plans that will aid in species recovery for each of them.
© Cengage Learning 2014

Figure 17.9 **Land use changes fragment habitat.** As housing developments are constructed, they permanently alter the landscape, creating fragmented islands of natural habitat cut off by new roads and the intrusion of human activities. Fragmentation makes it difficult for animals to move between the remaining islands of suitable habitat, creating the potential for local populations to go extinct.

In spite of this legal initiative, new species are regularly added to the list. By 2012, there were 1,392 threatened or endangered species in the United States (**Fig. 17.8**). The number of species listed represents a small fraction of those animals and plants in danger. The known species threatened with extinction may be 10 times higher than the listed species covered by the Endangered Species Act. Similar definitions are used by international organizations that monitor endangered and threatened species around the world. Listed species include familiar animals such as the polar bear, green sea turtle, Chinook salmon, black-footed ferret, and Canada lynx. Also included are lesser known species such as the key deer, a species found on only one small island along the south Florida Keys.

There is no specific threshold for when a species should be added to the endangered or threatened species list. The guidance provided to federal agencies requires them to complete a specific evaluation to list a given species when concerns are raised. In spite of the legal guidance, adding a particular species to the list is a long, controversial process. Wildlife biologists must establish that there is a near-term threat to the species' survival or its required habitat. For species determined to be in imminent danger of extinction, federal agencies or courts may intervene and take emergency action to prevent extinction from happening. Human activities that threaten organisms can be halted and the damage reversed if necessary. Emergency actions include halting construction of new dams or roads, stopping fishing, or banning the use of certain chemicals.

Preserving a species often means conserving or restoring large tracts of land

The conservation of biodiversity requires preserving habitats in a natural, undisturbed condition. There is little point in attempting to conserve a species or population if its habitat is in danger of disappearing. Conservation biologists approach this problem by working to preserve both the species and their habitats at the same time, relying on organizations such as zoos and federal agencies. Nonprofit organizations such as the Nature Conservancy also work to preserve critical habitat and endangered species in every state.

Habitat destruction and fragmentation represent a major threat to biodiversity. Habitat destruction occurs when humans alter an area through land conversion for agriculture or construction of roads, buildings, or other infrastructure. These activities alter the natural biodiversity by replacing it, killing or displacing the native species, and isolating the surviving populations into small islands of remaining habitat. Human activities also affect connections between natural spaces, making it difficult for organisms to find suitable habitats while moving between areas. *Habitat fragmentation* refers to a patchwork pattern of habitat disturbance that changes the landscape structure (**Fig. 17.9**). Fragmenting habitats are a major cause of species becoming threatened or endangered.

Habitat fragmentation creates problems of access to suitable habitats for all animals and potential conflicts with human activities. Throughout the eastern United States, large populations of white-tailed deer have been isolated because of housing and agricultural development. Deer must cross multiple roads and highways to move between suitable habitats. Deer on roads create major safety issues, causing more than one million car accidents, hundreds of deaths, and thousands of injuries each year (**Fig. 17.10**).

Once a habitat is degraded, corrective action or restoration is needed to restore biodiversity and natural processes. The degree of ecological restoration depends on the degree of disturbance, the area to be restored, the biodiversity remaining, the completeness of our knowledge of the ecosystem, and the amount of money available to fund it. Although complete restoration of a habitat is impossible, ecologists attempt to reestablish natural physical processes along with communities of native organisms that represent functioning food webs (**Fig. 17.11**). A range of restoration activities must take place in order for predisturbance biodiversity to return.

Figure 17.10 White-tail deer. Deer frequently move between habitat locations in search of food and shelter. When their paths cross human roads, a potentially dangerous mix is created, resulting in car–deer collisions that injure or kill thousands of humans and deer each year.

Figure 17.11 Restoring habitat is multidimensional. Habitat restoration requires more than assembling native plants and animals. Ecologists also restore water drainage patterns and soils to support the native mix of plants and animals being reintroduced. This photo shows planted trees covered by protective plastic tubes, a common device used to keep herbivores from browsing on them until they become well established.

Roads, fences, and similar structures can be formidable barriers to animal movement and causes of increased mortality for small and large animals alike. To reduce the stress and mortality of animals living in fragmented landscapes, wildlife biologists are developing innovative management solutions. Engineers have teamed with wildlife biologists to build specially designed wildlife overpasses and underpasses to safely permit animals to migrate across roadways (**Fig. 17.12**). These structures, planted with native vegetation and situated in known migration corridors, save animals and reduce the threat of human–animal collisions.

A common approach to reducing the impact of habitat fragmentation is the creation of habitat corridors that connect these areas and reduce the isolation of individual populations. Habitat corridors are important pathways that any organism can use to move between habitats that meet their needs. The Yellowstone-to-Yukon initiative is an example of a large-scale conservation effort designed to preserve significant patches of habitat strung along animal and bird migratory corridors, which will be available to many species besides grizzly bears.

Figure 17.12 Engineered corridors. Specially engineered overpasses and underpasses create safe, natural-appearing corridors that allow animals to easily cross busy highways. The corridors typically are planted with native plant species to make them more inviting and to provide suitable cover to the animals that use them. Large migratory animals such as deer, elk and bear use these structures to move from one area to the next without having to cross deadly highways.

- Conservation biology is about the study of Earth's biodiversity.

- The aim of conservation biology is to preserve both species and their habitats.

- Where large land tracts are not adjacent, corridors help create pathways between suitable habitat locations.

Figure 17.13 Snail darter.

1. Explain why governments use designations such as "endangered" when labeling populations of organisms.

2. Explain how the categories of endangered and threatened species differ in regard to the danger of extinction.

3. What is the relationship between biodiversity and ecosystem services in an undisturbed habitat?

4. What characteristics of an ecosystem need to be understood in order to undertake a successful restoration effort?

5. In 1976, a federal judge blocked the construction of a dam on the Tennessee River. The snail darter is a small (up to 3.5 inches long) fish native to waters of eastern Tennessee that feeds primarily on aquatic snails (**Fig. 17.13**). The court cited the fact that this project would alter the habitat of the river to the point of eliminating the snail darter. Was this protection warranted under the Endangered Species Act? Defend your answer based on the purpose of the Endangered Species Act.

17.3 Conserving Biodiversity Requires Efforts at Many Different Levels

Figure 17.14 Siberian tiger. Well adapted to the deep snows of Siberia, the Siberian tiger can easily speed across the snow to capture prey.

Extinction—elimination of a species, which may be local such as the western populations of Siberian tiger, or global as in the total elimination of passenger pigeons in the United States due to extensive hunting.

The endangered Siberian tiger is a large, powerful cat able to race across the deep snows of east Asia at speeds up to 30 mph (**Fig. 17.14**). Its distinctive orange and white coat with broad black stripes is easy to identify but difficult to see in its native habitat. The Siberian tiger is adapted to live in forests and hunt large herbivores. At one time Siberian tigers roamed much of Asia, but their populations have declined to critically low levels. Today they are found in only a tiny slice of their former range, and as few as 400 animals remain in the wild. In fact, all tigers in the western Asia populations are now extinct (**Fig. 17.15**). As you learned in Chapter 3, **extinction** is the eradication of a species, which may be local, such as the

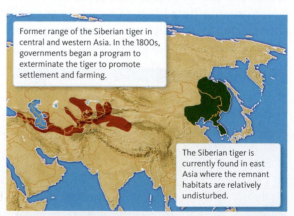

Former range of the Siberian tiger in central and western Asia. In the 1800s, governments began a program to exterminate the tiger to promote settlement and farming.

The Siberian tiger is currently found in east Asia where the remnant habitats are relatively undisturbed.

Figure 17.15 Present and former range of the Siberian tiger. The Siberian tiger once roamed much of Asia. Historic hunting records and recovered bones help us understand where they lived previously (brown area on left of map). Today, due to human activities they occupy a smaller and smaller area, seen in the dark green on the right-hand side of the map.
© Cengage Learning 2014

western populations of Siberian tiger, or global, such as the total elimination of U.S. passenger pigeons due to extensive hunting. Zoos play an active part in the global effort to save Siberian tigers by participating in captive breeding programs. More than 160 animals are being bred in captivity in an effort to ensure genetic variation and the survival of this species.

The dramatic reduction of the remaining tiger populations and the extinction of the western populations are the direct result of human activities, including habitat destruction, hunting, and elimination of the tigers' prey species. Ongoing human impacts, such as continued encroachments on its shrinking habitat, will reduce the wild population even further. There is hope, however. International organizations such as the United Nations and numerous nonprofit groups are focusing attention on preserving a range of habitats and organisms in order to achieve a significant reduction of the current rate of biodiversity loss at the global, regional, and national levels (**Fig. 17.16**). For example, as part of a global effort to protect ecosystems and the animals that inhabit them, the United Nations Environment Programme (UNEP) administers one of the largest conservation agreements: the Convention on International Trade in Endangered Species of Wild Fauna and Flora (CITES). This program provides international protection for organisms such as the Atlantic right whale by eliminating hunting and supporting the recovery of its populations.

In the following section, we'll examine the role that conservation groups play in preventing species from going extinct. Conservation groups, zoos, and aquariums around the world are networked together to save Siberian tigers and other species by monitoring genetic diversity and promoting captive breeding programs. Success in these programs, combined with habitat restoration, can lead to the reintroduction of these species and working ecosystems. Let's begin by learning more about the role of conservation in the survival of select species.

International networks of zoos and aquariums conserve endangered species

Siberian tigers are popular zoo attractions; observing large, powerful animals up close is exciting. Historically, zoos and animal parks were curiosities created by the rich for their personal entertainment. Exotic animals were collected with little thought of how their wild populations were affected or even how to properly care for them. Today, zoos and aquariums around the world lead efforts to preserve and enlarge the populations of specific endangered and threatened species (**Fig. 17.17**).

Many conservation groups and their programs are devoted to promoting reproduction in captivity. *Nongovernmental organizations* (NGOs) such as the World Wildlife Fund target the protection of animals and their habitats, creating opportunities to save endangered species. For instance, endangered animals such as rhinos are bred in captivity to sustain or increase their numbers (**Fig. 17.18**). International NGOs, including the World Association of Zoos and Aquariums, collaborate by registering their animals in a common database, which allows sharing of information on potential mates. Analysis of animal genetics and population data is used to develop a management plan designed to ensure the health of captive and wild populations. Managers try to minimize inbreeding and grow populations to sustainable levels, a process that has proven effective for more than 850 species or subspecies managed under cooperative conservation breeding programs.

© Secretariat of the Convention on Biological Diversity (SCBD)

2010 International Year of Biodiversity

Figure 17.16 Global efforts in conservation. The United Nations declared 2010 the year of biodiversity in an attempt to focus international attention and conservation emphasis on preserving Earth's species richness.

Figure 17.17 Siberian tiger cub. Zoos have played a critical role in maintaining the population of Siberian tigers. Intensive human intervention, such as bottle feeding cubs, helps ensure the long-term success of reproductive programs that maintain the species and its genetic diversity. Bottle feeding provides essential nutrients to cubs, helping them stay healthy and ensuring their survival.

Figure 17.18 Mating black rhinoceros. Zoos are important centers of controlled animal breeding programs, ensuring that genetically suitable mates produce healthy offspring to maintain or increase the population.

■ Range

Figure 17.19 California condor. With a wingspan of up to 10 feet, California condors are the largest land bird in North America. At one time they were common sights across much of the western United States. Hunting and habitat disturbance nearly wiped out this species, but today, through an intensive species recovery effort, they once again are being seen in the wild.

© Cengage Learning 2014

Not all species receive the same level of protection. Those considered attractive or charismatic get special attention for conservation. Reptiles, insects, and animals such as bats are considered less desirable or have negative cultural reputations and receive far less attention and funding. Large, appealing mammals such as pandas are used by conservation organizations to promote interest in preserving habitats and specific species. Although this may seem shortsighted, environmental groups use the popularity of engaging animals to achieve broader goals, such as obtaining funds for less well-known animals or those considered less attractive. By drawing public attention to the giant panda, for example, conservation groups have raised money and international awareness to support habitat and ecosystem preservation efforts in their native habitat.

Many species require assisted reproduction to maintain or increase their numbers

California condors once were common throughout much of the desert southwest. Although California condors are not as recognizable as pandas or tigers, as the largest North American land bird, their 10-foot wingspan draws considerable attention for their size and soaring ability (**Fig. 17.19**). Habitat destruction, climate change, and hunting combined with low birth rates and exacting mating behaviors resulted in the near extinction of this species. In 1987 the last 22 wild condors were captured, and an aggressive captive breeding program was started. Today this program still produces young that are released back into the wild, and although the recovery effort has increased the number of wild birds, there have been serious setbacks. About 40% of released condors die from lead poisoning as a result of eating lead bullets in dead animals, hitting power lines, and being attacked by golden eagles.

The captive breeding process is expensive and requires a great deal of human intervention. Condor eggs are removed from the mother soon after they are laid and carefully incubated (**Fig. 17.20**). Once hatched, handlers use puppets to hand feed the chicks and demonstrate nurturing parental behavior. During training, the young condors are taught to avoid human contact and deadly hazards such as power lines. This intensive care lasts for two years until the offspring are able to fend for themselves, and then they are released back into the wild.

Other species require far less intensive assistance. Captive panda breeding programs have been used to increase the numbers of pandas for decades. Pandas, which are native to China, number as few as 1,500 in the wild and roughly 250 in captivity. In the 1970s captive pandas were shipped between zoos and paired with mates to promote natural breeding, although once captured the pandas seemed to lose interest in reproduction. Fewer than 10% of male giant pandas engage in mating behaviors, and fewer than 30% of females conceive naturally. Poor success rates led to more invasive assistance; using artificial insemination, veterinarians insert the

a.

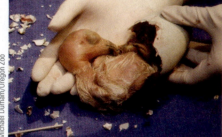

b.

Figure 17.20 Condor recovery requires intensive work. (a, b) The California condor species rescue and recovery effort is one of the most intense ever undertaken. Nearly extinct, the few remaining wild birds were captured and placed into a captured breeding program. Biologists incubate the eggs and, using a realistic looking hand puppet (b), they rear the young, teaching them the skills they need to survive on their own. This program has been a tremendous success, with animals released into the wild starting to reproduce naturally once again.

male's semen into the uterus of a reproductively active female (**Fig. 17.21**). This practice improved the successful reproduction rate to nearly 75%.

Genetic manipulation is sometimes used to assist in the targeted reproduction of specific endangered plants. This manipulation is designed to improve desirable physical characteristics, such as flowering, and resistance to disease and insect pests. The American chestnut tree once covered much of the United States east of the Mississippi River, accounting for nearly one of every four trees. In 1904 an invasive fungus called chestnut blight was accidentally introduced, and within 20 years almost three billion chestnuts were killed (**Fig. 17.22**). As a result, the plant and animal communities of eastern forests were dramatically altered.

The American Chestnut Foundation began to crossbreed a few remaining disease-resistant American chestnuts with a close cousin, the disease-resistant Chinese chestnut. Crossbreeding the two species mixes their traits, providing some of the American chestnut offspring with the disease-resistant characteristic of the Chinese species. After two decades of hybridization, the genetically altered American chestnut offspring possess the disease-resistant genes, providing a new genetic line that is able to survive the blight. Without human intervention, American chestnuts would have vanished from the landscape. With ongoing assistance in improving their disease resistance, someday these magnificent trees may once again dominate the eastern forests (**Fig. 17.23**).

Species recovery sometimes includes reintroducing species back into the wild

Preserving biodiversity may include the reintroduction of animals that were eliminated from their natural habitats in the past. Like the gray wolf, the black-footed ferret has been the focus of targeted reintroduction efforts during the past two decades (**Fig. 17.24a**). The black-footed ferret specializes in preying upon small mammals, including prairie dogs, ground squirrels, and rodents. Their numbers were dramatically reduced when their primary prey, prairie dogs, were systematically eliminated by ranchers who believed the burrow holes were hazardous to their cattle. Black-footed ferrets once were commonly found around prairie dog colonies from southern Canada to northern Mexico. By the early 1980s, hunting, habitat loss, and disease had reduced the population to only 50 members.

Effective restoration ecology requires that some level of a naturally occurring and functioning community be restored to the landscape. The reintroduction of the ferret was accomplished through an intensive captive breeding program at zoos along with the resurgence of prairie dog populations on federally protected lands in the west (**Fig. 17.24b**). Targeting reintroduction into native habitats with plentiful prairie dog populations helped ensure that ferrets had the opportunity to succeed in the wild. Today, the wild population exceeds 750, and nearly 300 additional ferrets are part of a successful captive breeding population. Although these projects are complex and require intensive management efforts, they provide evidence that endangered species management programs can succeed when endangered animals are reintroduced into their native habitats.

For animal reintroduction projects to be successful, it's important that wildlife biologists thoroughly understand the ecosystem and human activities. Reintroduction of a species is expensive, logistically challenging,

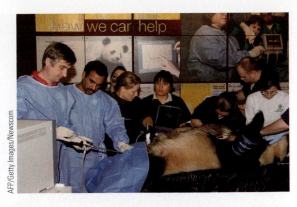

Figure 17.21 Panda artificial insemination. Pandas in captivity have poor reproductive rates. To improve reproductive success, veterinarians work with zoo staff to artificially inseminate females to increase the number of offspring.

Figure 17.22 Disease killed American chestnut trees. Once dominant in eastern forests, almost all of the chestnut trees in the U.S. were killed in just 20 years by the invasive disease chestnut blight.

Figure 17.23 Recovery of the American chestnut tree. Hybrid American chestnut trees have been bred to include disease-resistant genes from their Asian relatives.

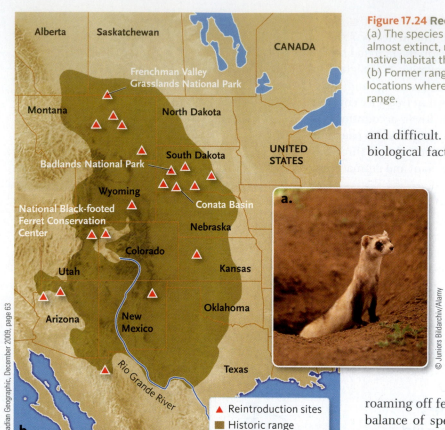

Figure 17.24 Recovery of the black-footed ferret.
(a) The species restoration program for the black-footed ferret, once almost extinct, now includes reintroducing this species back into its native habitat that once stretched from Canada to northern Mexico.
(b) Former range of the black-footed ferret. The red triangles indicate locations where ferrets have been reintroduced to areas of their former range.

▲ Reintroduction sites
■ Historic range

Canadian Geographic, December 2009, page 63

© Juniors Bildarchiv/Alamy

and difficult. It's important to know whether the physical and biological factors that caused the extinction of the species are still present. Factors such as habitat loss and degradation, illegal hunting, pollution, competition, and low population levels can cause the reintroduction to fail. To address these issues, scientists must complete comprehensive studies before beginning a reintroduction program.

By all accounts, the wolf reintroduction program in the GYE has been a success, and wolves were removed from the endangered species list in 2009. Within the conservation community, this was a highly controversial decision, and legal battles will continue for many years. Although partial protection remains in place, wolves that prey on livestock while they are roaming off federally protected lands can be killed. This delicate balance of species protection and conflicting human activities will continue to challenge wildlife biologists and land management agencies into the foreseeable future.

17.3 check + apply YOUR UNDERSTANDING

- Governments protect species and habitats by passing laws, educating the public, and setting aside suitable habitat.

- Zoos, aquariums, and nonprofits conserve the biodiversity of many endangered species through captive breeding programs.

- Assisted reproduction helps increase the numbers of many species managed in captivity.

- Under optimal conditions, reintroducing species back into the wild is part of a larger successful animal recovery program.

1. What is the role of zoos in preserving ecosystem biodiversity?

2. Describe two ways that humans can intervene in the reproductive cycle of an animal to help increase its numbers.

3. Why is being charismatic important if you are a threatened or endangered animal?

4. Many species of bats in the United States are in danger of extinction because of the harmful effects from a fungal disease called white nose syndrome, named for the white ring that grows around their snout (red arrows on photo) (**Fig. 17.25**). Since 2007 the disease has spread rapidly, killing more than a million bats. Because the syndrome is poorly understood, a number of biologists believe some bat species could become

CONTINUED

Figure 17.25 White nose syndrome. Red arrows point to the diseased tissues on the noses.

extinct due to the disease. Explain two techniques humans could use to prevent the species' extinction.

5. Many groups are lobbying for the reintroduction of the bison to the Great Plains (**Fig. 17.26**). They would like to restore the land to bison habitat and increase the numbers of animals to the millions. Make a case for why this would or would not succeed.

Figure 17.26 American bison.

17.4 Global Biodiversity Hotspots Are Important Reservoirs of Genetic Diversity

Stepping into the shade of the Amazon rainforest, you are quickly surrounded by a highly diverse curtain of life (**Fig. 17.27**). The remnants of this habitat give us an idea of the *species richness*—or number of species living there—that early biologists encountered as they surveyed these remote regions. In 1848, an English naturalist named Alfred Russell Wallace, the father of modern biogeography, traveled 2,000 miles up the Amazon from the Atlantic Ocean on a four-year quest to obtain new samples of animal life.

Wallace and his contemporaries observed that the unique conditions of the Amazon rainforest supported a vast amount of biodiversity absent in England and most other parts of the world. The large number of species provides us with clues about the development of biodiversity in the tropics (**Table 17.2**). Biologists have long recognized that species diversity increases as you move from the poles to the tropics, and tremendous efforts are aimed at better documenting and understanding the species richness contained there. Some evidence of this diversity is seen in a comparison between the state of Utah, located in a temperate zone, and that of the larger Amazonian basin (see **Table 17.2**). The Amazon is home to nearly 4 times as many birds and more than 35 times as many fish species.

The Amazon rainforest is just one of many *biodiversity hotspots,* locations where habitat conditions are able to support diverse numbers

Figure 17.27 Looking up from the floor of a rainforest. The Amazon rainforest is home to a significant portion of the biodiversity of life on Earth.

TABLE 17.2 Biodiversity of the Amazon Basin versus the State of Utah

Organism Type	Number of Species	
	Amazon Basin	State of Utah
Insects	2.5 million	Several thousand
Vascular plants	At least 40,000	2,000+
Mammals	427	134
Fish	3,000	83
Birds	1,294	367
Amphibians	428	17
Reptiles	378	57
Fungi	Unknown, perhaps 100,000+	Unknown, likely thousands
Microbes	Unknown, perhaps 1,000,000+	Unknown, likely tens of thousands

© Cengage Learning 2014

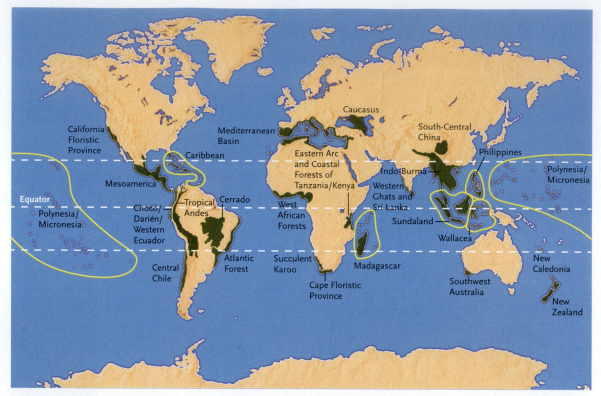

Figure 17.28 Global biodiversity hotspots. The green areas and circled island groups have been identified as key biodiversity hotspots. These hotspots hold especially high numbers of species, yet their combined area of remaining habitat covers less than 2% of Earth's land surface. This map shows the global distribution of areas of high biodiversity. These areas tend to be concentrated in warm, wet climates capable of supporting many ecological niches. A significant number of the world's species may occur in these rapidly disappearing areas.
© Cengage Learning 2014

of species. Most hotspots are located near the equator in warm, tropical climates that are associated with high rainfall (**Fig. 17.28**). Biodiversity hotspots represent large reservoirs of genetic variation, containing 44% of the world's plants and 35% of terrestrial vertebrates. Unfortunately, many of these same areas have undergone a dramatic transformation caused by human activities. Human development and habitat disturbance have reduced the size of these hotspots from nearly 12% to less than 2% of Earth's land surface. No one knows how many species have disappeared, but each extinct species diminishes the remaining biodiversity and ecosystem services of that area.

In the following section, we'll study the ongoing discovery of new species and examine two of the most important and well-studied regions with high levels of biodiversity: the Amazon rainforest and the Great Barrier Reef. Learning which ecosystem characteristics promote biodiversity and how humans can apply efforts to conserve these features will help us preserve the rich biodiversity for future generations and the health of our planet.

The Amazon rainforest is a very diverse ecosystem

The Amazon rainforest, located in the larger Amazon basin, is a unique biological treasure. The watershed, drained by the Amazon river, is huge, covering a region larger than 2.6 million square miles, more than half the area of all 50 U.S. states (**Fig. 17.29**). Spread over northern South America, the area contains many different types of ecosystems, each rich in biodiversity.

The Amazon basin contains over half of the remaining rainforests, and it is the largest and most species-rich tract of tropical rainforest in the world. The region is remote and difficult to navigate, so the species listed in **Table 17.2** represent a fraction of the total biodiversity contained there. The diversity of plant species is among the highest on Earth: a single area slightly larger than one-half square mile may contain as many as 75,000 species of trees and 150,000 of plants, with a biomass of nearly 2.5 billion pounds.

Why are rainforests so diverse? The year-round warm growing temperatures and the abundant base of producers, combined with plentiful resources and ecological niches, help sustain so much biodiversity. For example, the forest canopy may be subdivided into five or more distinct layers of plants, creating a multitude of unique habitat conditions. As we learned in Chapter 4, the greater the diversity in habitat conditions and geography, the greater the number of unique niches.

Unfortunately, the Amazon basin has undergone a significant amount of deforestation. Deforestation, which is the clearing of all trees and other vegetation from forests, results from human development for agriculture and animal grazing as well as settlement and logging. The amount of land cleared is significant and can be seen from space (**Fig. 17.30**). Given the level of species biodiversity, every square mile lost represents the extinction of many organisms. Deforestation tends to occur in nearly adjacent blocks, fragmenting the remaining habitat and isolating the remaining organisms. Ecologists are concerned about the effect deforestation in the rainforest will have on climate change. Amazonian rainforests account for 10% of the world's terrestrial primary productivity and 10% of the carbon sequestered in ecosystems, an important global resource for moderating the rise in atmospheric CO_2 levels and keeping global climate change in check.

The Great Barrier Reef is home to thousands of unique species

Hotspots also occur in marine ecosystems such as the Great Barrier Reef, which is the largest coastal reef system in the world. It stretches along Australia's eastern coastline for more than 1,500 miles, from the low-latitude tropics to temperate zones, and covers over 133,000 square miles, an area larger than the state of New Mexico (**Fig. 17.31**). The nearly 3,000 reefs and 900 islands in the chain are grouped into 70 bioregions, each containing a multitude of species and ecological niches (**Fig. 17.32**). Many species that live there are found nowhere else on Earth. As you look at **Table 17.3**, consider that the diversity of the species listed likely represents only a piece of the overall biodiversity of the region, with so many areas left to explore.

This collection of naturally fragmented habitats results from 20 million years of geologic processes, reef growth, and development. As in the rainforest, the tremendous biodiversity is the result of the partitioning of many ecological niches. As corals grow outward and upward they help to create a range of habitat conditions that provides cover for other animals and supports the survival of other species. Combined with abundant resources and favorable environmental conditions, reefs support large, rich, complex biological communities.

Figure 17.29 The Amazon basin. The Amazon basin covers much of the northern portion of the South American continent. The rainforest is just one of many ecosystems found across this extensive region drained by the Amazon River.
© Cengage Learning 2014

Astronaut Photography Database/NASA

Figure 17.30 Satellite photo showing habitat loss. The photo taken from space shows a several-square-mile area of the Amazon impacted by human activity. The roads and developments provide pathways into the forest, giving loggers access to the large amount of trees. Logging has greatly reduced the size of the rainforest.

Deforestation—the clearing of forests by humans for the development of agriculture, for animal grazing, and for settlement and logging. Deforestation is a primary cause of habitat fragmentation and reduction of biodiversity.

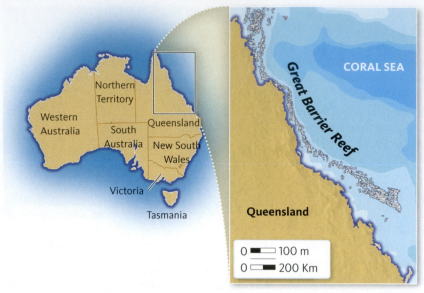

Figure 17.31 The great barrier reef. The reefs and small islands of the Great Barrier Reef stretch out along 1,500 miles of the east coast of Australia. This highly diverse landscape is home to tremendous biodiversity. The darker blue colors of the ocean indicate deeper areas, while the lighter blue are shallow areas.
© Cengage Learning 2014

Figure 17.32 Coral reefs. This photo, taken from an airplane, shows the irregular growth patterns of coral reefs. The range of depths, the interaction of ocean tides, and the range of physical habitats created by the reefs and islands provide ample niches to help support an immense amount of biodiversity.

TABLE 17.3 Biodiversity of the Great Barrier Reef

Organism Type	Estimated Number of Species
Whales, dolphins, porpoises	30
Vascular plants	1,500+
Corals	400
Sea snakes	15
Fish	1,500
Birds	215
Mollusks	5,000
Marine invertebrates	Unknown, perhaps thousands+
Microbes	Unknown, perhaps 1,000,000+

© Cengage Learning 2014

Each reef community occupies unique habitats created by the arrangement of the reefs and islands, the range of water depths, the variability of water salinity and temperatures, and the availability of nutrients. The fragmented landscape forms a remarkable amount of transition area between adjacent habitats, a feature called an *ecotone*. Ecotones contain a mix of environmental and biological characteristics from the habitats on each side, a feature that supports additional biodiversity. Boundaries between the beach or reef and the ocean possess physical characteristics of each habitat as the tide rises and falls. Terrestrial and aquatic organisms inhabit this area, as do other species successfully adapted to the ever-changing environmental conditions.

Although the possibility is remote, a wide range of threats are combining to put the health of this ecosystem in danger. Climate change threatens to alter the natural patterns of temperature and precipitation, potentially impacting species' abilities to survive there. One outcome of climate change—sea level rise—has already led to a decline in coral and seagrass communities. As the ocean levels rise, these communities are further submerged, lowering the amount of life-sustaining light that reaches them. Warmer-than-normal waters cause bleaching, killing the corals. Human activities, such as overfishing of commercial species, have had a dramatically negative effect on food webs and biological sustainability. Pollution is a persistent threat, and nutrient runoff from agricultural fields alters the sensitive chemical nutrient balance and the types of species that can survive there.

Biodiversity hotspots continue to yield new species

Our understanding of the living world grows each week with the discovery and cataloging of new species, most of which are found in biodiversity hotspots (**Fig. 17.33**). New species are found as biologists push further into remote terrestrial ecosystems such as jungles. Since 2000, wildlife biologists have discovered 25 new species of primates, mostly small lemurs and monkeys, and 30 new species of bats. Among the least explored regions of Earth, the world's oceans are believed to contain the greatest amounts of biodiversity.

To catalog the ocean's biodiversity, marine ecologists have increased their efforts to document and classify new species. Australian and Japanese waters are by far the most abundant; each has nearly 33,000 species. These regions have moderate climates and a wide variety of habitat availability. They are locations where deep currents that rise from the ocean floor carry with them abundant nutrients. A 10-year effort begun in the 1990s involved nearly 3,000 scientists who worked on locating and classifying new marine life. Their estimates indicated that 60% to 80% of ocean species are yet to be discovered.

Most species new to science are small; in fact, the vast majority would fit in the palm of your hand or would require a microscope to see. Consider that one teaspoon of soil alone may contain thousands of species of microbes. Microbes thrive in the oceans and may account for about half of the biomass on Earth. Although most have never been scientifically cataloged, they play a crucial role in ecosystem health. The abundance of microbes plays a critical role in most biogeochemical processes, accounting for almost half of global primary production and a major part of nutrient cycling. Identifying these new species will help us better understand their roles in working ecosystems and help governments justify their preservation.

In this chapter, we have learned about the role biodiversity plays in preserving healthy ecosystems and the conservation challenges we face. To preserve biodiversity, we need to retain intact, healthy ecosystems that provide critical habitat for communities of species. In some cases, human intervention is required to minimize the harm from invasive species or to assist threatened species and improve their reproductive success. This will require a deep understanding of how natural, undisturbed ecosystems function and the biology of the living things that compose them.

Figure 17.33 Newly discovered species. Each year scientists working in remote areas discover new species of plants and animals. (a) This Pinocchio-like tree frog species is among several new animals discovered in Indonesia's Foja Mountains. (b) The "dragon millipede" is one of more than 1,000 new species discovered around the Mekong River in southeast Asia over the last 10 years. Scientists suggest the millipede uses its bright color to warn predators of its toxicity. (c) A new species of brittlestar, a close relative of the sea star. This new species was discovered off the southern coast of Tasmania.

- Global biodiversity hotspots contain large numbers of habitat types and species.

- The Amazon basin in South America and the Great Barrier Reef in Australia are examples of regions containing thousands of unique species.

- Each year, scientists exploring biodiversity hotspots catalog new species.

1. How does species diversity change as you move from the poles toward the equator? What environmental factors contribute to this change?

2. Why is it important for wildlife biologists to continue to explore biodiversity hotspots to record species new to science?

3. What is the practical impact on biodiversity conservation from preserving one acre of an area in a biodiversity hotspot compared to an acre located in a less diverse area?

4. What is the general relationship between human activities and biodiversity in an ecosystem?

5. The temperate rainforests of the Pacific Northwest are home to many species (**Fig. 17.34**). Although separated by thousands of miles, this region shares several characteristics with the Amazon rainforest. Which of the following characteristics would you expect to be similar between the two regions—precipitation, high year-round temperatures, abundant resources, varying landscape that includes mountains and plains?

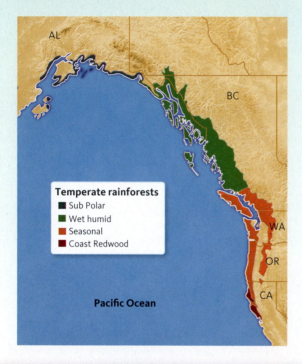

Temperate rainforests
- ■ Sub Polar
- ■ Wet humid
- ■ Seasonal
- ■ Coast Redwood

AL

BC

WA

OR

CA

Pacific Ocean

Figure 17.34 Distribution of temperate rainforests.
© Cengage Learning 2014

End of Chapter Review

Self-Quiz on Key Concepts

Conservation Biology and Biodiversity

KEY CONCEPTS: Conservation biology is the scientific study of the nature and status of Earth's biodiversity. Its aim is to protect species, their habitats, and larger ecosystems from excessive rates of extinction. Invasive species consume resources and alter the ecological balance, stressing or further endangering organisms found there. When populations are very low and a species is in danger of extinction, humans may intervene and protect organisms and their habitat. Biodiversity is critical to maintaining healthy ecosystems and the ecosystem services or the functions they play in biological and physical processes such as nutrient cycling and carbon sequestration.

1. Match each of the **conservation designations** with its description.

    Endangered species   a. a species that is likely to become significantly reduced in the near term throughout all or a large portion of its range

    Threatened species   b. a species that is in imminent danger of extinction across much or all of its range

2. It is likely that habitat fragmentation has lowered populations of migratory songbirds in the United States because:
   a. the remaining habitat patches likely lack food to sustain bird populations.
   b. the pairs of breeding birds cannot easily move from one habitat patch to another.
   c. small patches lack enough ecotone to support bird populations.
   d. female birds are unable to locate mates in small habitat patches.

3. The reintroduction of a species into a habitat must take into consideration which of the following factors?
   a. whether the habitat is in pristine, untouched condition
   b. whether there are humans living within a 100-mile radius
   c. whether there will be any competition for food
   d. whether the factors that caused the extinction are still present

4. Which of the following is not considered an ecosystem service provided by an intact habitat?
   a. cleaning water and air
   b. production of food
   c. ability to support housing developments
   d. maintenance of chemical cycles

5. The primary goal of creating wildlife corridors is to:
   a. provide habitat for reintroducing species.
   b. provide safe passage for cars and trucks passing through forests.
   c. connect fragmented habitats and provide safe passage for animals.
   d. improve the ability of predatory animals to hunt across roads.

Human Actions Help Preserve Many Species

KEY CONCEPTS: When species reach critically low population levels, humans may assist the reproductive success of a species through captive breeding programs. Endangered animals are isolated in zoos or parks where humans provide conditions that improve reproductive success. In a similar manner, hybridization of plants creates organisms able to survive in habitats where nonnative diseases or pests have been introduced or where a changing climate has altered environmental conditions.

6. Which of the following is a role not filled by zoos and aquariums in the support of species preservation?
   a. the conservation of endangered animal species
   b. the conservation of threatened animal species
   c. the development of species hybrids for sale to other zoos or parks
   d. increasing the numbers of rare species for potential reintroduction to the wild

7. Which of the following is a negative impact on biodiversity from climate change?
   a. increasing the availability of carbon dioxide for plant photosynthesis
   b. reducing the ability of invasive plants and animals to become established
   c. increased cooling of tropical waters brought about by changes to ocean currents
   d. altering the environmental characteristics of a habitat

8. Which of the following human interventions would have a positive effect on conserving biodiversity in a natural habitat?
   a. preventing the introduction of a nonnative disease
   b. removing a top-order predator considered a human pest
   c. adding nonnative plants to provide additional food for herbivores
   d. establishing a dedicated capture program to place animals in zoos

9. Invasive species have a *negative* effect on native organisms because:
   a. they lack barriers to population growth and displace native species.
   b. they increase the rate of nutrient cycling in a habitat.
   c. they decrease the rate of nutrient cycling in a habitat.
   d. they provide additional biological diversity and ecological niches.

10. Artificial insemination is a type of human intervention designed to:
    a. improve the natural reproductive rates in endangered animals.
    b. decrease the genetic variation found in a single species.
    c. improve the fertilization success rate for animals that have normally low natural rates.
    d. give zoos a supply of charismatic animals to attract visitors and generate money.

Biodiversity Hotspots

KEY CONCEPTS: Certain ecosystems, for example, rainforests and coral reefs, possess high concentrations of biodiversity. Biologists continue to explore these regions to catalog new species and document their role in ecosystem functioning as well as ecosystem services. Governments and conservation organizations often concentrate resources on protecting these relatively small areas that contain large numbers of threatened and endangered species.

11. Which of the following is likely to be considered a biodiversity hotspot?
    a. an Iowa cornfield where pesticides and herbicides are applied to promote corn yield
    b. a suburban neighborhood where fields have been converted to yards
    c. an uninhabited island located near the equator
    d. a high-plains desert located in Utah

12. Refer to **Figure 17.28**. Which of the following factors is important in supporting biodiversity hotspots?
    a. the east–west longitude
    b. temperature and moisture patterns
    c. whether the habitat is marine or terrestrial
    d. the availability of top-order predators to move between locations

13. The country of Brazil has set an ambitious goal that, by the year 2015, it will reduce deforestation rates in the Amazon rainforest by 80% of the rates recorded in 2005. Predict what effect this will have on ecosystem-wide biodiversity.
    a. It should rapidly increase by 2015.
    b. The rate of biodiversity loss should decrease by 2015.
    c. Loss of biodiversity should stop in 2015.
    d. Rates of biodiversity should continue to rise slowly beyond 2015.

Applying the Concepts

14. To promote the recruitment of marine life, many seaside communities are deliberately sinking ships to create man-made reefs (**Fig. 17.35**). What features does a sunken ship share with a natural reef that would attract biodiversity?

15. Since 1964, Congress has designated wilderness areas as places where the landscape is to be preserved in a manner that appears untouched by humans. How and why might biodiversity differ between a wilderness area and an equal-sized tract of land in an adjacent national forest where the landscape is managed for multiple uses?

Peter Leahy/Shutterstock.com

Figure 17.35 Sunken ships act as artificial reefs.

Figure 17.36 Brown and American white pelicans. (a) Brown pelican. (b) Biogeographic range of the brown pelican. (c) American white pelican. (d) Biogeographic range of the American white pelican.
© Cengage Learning 2014

16. The brown pelican and American white pelican are close relatives that have decidedly different habitat requirements (**Fig. 17.36**). The brown pelican is distinguished from the American white pelican by its coloration and its habit of diving for fish from the air, as opposed to cooperative fishing from the surface. Brown pelicans live on the coasts year-round, whereas American white pelicans migrate long distances between northern habitats where they breed in the summer and southern coastal areas where they overwinter. Speculate about which species might be more susceptible to problems with habitat fragmentation. Defend your answer.

17. Governments in many parts of the world that contain biodiversity hotspots are promoting an ecologically friendly type of tourism, called ecotourism, which attempts to minimize human impacts on the natural habitat. In developing countries, explain how a worldwide increase in ecotourism would influence the preservation of large tracts of land and biodiversity.

Data Analysis

Habitat Fragmentation Impacts Pollination and Seed Production

The destruction of natural habitats by humans is a major reason for the decreasing levels of biodiversity in regions dominated by agriculture. Bees are important pollinators of native plants as well as a wide range of commercial crops (**Fig. 17.37**). Habitat fragmentation affects the bees' foraging behavior and threatens their ability to move between their hives and locate suitable food sources of flower nectar and pollen grains. Native bees often nest in undisturbed grasslands that are isolated when agricultural development takes place, causing the bees to fly further to collect nectar and pollen—and pollinate the plants.

Figure 17.38 contains information on seed production for you to work with in answering Questions 1–4. The four separate graphs illustrate the number of seeds produced,

Figure 17.37 Honeybee.

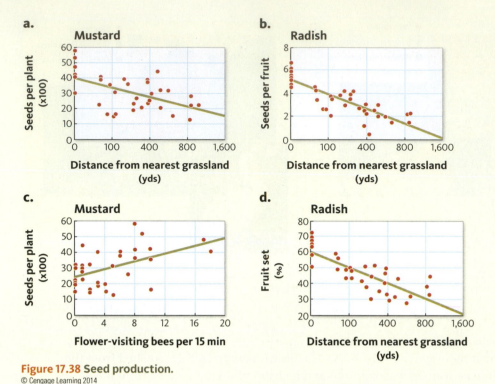

Figure 17.38 Seed production.
© Cengage Learning 2014

or fruit produced, by radish and mustard plants (y-axis) based on the distance from nearest grassland or number of bee visits (x-axis). Use the data provided and the statistically generated line for the data plots to answer Questions 1–4.

Adapted from Steffan-Dewenter I, Tscharntke T. 1999. Effects of habitat isolation on pollinator communities and seed set. *Oecologia* 121, 432–440.

Data Interpretation

18. Looking at Graph A, what is the effect on mustard seed production when the distance from the nearest grassland increases from 0 to 900 meters?

19. Graph B shows the same relationship for a different species—radish. Is the relationship between distance to grassland and seeds per plant the same?

20. The data in Graph C show the relationship between the number of flower visits (potential pollination events) over a 15-minute interval and the number of seeds produced per mustard plant. What happens to seeds produced per plant if you increase the number of visits from four to 16 over a 15-minute interval?

21. Graph D presents slightly different information. It shows the relationship between the percentage of fruits that are set (indicating pollination) for a radish plant as it relates to the distance to the nearest grassland bee habitat. To obtain a 50% fruit set success rate, how close does the nearest grassland need to be?

Critical Thinking

22. Crop production for radishes and bees is influenced by the distance the bees must travel. How might this relationship influence further fragmentation of the landscape by farmers?

23. To increase production, farmers often partner with beekeepers to place portable hives in their fields to increase pollination rates. What effect does this have on the conservation of natural habitat in these ecosystems? How about the impact on native bee biodiversity?

24. Consider the impact habitat fragmentation has on the pollination and seed production of native plant species that often are small in numbers and highly isolated and that may rely on specific species of native bees as pollinators. What are the possible consequences of losing these plant–pollinator interactions on the natural landscape?

Question Generator

Mussels Provide Ecosystem Services

Unique environmental conditions support different communities of mussels, which then provide different levels of ecosystem services. Mussels are sensitive to environmental conditions such as temperature, water flow, and nutrient availability, with each species responding to changes in slightly different ways. For instance, changes in factors such as temperature improve or reduce a mussel's ability to cycle nutrients or filter water.

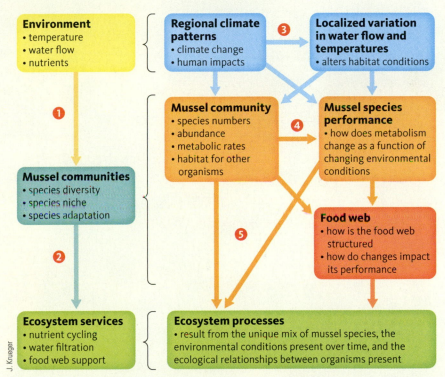

Richard Thorn/Visuals Unlimited/
Getty Images

Figure 17.39 Freshwater mussels.

Freshwater mussels perform an important function in streams and lakes. The mussel shells provide habitat for other organisms; they filter the water, cycle nutrients, and are an important component of the food web. For example, algae growing on mussel shells are eaten by herbivorous aquatic invertebrates (**Fig. 17.39**). The invertebrates attract small predatory fish, which in turn are eaten by larger fish. Mussel larvae must attach themselves to specific fish species as they begin to mature.

Their long life span and complex reproductive patterns make them very vulnerable to environmental damage, making them good indicators of ecosystem health. Currently, mussels are experiencing a global decline in abundance and biodiversity as a direct result of habitat damage by humans. The decline is cause for concern because ecologists recognize that an accelerated loss of mussels will result in an associated loss of ecosystem function.

Below is a block diagram that relates several aspects of the biology and ecology of freshwater mussels (**Fig. 17.40**). Use the background information along with the block diagram to ask questions and generate hypotheses. We've translated the components of the block diagram into several questions to get you started.

Figure 17.40 Factors relating to mussel abundance and biodiversity.

J. Krueger

Here are a few general questions to start you off.

1. How might increasing temperatures associated with climate change ultimately influence mussel biodiversity? (Arrow #1)

2. How would shifts in mussel biodiversity impact ecosystem services? (Arrow #2)

3. How do local climate patterns influence stream flow in a watershed? (Arrow #3)

4. What is the relationship between changes in the abundance of mussels and mussel metabolism? (Arrow #4)

5. How might the placement of a dam influence ecosystem services? (Arrow #5)

Use the block diagram to generate your own research questions and frame a hypothesis.

18 Patterns of Inheritance

18.1 The genetically unique human

18.2 Many traits are inherited in simple patterns

18.3 Some physical traits are expressed from more complex patterns of inheritance

18.4 Sex chromosomes contribute to determining gender and X-linked inheritance patterns

18.5 Domestication is the genetic modification of wild species

18.6 Genes control phenotype by controlling the synthesis of proteins

18.1 The Genetically Unique Human

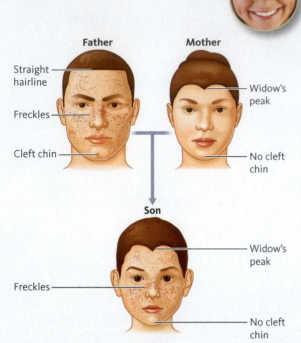

Figure 18.1 Inheritance of genetic traits. Genetic traits are passed from parents to their children. This drawing compares several facial features of a mother, father, and their son.
© Cengage Learning 2014

Labels: Father — Straight hairline, Freckles, Cleft chin; Mother — Widow's peak, No cleft chin; Son — Widow's peak, Freckles, No cleft chin

Figure 18.2 Unique features of organisms. If you look carefully, you can identify several differences in facial characteristics of these chimpanzees.

iStockphoto.com/Jeryl Tan

a.   b.

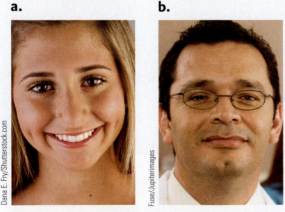

Dana E. Fry/Shutterstock.com   Fuse/Jupiterimages

Figure 18.3 Facial features determined by specific alleles. (a) Dimples. (b) Cleft chin.

It's likely that at some time in your life, you've compared yourself to your parents, grandparents, or siblings. Through this comparison, you can see your parents' specific genetic contributions to you. Physical features such as height, eye color, freckles, and hair color can be used to trace patterns of inheritance throughout your entire family (**Fig. 18.1**). Intuitively, we know that the flow and transmission of genetic information from our parents is not random but that it follows particular patterns. You are a mixture of half of your mother's DNA and half of your father's. Tracing through your family ancestry, you carry 25% of the same DNA as each of your grandparents, and 12.5% from each your great grandparents, and so on. These patterns of inheritance form the basis of the science of genetics.

Thousands of traits make up the human *(Homo sapiens)* phenotype. Some of them have simple patterns of inheritance that involve just one gene; others are far more complex and result from the interactions of several genes. Some simple traits are freckles, dimples, left- versus right-handedness, and eye, lip, and nose shape. More complex traits include height, eye color, and hair color. Some diseases, such as high cholesterol levels, cystic fibrosis, certain cancers, and Huntington disease, also are easily traced through family genetics.

Looking around campus it is easy to see each person's unique features. However, humans are not the only organisms that express individuality. Individual chimpanzees and gorillas have facial features as variable as those of humans, zebras have stripes that differentiate one zebra from another, and humpback whales have markings on the underside of the tail that distinguish individual whales (**Fig. 18.2**). In fact, all living things that arise from sexual reproduction are unique individuals. Let's look more closely at the human genetic profile.

The development of a human requires the complex coordination of 46 individual chromosomes that contain about 20,000 to 25,000 genes. These functional units of heredity, or *genes*, occupy only 5% of our genome and encode the information to produce traits such as hair color and hair texture. Through the processes of gene expression, different forms of genes, called *alleles*, code for different proteins that result in different traits, or *phenotypes*. For instance, the pigment protein melanin determines eye, skin, and hair color. Even though all humans share the same set of genes, through sexual reproduction and genetic recombination we are conceived as unique human beings (with the sole exception of identical twins). Each individual is a unique combination of alleles that interact with the environment before and after birth.

Let's examine your facial features, combining the perspectives of the artist, the geneticist, and the historian. The portrait artist first draws the shape of the head and then positions the eyes, nose, mouth, and ears. As the portrait's three-dimensional form develops, the artist highlights the underlying bone structures: the forehead, cheeks, and chin. With a drawing of the facial phenotype in place, the artist completes the portrait by applying paint to the canvas to indicate the colors of the skin, lips, eyes, and hair. To the geneticist, this snapshot in time is the result of activities of genes during the development and aging process of the face, head, and neck. Specific alleles dictate whether the ear lobes are attached or free, dimples are present, or the lips are thin, and these alleles can be traced to their particular location in the human genome (**Fig. 18.3**). Now, let's consider these traits from a historian's point of view. Many families investigate and document their family history and genealogy. First, think about which facial traits you have that are the same as your mother's or father's. Also, consider how your facial traits compare to those of your grandparents, siblings, aunts, and uncles. These traits have been passed down from generation to generation as part of your family heritage. Perhaps your family originally was from China. If so, it is likely that you have dark, straight hair and dark eyes with a distinct shape. Your genetics is part of your family history, and it enhances your understanding of your heritage.

Whether you are tracing traits at a family reunion or thinking in a larger context about where and when these genotypes and phenotypes first came about, a human family tree extends through millions of years. We are fascinated by both our short-term heritage as a family but also our ancestry as a species. The genetic history of our hominid ancestors is extensive, and we are beginning to better understand it. Take a moment and think about the origin of modern humans. Perhaps you've thought of questions such as the following: How long ago did modern humans develop? What features make us human? Where did humans originate? What does it mean to be closely related to chimpanzees?

The evolutionary history of humans is a complex subject involving several scientific fields, including archaeology, anthropology, and genetics. The current hypothesis on human origins is that *hominids,* which include humans and great apes, originated in Africa more than 6 million years ago, and branched off the evolutionary tree from chimpanzees (**Fig. 18.4**). Recall from Chapter 2 that this evolutionary branching means we share a common ancestor with chimpanzees but we are not descendants of chimpanzees. Although early hominids spent time living in trees, evidence from fossils strongly supports the hypothesis that they walked upright when they were on the ground. Another important evolutionary feature is the placement of the skull. On animals that walk upright, the skull is balanced on top of the erect body. Over time and throughout various regions, fossils indicate changes in bone thickness, overall human height, cranial space, and jaw and tooth structures (**Fig. 18.5**).

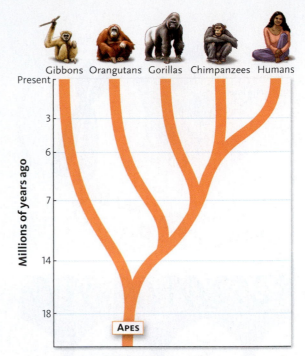

Figure 18.4 Hominid family tree. Humans originated in Africa more than 6 million years ago and share a common ancestor with chimpanzees.
© Cengage Learning 2014

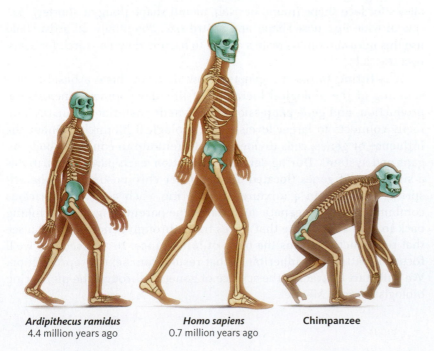

Ardipithecus ramidus
4.4 million years ago

Homo sapiens
0.7 million years ago

Chimpanzee

Figure 18.5 Hominids through time. Note changes in hip structure, skull placement, and overall height in this comparison of several hominids. Examine the skulls closely and note differences in overall size, cranial space, and jaw protrusion.
© Cengage Learning 2014

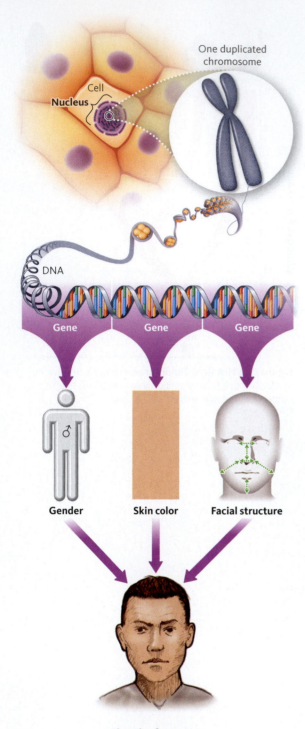

One duplicated chromosome

Cell

Nucleus

DNA

Gene   Gene   Gene

Gender   Skin color   Facial structure

Sketch of suspect

Figure 18.6 A facial profile. DNA can be used to establish rudimentary facial features.
© Cengage Learning 2014

Sexual reproduction—the creation of genetically different offspring from the fusion of male and female gametes. In this mode, half of the genetic material of the offspring comes from one parental gamete and the other half comes from the other parental gamete.

Just as with all other living things, as populations of humans responded and adapted to their environment, they evolved. For instance, in colder regions, humans had larger bones, shorter bodies, and carried more fat. In hot regions near the equator, humans had darker skin color and taller, leaner body builds. Through reproduction, the genes coding for these adaptations, or phenotypes, were passed on to many following generations. When two populations no longer interbred, a distinct species was formed. The human species *Homo sapiens* has lived for about 150,000 years and is distinguished from hominids by skeletal features including a high forehead, small nasal opening, distinctive chin, cylindrical rib cage, and narrow pelvis (see **Fig. 18.5**).

In this chapter on inheritance and genetics, we'll explain how patterns of inheritance arise and how traits are passed from parents to offspring. The essential concept is that genes inherited from parents determine physical traits, or, more simply put, genotype determines phenotype. These traits shape both short- and long-term evolutionary history. We focus on humans because they are most familiar, and we know much more about their patterns of inheritance than for other organisms. In this way this chapter aims to help you gain a deeper insight into your genetic past and future.

18.2 Many Traits Are Inherited in Simple Patterns

In addition to using genetics to trace ancestry, genetic technology is used in criminal cases. Imagine a scene where a suspect's DNA has been left behind but there were no eyewitnesses to the crime. What can the DNA evidence reveal? Today, scientists analyze DNA for patterns that may identify suspects in other crimes, but DNA evidence potentially could be used to create a mug shot. Specific genes code for facial structures, and DNA can be analyzed to identify gender, skin color, and facial structures. Because single genes determine many facial phenotypes, tests can identify alleles for face shape (round or oval), mouth shape (long or shorter), narrow or wide lips, nose shape, and nostril size. Potentially, an artist could use this information to create a sketch to narrow down a search for a suspect (**Fig. 18.6**)

It is fitting to discuss genetics now that you have a basic understanding of the biological hierarchy of life, development, meiosis, reproduction, and gene expression. The genetic material in our DNA directly connects to larger levels of the biological hierarchy. Under the influence of genes, cells divide and differentiate to construct body organs and systems. During sexual reproduction, each parent contributes a single set of genes (located on his or her chromosomes) to the offspring, resulting in a mixture of offspring, each containing various combinations of the genetic material of the parents (**Fig. 18.7**). Thinking back to your family tree that traces traits through generations, you see that reproduction forms the basis of inheritance. In this section, we'll focus on patterns of inheritance that result from sexual reproduction. We will start by tracing the science of genetics through the pioneering biologist Gregor Mendel.

Mendel developed a scientific understanding of the basics of inheritance

Gregor Mendel is one of the great historical figures in biology (**Fig. 18.8**). Although the inheritance of traits has been recognized for thousands of years, Mendel's work during the 1800s provided the deepest insights into this process. Prior to Mendel's work, it was believed that when two organisms reproduced, the parents' traits were simply blended in the offspring. This blending model predicted that a black male fox and a white female fox would produce gray offspring. However, because this did not always happen in nature, the mechanism for passing traits from generation to generation remained a mystery.

Mendel was the first person to provide an adequate scientific basis for how traits are inherited. Through his garden study of inheritance patterns of physical traits in peas, Mendel obtained clear-cut evidence that the blending model did not apply and that discrete factors, which today we call genes, are responsible for passing traits from one generation to the next. In the case of the fox (which Mendel did not study), his idea meant that the parent foxes would have discrete particles (genes) that are independent of each other and that their offspring would be either black or white. In nature, this is indeed the case. Examining simple physical traits such as coat color in foxes, flower color in peas, or dimples in humans allows us to see some of the patterns of inheritance in the natural world. Of course, there are many other traits, such as blood type or disposition to high cholesterol levels, which we do not see.

Mendel is considered the father of modern genetics because he developed the basic principles that govern inheritance during sexual reproduction. His work is an excellent example of the scientific process. In experiments conducted between 1856 and 1863 using approximately 29,000 pea plants, Mendel hypothesized that parents pass *discrete heritable factors*—what we now understand are genes—on to their offspring. Although many others were studying how traits were passed from generation to generation, Mendel designed rigorous experiments that allowed him to observe one trait at a time, and, as a result, he was able to mathematically support his experiments. His experiments led the way for scientists in the early 1900s to predict genetic outcomes based on mathematical probabilities.

Mendel discovered that each individual pea plant has two factors for each trait he studied and that these factors may or may not be identical. He also found that some of these factors are **dominant**, or preferentially expressed, whereas others, called **recessive**, are not. For instance, one trait he studied was purple and white flower color in pea plants. Through his experiments, Mendel was able to determine that when both the purple and white "factors" were present in the parents, the purple flower color was always expressed and, therefore, was the dominant form. He also discov-

Figure 18.7 Reproduction and the basis of inheritance. Notice the many similar physical traits of this family that were passed down from the parents to the children through sexual reproduction.

a.

b.

Purple flower (dominant)

White flower (recessive)

Figure 18.8 Gregor Mendel and the pea plant. (a) Mendel (1822–1884), the founder of our modern understanding of genetic traits. (b) The garden pea plant was Mendel's subject of study.
© Cengage Learning 2014

Dominant—in reference to alleles, the allele that is preferentially expressed; only one copy of the allele is needed for expression of the phenotype.

Recessive—in reference to alleles, the allele that is not preferentially expressed; two copies of the allele are needed for expression of the phenotype.

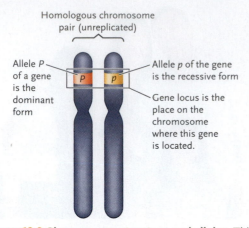

Homologous chromosome
pair (unreplicated)

Allele *P* of a gene is the dominant form

Allele *p* of the gene is the recessive form

Gene locus is the place on the chromosome where this gene is located.

Figure 18.9 Chromosome structure and alleles. This gene always occurs at the same location, or locus, on this chromosome. The two versions of the same gene are called alleles. On this homologous chromosome, the trait is heterozygous, P is dominant, and p is recessive.
© Cengage Learning 2014

ered that in order for the recessive white form to be expressed, two white factors were needed.

Modern scientists marvel at Mendel's work, especially considering DNA, meiosis, genes, and chromosomes had not yet been discovered. Today we understand Mendel's individual heritable factors to be genes, and the different forms of a gene are called alleles (Fig. 18.9). If the two alleles of the gene for any one trait are identical, the individual is considered homozygous for that trait. If the two factors code for different versions of a trait, the individual is considered heterozygous. In heterozygous individuals, the only allele expressed is the dominant form. The recessive allele is present in the genome, but the expression is hidden; that is, it does not have any noticeable effect on the appearance of the trait. When writing out the genetic condition, the alleles are represented by capital letters when dominant or lowercase when recessive. In the case of the flower color, if we use the letter "P" to represent the alleles, the purple flower is either homozygous (PP) or heterozygous dominant (Pp), and the white flower is recessive (pp). In the heterozygote flower (Pp), the purple allele (P) will mask the white (p) allele. In the case of the human face, dimples are inherited in a manner similar to that of flower color in peas. Dimples are a dominant trait, and the genotype is written as *DD* or *Dd.* A person with the genotype *dd* does not have dimples.

The garden pea plant used by Mendel was an especially good choice for studying genetic inheritance for several reasons. First, peas produce several pairs of contrasting traits that are easily observed, such as plant height, seed coat texture, seed color, flower position, and flower color. Mendel studied the inheritance of seven of these distinct characteristics, each with two well-defined traits (Fig. 18.10). Second, as a mathematician, Mendel recognized that he would need to sample a large number of seeds and plants to draw specific conclusions. He recognized that the pea plant has a short generation time and therefore produces large numbers of offspring in a short time. Third, recall from Chapter 1 that scientists use controlled experiments. These are tests in which all conditions, except for the variable under consideration, are controlled during the course of the study. In his experiments, Mendel could control each cross by pollinating the pea plant flowers by hand. This was tedious, time-consuming work, often another characteristic of scientific data collection. Finally, each pea plant flower contains both male and female reproductive parts and is capable of self-pollination. As a result, they can either self-pollinate or cross-pollinate with another plant. This charac-

Character	Traits crossed	Dominant	Recessive
Seed shape	round × wrinkled	round	wrinkled
Seed color	yellow × green	yellow	green
Pod shape	inflated × constricted	inflated	constricted
Pod color	green × yellow	green	yellow
Flower color	purple × white	purple	white
Flower position	axial (along stems) × terminal (at tips)	axial	terminal
Stem length	tall × dwarf	tall	dwarf

Figure 18.10 Mendel's model. Mendel used controlled experiments to study these seven genetic traits in garden peas. Notice that each trait follows a simple single-gene inheritance pattern and has a dominant and a recessive appearance.
© Cengage Learning 2014

teristic allowed Mendel to carefully control his experiments, crossing genetically identical individuals and creating pure breeding lines of plants. Pure breeding lines are beneficial for tracking purposes because both parents are homozygous for the trait of interest, and all of the resulting offspring have the identical traits as the parents.

Since Mendel's first experiments, many organisms such as fruit flies, corn, and bacteria have contributed to our understanding about inheritance (**Fig. 18.11**). Today, the organisms that scientists use in their studies are selected based on the size of their genome, how easily their genes can be manipulated, and how much information is known about the species. Both the fruit fly and corn are models for studying genetics because their genomes are well understood and they are easy to breed in the laboratory environment. The fruit fly genome is relatively small; it has only 170 million base pairs on eight chromosomes (four pairs), and about 15,000 genes. It may surprise you learn that corn has 20 chromosomes (10 pairs), 2.9 billion bases, and 32,000 genes; that's about 40% more genes than we have! Model organisms produce large numbers of offspring in a relatively short generation time, and crosses can be controlled with ease.

Figure 18.11 Other ideal models. The fruit fly (*Drosophila melanogaster*) and corn (*Zea mays*) are two model organisms commonly used in genetic research.

Alleles separate during the production of gametes

In addition to the inheritance of dominant and recessive alleles, Mendel also tested his pea plants to examine *how* alleles are inherited. His work led to the law of segregation, which states that each parent contributes one of the two alleles that are passed down to the offspring. This means that the members of each pair of alleles that code for a specific trait separate during meiosis when gamete-producing cells are made (**Fig. 18.12**). As each gamete (egg or sperm) is formed, it receives half of the total genetic complement of the parent, with only one of the versions of an allele. **Figure 18.12** illustrates how the alleles for flower color separate during meiosis of a heterozygous pea plant. In this case, each parent has an equal chance of contributing a dominant or a recessive allele to the offspring. During meiosis there is no preference for passing along dominant or recessive alleles. Many human traits work the same way. If your mother has dimples (DD) and your father does not (dd), you will inherit one allele (D) from your mother and one from your father (d), and in this case you will have dimples (Dd). If your parents both have dimples and you do not (dd), then your parents must be heterozygous for the trait (Dd).

Independent assortment leads to extensive genetic variation

Mendel's work also provided us with an understanding of how the inheritance of genes occurs on different chromosomes, called independent assortment. Mendel simultaneously observed two traits and determined that each of these traits is passed on independently of the other to offspring. He crossed homozygous plants having yellow, smooth seeds with homozygous plants having green, wrinkled seeds. If these traits were inherited together, the offspring would be identical to the parents (either yellow-smooth or green-wrinkled). Instead, Mendel observed that any combination of color and seed texture was possible. In this way, he concluded that

Allele—the different forms of a gene.

Homozygous—individual with two similar alleles for a single trait.

Heterozygous—individual with two different alleles for a single trait.

Law of segregation—each parent contributes one of the two alleles that are passed down to the offspring. The members of each pair of alleles that code for a specific trait separate during meiosis.

Independent assortment—the allele pairs of different traits separate during meiosis and pass on independently of each other to the offspring.

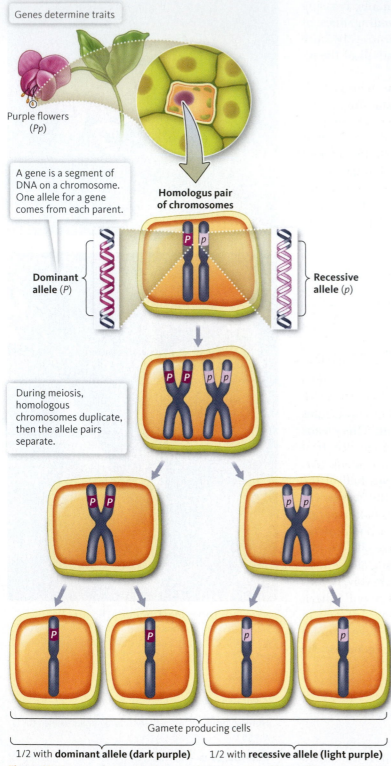

Genes determine traits

Purple flowers
(*Pp*)

A gene is a segment of DNA on a chromosome. One allele for a gene comes from each parent.

Homologus pair of chromosomes

Dominant allele (*P*)

Recessive allele (*p*)

During meiosis, homologous chromosomes duplicate, then the allele pairs separate.

Gamete producing cells

1/2 with **dominant allele (dark purple)**   1/2 with **recessive allele (light purple)**

Figure 18.12 Law of segregation. A brief summary of how alleles are inherited. Alleles from homologous chromosomes separate during meiosis when gamete-producing cells are made. In this case, the heterozygote parent passes on, through meiosis, either the dominant (purple flower) or recessive (white flower) allele.

© Cengage Learning 2014

these two traits are distributed independently of each other to offspring. Today we understand the mechanism behind this law and know that the distribution of individual chromosomes occurs randomly during meiosis. During meiosis, genes on different pairs of homologous chromosomes sort independently of each other, and this independent assortment is an important contribution to genetic variation in sexually reproducing organisms (**Fig. 18.13**). **Figure 18.13** summarizes both the behavior of chromosomes and the behavior of alleles during meiosis and highlights the law of segregation and independent assortment as they affect genetic behavior.

If we continue using the example of inheritance of human facial features, independent assortment states that because the genes for freckles and dimples are on separate chromosomes, they will be inherited independently of each other. This means that all genetic combinations are possible, leading to four potential phenotypic outcomes: freckles with dimples, no freckles with dimples, no freckles and no dimples, or freckles with no dimples.

Punnett squares model genetic outcomes for Mendelian inheritance

Understanding genetic processes allows us to predict the likelihood of genetic outcomes using the mathematical laws of probability. Probability is a valuable tool to use when humans develop new varieties of plants and livestock or when dealing with healthcare issues. Albinism, a complete or nearly complete lack of melanin pigment in the skin, hair, and eyes, is a visible trait that can occur in most mammals, birds, fish, reptiles, and amphibians (**Fig. 18.14**). It can even occur in plants, although as you would expect it is lethal, because these plants cannot perform photosynthesis. Because albinism is a recessive trait, it is possible to predict probable traits of offspring resulting from the mating of individuals who carry the recessive trait. Understanding the genetics of single gene traits allows breeders to accurately predict which plants or animals may have more desirable qualities, such as disease resistance or extreme environmental tolerances. It also can help explain and predict patterns of inheritance in human family lines.

The **Punnett square** is a graphic tool used by geneticists to show potential allelic combinations found in gametes and to predict the odds of each of the offspring genotypes occurring (**Fig. 18.15**). In other words, the Punnett square is a graphical representation of all of the potential combinations of genotypes that can occur in offspring, given the genotypes of their parents. Like the peas studied by Mendel, corn can have

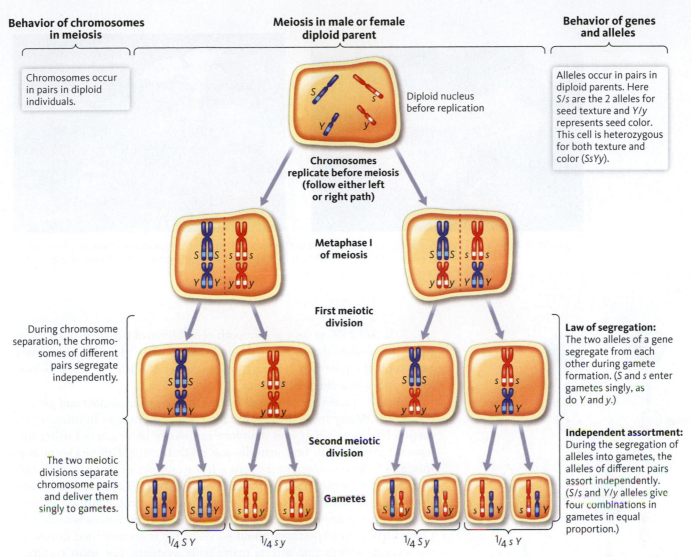

Behavior of chromosomes in meiosis

Chromosomes occur in pairs in diploid individuals.

During chromosome separation, the chromosomes of different pairs segregate independently.

The two meiotic divisions separate chromosome pairs and deliver them singly to gametes.

Meiosis in male or female diploid parent

Diploid nucleus before replication

Chromosomes replicate before meiosis (follow either left or right path)

Metaphase I of meiosis

First meiotic division

Second meiotic division

Gametes

¼ S Y   ¼ s y   ¼ S y   ¼ s Y

Behavior of genes and alleles

Alleles occur in pairs in diploid parents. Here *S/s* are the 2 alleles for seed texture and *Y/y* represents seed color. This cell is heterozygous for both texture and color (*SsYy*).

Law of segregation: The two alleles of a gene segregate from each other during gamete formation. (*S* and *s* enter gametes singly, as do *Y* and *y*.)

Independent assortment: During the segregation of alleles into gametes, the alleles of different pairs assort independently. (*S/s* and *Y/y* alleles give four combinations in gametes in equal proportion.)

Figure 18.13 Behavior of alleles and chromosomes during meiosis. This diagram shows how chromosomes, genes, and alleles behave during the production of gametes in meiosis. The gametes shown across the bottom illustrate how different combinations of alleles occur as a function of segregation and independent assortment, first described by Mendel.
© Cengage Learning 2014

smooth or wrinkled seeds. The dominant allele is smooth, designated "S"; the recessive allele is "s." Let's use a Punnett square to predict the outcome of a cross between a homozygous smooth-seeded plant and a homozygous recessive, wrinkled-seeded plant (see **Fig. 18.15**). First, write out the genotypes of this cross and determine what alleles each parent can possibly contribute to the offspring. In this case the cross is SS × ss. Next write the genotypes of each parent on the top and left edge of the grid. The first parent can only contribute an "S" allele and the second parent can only contribute an "s" allele. Now that the potential allelic contribution of each parent is shown on the outside edges of the Punnett square, complete the potential allelic combinations of the various gametes to make the potential allelic combinations of the offspring. If you add up the ratios, you have the predicted frequency of all of the potential genotypes among the offspring each time reproduction occurs between these two parental genotypes. In this case we have a 1 SS : 2 Ss : 1 ss genotypic ratio in percentages of 25%

Punnett square—a graphic tool used by geneticists to show potential allelic combinations found in gametes and to predict the odds of each of the offspring genotypes occurring; a graphical representation of all of the potential combinations of genotypes that can occur in offspring, given the genotypes of their parents.

Figure 18.14 Albinism. (a) An albino leopard frog found in Minnesota. The pink appearance occurs because of the double recessive allele (homozygous recessive) combination that eliminates the normal production of pigment in the skin and eyes. (b) An albino bean seedling that died by sundown.

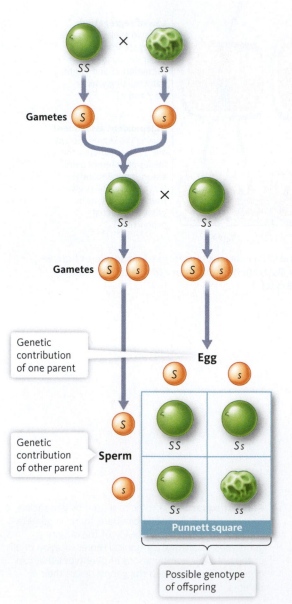

Figure 18.15 Punnett Square. This Punnett square predicts the outcome of crossing a homozygous smooth pea plant seed with a homozygous recessive wrinkled plant seed. There is a 75% chance the offspring seed will be smooth. The phenotypic ratio is 3 smooth seeds : 1 wrinkled and the genotypic ratio is 1 SS : 2Ss : 1ss.
© Cengage Learning 2014

SS : 50% Ss : 25% ss and a 3 smooth seed : 1 wrinkled seed phenotypic ratio. Remember that each of the internal squares in a Punnett square represents the potential combination of one set of alleles in gametes joining together in a single offspring.

Punnett squares are also used by healthcare professionals and genetic counselors. Many traits that lead to illnesses and diseases in humans are carried by single genes and therefore can easily be predicted using this low-cost graphic tool. Tay-Sachs disease, cystic fibrosis (CF), sickle-cell anemia, and lactose intolerance are just a few examples of single-gene conditions that can be predicted with a Punnett square. CF has a simple Mendelian recessive inheritance, and it is one of the most common genetic disorders of Caucasians. About one in 3,000 Caucasian babies is born with CF and has a life expectancy about 40 years. CF is an inherited disease of the mucous glands that affects many body systems. The most common signs and symptoms of the disorder include progressive damage to the respiratory system and chronic digestive system problems. Mutations to a single gene cause an improper salt balance in the cells that results in thick, sticky mucus. Researchers are focusing on ways to cure CF by correcting the defective gene or correcting the defective protein. About one in 25 Caucasians (more than 10 million Americans) is a carrier of this mutation. A genetic carrier is a person who carries the gene without expressing the trait and therefore is heterozygous for this trait.

Figure 18.16 shows a Punnett square used to examine a hypothetical problem involving the inheritance of CF. Imagine that your parents have been genetically tested and find that they both are carriers of the CF mutation. At every birth, what is the chance that the baby will inherit both recessive alleles and have CF? Looking at the Punnett square, we see that one of the four possible offspring combinations shows the double-recessive CF condition (the lower right square). Therefore, the answer is that your next brother or sister has a 25% chance of having CF. What is the chance that the baby will be a normal carrier? What about the odds that the baby will not carry the mutation? The Punnett square provides the answers. There is a 25% chance that your next brother or sister will have CF, a 25% chance he or she will not carry the mutation, and a 50% chance that he or she will be a heterozygous carrier and not have the disease.

This section has introduced you to the simple patterns of inheritance through basic genetics and reproduction. These patterns are a function of meiosis, and they are easily seen and predicted with Punnett squares. Although many traits are inherited in dominant and recessive patterns, some are passed on through patterns that are more complex. In the next section, we will examine other ways that traits are passed on from parent to offspring.

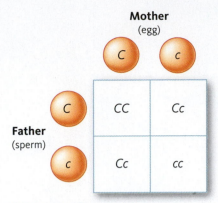

Figure 18.16 Cystic fibrosis. This Punnett square shows a cross between parents who are carriers (Cc) of the gene for cystic fibrosis. Since CF is recessive, offspring have a 25% of having CF and a 50% chance of being a heterozygous carrier who doesn't have the disease.
© Cengage Learning 2014

18.2 check + apply YOUR UNDERSTANDING

- In the process of reproduction, genetic material in the form of DNA is passed from the parents to the offspring. Asexual reproduction produces clones, and sexual reproduction introduces genetic variation in organisms.

- Gregor Mendel gathered the first evidence that genes are the distinct units passed on through generations resulting in the expression of traits. The law of segregation states that alternate forms of genes (alleles) separate during the formation of gametes and form new pairs during reproduction.

- During meiosis, genes on different pairs of homologous chromosomes sort independently of each other. This independent assortment is an important contribution to genetic variation in sexually reproducing organisms.

- Because several traits are inherited in simple patterns, genetic outcomes can be predicted using Punnett squares and family pedigrees.

1. Which method of reproduction, sexual or asexual, produces offspring that are clones of the parent? Which method requires more energy and resources? Why?

2. What insight did Mendel gain from observing one trait at a time in his studies of peas?

3. Roses are praised for their beautiful colors. Write a sentence that describes how the genotype of a rose determines its coloration.

4. Mendel used a scientific model that was very sound. Summarize why the pea plant was a successful model and speculate whether Mendel would have been just as successful using mice.

5. In camels, the trait for one hump is dominant to two humps. Create and fill out two practice Punnett squares crossing this single trait where the two parents are heterozygous and then where both parents are homozygous (one dominant and one recessive). In which cross is there a 100% chance that the offspring will be heterozygous for the trait?

18.3 Some Physical Traits Are Expressed from More Complex Patterns of Inheritance

Perhaps thinking back to the comparison you made of yourself to your parents, you can see where your freckles and dimples came from. However, you may be confused about other traits such as skin color and eye color. Mendel's laws of inheritance don't explain all patterns of inheritance that we see.

Although Mendel was a meticulous scientist, in several ways he was fortunate that he chose to study pea plant traits that were located on separate chromosomes and were clearly expressed in either the dominant or the recessive form. Scientists refer to traits that are expressed from a single gene and have one dominant and one recessive allele as simple or *Mendelian* traits. In humans, more than 18,000 Mendelian traits have been identified so far. Some familiar examples are cheek dimples, face freckles, cleft chin, and attached earlobes (**Fig. 18.17**). However, not all traits are expressed in this fashion, and other physical traits such as hair, skin, and eye color have a more complex pattern of inheritance. For instance, your height is determined by several genes, and your blood type is made up of two alleles drawn from a total of three alleles for this trait. In this section, we will briefly introduce you to some of these more complex patterns of inheritance.

Incomplete dominance produces intermediate forms

Mendel studied flower color in peas that was controlled in a simple dominant or recessive manner. However, today horticulturists rely upon another inheritance pattern to generate tremendous varieties of flower colors. In snapdragons and carnation flowers the visible phenotype is often an intermediate color between the dominant and recessive versions. The color property of the dominant gene is considered to show incomplete dominance rather than the familiar true dominance pattern. **Incomplete dominance** occurs when two alleles are needed to express a trait, but the dominant trait does not completely mask the recessive trait. When a red snapdragon (RR) is crossed with a white plant (rr), the offspring are heterozygous (Rr), but instead of being red, they are a blend of the two colors and are pink (**Fig. 18.18a**). An example of incomplete dominance in humans is hair texture. The dominant trait is curly hair and the recessive trait is straight hair. If you have wavy hair, you are heterozygous and

Incomplete dominance—occurs when two alleles are needed to express a trait, but the dominant trait does not completely mask the recessive trait.

Tongue roller

Widow's peak

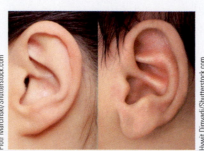
Attached ear lobe   Free ear lobe

Hitchhiker's thumb

Bent little finger

Figure 18.17 Common Mendelian traits in humans. These traits are simple and controlled by one gene.

express incomplete dominance. In horses, palomino color is inherited through incomplete dominance from chestnut and cremello parents (see **Fig. 18.18b**).

Multiple allele forms are expressed with codominance

So far we've learned about two forms of a single gene, dominant and recessive, in which only one trait is expressed. In some cases, when both alleles are present, both traits are expressed together. In some chickens, gray color is expressed from the presence of both black and white feathers. Roan cattle are a result of the expression of both red and white hairs. The expression of both the dominant and recessive traits in the phenotype of the organism is a genetic condition known as **codominance**. It may be helpful to note that the prefix "co-" means together.

For some traits, such as ABO blood type in humans, the situation it gets even more complicated. Some phenotypes are determined by *multiple alleles* (three or more) of a single gene. It is important to understand that although there may be three or more alleles within the population, each diploid individual will have only two of those alleles in its genotype. Humans, apes, and cows have four blood types (A, B, AB, and O) produced by three nonidentical alleles of a single gene (designated A, B, and O). Neither the A nor the B allele form is dominant over the other, and both may be expressed at the same time. The O blood type allele is not expressed as a codominant; rather, it acts as a recessive allele and is not expressed if an A or B form is present. The three alleles are written as I^A, I^B, and i. The relationship between the various genotypes and phenotypes is shown in **Table 18.1** and **Figure 18.19**. The A and B alleles each codes for a different carbohydrate that is embedded in the plasma membrane of a red blood cell. This carbohydrate acts as an identification marker to the immune system. During blood transfusions, the blood types must match properly or the blood can clump in the blood vessels, causing an extreme immune system reaction and potentially death. The immune system produces antibodies that recognize the appropriate molecular markers on our cells' surface and attacks those that don't match.

The genetics of the ABO blood type can also be used in forensic police work to eliminate suspects in specific crimes or to help solve paternity cases. Let's consider a complaint at the birth center of a local hospital. Assume the Smiths and the Johnsons think their babies were accidentally switched at the hospital before the babies were brought home. Baby Smith and Mr. Smith have type B blood, and Mrs. Smith has type O blood. Baby Johnson has type O blood, Mrs. Johnson has type AB, and Mr. Johnson has type B. By creating a simple Punnett square based on the blood types of

a.

Parents

Red (*RR*) × White (*rr*)

Offspring

Pink (*Rr*)

	R	R
r	Rr	Rr
r	Rr	Rr

b.

Cremello   Palomino   Chestnut

Fuse/Jupiterimages

Figure 18.18 Incomplete dominance. In incomplete dominance, a blended effect is seen. (a) When these homozygous red and white flowers cross, offspring are pink. (b) In these horses, palomino color is a result of a chestnut and a cremello parent.
© Cengage Learning 2014

TABLE 18.1 Blood Types of the Human ABO Blood Group

Blood Type	Carbohydrate Markers on Surface of Red Blood Cells	Antibodies Found in the Blood Plasma	Blood Types Accepted in a Transfusion
A	A	Anti-B	A or O
B	B	Anti-A	B or O
AB	A and B	None	A, B, AB, or O
O	None	Anti-A, anti-B	O

Adaptation of Russell Table 12.1 pg 248

Codominance—the expression of both the dominant and recessive traits in the phenotype of the organism.

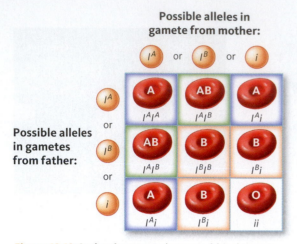

Possible alleles in gamete from mother:

I^A or I^B or i

Possible alleles in gametes from father:

I^A or I^B or i

	A I^AI^A	AB I^AI^B	A I^Ai
AB I^AI^B	B I^BI^B	B I^Bi
A I^Ai	B I^Bi	O ii

Figure 18.19 Codominance. In humans, blood type is expressed as codominance; three different alleles determine blood type. In the cross, all of the potential genotypes are represented.
© Cengage Learning 2014

the parents, you can figure out whether or not a mix-up occurred with the babies (**Fig. 18.20**). Recall that type B blood (the phenotype) occurs from the genotypes I^BI^B or I^BI^i. To solve this problem, set up a Punnett square using both of these possible genotypes (see **Fig. 18.20**). In this case, the Smiths could produce a baby with type O or B blood. The Johnsons' baby could be type A, AB, or B. Because "Baby Johnson" is type O, it appears the babies were switched at the hospital before the parents took them home.

It may be difficult for you to distinguish between incomplete dominance and codominance. In incomplete dominance one of the gene proteins is not functional; therefore, the traits blend together with only one allele fully expressed. For instance, in some flowers and horses, only one allele produces color; the other allele does not. In codominance, both alleles are equally expressed. Using flower color as an example, a snapdragon with red and white parents that shows incomplete dominance will be pink. However, other flower species with red and white parents that express codominance will be red and white spotted.

Some genes control more than one trait

In the human genome, it is very common for a single gene to control the expression of several phenotypic traits (**Fig. 18.21a**). When one gene controls two or more distinct and seemingly unrelated phenotypic traits, the

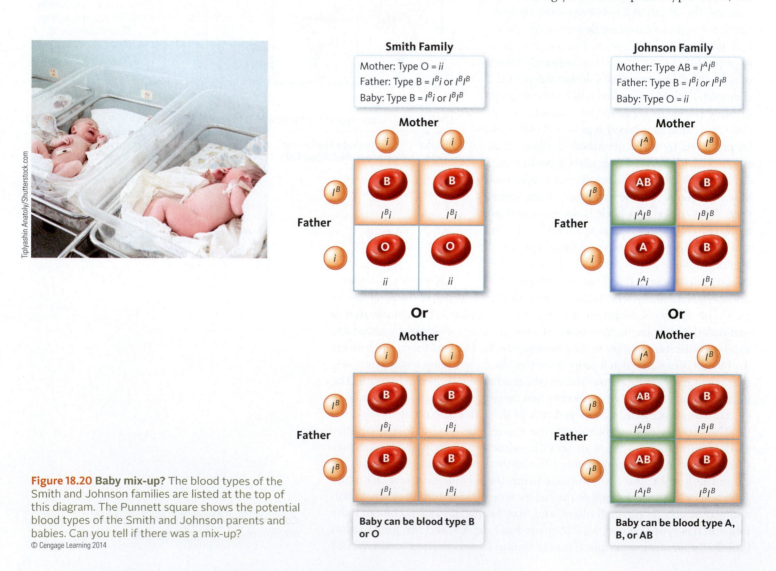

Figure 18.20 Baby mix-up? The blood types of the Smith and Johnson families are listed at the top of this diagram. The Punnett square shows the potential blood types of the Smith and Johnson parents and babies. Can you tell if there was a mix-up?
© Cengage Learning 2014

condition is called pleiotropy. In many cases, a single gene codes for a product that is subsequently used by different cell types performing different tasks. A gene may also be expressed at different stages of development. In either case, the result can be one gene that controls more than one trait. The lack of pigmentation seen in albinism and the abnormally shaped red blood cells in sickle-cell anemia are two examples of pleiotropy. As we discussed in Section 18.2, in albinism the recessive form of the gene results in a lack of pigmentation in skin, hair, and eyes (see **Fig. 18.21b**). In a similar manner, a large percentage of white cats with blue eyes are deaf. This is a pleiotropic effect in which the fur and eye color are associated with hearing loss. Sickle-cell anemia is also a result of two recessive alleles of one gene that codes for the hemoglobin molecule. The recessive form of this gene affects the normal shape of the red blood cell, causing it to have a crescent shape that restricts the flow of blood through tiny capillaries and vessels (**Fig. 18.22**). Sickle cells limit the amount of oxygen that reaches tissues, ultimately impacting the cardiovascular, respiratory, and immune systems.

From an evolutionary point of view, it is interesting to ask how an allele with such a negative effect as the sickle-cell trait remains in the gene pool rather than being selected against. In the case of sickle-cell anemia, heterozygote carriers offer some resistance to malaria. Other pleiotropic traits produce both beneficial and harmful effects. A gene that codes for hormones often is pleiotropic. For instance, in young human males testosterone improves reproductive fitness, but in adult males this hormone increases the risk for prostate cancer.

Some traits are controlled by several genes

Think about differences you see between people around you: small differences in skin, hair, and eye color, height, weight, length of arms and legs, and circumference of neck and head. In these characteristics, as in most human traits, the phenotype is controlled by many genes. This type of pattern is called polygenic inheritance (**Fig. 18.23**). In addition to genetics, these traits are influenced by a number of other factors, such as race, gender, nutrition, and amount of exercise. Because of multiple influences, these traits do not exhibit sharp distinctions in clear-cut classes; instead they vary along a continuous spectrum. For example, the additive effects of many alleles involved in the polygenic trait of human height result in a bell-shaped distribution (**Fig. 18.24**).

Most animals also express polygenic traits. In dogs, height, weight, character, working abilities, and some genetic defects are controlled through multiple genes. Because multiple genes control these traits, predicting outcomes is difficult. Dog breeders have a hard time ensuring that larger breeds do not have hip dysplasia, which is a polygenic trait. A

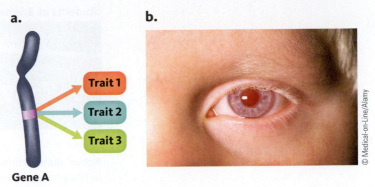

Figure 18.21 Pleiotropy. (a) In many organisms, one gene controls multiple traits. (b) As a result of a lack of pigment molecules in the cells, albinism affects three traits: hair color, skin color, and eye color.
© Cengage Learning 2014

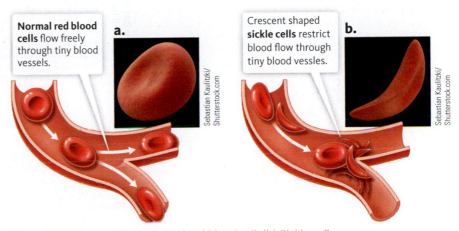

Figure 18.22 Sickle cell. (a) Normal red blood cell. (b) Sickle cell.
© Cengage Learning 2014

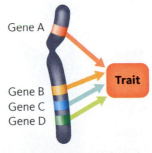

Figure 18.23 Polygenic traits. Polygenic traits are influenced by more than one gene.
© Cengage Learning 2014

Pleiotropy—genetic condition in which one gene controls two or more distinct and seemingly unrelated phenotypic traits.

Polygenic inheritance—genetic condition in which the phenotype is controlled by many genes.

Actual distribution of individuals in the photo according to height

Idealized bell-shaped curve for a population that displays continuous variation in a trait

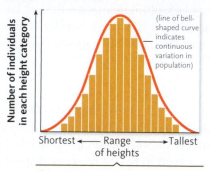

(line of bell-shaped curve indicates continuous variation in population)

Shortest ← Range → Tallest
of heights

Shortest ← Range → Tallest
of heights

If the sample in the photo included more individuals, the distribution would more closely approach this ideal.

Figure 18.24 Polygenic inheritance and human height. Human height typically is continuous in distribution due to polygenic inheritance. As a complex trait, height is influenced by hundreds of different genes and the environment. Over 80% of a population's variation is due to genetic factors. Environmental factors, such as diet, influence the other 20%.
© Cengage Learning 2014

Green

BBYY
BBYy
BbYY
BbYy

Blue

BByy
Bbyy

Yellow

bbYY
bbYy

White

bbyy

Figure 18.25 Polygenic inheritance and parakeets. Parakeet feather color results from the expression of two genes at once. The first gene codes for blue feathers (B) and the second gene codes for yellow feather color (y). A dominant allele results in the blue or yellow color. If dominant alleles for both colors are present, the bird shows green feather color.
© Cengage Learning 2014

simpler case of polygenic inheritance is feather color in parakeets, which is controlled by two genes. The first gene codes for blue feathers and the second gene codes for yellow feather color (**Fig. 18.25**).

Some traits are influenced by the environment

The environment surrounding cells is often an important factor in the expression of genetic traits. For example, the nutritional environment in which bones grow can result in shorter or taller individuals. Bone growth is greatly enhanced by vitamins, especially D and K, and minerals such as calcium, phosphorus, and magnesium. The hormonal environment plays an important role in the expression of sexual characteristics. It may surprise you that both men and women have genes for breast development and for beard growth; however, the expression of these genes is under the control of the sex hormones and therefore is expressed differently in males and females.

Physical aspects, including levels of light and temperature of the organism's environment, also influence the expression of some traits. The light environment (amount and intensity of sunlight) influences skin color, as well as vitamin D synthesis in humans and branching patterns in plants. Environmental temperature alters the expression of genes for fur color in snowshoe hares; cold temperatures change the brown fur to white. The temperature of egg incubation determines whether the alligator hatchling will be male or female. If the temperature is between 90°

and 93°F, the embryo will develop into a male; if the temperature is between 82° and 86°F, the embryo will become female. Intermediate temperatures result in a mixture of genders. Finally, soil pH influences flower color in hydrangeas, azaleas, and other plants. An acidic soil pH of 6 or below produces deep blue flowers, whereas a neutral pH of 6.8 or above yields pink flowers (**Fig. 18.26**).

In this section, we have illustrated the complex interactions between genes, alleles, environmental factors, and phenotypes. In many cases, patterns of inheritance are not simple. In the next section, we will examine one final exception to Mendel's simple patterns of inheritance by looking at inheritance patterns on sex chromosomes.

Figure 18.26 Environmental effects on genes. Conditions in the environment, such as soil acidity, can influence the expression of genes that control flower color in garden plants such as the hydrangea.

18.3 check + apply YOUR UNDERSTANDING

- Not all traits can be explained by Mendel's laws.

- Incomplete dominance produces blended, intermediate forms, whereas some traits are expressed with codominance due to multiple (more than two) allele forms.

- In pleiotropy, several traits are influenced by one gene, and polygenic traits are caused by the interaction of multiple genes.

- The environment also influences some patterns of inheritance.

Figure 18.27 Human eye color.

1. Define the terms *dominant* and *recessive* as they apply to a genetic trait. Would you expect to observe these forms of a genetic trait in all forms of life? Why?

2. Contrast the pleiotropic and polygenic patterns of inheritance.

3. A homozygous, recessive woman with her eyes spaced wide apart has a child with a homozygous, dominant man who carries the trait of eyes close together. Their child has eyes that are spaced in the middle (neither close together nor far apart). Which pattern of inheritance does the trait for eye spacing show? Using the letter "e", what are the genotypes of the mother and father?

4. Using three specific environmental factors, explain how these factors influence the phenotypes of plants. How do you think the same environmental factors influence the phenotypes of animals?

5. Below is a sampling of human eye color (**Fig. 18.27**). Is one or more than one gene necessary to express this trait? What type of inheritance pattern does this trait show? Explain your answers.

Sex Chromosomes Contribute to Determining Gender and X-Linked Inheritance Patterns

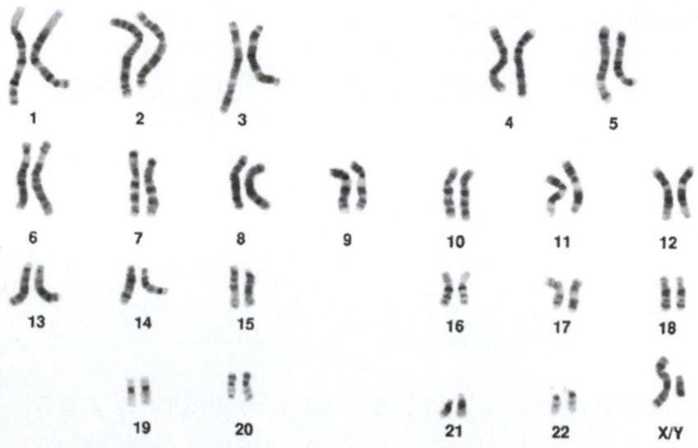

Figure 18.28 **Human karyotype.** A human male karyotype showing the relative sizes of the chromosomes. Note that the chromosomes are in an unduplicated condition. Chromosomes 1–22 are autosomes, and the X and Y are sex chromosomes.

Figure 18.29 **Sex determination.** Different animals have different ways to determine sex. Grasshoppers have 22 pairs of chromosomes plus an X chromosome; females have two X chromosomes and males have only one X chromosome. In bees, females are diploid (32 chromosomes) and males are haploid (16 chromosomes). Birds have 76 chromosome pairs plus Z and W chromosomes to determine sex. As reference, humans have 22 chromosome pairs plus the XX/XY sex-determining chromosomes.

© Cengage Learning 2014

Autosome—a chromosome that does not determine sex gender; in humans, chromosome pairs 1–22.

Sex chromosomes—chromosomes that determine sex gender; in humans, the X and Y chromosomes.

Recall from Chapter 11 that humans have 46 chromosomes, of which 23 are paternally derived and 23 are maternally derived. This full complement is referred to as diploid (2n = 46), in contrast to the gametes (sperm and eggs), which have only one set of chromosomes and are haploid (n = 23). These matched pairs of chromosomes are homologous to one another and contain different alleles for the same genes. The homologous chromosome pairs 1 through 22 are called autosomes. The X and Y chromosomes determine gender and are called sex chromosomes. Although normal diploid organisms have two homologous copies of each autosomal chromosome, they may or may not have sex-determining chromosomes as well. Humans have two different sex chromosomes that determine gender: the X and Y chromosomes. Human females have two homologous X chromosomes, whereas males have a nonhomologous set of one X chromosome and one Y chromosome that contain different genes.

In general, an organism's chromosomes are numbered from largest to smallest as a function of the number of base pairs on that chromosome. In humans, chromosome 1 has the largest number of base pairs, approximately 230 million, and contains about 4,200 identified genes. Your smallest autosome is chromosome 21 (not 22, although they are similar in size), a mere 47 million base pairs long with 446 genes (**Fig. 18.28**). The different sizes of chromosomes is one characteristic that scientists use to help identify them. It's interesting to note that chromosomes have been recognized and studied by cell biologists since the mid-1800s, but it wasn't until 1956 that scientists confirmed that the diploid number of chromosomes for humans was 46.

Sex chromosomes contribute to determining gender in animals

Plants and animals express gender, either a male or a female genotype. In some cases, an individual has both male and female structures. The pea plants studied by Mendel had flowers with both male and female parts on the same plant. Although this bisexual condition is common among living things, many species have separate male and female individuals. In addition to the gender-determining X and Y chromosomes, some organisms such as the alligator use environmental variables to determine sex, and some insects use social determinants. The vast biodiversity of plants and animals exhibit many different sex-determining systems.

The sex-determining system you are most familiar with is the XX/XY system that most mammals use to determine gender. More specifically, humans and most other mammals have a gene, called SRY, located on the Y chromosome that determines maleness. The SRY gene is responsible for the development of male gonads. If the Y chromosome is present, the individual develops testes and is a male; if the Y chromosome is absent, the individual does not develop testes and is a female. This implies that the female condition is the default gender in most mammals because it can arise with just the X chromosome. Insects, birds, and fish have different sex-determining systems (**Fig. 18.29**).

Some traits are carried on the X and Y chromosomes

Sex chromosomes also carry genes that are not related to sexual development. The human X chromosome is a large chromosome that carries about 1,000 genes. The much smaller Y chromosome carries only about 90 genes (see **Fig. 18.28** for size comparison). Any gene located on either sex chromosome is called **sex-linked**. Females carry two alleles for all genes carried on the X chromosome. Males carry only one allele (because they carry only one X chromosome). Males inherit the X chromosome from their mother and the Y chromosome from their father. Hemophilia, red–green color blindness, and Duchenne muscular dystrophy are human recessive disorders passed down mostly from mother to son on the X chromosome.

Fruit flies also use the XX/XY system to determine sex. One of the traits carried on the X chromosome is eye color. Red eye color is dominant and white is recessive (**Fig. 18.30**). Patterns of inheritance for eye color in fruit flies can be tracked (**Fig. 18.30c**). Although females can have white eyes, the trait is mostly expressed in males. This is especially true for recessive traits. Because males only carry one X chromosome, if they inherit one recessive allele, they express the trait. Females, however, need to inherit two recessive alleles to express the phenotype. Another characteristic of an X-linked trait is that all daughters of an affected male are carriers of the disorder. In contrast, sons from an affected male will never carry the father's recessive allele because they inherit their X chromosome from their mother (**Fig. 18.31**).

Pedigrees are tools to track patterns of inheritance in families

Another tool that geneticists use to track patterns of inheritance is the **pedigree**. Pedigrees are a type of family tree that allows you to track the history of a particular trait over several generations (**Fig. 18.32**). Pedigrees can be helpful in determining if a trait is an autosomal, X-linked, dominant, or recessive trait. Like Punnett squares, pedigrees are especially useful in human genetic counseling when a family member is known to carry an undesirable or lethal gene. Animal and plant breeders also make use of pedigree charts to track physical traits in their breeding lines.

Let's look at an example of how we can use a pedigree to track red–green color blindness in a family (see **Fig. 18.32**). In the Caucasian population, approximately 8% of men and 0.5% of women have the recessive trait for red–green color blindness. Affected individuals have normal

a.

Carolina Biological Supply Co/Visuals Unlimited, Inc.

b.

Carolina Biological Supply Co/Visuals Unlimited, Inc.

Figure 18.30 Eye color phenotypes in *Drosophila*. (a) The normal "wild-type" red color. (b) The mutant white eye color caused by a recessive allele from a sex-linked gene on the X chromosome. (c) Three generations of *Drosophila* flies. You can see by the last generation that segregation of the trait for white eyes is a sex-linked trait on the X chromosome.
© Cengage Learning 2014

c. Red-eyed female × white-eyed male

Parental generation

Red eyes (wild type) ♀ × White eyes ♂

X^{w^+} X^{w^+} X^w Y

1st generation

Red eyes ♀   Red eyes ♂

w^+ / w w^+ / (Y)

2nd generation

Sperm: w^+, w (Y)

Eggs: w^+, w

	w^+ (sperm)	Y (sperm)
w^+ (egg)	w^+ / w^+	w^+ / Y
w (egg)	w^+ / w	w / Y

All red-eyed females   ½ red-eyed, ½ white-eyed males

³⁄₄ red eyes : ¹⁄₄ white eyes

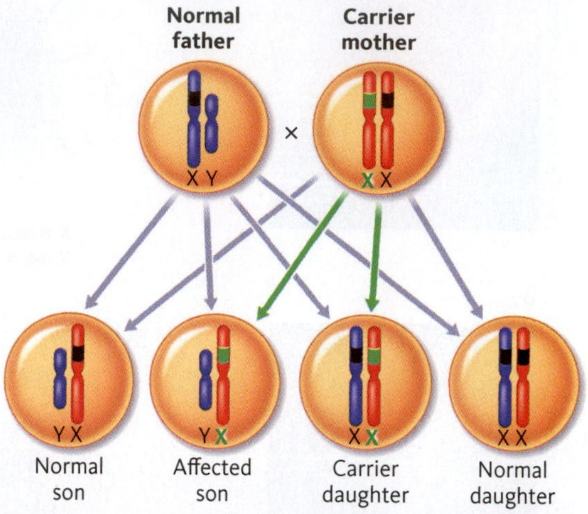

Figure 18.31 Inheritance of red–green color blindness: an X-linked recessive trait

Normal father

Carrier mother

X Y × X X

Normal son — Y X

Affected son — Y X

Carrier daughter — X X

Normal daughter — X X

Figure 18.31 X-linked inheritance. Males receive their X chromosome from their mother; therefore, they are more likely to inherit X-linked recessive traits.
© Cengage Learning 2014

Inheritance of red–green color blindness: an X-linked recessive trait

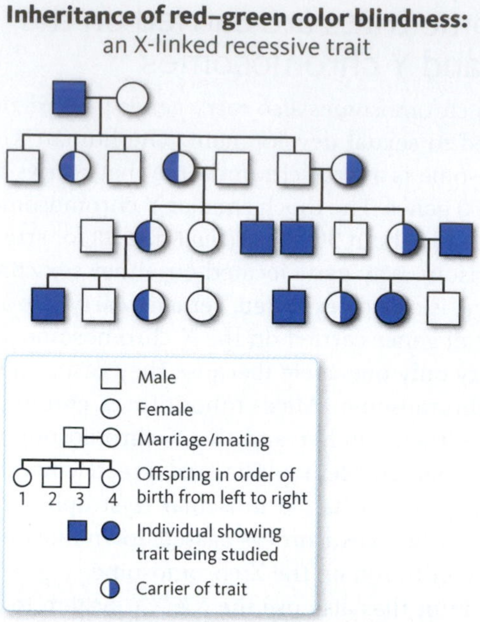

Figure 18.32 Inheritance of red–green color blindness. This family pedigree shows that males are more likely to express the recessive condition than are females.
© Cengage Learning 2014

	Male
	Female
	Marriage/mating
1 2 3 4	Offspring in order of birth from left to right
	Individual showing trait being studied
	Carrier of trait

Figure 18.33 A simple test for red–green color blindness. Can you see the number in the colored background?

vadim kozlovsky/Shutterstock.com

	X^N	X^n
X^N	$X^N X^N$ Female normal	$X^N X^n$ Female carrier
Y	$X^N Y$ Male normal	$X^n Y$ Male affected

Figure 18.34 Color blindness Punnett square. Normal vision is noted as dominant (NN or Nn) and color blindness as recessive (nn). What is the takeaway from this figure? There is a 50/50 chance that a male will be born affected by color blindness.
© Cengage Learning 2014

visual acuity (sharpness); however, they have a difficult time distinguishing between shades of red and green. For example, someone with red–green color blindness can't visualize the number in **Figure 18.33** because the red–green color-sensing cells of the retina are defective. Assume a female with normal vision who carries red–green color blindness and a male with normal vision have children. A Punnett square predicts the potential genotypes of the offspring of this couple (**Fig. 18.34**). In order to clearly track these traits through several generations in a family tree, standardized symbols are used (see **Fig. 18.32**). Basic Mendelian principles of inheritance and probability are used to interpret pedigrees. Understanding the phenotype of a specific genetic condition and the pattern of inheritance of that trait allows you to associate a genotype with most individuals in the family tree. This information can be used to better understand the trait and to predict the genotype and phenotype of potential offspring. It is particularly important to determine the probability of a genetic defect in a potential child or of a specific trait when breeding domesticated plants and animals.

In the last several sections, we introduced both simple and complex patterns of inheritance and concluded with sex-linked patterns. You have seen Punnett squares and pedigrees put into action to trace patterns of inheritance. In the next section, you will learn how breeders have identified and tracked desirable traits to create agricultural and other products of interest.

18.4 check + apply YOUR UNDERSTANDING

- In some organisms, sex chromosomes determine gender. Most mammals, including humans, use the XX/XY system.

- Sex-linked genes have distinct patterns of inheritance. Females carry two X chromosomes, and males carry one X chromosome and one Y chromosome.

- Pedigrees are tools that are used to track autosomal, X-linked, dominant, or recessive traits.

1. Humans have 23 pairs of chromosomes for a total of 46. How many of the 46 chromosomes are autosomes? What does the term *autosome* mean?

2. Which type of chromosome and system determine gender in most mammals?

3. In humans, why would more diseases found on X chromosomes appear in males than in females even though females have twice as many X chromosomes as men?

4. Refer to the human karyotype shown in **Figure 18.28** and predict whether chromosome 9 or 13 carries the most genes. Explain your prediction.

5. Examine the pedigree for Huntington disease (**Fig. 18.35**). Is this disease passed on an autosome or is it X-linked? Explain whether it is dominant or recessive.

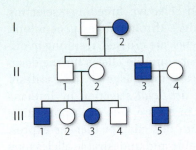
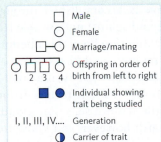

Figure 18.35 Huntington's pedigree. Pedigree of family carrying the alleles for Huntington disease.
© Cengage Learning 2014

18.5 Domestication Is the Genetic Modification of Wild Species

In his prize-winning book *Guns, Germs, and Steel,* Jared Diamond examines the history of human societies from small bands of hunter–gatherers to the beginnings of food production and agriculture. According to Diamond, one of the forces driving this transformation was the domestication of plants and animals. Hunter–gatherer societies could only support small populations because of the energy expended to find the small number of edible wild animal and plant species worth hunting or gathering.

Most of the wild plants in any region are poisonous, unpalatable, tedious to prepare, low in nutrition, or difficult to obtain. However, when humans select and grow a few species of domesticated plants and animals, an area of land can support a larger number of people. With the domestication of plants and animals, human populations grew and societies changed. Today we live in a complex society dependent on farmers who, in turn, depend on breeders to apply their knowledge of reproduction and genetics to produce plants and animals that yield high-quality foods, strong building materials, good-natured companions, and beautiful flowers.

How is domestication accomplished? How does a wild species genetically change into a domesticated species? What are some of the challenges and problems with the domestication process? In this section, we will examine the domestication of a wild plant and an animal species and apply simple Mendelian patterns of inheritance to show the genetic consequences of inbreeding.

Domestication involves artificial selection of traits

Domestication is accomplished through selective breeding—selecting for desirable and against undesirable traits—through many generations. Individuals exhibiting favorable phenotypes are crossed, resulting in offspring with an increased likelihood of carrying the alleles for the desirable traits and a decreased likelihood of carrying alleles for the less desirable traits. Domestication results from *artificial* selection by humans rather than the natural selection found in nature. The concept is fairly straightforward; however, you know from this chapter that the expression of genes can be very complex. Desirable and undesirable alleles may occur on the same chromosome or be carried on a sex chromosome; the environment can play a role in the expression of phenotype; and finally there is the element of chance.

Almonds are plant seeds rich in carbohydrates, proteins, and oils, making them an especially nutritious and energy-dense food (**Fig. 18.36**). In the wild, almond species are bitter and unpalatable because they contain the compound amygdalin. When eaten, amygdalin breaks down into the highly toxic chemical cyanide and thus serves to defend the seed by deterring predators, who learn to avoid it.

The production of amygdalin is inherited as a single gene. A simple mutation in this gene changes a bitter-tasting almond into a nonbitter

Figure 18.36 The almond tree, fruit, and seed.

one. Traits with simple patterns of inheritance make the almond and other species excellent candidates for domestication. It is believed that around 3000 B.C., hunter–gatherers discovered trees that by random mutation produced nonbitter almonds, and then through trial and error they selected trees that produced the best-tasting almonds. Today, in addition to selecting against bitter taste, almond breeders select for sweetness, which is transmitted as a dominant allele. The domestication of almonds illustrates how a wild, poisonous plant can become domesticated into a cultivated, sweet-tasting product.

Characteristics of wild animals can be produced relatively quickly

Now let's look at an animal experiment that was established to gain insight into the domestication of animals. The Russian scientist Dmitri Belyaev ran a 40-year selective breeding experiment to investigate the genetic processes that occur when an animal becomes domesticated from its wild relative. He attempted to domesticate the silver fox, a member of the dog family *Canidae,* because it was easy to acquire and maintain in captivity. With a short gestation period of 52 days and an average litter size of five, the silver fox was relatively easy to breed.

The breeders began by selecting for a single behavioral trait: low fear of humans. Eventually, they selected for foxes that exhibited positive responses to people. After 10 generations they successfully produced foxes that not only showed no fear of humans but also exhibited several behavioral characteristics similar to those of domesticated dogs. They wagged their tails, affectionately licked their human caretakers, and produced doglike vocalizations (**Fig. 18.37**). Surprisingly, the behavioral changes were accompanied by physical changes to the fox phenotype. Domesticated foxes inherited genes for spotted coats, somewhat floppy ears, and curled tails. Also, the animals' skull and jaw shape differed noticeably from those of the original wild foxes first bred in the 1950s.

By selecting for a docile disposition, breeders inadvertently selected silver foxes with lower levels of adrenaline. Today scientists are still trying to determine the genetic connection between low adrenaline levels and the various physical changes that occurred during the breeding experiment. To date, 40 genes have been identified that differ between the domesticated and wild fox. The function of these genes is still unclear. This experiment illustrates how phenotypic and behavioral characteristics of wild animals can be selected for and bred in a relatively short period of time to produce a dramatically different animal, one that is domesticated. The domestication of the silver fox also illustrates the complex patterns of inheritance. In this case, the domestication process did not take many generations; however, determining the effects of the 40 genes on various traits will take some time.

Inbreeding increases the chance of genetic disease

The domestication of wild plants and animals involves making crosses between different individuals, called *outcrossing,* and crosses between closely related individuals, called *inbreeding.* Over time, these two breeding strategies produce opposite results in a population. Outcrossing

a.

b.

Figure 18.37 Silver fox. (a) The domesticated silver fox developed many physical and behavioral characteristics similar to those of domesticated dogs. (b) The silver fox looks dramatically different than its wild cousin, the red fox.

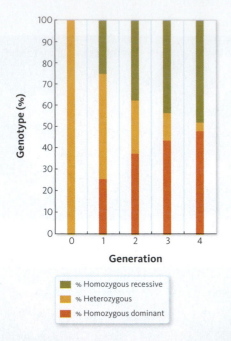

Figure 18.38 Inbreeding depression. Inbreeding for a heterozygous trait can shift the genotype of a population quickly. After four generations using self-fertilizing plants, 47% of the population will be homozygous dominant (orange bar), 6% will be heterozygous (yellow bar), and 47% will be homozygous recessive (green bar) for this particular trait. Some homozygous recessive alleles are harmful, reducing the health and vigor of the population.
© Cengage Learning 2014

brings together different combinations of alleles and thus increases genetic variability, whereas inbreeding decreases genetic variability.

Humans attempting to domesticate a wild plant or animal use inbreeding to produce genetically identical or *homozygous* individuals for the desired trait. If a breeder starts with a single self-fertilizing individual that is heterozygous for a particular trait and then inbreeds the offspring for four generations, the proportion of the population that will be homozygous dominant is 47%, with 6% being heterozygous and 47% homozygous recessive for this trait (**Fig. 18.38**). Even though this example represents an extreme case of inbreeding, it demonstrates that continued mating of closely related individuals within a line or family increases the chance of bringing together hidden recessive alleles that previously were masked in the heterozygous condition. Some recessive alleles are harmful when homozygous; therefore, inbred populations often have reduced health and vigor, a condition called inbreeding depression. In humans, studies have shown that inbreeding also increases the risk of birth defects and miscarriages and is associated with higher infant mortality. These consequences have led states to enact laws prohibiting marriage between blood relatives as close as first cousins.

Inbreeding depression is evident in the intense breeding of show dogs, beef and dairy cattle, racehorses, and high-value crops. Prize-winning dogs meet specific breeding standards, which are written descriptions of the ideal phenotype of the breed. You may be familiar with the short legs of dachshunds, squashed faces of pugs, and spotted coats of Dalmatians. Intense inbreeding to meet the breed standard has led to breeds with characteristic health problems, which include growth and developmental abnormalities, reproductive or behavioral disorders, and degenerative diseases. For example, Dalmatian breeders who have selected for increased spotting of the coat inadvertently increased the occurrence of painful and potentially deadly kidney stones (**Fig. 18.39**). In Labrador retrievers, selecting for coat color has resulted in a specific type of blindness in some animals. In bassett hounds, an autosomal recessive trait causes blood disorders.

Problems associated with inbreeding also plague plant breeders. Selective breeding for one trait may cause a tradeoff for another. Breeders discovered that the trait for disease resistance in wheat is linked to genes that control seed head development and crop yield. Thus, gains in disease resistance to fungal rusts were offset by declines in yield. To combat inbreeding depression, breeders return to the place of origin to find plants that possess greater genetic variation and through natural selection often have genes that confer resistance to disease. For instance, corn breeders have tapped the genetic resources of *Zea diploperennis,* an ancient relative of modern-day corn, which contains genes that make it resistant to corn viruses (**Fig. 18.40**). In the final section of this chapter, we will explore more closely how the genetic makeup of organisms influences physical traits.

Figure 18.39 Inbreeding depression can cause health conditions. Dalmatians exhibit problems with inbreeding depression. Kidney stones, which can be very painful and sometimes deadly, increase due to breeding for specific spot patterns.

a.　**b.**

Figure 18.40 Corn breeding. (a) Modern corn is backcrossed with its ancestor (b) *Zea diploperennis* to create offspring that have resistance to diseases that modern varieties of corn have lost due to inbreeding depression.

Inbreeding depression—reduced health and vigor as a result of breeding related individuals.

- The principles explained by Mendel and Darwin have been used for thousands of years to domesticate plants and animals.

- By selecting for desired traits, the phenotype of an organism can be manipulated for a desired outcome.

- Continued inbreeding in a population increases the likelihood of expressing harmful recessive alleles, which leads to reduced health and vigor.

1. What is meant by artificial selection? How does this term apply to the domestication process?

2. How does inbreeding depression work against a breeder's attempt to improve the odds of a favorable phenotypic outcome?

3. Compare two plants of the same species, one that consistently self-fertilizes and one that consistently outcrosses. Predict which of the two plants will produce the most genetically similar seeds. Which will produce the larger, more vigorous seeds?

4. There are several populations of northern flying squirrels in the United States. One of the populations lives in Pennsylvania and is small; the other is quite a bit larger living in South Dakota. Is one of these populations more susceptible to inbreeding depression? Explain your answer.

5. What contribution to the field of genetics of domestication did Dmitri Belyaev and his colleagues make? How might these contributions be applied to questions regarding human genetics?

18.6 Genes Control Phenotype by Controlling the Synthesis of Proteins

Whether we are referring to the domestication of plants and animals, the structure of your face, or the patterns of inheritance of a rare disease, all of these phenotypes are the result of interactions of genes. In our own bodies, our phenotype is the result of developmental and physiological processes that are directed by the activities of our genes—our genotype. To take this concept further, let's look at the phenotype that involves the presence or absence of skin pigmentation (**Fig. 18.41**). The lack of pigmentation is called albinism; the presence of some pigmentation is the normal condition. The allele for normal skin pigment, dominant (A), is expressed over the recessive albino allele (a) and is under the control of a single gene that shows a simple Mendelian pattern of inheritance. It's important to distinguish this trait from the relative lightness or darkness of skin color, which in humans is controlled by the action of many other genes.

Within the lower levels of the skin, specialized cells produce the brown pigment melanin through a series of chemical reactions (**Fig. 18.42**). By controlling the steps in these reactions, genes ultimately control the synthesis of melanin and therefore control the presence or absence of skin color. In the heterozygous condition *(Aa)*, a person's skin cells have one functioning allele and one nonfunctioning allele and are capable of producing the enzyme to complete the chemical synthesis of melanin. However, in the recessive homozygous condition *(aa)*, skin cells lack a functioning gene or, to be more precise, a functioning allele

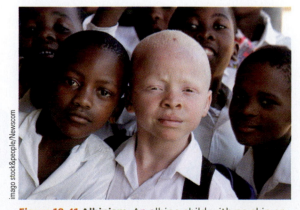

Figure 18.41 Albinism. An albino child with no skin or eye pigmentation contrasts sharply with darkly pigmented classmates. The albino child displays many of the same facial traits as the other children. At a glance, the only difference seems to be skin color controlled by a single gene.

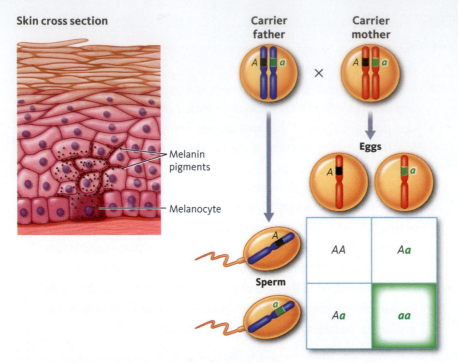

Skin cross section

Melanin pigments

Melanocyte

Carrier father × **Carrier mother**

Eggs

Sperm

	AA	Aa
	Aa	aa

Figure 18.42 Genetic control of skin pigmentation. Pigment-producing cells of the lower skin layer produce the pigment, melanin. Albinism is a result of inheriting two recessive alleles (aa) that prevent melanin production. This Punnett square predicts the outcome of two carriers (Aa) that reproduce.
© Cengage Learning 2014

Structural proteins—proteins that give body parts form and structure, including hair, muscle, skin, tendons, cartilage, and connective tissues.

Functional proteins—proteins that protect, communicate, transport oxygen, or catalyze chemical reactions; examples are antibodies, hormones, hemoglobin, and enzymes.

capable of directing the production of melanin. Because all body cells contain the same genetic information or have the same genotype, all body cells contain two copies of this faulty allele. The condition arises when a person inherits two recessive alleles for this gene from his or her parents and consequently produces no melanin and has no pigmentation in the iris of the eyes or skin, a classic case of albinism. In the last section of this chapter, you will learn how the genotype of an organism is translated into its phenotype.

Genes are expressed as proteins

Let's take the albinism example one step further and ask, "What makes the recessive allele incapable of directing the production of melanin?" From Chapter 7 the answer is that *genes control all traits, including melanin formation, by directing the synthesis of a large number of proteins.* An overview of this process is shown in **Figure 18.43**, which points out some of the traits associated with the human face, such as facial muscles, cartilage composing the nose and ears, the underlying connective tissue, and the blood pigment hemoglobin that colors lips red. The bodies of humans, like those of all animals, are composed chiefly of **structural protein**. Hair, muscle, skin, and connective tissues are composed of these structural proteins. By directing the building of proteins, genes direct facial structures. Proteins such as antibodies, hormones, hemoglobin, and enzymes are examples of **functional proteins**, which protect, communicate, transport oxygen, or catalyze chemical reactions. If the genes that produce some of these proteins become dysfunctional, the result is usually a functional or biochemical disorder such as sickle-cell anemia, diabetes, or albinism. The patterns of inheritance for many of the human traits we introduced in this chapter are summarized in **Table 18.2**.

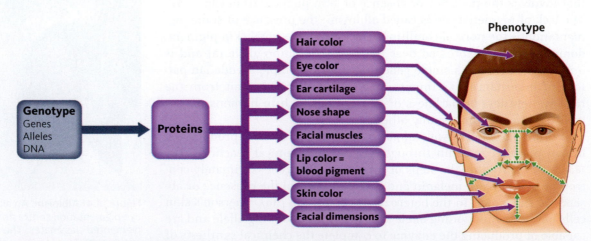

Phenotype

Genotype
Genes
Alleles
DNA

Proteins

Hair color
Eye color
Ear cartilage
Nose shape
Facial muscles
Lip color = blood pigment
Skin color
Facial dimensions

Figure 18.43 Genes control phenotype. Genes direct the formation of facial structures during the developmental and aging processes by controlling the production of a large variety of proteins.
© Cengage Learning 2014

TABLE 18.2 Patterns of Inheritance in Humans

Trait	Pattern of Inheritance
Albinism—presence of melanin	Single gene—recessive
Eye color	Polygenic
Red–green color blindness	X-linked gene
A and B blood type	Single gene—codominance
Cystic fibrosis	Single gene—recessive
Ability to digest lactose	Single gene—dominance
Tay-Sachs disease	Single gene—recessive

© Cengage Learning 2014

Even though we have focused mostly on human traits, it's important to see that all organisms exhibit the same simple and complex patterns of inheritance and appear to us as expressions of their underlying genotypes. The natural world is a landscape of genetic expression. The skin secretions of a poison dart frog, the camouflage coloration in toads, the leaf shape of an oak, the antibiotic resistance of infectious bacteria, and the green color of the forest all have a genetic basis in DNA. All are determined by the sequence of nucleotides that form genes. In fact, it could be concluded that this entire text is about the phenotypes of living things as shaped by selective forces on their genotypes.

18.6 check + apply YOUR UNDERSTANDING

- Genes control the expression of traits by controlling protein synthesis.

- The genotype (genetic makeup) of an organism determines the phenotype (physical traits) of the organism.

1. Explain why all body cells contain the same genotype.
2. What percentage of sperm cells with the recessive allele for albinism will an albino man produce?
3. Genes control traits by directing the synthesis of which class of macromolecules?
4. Collagen fibers are strong protein fibers in bone and connective tissue. Is collagen considered a structural or a functional protein?
5. Enzymes are a type of functional protein. How are enzymes used in fat metabolism?

End of Chapter Review

Self-Quiz on Key Concepts

The Basis of Inheritance

KEY CONCEPTS: Sexual reproduction introduces genetic variation in organisms. Gregor Mendel gathered the first evidence that genes are the distinct units passed on through generations resulting in the expression of traits. Alternate forms of genes (alleles) separate during the formation of gametes and form new pairs during reproduction. Because many traits are inherited in simple patterns, genetic outcomes can be predicted using Punnett squares and family pedigrees.

1. Match each of the following **gene terms** with its characteristic, description, or example.

    Gene
    Allele
    Heterozygous
    Homozygous
    Dominant allele
    Recessive allele

    a. two different alleles for a particular gene
    b. alternate forms of a gene
    c. has no noticeable effect unless two alleles are present
    d. the allele that is preferentially expressed, even when only one copy is present
    e. two of the same alleles for a particular gene
    f. DNA sequence coding for a protein product

2. Match each of the following **genetic terms** with its characteristic or description.

    Meiosis
    Independent assortment
    Law of segregation

    a. process of gamete formation
    b. describes the separation of *alleles* during meiosis
    c. describes the behavior of *chromosomes* during meiosis

3. Which of the following statements regarding sexual reproduction is not true?
    a. Meiosis and fertilization increase genetic variation.
    b. Each parent contributes half of the genetic material to the resulting offspring.
    c. Sexual reproduction costs the organism less energy to produce large numbers of offspring.
    d. Sexual reproduction includes the fusion of gametes and genetic information.
    e. Sexual reproduction includes meiosis (gamete formation) and the fertilization of gametes.

4. Albinism is a recessive trait. What is the outcome of a heterozygous normal-colored male crossed with an albino female?
    a. all albino
    b. half albino and half normal
    c. all normal
    d. 75% normal and 25% albino
    e. 75% albino and 25% normal

More Complex Patterns of Inheritance

KEY CONCEPTS: Not all traits can be explained by Mendel's laws. Incomplete dominance, codominance, multiple genes, and the environment also influence patterns of inheritance.

5. Match each of the following **patterns of inheritance** with its characteristic, description, or example.

    Incomplete dominance
    Codominance
    Pleiotropy
    Polygenic

    a. One gene affects many phenotypic traits.
    b. Multiple genes control the trait expressed.
    c. Trait expressed is a blend of the dominant and recessive alleles.
    d. More than two alleles for the same gene control the trait expressed.

6. A mother has type A blood, and a father has type B blood. They have two children, one with type AB blood and the other with type O. What are the genotypes of the parents?

   a. $I^A I^A$ and $I^B I^B$

   b. $I^A I^i$ and $I^B i$

   c. $I^A I^i$ and $I^B I^B$

   d. $I^A I^A$ and $I^B i$

   e. none of the above

7. Andalusian chickens show incomplete dominance in feather color. A cross between a black bird and a white bird results in a gray chicken. What ratio of offspring is expected when a black homozygote crosses with a white homozygote?

   a. 1 black : 2 gray : 1 white

   b. all gray

   c. 3 black : 1 white

   d. all black

   e. 1 black : 1 gray : 1 white

Sex Chromosomes and X-Linked Inheritance Patterns

KEY CONCEPTS: In some organisms, sex chromosomes determine gender. Sex-linked genes (genes carried on sex chromosomes) have distinct patterns of inheritance; females carry two X chromosomes and males carry one X and one Y chromosome.

8. Match each of the following **chromosome terms** with its characteristic, description, or example.

Linked	a. in humans, chromosomes 1–22
X-linked	b. genes that do not show independent assortment
Sex chromosomes	c. genes or traits associated with the X chromosome
Autosomes	d. chromosomes involved in sex determination

9. Characteristics of X linked recessive traits include which of the following?

   a. In order to express the recessive trait, females must inherit two copies of the allele.

   b. In order to express the X-linked recessive trait, males only need to inherit one copy.

   c. All daughters from an affected father will carry the trait.

   d. Sons from an affected male will never carry the recessive allele.

   e. All of the above statements are true.

Artificial Selection of Traits for Domestication

KEY CONCEPTS: The principles explained by Mendel and Darwin have been used for thousands of years to domesticate plants and animals. By selecting for desired traits, the phenotype of an organism can be manipulated for a desired outcome.

10. Inbreeding depression is the result of:

    a. the unintentional concentration of undesirable traits while concentrating desired traits.

    b. inheritance patterns from linked genes.

    c. the loss of alleles from the gene pool.

    d. continuous selective breeding in small populations.

    e. all of the above.

11. Like the domestication of many animals, the first individuals of the silver fox were selected based on

    a. their interactions with humans (low fear).

    b. physical traits such as size, speed, and coat color.

    c. growth rate.

    d. none of these traits.

    e. all of these traits.

Genotype to Phenotype

KEY CONCEPTS: Genes control the expression of traits by controlling protein synthesis. The genotype (genetic makeup) of an organism determines the phenotype (physical traits) of the organism.

12. Match each of the following **gene expression terms** with its characteristic, description, or example.

Genotype   a. protein such as hormones, antibodies, and enzymes
Phenotype   b protein such as hair, muscle, and connective tissues
Structural protein   c. genetic makeup of an organism
Functional protein   d. physically expressed traits

Applying the Concepts

13. Freckles are a dominant trait. Two parents, both with freckles, give birth to a child without freckles. What are the genotypes of the parents?

14. The Joneses are afraid there has been a mix-up at the birthing center of the hospital. Mr. Jones has type A blood and Mrs. Jones has type O blood. Use a Punnett square to predict what type of blood Baby Jones will have.

15. Which of the pedigrees below (**Fig. 18.44**) represents the inheritance pattern of a dominant trait? A recessive trait? Use the letter "A" to represent the dominant allele and "a" to represent the recessive allele and fill in the genotypes of as many individuals as you can.

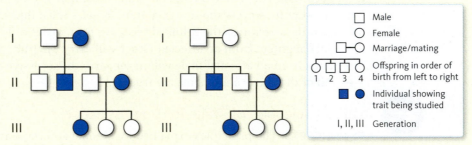

Figure 18.44 Pedigree analysis.
Copyright © 2000. Phillip McClean. http://www.ndsu.nodak.edu/instruct/mcclean/plsc431/mendel/mendel9.htm

16. The following pedigree (**Fig. 18.45**) represents the inheritance of an X-Linked recessive trait.
    a. Complete the genotype for each of the family members.
    b. Why is there never any male-to-male transmission?
    c. If a mother has the trait, what is the likelihood that her sons will have the trait? Why is this?

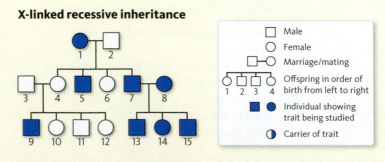

Figure 18.45 Pedigree of an X-linked recessive inheritance.
© Cengage Learning 2014

Data Analysis

Pedigree Analysis of Cystic Fibrosis

In Section 18.2 you learned about cystic fibrosis (CF), the fatal genetic disorder carried on chromosome 7 (**Fig. 18.46**). This inherited disease creates a thick, sticky mucus that affects many body systems, including the respiratory and digestive systems. You also were introduced to the use of pedigrees to trace family traits. Imagine you are part of this family and you would like to understand your chances of carrying the gene that causes CF. Analyze the pedigree on page 576 (**Fig. 18.47**) and answer the questions that follow. It will be helpful to write out some Punnett squares to help with your analysis.

Cystic fibrosis is the most common fatal hereditary disorder affecting Caucasians in the United States. It also is the most common cause of chronic lung disease in children and young adults in the United States, affecting one in 2,000.

Mucus builds up and blocks pathways of digestion and absorption; it also increases rates of respiratory infections. CF affects many organ systems.

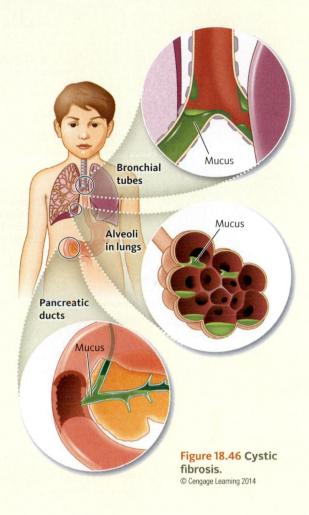

Bronchial tubes

Mucus

Mucus

Alveoli in lungs

Pancreatic ducts

Mucus

Figure 18.46 Cystic fibrosis.
© Cengage Learning 2014

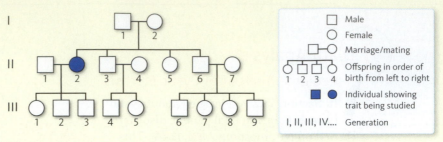

Figure 18.47 Pedigree analysis of cystic fibrosis.
© Cengage Learning 2014

Data Interpretation

17. Complete the following table, which summarizes some of the characteristics of the common patterns of inheritance.

Pattern	Number of Allele Copies Needed for Expression	Are Parents of Affected Child Usually Affected (Y/N)?	Equally Likely To Be Expressed in Both Sexes (Y/N)?	Transmitted Father to Son (Y/N)?	Is Expressed Trait Likely to Skip a Generation (Y/N)?
Autosomal dominant					
Autosomal recessive					
X-linked dominant					
X-linked recessive					

18. What pattern of inheritance does this disorder follow?

19. Using the letter "C" to represent the CF gene and *normal, carrier,* and *cystic fibrosis,* complete the table to the right.

Family Member	Genotype	Phenotype
I-1		
I-2		
II-1		
II-2		
III-1		

Critical Thinking

20. If the original parents (I-1 and I-2) had another child, what would be the likelihood that the child would have cystic fibrosis?

21. What are the possible genotypes for III-3 and III-4? Is there any chance they could have a child with cystic fibrosis? Explain your answer.

22. Imagine III-1 is expecting a child. The couple decides to go to genetic counseling to determine the likelihood of their child having cystic fibrosis. Although expensive, genetic testing for cystic fibrosis is available. As the genetic counselor, what would you recommend to the family?
   a. Should III-1 be tested for the gene?
   b. Should the baby's father be tested?
   c. If the father is heterozygous, what is the likelihood of the child being affected?

23. What if the father is homozygous dominant?

Question Generator

Inheritance Patterns in Breast and Ovarian Cancers

In the 1990s, biologists identified two genes associated with breast and ovarian cancer. These genes, named "breast cancer 1" (BRCA1, pronounced "brak-uh") and BRCA2, are mutated forms of normal genes that suppress tumor development by controlling cell growth and cell death. They are located on chromosomes 17 and 13, respectively (**Fig. 18.48**). The BRCA mutations are transmitted in an autosomal dominant pattern in a family. Because each body cell is diploid, every person has two genes, one inherited from each parent. When a person has inherited one altered or mutated copy of either the BRCA1 or BRCA2 gene, his or her risk for various types of cancer increases.

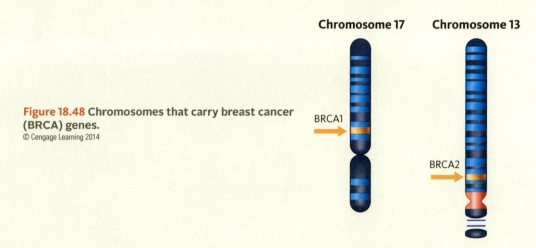

Figure 18.48 Chromosomes that carry breast cancer (BRCA) genes.
© Cengage Learning 2014

Below is a block diagram that relates cancer genes, patterns of inheritance, and risk of developing breast or ovarian cancer (**Fig. 18.49**). Use the background information along with the diagram to ask questions and generate hypotheses. We've translated the components of the block diagram into several questions to get you started.

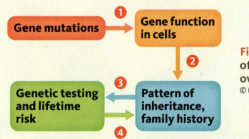

Figure 18.49 Factors relating cancer genes, patterns of inheritance, and the risk of developing breast or ovarian cancer.
© Cengage Learning 2014

Here are a few questions to start you off.

1. How does the gene's mutation affect its function in cells? (Arrow #1)

2. What is the relationship between how the gene functions and the way it is inherited in families? (Arrow #2)

3. What are my chances of having a BRCA mutation if my mother or father has it? (Arrow #3)

4. How does the autosomal dominant pattern influence my lifetime risk for breast cancer? (Arrow #4)

5. Why do some people who inherit a BRCA mutation never get breast cancer? (Arrow #4)

Use the diagram to generate your own research question and frame a hypothesis.

19 Biotechnology and Genetics

19.1 Biotechnology and corn—an ancient and modern science

19.2 Biotechnology is the engineering of biological systems

19.3 Genetic engineering techniques are used to modify genes and organisms

19.4 Biotechnology is changing the face of health care

19.5 DNA profiling is used to establish genetic relationships

19.1 Biotechnology and Corn—An Ancient and Modern Science

The Cerro Bau'l mesa, a Peruvian landmark, sits 8,000 feet above sea level and contains ruins of structures built by the Wari people between 600 and 1000 A.D. (**Fig. 19.1**). Four hundred years before the Inca rose to dominance, the Wari settled this barren mesa and built a city of temples, gathering halls, and residences for 1,000 people. To support religious and diplomatic ceremonies, a select group of high-status women ran an industrial-scale brewery, churning out hundreds of gallons of corn-based beer per week. For unknown reasons, the brewery was ritually burned to the ground 1,000 years ago when the Wari abandoned the settlement and mysteriously moved to another location.

Archaeological remains show this brewery used several rooms to mill and ferment sprouted corn kernels to make a beer called *chicha.* The brewery contained multiple fire pits to boil the mixture, which was later transported to the fermentation area and placed in large vats to age for several days. The brewing process they used is very similar to the one used today, except that today's brewers ferment with yeast, whereas the Wari chewed the boiled pulp, using enzymes in their own saliva to start the fermentation. Even before the Wari began brewing beer, Egyptians and Babylonians half a world away were brewing beer during the third century B.C. It is fascinating to think that a few thousand years ago this elaborate brewery practiced biotechnology!

At a broad level, biotechnology refers to scientific processes that use living organisms or their characteristics to study and make products for human use. Today, biotechnology includes forensics, paternity testing, commercial production of medicines, and genetic engineering of plants and animals for food production. Although they may not have understood the science behind it, the Wari people used biotechnology for fermentation and food production. Their successors continued to experiment with ways of improving food production. DNA studies suggest that improving corn crops by purposefully selecting for various traits led to 35 different varieties in Peru before 1500. To date, thousands of varieties of corn have been cultivated for various characteristics and climates throughout the world.

Zea mays, or maize as it is known around the world, was domesticated more than 8,000 years ago in Mexico from a wild grass known as *teosinte* (**Fig. 19.2**). Maize has been genetically manipulated for thousands of years to produce a food plant that has enormous global economic importance. Today it is the largest crop in America, and world exports continue to increase due to the growing demand for meat products and ethanol (biofuels). By 2013, the United States is projected to increase its production to 12 billion bushels, which is 70 percent of the global market share (**Fig. 19.3**).

Over the last 50 years in the United States, corn yield has doubled from 70 to more than 150 bushels per acre due to improved farming techniques (agronomy), breeding programs, and biotechnology. Genetic analysis, manipulation, and selection have both improved the stress tolerance of corn and allowed it to be grown with less fertilizer. Corn hybrids can now better tolerate cold, drought, pests, disease, herbicides, and high-density plantings. Yield improvements are also a result of the hybrids that produce more kernels per ear that fill in more efficiently (**Fig. 19.4**).

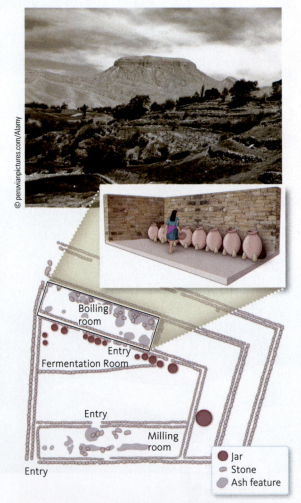

© peruvianpictures.com/Alamy

Figure 19.1 Wari brewery on Cerro Bau'l mesa. Cerro Bau'l mesa, located in Peru, was home to one of the Wari settlements between 600 and 1000 A.D. This ancient brewery is an example of biotechnology!

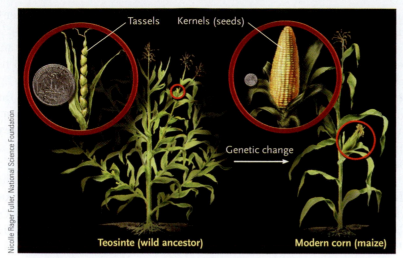

Figure 19.2 **Comparison of ancient and modern corn.** Corn's wild ancestor is a grass called teosinte (left). These plants have the same number of chromosomes with genes in similar arrangements. Through selective breeding, plant geneticists have increased the size of the corn cob, with more rows of kernels. Five areas of their genomes cause the major differences between teosinte and modern corn (right).

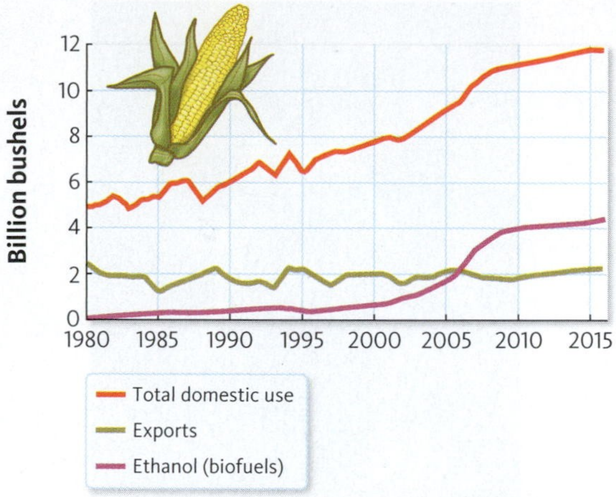

Figure 19.3 **U.S. corn production, 1980–2015 (projected).** By 2013, it is estimated the U.S. will produce 70% of the world's corn. While U.S. exports have remained steady, worldwide exports continue to grow. Increasing biofuel and feed production for livestock drive demand.
© Cengage Learning 2014

As a response to the demands of feeding an ever-increasing world population, biotechnology is also being used to increase the nutritional value of food. Recently corn has been bioengineered so that it is fortified with three essential vitamins frequently missing in the diets of millions of people in developing countries. Scientists introduced five genes from other organisms into a variety of corn commonly grown in Africa. These genes increase levels of beta-carotene, vitamin C, and folate, compounds which are known to prevent several diseases afflicting the people of Africa. This hybrid is not yet available and is now being tested through traditional breeding programs to select for the plants that are best adapted for specific regions and environmental conditions.

In this chapter, we will discuss the important roles that biotechnology plays in our living world. We will look at the engineering of new crops, food products, and medicines and discuss the ethics of these inventions. We also will look at how biotechnology is used as an identification tool in both the healthcare and justice systems.

Figure 19.4 **Yield improvements from biotechnology and agronomy practices.** The bottom cob has less fill, or kernels per ear, than the top cob, which is completely filled in.

19.2 Biotechnology Is the Engineering of Biological Systems

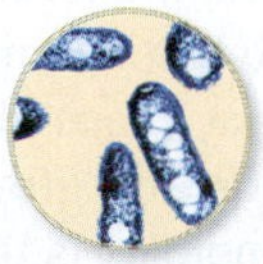

What do food, fiber, feedstock, fuel, and pharmaceuticals have in common? Products in each of these categories hold an important place in the global economy and play a role in the well-being of all the people in the world. Each of these industries also relies heavily on biotechnology to produce these valuable products. As the name implies, **biotechnology** is an interaction between *technology* and *biological systems*. Technology is defined as the application of science for industrial or commercial uses. In other words, biotechnology is a scientific field that uses biological agents,

Biotechnology—a scientific field that uses biological agents, such as microorganisms or cellular components, in a controlled manner to solve problems or to make products that have beneficial uses.

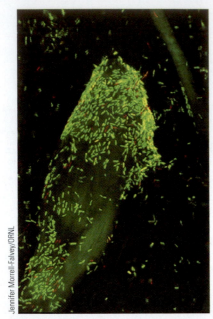

Figure 19.5 Bacteria and Biofuels. Researchers are studying the use of microbes to produce cheaper biofuels. This species of bacteria can break down cellulose and switchgrass. The products are then used by the bacteria during their metabolic pathways, producing ethanol. In this photo, scientists have added a fluorescent dye to monitor these bacteria.

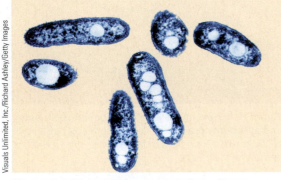

Figure 19.6 Bioengineered plastic from plants. Recently, scientists have genetically modified bacteria and plants, inserting genes into the DNA of these organisms that cause them to produce granules of biodegradable plastic inside their cells. This bioengineered plastic is secreted by the plant or harvested from the cells to make the final product that can be used by the plastic industry.

Genomics—the study of all the genes of an organism, examining both the DNA sequences and the interactions of these genes with each other and with the organism's environment.

such as microorganisms or cellular components, in a controlled manner to solve problems or to make products that have beneficial uses (**Fig. 19.5**).

The study of biotechnology began with selective breeding programs designed to domesticate crops and animals, as well as from the use of microorganisms to make beer, wine, bread, and cheese. Many centuries ago people learned that, with the help of microbes, fruit juices ferment into wine, milk solidifies into cheese or yogurt, and yeast helps make products such as bread and beer. Today, agriculture, waste recycling, food processing, and energy production all use biotechnology to thrive. It also plays an integral role in medicine and forensics.

Let's look at an example of how biotechnology affects our lives today. Plastics are the third largest manufacturing industry in the United States, with production facilities in every state. Although the economic impact of plastics manufacturing is significant, from an environmental perspective plastic is difficult to recycle or dispose of because of how long it takes these products to degrade in a landfill. Plastics account for 5 percent of the nation's consumption of petroleum and natural gas, and only about 5 percent of all plastics manufactured are recycled. Consider for a moment the number of items within your reach that are made of plastic or have plastic components. More than one trillion pounds of plastic waste sits in U.S. landfills. Even though plastic products play an important and valuable role in our lives, the environmental costs create short- and long-term problems. This is where biotechnology comes on the scene.

Making plastics from biological systems solves many of these problems. One of the first plastics invented, celluloid, was made in the late 1800s from the cellulose molecules found in plants. Today polymers for plastics are produced from corn, soybeans, potatoes, bacteria, algae, and other natural resources (**Fig. 19.6**). Unlike petroleum-based synthetic plastics, plant-derived substances are naturally biodegradable; they also are renewable. In this section, we will introduce you to the ever-expanding field of biotechnology, focusing on its history, its power, and some of the ethical dilemmas created as scientists manipulate the DNA of microbes, plants, and animals.

The study of genomes provides important tools for biotechnology

Whether the goal is to make a new renewable plastic, a pharmaceutical, or an improved food crop, the ability to engineer a biological system means that you have to understand how the system works. Often, scientists start by understanding the genome of the organism to be manipulated. Throughout this book, we have emphasized the role that genes play, individually and collectively, in determining structure, directing growth and development, and controlling biological functions. **Genomics** is the study of all the genes of an organism, examining both the DNA sequences and the interactions of these genes with each other and the organism's environment. The goal of this field is to gain an understanding about fundamental biological processes and to use this information to predict how living systems operate at the molecular, cellular, community, and population levels. For instance, scientists use genomics to characterize and manipulate the gene inserted into plants to produce products such as plastic.

The first step in understanding an organism's genome is to decipher the genetic code, or to *sequence* it, meaning to identify the order and location for each individual base of the DNA code. Scientists around the world have documented the genomes of more than 1,000 species, a costly and time-consuming process. **Table 19.1** lists some genomic information for several organisms, including humans, corn, and *Escherichia coli* bacteria. With this information, scientists can map the connections between genes and the traits or diseases they control. Sequencing was one of the first steps used by scientists to isolate the plastic-making gene in bacteria and insert it into a plant.

Another example of the partnership between genomics and biotechnology is improvements to food crops. It is estimated that 40 to 50 percent of the world suffers from diseases caused by vitamin and mineral deficiencies. Using biotechnology, scientists have learned how to fortify crops through the introduction of genes that turn on metabolic pathways to produce key micronutrients. These modifications would not be possible without information first provided by genomics. Recall that genes carry the DNA code and express it in the form of proteins that function as enzymes along metabolic pathways that make vitamins and nutrients as products. In order to make corn that is fortified with beta-carotene, vitamin C, and folate, scientists use genomics to identify genes of interest and develop a plan to introduce the genes into the corn plant (**Fig. 19.7**). Once the genes are inserted into the corn's genome, the corn makes these vitamins and nutrients. Like all scientific endeavors, this engineering process leads to other questions that genomics can answer. How will these genes interact with the rest of the plant's genome? Are there any environmental factors that influence the growth, development, and expression of these new genes? By understanding the chromosomal location of the genes, the function of the gene products, and how certain gene products are influenced by the environment, new technologies can be developed to make use of living systems for carbon storage and cycling, bioenergy production, sustainable agriculture, and environmental waste cleanup (remediation). We will explore in detail how organisms are engineered with modified genes in Section 19.3.

Biotechnology is also important because it provides a set of tools for other areas of biology. For example, the techniques and outcomes of genomics are also used to analyze and verify evolutionary relationships. The analysis of evolutionary relationships is emerging as a critical means for determining the function and interactions of genes within a genome. Molecular evolutionists compare the genomes of various organisms to analyze the number of changes that DNA sequences undergo through the course of evolution. Using this information, researchers can determine functionally important regions within genes and construct a molecular timescale of species evolution.

Many scientists compare the completion of the sequencing of the human genome in 2003 to achievements such as space flight. They believe it will be the most significant discovery of the 21st century and perhaps of all time. The knowledge and technologies developed from this information will improve human welfare and increase our life span through new diagnostics, treatments, and cures and preventions of all human diseases. The knowledge from the field of genomics will also enable scientists to more quickly manipulate genomes to enhance the world we live in.

TABLE 19.1 Genomic Information of Several Organisms

Organism	Number of Base Pairs	Number of Genes (Approximate)	Density of Genes (Approximate)	Number of Chromosomes
Human	3.2 billion	~25,000	1 gene per 100,000 bases	46
Mouse	2.6 billion	~25,000	1 gene per 100,000 bases	40
Fruit fly	137 million	13,000	1 gene per 9,000 bases	8
Corn	2.4 billion	~50,000	1 gene per 48,000 bases	10
Yeast	12.1 million	6,000	1 gene per 2,000 bases	32
Escherichia coli bacteria	4.6 million	3,200	1 gene per 1,400 bases	1

© Cengage Learning 2014

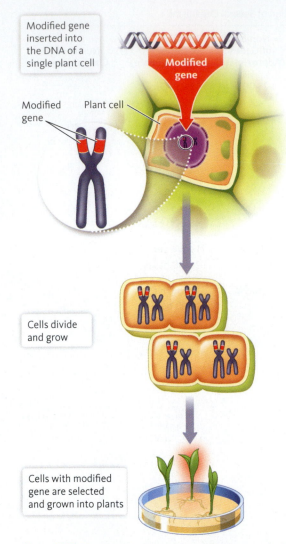

Modified gene inserted into the DNA of a single plant cell

Modified gene

Plant cell

Cells divide and grow

Cells with modified gene are selected and grown into plants

Figure 19.7 The introduction of modified genes into plants. This simplified illustration shows how genes are introduced into plants. The modified gene (red) is inserted into the plant's DNA, which is then inserted into a plant cell. As this cell divides, the modified DNA is passed on to new cells. Through culturing techniques, the plant with the modified genome is selected for and grown.
© Cengage Learning 2014

Bioethics—field of study that examines legal, social, and moral standards as they apply to biotechnology and medicine.

Genetically modified (GM) food—crops that are genetically altered with genes from another organism.

As biotechnology progresses, new safety and ethical issues arise

As you continue with this chapter, learning more about what biotechnology is and the amazing problems it can solve, it is important that you also consider the safety and ethics of this field. Along with providing new and powerful knowledge, the field of biotechnology brings with it a wide range of controversial issues. As new genetic technologies are developed, agencies, researchers, and the public openly discuss whether new research, techniques, treatments, and products meet ethical standards. Ethics defines a code based on professional, national, cultural, and religious beliefs that society honors. Scientists and bioengineers are trained to answer objective questions, but deciding whether something is right or wrong is a subjective question for society to decide. The field of **bioethics** examines legal, social, and moral standards as they apply to biotechnology and medicine. Many bioethics committees around the world work together with the public to determine procedures and policies for new biotechnological inventions.

In the United States, agencies that regulate food, medical, and environmental policies also regulate the biotechnology industry. These agencies include the U.S. Food and Drug Administration (FDA), the Environmental Protection Agency (EPA), and the U.S. Department of Agriculture (USDA). Depending on how the product may be used, one or more of these agencies will oversee the process. These agencies help to determine the safety and efficacy of new medicines, foods, and diagnostic tests.

Crops that are genetically altered with genes from another organism are referred to as **genetically modified (GM) foods**. Despite the federal and international regulation of GM foods, consumers are concerned about the long-term effect of these GM organisms from both human health and environmental perspectives. Inserting another organism's genes into a corn or tomato plant seems unnatural to consumers and causes them to pause and consider the product's safety. Another concern relates to allergic reactions to foods. Food allergies are caused by a reaction to certain proteins, and because GM organisms express proteins they do not normally express, there is a concern that these proteins in their new environment will cause allergic responses. As long as the gene inserted into the plant or animal did not originally cause allergic reactions, the modified organism should not either. Although it is understandable that consumers are concerned, scientific studies and government agencies do classify GM foods as safe.

Another concern with GM crops relates to the potential impact on the environment. Although GM crops greatly reduce the need for pesticide applications, herbicide use, and soil erosion, there are other environmental concerns. As with all organisms, natural selection favors organisms that can adapt to the current environment. As pests and weeds adapt to the GM crops, we will unintentionally be selecting for organisms that are most resistant to the modified genome. In order to reduce this effect, regulatory agencies establish guidelines. For instance, for GM corn, the EPA requires that farmers plant at least 20 percent of their crop using conventional seeds and up to 50 percent in areas where cotton is cultivated. Mandates such as this show the government's concern about GM crops causing harm to other crops or the environment. In initial studies, pests, weeds, and other plants in fields planted in this manner have developed little to no resistance to these new technologies.

Although the issues with GM crops focus on safety and practical concerns, other technologies, such as human genetic testing, also pose ethical questions. For example, is it right or wrong to perform genetic tests on a fetus? Under what circumstances should these tests be performed? Who should be able to gain access to the test results? If parents learn that their unborn child has an untreatable defect, what are the acceptable options? Now consider the use of human embryos in a laboratory. Under what circumstances is research using these embryos acceptable? How should embryonic stem cells be used? Similar questions can be asked of reproductive or therapeutic cloning. Although we have been manipulating nature by creating hybrid plants and animals for thousands of years, at some point we need to ask ourselves: have we gone too far in altering living things? As you progress through the rest of the chapter, keep these issues in mind as you learn more about the powerful field of biotechnology.

19.2 check + apply YOUR UNDERSTANDING

- Biotechnology is both a modern and an ancient science.

- Biotechnology combines engineering solutions and biological systems to produce food, medicine, and other useful products.

- Genomics, the study of genes, gene products, and gene interactions, is an important field that provides foundational information for the invention of biotechnology innovations.

- Along with powerful knowledge, the field of biotechnology brings with it controversial and ethical issues.

1. List several examples of the use of biotechnology in the food industry.
2. How does genomics enhance biotechnology?
3. What are some ethical concerns of genetic testing?
4. Life insurance companies make money when they receive payments for as long as possible before a person dies and then they have to pay the family. How will genetic testing for specific diseases affect this business? How might these genetic tests be a cost savings for the insurance company?
5. Bioremediation is the use of bacteria to clean up environmental wastes such as oil spills and heavy metals. How might biotechnology help provide a solution to this emerging waste disposal industry? Devise a series of steps that would solve the problem. (Hint: use the plastics example to devise your plan.)

19.3 Genetic Engineering Techniques Are Used to Modify Genes and Organisms

The nutritionally fortified corn crops, pest-resistant potatoes, and plants and bacteria used to produce plastic were the result of taking a gene from one organism and inserting that sequence into the DNA of another species. This process, called genetic engineering, is the genetic manipulation of an organism to create the *genetically modified organisms (GMOs)* we have introduced. You may hear GMOs referred to as being transgenic and the processes as *recombinant DNA technology* because genes are transferred from one organism to another (**Fig. 19.8**).

The genetic engineering of new products and therapies shows a nearly infinite amount of potential to improve the lives of humans. Many drugs manufactured today are proteins produced and purified from bacteria. Corn and switchgrass plants that produce plastic are genetically engineered with a gene from bacteria that naturally makes plastic. In addition to these ex-

Genetic engineering—the genetic manipulation of taking a gene from one organism and inserting that sequence into the DNA of another species.

Courtesy of Dennis Gonsalves, Cornell University

Courtesy of Dennis Gonsalves, Cornell University

Figure 19.8 Non-transgenic vs. transgenic tomato plants. The transgenic tomato plants on the top have been genetically modified to be virus resistant. Those on the bottom were not modified and have a reduced yield. While the plants look the same, the yield of the virus-resistant plants is much higher.

Vector—a carrier. In biotechnology, a carrier of genetic information to another organism (for example, a plasmid). In infectious diseases, a carrier of an infectious agent to another organism.

amples, genetic engineering can produce new characteristics or phenotypes of a variety of organisms. For instance, tomatoes, strawberries, pineapples, sweet peppers, and bananas have been modified with a gene that delays ripening, allowing the fruit to stay fresh longer. This change to the plant's phenotype enables farmers to harvest an entire crop simultaneously and allows for a longer transport time to market. Another valuable use of genetic engineering is gene therapy, where defective or missing genes that result in diseases are corrected through the addition of normal genes to potentially treat genetic disorders. At this point in this new scientific revolution, there seem to be endless applications for genetic engineering. In this section, we will introduce you to how organisms are genetically engineered to produce useful proteins, new plant phenotypes, and gene therapy products.

Genetic engineering produces drugs and gene therapies

Many living organisms produce compounds that have therapeutic value for us (**Fig. 19.9**). Genetic engineering is used to produce products in large, safe quantities—a benefit over traditional methods of pharmaceutical drug production. Before modern biotechnology techniques were developed to capitalize on these medicinal products, animals and plants were sacrificed in order to obtain small amounts of the drug. Today, genetic engineering provides cheaper and quicker ways to tap into natural diversity, and, as a result, scientists are investigating many plants and animals as sources of new medicines. **Table 19.2** lists several pharmaceutical products that currently are produced from genetically engineered organisms. Since the first genetically engineered drug was approved in 1982 (human insulin), more than 200 new therapies and vaccines have been approved, with several hundred more in clinical trials.

We'll start our exploration of genetic engineering with an overview of the process and gradually add details as we progress throughout this section. Genetic engineering requires three elements: the gene to be transferred, a host cell into which the gene is inserted, and a carrier called a **vector** to transfer the gene to the new organism. The first step involves gathering genetic information, such as identifying specific genes and their protein products. For instance, the gene that makes insulin, which controls blood sugar, has been identified. The second step involves isolating this gene segment. The third step inserts the gene into a vector that will physically carry the gene into a new host cell. Because there are so many different applications of genetic engineering, there are many variations in vectors and host cells.

a.

Le Do/Shutterstock.com

b.

Rusty Dodson/Shutterstock.com

Figure 19.9 Genetic engineering produces drugs. (a) The anticoagulant warfarin was originally identified in the sweet woodruff plant. (b) The saliva of the Gila monster lizard contains a protein that helps to sustain blood glucose levels in diabetics.

Sources of common vectors include bacteria, viruses, and yeasts.

Plasmids, from bacteria, are often used as a vector to transfer genes from one organism to another because they have two very important characteristics. A *plasmid* is a double-stranded circular DNA molecule that is separate from the bacterial chromosome (**Fig. 19.10**). Plasmids can pass readily from one cell to another, even when the cells are from different species far apart on the evolutionary tree. In other words, plasmids can transfer genes between humans and bacteria and vice versa. Thus, a human gene can be inserted into a plasmid, and the plasmid can then be introduced into a host (such as bacteria) for a biotechnological application. Plasmids have a unique characteristic in that they can replicate themselves. Often vectors are inserted into bacteria because bacteria reproduce rapidly. One bacterium carrying a plasmid with a human gene can produce millions of copies of the gene of interest in a short period in this process known as *gene cloning.* Let's examine how genetic engineering has improved the production of human insulin over the years.

In type 1 diabetes, the immune system attacks and destroys the cells of the pancreas that make insulin. This information was not known before the 1920s, and people who developed type 1 diabetes died within a year. In the 1920s, more was learned about the disease, and insulin was extracted and purified from fetal calf pancreas glands. However, this process produced only small amounts of purified hormone. Later, pancreas glands from swine and cattle slaughtered for food were used, but these processes took about 8,000 pounds of animal pancreas glands to produce one pound of insulin. In the 1950s and 1960s, the insulin protein was characterized, and in the early 1980s, genetic engineering techniques using a vector (plasmid) allowed the insulin gene to be introduced into a strain of *E. coli* bacteria commonly found in the human intestine. Today, genetic engineering produces abundant quantities and, more importantly, produces chemically identical human insulin rather than a swine- or cattle-based hormone that is less effective.

How is genetically engineered insulin made? Following the three steps shown in **Figure 19.11**, you can see the healthy gene that makes human insulin is isolated from pancreatic cells (step 1) and inserted into a plasmid vector (step 2). The vector is then introduced into *E. coli* bacteria, which naturally live in the human gut. The host bacteria take up the vector and are cultured in the lab, where they grow and reproduce efficiently. Inside each bacterium, the human insulin protein is made (step 3). The bacteria are then harvested, and the insulin is purified and processed to be used by the diabetic. Although many genetic engineering techniques use bacteria or yeast as a "living factory" to produce the gene product of interest, these microbes do not have the mechanisms needed to make many complex proteins properly folded and assembled in the correct format for eukaryotes. In these situations, animals or cells cultured in a lab are used.

The second application of genetic engineering that we will discuss is gene therapy. More than 4,000 known diseases, such as cystic fibrosis, severe com-

TABLE 19.2 Genetically Engineered Products.

Product	Use	Host organism
Insulin	Human hormone used to treat diabetes	Bacteria/yeast
Human growth hormone	Human growth hormone used to treat dwarfism	Bacteria
Bovine growth hormone	Bovine growth hormone used to increase milk yield of cows	Bacteria
Antithrombin	Anti–blood clotting agent used in surgery	Goats
Penicillin	Antibiotic used to kill bacteria	Fungi/bacteria
Vaccines	Hepatitis B antigen for vaccination	Yeast
Tissue plasminogen activator	Enzyme used to dissolve blood clots and administer for heart attacks	Hamster cell cultures
Interleukin-2	Interleukin-2 used to treat some cancers and immune deficiencies	Bacteria

© Cengage Learning 2014

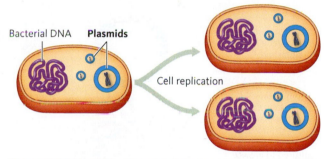

Figure 19.10 Plasmids. Plasmids from bacteria are often used as vectors because they can transfer genes between species and replicate quickly.
© Cengage Learning 2014

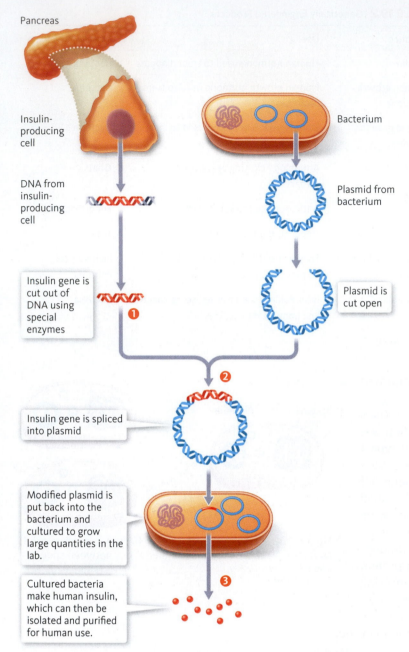

Pancreas

Insulin-producing cell

DNA from insulin-producing cell

Insulin gene is cut out of DNA using special enzymes ❶

Insulin gene is spliced into plasmid ❷

Modified plasmid is put back into the bacterium and cultured to grow large quantities in the lab.

Cultured bacteria make human insulin, which can then be isolated and purified for human use. ❸

Bacterium

Plasmid from bacterium

Plasmid is cut open

Figure 19.11 Genetic engineering of insulin. This process shows how the human insulin gene is inserted into a plasmid and placed in a bacterium for the manufacture of insulin. A healthy insulin gene is isolated from pancreatic cells ❶ and inserted into a plasmid vector ❷. The vector is then introduced into *E. coli* bacteria, which take up the vector. These bacteria are cultured and grown in the lab. Inside each bacterium, the human insulin protein is made ❸. This protein can be purified and then delivered pharmaceutically to a diabetic patient.
© Cengage Learning 2014

Gene therapy—a technique that inserts a normal gene into the genome, replacing the defective gene that causes disease.

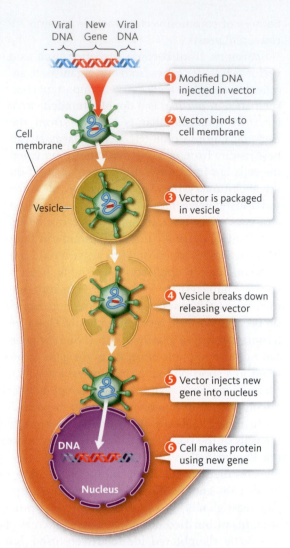

Viral DNA New Gene Viral DNA

❶ Modified DNA injected in vector

❷ Vector binds to cell membrane

Cell membrane

Vesicle

❸ Vector is packaged in vesicle

❹ Vesicle breaks down releasing vector

❺ Vector injects new gene into nucleus

DNA

Nucleus

❻ Cell makes protein using new gene

Figure 19.12 Viral vectors are used to deliver modified genes in gene therapy for diseases caused by single-gene defects. Viral vectors are modified by removal of the harmful viral genes and the insertion of the therapeutic gene (step 1). The virus then enters the human cell (steps 2–3). The cell naturally defends itself and digests the vesicle used to engulf the virus (step 4). The exposed virus survives and uses its molecular machinery to infect the nucleus (step 5). Now that the therapeutic gene is delivered to the patient's cells, they can produce the protein of interest (step 6).
© Cengage Learning 2014

bined immunodeficiency (SCID), multiple sclerosis, muscular dystrophy, and some forms of blindness, are caused by a defect in a single gene. Single-gene defects that cause a defective protein or enzyme are candidates for **gene therapy**, which is a technique that inserts a normal gene into the genome, replacing the defective gene that causes disease. This new gene produces a working enzyme or protein that replaces or inactivates the faulty one.

The most common technique for gene therapy delivers a normal gene into the genome of the target organism using a modified virus as the vector (**Fig. 19.12**). This viral vector is modified by removal of its harmful viral genes and is used to deliver the therapeutic gene to the patient's cells. Viruses are chosen as a vector because they easily attach to human cells and insert their genome inside. As the newly infected cell replicates, the newly altered genome with the therapeutic gene is produced in each new generation of cells.

Because gene therapy is still in an experimental stage of development, it is used only on diseases with no other cure. SCID is a rare disease that leaves newborns with little or no immune system to defend against foreign invaders such as bacteria, viruses, and fungi. The primary cells of the immune system develop in the bone marrow and later mature into specialized cells called B-lymphocytes and T-lymphocytes. These two cell types are widespread throughout the body and are ready to identify and destroy foreign invaders to protect the body from infection. A deficiency in both T- and B-lymphocyte production and function results in a weakened immune system. Children with SCID often get many infections and as a result fail to grow and develop properly. If untreated, children born with SCID who are not isolated in germ-free plastic bubbles die in their first year (**Fig. 19.13**).

Gene therapy has been used to successfully treat children with this disorder. **Figure 19.14** shows the steps in the process, in which a GM virus is used to deliver a healthy gene into bone marrow cells taken from the child (steps 1–3). The virus uses its replication mechanisms to incorporate the gene into the bone marrow cells (steps 4 and 5). These transformed cells are then placed back into the patient's bloodstream to produce a functional gene product (steps 6 and 7) (see **Fig. 19.14**). Because these bone marrow cells reproduce, each new generation of cells is programmed to make the missing gene product, and the child exhibits normal immune function.

As you can imagine, because gene therapy is a relatively new medical technology, it comes with risks and ethical concerns. Because it is difficult to evaluate for long-term effects, the true test comes when patients are monitored after they have received gene therapy treatments. Gene therapy has

LAURENT/BSIP/age fotostock

Figure 19.13 SCID. Children with SCID are isolated from a world where harmless germs could kill them. Within a minute of being born, the baby is placed in a plastic bubble that will protect him or her from getting infections. If untreated and not kept in plastic bubbles, children with SCID die in their first year.

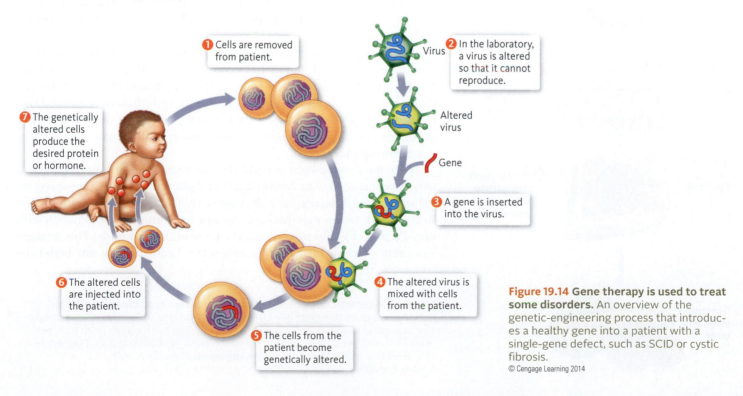

1 Cells are removed from patient.

Virus

2 In the laboratory, a virus is altered so that it cannot reproduce.

Altered virus

Gene

3 A gene is inserted into the virus.

4 The altered virus is mixed with cells from the patient.

5 The cells from the patient become genetically altered.

6 The altered cells are injected into the patient.

7 The genetically altered cells produce the desired protein or hormone.

Figure 19.14 Gene therapy is used to treat some disorders. An overview of the genetic-engineering process that introduces a healthy gene into a patient with a single-gene defect, such as SCID or cystic fibrosis.
© Cengage Learning 2014

D.D. Hardee/USDA

Figure 19.15 GM vs. non-GM cotton. Results of insect infestation on Bt (left) and non-Bt (right) cotton bolls.

been successful in reversing SCID, blindness, and some other conditions, but in some cases, patients have developed specific forms of cancer, and gene therapy treatments for these diseases are under additional scrutiny. The complications in gene therapy are associated with the long-term effects of the virus that is used as a vector. As our knowledge increases, gene therapy delivery systems and long-term survival will improve.

In addition to questions regarding the safety and ethics of human clinical trials using these techniques, other ethical questions arise. Who should pay for these experimental treatments: insurance companies or the family? Who has access to these treatments? Which diseases should be targeted? Will people abuse this technology to fashion individuals who meet certain cultural or intellectual standards? Again, as technologies emerge into new territories, more questions are generated than are answered.

Genetic engineering is used to produce new characteristics for crop production

Just as bacteria are manipulated to produce medicines, agricultural crops can be modified to take on new characteristics related to nutrition and farm productivity. Most frequently, crops are modified for either pest or herbicide resistance, which increases yield or efficiency on the farm. Organic farmers have sprayed their crops with the soil bacterium *Bacillus thuringiensis* (Bt) for more than 50 years because it makes a natural insecticide that is effective against specific crop-eating insects. "Bt crops" are some of the most popular food crops planted around the world for their resistance to certain insect pests (**Fig. 19.15**).

With the capability to sequence, study, and identify genes, scientists were able to isolate the Bt-specific gene that produces the protein that acts as a pesticide on insect larvae (**Fig. 19.16**). Using a plasmid vector, this gene has been successfully inserted into soybeans, corn, potatoes, and cotton so that these plants can produce the insect toxin. With the plant "naturally" producing the toxin, the need to spray for certain pests is reduced. These crops save the farmer money and reduce the farmer's exposure to potentially toxic chemicals.

Another group of important GM crops is modified to be resistant to herbicides. The best-selling herbicide contains glyphosate, which inhibits important metabolic pathways in many plants. Scientists identified a gene that is resistant to this molecule and ultimately created GM glyphosate-resistant soybeans, corn, sugar beets, and canola. With the herbicide resistance, farmers can spray fields to eliminate weeds without harming their crops.

With the development of pesticide- and herbicide-resistant crops, scientists are concerned with nearby insect and plant populations becoming resistant, just as humans have developed resistance to certain antibiotics. Regulatory agencies mandate that Bt crops be planted in combination with non–Bt crops to help prevent insecticide resistance in pests. This strategy has been effective in managing resistance. Unfortunately, for herbicide-

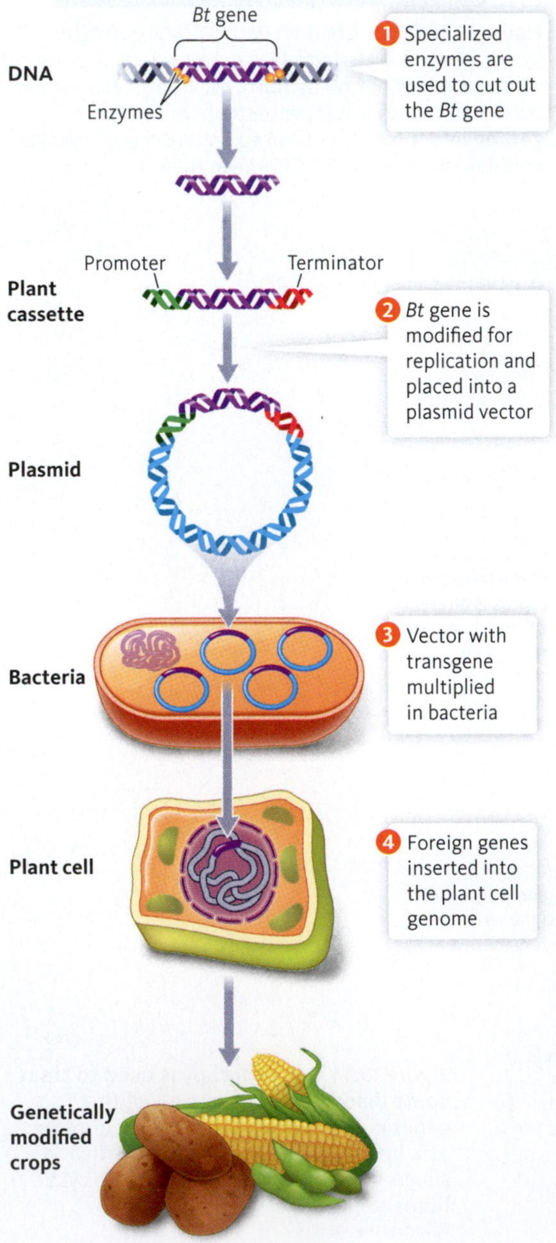

Bt gene

DNA
Enzymes

1 Specialized enzymes are used to cut out the *Bt* gene

Promoter   Terminator

Plant cassette

2 *Bt* gene is modified for replication and placed into a plasmid vector

Plasmid

Bacteria

3 Vector with transgene multiplied in bacteria

Plant cell

4 Foreign genes inserted into the plant cell genome

Genetically modified crops

Figure 19.16 Genetic engineering of crops for insect resistance. Specialized enzymes are used to isolate and insert the Bt gene into a plasmid (steps 1 and 2). The plasmid is introduced into bacteria (step 3). The bacteria are used to infect plant cells, and the DNA is transferred to the plant genome (step 4). The genetically modified plants grow and exhibit the insect-resistance trait.
© Cengage Learning 2014

resistant crops, the story isn't as hopeful. For instance, if a certain population of weeds has been exposed repeatedly to one particular herbicide, the surviving weeds will have adapted a resistance to the chemical. When these weeds reproduce, they pass their resistant genes on to the next generation, creating "super weeds" that are even more difficult to kill. This dilemma is one factor that makes the public, the government, and the biotechnology industry apprehensive about some of these new technologies.

19.3 check + apply YOUR UNDERSTANDING

- The genetic engineering of new products shows incredible potential to improve the lives of humans by producing cheaper, more effective products in a relatively quick manner.

- Genetic engineering requires three elements: the gene to be transferred, a host cell into which the gene is inserted, and a vector to transfer the gene to the new organism.

- Gene therapy is a technique that inserts a normal gene into the genome, replacing a defective gene that causes disease.

- Although their safety and efficacy are often debated, many crops are genetically modified for either pest or herbicide resistance to increase yield or efficiency on the farm.

1. Genetically modified organisms are often referred to as transgenic. Explain what transgenic means.

2. It is easier to genetically engineer microbes, but sometimes animals must be used to produce the product of interest. Why is this?

3. What is a plasmid, and how is it used in genetic engineering?

4. List an advantage and a disadvantage to using human viruses as vectors in gene therapy.

5. By the end of the 21st century, some studies estimate that global food production will need to double. Among cereal crops, rice is unique in that it is the only plant naturally resistant to fungal rusts that sometimes wipe out wheat, maize, barley, and oats. Describe the process scientists might use to genetically engineer these cereal grains to improve crop productivity.

19.4  Biotechnology Is Changing the Face of Health Care

Biotechnology is being used to find cures and improve treatments for diseases such as diabetes. About 8 percent of people in the United States have diabetes. Overall, the risk of death among people with diabetes is two times higher than it is for people without the disease. In the United States, diabetes is one of the top 10 causes of death. There are three main kinds of diabetes: type 1, type 2, and gestational. Although each form of the disease has certain characteristics, diabetes is always marked by high levels of blood glucose caused either by a lack of insulin production or by the cells' inability to use insulin properly to break down glucose. Diabetes also leads to high incidences of heart disease, elevated blood pressure, stroke, kidney disease, blindness, and nervous system diseases.

Metabolic disorders such as diabetes hinder the way the body uses and stores digested food. Most of the food people eat is eventually broken down into glucose, the main source of fuel for the body. Specialized cells in the pancreas release insulin into the bloodstream. The insulin then helps glucose enter liver, fat, and muscle cells, where it can be stored or

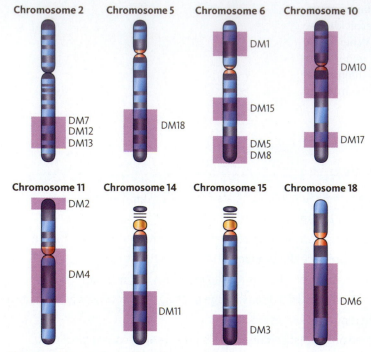

Chromosome 2
DM7
DM12
DM13

Chromosome 5
DM18

Chromosome 6
DM1
DM15
DM5
DM8

Chromosome 10
DM10
DM17

Chromosome 11
DM2
DM4

Chromosome 14
DM11

Chromosome 15
DM3

Chromosome 18
DM6

Figure 19.17 Genetic markers for type-1 diabetes.
Genetic markers are DNA sequences that indicate an increased risk of developing a specific disease or disorder. On the chromosomes shown, genetic markers are highlighted in purple and numbered as they were discovered (DM1 = diabetes marker 1, DM2 = diabetes marker 2, etc.)
© Cengage Learning 2014

Genetic markers—DNA sequences that indicate an increased risk of developing a specific disease or disorder. Genetic markers have been identified for many human diseases, including colon, prostate, and breast cancer, heart disease, Alzheimer disease, type 1 diabetes, and Parkinson disease.

used for cellular energy. When people eat food, a healthy pancreas automatically produces and releases the proper amount of insulin to move glucose from blood into cells. In people with diabetes, glucose builds up in the blood, overflows into the urine, and passes out of the body. Thus, the body loses its main source of fuel and switches to using fat for energy, even though the blood contains large amounts of glucose. As a result, people with diabetes have an increased appetite, thirst, and urination; they also suffer from blurry vision, and extreme fatigue. Current diabetes research focuses on the use of biotechnology to find a cure for this disease. In this section, we will learn more about how biotechnology is changing the face of health care in treating diseases such as diabetes.

The application of genomics enhances health care

Today, genomics plays a large role in the diagnosis and treatment of diseases. Genomic studies have provided much of the current information regarding diabetes. In type 1 diabetes, the pancreas does not produce insulin. In type 2 diabetes, the body's cells are resistant to insulin. Type 2 diabetes typically is associated with age, obesity, and inactivity. Although the mechanisms are still not fully understood, scientists believe type 1 diabetes is caused by genetic, environmental, and immune factors, whereas type 2 diabetes is caused by genetic and environmental factors. Because genetics is associated with both types of diabetes, understanding more about the interactions of the specific genes that trigger insulin production, insulin resistance, or the death of insulin-producing cells will help medical researchers predict, treat, and cure diabetes.

One tool researchers use is **genetic markers**, which usually are DNA sequences that indicate an increased risk of developing a specific disease or disorder. Genetic markers have been identified for many human diseases, including colon, prostate, and breast cancer, heart disease, Alzheimer disease, and Parkinson disease. Genetic markers have also been identified for type 1 diabetes, and it is now possible to screen relatives of people with type 1 diabetes to determine whether those relatives also are at risk. **Figure 19.17** shows more than a dozen genetic markers that play a role in the susceptibility to type 1 diabetes. From this diagram you can see the complexity of this disease. Fifteen markers (genes) on fourteen chromosomes influence this disease. Once people are identified as high risk, they can benefit from early medical intervention. These genetic markers will be especially helpful for diseases such as diabetes, cancer, and heart disease, for which lifestyle modifications can prevent or delay the onset of disease.

A genetic profile is a collection of information about an individual's genes and includes a person's genetic markers. Today many tools are available for examining a patient's genetic profile, family history, and environmental exposures to assess a disease. Genetic profiles are also being used to assess the response to drugs. For instance, one of the most widely prescribed medications to prevent blood clots (warfarin) has several known adverse side effects. The use of genetic markers in the genes that metabolize this medication allow for more accurate dosing that takes into account the age, gender, weight, and genetic profile of the patient. This informa-

IMAGEMORE/age fotostock

Figure 19.18 Using genomics to advance drug development. The *Artemisia annua* plant is one of the many plants that contain useful chemicals that can be harvested for medicines. This plant produces a drug used to treat malaria. Scientists have identified the genes that increase the amount of the chemical and the genes that increase the size of the plant to improve access to this drug.

tion not only is being used to help treat patients; it also can be used to design more effective drugs.

Not only is genomics being used to access a patient's response to a particular treatment; it is also being used to advance drug development. The more scientists understand about genes and their interactions in the body, the more quickly drugs can be developed and improved. For example, one of the most effective treatments of malaria uses a chemical obtained directly from the plant *Artemisia annua* (**Fig. 19.18**). Because only about 1 percent of the plant can be converted into the drug, many plants must be grown. Scientists have identified the genes that increase the amount of the chemical and the genes that increase the size of the plant. With these discoveries, scientists can boost production of this key drug used to treat malaria. Also with this information, scientists will soon be able to genetically engineer yeast, bacteria, or other cultures to synthesize the chemical in larger quantities. By studying how various cells respond to this drug, scientists believe it has potential for treating other infectious diseases and some forms of cancer.

Stem cells offer new potential for studying and treating disease

Particular cells in the body called **stem cells** are important for living organisms. Stem cells have the remarkable potential to develop into many different cell types in the body during early life and growth (**Fig. 19.19**). Recall from Chapter 7, stem cells are different from other cells in two important characteristics. First, they are unspecialized cells capable of renewing themselves through cell division. Second, under the proper conditions, they can be coaxed to become tissue- or organ-specific cells with specialized functions. In some organs, such as the intestine and bone marrow, stem cells regularly divide to repair and replace worn out or damaged tissues. When a stem cell divides, each new cell has the potential to either remain a stem cell or become another type of cell with a more specialized function, such as a muscle, brain, or red blood cell. In other organs, however, such as the pancreas and the heart, stem cells only divide under special conditions.

The two main sources for stem cells are embryonic stem cells and adult stem cells. Within the first week of development following fertilization, the embryo consists of a few hundred cells contained in a ball-shaped structure. Two hundred **embryonic stem cells** line this hollow mass and develop into all of the various types of cells in the body, including all of the many specialized cell types and organs such as the heart, lung, skin, and other tissues (see **Fig. 19.19**). In some adult tissues, such as bone marrow, muscle, and brain, discrete populations of **adult stem cells** continue to generate replacements for cells that are lost through normal wear and tear, injury, or disease. Compared to embryonic stem cells, adult stem cells are limited in the types of cells they can become. Embryonic stem cells have more versatility, with the potential to develop into almost every cell type in the body, so they are more valuable for research and therapeutic use.

Using stem cell technology, scientists study human development, diseases, and disease treatments. They also generate new tissues to replace

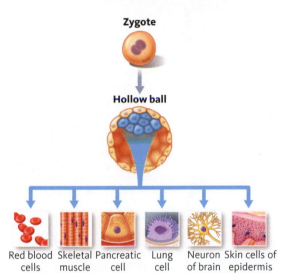

Zygote

Hollow ball

Red blood cells | Skeletal muscle | Pancreatic cell | Lung cell | Neuron of brain | Skin cells of epidermis

Figure 19.19 Embryonic stem cells. Stem cells give rise to all of the various types of cells in the body. Embryonic stem cells are taken from a one-week-old embryo and, because of their potential to develop into many different cells, tissues, and organs, they are more valuable for research and therapeutic use.
© Cengage Learning 2014

Stem cells—unspecialized cells that have the potential to develop into many different cell types and tissues in the body during early life and growth.

Embryonic stem cells—undifferentiated cells derived from the inner cell mass during the first week of embryonic development; develop into all of the various types of cells and organs such as the heart, lung, skin, sperm, eggs, and other tissues.

Adult stem cells—undifferentiated cells derived from nonembryonic cells. Through cell division, these cells continue to generate replacements for cells that are lost through normal wear and tear, injury, or disease.

damaged or lost tissues. The hope is that one day stem cells will reverse diseases such as diabetes, Parkinson disease, and even heart disease. Specifically for diabetes, if scientists can trigger stem cells to become insulin-producing cells in the pancreas, then those cells could be introduced into someone with type 1 diabetes to ultimately cure their disease. Alternatively, if someone has tissue damage from heart disease or a stroke, stem cells could regenerate and replace the damaged tissue; the same could be true for people with spinal cord injuries or degenerative diseases such as Parkinson disease and Alzheimer disease.

One advantage of stem cell therapy is the elimination of potential rejection responses common in nearly all organ transplants, even between the same species. Because the human immune system is designed to attack cells it does not recognize as its own, white blood cells try to destroy foreign cells that do not show the correct cell surface marker proteins. **Therapeutic cloning** is a technique that produces stem cells genetically matched to the patient by making embryonic stem cells from their own adult tissues. As the name implies, this type of cloning is for therapeutic purposes. It is aimed at generating cloned embryos, which provide embryonic stem cells for the repair of damaged or defective tissues. Unlike what you see in the movies, this process does not make an identical clone of a human.

Figure 19.20 shows an experimental approach using therapeutic cloning to cure type 1 diabetes. The goal of this treatment is to generate functional insulin-producing cells in the pancreas. First, medical researchers isolate the diabetic patient's DNA from a sample of skin cells. Next, they inject this DNA into a donor egg cell that has been emptied of its own genetic contents. The embryo grows in a culture containing the nutrients necessary to promote cell division. After about a week, researchers harvest the embryonic stem cells from the lining of the mass and use growth factors to coax them into developing into insulin-producing cells. Finally, doctors inject millions of insulin-producing cells back into the patient to reverse the effects of diabetes. Researchers have developed therapeutic cloning models using skin

Therapeutic cloning—a technique that produces stem cells genetically matched to the patient by making embryonic stem cells from his or her own adult tissues; aims to generate cloned embryos, which provide embryonic stem cells for repair of damaged or defective tissues.

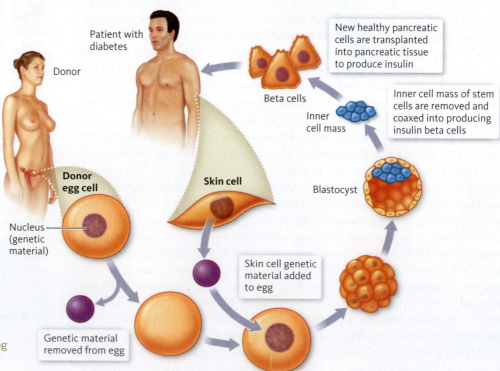

Figure 19.20 Therapeutic cloning. In this experimental method, adult stem cells are used to create healthy insulin-producing cells to treat diabetes. Researchers have developed therapeutic cloning models using skin cells for more than 20 diseases.
© Cengage Learning 2014

Patient with diabetes

Donor

New healthy pancreatic cells are transplanted into pancreatic tissue to produce insulin

Beta cells

Inner cell mass

Inner cell mass of stem cells are removed and coaxed into producing insulin beta cells

Donor egg cell

Skin cell

Blastocyst

Nucleus (genetic material)

Skin cell genetic material added to egg

Genetic material removed from egg

cells for more than 20 diseases, including Parkinson disease, Down syndrome, SCID, Huntington disease, and muscular dystrophy.

Research on stem cells continues to advance our knowledge about how an organism develops from a single cell and how healthy cells can replace damaged cells in adult organisms. Stem cell research is one of the most fascinating areas of present-day biology, but this research also raises scientific and ethical questions. The most highly debated issue with stem cells is the source of embryonic stem cells, which arise from fertilized eggs remaining from *in vitro* procedures (manual fertilization of a sperm and an egg in a laboratory dish). Although patients donate these embryos for research with informed consent, some people argue that because the embryos represent a potential life, destroying them for research is wrong. Innovative research includes new techniques that genetically reprogram some specialized adult cells, such as those found in the skin, to have characteristics of embryonic stem cells, providing hopes that these cells will be as powerful as embryonic stem cells.

19.4 check + apply YOUR UNDERSTANDING

- Biotechnology and genomics play a large role in finding cures, improving treatments, and advancing drug development for diseases. Genetic markers, stem cells, and therapeutic cloning are some of the important tools for studying and treating disease.

- Genetic markers indicate an increased risk of developing a specific disease or disorder.

- Stem cells are unspecialized cells capable of renewing themselves through cell division, and, under the proper conditions, they can be induced to become specific cells with specialized functions. Using stem cell technology, scientists study human development, diseases, and disease treatments as well as generate new tissues to replace damaged or lost tissues.

- Therapeutic cloning generates cloned embryos, which can then provide embryonic stem cells for the repair of damaged or defective tissues.

- Research on embryonic stem cells raises ethical concerns, posing the question of when life begins.

1. What are the two characteristics that make stem cells unique?
2. Describe how genetic markers and genetic profiles are being used to personalize medicine (assign individual treatment).
3. Genetic markers are an important research and diagnostic tool in the biotech toolbox. In your own words, explain how genetic markers are useful in the diagnosis of disease.
4. Although the technology to clone a pet exists, present a biological argument (genetic and behavioral) as to why this technology cannot replace a pet.
5. Some muscular dystrophies are inherited diseases that involve muscle weakness and degeneration of muscle tissue over time. Explain how the process of therapeutic cloning might be used to cure muscular dystrophy.

19.5 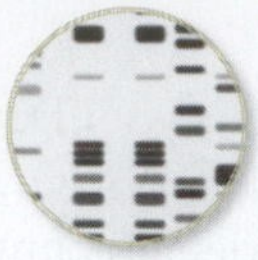 DNA Profiling Is Used to Establish Genetic Relationships

Biotechnology is important in establishing genetic relationships, not just from a healthcare perspective but also from evolutionary, forensic, and conservation biology points of views. For instance, tools from biotechnology are being used to combat elephant poaching in Africa. Ivory, illegally taken from elephant tusks, serves as the raw material for trinkets, gun handles, and jewelry. Although elephant hunting is highly regulated around the world, it may surprise you that this business continues to flourish (**Fig. 19.21**). In 2008, conservation biologists estimated the African elephant population at 470,000, with a death rate from illegal hunting of 8 percent throughout Africa. With mortality rates this high, large herds of elephants could be extinct by 2020. Efforts in wildlife crime investigations and increasing public awareness of the elephants' plight are critical to conserving this animal species.

Until recently, ivory poaching was a high-profit, low-risk endeavor. Elephants live in remote areas with few game wardens to protect them, making the illegal killing of these animals easy and nearly risk-free. Biotechnology is being used to change this tragic situation. Wildlife biologists use DNA to establish a profile to trace the ivory back to the elephant population it came from. This genetic or **DNA profile** determines the genotype (gene makeup) of an organism at several highly variable sites in the genome and can be gathered from dung, saliva, blood, fur and hair, skin, and other tissues. After law enforcers seize a shipment of ivory, scientists grind up a piece of the tusk to extract and analyze the DNA. They compare the DNA sequence from the tusk to DNA gathered from the dung of wild elephants. Different populations of elephants have slight variations in their DNA, and using this information, scientists are able to pinpoint the population the poached material came from. Because poachers target large elephant herds, DNA sampling helps to focus law enforcement on areas with significant poaching activity. It also raises a country's responsibility to prevent poaching within its borders. In the following section, we will explore how the genetic profile is used to identify an individual by using his or her unique DNA makeup and how it has dramatically improved wildlife management, criminal investigation, and forensic science.

DNA profile—a genetic fingerprint established by analyzing several highly variable sites in the genome.

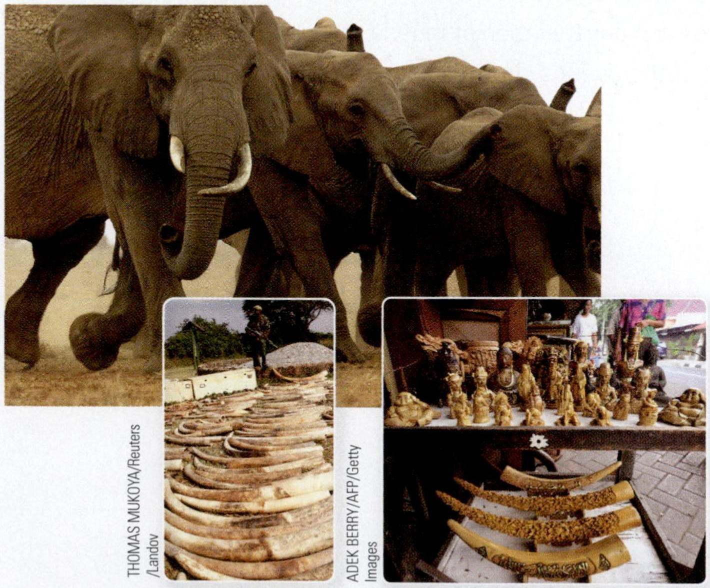

Steffen Foerster Photography/Shutterstock.com

THOMAS MUKOYA/Reuters /Landov

ADEK BERRY/AFP/Getty Images

Figure 19.21 Elephant poaching for ivory. Elephants are poached for their ivory tusks, which are sold on the black market for trinkets, jewelry, and other products. DNA technology is now being used to help catch poachers by connecting ivory sold in markets to specific animals killed in protected reserves.

Wildlife managers use information on genetic diversity in conservation efforts

As you recall from Chapter 17, conservation biologists are concerned with preserving species, biodiversity, and ecosystems. One measure of biodiversity is the inherent genetic diversity at both the species and population levels. Recall that the greater the genetic variation of a population, the more successfully it is able to adapt to environmental changes to survive. Decreased genetic variation often leads to problems from inbreeding and reduced population fitness. Genomics, genetic markers, and genetic profiling are used to measure genetic diversity and to determine strategies for conservation efforts. In populations where there is little genetic variation but other populations exist, wildlife managers introduce new genetic stock to help increase genetic variation and the fitness of that group.

One example of using DNA profiling for conservation efforts is the plan for protecting the endangered Florida panther (**Fig. 19.22**). These panthers require about 200 square miles of territory for reproductive purposes, and at one time the Florida panther could be found throughout the southeast United States. Now, however, inbreeding, habitat loss, and severe habitat fragmentation threaten the species. At this point, fewer than 100 individuals remain in the southern tip of Florida. With such a small population, researchers devised a plan to introduce several panthers from Texas, where population numbers are good, into the Florida panther population. From a genetic and public point of view, mixing these two subspecies is controversial, but conservation biologists believed it was the only way to save the Florida species from extinction. Since this genetic introduction, the wild Florida panther population has increased from 30 to 80 individuals over a period of about 10 years.

Conservation efforts also include genetic analysis from preserved specimens. Scientists use genetic information obtained from fossils and other preserved specimens to understand how a species has adapted over time to various environmental changes. The greater prairie chicken population has declined dramatically in Illinois (**Fig. 19.23**). Studies of museum specimens indicate that certain genes helpful in reproducing are missing in this particular population today. Scientists hope that the reintroduction of these genes from other populations in Kansas, Nebraska, and Minnesota will help the population in Illinois rebound. So far, follow-up genetic studies indicate the genetic diversity of the population is improving, but habitat loss is still a factor that limits the success of this population.

The illegal trade of poached ivory across international borders highlights another challenge for conservation biologists. Wildlife crimes—for example, animal poaching, trafficking of endangered species, the illegal sale of wild meat, and the use of endangered species in products—need to be managed between countries with competing interests. Molecular methods for detecting such wildlife crimes are proving useful in deterring these activities. For example, law enforcers used DNA analysis in a high-visibility international case to detect the unlawful importation and use of whale meat in Japan for sushi. Another case in the United States involved an arrest based on DNA evidence comparing remains left on Clint Eastwood's property to the taxidermy mount of a trophy-quality large mule deer. The court fined the poacher several thousand dollars and suspended his hunting license for three years. This case was used to increase awareness and raise funds to invest in a new crime lab that has since used DNA evidence to break elk-poaching rings, round up sturgeon caviar collectors on the Sacramento River, and investigate the mysterious drowning of several hundred immature terns in California harbors.

Forensic scientists use DNA profiles as courtroom evidence

Just as DNA profiles are used to detect illegal activities involving wildlife, they are also used as evidence at other crime scenes. **Forensic science** uses scientific techniques to analyze crime scenes, identify accident vic-

Figure 19.22 Florida panther. DNA profiling is used to monitor and improve the genetics of populations of this endangered panther.

Figure 19.23 Greater prairie chickens. Female (left) and male (right) greater prairie chicken museum specimens were used for DNA profiling to compare current population genetics with older populations.

Forensic science—field that uses scientific techniques to analyze crime scenes, identify accident victims, and establish paternity in child support cases.

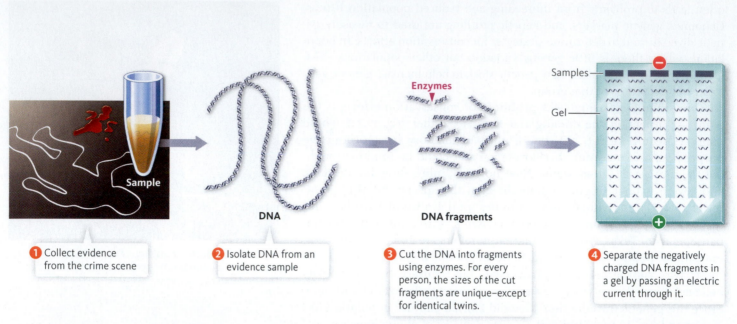

Figure 19.24 Process of analyzing forensic DNA evidence. After forensic evidence is collected, DNA is isolated and cut into smaller fragments so that each fragment can be separated and analyzed with 13 different probes. The evidence at this crime scene matches the DNA from Suspect #2 because the DNA matches at 3 bands. The probability of its being someone else in the general population is incredibly low.
© Cengage Learning 2014

1 Collect evidence from the crime scene

2 Isolate DNA from an evidence sample

3 Cut the DNA into fragments using enzymes. For every person, the sizes of the cut fragments are unique—except for identical twins.

4 Separate the negatively charged DNA fragments in a gel by passing an electric current through it.

tims, and establish paternity in child support cases. As with criminal evidence, scientists compare the DNA profile of the suspect to the DNA evidence left at the crime scene, often in the form of hair, blood, semen, saliva, or skin. DNA profiles have also been used to establish the innocence of individuals convicted of crimes they did not commit.

The first step is to collect DNA evidence in the form of cells from a crime scene. Because a DNA profile is the same for every cell, tissue, and organ of a person, any type of cell can be used. In addition, because only a few cells are required, it is increasingly difficult for someone to leave a crime scene without leaving a piece of incriminating biological evidence behind. DNA even can be found on physical evidence that is decades old. However, several factors can affect DNA, including heat, sunlight, moisture, bacteria, chemicals, and mold. Therefore, not all DNA evidence will result in a usable DNA profile. Furthermore, just like standard fingerprints, DNA testing cannot determine when the suspect was at the crime scene or for how long.

Figure 19.24 shows the steps in creating a DNA profile. Scientists use different sampling and analysis techniques based on the type of sample collected. Sometimes the DNA evidence is so small that scientists mimic the DNA replication process to duplicate (or amplify) the sample many times in order to have enough sample to analyze. In the case shown in **Figure 19.24**, the DNA is isolated and cut into smaller fragments so that each fragment can be separated and analyzed. Specialized enzymes are used to cut the DNA at specific sequences so that the fragment profiles can be created and compared.

To identify individuals based on their DNA profile, forensic scientists scan 13 highly specific DNA loci (or regions) on various chromosomes, which are known to vary from person to person (**Figs. 19.24** and **19.25**).

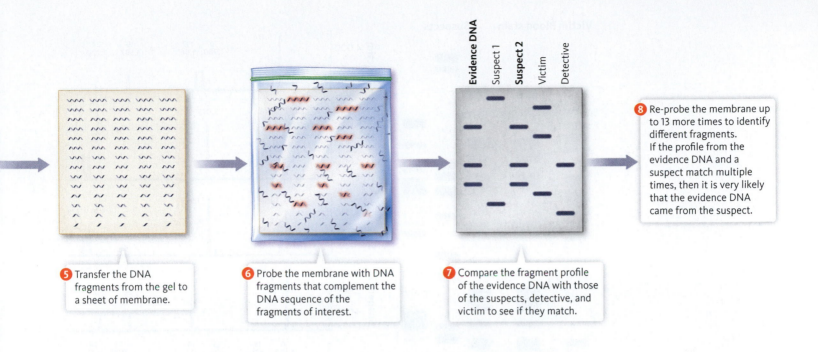

5 Transfer the DNA fragments from the gel to a sheet of membrane.

6 Probe the membrane with DNA fragments that complement the DNA sequence of the fragments of interest.

7 Compare the fragment profile of the evidence DNA with those of the suspects, detective, and victim to see if they match.

8 Re-probe the membrane up to 13 more times to identify different fragments. If the profile from the evidence DNA and a suspect match multiple times, then it is very likely that the evidence DNA came from the suspect.

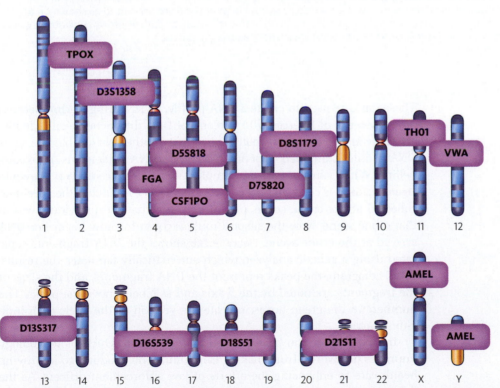

Figure 19.25 Standard sites of analysis for DNA evidence. In addition to analysis of the X and Y chromosomes, 13 standard loci are used to analyze DNA crime evidence. These loci are standardized throughout the world and are non-coding regions of DNA. Since the 13 loci are located on different chromosomes, they assort independently and decrease the likelihood that other people would have the same assortment of chromosomes, or the same profile.

© Cengage Learning 2014

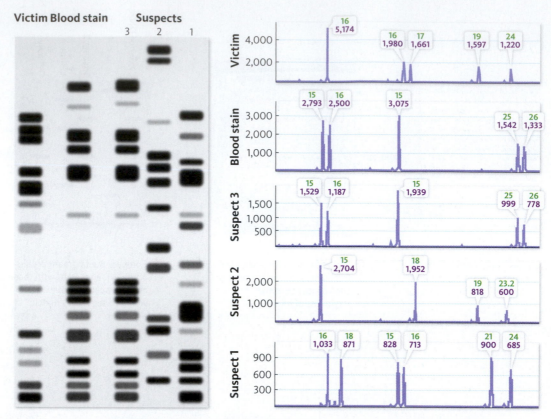

Figure 19.26 DNA profiles and crime investigation. Both of these diagrams show the DNA profile of suspects of a crime. Today most profiles are analyzed by computer and are presented as shown on the right, with the peaks representing a DNA fragment. In both cases you can gather information about which suspects were at the scene of the crime. Which suspect would you pursue?

© Cengage Learning 2014

These data are used to create a DNA profile, or DNA fingerprint. Because only one-tenth of 1 percent (0.1 percent) of DNA differs between individuals, there is an extremely small chance that another person has the same DNA profile. Combined with other evidence, DNA evidence is quite compelling. When you compare the samples from the three suspects shown in **Figure 19.26a**, it is clear that the DNA from Suspect 1 matches the specimen collected at the crime scene, providing proof that the individual was at that crime scene. Now the suspect must account for how his or her DNA arrived at the crime scene. **Figure 19.26b** shows the DNA fragments separated using a genetic analyzer, which automatically tabulates the results. In this diagram, the peaks represent the DNA fragments, and the sizes of the fragments are noted by the Y-axis and as labels above the peaks. The numbers in the green boxes indicate the version of the allele each individual carries.

In this chapter you have learned many ways in which biotechnology improves our world: from making important food products to improving health care to enhancing the efforts of law enforcement officers. As the debate continues on the long-term health and environment effects from GM organisms, you now should have a better understanding of both the ethics and technology behind this science. This knowledge will help you understand and shape public policy as it relates to biotechnology, medicine, and agriculture. Although the list of biotechnology applications is endless, we hope you have gained an appreciation of the interaction of genomics, biology, and technology.

19.5 check + apply YOUR UNDERSTANDING

- DNA profiling is used to establish genetic relationships.

- Information from DNA profiles is used for convicting criminals, proving innocence, and conserving wildlife.

1. What are three techniques used to establish genetic relationships?
2. List five sources of DNA evidence that forensic scientists collect to investigate a case.
3. DNA profiling is often referred to as DNA fingerprinting. Explain why DNA is referred to as a fingerprint.
4. Imagine you are on a jury to determine the guilt or innocence of a suspect for murder. What would hold more weight: a DNA profile using 8 loci or one using 20 loci? What are the advantages and disadvantages of using 8 or 20 loci?
5. DNA profiling is also used for paternity testing. Explain how this technique could be used to identify a child's father.

End of Chapter Review

Self-Quiz on Key Concepts

The Science of Biotechnology

KEY CONCEPTS: Biotechnology is both a modern and an ancient science that uses biological systems to make useful products. Biotechnology combines engineering solutions and biological systems to produce food, medicine, gene therapy, stem cells, and forensic evidence. Genomics, the study of genes, gene products, and gene interactions. is an important field that provides foundational information for the invention of biotechnology innovations.

1. Match each of these **fields of biotechnology** with its description or example.

   Biotechnology   a. the genetic manipulation of an organism
   Genomics   b. the study of genes, gene products, and their interactions
   Genetic engineering   c. a technique that inserts a normal gene into the genome
   Gene therapy   d. the application of engineering to biological systems to make desirable products

2. Which of the following is *not* an application of genetic engineering in plants?
   a. to improve the nutritional quality of food
   b. to improve crop yields
   c. to make insect-resistant crops
   d. All of these are applications of genetic engineering of plants.

3. By knowing the sequence of the entire human genome, scientists:
   a. now understand how every gene functions.
   b. can map all human illnesses.
   c. can use this information for understanding individual genes.
   d. can understand how gene therapy works.

Tools and Applications of Biotechnology

KEY CONCEPTS: Genetic engineering, DNA cloning, genetic markers, stem cells, and DNA profiling are some of the important tools and techniques used in biotechnology. These tools are improving medicine, enhancing agriculture, personalizing medicine, convicting criminals, proving innocence, and conserving wildlife.

4. Match each of these **biotechnology tools** with its description or example.

Genetic markers a. undifferentiated and unspecialized cells
Stem cells b. produces stem cells from adult tissues
DNA profiling c. DNA sequences used to identify individuals
Therapeutic cloning d. DNA fingerprint of an individual

5. Plasmids that are used for genetic engineering are:
   a. double-stranded DNA molecules from bacteria.
   b. double-stranded RNA molecules from bacteria and viruses.
   c. naturally occurring in all eukaryotes.
   d. unique, single-stranded DNA molecules.

6. Embryonic stem cells are a powerful research tool because:
   a. they are the only source for therapeutic cloning.
   b. they can be induced into any cell type.
   c. they are easy to manipulate.
   d. they are easy to obtain.

Ethical Concerns of Biotechnology

KEY CONCEPTS: Along with powerful knowledge, the field of biotechnology brings with it controversial and ethical issues. Research on embryonic stem cells poses the question of when life begins. Genetic testing considers the rights of privacy and confidentiality. The use of genetically modified organisms raises safety and environmental concerns. As biotechnology expands, society will have to make difficult ethical decisions.

7. Match each of the **types of stem cells** with its description or example.

Embryonic stem cells a. undifferentiated cells from the skin and other organs
Adult stem cells b. unspecialized cells that give rise to all the types of cells in the body

8. Which government agencies help to determine the safety and efficacy of new biotechnology products?
   a. FDA
   b. USDA
   c. EPA
   d. All of the above

9. One concern with gene therapy is:
   a. it has never been used before.
   b. the outcome is not predictable.
   c. it has not been used to successfully treat any human diseases.
   d. it is difficult to test for long-term effects in humans.

Applying the Concepts

10. Ten years after the 2001 attack on the World Trade Center, medical examiners have identified 60 percent of the remains collected, and they have confirmed the identity of 60 percent of the victims. Explain how forensic scientists use DNA profiling to identify the victims. Speculate why 40 percent of the victims and collected remains would still be unidentified.

11. In terms of increasing food production, industrial countries are driven by market demands and profit. On the other hand, developing countries face urgent needs to alleviate poverty, hunger, and starvation. Most governments in sub-Saharan Africa have long-term policy goals to attain food self-sufficiency. Although it is acknowledged that biotechnology can help support national policy, there is still widespread resistance among Africans to incorporate

biotechnology. Investigate this issue and present the benefits and drawbacks of using biotechnology to improve food production in this region.

12. The onset of some diseases, such as Huntington disease, cancer, and familial hypercholesteremia, occurs later in life. Genetic testing is available for these diseases, allowing for the diagnosis of these adult-onset disorders before symptoms occur. What are the emotional and medical effects of a positive test for a condition that causes premature death? Is it appropriate to offer such testing if there is no cure to the fatal condition? What type of discrimination might someone with a positive diagnosis face?

Data Analysis

DNA Profiling at a Crime Scene

One common technique for creating DNA profiles examines slight gene sequence differences called *gene variants,* which all individuals carry, referred to as STRs (short tandem repeats). **Figure 19.26b** shows DNA profiles of samples from blood collected at the scene of a crime and from four suspects. These data show alleles from three loci: D3S1358, vMA, and FGA. The numbers in the green boxes indicate the version of the allele each individual carries.

From Thompson WC, Ford S, Doom T, Raymer M, Krane DE. 2003. Evaluating forensic DNA evidence: essential elements of a competent defense review. Part 1. The Champion 27(3):16–25.

Data Interpretation

13. Based on the above results, can any of the suspects be excluded as the one who committed the crime?

14. Do the data indicate that one of these suspects was at the crime scene and should be investigated further?

Critical Thinking

15. Although the graphical data are quick to interpret, examine the graphs more closely and note the allele version of each sample listed in **Table 19.3**. The blood sample is outlined for you.

16. In **Table 19.3**, why are there always two numerical values for each gene locus?

17. If you are the defense attorney, what are some potential sources of error you will want the jury to consider?

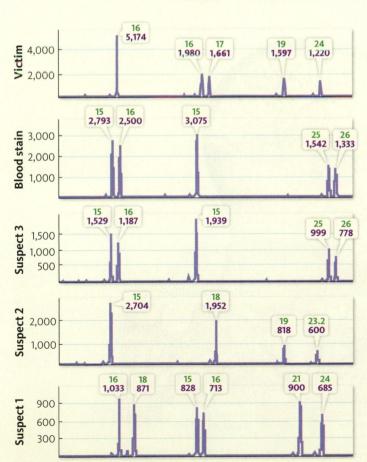

Figure 19.26b DNA profile from crime scene evidence.
© Cengage Learning 2014

TABLE 19.3 Alleles for Analyzed Gene Loci

Sample	Locus D3S1358	Locus vWA	Locus FGA
Blood	15, 16	15	25, 26
Suspect 1	16, 18	15, 16	21, 24
Suspect 2			
Suspect 3			
Victim			

© Cengage Learning 2014

18. If you are the prosecuting attorney, how can you make this analysis more powerful in the courtroom?

19. It is important to know how variable a particular locus is and how frequently a particular allele occurs in the population. Why would it be important to understand the frequency of the alleles?

Question Generator

Parkinson Disease and Gene Therapy

Parkinson disease is a degenerative brain disorder that affects 1 in 100 people over the age of 60 years, as well as a smaller number of younger people. The disease is caused by a combination of genetic and environmental factors and results in a loss of motor skills, tremors, loss of balance, muscle stiffness, and slow movement. The *substantia nigra* is a region of the brain that produces dopamine, a neurotransmitter responsible for smooth, coordinated muscle movements (**Fig. 19.27**). When 80 percent of the neurons in this region of the brain die or are impaired, symptoms of Parkinson disease develop. Medicines that replace or mimic dopamine help to alleviate the primary symptoms of Parkinson disease.

Researchers are hopeful that stem cell research and gene therapy will be used in the future to help the body make healthy dopamine-secreting neurons in this region of the brain.

Experimental approaches use gene therapy techniques utilizing specific viral vectors to deliver three genes involved in dopamine synthesis into the *substantia nigra.* To date, these clinical trials have been encouraging for patients with Parkinson disease.

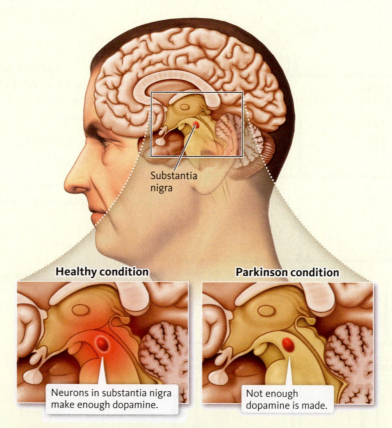

Substantia nigra

Healthy condition

Neurons in substantia nigra make enough dopamine.

Parkinson condition

Not enough dopamine is made.

Figure 19.27 Parkinson disease and the substantia nigra, the dopamine-producing region of the brain.
© Cengage Learning 2014

Below is a block diagram (**Fig. 19.28**) that relates several aspects of the biology of Parkinson disease and gene therapy. Use the background information along with the diagram to ask questions and generate hypotheses. We've translated the components of the block diagram into several questions to get you started.

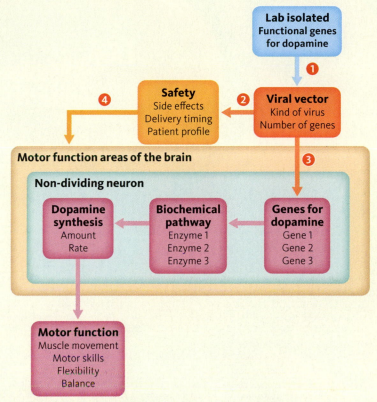

Figure 19.28 Factors relating to Parkinson disease and gene therapy.
© Cengage Learning 2014

Here are a few questions to start you off.

1. How do researchers get functional genes into a viral vector? (Arrow #1)

2. What are the safety concerns of using a viral vector? (Arrow #2)

3. What is the relationship of the viral vector to the genes for dopamine? (Arrow #3)

4. How do researchers deliver the viral vector precisely to the local region of the brain? (Arrow #4)

Use the diagram to generate your own research question and frame a hypothesis.

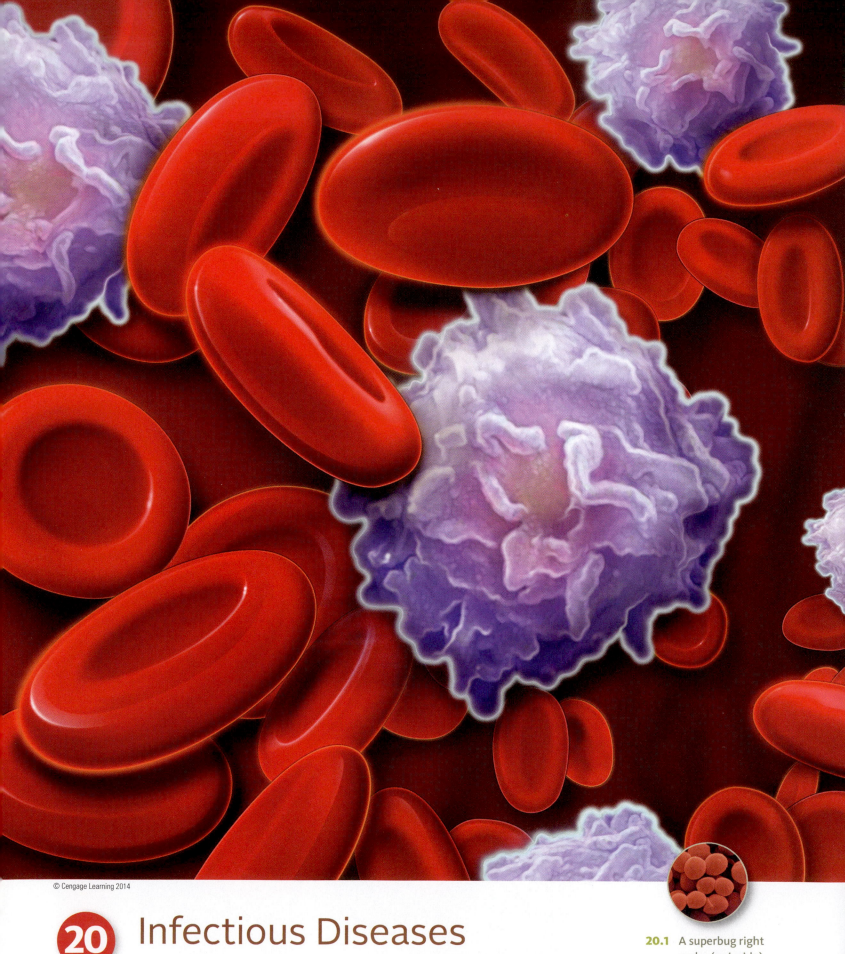

20 Infectious Diseases and the Body's Responses

20.1 A superbug right under (or inside) your nose

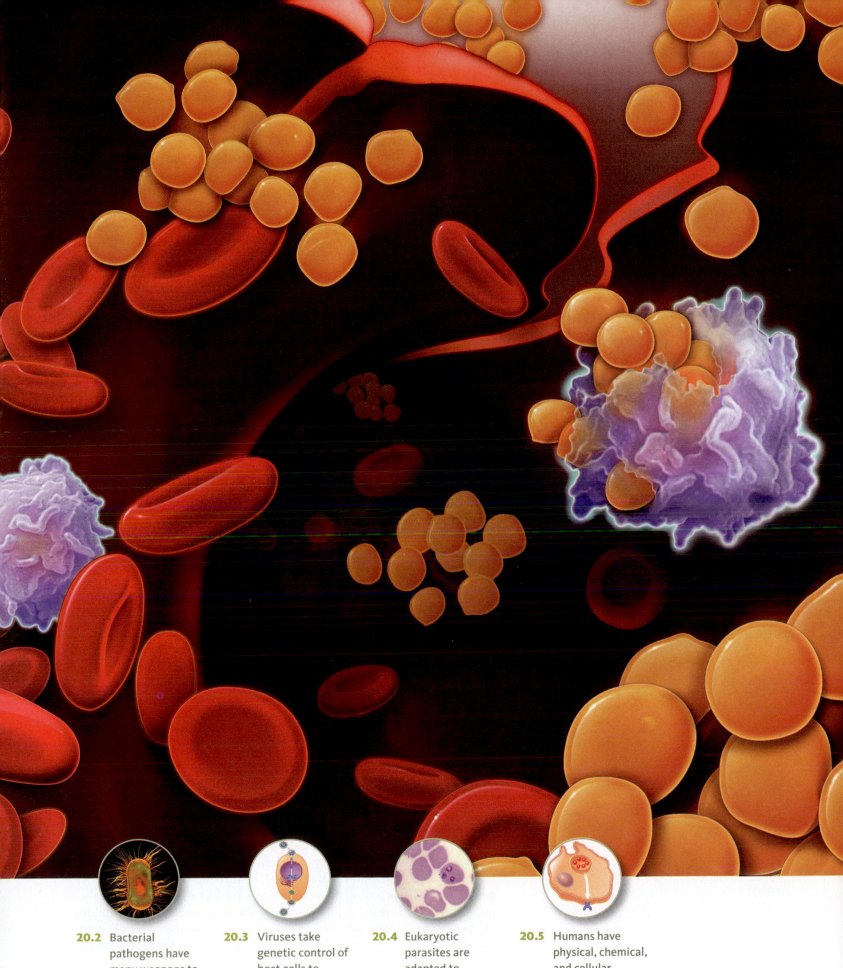

20.2 Bacterial pathogens have many weapons to infect the body and cause disease

20.3 Viruses take genetic control of host cells to produce new viruses

20.4 Eukaryotic parasites are adapted to different hosts and different habitats

20.5 Humans have physical, chemical, and cellular defenses against pathogens

20.1 A Superbug Right Under (or Inside) Your Nose

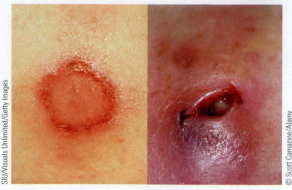

Figure 20.1 MRSA Infection. An infection caused by MRSA at first appears similar to a spider bite (left). As the infection progresses, the infected area becomes more inflamed and fills with pus (right).

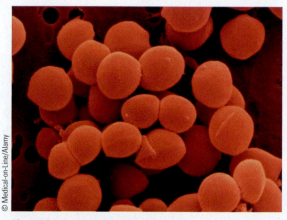

Figure 20.2 The superbug MRSA. This color-enhanced micrograph shows *Staphylococcus aureus*, the bacteria that causes MRSA (methicillin-resistant *Staphylococcus aureus*). *Coccus* refers to the sphere-like shape of the bacteria.

It all started with what appeared to be a spider bite on his elbow. The sore was red, slightly tender to the touch, and swollen. It also had a small amount of pus and other fluids (**Fig. 20.1**). The high school wrestler didn't give it much thought until three days later when the sore hadn't started healing and the swelling had gotten worse. He visited the clinic and was prescribed penicillin, an *antibiotic* that inhibits the growth of bacteria. Several days later, the swelling had increased to the size of a golf ball. What started as a minor-looking sore had developed into a deeper, more serious infection caused by a common strain of bacteria that lives on the skin. This severe infection is caused by a pathogen, resistant to many antibiotics, called MRSA, which stands for methicillin-resistant *Staphylococcus aureus* (**Fig. 20.2**). A *pathogen* or infectious agent—sometimes called a germ—is a microbe or microorganism such as a virus, bacterium, or fungus that causes disease.

MRSA is one of several strains of the bacterial species *S. aureus,* or "staph" for short. Staph is a common inhabitant of the human body along with many other microbes that are part of the body's *normal flora,* which are harmless and usually beneficial microbes that live on the skin and inside the respiratory, digestive, and reproductive tracts. Staph's ecological niche is the warm, moist entrance to the nose. From there it spreads to other areas of the body.

Everyone is susceptible to bacterial, viral, and fungal skin infections, which can be transferred from one person to another through body secretions, blood, skin flakes, and wastes. Wrestlers and other athletes in contact sports are especially at risk because of repeated contact that can cause open wounds. MRSA bacteria can spread through sweaty skin-to-skin contact with an infected person or through frequently touched areas such as wrestling mats, treatment tables, locker room benches, and gym equipment. The bacteria then may enter the blood and body tissues through small cuts or skin abrasions. The bacteria are common; in fact, 1 in 100 people carries the MRSA strain of bacteria on his or her body without showing symptoms or getting sick. MRSA has been reported among people living in close contact with each other, such as prisoners and military recruits.

Although most MRSA skin infections aren't serious, about 6 percent of cases can be life-threatening. In the case described here, the 17-year-old wrestler eventually died of blood poisoning caused by bacterial toxins. MRSA is considered a "superbug" because it is resistant to many antibiotics, including penicillin and methicillin. MRSA infections currently are treated with other types of antibiotics, but there have been reports that some MRSA strains are becoming resistant to these drugs as well. As a result, MRSA is sometimes referred to as multidrug-resistant *Staphylococcus aureus.*

Antibiotic resistance is a critical healthcare challenge around the globe. The MRSA strain has resistance genes that make a protein that binds to methicillin and inactivates it. Bacteria reproduce rapidly and can transfer genes from one bacterium to another, so a harmless strain can quickly acquire the genes to resist antibiotics and cause disease. From an evolutionary perspective, antibiotics act as a strong force of natural selection. Antibiotic drugs kill bacteria lacking the resistance gene; the few resistant bacteria reproduce and pass their resistance genes rapidly to the next generation (Fig. 20.3). In other words, bacterial pathogens can evolve at much faster rates than humans can develop new antibiotic drugs. For this reason, some physicians are less likely to prescribe antibiotics for their patients who may not need them.

The skin and membranes lining the body's openings are very effective barriers against the entry of pathogens such as MRSA. However, if the skin or the mucous membranes are damaged or punctured, staph can gain entry into the blood and become a pathogen, or a microbe capable of causing *infectious disease.* As a pathogen, staph bacteria have the ability to multiply, or infect, the particular site for the purpose of using the host as a source of nutrients or replication. Ultimately, any infection may cause disease if the following conditions are met: a capable strain of a pathogen makes contact with a susceptible host, and the environmental conditions within the body are favorable for the pathogen.

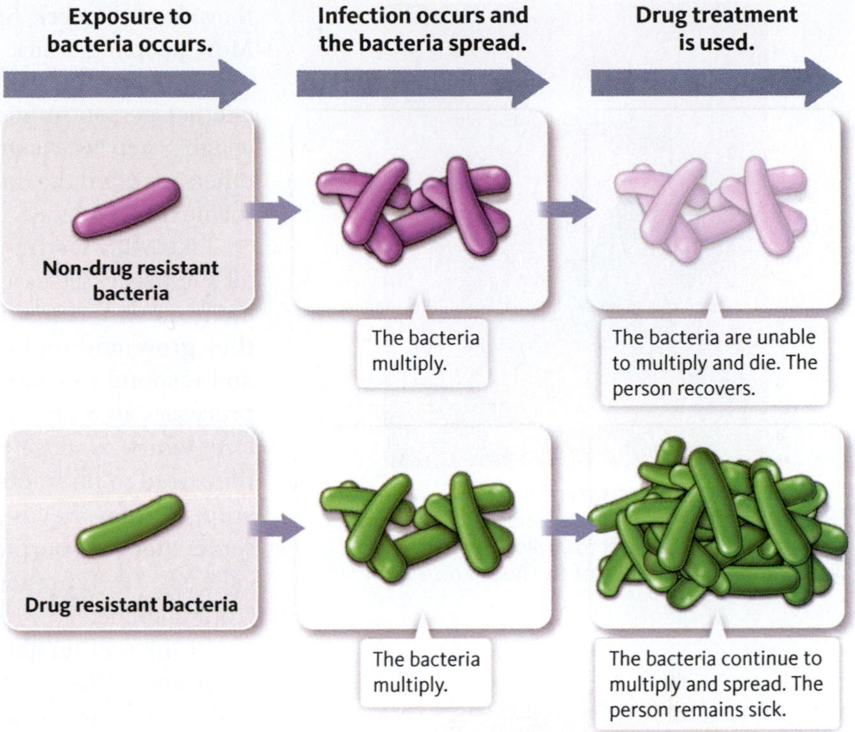

Disease is marked by the appearance of various symptoms, such as a rash, diarrhea, or fever, which result from the pathogen interacting with the host's defense. In MRSA infections, the redness and swelling of the sore is the body's inflammatory response to the bacterial infection. Staph bacteria are not defenseless; in fact, once staph gains access to the other tissues, it is a powerful bacterial pathogen that can inhibit antibiotics, make potent toxins, kill white cells, and hide from the body's surveillance system.

In the chapter ahead, we'll investigate the battle between your body and numerous infectious agents and *parasites* that invade the body to obtain nutrients or as a means of multiplication. Some parasites, usually those that are microorganisms, are called *pathogens,* and in this chapter we use the terms to mean the same thing. Let's begin by exploring the enormous variety of parasites and outline the complex methods the body uses to defend itself. Your day-to-day health is a testament to the effectiveness of your body's defenses. Many important concepts at the cellular and molecular levels will help you understand the medical news reported in the mass media and how to prevent infectious disease in your future.

Exposure to bacteria occurs.	Infection occurs and the bacteria spread.	Drug treatment is used.

Non-drug resistant bacteria

The bacteria multiply.

The bacteria are unable to multiply and die. The person recovers.

Drug resistant bacteria

The bacteria multiply.

The bacteria continue to multiply and spread. The person remains sick.

Figure 20.3 Resistant bacterial populations. Antibiotics exert strong selection pressure on bacterial populations in or on the body. Nonresistant strains are eliminated, whereas resistant cells multiply and spread.
© Cengage Learning 2014

20.2 Bacterial Pathogens Have Many Weapons to Infect the Body and Cause Disease

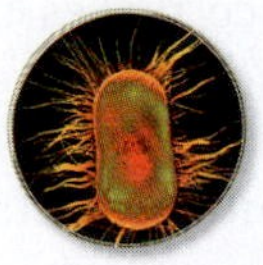

Although it's unlikely that you'll be diagnosed with a MRSA infection, most of us have had strep throat at one time or another (Fig. 20.4). Strep throat is caused by a pathogenic strain of the bacterium *Streptococcus pyogenes.* The symptoms of strep throat include swollen lymph nodes on

Infection (infect)—a reference to the ability of pathogenic organisms to enter a host through wounds or openings to the body and cause disease symptoms.

Disease—a condition affecting an organism, most often associated with specific symptoms such as physical ailments. Diseases are often caused by infectious agents, which include viruses, bacteria, fungi, protists, and multicellular parasites.

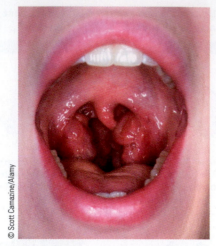

Figure 20.4 A common strep throat infection.
Swollen, bright red tissue at the back of the throat is characteristic of strep throat, along with swollen lymph nodes on the side of the neck and a fever.

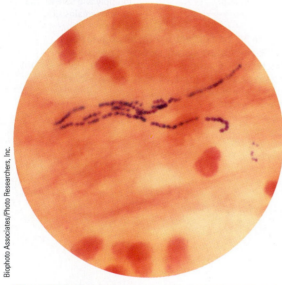

Figure 20.5 The pathogen *Streptococcus pyogenes* causes strep throat. Strep bacteria divide to form long chains of cells. Each chain consists of a dozen cells attached end to end. The larger, reddish-pink cells are epithelial cells from the throat.

Prokaryotic—small, simple cells that lack organized compartments; includes bacteria and archaea.

Cell wall—protective outer structure that surrounds bacteria; composed of molecules that are highly cross-linked to form a strong, rigid structure. The wall determines the bacteria's structure and shape and protects it from expanding and bursting.

the side of the neck, bright red tissue at the back of the throat, and a fever. Most people get sick three to five days after exposure to the pathogen, which is transferred from one person carrying the infectious bacterium to another person in microscopic airborne droplets carried in a sneeze or a cough. Strep is a *communicable* disease, meaning it is easily transmitted to others, so often the doctor asks you to stay home from class and to rest for the next several days until you recover.

Like MRSA, strep bacteria are single-celled organisms that carry out all the processes essential for life. They exchange materials with their environment, transform nutrients into energy, build new structures as they grow, and replicate their genetic information as DNA. They sense and respond to changes in their environment and perform all of these processes in a structure only about a few cubic micrometers (μm) in size, which is close to 2/10,000 of an inch. Bacteria are so small that they need to be stained in order for you to even see them with a microscope. When viewing strep bacteria under high power, you will observe that the purple-stained, round-shaped bacteria form chains of cells (**Fig. 20.5**). In fact, the word *strep* means "chain" and *coccus* means "round shape."

In the section ahead, we'll explore how bacterial pathogens such as strep and MRSA overcome our bodies' defenses and infect our skin, brain, eyes, ears, nasal sinuses, lungs, and urinary and reproductive systems. It's important to understand that only a few bacterial species live as human pathogens. The overwhelming majority of bacteria that we encounter each day either are beneficial or neutral in our lives.

Bacteria are small, prokaryotic cells surrounded by a rigid wall

In order to see the internal structure of a bacterial cell, microbiologists use a high-powered electron microscope able to magnify bacteria several thousands of times. At this magnification, you see that a bacterial cell has a much simpler structure and organization than the larger human white blood cell that attacks it (**Fig. 20.6**). Recall from Chapter 1 that bacteria are **prokaryotic** in their organization and structure, which means that, unlike *eukaryotic* cells, they lack a nucleus surrounded by a membrane. Bacterial cells also lack other internal structures, including organelles such as mitochondria, endoplasmic reticulum, or Golgi complexes (**Table 20.1**)

Many of the metabolic processes that take place inside eukaryotic cells occur on the inner or outer surface of the plasma membrane of the bacteria. Proteins are produced on organelles called *ribosomes.* However, bacterial ribosomes are smaller than eukaryotic ribosomes and have a slightly different molecular structure. This difference makes bacterial ribosomes an excellent target for antibiotics that stop the protein synthesis and growth of bacteria in the body. Bacteria such as MRSA become resistant to antibiotics through gene transfers and mutations that alter metabolic pathways and cellular responses that allow the bacteria to adapt and become resistant to antibiotics. Excessive use of antibiotics can accelerate the occurrence of resistance to antibiotic drugs in a bacterial population.

Bacteria have **cell walls** made of highly cross-linked molecules that form a strong, rigid structure. The wall determines the bacteria's structure and shape and protects the bacteria from expanding and bursting

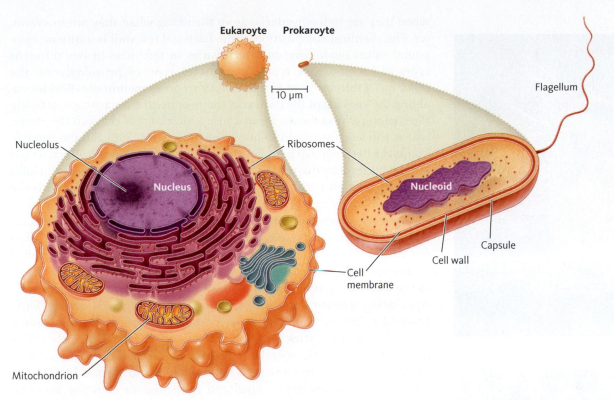

Eukaroyte   Prokaroyte

Flagellum

Nucleolus

Ribosomes

Nucleus

Nucleoid

Capsule

Cell wall

Cell
membrane

Mitochondrion

10 µm

Figure 20.6 Comparison of prokaryotic and eukaryotic cell structures. Bacteria are much smaller and their prokaryotic cell structure is much less compartmentalized than the eukaryotic human white blood cell that attacks them. Prokaryotic cells are surrounded by a rigid cell wall and lack a nucleus and other internal organelles. Many prokaryotes have a flagellum for mobility and a sticky capsule that helps them adhere to surfaces (like human tissue). Both cell types have a plasma membrane, ribosomes, cytoplasm, and DNA.
© Cengage Learning 2014

TABLE 20.1 Bacterial (Prokaryotic) and Human (Eukaryotic) Cell Structures

Cellular Processes	Bacterial Cell (Prokaryotic)	Human Cell (Eukaryotic)
Movement of cell	Flagella (some)	Flagella (sperm)
Attachment or outer wall layer	Capsule (some), cell wall	No wall present
Controls transport of substances in/out of cell	Plasma membrane	Plasma membrane
Energy (ATP) generation, cellular respiration	Plasma membrane	Mitochondria
Lipid synthesis	Plasma membrane	Smooth endoplasmic reticulum
Packaging	Cytoplasm	Golgi apparatus
Secretion	Cytoplasm	Vesicles
Internal digestion	Cytoplasm	Lysosomes
Proteins synthesis	Ribosomes (small) in cytoplasm	Ribosomes (larger) in rough endoplasmic reticulum and cytoplasm
Genetic element(s)	Circular chromosome in nucleoid region, plasmid(s)	Linear chromosomes (23 pairs) in nucleus
Genetic material (bases) Genes	DNA (A, T, G, C) ~500–5,000 genes	DNA (A, T, G, C) 20,000–25,000 genes

© Cengage Learning 2014

a.

Dr. Terry Beveridge/Visuals Unlimited/Getty Images

b.

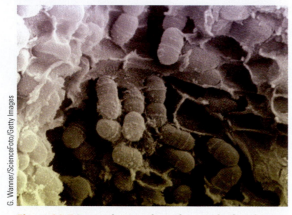

G. Wanner/ScienceFoto/Getty Images

Figure 20.7 Bacteria attach to the particular tissues they attack. Bacteria attach to body surfaces (a) by using long protein threads or (b) by secreting sticky substances that create biofilms. Dental plaque on teeth is an example of a biofilm.

Capsule—outer layer of carbohydrates and proteins secreted by some bacteria that helps them stick to surfaces. For example, strep bacteria have a capsule that helps them stick to the epithelial cells of the throat and prevents them from being washed away into the stomach during swallowing.

when they are in freshwater or from shrinking when they are in saltwater. The chemical composition of the bacterial cell wall is a unique compound called *peptidoglycan,* which varies in thickness in two different types of bacteria. One type has a thicker layer of peptidoglycan; the other has a thinner layer in addition to an outer membrane. Besides enabling the viewing of the bacteria's shape under the microscope, staining the bacteria allows the doctor to know which kind of cell wall they have, which will suggest the antibiotic to be prescribed. For example, penicillin blocks the chemical pathway that makes peptidoglycan, which would kill the bacteria by disrupting cell wall construction. Although penicillin works against all bacteria, it's most effective against bacteria with the thick peptidoglycan layer.

Bacteria secrete substances or make structures that help them attach to the particular tissues they attack. Some bacteria or even different strains secrete an outer layer of carbohydrates and proteins called a capsule. Strep bacteria, for example, have a capsule that allows them to stick to the epithelial cells of the throat and prevents the bacteria from being washed away into the stomach during swallowing. Thin, fingerlike protein filaments project from the surface and help some bacterial species to stick to solid surfaces (**Fig. 20.7a**). Once anchored in place, many bacteria secrete a protective, slimelike substance that forms a layer, or *biofilm* (**Fig. 20.7b**). The ability to form a biofilm appears important for the survival and transmission of many bacterial pathogens such as staph.

Bacteria may be simple in their structure, but metabolically they are very complex. Bacteria, including pathogenic ones, can be identified using the differences among their structures, molecular compositions, and metabolic capabilities. Appropriate antibiotics can then be used that will inhibit the bacteria's growth and reproduction and thus prevent disease.

Bacteria absorb nutrients and multiply during an infection

Nutritionally, bacteria require many of the same molecules that humans do. They need a source of energy along with carbon, hydrogen, oxygen, and nitrogen to build their cellular structures such as phospholipid membranes, DNA molecules, and proteins. They need minerals for enzymes and metabolic pathways. The function of a nutrient is similar whether in a bacterium or a blue whale. For example, both organisms use glucose to fuel cellular respiration and ATP production. Similarly, pathogens obtain their nutrients by breaking down human cells and tissues to acquire carbohydrates, lipids, proteins, and nucleic acids to build their own bodies and fuel their growth and reproduction.

In the same way that nutrients in food items fuel the growth of blue whales and humans, they also fuel the growth of bacteria. On an individual cell level, growth is an increase in the size of the bacterium. Growth is achieved by building proteins and ribosomes, absorbing water, and enlarging the cell wall so that the cell reaches a particular volume. As a cell *increases* in size, the amount of surface area relative to its volume is *reduced.* As the cell continues to grow, materials entering and leaving through the cell's surface can no longer keep up with the demand for resources needed to support the growing cell. One solution to this problem is for the cell to divide into two. Therefore, when microbes grow, they also rapidly divide so that the number of individual

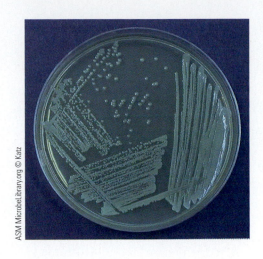

Figure 20.8 Bacterial colonies grown in a lab. Under the favorable conditions of an agar plate, bacteria divide and grow into discrete colonies. One colony arises from one bacterium. The plate shows *Staphylococcus aureus*.

cells increases. Thus, when we discuss microbial growth, we are referring to the growth of the *population* of bacteria and the number of cells that are produced.

During the early stages of an infection, microbial populations rapidly grow and multiply. This is the case when MRSA bacteria enter the bloodstream, where nutrients are abundant and competition is low. We see this at the doctor's office or in a lab when bacteria swabbed onto an artificial growth medium (such as an agar plate) reproduce and multiply into visible *colonies* containing billions of cells in as few as 24 hours (**Fig. 20.8**).

As with any cell, before a bacterium divides, it must first replicate its DNA. The bacterial chromosome forms a circle, and during replication enzymes place nucleotide bases into their proper places and a new circular DNA molecule is created (**Fig. 20.9**). Some bacterial species also contain other small pieces of circular DNA called plasmids (see **Fig. 19.10** and **Fig. 20.9**). During cell division, plasmids replicate independently and are inherited by the daughter cells. Plasmids carry genes for a variety of proteins, including toxins, that bind to and halt antibiotics from functioning, providing a type of antibiotic resistance. It is important to note that plasmids can be exchanged between different bacteria. In effect, any bacterial cell can receive plasmid genes from another bacterium, making it capable of causing disease or becoming resistant to antibiotic treatment.

Bacterial pathogens advance through steps in the infectious disease cycle

Infectious diseases caused by bacteria and other pathogens follow a series of steps called the disease cycle (**Fig. 20.10**). Although not all diseases progress through the steps in the same order, the disease cycle provides a framework for understanding how pathogens cause infection and disease. Some pathogens such as MRSA infect the body and persist for long periods of time; others such as the bacteria that cause cholera have a rapid "get in and get out" strategy. Let's trace the steps in the cycle taken by the fast-acting cholera bacterium. Cholera is a disease that is uncommon in the United States because our water supply is clean. However, in the developing world, cholera outbreaks are more common and cause severe diarrhea that has killed millions of people. We'll begin at the source: the disease *reservoir*.

Cholera bacteria live in water reservoirs contaminated by sewage and human waste. Infection begins when the pathogens are transmitted from the contaminated reservoir and enter the body. Cholera bacteria enter the body when they are swallowed into the digestive system. Most bacteria are killed by the strong acids of the human stomach, but some survive passage through the stomach and enter the small intestine to cause a cholera infection.

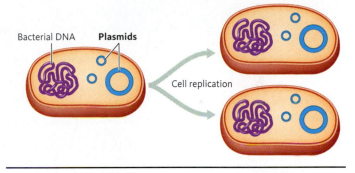

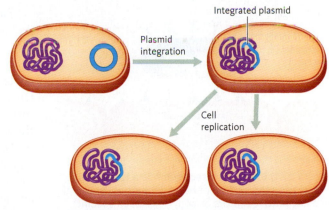

Figure 20.9 Plasmids. Bacteria have DNA in their large, circular chromosome as well as in small plasmids. When a bacterium divides, it replicates its DNA and plasmids (top figure). In some cases the genes of a plasmid are integrated into the chromosome (bottom figure). Plasmids often carry genes that code for toxins and antibiotic-resistant proteins.
© Cengage Learning 2014

Plasmids—small pieces of circular DNA found in bacteria. Plasmids often code for a small collection of proteins important to bacterial infection or survival.

Disease cycle—infectious diseases caused by bacteria and other pathogens follow a series of steps called the disease cycle. Not all diseases progress through the steps in the same order. The disease cycle provides a framework for understanding how pathogens cause infection and disease.

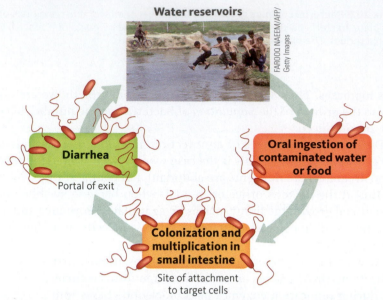

Water reservoirs

FAROOQ NAEEM/AFP/Getty Images

Diarrhea
Portal of exit

Oral ingestion of contaminated water or food

Colonization and multiplication in small intestine
Site of attachment to target cells

Figure 20.10 The disease cycle of cholera. Worldwide, 3–5 million cases of cholera are reported each year and about 120,000 of these people die from the infection. Cholera bacteria are transmitted from a water reservoir into the body, where they target cells of the small intestine (site of attachment) and cause severe dehydration from diarrhea. The bacteria are expelled through diarrhea (portal of exit) and deposited back into a water reservoir for the cycle to continue.
© Cengage Learning 2014

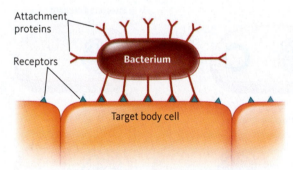

Attachment proteins

Receptors

Bacterium

Target body cell

Figure 20.11 Bacterial attachment to a target cell. Proteins on the surface of bacterial cells fit like a lock and key onto receptors of body cells. Each specific pathogen has specific target cells to which it attaches.
© Cengage Learning 2014

Each specific pathogen generally infects a specific tissue in the body—its target cell. The essential first step is *attachment* of the pathogen to a host target cell (**Fig. 20.11**). Strep bacteria attach to cells lining the throat, MRSA attaches to epithelial cells of the skin and respiratory tracts, and cholera bacteria attach to the epithelial cells that line the small intestine. Once attached, the bacterial cells can then attack and gain resources necessary for growth and survival; as a result, the bacteria are able to spread throughout the tissue.

Once attached and colonized, the bacteria initiate the disease process when they *damage* or destroy host cells and the underlying connective tissue. Unlike viruses that work *inside* host cells, bacterial pathogens live and act *outside* host cells in the blood and body tissues. Many bacteria secrete enzymes and toxins into their environment that can alter cell structure or function. For instance, strep bacteria make an enzyme that breaks down red blood cells. Cholera bacteria produce a toxin that does not directly damage the cell's *structure;* instead, it alters the host cell's *function.* The cholera toxin activates enzymes that pump water and electrolytes such as sodium and potassium ions from the blood into the intestine. The result is a profuse, watery diarrhea in which the water composing a person's blood is physically pumped into the intestines and out of the body. Water loss in the diarrhea ultimately reduces blood pressure, and the person goes into shock and may die within two to three hours if the fluid and electrolytes aren't quickly replaced.

The diarrhea carries the cholera bacteria out through a *portal of exit.* Human wastes find their way back into the drinking water reservoir, and the disease cycle continues into a new host. This cycle of events illustrates how diseases such as cholera are able to rapidly spread through communities with unsanitary conditions and contaminated drinking water.

20.2 check + apply YOUR UNDERSTANDING

- Bacteria are small, prokaryotic cells surrounded by a rigid cell wall that determines the shape of the cells. They lack a nucleus surrounded by a membrane and other organelles such as mitochondria, endoplasmic reticulum, or Golgi complexes.

- Although most bacteria that we encounter each day are either beneficial or neutral in our lives, some bacteria cause disease. Many species have the ability to protect themselves and cause disease by producing toxins that cause infection and disease and chemicals that inhibit antibiotics taken to fight the infection or disease.

- Infectious diseases caused by bacteria and other pathogens follow a series of steps called the disease cycle. The ability of any pathogen to cause infection depends on the genes it carries and the proteins it expresses.

1. What are two functions of the bacterial cell wall?
2. What is a plasmid, and why is it important for pathogens?
3. Bacteria require a source of energy and a source of raw materials to grow. List four chemical elements that all bacteria need and what structures they build with them. Would bacteria require a supply of phosphorus (P)?
4. Does infection always lead to disease? Can you have disease without infection? Explain and give an example.
5. Gonorrhea is a sexually transmitted disease with a characteristic incubation period of two to seven days, which is the elapsed time between a person's exposure to the pathogen during sex and the onset of symptoms such as painful urination and a discharge (**Fig. 20.12**). (a) Refer back to the disease cycle shown in **Figure 20.10** and determine which of the stages of the disease cycle likely occur during the incubation and illness stages. (b) Which part of the disease cycle is blocked by using a condom? (c) Which part of the disease cycle is blocked by antibiotics?

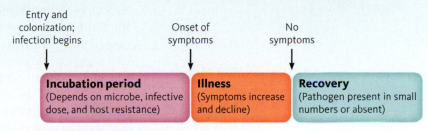

Entry and colonization; infection begins → **Incubation period** (Depends on microbe, infective dose, and host resistance)

Onset of symptoms → **Illness** (Symptoms increase and decline)

No symptoms → **Recovery** (Pathogen present in small numbers or absent)

Figure 20.12 Disease stages of gonorrhea.
© Cengage Learning 2014

20.3 Viruses Take Genetic Control of Host Cells to Produce New Viruses

You may be unaware, but herpes is a common virus that you have a very good chance of being exposed to and carrying throughout your lifetime. About 80 percent of adults carry the virus that causes oral herpes, an infection characterized by highly contagious cold sores and fever blisters on the outer and inner surfaces of the mouth (**Fig. 20.13**). As the name implies, most people mistakenly associate these "cold sores" with cold symptoms rather than with oral herpes, which is spread by the herpes simplex virus-1 (HSV-1). This virus spreads mostly through direct contact with skin or saliva, but it also can be transmitted through sexual contact of the genitals through oral sex. Although most cases of genital herpes do not come from HSV-1, this particular virus causes 30 percent of genital herpes cases in the United States. Another herpes virus known as herpes simplex virus-2 (HSV-2) causes most forms of genital herpes.

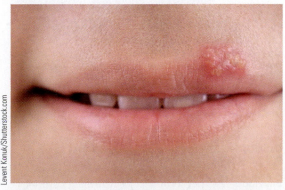

Figure 20.13 This cold sore is a symptom of oral herpes. Eighty percent of adults are infected with the virus that causes oral herpes (HSV-1).

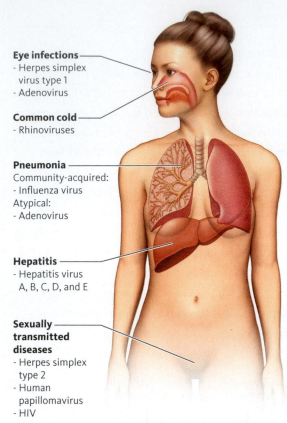

Eye infections
- Herpes simplex virus type 1
- Adenovirus

Common cold
- Rhinoviruses

Pneumonia
Community-acquired:
- Influenza virus
Atypical:
- Adenovirus

Hepatitis
- Hepatitis virus A, B, C, D, and E

Sexually transmitted diseases
- Herpes simplex type 2
- Human papillomavirus
- HIV

Figure 20.14 Viruses infect all parts of the body.
Many infectious diseases are caused by viruses that can infect almost any part of the body.
© Cengage Learning 2014

a.
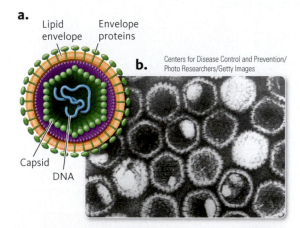

Lipid envelope
Envelope proteins

b.

Centers for Disease Control and Prevention/ Photo Researchers/Getty Images

Capsid

DNA

Figure 20.15 General structure of the herpes virus.
(a) Like many viruses, the herpes virus has a DNA core, a capsid, and an envelope made of lipids and proteins. (b) Electron micrograph of the herpes virus.
© Cengage Learning 2014

Viruses—intracellular parasites composed of a nucleic acid core (DNA or RNA) wrapped with a protein coat.

Capsid—a protective protein coat that covers an individual virus.

Envelope—another form of protective outer covering found on some viruses. The envelope is comprised of a glycolipid coat rich in proteins.

The initial infection of oral herpes usually occurs with a simple and playful kiss to an infant who has an undeveloped immune system. In the case of HSV-1, the virus usually infects and damages the epithelial cells of the lips. Entrance to the host can also occur through the mouth, eyes, or genital openings, or through wounds and damaged cells.

There is no cure for herpes. Once you contract the virus, you remain infected for the rest of your life. The symptoms of the disease may go away, but the infection remains. The virus has a unique adaptation in that it also infects neurons where it "hides out" until stress or hormones act on the neurons to "awake" the virus. When the virus "awakens," the infected person will experience a recurrence, or outbreak, of cold sores. Despite ongoing research efforts, an effective vaccine has yet to be developed. Antiviral medication minimizes outbreaks by preventing the virus from replicating its genome. Topical creams are available that shorten the healing time and duration of symptoms by reducing inflammation.

Diseases caused by viruses are commonly experienced throughout our lifetimes. Many common infectious diseases, such as chicken pox, the common cold, hepatitis, herpes, mononucleosis, measles, mumps, polio, rabies, and West Nile fever, are caused by viruses (**Fig. 20.14**). Infectious diseases caused by viruses have shaped historical events throughout the world. Some experts estimate that smallpox, measles, and the flu accounted for 90 percent of the deaths among the Native American population after the Europeans settled the New World. Centuries later, the influenza virus caused the Spanish flu epidemic of 1918, which killed an estimated 50 million people worldwide, about 3 percent of the total population. Even today, the human immunodeficiency virus (HIV) has killed more than 25 million people globally. In the section ahead, we'll explore the physical and genetic characteristics of viruses that lead to human infectious diseases and ways to exploit these characteristics to combat these viruses.

Viruses are nonliving particles that rely on host cell machinery to replicate

Viruses are not living organisms—they are not made of cells, nor can they reproduce on their own. As nonliving particles, viruses have several distinguishing characteristics. By definition, **viruses** are intracellular parasites composed of a nucleic acid core (DNA or RNA) wrapped with a protein coat called a **capsid**. Many animal viruses also have an outer **envelope**, a lipid coat rich in proteins (**Fig. 20.15**). Viruses do not have ribosomes and the necessary internal structures necessary to make proteins and replicate their own nucleic acid. They use the host cell's machinery for these processes. Despite appearing as a living entity, outside of its host cell a virus is just a lifeless particle.

Viruses display a wide variety of shapes and sizes, which are determined to a great extent by the surrounding capsid and envelope (**Fig. 20.16**). In general, viruses are 100 times smaller than bacteria and cannot be seen with a light microscope. Because they use the host cell to replicate, their genomes are small. Viral genomes vary in size but code for anywhere from 4 to 100 different proteins. For example, the virus that causes the flu codes for only 14 proteins, whereas the human cell that the virus infects codes for 20,000 proteins.

Viruses are extremely diverse, infecting bacteria, fungi, plants, and animals. In fact, viruses are parasites on nearly every known organism. More than 5,000 virus species have been identified and classified into species, genera, and families based on the type of nucleic acid (RNA or DNA)

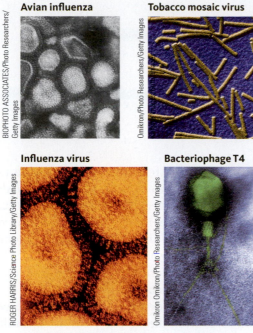

Avian influenza (BIOPHOTO ASSOCIATES/Photo Researchers/Getty Images)

Tobacco mosaic virus (Omikron/Photo Researchers/Getty Images)

Influenza virus (ROGER HARRIS/Science Photo Library/Getty Images)

Bacteriophage T4 (Omikron Omikron/Photo Researchers/Getty Images)

Figure 20.16 Virus diversity. Viruses display a wide variety of shapes and sizes, primarily based on their surrounding capsid and protein coat. They can infect many different hosts, including animals, plants, fungi, and bacteria.

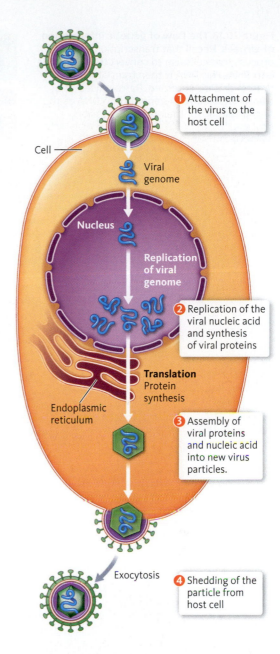

Cell

Viral genome

Nucleus

Replication of viral genome

Translation Protein synthesis

Endoplasmic reticulum

Exocytosis

1 Attachment of the virus to the host cell

2 Replication of the viral nucleic acid and synthesis of viral proteins

3 Assembly of viral proteins and nucleic acid into new virus particles.

4 Shedding of the particle from host cell

Figure 20.17 Virus replication. The viral replication process includes four main steps: (1) attachment of the virus to the host cell, (2) replication of the viral nucleic acid and synthesis of viral proteins, (3) assembly into a virus particle, and (4) shedding of the particle from the host cell.
© Cengage Learning 2014

they contain, the presence and shape of their outer envelope, and the way they replicate.

Viruses are host-specific parasites that rely on a host cell to provide habitat, raw materials, and energy for replication. Because the virus robs the host cell of its resources, the host cell is harmed, often resulting in cell death. The process in which viruses create new copies of themselves is called the *viral replication cycle* (**Fig. 20.17**). This cycle includes four main steps: attachment, replication, assembly, and shedding. Let's trace these steps using the herpes simplex virus that we introduced in the section opener infecting an epithelial cell (see **Fig. 20.17**).

First, the virus enters into the host cell. The envelope proteins on the virus *attach* to cell membrane proteins specific to the host cell. For example, the HSV-1 envelope proteins specifically bind to proteins found on mucous membranes of human epithelial tissue. Slight pH and temperature changes brought about by contact with the host cell activate enzymes that degrade the viral capsid, leading to release of the viral genetic content into the host cell.

The second step occurs once the virus is inside the cell, where it begins to *replicate.* The DNA of the herpes virus enters the host cell's nucleus, where it uses the cell's enzymes, nucleotides, ATP, and other raw materials to replicate and then transcribe the viral DNA into viral RNA. The RNA leaves the nucleus and attaches to ribosomes in the endoplasmic reticulum, where it is translated into viral proteins. Along with cellular proteins from the host's genes, these new proteins regulate genes to produce many copies of its genome as well as the outer envelope proteins needed for infecting new cells. The replication process provides the directions, enzymes, structural proteins, and other raw materials for the third step in the cycle: *assembly.* The particles now have all the instructions and proteins necessary to assemble themselves into new viruses.

Figure 20.18 The flow of genetic information of viruses. Recall that transcription is the process that cells use to convert the DNA code into RNA. This RNA is then translated into a protein that is expressed. Retroviruses transcribe their RNA genome into DNA as the beginning part of their gene expression process.

© Cengage Learning 2014

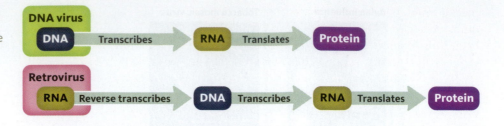

As the cell dies, the newly created virus particles are released in the fourth and final step of the cycle, a process called *viral shedding.* Once shed, these particles go on to infect the next set of host cells. One reason HSV-1 is so prevalent is that the virus is transmitted during viral shedding, which often occurs when there are no noticeable symptoms of an infection or outbreak. So how will you know that your sex partner is infected and can pass herpes to you? Unfortunately, without visible symptoms, you may have no idea that you are being infected.

During this cycle, the virus depletes the host cell's resources and the cell dies. The cold sores from the herpes virus are a result of epithelial cell damage and death. In some other diseases, the damage is severe enough that organ failure can occur. During a herpes outbreak, the virus is actively replicating, shedding, and killing host cells. The result is a painful blister that takes several days to heal. Between outbreaks, the virus is still alive, but it enters a dormant stage where it is inactive. During this stage, the infected cells carry out their normal functions without showing signs of infection. In the case of herpes, the virus enters specific sensory neuron cells, where it turns off much of its gene expression. When the host's immune system weakens from stress, fever, or hormonal changes, gene expression is turned back on, and the virus is activated and begins its replication cycle again. Not all viruses have a dormant phase, but many do. Epstein-Barr virus, the hepatitis C virus, and the human papilloma virus (HPV) are a few examples of dormant viral infections. It is often during these dormant periods that the virus is unintentionally spread from person to person.

The herpes virus is a DNA virus and follows the flow of genetic information similar to human cells: DNA to RNA to protein. However, many viruses instead contain RNA as their genome (**Fig. 20.18**). **Retroviruses**, such as HIV, are a class of RNA viruses that use their own special enzymes to direct the replication of their RNA genomes in the host nucleus. Retroviruses use their own enzymes to reverse the normal path of genetic information. They first convert the RNA into a DNA molecule that then enters the nucleus, where it is inserted into the host cell's genome (**Fig. 20.19**). Now that the viral DNA is in the host chromosome, it will replicate with each cell division. The virus has set up a situation where not only will the host cell machinery be used for gene expression, but, also with each cell division, the viral genome will be replicated. Retroviruses are an example of how adapted viruses are to their specific human hosts!

One final way that viruses rely on their host cell is through the viral envelope that dictates how the virus infects cells. Many human viral pathogens have an envelope that forms using membranes from the host cell. For instance, as with other DNA viruses, the herpes envelope forms from the host cell's nuclear membrane. Other virus envelopes form from the cell's

Retroviruses—a class of RNA viruses, such as HIV, which use their own special enzymes to direct the replication of their RNA genomes in the host nucleus. Retroviruses use their own enzymes to reverse the normal path of genetic information. They first convert the RNA into a DNA molecule, which then enters the nucleus where it is inserted into the host cell's genome.

plasma membrane, endoplasmic reticulum, or Golgi apparatus. The outer envelope of HIV is made from the plasma membrane of the host. This feature allows viruses to go unidentified in the body, thus evading the immune system that is on the lookout for foreign particles. It is also an ingenious way for the virus to fool the host cell into allowing it to pass through the plasma membrane and enter its interior. Influenza A has several characteristics, including the structure of its envelope, that make it successful at causing infection. Because this virus causes the most cases of flu, we will next look at the structure, replication, and adaptations of this virus.

Viruses are difficult to treat because they mutate and evolve

Each year between 250,000 and 500,000 people around the world die of the flu. From three to five million people suffer from severe flu symptoms, which include a fever, cough, headache, muscle aches, fatigue, runny nose, and sore throat. The flu is mainly caused by two variants, or strains: influenza A and influenza B. Most healthy adults can spread the virus from one day before their symptoms arise to one week after they become sick, an important reason why the virus can spread so rapidly within a population. Although public awareness, sanitation, and vaccination greatly reduce the concern over influenza, the flu virus has ways of eluding these controls.

Different strains of the same virus are able to infect dissimilar species. The swine flu, a viral strain that normally infects pigs, causes hundreds of deaths each year in the U.S. Based on the proteins of the envelope, some strains of influenza A target pigs, whereas other strains target birds. Most viruses only infect one particular species; however, influenza is different in that some strains can mutate enough that the new viral envelope is able to cross over and infect multiple species. This is a public health concern because it is through these crossovers that we may be exposed to new strains to which we have no immunity (**Fig. 20.20**).

As a virus attacks, the immune system reacts to prevent reinfection by blocking the virus from entering the cell. The virus adapts by mutating and changing the surface proteins on its envelope, allowing it to gain entry into the host cell. The influenza virus is able to quickly evolve. As an RNA virus, replication occurs much faster with a higher rate of mutation because it does not include a proofreading step. Each round of influenza replication typically produces at least one mutation. If this mutation allows the virus to gain entrance into a cell to replicate, then natural selection favors this new strain, allowing the virus to evolve (**Fig. 20.21a**).

Another way the influenza virus evolves is through genetic recombination. If two strains of the influenza A virus infect a cell simultaneously, often a new genetic combination is produced (**Fig. 20.21b**). Each of the worldwide influenza epidemics has been caused by a strain newly introduced to humans and transmitted from another species. Fortunately, these dangerous new combinations are infrequent. When the genome of the virus is reshuffled, it still needs to be able to perform its job to help the virus replicate and propagate itself. If the combination is not just right, the virus particle is not able to survive. Because of the high rate of mutation and genetic recombination,

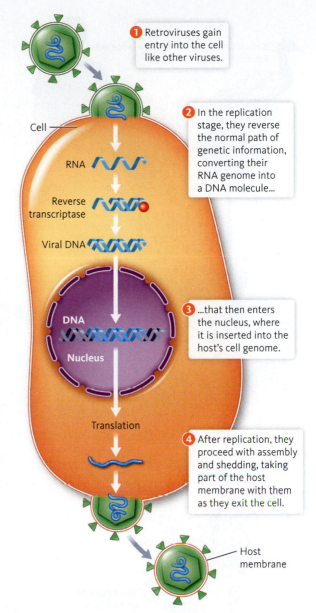

① Retroviruses gain entry into the cell like other viruses.

Cell

RNA

② In the replication stage, they reverse the normal path of genetic information, converting their RNA genome into a DNA molecule...

Reverse transcriptase

Viral DNA

DNA

③ ...that then enters the nucleus, where it is inserted into the host's cell genome.

Nucleus

Translation

④ After replication, they proceed with assembly and shedding, taking part of the host membrane with them as they exit the cell.

Host membrane

Figure 20.19 The retrovirus replication cycle.
① Retroviruses gain entry into the cell like other viruses do. **②** In the replication stage, they use specialized enzymes that they carry to reverse the normal path of genetic information, converting their RNA genome into a DNA molecule **③** that then enters the nucleus, where it is inserted into the host cell's genome. **④** After replication, they proceed with assembly and shedding, taking part of the host membrane with them as they exit the cell.
© Cengage Learning 2014

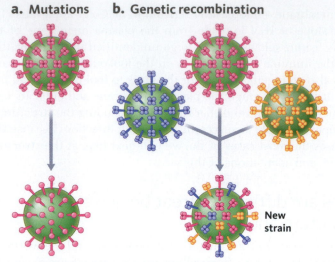

a. Mutations   **b. Genetic recombination**

New strain

Figure 20.21 Genetic recombination of the influenza virus. Mutations in the influenza virus recombine, creating new strains. Both mutations (a) and recombination of strains (b) create new strains of the influenza virus. Each of the worldwide influenza epidemics has been caused by a new strain introduced to humans and transmitted from another species. Due to the high mutation rate, new strains are common, and a new flu vaccination is necessary each year.
© Cengage Learning 2014

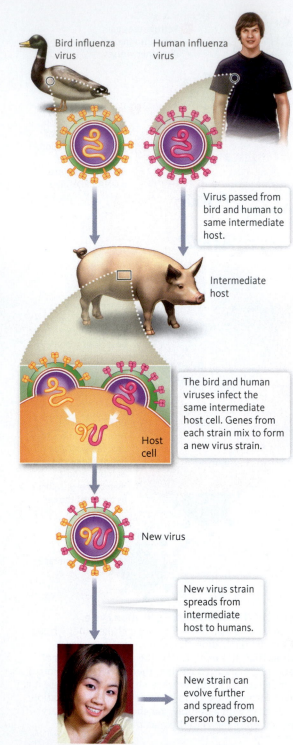

Bird influenza virus

Human influenza virus

Virus passed from bird and human to same intermediate host.

Intermediate host

The bird and human viruses infect the same intermediate host cell. Genes from each strain mix to form a new virus strain.

Host cell

New virus

New virus strain spreads from intermediate host to humans.

New strain can evolve further and spread from person to person.

Figure 20.20 Influenza strains can mutate and infect different species. The mechanisms by which the influenza virus can alter its genome and infect multiple species.
© Cengage Learning 2014

various strains circulate each year. As a result, each year a new flu vaccination is necessary.

Antiviral medicines lessen the effect of viral diseases

When you get the flu, antiviral drugs are sometimes used to shorten the effects of the viral infection by preventing viral replication. With their capsids, envelopes, and genomes, viruses are completely different in structure and function from bacteria, which have cell walls and more sophisticated components. Because of these fundamental differences, antibiotics are not effective in treating viral infections. Antiviral medicines inhibit the viral cycle by targeting various points of the virus replication cycle (**Fig. 20.22**). Some of these targets are generic and can be used on several viruses; others are specific to the virus. Developing antiviral drugs is complicated because viruses use host cells to replicate, and it is difficult to target a virus without harming the host. The high mutation rate of RNA viruses and retroviruses adds complexity to drug development.

Some viruses transform bacteria into disease-causing agents

Viruses play another important role in human infectious disease. Viruses also infect bacteria, causing some of the infected bacteria to transform from harmless to infectious agents capable of causing disease. The bacterium that causes cholera exists in both a harmless form and a pathogenic form. The deadly form results when a bacterial virus infects the bacterium. As part of the viral replication cycle, the virus inserts some of its DNA into the bacterial genome. The inserted gene codes for the production of the deadly cholera toxin. Once the viral gene is inserted into the bacterial genome, it is passed on to the next

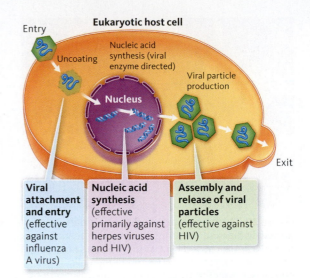

Figure 20.22 Drug targets for viruses. Antiviral medicines inhibit the viral cycle by targeting various points of the virus replication cycle.
© Cengage Learning 2014

Entry
Eukaryotic host cell
Uncoating
Nucleic acid synthesis (viral enzyme directed)
Viral particle production
Nucleus
Exit

Viral attachment and entry (effective against influenza A virus)

Nucleic acid synthesis (effective primarily against herpes viruses and HIV)

Assembly and release of viral particles (effective against HIV)

generation as the bacterium reproduces. These types of gene insertions cause diphtheria, scarlet fever, and food poisoning from some botulinum and staph bacterial toxins.

A number of animal viruses are known to cause cancer. For example, cervical cancer has been traced to HPV. In several cases, the mechanism of disrupting the cell cycle is similar to the gene transfer that occurs when some viruses infect bacteria. Insertion of the viral genome disrupts the cell cycle, causing uninterrupted cell division. The hepatitis B and hepatitis C viruses can cause cancer in humans through this type of gene transfer.

20.3 check + apply YOUR UNDERSTANDING

- Viruses are nonliving particles that take genetic control of host cells to make new viruses.

- Viruses are extremely diverse pathogens that infect bacteria, fungi, plants, and animals and cause many common infectious diseases.

- These intracellular parasites rely on a host cell to provide habitat, raw materials, and energy for replication. Most viruses are host-specific, infecting one particular species. However, some mutate their genetic information to infect other closely related species.

- Antiviral drugs are sometimes used to shorten the effects of a viral infection by blocking steps in the viral replication process.

1. Explain how viruses are nonliving intracellular parasites.

2. What are the two structural components that all viruses have? What is a third structure that most animal viruses have?

3. Why is it advantageous to the virus to use the host's cell membrane to form its outer envelope?

4. Imagine that you are a lab technician processing samples from a local stream that is used for drinking water. Your lab test results concern you because they indicate an abnormally high amount of bacteria viruses. Even though you know that bacteria viruses only infect bacteria, why are you concerned?

5. HPV is a sexually transmitted virus that is associated with cancer. Gardasil® is a recommended vaccine that reduces the rates of cancer. This vaccine contains HPV-like particles that activate the immune system, protecting vaccine recipients from infection. However, it does nothing to treat an existing infection. Why do you think this vaccine is administered to girls before they become sexually active? HPV shows stability. Based on what you learned about how viruses evolve, is HPV a DNA virus or an RNA virus?

20.4 Eukaryotic Parasites Are Adapted to Different Hosts and Different Habitats

Imagine living in the tropics of Africa and awakening in the middle of the night with a fever, headache, and chills. You are scared because you know these are the first symptoms of malaria, a disease that has caused the death of some of your friends and family members. You remember watching

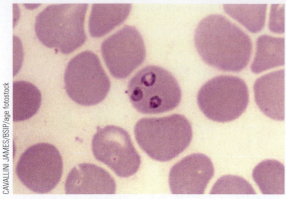

Figure 20.23 Malaria—an important human parasitic disease. The most severe and deadly cases of malaria are caused by a protist species known as *Plasmodium falciparum*. The red blood cells in the center are infected by parasites in an early stage of development.

your friends go through cycles of fever, painful muscle aches, and nausea that left them exhausted and weak. You drift back to sleep, hoping that if you have malaria, you can get to the doctor soon to minimize the spread of this infection throughout your body.

Malaria is considered to be the most important human parasitic disease, with about half of the world's population at risk of infection. Each year, nearly one million people die of this preventable and curable disease. The highest number of deaths occur among children younger than five years living in Africa.

Malaria is caused by several species of protist in the genus *Plasmodium* (**Fig. 20.23**). This single-celled eukaryotic parasite is spread between humans and the mosquitoes that flourish in the warm, rainy seasons of the tropics. The parasite has a complex life cycle. It relies on both its mosquito and human hosts during its various stages of development and reproduction (**Fig. 20.24**). As with many infectious diseases, transmission of the disease relies on a *vector* that carries the parasite from one host to the next. The mosquito vector for malaria not only

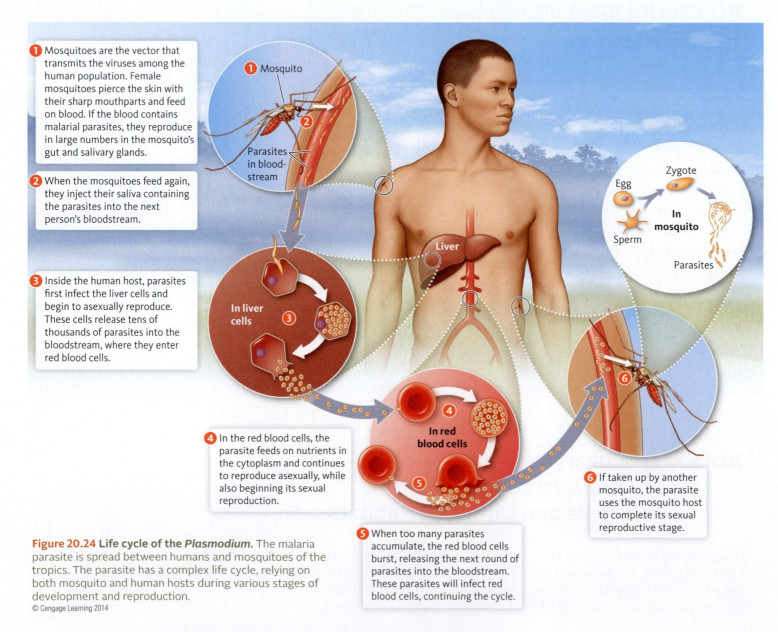

1 Mosquitoes are the vector that transmits the viruses among the human population. Female mosquitoes pierce the skin with their sharp mouthparts and feed on blood. If the blood contains malarial parasites, they reproduce in large numbers in the mosquito's gut and salivary glands.

2 When the mosquitoes feed again, they inject their saliva containing the parasites into the next person's bloodstream.

3 Inside the human host, parasites first infect the liver cells and begin to asexually reproduce. These cells release tens of thousands of parasites into the bloodstream, where they enter red blood cells.

4 In the red blood cells, the parasite feeds on nutrients in the cytoplasm and continues to reproduce asexually, while also beginning its sexual reproduction.

5 When too many parasites accumulate, the red blood cells burst, releasing the next round of parasites into the bloodstream. These parasites will infect red blood cells, continuing the cycle.

6 If taken up by another mosquito, the parasite uses the mosquito host to complete its sexual reproductive stage.

Figure 20.24 Life cycle of the *Plasmodium*. The malaria parasite is spread between humans and mosquitoes of the tropics. The parasite has a complex life cycle, relying on both mosquito and human hosts during various stages of development and reproduction.
© Cengage Learning 2014

transfers the virus; it also is an important host in the life cycle of the *Plasmodium.* This protist relies on the mosquito for part of its sexual development (see **Fig. 20.24**).

The interactions between the mosquito, *Plasmodium,* and humans illustrate that hosts and parasites have strong evolutionary ties to one another. As a parasite interacts with its host, the host's body defends itself, and the parasite adapts to these responses with countermeasures of its own. The use of two hosts is an adaptive advantage for the malaria-causing parasite. It shortens the time spent in one particular host and lessens the host's ability to adapt and respond. If the parasite stayed solely inside either the body of the human or that of the mosquito, the immune system of that host might eventually defeat the attacking species. *Plasmodium* also increases its fitness through a dispersal adaptation that allows it to travel easily to another host with a new set of resources. The parasite has one last advantage in this evolutionary arms race. Generally, parasites are adapted to hosts with a longer life span. The parasite produces more offspring to perpetuate the species in a shorter period, allowing the parasite to adapt more quickly to the host.

Malaria has been difficult to control because of the parasite's ability to adapt. The use of preventive drugs is challenging because, through natural selection, the parasite has evolved resistance to each of these drugs. For this reason, vaccine development has been unsuccessful. The best disease control programs involve either eliminating the mosquitoes that carry the parasite or preventing them from biting and transmitting the disease. Although the insects are becoming pesticide resistant, some progress has been made in controlling their populations by applying pesticides and by draining standing water where they lay eggs. Another effective control measure is the distribution and use of insecticide-treated bed nets, which have been shown to reduce the death of young children due to malaria by 20 percent (**Fig. 20.25**).

Unlike viral and bacterial pathogens, the malaria pathogen is eukaryotic. Eukaryotic parasites, which include fungi, worms, and protists, are common and remarkably diverse. They possess a wide range of adaptations that allow them to enter, live, and obtain nutrients from many different places in the human body. Ticks, lice, leeches, and fungi (such as athlete's foot) live on the outside of the body, whereas tapeworms and most protists live inside the host in the blood, intestines, or other tissues. In the section ahead, we will look at the adaptations and life cycles of different types of eukaryotic parasites that allow these organisms to survive and reproduce in different environments and, ultimately, in the human body habitat.

Figure 20.25 Malaria bed net. To help prevent malaria, insecticide-treated bed nets are used at night when mosquitoes are most active. The net is a physical barrier with an insecticide that lasts up to three years and both kills and repels mosquitoes. These bed nets are one of the most effective (and inexpensive) malarial treatments.

Most human fungal infections are localized to the skin, hair, and nails

In Chapter 15, you learned that some fungi specialize in degrading tissues of living organisms such as plants and humans. Fungi cause athlete's foot, sinusitis, skin diseases, and vaginal infections. In general, pathogenic fungi tend to cause infections on the surface of the body in the hair, skin, and nails, where they utilize keratin protein as a source of nutrients and energy. These fungi are adapted to penetrate animal cells to gain access to nutrients. They grow in the upper layer of the skin, causing inflammation and tissue damage.

Maybe you are familiar with the itching and burning sensation from one of the most common fungal skin diseases: ringworm, athlete's foot, and jock itch. This fungus is contagious and can easily spread by contact between individuals or through other items such as shower floors, pets, or tanning beds. Wrestlers commonly contract ringworm, and they are not allowed to participate in the sport when they have an infection because of

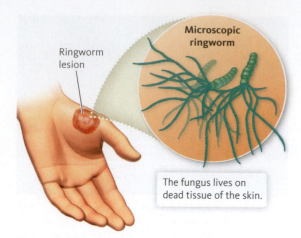

Microscopic ringworm

Ringworm lesion

The fungus lives on dead tissue of the skin.

Figure 20.26 Ringworm is a fungal disease localized to the skin. The infection is not caused by a worm but rather by several different types of fungi that form a ring-like pattern as they grow.
© Cengage Learning 2014

a.

© Scott Camazine/Alamy

b.

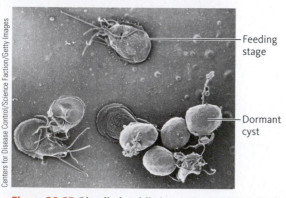

Centers for Disease Control/Science Faction/Getty Images

Feeding stage

Dormant cyst

Figure 20.27 *Giardia lamblia* is a common intestinal parasite. (a) *Giardia* is a protozoan that causes intestinal pain and diarrhea. (b) The two stages of the *Giardia* life cycle: the feeding stage (top center) and the dormant cyst stage (lower right).

Protozoans—unicellular, heterotrophic protists that must acquire nutrients from their environment. Some protozoa are also infectious disease agents in the form of parasites.

the contagious nature of the fungi. Despite its name, ringworm is not caused by a worm but rather by several different types of fungi. It is actually named for the pattern of inflammation created as the fungus grows outward (**Fig. 20.26**). Although they are irritating, most fungal pathogens are easily treated with antifungal medicines and are not considered a significant health threat.

Many protozoans and parasitic worms have complex life cycles

The malaria-causing *Plasmodium* belongs to a group of unicellular, heterotrophic protists referred to as **protozoans**. As heterotrophs, they must acquire nutrients from their environment. From an ecological perspective, protozoans serve an important role in the food chain, eating bacteria and algae that accumulate in aquatic and soil habitats. Some protozoa are also infectious disease agents in the form of parasites.

Protozoans can cause life-threatening diseases such as amoebic dysentery, African sleeping sickness, and malaria. Protozoan diseases are often found in areas of overpopulation and poverty, where there is a lack of proper sanitation, safe water, or sewage disposal. Many protozoan infections are spread when water contaminated with both sewage and protozoans is used for bathing or preparing food. In this way, they follow a fecal portal of exit and a mouth portal of entry: a fecal–oral disease. Protozoan infections can be successfully treated with drugs, but reinfection often occurs if the drinking water supply is not kept free from the disease-causing organisms. In the developing world, this cycle of treatment and reinfection can continue for years.

The protozoan *Giardia lamblia* is the most common intestinal parasite worldwide (**Fig. 20.27a**). Perhaps you have camped in the wilderness and boiled your drinking water to kill this parasite to prevent sickness. *Giardia* is found in soil, food, or water that has been contaminated with feces containing the parasite.

Like many protozoans, *Giardia* has a two-stage life cycle. The actively feeding form marks one stage, and dormancy dominates the second stage. Under adverse environmental conditions, *Giardia* forms a structure known as a **cyst**, which has a hard outer surface that protects it from desiccation (see **Fig. 20.27b**). Like a plant seed or fungal spore, this adaptation enables *Giardia* to live outside of the body for several months when food, water, or oxygen levels are low. It is also an effective means of dispersal, providing protection so that *Giardia* can survive the journey from one host to the next. The cyst is so durable that this protozoan can survive in chlorinated city water sources and swimming pools as well as in very cold mountain streams.

When you think of human infectious diseases, you usually don't think of worms, but on a global scale, parasitic worms cause most human infectious diseases. Like protozoan parasites, parasitic worms have complex life cycles. Roundworms, pinworms, tapeworms, and hookworms are common examples of parasitic worms (**Fig. 20.28**). They probably are more common than you realize: one in five people have roundworms that can grow to 8 inches long! They are especially abundant in tropical and subtropical regions of the world.

Parasitic worms typically live in the digestive tract, feeding from the host and causing weakness and disease. Some species of worms inhabit the blood and other tissues of the host. Parasitic worms do not kill their hosts but typically disrupt the hosts' ability to absorb nutrients, leaving them with abdominal discomfort, diarrhea, and no appetite.

Although there are some similarities in the life cycles of *Plasmodium* and parasitic worms, as multicellular organisms, parasitic worms have more complex reproductive systems and life cycles involving multiple habitats and hosts. Most parasitic worms involve **intermediate hosts** for the development of their *larval* stages and a second host for the *adult* stages of life. During the life cycle of a common tapeworm, like that of many worms, cysts are produced that form within the muscle tissue and develop around the larvae (**Fig. 20.29**). As in the protozoan life cycle, these cysts aid in dispersal and protection.

Tapeworms are most widespread in areas of poor sanitation. Infection is easily prevented with proper cooking techniques and is easily treated with specialized medications once diagnosed. For treatment, an oral drug that targets only tapeworms is used. The medicine is lethal to tapeworms on contact and acts by interfering with steps of aerobic respiration in their mitochondria. As in all respiring organisms, once aerobic respiration is shut down, the organism must be able to survive without oxygen or it will die. The drug targets an enzyme specific to worms and, therefore, does not harm human cells. This chemical compound kills the tapeworms by ultimately depleting them of energy.

It may puzzle you that given the availability of advanced medicines, millions of people in undeveloped nations die each year from parasitic worms. Although we have made progress in understanding and treating parasitic diseases, infections continue to increase around the world. One reason is that worms, as well as protozoa, are difficult to study in the laboratory. Both require the invasion of a suitable host to complete all or part of their life cycle; therefore, they cannot be studied independently. As mentioned earlier, it is difficult to treat these parasites without harming the host. Although these problems are factors in combating infectious disease, the largest reason for widespread problems of infection is lack of proper sanitation, clean water sources, proper hygiene, and available medicines.

Figure 20.28 Parasitic worms cause most human infectious diseases. Parasitic worms are common in humans, dogs, and farm animals. Examples include (a) the pinworm, (b) *Ascaris* roundworm, and (c) hookworm.

Cyst—a dormant stage of the life cycle of some parasites that consists of a structure which has a hard outer surface that protects it from desiccation and helps with dispersal.

Intermediate hosts—required host necessary for the completion of particular life stages of some parasitic organisms such as tapeworms. Most parasitic worms involve one host for the development of their *larval* stages and a second host for the *adult* stages of life.

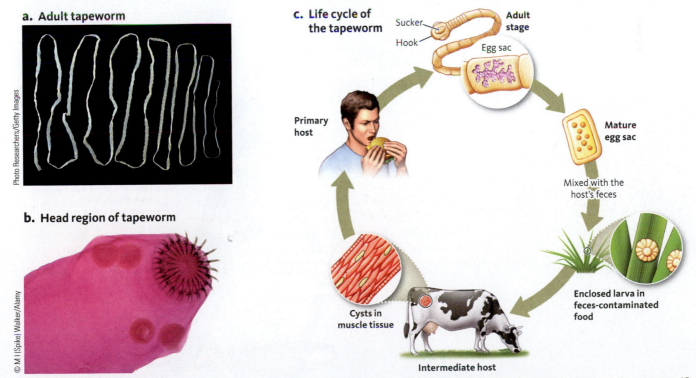

a. Adult tapeworm

b. Head region of tapeworm

c. Life cycle of the tapeworm

Sucker

Hook

Adult stage

Egg sac

Primary host

Mature egg sac

Mixed with the host's feces

Enclosed larva in feces-contaminated food

Intermediate host

Cysts in muscle tissue

Figure 20.29 The parasitic tapeworm. There are many species of tapeworms that live in the digestive tract. Most species are specific and feed on one particular host. (a) This beef tapeworm can grow up to 25 feet long in the human intestine. (b) Head region of a tapeworm with hooks for attachment. (c) Life cycle of the tapeworm, which uses an intermediate host and multiple habitats.
© Cengage Learning 2014

Now that you have been introduced to bacterial, viral, and eukaryotic pathogens, in the final section of the chapter we will explore how the body defends itself against these invaders. These parasites have evolved along with the immune system. The constant adaptations of both the host and the pathogen have allowed them to survive.

20.4 check + apply YOUR UNDERSTANDING

- Eukaryotic parasites, which include fungi, worms, and protists, are common and remarkably diverse. They have adaptations that allow them to enter, live in, and obtain nutrients from many different places in the human body, from the outside layers of the skin to the lining of the intestine.

- Many protozoans and worms have a two-stage life cycle, and some infect two or more hosts in their life cycle. Cysts are a dormant life stage that enables these parasites to survive unfavorable conditions outside of the body and become transmitted to a new host.

- Infectious diseases caused by eukaryotic parasites are often found in areas of overpopulation and poverty, where there is a lack of proper sanitation, safe water, or sewage disposal.

1. What structural adaptations do pathogenic fungi have to obtain their resources to live on human skin? What specific nutrients of the skin do they utilize?

2. Parasitic worms do not kill their hosts. How is this advantageous to the tapeworm?

3. Why is it more difficult to develop drugs against eukaryotic parasites than to bacterial parasites?

4. Parasites have many adaptations that enhance their survival and reproduction. During the life cycle of *Plasmodium falciparum,* infected people release an airborne chemical attractant that lures mosquitoes. Refer back to **Figure 20.24** and predict at which stage the parasite might release this chemical to increase survival.

5. Throughout this chapter, we have shown you examples of the life cycles of various organisms. **Figure 20.30** shows the life cycle of a liver fluke, which is another common parasitic worm. Three separate host organisms are shown in the figure. Identify these hosts as either the intermediate host or the final host.

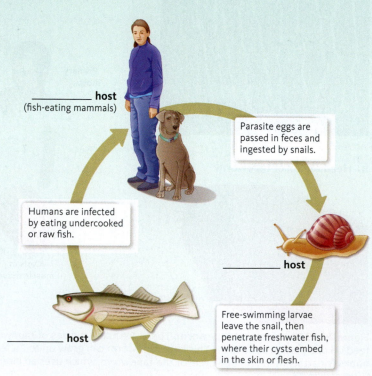

_____ host
(fish-eating mammals)

Parasite eggs are passed in feces and ingested by snails.

Humans are infected by eating undercooked or raw fish.

_____ host

Free-swimming larvae leave the snail, then penetrate freshwater fish, where their cysts embed in the skin or flesh.

_____ host

Figure 20.30 Liver fluke life cycle.
© Cengage Learning 2014

20.5 Humans Have Physical, Chemical, and Cellular Defenses against Pathogens

Puncture your skin on rusty barbed wire, and the bacteria present on the surface of the wire get direct access to your bloodstream. The bacteria living on the rust might include *Clostridium tetani,* the pathogen that causes lockjaw or tetanus (**Fig. 20.31**). When the skin is punctured, bacteria can enter into the body, where they grow, reproduce, and begin to produce a toxin. The bacteria don't spread far from the initial wound, but the toxin does. The toxin is able to enter neurons, and as the nervous system is affected, the toxin spreads throughout the body, first affecting muscles and nerves of the neck and jaw—hence the name lockjaw. In the case of tetanus, it is not the spread of the bacteria that causes disease and damage but rather the spread of the bacterial toxin throughout the nervous system.

Children typically are vaccinated against diphtheria, pertussis, and tetanus with a DPT vaccine given between the ages of two months and five years. Vaccines are designed to "educate" the body's immune system to recognize and eliminate a particular threat based on its specific molecular structure. The DPT vaccine contains a chemically treated version of the toxin that stimulates the immune system to be on the lookout for the tetanus toxin should it be encountered in the future.

This chapter began by presenting some infectious agents and how they cause disease. Looking at several agents sets the stage for understanding what the body must deal with: a large variety of pathogens and parasites of various sizes, molecular composition, and strategies. For instance, viruses pirate the genetic machinery of cells, whereas other pathogens produce enzymes and toxins that damage cells. In the section ahead, we will explore the three lines of defense between a pathogen and body cells (**Fig. 20.32**). The human body has a very strong set of defenses, ranging from thick physical barricades to an army of chemical and cellular agents that have been trained to detect and eliminate the pathogens and parasites we share the world with.

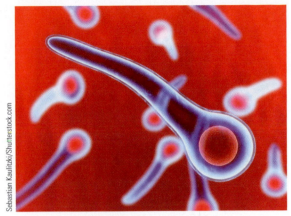

Figure 20.31 *Clostridium tetani,* **the pathogen that causes lockjaw or tetanus.** Tetanus bacteria live in low-oxygen conditions, such as those that exist in the soil and in rust. The dormant tetanus bacteria resemble tennis rackets.

Vaccines—a biological preparation that introduces a killed or inactivated virus or bacterial antigen into the body to "educate" the body's immune system to recognize and eliminate a particular threat based on its specific molecular structure.

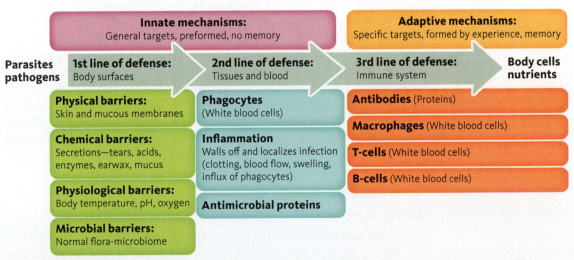

Figure 20.32 The three lines of defense between a pathogen and body cells. To gain nutrients or replicate successfully inside the body, parasites must breach three lines of defense: body surfaces, nonspecific defenses in the tissues and blood, and adaptive and specific defenses of the immune system.

© Cengage Learning 2014

The skin and mucous membranes are effective barriers against infection

Healthy skin provides a formidable physical barrier to most pathogens and parasites. Three characteristics make skin an effective barrier. First, epithelial tissue is composed of closely packed cells that function like a brick wall. Cells are tightly bound to other cells by intercellular bridges of proteins. Second, the outer layer of skin cells, or the epidermis, is relatively short-lived and sheds from the body as cells beneath it divide and grow. As the skin sheds, it continually removes microbes from the surface. Third, the outer skin surface is relatively resource-poor and inhospitable to microbial growth. Microbes of the body's normal flora are adapted to these conditions and generally outcompete pathogens that recently arrived on the scene. One critical aspect of skin is how dry it is. Bacteria need moisture for metabolic reactions and reproduction. This explains why populations of bacteria flourish in moist areas of the body, such as the feet, groin, and armpits.

Mucous membranes consist of epithelial tissues and their underlying connective tissue that line the body cavities and open to the external environment. Examples are the mouth, nose, eyes, ears, anus, urethra, and vagina. These linings represent an enormous surface area where pathogens frequently make their first contact with the body. Some mucous membranes have glands that secrete a thick, sticky fluid called mucus. Mucus is a sugar-protein molecule that keeps the cells moist and prevents bacteria from attaching to the epithelium. This coating is effective against most, but not all, pathogens. Cholera bacteria, for example, are adapted to penetrating the mucous layer in the intestines.

Body surfaces are also protected by mechanisms that physically wash away microbes and reduce colonization. Tears pass from the upper eyelid to the corner of the eye near the nose, where they drain into the nasal cavity. Saliva continually washes microbes from the mouth into the stomach, and urine washes microbes from the urinary tract. Perspiration, tears, and saliva contain a potent antibacterial enzyme that breaks down the peptidoglycan found in the cell wall of bacterial pathogens. Chronic health conditions that cause reduced tear or saliva secretions are major factors that weaken defenses and predispose the body to infections of the eyes and mouth.

The pH of the stomach is one of several environmental factors that determine whether infection and disease occur. Like all organisms, pathogens are adapted to a specific range of conditions that compose their ecological *niche*. Many pathogens infect hosts with a particular body temperature or tissues with a particular level of oxygen. For example, bacteria that cause dental cavities only live under low-oxygen conditions in the mouth, and anthrax bacteria that infect humans do not infect birds because of birds' high body temperature. These barriers are called *physiological* barriers to infection. They, along with the physical, chemical, and microbial barriers, make very effective surface barriers.

The second line of defense includes phagocytosis and inflammation

When the skin is broken, microbes living on the skin, such as staph bacteria, are among the first pathogens to enter the cut and invade deeper tissues. When this happens, the second line of defense against infection is rapidly activated. This line of defense involves white blood cells and blood proteins that possess three important characteristics. First, they act against any cel-

Epidermis—the upper or outer layer of the two main layers of cells that make up the skin. The epidermis is mostly made up of flat, scalelike cells. The deepest part of the epidermis also contains melanocytes that produce melanin, which gives the skin its color.

Mucous membranes—epithelial tissues and their underlying connective tissue that line the body cavities and open to the external environment. Examples are the mouth, nose, eyes, ears, anus, urethra, and vagina. These linings represent an enormous surface area where pathogens frequently make their first contact with the body.

Mucus—a thick, sticky fluid secreted by some mucous membranes that have glands. For example, the nasal passages secrete mucus that varies in consistency depending on whether you are congested or have a runny nose. Mucus is a sugar-protein molecule that keeps the cells moist and helps create a barrier to infection.

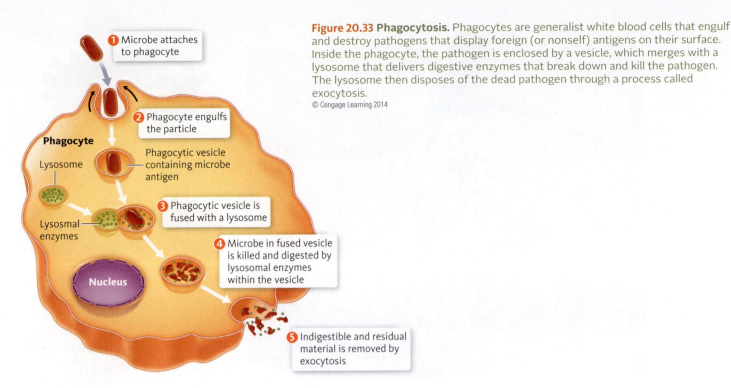

Figure 20.33 Phagocytosis. Phagocytes are generalist white blood cells that engulf and destroy pathogens that display foreign (or nonself) antigens on their surface. Inside the phagocyte, the pathogen is enclosed by a vesicle, which merges with a lysosome that delivers digestive enzymes that break down and kill the pathogen. The lysosome then disposes of the dead pathogen through a process called exocytosis.
© Cengage Learning 2014

❶ Microbe attaches to phagocyte

❷ Phagocyte engulfs the particle

Phagocyte

Lysosome

Phagocytic vesicle containing microbe antigen

Lysosmal enzymes

❸ Phagocytic vesicle is fused with a lysosome

❹ Microbe in fused vesicle is killed and digested by lysosomal enzymes within the vesicle

Nucleus

❺ Indigestible and residual material is removed by exocytosis

lular or chemical substance that is recognized as foreign to your body, or *nonself*. Microbes have specific molecules on their surface that the body detects as being "out of place." These molecules are called **antigens**. Proteins that make up the viral capsid and envelope, bacterial cell walls, or the cuticle of a hookworm parasite are examples of antigens. Second, the members of the second line of defense are *preformed*, which means that they are in place and ready to go if called. Third, this set of defenses is *nonspecific* and targets general threats. Essentially it is geared to eliminate any infectious agent or toxin it encounters. It has no memory of past infections and does not target specific pathogens, such as strep or staph bacteria.

Cells and molecules of the immune system patrol the body for signs of antigens and infection. A special group of white blood cells called **phagocytes** recognize, ingest, and kill microbes that display nonself antigens on their surface (**Fig. 20.33**). One type of phagocyte, called a **macrophage**, is able to capture and ingest as many as 100 bacteria at a time. Some phagocytes are attached to specific tissues in the liver, spleen, bone marrow, and lungs, where they filter microbes from the bloodstream. Other types of phagocytes circulate throughout the body on the lookout for antigens. The ingested microbe is killed by powerful digestive enzymes held in the phagocyte's lysosome (**Fig. 20.34**).

In addition to white blood cells, the body produces a number of defensive proteins that protect us. A special group of 20 different antimicrobial proteins is always circulating in the blood. These proteins will attach to any antigen they encounter. In the case of an invading bacterium, these antimicrobial proteins bind to particular sugars or proteins on the bacterial cell wall. One of three scenarios results from this binding: (1) the protein kills the bacterium directly, (2) it attracts phagocytes to come and ingest the bacterium, or (3) it releases chemical signals to intensify the inflammatory response.

The **inflammatory response** is triggered by infection or tissue damage and injury (**Fig. 20.35**). It functions to "wall off" an infection and prevent the spread of pathogens to other tissues. The inflammatory

Antigens—specific molecules on the surface of microbes that our body is able to detect as being "out of place." Examples of antigens are proteins that make up the viral capsid and envelope, bacterial cell walls, and the outer cuticle of hookworm parasites.

Phagocytes—a special group of white blood cells that are able to recognize, ingest, and kill microbes that display nonself antigens on their surface.

Macrophage—one type of phagocyte that is able to capture and ingest and destroy as many as 100 bacteria at a time.

Inflammatory response—a fundamental, nonspecific type of response by the body to disease and injury characterized by the classic signs of pain, heat (localized warmth), redness, and swelling. Inflammation is a key part of the body's defense system, providing an essential protective response by the body's system of self-defense. Common occurrences such as a mosquito bite, a splinter, a virus infection, a bruise, or a broken bone can trigger an inflammatory response and dispatch cells and chemicals to the site to repair the damage.

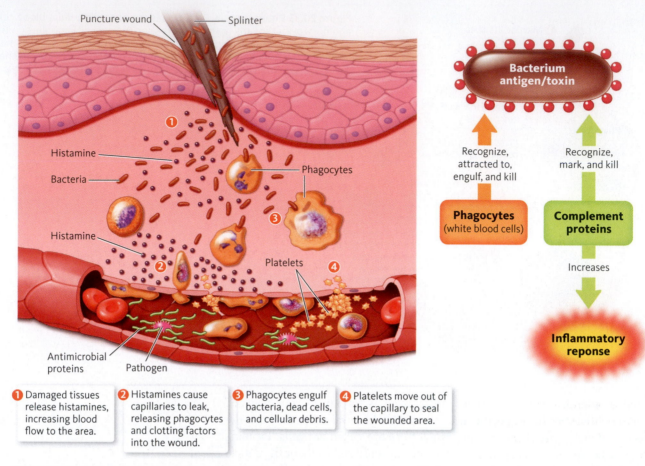

Puncture wound — Splinter

Histamine

Bacteria

Histamine

Phagocytes

Platelets

Antimicrobial proteins   Pathogen

Bacterium antigen/toxin

Recognize, attracted to, engulf, and kill

Recognize, mark, and kill

Phagocytes (white blood cells)

Complement proteins

Increases

Inflammatory reponse

❶ Damaged tissues release histamines, increasing blood flow to the area.

❷ Histamines cause capillaries to leak, releasing phagocytes and clotting factors into the wound.

❸ Phagocytes engulf bacteria, dead cells, and cellular debris.

❹ Platelets move out of the capillary to seal the wounded area.

Figure 20.34 Phagocytosis and inflammation. Phagocytes squeeze through capillaries to attack pathogens in the tissue spaces of the body before they enter the bloodstream. Within the blood vessels, antimicrobial proteins attach and signal an even greater inflammatory response.
© Cengage Learning 2014

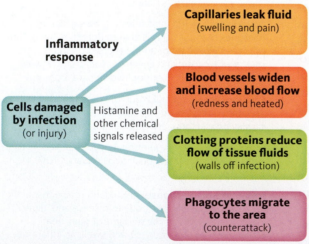

Inflammatory response

Cells damaged by infection (or injury)

Histamine and other chemical signals released

Capillaries leak fluid (swelling and pain)

Blood vessels widen and increase blood flow (redness and heated)

Clotting proteins reduce flow of tissue fluids (walls off infection)

Phagocytes migrate to the area (counterattack)

Figure 20.35 Inflammatory responses are triggered by infection or tissue injury. Damaged cells produce chemical signals that initiate four actions that comprise the body's inflammatory response. These actions function to wall off and overpower the infection.
© Cengage Learning 2014

response begins when injured cells release a chemical alarm signal. *Histamine* and other chemicals cause blood vessel walls to separate and leak fluid into the surrounding tissue spaces, resulting in the four cardinal signs of inflammation (see **Fig. 20.35**). Proteins in the leaked fluid make many small blood clots that slow the flow of tissue fluids around the area. This causes the infected area to *swell, which pushes on pain receptors.* Blood vessels dilate, or widen, to increase the flow of blood to the area, causing the inflamed area to appear *red and warm.* Phagocytes engulf the pathogen and clear the area of cellular damage and debris.

The third line of defense eliminates and remembers the specific pathogens it encounters

When the first and second lines of defense fail to prevent invasion and infection, the body has a powerful third line of defense: the **adaptive immune system**. This line of the body's defenses involves specialized defensive cells and molecules, which have the ability to recognize antigens at a very fine level of detail. This "high definition" allows the body to recognize and respond to a *specific* invader, such as tetanus, strep, or staph bacteria. Remarkably, once the body has been exposed to a pathogen, it will remember the pathogen's pattern of molecular antigens and be ready to protect the body in the future against the same threat. The encounter has been imprinted on the immune system's *memory*, which began with the first microbes you encountered on the day of your birth and continues to today. This is the reason why you do not contract the same infectious disease from the same strain of pathogen a second time. In other words, your body becomes *adapted* and protected through experience. If this is true, then why do we catch a cold many times during our lifetime? The virus that causes the common cold occurs in more than 100 different strains, and after we catch a cold, we then become immune to the specific strain of cold virus that infected us.

The adaptive immune system consists of organs and tissues of the lymphatic system and a group of white blood cells called **lymphocytes** (**Fig. 20.36**). At this moment, more than 10 trillion lymphocytes are patrolling the blood and lymphatic fluids of your body. There are many different types of lymphocytes, each of which perceives different antigens and performs different immune system functions. Like all white blood cells, lymphocytes begin their life in the red marrow of bones. Lymphocytes called **B-cells** remain in the *bone marrow* or migrate to the spleen, where they mature and learn to recognize specific antigens. B-cells specialize in eliminating parasites and pathogens such as staph and strep bacteria that attack cells in the blood and tissues.

A second group of lymphocytes, called **T-cells**, are "educated" in the *thymus gland* (**Fig. 20.36a**). Here they learn to make the sharp distinction between molecular markers that say "self" from antigens that indicate "nonself," or are foreign to the body. Unlike B-cells, T-cells probe and detect markers on cell surfaces and specialize in identifying and eliminating *virus-infected cells*. Let's now look at the two arms of the adaptive immune system, one that specializes in eliminating threats outside and another in eliminating pathogens that infect the body inside cells.

Attacking and destroying specific microbes in body fluids

Suppose a staph bacterium invades the body fluids through a wound and is engulfed by a macrophage in a lymph node. The macrophage not only destroys the pathogen but also displays parts of the staph's cell wall antigens on its surface, where it can be recognized by one of

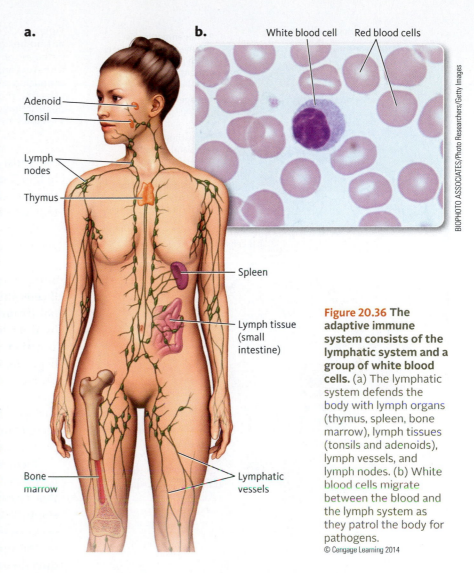

a.

Adenoid
Tonsil
Lymph nodes
Thymus
Spleen
Lymph tissue (small intestine)
Bone marrow
Lymphatic vessels

b. White blood cell   Red blood cells

BIOPHOTO ASSOCIATES/Photo Researchers/Getty Images

Figure 20.36 The adaptive immune system consists of the lymphatic system and a group of white blood cells. (a) The lymphatic system defends the body with lymph organs (thymus, spleen, bone marrow), lymph tissues (tonsils and adenoids), lymph vessels, and lymph nodes. (b) White blood cells migrate between the blood and the lymph system as they patrol the body for pathogens.
© Cengage Learning 2014

Adaptive immune system—a powerful third line of defense for the body that includes a wide range of specialized white blood defensive cells and molecules, which have the ability to recognize antigens at a very fine level of detail.

Lymphocytes—a group of white blood cells that always patrol the blood and lymphatic fluids of your body. There are many different types of lymphocytes, each of which perceives different antigens and performs different immune system functions.

B-cells—a specialized group of lymphocytes that remain in the *bone marrow* or migrate to the spleen, where they mature and learn to recognize specific antigens. B-cells specialize in eliminating parasites and pathogens such as staph and strep bacteria, which attack cells in the blood and tissues.

T-cells—a group of lymphocytes, which are "educated" in the *thymus gland* and learn to make the sharp distinction between molecular markers that say "self" from antigens that say "nonself," or are foreign to the body. Unlike B-cells, T-cells probe and detect markers on cell surfaces and specialize in identifying and eliminating *virus-infected cells*.

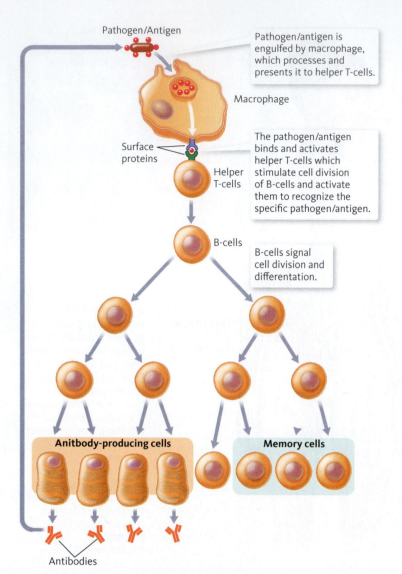

Pathogen/Antigen

Pathogen/antigen is engulfed by macrophage, which processes and presents it to helper T-cells.

Macrophage

Surface proteins

The pathogen/antigen binds and activates helper T-cells which stimulate cell division of B-cells and activate them to recognize the specific pathogen/antigen.

Helper T-cells

B-cells

B-cells signal cell division and differentiation.

Anitbody-producing cells

Memory cells

Antibodies

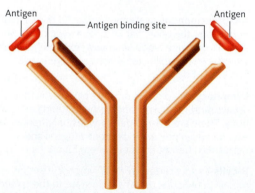

Antigen

Antigen binding site

Antigen

Figure 20.38 Antibody structure. Once attached, antibodies cause the destruction of the pathogen/antigen by marking it for destruction with antimicrobial proteins and phagocytes.
© Cengage Learning 2014

Helper T-cells—a specialized class of white blood cells that can work in concert with macrophages to destroy pathogens. As the name implies, helper T-cells *help* in the function and control of virtually all immune system activities.

Figure 20.37 During development, B-cells specialize to produce antibodies or memory cells. The initial encounter between a pathogen and the immune system triggers B-cells to divide and form two antigen-specific populations of cells: 1) antibody-producing cells and 2) memory cells. Specific antibodies then attach to the antigens on the pathogen and mark them for destruction.
© Cengage Learning 2014

the most important cells in the immune system: a helper T-cell. As the name implies, helper T-cells *help* in the function and control of virtually all immune system activities.

Figure 20.37 illustrates the series of events that begins when a macrophage engulfs a staph bacterium and displays it to a helper T-cell. The helper T-cell binds to the antigen on the macrophage, which in turn triggers B-cells into action. In addition to strong chemical signals from a helper T-cell, B-cells are activated by directly binding to antigens on the staph cell wall (**Fig. 20.37**). Together, these stimuli trigger the B-cell's nucleus to begin rapid mitosis and cell division. The result of this *primary immune response* is an army of antigen-specific B-cells that all express the exact same receptors at their surfaces. In other words, their education has been completed, and they are now employed in the body to perform a specific job.

Most B-cells go on to differentiate into a special cell that makes antibodies. **Antibodies** are Y-shaped proteins that circulate in the blood and bind specifically to the staph antigen they recognize (**Fig. 20.38**). This recognition occurs when the antigen binds to the specific binding site on the tip of the antibody molecule. The antibodies act directly on the staph bacteria by marking it for destruction with antimicrobial proteins and phagocytes, which in turn amplifies the inflammatory response. Antibodies also bind many antigens together into a large mass of bacteria, which is then "cleaned up," or cleared, by phagocytes attracted to the area. This hypothetical battle has involved the interaction between the second and third lines of defense and has resulted in swollen and tender lymph nodes. Your body has antibodies specific to every pathogen or antigen it has encountered during your life and can respond to eliminate those same threats in the future.

In addition to antibody-producing cells, another population of cells—**memory cells**—is made from the initial contact between B-cells and an antigen (see **Fig. 20.37**). Memory cells express receptors for the antigen just like other members of the population, but they circulate in the body for years or even decades. If the same antigen is encountered in the future, antigen-specific memory cells are able to respond rapidly to overwhelm the bacterium before it has a chance to cause an infection and disease.

Memory cells respond to an attack with a *secondary immune* response. This is the basis for **vaccination**, which introduces a killed or inactivated virus or bacterial antigen into the body to produce a population of memory cells. Vaccination activates your immune system in advance of you coming into contact with a known threat. The secondary response occurs faster and to

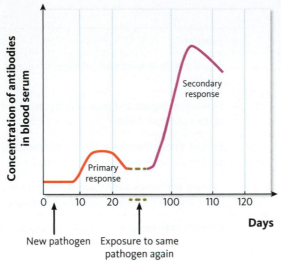

Figure 20.39 **Adaptive nature of B-cells.** Antibodies against a pathogen are made faster and in greater concentration after a pathogen causes a first-time infection.
© Cengage Learning 2014

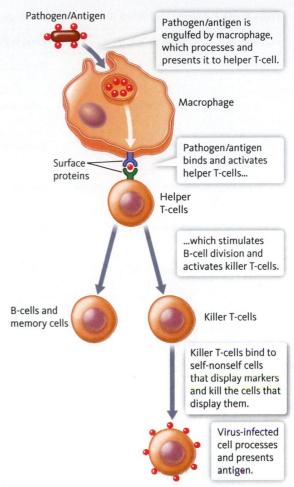

Figure 20.40 **The central role of the helper T-cell.** Helper T-cells coordinate the counterattack by signaling specific killer T-cells to attack the pathogen and by signaling B-cells to divide into an army of antibody-producing cells and B memory cells.
© Cengage Learning 2014

a greater level than the primary response (**Fig. 20.39**). A vaccine for measles (virus), mumps (virus), polio (virus), and tuberculosis (bacteria) triggers a primary immune response and the subsequent antibodies and memory cells against each of these pathogens. Once again, each vaccine introduces its specific antigen into the body, which selectively activates a small population of memory cells to divide and produce a large population of cells dedicated to eliminating the specific antigen. In some cases, a *booster shot* is needed to maintain a necessary level of protection. For instance, a booster shot against tetanus should be given every 10 years.

The immune response takes several days and sometimes weeks to develop a population of cells and antibodies large enough to fend off an infection. In some cases, such as tetanus, the pathogen can harm or kill a person before the immune system has built up the body's immune defenses. To counteract the toxin, a person is given a large dose of the antibodies specific to the toxin directly into the blood. This type of immunity does not last as long as the immunity provided by the body's own antibodies, when memory cells are made. Now let's turn to one of the most challenging scenarios your immune system faces: detecting and eliminating pathogens that get inside body cells.

Attacking and destroying infected cells

Many types of pathogens and parasites invade the body fluids and then live *inside* body cells. This list includes all viruses, bacteria that cause tuberculosis and meningitis, and protozoa that cause malaria and sleeping sickness. When these pathogens attack, antibodies are not effective because antigens are hidden inside cells. However, antibodies can protect the body against the spread of the disease outside of cells as well as against future infections. If antibodies and B-cells are ineffective, then how does the body cope with these intracellular pathogens? The answer is T-cells.

In addition to activating B-cells, the helper T-cell sends out signals that stimulate other T-cells to divide and produce a large population of specialized killer T-cells (**Fig. 20.40**). Unlike B-cells that only detect nonself anti-

Antibodies—a group of Y-shaped proteins that circulate in the blood and bind specifically to antigens. This recognition occurs when the antigen binds to the specific binding site on the tip of the antibody molecule.

Memory cells—cells made from the initial contact between B-cells and an antigen. Memory cells express receptors for the antigen, but they circulate in the body for years and even decades. If the same antigen is encountered in the future, memory cells respond rapidly to overwhelm the bacterium before it has a chance to cause an infection and disease.

Vaccination—typically, an injection that contains materials specifically designed to activate your immune system in advance of you coming into contact with a known pathogen.

Killer T-cells—a class of white blood cells that work in concert with helper T-cells. Killer T-cells directly kill infected cells before the pathogen has had a chance to replicate. Some T-cells become memory cells that circulate in the blood for years, waiting for the pathogen or antigen to reappear.

TABLE 20.2 Summary of White Blood Cells in Vertebrates

Component	Type of Cell	What It Perceives	Function(s)
Phagocytes	Phagocytes	General nonself antigens	Engulf and destroy nonself pathogens, bacteria, and cellular debris.
	Macrophages	General nonself antigens	Destroy pathogens and process and present antigens on surface to helper T-cells.
Lymphocytes	B-cell	Specific nonself antigens	Receive signals from helper T-cells to divide and multiply; attach to specific antigens and multiply.
	Antibody-producing cell	Specific nonself antigens	Produce antibodies that attach to and destroy objects that express specific antigens; enhance inflammatory response.
	Memory B-cell	Specific nonself antigens	Circulate with receptors that will bind to specific antigens; divide and multiply during secondary immune response.
	Helper T-cell	Specific self and nonself antigens	Bind to macrophages that have engulfed and processed antigens; signal B-cells to multiply and killer T-cells to attack.
	Killer T-cell	Specific self and nonself antigens	Receive signals from helper T-cells. Directly attack infected cells that display both self and specific nonself antigens on their surface.

© Cengage Learning 2014

gens, T-cells look for cells that express a combination of markers that includes *both* self and specific nonself antigens. B-cells *indirectly* kill pathogens by making antibodies, whereas killer T-cells *directly* kill infected cells before the pathogen has had a chance to replicate. Like B-cells, each killer T-cell expresses a single type of receptor on its surface. Some T-cells become memory cells that circulate in the blood for years, waiting for the pathogen or antigen to reappear in the future.

One very damaging strategy used by some viruses is to directly invade immune cells and suppress the body's ability to defend itself. In acquired immunodeficiency syndrome (AIDS), HIV attacks helper T-cells. By striking at the central regulatory cell of the immune system, HIV prevents the body from mounting an effective immune response. This leaves the body susceptible to infection from a wide variety of other pathogens, as well as cancer cells, which can then remain undetected in the body.

The human immune system isn't unique. Vertebrates such as fish, amphibians, reptiles, and mammals all share the same kinds of immune cells, or white blood cells (**Table 20.2**). The functions of only a small number of cells that participate in the immune response are highlighted in **Table 20.2**. Many more types of cells have evolved as part of an immensely complex and effective defense against pathogens. In turn, pathogens have evolved countermeasures that allow them to persist and survive in human hosts. The constant attack and counterattack have been shaped by changes in parasite and host genomes through natural selection driven by sexual reproduction and mutation. Pathogens that mutate are much more difficult for the host to recognize, and, in turn, changes in the host genome create new barriers and defenses against disease.

In this chapter we have surveyed various pathogens that cause infectious diseases in humans. We have examined the responses the body has to combat these pathogens. Infectious disease results from the interaction of a pathogen and a susceptible human host in an environment favorable for the pathogen. The infectious disease cycle begins when the pathogen is transmitted from a reservoir through a portal of entry, attaches to a target tissue, and multiplies to cause infection. Host defenses counterattack through three lines of defense: surface barriers, white blood cells, and molecules comprising the nonspecific and adaptive immune systems.

20.5 check + apply YOUR UNDERSTANDING

- The human body has a very effective set of defenses, ranging from thick physical barricades to an army of chemical and cellular agents that have been trained to detect and eliminate pathogens foreign to the body.

- Body defenses include the skin and mucous membranes that support the first line of defense. The second line of defense involves white blood cells and special blood proteins. Antimicrobial proteins are a part of the inflammatory response, which prevents the spread of pathogens to other tissues. The final line of defense is the adaptive immune system, which recognizes, eliminates, and remembers the specific pathogens it encounters.

1. What is an antigen? Name two antigens that commonly challenge the body.

2. Describe three characteristics of the second and third lines of defense. Explain.

3. Place each of the following molecules and cells into its correct column (does it attack pathogens outside or inside body cells?) and row (is it a molecule or cell?) in the table below:

   antibodies   helper T-cells
   antibody-producing cell   histamine
   B-cells   killer T-cells
   complement proteins   memory cells
   enzymes   mucus
   gastric juice

	Recognize and Attack Pathogens Outside of Cells	Recognize and Attack Pathogens Inside Cells
Molecules		
Cells		

4. A friend has been sexually active for many years and is concerned about HIV or herpes virus. A blood test reveals the presence of antibodies to the herpes simplex virus but not to HIV. What does this mean in terms of his exposure to pathogens?

5. Predict the consequence to a person if a virus infects the thymus during early fetal life, when his or her immune system is still developing.

End of Chapter Review

Self-Quiz on Key Concepts

Bacterial Pathogens

KEY CONCEPTS: Bacteria are small prokaryotic cells surrounded by a rigid cell wall that determines the shape of the cell. They lack a nucleus surrounded by a membrane and other organelles such as mitochondria, endoplasmic reticulum, or Golgi complexes. Although most bacteria that we encounter each day are either beneficial or neutral in our lives, some cause disease. Many species have the ability to protect themselves and cause disease by producing toxins and chemicals that inhibit antibiotics used to prevent infection or disease. Infectious diseases caused by bacteria and other pathogens follow a series of steps called the disease cycle. The ability of any pathogen to cause infection depends on the genes it carries and the proteins it expresses.

1. Match each of the following **bacterial structures** with its characteristic, description, or example.

   Capsule
   Cell wall
   Toxin
   Plasmid

   a. a substance that changes host cell function
   b. a substance or structure for attachment
   c. a small piece of circular DNA
   d. the part of the cell that provides structure, shape, and protection

2. Place these steps in the infectious disease cycle in the correct sequence.
   ____ Entry into host
   ____ Exit from host
   ____ Reservoir holding the pathogen
   ____ Infection of host cell
   ____ Attachment to cell
   ____ Disease state
   ____ Cellular damage

3. There is a difference between being infected with a pathogen and having a disease. Select the statements below that are correct.
   a. A carrier without any symptoms is infected but not diseased.
   b. A carrier without any symptoms is infected and diseased.
   c. A person with symptoms is infected and diseased.
   d. A person with symptoms is not infected but has the disease.

Viral Pathogens

KEY CONCEPTS: Viruses are nonliving particles that take genetic control of host cells to make new viruses. Viruses are extremely diverse pathogens that infect bacteria, fungi, plants, and animals, causing many common infectious diseases. These intracellular parasites rely on a host cell to provide habitat, raw materials, and energy for replication. Most viruses infect one particular species; however, some mutate their genetic information to infect other closely related species. Antiviral drugs are sometimes used to shorten the effects of a viral infection by blocking steps in the viral replication process.

4. Match each of the following **virus structures** with its characteristic, description, or example. Each term on the left may match more than one description on the right.

   Capsid
   Envelope
   Nucleic acid core

   a. contains both proteins and lipids, often made from host cell membranes
   b. protein coat that surrounds the genome
   c. can be RNA or DNA
   d. provides structure, shape, and protection

5. Which characteristic best describes viruses?
   a. eukaryotic
   b. prokaryotic
   c. noncellular
   d. prokaryotic and noncellular

6. Which statement best describes a virus replication sequence?
   a. replication, attachment, assembly, and shedding
   b. shedding, attachment, replication, and assembly
   c. shedding, assembly, replication, and attachment
   d. attachment, replication, assembly, and shedding

Eukaryotic Parasites

KEY CONCEPTS: Eukaryotic parasites, which include fungi, worms, and protists, are common and remarkably diverse. They have adaptations that allow them to enter, live in, and obtain nutrients from many different places in the human body, from the outside layers of the skin to the lining of the intestine. Many protozoans and worms have a two-stage life cycle, and some infect two or more hosts in their life cycle. Cysts are a dormant life stage that enables these parasites to survive unfavorable conditions outside of the body and become transmitted to a new host. Infectious diseases caused by eukaryotic parasites are often found in areas of overpopulation and poverty, where there is a lack of proper sanitation, safe water, or sewage disposal.

7. Match each of the following **eukaryotic parasites** with its characteristic, description, or example. Each term on the left may match more than one description on the right.

   Protozoan   a. unicellular
   Fungus   b. multicellular
   Worm   c. uses cysts as part of its life cycle
        d. uses intermediate hosts to complete its life cycle
        e. often uses more than one host to complete its life cycle

8. Pathogens that cause human disease include:
   a. bacteria and viruses
   b. fungi, protozoans, and worms
   c. protozoans and worms
   d. bacteria, viruses, fungi, protozoans, and worms

9. Some eukaryotic parasites use a cyst as part of their development. This structure is important for: (select all that apply)
   a. dormancy when nutrients and resources are low.
   b. feeding and growing.
   c. aiding in dispersal to a new host.
   d. providing a protective coat to survive outside of the host.

The Human Body's Defenses against Pathogens

KEY CONCEPTS: The human body has a very effective set of defenses, ranging from thick physical barricades to an army of chemical and cellular agents that have been trained to detect and eliminate pathogens foreign to the body. These defenses include the skin and mucous membranes, which provide the first line of defense. The second line of defense involves white blood cells and special blood proteins that counterattack pathogens outside cells, in the blood and extracellular fluids, and inside cells, where pathogens take refuge and hide. Antimicrobial proteins are part of the inflammatory response that prevents the spread of pathogens to other tissues. The final line of defense is the adaptive immune system, which recognizes, eliminates, and remembers the specific pathogens it encounters. This arm of the body's defenses attacks and destroys microbes in body fluids and infected cells.

10. Match each of the following **lines of defense** with its characteristic, description, or example. Each term on the left may match more than one description on the right.

First line of defense
Second line of defense
Adaptive immunity

a. skin and mucous membranes
b. secretions such as tears and mucus
c. changes in body temperature, pH, and oxygen levels
d. antimicrobial proteins
e. macrophages
f. inflammation
g. antibodies
h. phagocytes
i. T-cells and B-cells

11. Which of the following are characteristics of antibodies?
    a. part of the adaptive immune system
    b. made by B-cells
    c. circulate in the blood and fluids and bind specifically to antigens
    d. all of the above

12. Which of the following are included in adaptive immunity responses?
    a. antibodies, macrophages, T-cells, and B-cells
    b. physical and chemical barriers
    c. antibodies
    d. T-cells and B-cells

13. Many times multiple doses of a vaccine are given to prevent disease. A second vaccine dose initiates a response that:
    a. lasts longer than the first.
    b. makes antibodies more quickly and in greater amounts than the first vaccination.
    c. activates a small population of memory cells.
    d. all of the above

Applying the Concepts

14. The brain and spinal cord are nourished and cushioned by circulation of a fluid called cerebrospinal fluid. This fluid normally is sterile and flows between membrane layers surrounding the brain called meninges. However, in the disease meningitis, the cerebrospinal fluid becomes infected with pathogens, usually caused by either a *viral* or a *bacterial* infection. Knowing whether a virus or a bacterium causes meningitis is important because the severity of illness and the treatment differ. Which of the following components of the body's defenses would you expect to find in the cerebrospinal fluid if it is a *bacterial* infection? (a) macrophages, (b) killer T-cells, (c) B-cells, (d) antimicrobial proteins.

15. Dental plaque is actually a type of biofilm formed by bacteria colonizing and attaching themselves to the surface of the teeth. If not removed within 48 hours, the plaque begins to harden into tartar, which is rock hard and difficult to remove. Bacteria in dental plaque live in a protected environment with little oxygen and produce lactic acid that can lead to gum disease and tooth decay. (a) How does forming a biofilm adapt oral bacteria to survive in the mouth? (b) Where do these bacteria acquire the nutrients to provide energy and raw materials to grow and reproduce? (c) Which environmental factors in the mouth favor the growth of oral bacteria? (d) Which metabolic process do oral bacteria use to produce energy? (e) Explain how brushing and flossing prevent tooth decay.

16. Hookworms are internal parasites of humans, cats, and dogs. The life cycle is shown in **Figure 20.41**. For each of the following prevention and treatment methods, determine where in the life cycle (A, B, C, D) of the parasite the treatment works: (a) using a deworming drug, (b) cleaning up dog feces from a kennel or yard, (c) having children wash their hands after playing and eating, (d) preventing the cat from using the sandbox as a litter box, (e) treating pregnant dogs with deworming drugs, (f) not walking barefoot in areas where people walk their dogs.

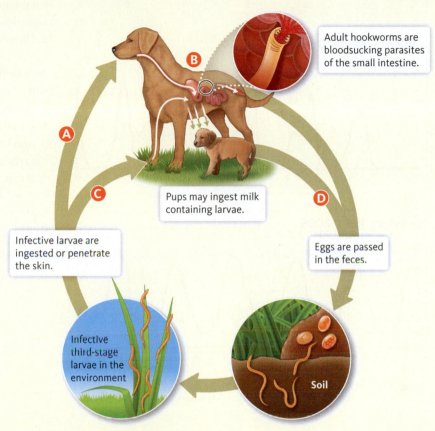

Adult hookworms are bloodsucking parasites of the small intestine.

Pups may ingest milk containing larvae.

Eggs are passed in the feces.

Infective larvae are ingested or penetrate the skin.

Infective third-stage larvae in the environment

Soil

Figure 20.41 Hookworm life cycle.
© Cengage Learning 2014

Data Analysis

The Flu Season and Seasonal Cycles of Infectious Diseases

Influenza spreads around the world in seasonal epidemics, resulting in the deaths of between 250,000 and 500,000 people every year. The flu is caused by the influenza virus, which appears in populations as three major types: A, B, and C. The A type, responsible for many large outbreaks, has at least 10 different strains (**Fig. 20.42**). Each season as many as five different strains break out and spread by aerosol droplets of saliva from one person to another. The cycle is so regular that outbreaks can be predicted and even tracked from one region to the next.

Figure 20.42 Influenza virus.
© Cengage Learning 2014

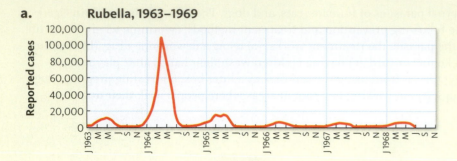

a. Rubella, 1963–1969

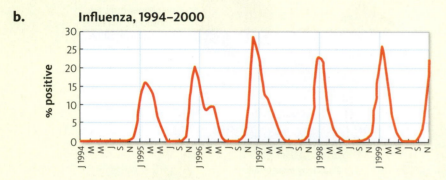

b. Influenza, 1994–2000

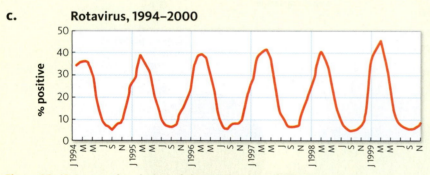

c. Rotavirus, 1994–2000

Figure 20.43 Seasonal variation of three infectious diseases.
Dowell SF. 2001. Seasonal variation in host susceptibility and cycles of certain infectious diseases. Emerg Infect Dis 7, 369–374.

Many other infectious diseases also show an epidemic, or a rapid increase in the number of people infected (**Fig. 20.43**). The graphs in **Figure 20.43** plot data on a bimonthly basis for three viral diseases: German measles (rubella), the flu (influenza), and rotavirus. Rubella is reported as the number of reported cases from 1963 to 1969, whereas the influenza and rotavirus graphs show the percent of specimen samples that tested positive for these two pathogens from 1994 to 2000. All three diseases are highly contagious, with symptoms that develop rapidly but last for only a short time.

Rotavirus is a virus that infects the intestinal tract of almost all young children by age 5 years and is a common cause of severe diarrhea among infants and young children. A strain called rotavirus A accounts for greater than 90 percent of cases and is endemic, or is constantly present in all populations worldwide.

Adapted from Dowell SF. 2001. Seasonal variation in host susceptibility and cycles of certain infectious diseases. Emerg Infect Dis 7, 369–374.

Data Interpretation

17. In which season(s) do these three viral outbreaks occur?

18. Are the infection peaks for rotavirus consistent from year to year? Is the length of the rotavirus season also consistent?

19. How long does the flu season last in the United States?

20. Which of the three pathogens shows a large epidemic? How many more people were infected during this outbreak than in a regular, smaller outbreak?

Critical Thinking

21. Would the months be the same in the northern and southern hemispheres? How does this affect the infection rates of people worldwide?

22. Which environmental factors do you think play a role in the seasonal cycles of these viral diseases?

23. How would the seasonal pattern of a disease that is vectored by a mosquito be more difficult to explain?

24. If people in the population develop immunity to the flu, why is there a new outbreak every year? Is it possible to get a flu shot and still get influenza during flu season?

Question Generator

Macrophages Fail to Kill Pathogenic Strains of Tuberculosis Bacteria

Macrophages are white blood cells that roam throughout tissue spaces or are attached to tissues. They recognize, attack, destroy, and engulf invading bacteria, viruses, and other injurious agents. A macrophage is capable of engulfing as many as 100 bacteria at a time.

The infectious agent for tuberculosis is the bacterium *Mycobacterium tuberculosis.* It is predominantly transmitted through aerosol droplets containing one to three bacteria. Once they gain access to the new host's lungs, nonpathogenic strains are engulfed and killed by the macrophages that line the walls of the lungs. Some pathogenic bacteria are engulfed, while others are not. Those that are engulfed are able to survive and even multiply within the macrophage (**Fig. 20.44**). Therefore, the bacteria are able to hide within the very agents that are specialized to kill them.

Below is a block diagram of the steps involved in phagocytosis (**Fig. 20.45**) that relates to the interaction between the pathogenic strains of *M. tuberculosis* bacteria inside macrophages. Use the background information along with the block diagram to ask questions and generate hypotheses. We've translated the components of the block diagram into several questions to get you started.

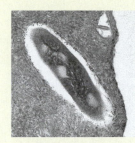

Figure 20.44 Tuberculosis in a lysosome.
Reprinted from Cell Host & Microbe, Volume 10, Issue 3, Hélène Botella, Pascale Peyron, Florence Levillain, Renaud Poincloux, Yannick Poquet, Irène Brandli, Chuan Wang, Ludovic Tailleux, Sylvain Tilleul, Guillaume M. Charrière, Simon J. Waddell, Maria Foti, Geanncarlo Lugo-Villarino, Qian Gao, et al., "Mycobacterial P1-Type ATPases Mediate Resistance to Zinc Poisoning in Human Macrophages", pp248–259., 2011, with permission from Elsevier.

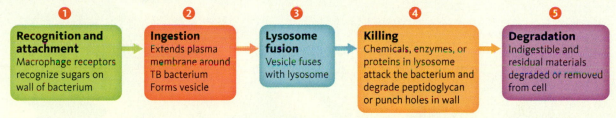

① Recognition and attachment
Macrophage receptors recognize sugars on wall of bacterium

② Ingestion
Extends plasma membrane around TB bacterium
Forms vesicle

③ Lysosome fusion
Vesicle fuses with lysosome

④ Killing
Chemicals, enzymes, or proteins in lysosome attack the bacterium and degrade peptidoglycan or punch holes in wall

⑤ Degradation
Indigestible and residual materials degraded or removed from cell

Figure 20.45 Factors relating to phagocytosis, pathogenic strains of *M. tuberculosis* bacteria, and macrophages.
© Cengage Learning 2014

Here are a few questions to start you off.

1. What is the difference between the sugar molecules on the pathogenic and the nonpathogenic strains? (Block #1)

2. How does the capture by a vesicle affect a tuberculosis bacterium? (Block #2)

3. How long does it take from attachment to fusion? Is there a time delay in pathogenic strains? (Block #3)

4. What kinds of chemicals do macrophages use to kill pathogenic and nonpathogenic strains? (Block #4)

5. Are there any signs of degradation in pathogenic bacteria? Is it successful at all? (Block #5)

Use the diagram to generate your own research question and frame a hypothesis.

21 The Biology of Chronic Disease

21.1 Early humans were adapted to their environment; modern humans are not

21.2 Chronic diseases result from genetic and behavioral responses

21.3 Cardiovascular disease is caused mostly by behavioral factors

21.4 Mutations that alter cell-cycle regulation cause cancer

21.1 Early Humans Were Adapted to Their Environment; Modern Humans Are Not

Figure 21.1 Early humans compared to modern humans. Artistic portrait of early humans (the oldest humans known from 1.8 million to 143,000 years ago) compared to an American couple today.

Take a moment to imagine how your *Homo sapiens* relatives lived thousands of years ago when early humans first migrated to the Americas (**Fig. 21.1**). Those ancestors were adapted to a life of intense physical activity. In order to survive they hunted and gathered their food, defended themselves against large mammals and other dangers, and sought shelter. Because there was no established agriculture, they spent their days in search of animals and plants that made up their diet. Meat was their source of protein, and fruit and vegetables were their main source of carbohydrates. Studies show these hunter–gatherer ancestors were generally healthy, fit, and largely free of the chronic diseases common today. They were healthy and fit in part because they were accustomed to strenuous physical activity and a diet much richer in fiber, vitamins, and minerals than that provided by our modern diet (**Table 21.1**).

Why are we looking at the lifestyles and dietary habits of our hunter–gatherer ancestors? A partial answer comes when we consider the following timeline. If we estimate that 100,000 generations of people were hunter–gatherers, then we can say that only about 500 generations have depended on agriculture, only 10 generations have lived since the start of the Industrial Age, and only 2 generations have grown up with highly processed fast foods. The evidence provided by fossils, other preserved specimens, and DNA and protein sequence analysis tells us that the genes we carry today are 99.5 percent similar to those of the hunter–gathers. This means that our bodies are adapted to an ancestral diet and lifestyle that, in general, we no longer follow.

From a biochemical point of view, an optimally functioning human body requires more than 50 nutrients. When that body has either too few

TABLE 21.1 Comparison of Early Human Diet to Modern Diet

	Early Human Diet	Modern Human Diet
Protein	Very lean	Fatty
Carbohydrates	Vegetables	Grains/refined sugars
Fats	Balanced intake	Proinflammatory
Animal/fish	65% of diet	15% of diet
Vegetables/fruit	100 different plants	Small selection
Fiber	100 grams/day	20 grams/day
Vitamins/minerals	High intake	Low intake
Grains	None	Substantial
Dairy	None	Substantial
Pressed oils	None	Substantial
Trans fatty acids	Negligible	Substantial
Alcohol	None	3% of overall calories

© Cengage Learning 2014

or too many of these nutrients, genes cannot do their jobs, and cell structure and function are compromised. When hunter–gatherer societies transitioned to an agricultural, grain-based diet, their general health deteriorated. Studies of partial skeletons reveal that this dietary change resulted in shorter life spans, higher childhood mortality, and a higher incidence of osteoporosis, rickets, and other diseases caused by mineral and vitamin deficiencies.

In the 21st century we have continued to modify our diet and lifestyle, with the outcome that many chronic diseases have become commonplace. Chronic diseases, which include heart disease, diabetes, cancer, Alzheimer disease, obesity, and arthritis, are long-term diseases that affect a person's function for three or more months. At a national level there is great concern regarding the overall declining health of Americans. Many factors contribute to our current health conditions, and obesity is one of our top medical concerns.

Humans are integrated biological systems with highly ordered parts that are interdependent and interactive. These component parts are dynamic, constantly changing, and adapting to environmental inputs and conditions. When one part fails to communicate or connect with another part, the system starts to break down. Obesity is a condition that negatively affects many parts of the system, leading to a cascade of problems at the hormonal and cellular levels. In addition to decreased life expectancy, the risks of suffering from other preventable and chronic diseases, including cardiovascular disease, hypertension (high blood pressure), diabetes, and some cancers, increase as obesity levels increase.

Preventing chronic disease requires a closer look at the major conditions that affect us—namely, heart disease and stroke, cancer, diabetes, arthritis, respiratory diseases, and obesity (Fig. 21.2). Although each of these diseases has a genetic component, the growing prevalence of the disease is simply the body's poor response to our modern environment. These responses take place throughout the entire hierarchy of organization, from cells to tissues and organs to organ systems, all the way up to the scale of human populations. In this chapter, we will apply and integrate many of the biological concepts you have learned throughout this textbook to examine the underlying causes of a few of our nation's top health concerns. From among the many chronic diseases affecting the health of the human population, we focus on those diseases that can be directly impacted by you and your lifestyle decisions. The earlier in your adult life that you make healthy decisions, the better your chances that you can prevent the high cost and suffering associated with these conditions.

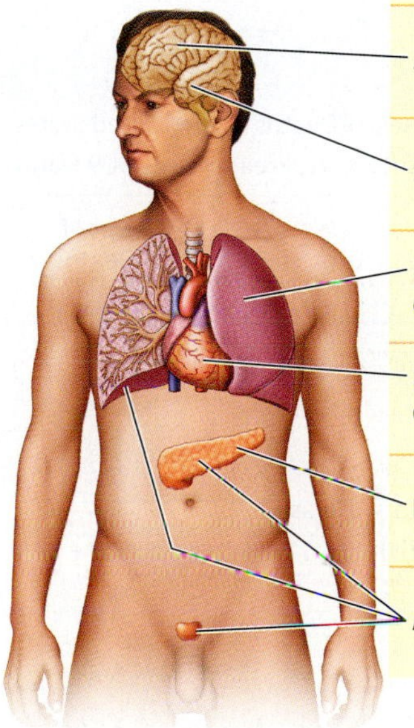

Cause of death	% of U.S. deaths
Alzheimer's	3%
Stroke	5%
Respiratory disease	5%
Heart disease	25%
Diabetes	3%
All cancers	23%

Figure 21.2 Chronic diseases. Chronic diseases such as obesity, diabetes, heart disease, and cancer account for greater than 70 percent of the deaths in the United States. These diseases are based on genetic factors, which cannot be modified, but also are greatly influenced by lifestyle factors, which are modifiable.
© Cengage Learning 2014

21.2 Chronic Diseases Result from Genetic and Behavioral Responses

Chronic diseases affect people of all ages. Maybe you or a friend has asthma, one of your parents has high blood pressure, or an elderly relative has Alzheimer disease. Whereas some chronic disease symptoms can be managed, usually a chronic illness does not go away. Worldwide, chronic diseases are the top cause of death, accounting for 60 percent of all deaths. In fact, twice as many people die of chronic illnesses as they do of nutritional

Chronic disease—an illness or medical condition that lasts over a long period and sometimes causes a long-term change in the body that affect a person's function for three or more months. Examples are heart disease, diabetes, cancer, Alzheimer disease, obesity, and arthritis.

deficiencies and all infectious diseases, including HIV/AIDS, tuberculosis, and malaria combined. Another startling statistic is that worldwide deaths from chronic diseases are projected to increase by 15 to 20 percent over the next 10 years. Chronic diseases represent an epidemic that demands the attention of individuals, governments, scientists, and healthcare providers around the world.

A question you may have is "Why are chronic diseases so prevalent today?" One possible reason is that we are no longer well adapted to our current environment. Throughout the text, we've presented examples of how organisms are adapted to their environment, including those of our ancestors described in Section 21.1. In general, natural selection favors genetic mutations and combinations that produce traits that are adaptive in nature, that is, traits that increase fitness and promote survival and reproduction. However, along the evolutionary pathway some mutations occur that decrease fitness, survival, and reproduction. These are considered *maladaptive.* The same can be said of behaviors and lifestyle choices: some enhance fitness, whereas others result in diseases such as asthma, high blood pressure, and cancer. From a biological point of view, *maladaptation* is not a term used to judge other people's behaviors or to blame them. Rather, it means that humans, like all organisms, are not perfectly adapted to their environment. In this section, we will explore chronic diseases that result from genetic and behavioral responses to our environment.

Chronic diseases are prevalent, costly, and often preventable

In the United States, chronic diseases such as heart disease, cancer, and diabetes are the leading causes of death and disability (**Table 21.2**). It's important to note that although the majority of people with chronic conditions are younger than 65 years, the likelihood of having a chronic condition increases as you age. Also, as life expectancy increases, more people will be living with chronic illnesses. For example, hypertension (high blood pressure), the most common chronic condition, affects a greater percentage of older people than younger people. Biologically, as we age our blood vessels stiffen, raising the blood pressure in the body as the force of blood through the circulatory system causes an inflammatory response in the blood vessels. As the blood vessels are exposed to this inflammation throughout our lifetime, the tissue increasingly thickens and loses flexibility. In addition, as we age, lifestyle factors change. We may eat less nutritiously and become less active, increasing the risk of hypertension.

Sometimes a chronic disease results from inherited genes, but more often the cause is a lifestyle factor, such as eating a poor diet, using a harmful substance such as tobacco, or being physically inactive. In fact, the Centers for Disease Control and Prevention estimate that 80 percent of cases of heart disease and type 2 diabetes and 40 percent of cases of cancers could be prevented by exercising, consuming a proper diet, and avoiding tobacco. Although chronic diseases are among the most common and costly health problems, they are also among the most preventable.

Adopting healthy behaviors, such as eating recommended amounts of nutritious foods, being physically active, and avoiding tobacco use, can prevent or control the devastating effects of these diseases. Behaviors and genetic factors that increase the chance of developing a disease are called **risk factors**. Previous research indicates that *modifiable* risk factors are responsible for a large number of premature deaths in the United States.

TABLE 21.2 Top Causes of Deaths in the United States (National Vital Statistics Reports, Preliminary 2009 Data)

Cause of Death	Percent of U.S. Deaths
Heart attack	25%
Cancer	23%
Stroke	5%
Chronic lower respiratory disease (emphysema, bronchitis)	5%
Accidents	5%
Diabetes	3%
Alzheimer disease	3%
Pneumonia and influenza	2%
Other causes	29%

Kochanek KD, Xu JQ, Murphy SL, et al. 2011. Deaths: preliminary data for 2009. National Vital Statistics Reports, Volume 59, Number 4. Hyattsville, MD: National Center for Health Statistics.

Risk factors—behaviors and genetic factors that increase the chance of developing a disease. Modifiable risk factors are those that involve diet or other behaviors, which individuals can influence. Examples include smoking, diet, and certain forms of high blood pressure.

The number of deaths attributed to various modifiable risk factors are shown in graphical form in **Figure 21.3**. Notice that each bar also correlates the risk factor to the number of deaths from each of the top chronic diseases. Smoking, obesity, and high blood pressure are the three leading causes of death in the United States. All three can be prevented through behavior, diet, and physical activity.

Chronic diseases are costly. More than 75 percent of the nation's healthcare dollars are spent on people with chronic illnesses. This number will undoubtedly increase as a larger proportion of our population ages and the number of people with chronic conditions grows. In fact, the number of people who live to be older than 65 years is expected double over the next 25 years. People with chronic conditions, particularly those with multiple chronic conditions, are the heaviest users of healthcare services.

Chronic diseases are not just a financial burden; they affect people physically, mentally, and socially. Some chronic conditions are highly disabling. For example, diabetes may not disable a person immediately, but if it is not treated early and effectively, it often leads to complications such as loss of limb circulation or blindness later in life. After a heart attack or stroke, some people return to their former levels of daily activity; others do not. Some individuals with chronic conditions live full, productive, and rewarding lives; for others, isolation, depression, and physical pain are the consequences of severe chronic illness.

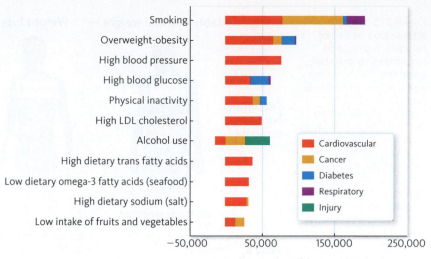

Deaths attributable to individual risks in both sexes

Figure 21.3 Number of deaths associated with various risks in the United States. Each risk is broken down into causes of death. Cardiovascular disease, cancer, and diabetes are most frequently associated with these risk factors. Notice that alcohol is protective of cardiovascular disease to a certain point.
© Cengage Learning 2014

Obesity is a response to our diet and physical activity levels

Let's now take a closer look at obesity. Obesity is a global epidemic that is the fastest growing health challenge the United States has ever faced, with two-thirds of Americans overweight (**Fig. 21.4**). Obesity is a complex disease that results from a caloric imbalance caused by genetic, behavioral, and environmental factors. It is considered a *preventable disease*—a disease that has an available cure or vaccination—so people should not get it or have major problems with it. Our current culture promotes increased caloric intake, consumption of nonhealthy foods, and an inactive lifestyle, all of which contribute to obesity. In 2010, medical costs attributed to obesity were almost 10 percent of all medical spending and projected to be 20 percent by 2018. Obesity is a global health problem that has significant medical, political, behavioral, social, and biological implications.

From a biological standpoint, genetics plays a minor role in obesity. However, other biological responses in obese people alter hormone levels to encourage the accumulation of body fat. Behaviors such as overeating, eating large amounts of

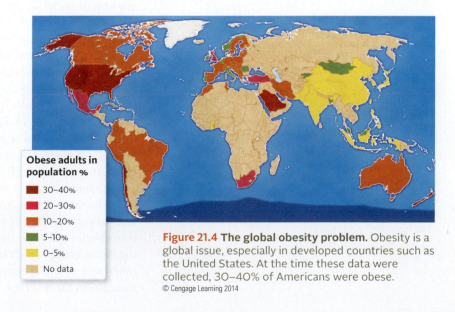

Figure 21.4 The global obesity problem. Obesity is a global issue, especially in developed countries such as the United States. At the time these data were collected, 30–40% of Americans were obese.
© Cengage Learning 2014

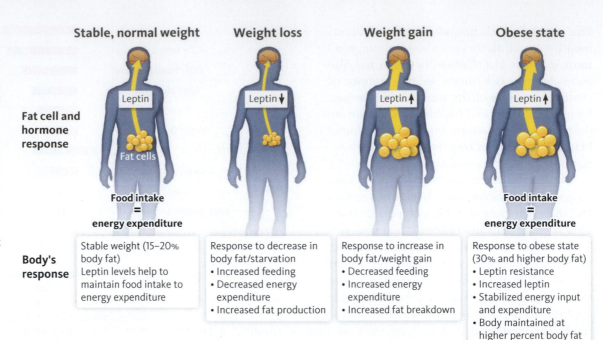

Figure 21.5 The interaction between fat cells, leptin, and responses to maintain body fat. The hormone leptin works to maintain body fat at 15 to 20 percent. Leptin levels fluctuate based on fat levels. However, once fat levels reach an obese state, the body ignores these chemical signals and resets itself to maintain an obese state rather than the normal 15 to 20 percent fat range.
© Cengage Learning 2014

Stable, normal weight

Weight loss

Weight gain

Obese state

Fat cell and hormone response

Leptin

Leptin ↓

Leptin ↑

Leptin ↑

Fat cells

Food intake = energy expenditure

Food intake = energy expenditure

Body's response

Stable weight (15–20% body fat)
Leptin levels help to maintain food intake to energy expenditure

Response to decrease in body fat/starvation
• Increased feeding
• Decreased energy expenditure
• Increased fat production

Response to increase in body fat/weight gain
• Decreased feeding
• Increased energy expenditure
• Increased fat breakdown

Response to obese state (30% and higher body fat)
• Leptin resistance
• Increased leptin
• Stabilized energy input and expenditure
• Body maintained at higher percent body fat

fat, and physical inactivity reset the body's homeostatic feedback loops that regulate appetite and body fat distribution, resulting in a tendency for a person to gain weight. Recent genetic studies show that metabolism and physical activity closely interact. Because our human genes were selected for under certain environmental conditions, a change in environment can promote faulty traits that include disease. The good news is that, with current technology and knowledge, many obesity-related chronic illnesses can be reversed if people are willing to make some lifestyle and dietary changes.

Behaviors that lead to obesity influence chemical reactions in excess fat stored at the cellular and tissue levels. When you consume too few or too many calories, or when you consume a diet high in saturated fats, the body responds by adjusting your appetite, metabolism, and fat storage. Specialized fat cells called *adipose* cells produce and secrete the hormone *leptin,* which is a protein that travels through the circulatory system to all parts of the body. The body uses this hormone as a chemical signal to determine the amount of body fat present. Endocrine and dietary factors regulate leptin expression. This feedback loop was an important adaptation in the survival of our ancestors because food was not always in steady supply.

Under normal circumstances, leptin is involved with inhibiting appetite and ultimately with balancing energy uptake and expenditure. The amount of leptin in the body is proportional to the number of adipose cells; therefore, obese people have large amounts of circulating leptin. When the amount of leptin reaches a certain level, overweight or obese individuals develop a resistance to it, and their appetites are not suppressed efficiently. The result is an increase in energy intake, adding "fuel to the fire." A diet high in saturated fats also induces leptin resistance.

The interaction between fat cells, leptin, and responses to maintain body fat is shown in **Figure 21.5**. The body releases leptin to levels that maintain a normal state of 15 to 20 percent body fat. If fat reserves drop when the body is in the normal state, the body responds to the starvation mode by decreasing leptin levels, which increases the appetite to promote fat storage. On the other side of the spectrum, when the body senses weight gain through an increase in fat reserves, high levels of leptin decrease the appetite and increase the use of fat for metabolic energy. Like all homeostatic controls, the body prefers to maintain body fat in a narrow range. The fourth scenario is the response in obese people when the range for percent body fat is outside of the normal range, that is, from 30 to 35 percent (see **Fig. 21.5**). In the obese state, the body is resistant to leptin and ignores the hormonal signals. The system is "recalibrated" so that signals balance food intake and energy expenditure. The situation gets even more complicated when you consider that despite what signals the body sends, we often choose to ignore the signals and eat more than our bodies tell us to.

Another cellular response in obese people is inflammation caused by an increased number of macrophages in adipose tissue. Recall from Chapter 20 that macrophages play a vital role in the immune system. Although the inflammatory response to pathogens is a vital and positive part of the body's defenses, sometimes inflammation, when directed at the body's own tissues, has negative consequences. Additional macrophages and inflammation increase the chances of developing arthritis, some types of cancers, and cardiovascular disease. Inflammation can interfere with normal metabolic pathways, causing more stress on the body and leading to disease.

21.2 check + apply YOUR UNDERSTANDING

- Chronic diseases are recurring, long-term health problems that are the leading cause of death. They are prevalent and represent significant costs to society, but many are preventable.

- Obesity is a response to our current diet and physical activity levels. Maladaptive responses are seen at all levels of biological hierarchy— from the cellular and tissue levels to the population level.

1. What is a modifiable risk factor?
2. List five modifiable risk factors that are leading causes of death in the United States.
3. Explain the role of adipose tissue in the maintenance of body fat.
4. Why is it difficult to predict the outcome of a person's chronic disease?
5. Bronchiectasis is a disease marked by irreversible widening of the lung passageways. It is associated with infections and inflammation of the passageways that cause the epithelial layer to be replaced with fibrous scar tissue. Would you characterize bronchiectasis as a chronic disease? Defend your answer.

21.3  Cardiovascular Disease Is Caused Mostly by Behavioral Factors

Your heart is about the size of your fist and is made of cardiac muscle that contracts involuntarily. In an average lifetime, the human heart beats almost 100,000 times per day and close to 38 million times per year. With each contraction, the heart sends blood along the 100,000-mile-long network of blood vessels to supply the body with oxygen and nutrients and to carry away carbon dioxide and waste products. In addition to delivering oxygen and removing wastes, your heart and blood distribute nourishment from the digestive system and hormones from glands. Because the heart lies at the center of the blood-delivery system, it is the engine of human life.

Considering the essential functions of the heart, it seems logical to take good care of it. In spite of this, the incidence of heart disease continues to rise, largely due to poor diet and lifestyle choices. Each year heart disease kills more Americans than cancer does, and it remains the number one preventable cause of death. A heart attack can strike at any time and without warning. In minutes, it can irreparably damage the heart, our body's hardest-working muscle. For many, a heart attack is the first sign of heart disease.

Heart disease is a broad family of conditions that includes heart rhythm problems (known as arrhythmias), congenital heart defects, and cardiovascular diseases that involve the heart as well as the blood vessels. In the following section, we will focus on the three most common cardiovascular diseases: heart attack, stroke, and high blood pressure. As with the other chronic diseases, both modifiable and unmodifiable factors influence heart disease.

Heart attacks and strokes are associated with high blood pressure and atherosclerosis

As a refresher from Chapter 9, **Figure 21.6** shows the relationships between the various blood vessels associated with the cardiovascular system. Notice that the larger arteries take blood away from the heart and branch into smaller and smaller vessels, eventually becoming thin-walled capillaries. Through the capillaries oxygen, nutrient, and waste exchanges occur between the blood and tissues. From the capillaries, the blood, now deoxygenated, passes into vessels that eventually widen and connect with the veins to return the blood to the heart.

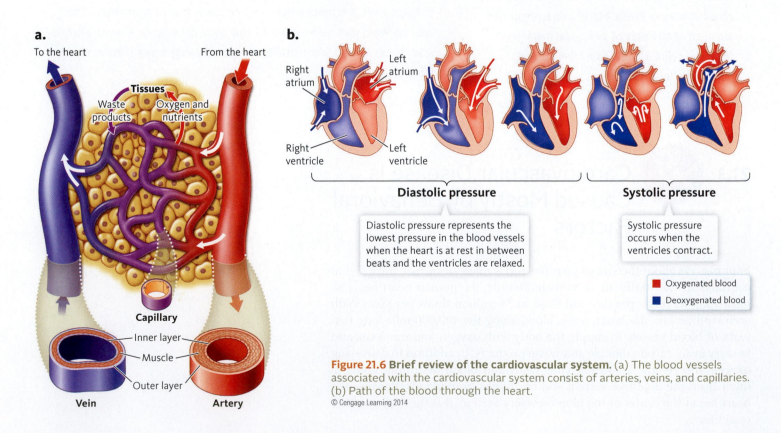

Figure 21.6 Brief review of the cardiovascular system. (a) The blood vessels associated with the cardiovascular system consist of arteries, veins, and capillaries. (b) Path of the blood through the heart.
© Cengage Learning 2014

Figure 21.6 illustrates an important characteristic of the arteries: they have thicker walls than the veins do. This characteristic helps the arteries to control blood flow by maintaining blood pressure. Because the arteries are closest to the heart, the blood pressure is highest in the arteries closest to the heart. Under ideal circumstances, blood flows through this circuit at a fairly constant pressure. **Blood pressure** is the force of blood pushing against artery walls and is created by contraction of the heart muscles. Your blood pressure is recorded as two numbers, such as 120/80. The top number is the *systolic pressure,* which is the maximum pressure in the arteries when the heart beats. The bottom number is the *diastolic pressure,* which represents the lowest pressure in the blood vessels when the heart is at rest between beats. Referring specifically to the chambers of the heart, systolic pressure occurs when the ventricles contract, and diastolic pressure occurs when they relax (see **Fig. 21.6b**).

Physiologically, blood pressure monitors the movement of blood in the body. It senses blood loss, stress levels, and changes in body position. Blood pressure is one of the vital signs that doctors monitor as an indicator of health, along with pulse, temperature, and respiratory (breathing) rate. Deviations from normal blood pressure levels may indicate heart malfunction or a problem with circulation. **High blood pressure (hypertension)** means the pressure in your arteries is regularly above the normal range. Healthcare professionals define high blood pressure as a consistently elevated pressure of 140 systolic or higher and/or 90 diastolic or higher. Because high blood pressure makes the heart work harder and damages blood vessels, people with high blood pressure are susceptible to hardened arteries, stroke, heart diseases, kidney failure, headaches, and other health problems. For instance, if the blood vessels in the kidneys are damaged, they may stop removing wastes and extra fluid from

Blood pressure—the force of blood pushing against artery walls created by contraction of the heart muscles. Your blood pressure is recorded as two numbers, such as 120/80. The top number is the *systolic pressure,* which is the maximum pressure in the arteries when the heart beats. The bottom number is the *diastolic pressure,* which represents the lowest pressure in the blood vessels when the heart is at rest between beats.

High blood pressure (hypertension)—a physical condition in which the blood pressure in your arteries is regularly above the normal range. Healthcare professionals define high blood pressure as a consistently elevated pressure of 140 or higher systolic and/or 90 or higher diastolic.

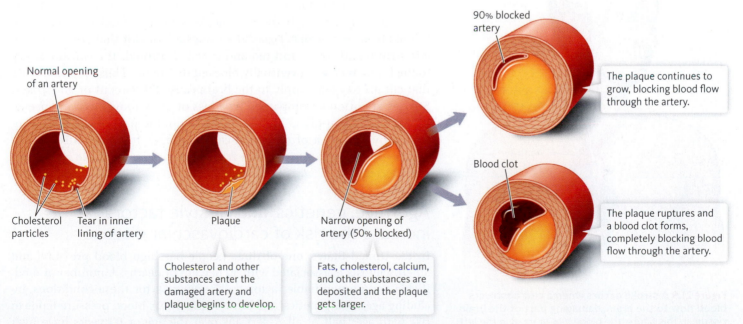

Normal opening of an artery

Cholesterol particles

Tear in inner lining of artery

Plaque

Cholesterol and other substances enter the damaged artery and plaque begins to develop.

Narrow opening of artery (50% blocked)

Fats, cholesterol, calcium, and other substances are deposited and the plaque gets larger.

90% blocked artery

The plaque continues to grow, blocking blood flow through the artery.

Blood clot

The plaque ruptures and a blood clot forms, completely blocking blood flow through the artery.

Figure 21.7 Atherosclerosis and heart disease. High blood pressure is associated the buildup of plaques that leads to atherosclerosis and the hardening of the arteries. Most heart attacks and strokes are caused by atherosclerosis.
© Cengage Learning 2014

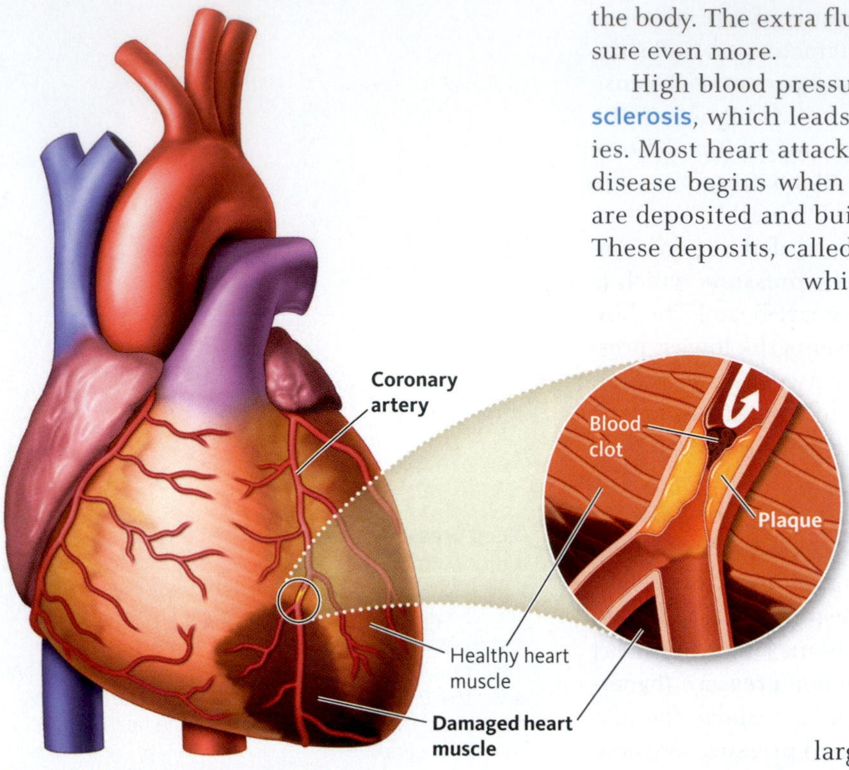

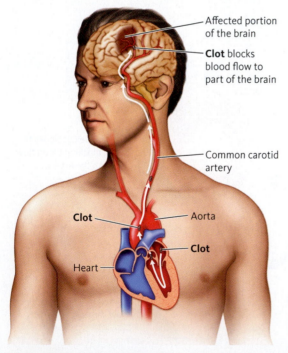

Figure 21.8 Coronary heart disease. The muscle tissue downstream of the blocked artery is deprived of oxygen and damaged.
© Cengage Learning 2014

Coronary artery

Blood clot

Plaque

Healthy heart muscle

Damaged heart muscle

Affected portion of the brain

Clot blocks blood flow to part of the brain

Common carotid artery

Clot

Aorta

Clot

Heart

Figure 21.9 A stroke occurs when a clot obstructs blood flow to the brain, damaging part of the brain tissue. In this figure, the blood clot started in the left atrium, ruptured, and traveled through the carotid artery to the brain, where it eventually blocked the flow of blood.
© Cengage Learning 2014

the body. The extra fluid in the blood vessels may then raise blood pressure even more.

High blood pressure is associated with a disease known as **atherosclerosis**, which leads to a condition known as hardening of the arteries. Most heart attacks and strokes are caused by atherosclerosis. This disease begins when fats, cholesterol, calcium, and other substances are deposited and build up on the inner walls of the arteries (**Fig. 21.7**). These deposits, called *plaques,* cause the arteries to become narrower, which significantly reduces blood flow. When flow of blood is reduced, the muscle serviced by that artery receives less oxygen and nutrients. Through action of a feedback loop, atherosclerosis is a self-accelerating process. The intense pressure of the blood forced against the artery wall damages the artery and causes inflammation, which in turn increases the rate of atherosclerosis. Atherosclerosis also increases blood pressure as it narrows the arteries.

The high rates of metabolic activity in the heart and brain require a disproportionately large amount of oxygenated blood. In fact, 10 percent of the body's blood is delivered to the heart through the left and right coronary arteries; 30 percent of the blood goes to the brain. When the coronary artery supplying the heart muscle with oxygen is obstructed with plaque, the result is damage to a portion of the heart muscle (**Fig. 21.8**). Heart attacks occur when localized muscle tissues die and the heart loses some of its function. This type of heart disease is called *coronary heart disease.*

Plaques can rupture and cause blood clots that block blood flow. The harm caused depends on where the clot lodges in the circulatory system. If a blood clot blocks a blood vessel that supplies the heart, it causes a heart attack. If the clot blocks a blood vessel that feeds the brain, it causes a **stroke**. **Figure 21.9** shows a blot clot that started in the left atrium and then ruptured and traveled through the carotid artery to the brain, where it eventually blocked the blood. These types of clots that cut off oxygen supply to the brain cause 80 percent of all strokes. Strokes are often accompanied by a loss of body movement or speech. When oxygen supply is limited in the brain, the part of the body controlled by that area of the brain can no longer function. Strokes can also be caused by rupture of the fragile arteries in the brain as a result of hypertension.

Age, race, genetics, and lifestyle factors increase the risk of cardiovascular disease

In the United States, one of three people has high blood pressure, and atherosclerosis is associated with 30 percent of deaths. A number of modifiable and unmodifiable factors raise your risk for these conditions, including age, race, genetics, and lifestyle (**Fig. 21.10**). Blood pressure tends to rise with age; half of all Americans over the age of 60 years have high blood pressure. High blood pressure can affect anyone, but it occurs more frequently in African-American adults than in Caucasians or Hispanic Americans. African-Americans not only develop high blood pressure

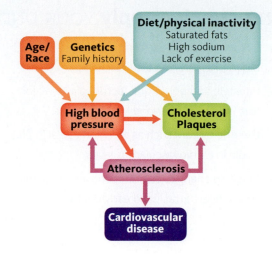

Figure 21.10 Risks of cardiovascular disease. The arrows in this diagram point to the complexity of interactions between unmodifiable and modifiable risk factors for cardiovascular disease.
© Cengage Learning 2014

sooner, they also have more severe cases and higher rates of death due to blood pressure complications.

Family history also plays a role in high blood pressure and atherosclerosis. Medical researchers have identified more than a dozen genes linked to high blood pressure, and several genes have also been linked to high cholesterol levels that cause atherosclerosis. We will examine in closer detail the most common form of inherited high cholesterol that starts at birth and causes heart attacks at an early age. A mutation on chromosome 19 that codes for the LDLR gene provides instructions for making a protein called a low-density lipoprotein receptor, which causes the disease. This receptor binds to the primary carriers of cholesterol in the blood, called low-density lipoproteins (LDLs) (**Fig. 21.11**). These receptors play a critical role in regulating cholesterol levels by removing LDLs from the bloodstream. Some LDLR mutations reduce the number of LDL receptors produced within cells, whereas others disrupt the receptors' ability to remove LDLs from the bloodstream. As a result, people with mutations in this gene have very high levels of blood cholesterol, leading to an increased risk for heart disease.

This type of high cholesterol caused by LDLR mutations is inherited in an autosomal dominant manner. Recall that this means that one copy of the mutated gene is enough to inherit the disease. A parent who carries a mutated gene has a 50 percent chance of passing on this gene sequence to each of his or her children. One in 500 individuals is a heterozygote, carrying one mutated gene copy causing the condition. More rarely, a person inherits the gene mutation from both parents, making him or her genetically homozygous. Individuals who are homozygous have a much more severe form of the disease, with heart attack and death often occurring before age 30 years.

Although you are not in control of your genetics, age, or race, you can focus on the lifestyle factors that heavily influence cardiovascular health. A diet similar to that of our hunter–gather ancestors—that is, a diet low in sodium, alcohol, and saturated fat—reduces your risks, as does a diet high in potassium from fruits and vegetables. Sodium and potassium are important for regulating fluid levels in your body. When sodium levels are high and potassium levels are low, the body retains water. With excess fluid in the blood, the blood volume increases, raising blood pressure. A diet low in saturated fats is important because excess fat and cholesterol build up to form plaques. Maintaining proper body weight and exercise levels also is important to maintain the blood circuit and the health of the heart. Smoking and stress are two other more important factors that increase blood pressure and hardening of the arteries. These recommendations are based on current scientific knowledge, but it is easy to picture our hunter–gatherer ancestors living in this manner.

Combined with obesity, signs of cardiovascular disease are now present in many adolescents. As these children reach adulthood, cardiovascular diseases will affect younger and younger adults. In order to stem the tide of cardiovascular disease, prevention is required. It starts with education and an awareness of how you can change your diet, stress levels, and activity levels to maintain a healthy heart. It is important that you monitor your weight and your blood pressure and blood cholesterol levels. Heart disease not only is preventable, but it is reversible if it is addressed early enough.

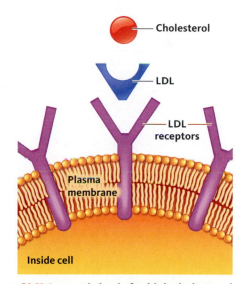

Figure 21.11 A genetic basis for high cholesterol. Some gene mutations reduce the number of LDL receptors produced within cells, whereas others disrupt the receptors' ability to remove LDLs from the bloodstream. People with these mutations have very high levels of blood cholesterol, leading to an increased risk for heart disease.
© Cengage Learning 2014

Atherosclerosis—a medical condition of reduced blood flow caused by buildup of plaques on the inner walls of the arteries. This condition is associated with high blood pressure and leads to hardening of the arteries. Most heart attacks and strokes are caused by atherosclerosis.

Stroke—a medical condition caused by blockage of a blood vessel that feeds the brain. These blockages can cut off oxygen supply to the brain, accounting for 80 percent of all strokes. These types of strokes often are accompanied by a loss of body movement or speech. When oxygen supply is limited in the brain, the part of the body controlled by that area of the brain can no longer function.

- Common cardiovascular diseases include high blood pressure and atherosclerosis.

- Lifestyle and behavioral factors play a major role in this number one killer of adults.

- Genetic factors such as race and age also play a role in cardiovascular diseases.

- Lifestyle changes can dramatically lower the risk of heart attack and early death.

1. What are several lifestyle factors that you can incorporate to help prevent cardiovascular disease?

2. What is the relationship between atherosclerosis, plaques, and strokes?

3. Explain what is meant by "atherosclerosis is a self-accelerating process."

4. Imagine that a family member just told you his blood pressure was 160/110. What are the consequences to his cardiovascular health if his blood pressure remains untreated? What advice would you give to him to improve his condition?

5. Predict how blood pressure responds under each of the following conditions: (a) at rest in bed, (b) while standing upright, and (c) during severe blood loss after an auto accident.

21.4  Mutations That Alter Cell-Cycle Regulation Cause Cancer

As you walk around campus, count the number of people smoking cigarettes. National statistics indicate that one in five college students smokes. Many smokers start their habit as teenagers, but due to nicotine addiction, they have a hard time quitting despite the known harmful health effects. From 80 to 90 percent of lung cancers occur in smokers.

Whether you are breathing fresh air or secondhand smoke or are inhaling from a cigarette, air passes through your nose, down your trachea or windpipe, and into the lungs, where it spreads through your bronchial tubes to interact with the blood carried in the capillaries (**Fig. 21.12**). Most lung cancers begin in the epithelial cells lining this pathway. These cells are vulnerable to **carcinogens**, which are agents that cause genetic mutations leading to uncontrolled cell division and cancer. Tobacco smoke, ultraviolet (UV) light, and many pesticides are familiar carcinogens.

Lung cancers, like most cancers, are named for the cell type in which they develop. Although we often think of **cancer** as a single disease, in fact it is a family of diseases that have one thing in common: the uncontrolled growth of abnormal cells. One feature that makes many cancers deadly is their ability to spread to various parts of the body through the blood or lymph fluids. Because different types of cells perform specific functions in the body, each cancer has a different set of characteristics. Regardless of the type of cancer, it is a painful (and expensive) disease that causes the body to break down at the cellular level and stop functioning properly. The genetic mutations negatively influence fitness, and the disease can be considered another maladaptive response to our environment.

Symptoms of lung cancer result from the lung tissue not being able to act as a moist, flexible respiratory surface for gas exchange. The lungs produce a sticky mucus that acts as a lubricant in the respiratory pathway, trapping unwanted particles. The cilia in the epithelial cells of the larger airways help with this process. When particles weigh down the cilia, they are unable to move the mucus through the respiratory system. Difficulty

Carcinogens—agents that cause genetic mutations leading to uncontrolled cell division and cancer. Tobacco smoke, ultraviolet light, and many pesticides are familiar carcinogens.

Cancer—a family of diseases that have one thing in common: the uncontrolled growth of abnormal cells. One feature that makes many cancers deadly is their ability to spread to various parts of the body through the blood or lymph fluids. Because different types of cells perform specific functions in the body, each cancer has a different set of characteristics.

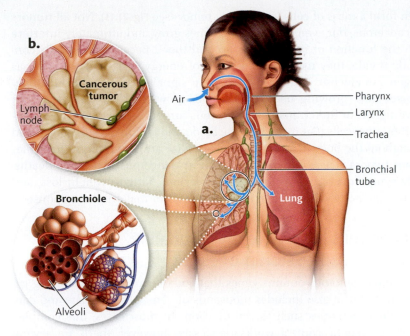

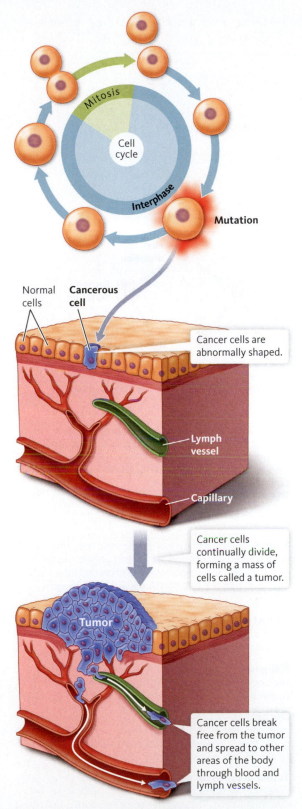

Figure 21.12 Most lung cancers begin in the epithelial cells lining the respiratory pathway. (a) Pathway of air through the respiratory system. (b) A cancerous tumor in the lung that is spreading throughout the body via the lymph system (shown in green).
© Cengage Learning 2014

breathing, wheezing, shortness of breath, chest pain, and coughing up blood all are signs of lung cancer.

In the following section, we will explore the cellular, genetic, and environmental factors of cancer. We have chosen to focus on three different types of cancer—lung, skin, and breast cancer—because they are some of the most frequently diagnosed and preventable cancers. We will continue to emphasize that some cancers are preventable and that decisions you make *today* greatly influence your risk factors of dying from cancer.

Cancer cells divide continuously without control

Cancer is a disease caused by disrupted genetic regulation and expression. Cancer cells arise from DNA mutations that disrupt normal cell cycle and function (**Fig. 21.13**). The result is a loss of cellular control with genetic defects at all levels of hierarchy in the human body. Recall that normal cells divide in a highly regulated and orderly process that builds new tissues during growth and replaces injured cells. Chemicals from outside and inside the cell signal the growth and differentiation processes. When a normal cell acquires a number of mutations that alter its ability to properly regulate its own growth and division, it becomes transformed into a cancer cell. With each cell cycle the cancer cell (blue cell in **Fig. 21.13**) passes on these mutations and then starts to spread throughout the body. One of the cells escapes from the tumor and enters into the bloodstream. This cell can establish itself in a new location and proliferate.

On the tissue level, cancer cells generally arise in continually dividing tissues such as epithelial and connective tissues. Nerve and muscle tissues rarely become cancerous because they do not divide once they differentiate and mature. Most normal cells divide and are neatly organized to create the tissue layer, but cancer cells appear chaotic and disorganized and

Figure 21.13 Cancer at the cellular level. Cancer results from a loss of control of the cell cycle (top). Cancer cells behave differently than normal cells. They form a tumor and can break off, spreading to other areas (bottom). Many characteristics of cancer cells are shown.
© Cengage Learning 2014

often form a mass of cells known as a **tumor** (see **Fig. 21.13**). Not all tumors are cancerous, but even noncancerous ones grow and ultimately interfere with the function of normal cells around them. Because these cells constantly divide, they use a greater share of energy and resources such as oxygen and glucose. Tumors also send out signals to build a blood supply to meet their growing needs. As tumors grow, they can interfere with nerve signals or fluid transport. Cancerous tumors can **metastasize**, meaning they shed cells that move into the blood and lymph system and spread throughout the body. Cancers that metastasize are also referred to as **malignant tumors**. These tumors are the most dangerous and cause deaths. Early symptoms of cancer include skin lesions, fatigue, weight loss, and fever as the body's response to fight off the abnormal cellular responses.

Cancer cells have defective signaling and gene-expression processes

The human genome contains genes that code for proteins producing genetic traits, but it also includes thousands of genes that regulate and control cellular activities such as cell division. In most cases, this complex system of genetic control works flawlessly; however, there are several cases where this system goes awry, such as in cancer.

Recall from Chapter 7 that the cell cycle relies on proper levels of transcription (DNA to RNA) and translation (RNA to protein) of certain genes. The genome contains normal genes that regulate cell growth and differentiation. One way to think of cell-cycle regulation is to view it like traffic regulation by a series of light signals at intersections along the road. For example, a green traffic light signal causes the cell to move forward through the cell cycle and divide. In many forms of cancer, mutations cause the traffic light to remain constantly green, with the cell always moving forward in the cell cycle and dividing. The mutated forms of these "green light" genes are called **oncogenes** (the prefix *onco-* means tumor). Another family of genes involved in regulating transcription, DNA repair, and cell-to-cell communication are **tumor suppressor genes**, which serve two functions: to prevent cell division and to lead to cell death. In this analogy, these genes act as the red light that brings traffic to a halt. A third gene family associated with cancers are **DNA repair genes**, which, as their name implies, function to detect and repair DNA damage from mutations or other damage to DNA. The products from these three gene families work together to regulate the movement of the cell through its normal cell cycle, keeping the cell dividing at a rate that is appropriate for its role in the body.

For a normal cell to be transformed from a healthy cell to a cancer cell, several gene mutations must occur (**Fig. 21.14**). Cancer usually results from mutations in the cell-cycle regulator (green light stays on), the tumor suppressor genes (red light stays off), and the DNA repair genes (mutations remain undetected). This combination of mutations permits the cell to cycle rapidly and divide uncontrollably. Mutations typically start in single cells, which pass the abnormality along to each of the daughter cells during cell division. In order for these mutations to accumulate, they must pass through many generations of cell division (see **Fig. 21.14**). For this reason, cancer is more common in older than in younger individuals. This multistep process can proceed rapidly or can take more than a decade, depending on the tissue and cancer type. Regardless, the collection of mutations causes the cells to grow, divide, and develop into a tumor.

What causes the mutations that cause tumors and cancers to develop? In addition to mutations that naturally occur as a part of the cell cycle,

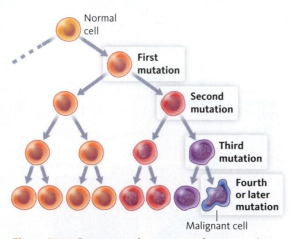

Figure 21.14 Gene mutations accumulate over time. Several gene mutations must occur and accumulate over many cell generations for cancer to develop. This is why cancers are more common in tissues that constantly divide.
© Cengage Learning 2014

Tumor—an abnormal mass of cells resulting from uncontrolled cell division. Not all tumors are cancerous, but even noncancerous ones grow and ultimately interfere with the function of normal cells around them.

Metastasize—the shedding of cancerous cells from tumors that move into the blood and lymph system to spread throughout the body.

Malignant tumor—a tumor that spreads throughout the body (metastasizes); the most dangerous type of tumor that can cause deaths.

Oncogenes—a family of genes that transform normal cells into tumor cells. Mutated forms of "green light" genes found in many forms of cancer that cause cells to always move forward in the cell cycle and divide.

Tumor suppressor genes—a family of genes involved in regulating transcription, DNA repair, and cell-to-cell communication. These genes serve two functions: to prevent cell division and to lead to cell death.

DNA repair genes—a gene family associated with cancers, which as their name implies, function to detect and repair DNA damage from mutations or other damage to DNA.

some mutations are inherited. Other mutations are caused by a wide variety of carcinogens in the environment. Viruses cause other mutations, such as those that cause cervical cancer. Carcinogens include ionizing radiation, tobacco, cigarette smoke, UV radiation, pesticides, and other chemicals. The tobacco smoke inhaled by smokers contains more than 60 chemical carcinogens (**Fig. 21.15**). Many cancers associated with carcinogens can be prevented by minimizing exposure to the known risk factors, making this an effective means of avoiding the disease.

Diet modifies the risk of cancers

Our caveman ancestors didn't live long enough for mutations from environmental factors to accumulate to a point where they triggered the onset of cancer. It is hypothesized that if our ancestors had lived longer, their rates of cancer still would be lower than ours are today because of their active lifestyle, high consumption of fruits and vegetables, and low consumption of sodium. The nutrients in our diet influence cellular processes. When the nutrients received by our body are out of balance, control over cellular processes can be lost. In other words, the nutrients we consume can either prevent or promote cancer and other chronic diseases.

Cancer can be caused by a variety of factors and may develop over a number of years. Some risk factors, such as diet and activity level, can be controlled. Choosing the right balance of healthy behaviors and avoiding exposure to certain environmental carcinogens can help prevent the development of cancer. From both a personal point of view and a biological perspective, understanding the factors that regulate gene expression provides you with a better understanding of how cells operate and how cancer can develop. Later in the chapter, we will turn your attention to the influence of heredity (genetics) on certain cancers.

Figure 21.16 shows the relationship between the cell cycle and some nutrients common in the food that you eat. Zinc, magnesium, selenium, vita-

Figure 21.15 Some of the many carcinogens in a cigarette. Over 60 carcinogens are inhaled from cigarettes. These carcinogens are a leading cause of mutations that disrupt the cell cycle and lead to cancer.
© Cengage Learning 2014

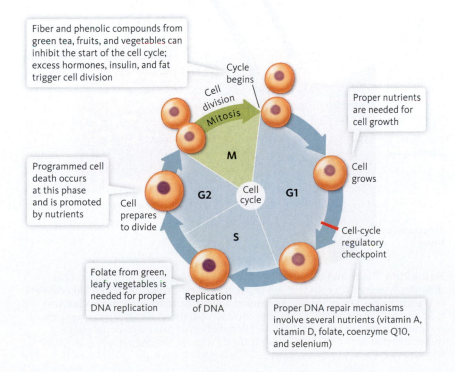

Fiber and phenolic compounds from green tea, fruits, and vegetables can inhibit the start of the cell cycle; excess hormones, insulin, and fat trigger cell division

Cycle begins

Cell division

Mitosis

Proper nutrients are needed for cell growth

M

Cell grows

Programmed cell death occurs at this phase and is promoted by nutrients

Cell prepares to divide

G2

Cell cycle

G1

Cell-cycle regulatory checkpoint

S

Folate from green, leafy vegetables is needed for proper DNA replication

Replication of DNA

Proper DNA repair mechanisms involve several nutrients (vitamin A, vitamin D, folate, coenzyme Q10, and selenium)

Figure 21.16 Nutrition influences on the cell cycle. Vitamins and minerals are required for cell growth, DNA repair, and DNA replication. Depending on the cell type and where it is in its cycle, nutrients play an important role in initiating and terminating the cell cycle.
© Cengage Learning 2014

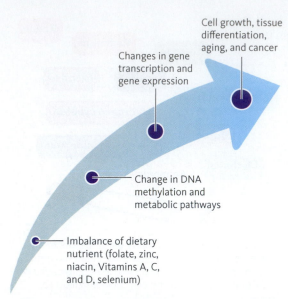

Cell growth, tissue differentiation, aging, and cancer

Changes in gene transcription and gene expression

Change in DNA methylation and metabolic pathways

Imbalance of dietary nutrient (folate, zinc, niacin, Vitamins A, C, and D, selenium)

Figure 21.17 Proper nutrients are needed for normal gene transcription and regulation. When nutrients are lacking or out of balance, cancer can occur. Changes in DNA structure and metabolic pathways lead to changes in transcription and gene expression, which can lead to cell growth, tissue differentiation, aging, and cancer.
© Cengage Learning 2014

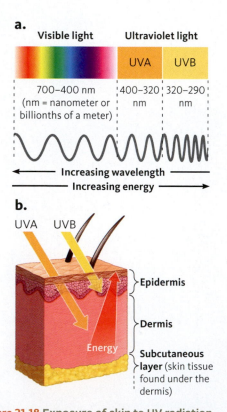

a.

Visible light	Ultraviolet light	
	UVA	UVB
700–400 nm (nm = nanometer or billionths of a meter)	400–320 nm	320–290 nm

← Increasing wavelength
Increasing energy →

b.

UVA UVB

Epidermis

Dermis

Energy

Subcutaneous layer (skin tissue found under the dermis)

Figure 21.18 Exposure of skin to UV radiation. (a) Spectrum of visible and high-energy UV light to which your skin is exposed. (b) UVA rays penetrate the skin more deeply, causing damage to the deeper layers of the skin. This damage introduces structural changes to DNA that impact the cell cycle and can cause skin cancer.
© Cengage Learning 2014

min A, vitamin C, vitamin D, and folate are just a few of the vitamins and minerals important for DNA maintenance activities (**Fig. 21.17**). These nutrients play a molecular role as substrates or cofactors needed in enzymatic reactions in the metabolic pathways that regulate DNA synthesis and repair and the expression of genes. Let's now apply the basic concepts of the cell cycle, mutations, and gene expression to two particular kinds of cancers: skin and breast cancer.

Radiation causes mutation and skin cancer

Whether it's the first warm day in the springtime, a summer day at the beach, or a routine trip to the tanning bed, you probably can picture yourself catching a few rays to achieve a glowing tan. What you may not realize is that each time you expose yourself to this type of UV radiation, you increase your risk for skin cancer. Most skin cancers are associated with long-term exposure to sunlight, tanning beds, and high-energy UV wavelengths (10–400 nm), which induce structural changes to DNA that impact the cell cycle. Sunlight comprises many types of electromagnetic radiation, including the visible light, UVA, and UVB rays (**Fig. 21.18**). UVA rays penetrate the skin more deeply and as a result cause deeper skin damage (**Fig. 21.18b**). Because tanning beds primarily use UVA rays, their use doubles your risk for most skin cancers. Given the potential for severe health problems, their popularity may surprise you, and many people believe the government should more tightly regulate the use of tanning beds.

To better understand skin cancer, let's start by reviewing the anatomy of your skin, which is the largest organ of your body covering and protecting your internal organs from injury. Skin is composed of three layers (**Fig. 21.19**). The outermost layer, the *epidermis,* consists mainly of epithelial cells and is always dividing at the basal layer. The epidermis is attached to the deeper *dermis* layer, which secretes collagen and elastic

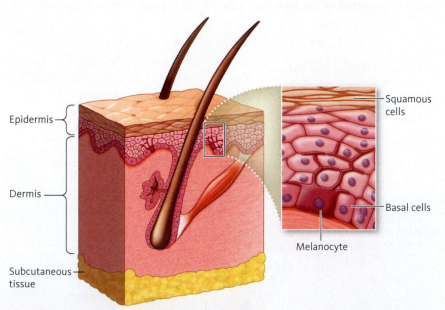

Squamous cells

Epidermis

Dermis

Basal cells

Subcutaneous tissue

Melanocyte

Figure 21.19 Tissue and cell layers of the skin. The three layers of skin contain different types of skin cells that are prone to different types of cancer. Skin cancers in the squamous layers are more common and easier to treat, whereas cancers that originate in melanocytes are less common but more deadly. The extensions on the melanocyte cell are stimulated by UV radiation to make additional pigment molecules that act as filters to prevent UV damage to the deeper layers of skin.
© Cengage Learning 2014

fibers, giving the organ its strength and stretching ability. The skin's base is a fatty *subcutaneous* layer, which insulates and cushions the body while also serving as fuel reserve in case of food shortage. Because skin cancers typically develop in the epidermis, they often are clearly visible and among the easiest to detect in their early stages (see **Fig. 21.19**).

Due to exposure to UV radiation, skin cancer starts in the outer epidermal layer of the skin, which has three cell types (see **Fig. 21.19**). Skin cancers are named after these cell types (squamous, basal, melanocytes) and are categorized into two main types: melanoma and nonmelanoma cancer. Nonmelanoma is the most common skin cancer, with 90 percent of these cancers associated with UV radiation from the sun. Skin cancers formed in the basal and squamous cells are grouped together as nonmelanoma skin cancers because they tend to act very differently from melanomas. Nonmelanoma cancers are *unlikely* to metastasize because the blood and lymph vessels are located deeper in the tissue. Melanoma is not that common, but it is more serious because of its ability to spread quickly throughout the body.

Melanoma is caused by mutated melanocyte cells that are in the deepest layer of the epidermis. Melanocytes are a type of pigment cell that produces skin, eye, and hair color. These cells cluster together, forming benign growths called moles (**Fig. 21.20**). Most moles are harmless and slowly change over time. However, under the right conditions they can develop into the most dangerous kind of skin cancer—malignant melanoma.

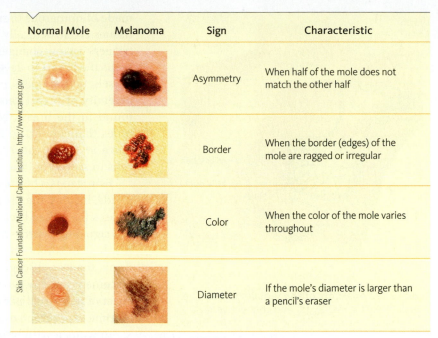

Normal Mole	Melanoma	Sign	Characteristic
		Asymmetry	When half of the mole does not match the other half
		Border	When the border (edges) of the mole are ragged or irregular
		Color	When the color of the mole varies throughout
		Diameter	If the mole's diameter is larger than a pencil's eraser

Skin Cancer Foundation/National Cancer Institute, http://www.cancer.gov

Figure 21.20 Signs of skin cancer. Skin cancer is more easily detected than other cancers because changes are visible on the skin. Changes in mole size, symmetry, color, shape, and texture are signs of potential skin cancer.
© Cengage Learning 2014

A combination of heredity and environmental factors cause some cancers

We have emphasized that most cancers are caused by mutations in several genes that regulate the cell cycle. Therefore, when cancer is hereditary, usually one mutated copy is inherited and other mutations accumulate over time. Hereditary cancers—those that have developed because of a gene mutation passed down from a parent to a child—account for only 5 to 10 percent of all cancers. The most common hereditary cancers are breast, ovarian, prostate, and colon cancer.

Mutations in one family of genes, called *BRCA* genes, are strongly associated with about 5 to 10 percent of breast and ovarian cancers (**Fig. 21.21**). There is also a correlation between prostate cancer patients and mutated *BRCA* genes. These genes help to control cell multiplication, and they repair damage to DNA. When the genes are mutated, they are unable to do their job, and cancer develops.

One in eight women in the United States will be diagnosed with breast cancer, making it one of the most common cancers. Breast cancer also occurs in men, but because men have less breast tissue, breast cancer is 100 times less common than in women. Breast cancer is one of the oldest known forms of cancers in humans, with detailed descriptions dating back 3,600 years to ancient Egypt.

So, what exactly is breast cancer? To better understand the disease, we'll start with the anatomy of the breast, which is a milk-secreting gland

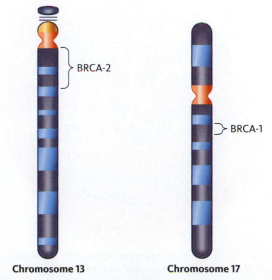

Chromosome 13   Chromosome 17

Figure 21.21 Mutated BRCA genes are inherited and associated with many cancers. These tumor-suppressor genes are involved in cell-cycle control and DNA repair. When the genes are mutated, they are unable to do their job, and cancer develops. Two different BRCA genes on chromosomes 13 and 17 are shown here.
© Cengage Learning 2014

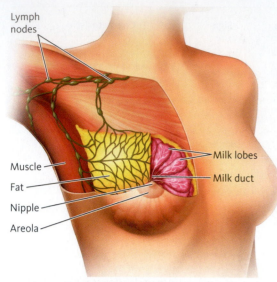

Figure 21.22 Anatomy of the female breast. The female breast is mostly fat and connective tissue enclosing the milk glands and duct system. Tumors and changes in breast tissue can be detected through mammograms (x-rays) and clinical and breast self-exams.
© Cengage Learning 2014

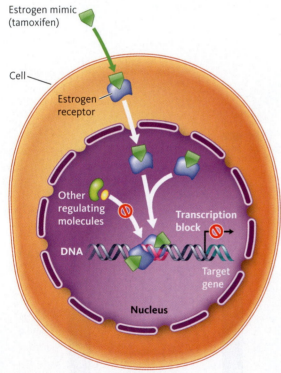

Figure 21.23 Blocking estrogen receptors to inhibit cell division and treat breast cancer. A drug that mimics estrogen passes through the cell membrane and binds to estrogen receptors. Under normal circumstances, other regulating molecules bind to the complex to initiate transcription, but because the molecule is not actually estrogen, proper gene expression is halted, which then stops cell division.
© Cengage Learning 2014

made of numerous tissue types (**Fig. 21.22**). Prior to puberty, human breast tissues are structurally identical in men and women. In females, puberty triggers many changes, including the production of estrogen, which helps the breast to develop. Inside the breast is a network of ducts, blood vessels, lymph nodes, connective tissue, adipose tissue, and muscles. The diversity of breast cancer types is related to the many types of tissues found in the breast. Most breast cancers begin in the breast ducts or tubes that connect to the nipple. Lymph nodes are of particular importance in breast cancer because they can serve as a pathway for cancer cells to spread to other parts of the body.

Estrogen is a hormone-signaling molecule that triggers specific cells to divide, helping the breast tissue grow. Increased exposure to estrogen increases the risk of breast cancer because it disrupts the balance of normal cell signaling. Estrogen levels can become out of balance in several ways. Alcohol consumption increases levels of estrogen. The more alcohol you drink, the less efficiently the liver can metabolize estrogen. Hormone replacement therapy, hormonal-based birth control drugs and devices, early onset of menstrual periods, and late onset of menopause also increase the lifetime exposure to estrogen.

In women with breast cancer, there is a distinct relationship between estrogen levels and estrogen *receptor* levels. Recall from Chapter 7 that hormones bind to cellular receptors on specific target tissues and in turn cause various genes to turn on or off (see **Fig. 7.24a**). Cells with estrogen receptors grow and multiply when estrogen attaches to the receptors. Although the mechanism is not fully understood, about 75 percent of female patients with breast cancer show a positive relationship between estrogen levels and receptors. Once breast cancer is established, estrogen serves as a signaling molecule causing breast cancer cells to grow and divide at high rates.

One common treatment of breast cancer consists of blocking the estrogen receptors to inhibit cell division. **Figure 21.23** shows the mechanism behind these treatments. A drug (tamoxifen) that mimics estrogen diffuses through the cell membrane and binds to estrogen receptors. The drug mimics estrogen enough to bind the receptor, but it cannot initiate the normal cell response. The complex prevents normal transcription (DNA to mRNA) and ultimately prevents excess cell division (see **Fig. 21.23**). Without the drug, estrogen would bind to the receptors, and transcription would result in cell division.

Cancer treatments remove the tumor or disrupt the cell cycle

Let's close this discussion on cancer by looking at the biological basis of several cancer treatments. Because each cancer and cancer patient is different, cancer treatment is always tailored to the patient's age, health, gender, and disease stage. Success often depends upon early detection and treatment. Many late-stage cancers are extremely difficult to treat. Cancer treatment often involves a number of simultaneous medical procedures, which may include surgery, radiation therapy, or drug treatment such as chemotherapy. Because cancer refers to an entire family of diseases affecting all types of tissues, a single treatment or cure likely will never be developed.

Let's take a look at some of the most common treatments of cancer. In some cases, the disease can be cured by removing the existing tumor. In these cases, the cells carrying the mutations are removed from body. In patients with breast cancer, surgery is often used to remove just the tumor (called a *lumpectomy*), or the entire breast may be removed in a *mastectomy*. The most common treatment of skin cancers is removal of the tumor and damaged tissue through surgery. External radiation therapy is also used as a treatment option for many cancers. Radiation therapy involves the targeted use of ionizing radiation to kill cancer cells and shrink tumors (**Fig. 21.24**). Radiation can be given externally using a focused beam of energy, or internally using a small radioactive capsule that is placed next to the tumor. Radiation damages both cancerous and normal cells, but normal cells usually have the ability to recover, whereas cancer cells die off. The goal of radiation therapy is to kill off as many cancer cells as possible while limiting the damage to healthy tissues. Surgery and radiation therapy are used in the same way to treat many other cancers, including prostate cancer.

If the cancer has metastasized to other sites, complete removal by surgery is impossible, and other treatments that target the cancer at the genetic level are used. These treatments are linked to gene expression and typically interfere with the disruptive nature of cancer cell division. Chemotherapy involves treating cancer patients with drugs that destroy cancer cells by interfering with the processes that control the cell growth and multiplication phases of the cell cycle. Chemotherapy is delivered through the blood to reach cancer cells wherever they have spread. Because these drugs are most effective against rapidly dividing cells, normal cells such as those of the intestinal tract, inside of the mouth, and hair follicles also are harmed. This is the reason why cancer patients undergoing chemotherapy have difficulty digesting foods and often lose much of their hair. Because cancer cells divide at a fast rate, mutate, and lack proper DNA repair mechanisms, they can become resistant to drug treatments. To combat this drug resistance and because many drugs work better in combination, chemotherapy often involves the use of two or more drugs.

Table 21.3 lists several categories of chemotherapy drugs along with their cellular mechanisms. Paclitaxel (Taxol) is a drug commonly used to treat breast and prostate cancer. It is also used for treatment of ovarian, lung, bladder, and some other cancers. Paclitaxel belongs to a class of chemotherapy drugs called *plant alkaloids* and is made from the bark of the Pacific yew tree. It also has been found on certain species of fungi that live on the bark of these trees. The periwinkle plant, the Asian happy tree, and the May apple plant also produce valuable chemotherapeutic compounds (**Fig. 21.25**). Plant alkaloids are cell cycle specific, which means they attack the cancer cells during various phases of division. Paclitaxel interferes with cell division by disrupting microtubules within the cancer cell. Remember that during division, microtubules attach and move chromosomes into the daughter cells. Disrupting the microtubule's normal function leads to cell death (**Fig. 21.26**).

In Chapter 19, you learned how biotechnology and genomics play a large role in finding cures, improving treatments, and advancing drug development for diseases. Cancer drug development involves finding products that will disrupt the signaling pathways of cancer cells and help to restore the signaling pathways and cell division of normal cells. In the case of paclitaxel, both healthy and unhealthy cells are destroyed. The future of cancer drugs is more direct targeting of cancer cells. The analogy is that drugs such as paclitaxel act like a bomb and destroy everything in their path; future drugs will behave like a gun and have a much more specific target. Tamoxifen, the

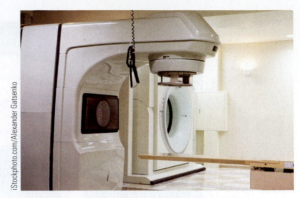

Figure 21.24 Radiation therapy. This radiation device delivers a targeted high energy beam that kills cells and shrinks tumors.

TABLE 21.3 Classes of Chemotherapeutic Drugs

Chemotherapy Drug Class	Cellular Mechanism
Alkylating drugs	Attack DNA
Antimetabolites	Prevent DNA synthesis
Antitumor antibiotics	Prevent DNA synthesis and interfere with other cellular proteins
Plant alkaloids	Prevent cells from dividing normally
Steroid hormones	Slow the growth of some cancers by interfering with hormones that trigger cell growth and replication

© Cengage Learning 2014

Radiation therapy—a medical treatment for cancer that involves the targeted use of ionizing radiation to kill cancer cells and shrink tumors. Radiation can be given externally using a focused beam of energy or internally using a small radioactive capsule that is placed next to the tumor. Although radiation damages both cancerous and normal cells, normal cells usually have the ability to recover, whereas cancer cells die off.

Chemotherapy—a medical treatment for cancer that involves treating patients with drugs that destroy cancer cells by interfering with the cell growth and multiplication phases of the cell cycle. Chemotherapy is delivered through the blood to reach cancer cells wherever they have spread. Because these drugs are most effective against rapidly dividing cells, normal cells such as those of the intestinal tract, inside of the mouth, and hair follicles also are harmed.

Figure 21.25 Some plants produce chemotherapy drugs. These plants produce cell-cycle–specific compounds that disrupt cell division. Left to right: Pacific yew tree, periwinkle plant, Asian happy tree, and May apple plant.

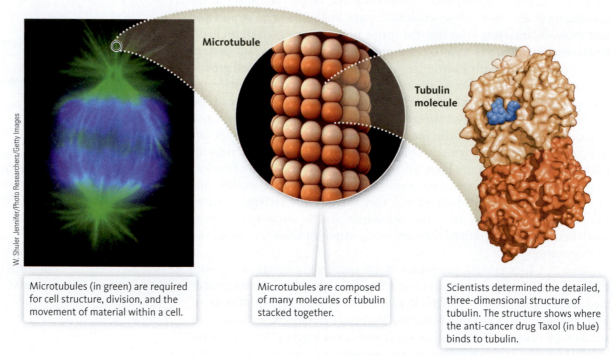

Microtubules (in green) are required for cell structure, division, and the movement of material within a cell.

Microtubules are composed of many molecules of tubulin stacked together.

Scientists determined the detailed, three-dimensional structure of tubulin. The structure shows where the anti-cancer drug Taxol (in blue) binds to tubulin.

Figure 21.26 Mechanism of paclitaxel (Taxol®) that causes cell death. Microtubules (green) are critical for cell division and normal cell function. To prevent cancer cells from dividing, paclitaxel (Taxol) is one chemotherapeutic drug often used. This compound (blue) binds to the tubulin molecules that make microtubules, preventing normal function during cell division and ultimately resulting in cell death. Taxol cannot distinguish between normal and cancer cells, so both are prevented from dividing, leading to unpleasant side effects from the drug.
© Cengage Learning 2014

estrogen-mimic drug, is an example of a targeted therapy (see **Fig. 21.23**). As researchers learn more about cancer-causing gene mutations and changes in gene expression in cells, the goal is to develop individualized drug treatments that specifically target these changes.

In this final chapter of your text, we have applied numerous biological concepts to a discussion of risk factors, disease, and your health. As a college student, these chronic diseases may not seem important to you. Whether it is you or a family member or close friend, you will be affected either directly or indirectly by some of the diseases discussed in this chapter. It is our hope that you apply the biology you have learned to better understand these diseases and to take the necessary steps to live a long and healthy life.

21.4 **check + apply** YOUR UNDERSTANDING

- Cancer refers to a family of related diseases associated with uncontrolled cell division.

- DNA mutations that affect the cell cycle and cellular controls are leading causes of cancer. Few cancers are inherited; most are related to lifestyle or exposure to carcinogenic toxins.

- Cancer treatments take on many forms, and the effectiveness of medical treatments is continually improving.

1. How is cancer a disease of disrupted genetic regulation and expression?
2. If cancer forms in the breast, how does it spread throughout the body?
3. Why do people with inherited mutations have a greater chance of developing certain cancers?
4. Summarize the activity and role of cell-regulator genes, oncogenes, tumor suppressor genes, and DNA repair genes in normal cell division and cancerous cell division.
5. Antiangiogenic drugs are one type of cancer drug we did not discuss. These drugs slow the growth of blood vessels that feed various cells. Apply your knowledge of how tumors grow to suggest how these drugs might work.

End of Chapter Review

Self-Quiz on Key Concepts

Chronic Disease Is Prevalent and Preventable

KEY CONCEPTS: In the United States, chronic diseases such as heart disease, cancer, and diabetes are the leading causes of death and disability. Modifiable behaviors, such as eating nutritious foods, being physically active, and avoiding tobacco use can prevent or mitigate the devastating effects of these diseases. These behaviors decrease risk factors for a large number of chronic diseases that result in premature deaths. Although each of these illnesses has a genetic component, the disease is simply a maladaptive response to our environment.

1. Match each of the following **health terms** with the best definition or example.

   Chronic disease   a. genetics, age, gender, and race
   Maladaptation   b. a factor that increases the risk of disease
   Unmodifiable risk factor   c. a response that decreases survival and reproduction
   Modifiable risk factor   d. a long-term disease that often is preventable
   Risk factor   e. preventable factors such as high-sodium diet, smoking, and physical inactivity

2. Modifiable risk factors that increase the risk of many chronic diseases include:
   a. age, gender, and race.
   b. age, gender, and diet.
   c. diet, physical inactivity, and gender.
   d. diet, physical inactivity, and other behaviors.

3. Obesity is a risk factor for which of the following chronic diseases?
   a. cancer
   b. cardiovascular disease
   c. diabetes
   d. all of the above

Cardiovascular Disease and Modifiable Lifestyle Factors

KEY CONCEPTS: Cardiovascular disease, caused mostly by lifestyle or behavioral factors, is the number one killer in the United States. Most heart attacks and strokes are caused by atherosclerosis, which is associated with high blood pressure. Although some factors such as genetics, age, or race are unmodifiable, or beyond your control, you can focus on the lifestyle factors that heavily influence cardiovascular health. Prevention starts with your education about and awareness of how you can change your diet, stress levels, and activity levels to maintain your weight, blood pressure, and blood cholesterol levels.

4. Match each of the following **cardiovascular conditions** with its characteristic, description, or example. Each term on the left may match more than one description on the right.

Atherosclerosis
High blood pressure

a. excessive pressure on the blood vessel walls
b. a vital sign that indicates heart health and function
c. associated with hardening of the arteries
d. associated with high-cholesterol diets

5. Which one of the following statements is correct?
   a. Heart attacks are caused by a blocked artery in the coronary circuit, and a stroke is caused by a blocked artery to the brain.
   b. Heart attacks and strokes are caused by a blocked artery in the coronary circuit.
   c. Heart attacks and strokes are caused mainly by genetic factors.
   d. Heart attacks and stroke are caused mainly by age.

6. High cholesterol is a risk factor for cardiovascular disease. Cholesterol is considered:
   a. a dietary factor.
   b. a genetic factor.
   c. both a dietary and a genetic factor.
   d. neither a dietary nor a genetic factor.

Cancer and Gene Mutations

KEY CONCEPTS: Cancer is a disease of genetic regulation and expression. Cancer cells arise from DNA mutations that disrupt normal cell cycle and function. The result is a loss of cellular control with profound genetic defects at all levels of hierarchy in the human body. Many cancers are closely associated with carcinogens and can be prevented by minimizing exposure to these agents. Although some cancers are hereditary, most cancers are not due to a single inherited change in a gene but rather to mutations in several genes that regulate the cell cycle, attachment, and movement to new sites. Cancer treatment often involves a number of simultaneous medical procedures that may include surgery, radiation therapy, or drug treatment such as chemotherapy.

7. Match each of the following **cancer terms** with its characteristic, description, or example.

Hereditary cancer
Tumor
Malignant tumor

a. cancer caused by excess cell growth and division
b. cancer that has spread to another location
c. cancer that is passed on from parents to offspring

8. The normal function of the family of *BRCA* genes is to help prevent uncontrolled cell division. Which family of genes do the *BRCA* genes belong to?
   a. oncogenes
   b. tumor suppressor genes
   c. DNA repair genes
   d. none of the above

9. Examples of carcinogens include:
   a. UV radiation from tanning beds.
   b. cigarette smoke.
   c. alcohol.
   d. all of the above.

Applying the Concepts

10. Using **Figure 21.27**, assess the risk of developing cardiovascular disease and diabetes for each of the following people. How many high-risk factors does each person have? Using a scale of high, moderate, or low, what do you think is the overall risk for the person developing heart disease and diabetes? Justify your answers. (a) a 60-year-old male construction worker who eats fast food every day for lunch; (b) a 50-year-old female CEO who drinks soda and eats sweets to get her through each stressful day; and (c) a 20-year-old college student who eats pizza, hamburgers, and other fast food each night for dinner before playing basketball each evening.

11. Mutations of the *BRCA1, PTEN,* and *ERBB-2* genes increase the risk of developing breast cancer. **Table 21.4** lists the function of these genes. Classify each gene as either an oncogene or a tumor suppressor gene (TSG). List key words from the description that support your classification. It may be helpful to think of the green/red light analogy for your classifications.

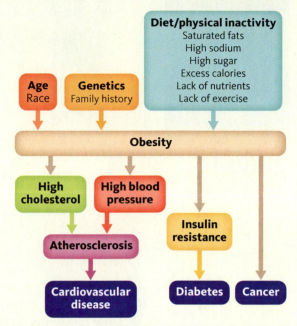

Figure 21.27 Interactions of risk factors, obesity, and chronic diseases.
© Cengage Learning 2014

TABLE 21.4 Breast Cancer Genes

Gene Mutation	Function	Oncogene or TSG?
BRCA-1	Encodes a protein that is directly involved in repairing damaged DNA	
PTEN	An enzyme that acts as part of a chemical pathway that signals cells to stop dividing and triggers cells to undergo a form of programmed cell death	
ERBB-2	One member of a family of genes that provide instructions for producing growth factor receptors (growth factors are proteins that stimulate cell growth and division)	

© Cengage Learning 2014

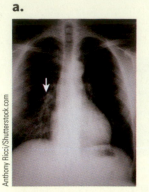

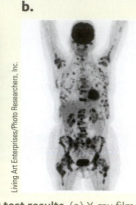

Figure 21.28 Lung cancer test results. (a) X-ray film. (b) PET scan.

12. Your aunt tells you that she has just been diagnosed with lung cancer. The oncologist has two images: a chest X-ray film showing a tumor in the left lung (arrow) and a PET scan, which highlights specific areas (in black) where tumors are located (**Fig. 21.28**). Considering surgery, radiation therapy, and chemotherapy, what treatment option would you recommend to your aunt given the information you have? Choose any of these treatments that you believe apply and provide support for your choice(s).

Data Analysis

Marketing, Cigarettes, and Lung Cancer

Lung cancer was a rare disease in the United States before widespread increases in cigarette smoking toward the end of the 1800s. Lung cancer tumors comprised about 1 percent of all cancers in the late 1870s but rose to nearly 15 percent by the early 1900s. Medical reports reveal that the incidence of lung cancer had increased from less than 1 percent in 1852 to greater than 5 percent by 1950. By then the terrible health effects of cigarettes started to become widely known, and concerns about the link to lung cancer were voiced around the globe. Although extensive medical research over the last 70 years has clearly shown the relationship between tobacco use and cancer, tobacco use in the United States remains high. Public education about the increased health risks has helped; smoking rates in the United States dropped by half from 1965 to 2006. Although cigarette smoking has leveled off or declined in developed nations, it continues to increase in some developing countries. In the developing world, tobacco use is increasing by more than 3 percent per year. Worldwide, cigarette sales are a big business, with more than five trillion cigarettes produced each year. More than one billion people, or greater than 16 percent of the world's population, use tobacco.

Tobacco ads deliberately target certain groups of people. Joe Camel was a popular icon for many years, until the Federal Trade Commission banned the marketing campaign in the mid-1990s because the cartoon character targeted children (**Fig. 21.29a**). In fact, Joe Camel was as popular as Mickey Mouse with children. Greater than 90 percent of six-year-olds recognized this character and associated him with cigarettes. Other tobacco companies targeted women by associating empowerment with smoking cigarettes (**Fig. 21.29b**). Cigarette companies attract students into the addictive habit by hosting music events and bar parties in college towns.

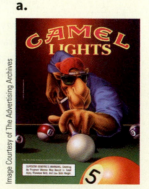

Figure 21.29 Tobacco advertisements. (a) Joe Camel, a popular ad in the 1980s. (b) An ad targeting women.

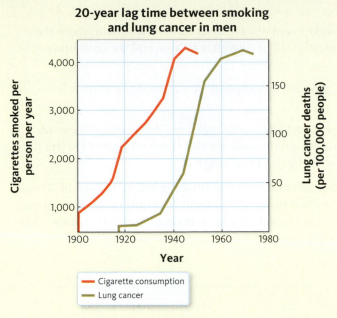

20-year lag time between smoking and lung cancer in men

Legend:
- Cigarette consumption
- Lung cancer

Figure 21.30 Twenty-year lag time between smoking and lung cancer.
© Cengage Learning 2014

Data Interpretation

13. Using **Figure 21.30**, what is the cigarette consumption rate for a lung cancer death rate of about 100 per 100,000 people in 1950?

14. What happens to the lung cancer death rate as cigarette consumption rates go from 1,000 to 2,000 cigarettes per year? What is the daily consumption of cigarettes for a person who smokes 3,000 cigarettes per year?

15. Is the lag time between consumption and onset of lung cancer consistent for all rates of consumption? Explain.

16. **Table 21.5** shows a trend in the 1990s among college smokers. What is the overall trend comparing 1993 to 1997? In which two groups did the rate of smoking increase the most? Compare the changes in smokers of the male and female populations. What is striking about these data? Summarize the 1997 results comparing (a) males to females, (b) percentage of smokers by race, and (c) percentage of smokers by college class.

TABLE 21.5 Smoking Prevalence in College Students from 1993–1997*
Change in Smoking Prevalence Among Subgroups of College Student Characteristics, 1993 vs. 1997

Student Characteristics	Smoked in Past 30 Days, %		
	1993 (N = 15,103)	1997 (N = 14,521)	Increase %
All students	22.3	28.5	27.8
Male	22.3	27.5	23.4
Female	22.3	29.2	31.0
Hispanic	22.7	25.4	12.0
Non-Hispanic	22.2	28.8	29.9
White	23.2	30.4	31.2
African American	9.6	13.7	42.7
Asian/Pacific Islanders	18.3	22.4	22.5
Other	24.7	26.3	6.4
Aged < 24 years	22.4	29.0	29.8
Aged 24 years	21.8	25.8	18.5
Freshmen	24.3	31.2	28.4
Sophomores	24.2	29.2	20.7
Juniors	22.2	29.4	32.4
Seniors	20.8	25.3	21.6

*1993 30-day smoking rate is used as baseline for comparison with 1997 rate.
Data are based on past 30-day cigarette use.
Data source: Henry Wechsler, Nancy A. Rigotti, Jeana Gledhill-Hoyt, and Hang Lee. Increased Levels of Cigarette Use Among College Students: A Cause for National Concern, *JAMA*, Nov. 1998; 280: 1673–1678.

Critical Thinking

17. Based on your understanding of the development of cancerous tumors, speculate about the reason for the lag time between the consumption of cigarettes in men and the development of lung cancer.

18. Predict whether the same lag time and general consumption trends would be similar for women. Explain your reasoning.

19. Closely examine the ads shown in **Figure 21.29**. What social messages does each of these ads send to the consumer? What gender and age-group audience does each ad target? Do you think that these ads accurately portray the social effects of smoking?

20. Current estimates indicate that 90 percent of smokers begin smoking before the age of 20 years. Smoking among college students has declined since 1997 but still is a concern. In 2008, about 22 percent of 18- to 24-year-olds smoked. One recent trend is smoking tobacco from pipes. List several reasons why high school and college students might start smoking.

Question Generator

Treating Breast Cancer

Hormones play a role in the development of breast cancer. During the reproductive period of her life, a woman's ovaries produce estrogen, a steroid hormone that promotes the growth and development of bones, muscles, and breast tissues. During menopause, the ovaries degenerate, causing estrogen production to decline. Other factors, such as obesity and hormone treatments, also influence estrogen levels. Estrogen is an important cell-signaling molecule that triggers gene expression, DNA synthesis, and the cell cycle (**Fig. 21.31**). If these signals are out of balance, uncontrolled cell growth and division can lead to breast cancer.

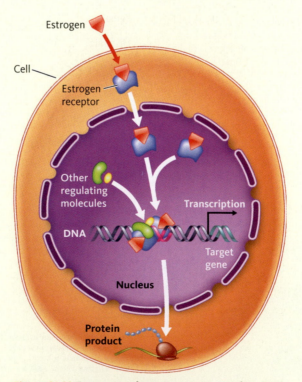

Figure 21.31 Estrogen triggers gene expression, DNA synthesis, and the cell cycle.
© Cengage Learning 2014

Below is a block diagram (**Fig. 21.32**) that relates several aspects of hormones, cellular mechanisms, and breast cancer. Use the background information along with the concept map to ask questions and generate hypotheses. We've translated the components of the block diagram into several questions to get you started.

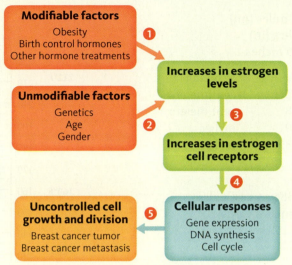

Figure 21.32 Factors relating estrogen, cellular mechanisms, and breast cancer.
© Cengage Learning 2014

Here are a few questions to start you off.

1. What type of modifiable factors might increase estrogen levels? (Arrow #1)

2. What type of unmodifiable factors might increase estrogen levels? (Arrow #2)

3. How do the levels of estrogen influence cell receptors? (Arrow #3)

4. How do increases in cell receptors influence cellular responses? (Arrow #4)

5. Which cellular responses affect breast cancer? (Arrow #5)

Use the diagram to generate your own research question and frame a hypothesis.

Appendix A
Units of Measure

Length

1 kilometer (km) = 0.62 miles (mi)
1 meter (m) = 39.37 inches (in)
1 centimeter (cm) = 0.39 inches

To convert	multiply by	to obtain
inches	2.25	centimeters
feet	30.48	centimeters
centimeters	0.39	inches
millimeters	0.039	inches

Area

1 square kilometer = 0.386 square miles
1 square meter = 1.196 square yards
1 square centimeter = 0.155 square inches

Volume

1 cubic meter = 35.31 cubic feet
1 liter = 1.06 quarts
1 milliliter = 0.034 fluid ounces = 1/5 teaspoon

To convert	multiply by	to obtain
quarts	0.95	liters
fluid ounces	28.41	milliliters
liters	1.06	quarts
milliliters	0.03	fluid ounces

Weight

1 metric ton (mt) = 2,205 pounds (lb) = 1.1 tons (t)
1 kilogram (kg) = 2.205 pounds (lb)
1 gram (g) = 0.035 ounces (oz)

To convert	multiply by	to obtain
pounds	0.454	kilograms
pounds	454	grams
ounces	28.35	grams
kilograms	2.205	pounds
grams	0.035	ounces

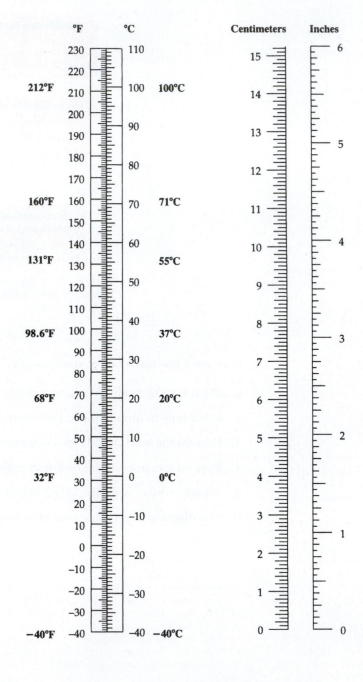

Temperature

Celsius (°C) to Fahrenheit (°F):

$$°F = 1.8 \ (°C) + 32$$

Fahrenheit (°F) to Celsius:

$$°C = \frac{(°F - 32)}{1.8}$$

	°C	°F
Water boils	100	212
Human body temperature	37	98.6
Water freezes	0	32

Appendix B
Answers to End of Section and End of Chapter Questions

CHAPTER 1

1.2 Check + Apply Your Understanding

1. The cell is the smallest level of hierarchy that is able to operate on its own—living and reproducing.
2. As animals, krill would be placed in the broad category of consumers.
3. Myoglobin is a molecule, and a whale's muscle tissue is composed of many types of specialized cells.
4. This occurs as a function of their ability to sense and respond to changes in their environment, a fundamental characteristic of life.
5. A close examination of their DNA would likely confirm their close relationship. Investigating their common ancestry through the fossil record will also provide evidence.

1.3 Check + Apply Your Understanding

1. An **atom** consists of three main subatomic particles—**protons,** neutrons, and electrons. Depending on the number of protons, atoms come in a variety of forms called **elements.** When two atoms bond together they form a **molecule.**
2. Biomolecules include carbohydrates, lipids, proteins, and nucleic acids. The three elements common to these four groups are C, H, and O.
3. Water sticks together in droplets and absorbs large amounts of heat because of the attraction of one molecule of water to another. This attraction occurs between the relatively negatively charged oxygen atom of one water molecule to the relatively positively charged hydrogen atoms on the adjacent molecules.
4. Chitin is constructed from four different elements—carbon, hydrogen, oxygen, and nitrogen—that total 26 atoms.
5. Paramecium is a eukaryotic cell because it is composed of several separate specialized compartments

such as the food vacuole and the nucleus. It is more complex than a bacterium since it includes a mouth and digestive organelles called food vacuoles.

1.4 Check + Apply Your Understanding

1. Darwin proposed that all organisms share a common ancestor and therefore share important yet common characteristics.
2. We are able to observe evolutionary changes in organisms such as bacteria over a much shorter period of time due to the fact that bacteria reproduce so quickly. Large, complex organisms that take far longer to reproduce might not exhibit evolutionary changes for thousands of years.
3. Evolutionary changes are not a matter of need—but instead relate to differential survival and reproduction based on adaptations to the environmental conditions that exist. Blind cavefish did not need to lose their eyesight—instead, those that lost their eyesight were in fact better adapted to the new environment.
4. DNA carries the code or instructions for the cell in gene segments. These genes are used to make products in the form of proteins.
5. Evolution and fitness relate to the environment of a certain population. If an organism is more suited to this environment, it will survive and pass its genes on. The genes accumulate in the population's gene pool. Evolution does not mean perfection—if an organism is good enough to survive and reproduce, it has accomplished its needs for the species to exist.

1.5 Check + Apply Your Understanding

1. Nonexperimental, descriptive methods; and experimental, manipulative methods.
2. Observable characteristics, the fossil record, embryonic developmental

patterns, and molecular data.
3. I would explain that he or she is misunderstanding the definition of theory and that a theory is a well-established explanation supported by years of evidence and experimentation. I would also add that several lines of evidence support the theory of evolution.
4. At the bottom—these layers are the oldest sediment, while the top of the cliffs are newer layers.
5. Since the plant roots live underground, it is more difficult to observe and collect data.

Chapter 1
End of Chapter Questions

1.

consumer	e.
producer	c.
decomposer	a.
eukaryote	b.
prokaryote	d.

2.

DNA	c.
genome	d.
chromosome	a.
gene	b.

3.

atom	e.
molecule	a.
cell	b.
population	d.
biosphere	c.

4. c.
5. d.
6.

theory	d.
hypothesis	c.
experiment	b.
scientific method	a.

7. d.
8. b.
9.

natural selection	c.
evolution	b.
adaptation	a.

10. d.
11. There are two key elements to consider. First is that other scientists examined the work of Woo-suk Hwang and could not verify his results, which resulted in the retraction of his work—this shows the power of the peer review process. Second, this story illustrates the "human" na-

ture of scientists who feel pressure to provide positive results at the conclusion of expensive and time-consuming research—an element that sometimes compels researchers to fake their results.
12. The old saying "if it sounds too good to be true—it must not be true" represents a potential start to the critical thinking process. Bold statements of "incredible" weight loss from taking a single product are contrary with our own experiences involving calorie consumption, exercise regimens, and weight loss or gain. We understand from our own experience that these elements are the key components to consider—rather than a quick commercial solution.
13. A basic understanding of the role of biodiversity in the normal functioning of any ecosystem tells us that the more we know about ecological interactions among organisms, the better we are able to recognize ecological problems and potentially take actions to conserve species. Although less charismatic organisms like earthworms draw less attention, they play as significant a role as larger, more familiar organisms, like blue whales.
14. This story illustrates the power of the fact that in science, there is not a quest for one "truth"; instead, science is a continual journey that seeks to find the most compelling evidence in support of a line of thinking, or hypothesis. In this regard, the "new information" does not invalidate the other science investigations in this area—it adds to the long line of evidence and helps to improve our understanding of how and when humans evolved the capacity to walk upright.
15. Approximately 1,300 whales were harvested in 2006 by Japan.
16. Between 2002 and 2006 there was an increase of

roughly 600 whales harvested per year—a nearly 100% increase in numbers.
17. According to marinebio.org (accessed Oct. 11, 2010), "The current world population for the species is estimated to be around 60,000 individuals (6,000–8,000 in the North Pacific, 12,000 in the North Atlantic, and approximately 40,000 in the Southern Hemisphere) or about 30%–35% of their original population before whaling." Estimating the population of humpback whales is an equally difficult task as it is for blue whales; understanding this, the numbers represent the best estimates based on sampling the population.
18. It is assumed that the harvesting of whales by Japan is leading to a better understanding of the whale's anatomy, physiology, and developmental patterns. Some information on gender ratios and reproduction can be obtained by simply counting the numbers of males vs. females harvested and the estimates for ages at which they were harvested. The relatively high number of whales harvested (1,000) is considered by many international conservation organizations to be of limited value to our understanding of the species.
19. Some argue that ongoing harvesting provides us additional information each year that helps us better understand the species and develop conservation efforts to maintain or increase their numbers. Against harvesting any of these whales runs counter to the conservation effort—especially in light of the decades of unregulated hunting that caused a significant loss of the population. The cultural significance of whales to the Japanese outweighs any potential, marginal impact to the species.
20. Whales are an important part of the history and

culture of Japan—this historical right to harvest whales cannot be erased by some international consensus that it is not right. Japan maintains that scientific harvesting of whales benefits our understanding of the species and improves our ability to develop and implement true conservation efforts.

21. The international moratorium is a just and positive move by the international community to conserve a species that was subject to unregulated hunting for so long. This action provides hope to the tribe that this culturally important species will be available for future generations.

Potential Student-Generated Questions

1. Would molecular studies help to clarify the evolutionary relationships among these three species?
2. If the elephant is more closely related to the hyrax, what happened to the elephant's lower incisors?
3. What other anatomical features do these animals share?
4. Are habitat, diet, and social behaviors important in determining evolutionary relatedness?

CHAPTER 2

2.2 Check + Apply Your Understanding

1. A species is a group of organisms that can breed and produce viable offspring.
2. Fungi.
3. No—these populations do not interbreed, so they are considered separate species.
4. The branching of plants occurred before animals.
5. A—Crocodiles and mammals share an additional characteristic in the amniotic egg.

2.3 Check + Apply Your Understanding

1. Animal characteristics include the following: multicellular, eukaryotic, heterotrophic, sexual reproduction (for most animals), quickly respond to their environment, ingest nutrients.
2. DNA analysis.
3. Reproduction.

4. Chordates.
5. Based on this information, it could belong to the Animal Kingdom or the Protist Kingdom. One could determine its classification based on DNA analysis. (This is a hydra, which is an animal.)

2.4 Check + Apply Your Understanding

1. Cyanobacteria help to cycle carbon through photosynthesis.
2. They help to produce commercial food products and medicines.
3. Similar—grow and reproduce asexually like bacteria, prokaryotic, one circular DNA chromosome in the cytoplasm and have cell walls; different—ribosome structure, composition of the cell wall.
4. Yes—archaeans can live in these extreme environments.
5. Since protists are eukaryotes and cyanobacteria are prokaryotes, the cellular structure can be used to determine differences. Generally, prokaryotes are smaller and less complex than eukaryotes. The cyanobacteria will not have any membrane-bound organelles, such as a nucleus, vacuole, or chloroplasts. The cyanobacteria will also be smaller.

Chapter 2
End of Chapter Questions

1.
lineage	b.
species	d.
shared characteristics	a.
evolutionary tree	c.

2. c.
3. b.
4.
protists	a., b., c.
fungi	c.
plants	a.
animals	b.
prokaryotes	a., c.
autotrophs	a.
heterotrophs	b., c.

5. e.
6. c.
7. Since the Protist Kingdom represents a taxonomic "grab bag," laboratory investigations into the specific genetic differences might move groups from the Protist Kingdom into new kingdoms—or potentially new domains.

8. Yes, it would appear from their physical appearance that these two wolf species are closely related; however, this is not the best evidence to use to determine relatedness. Although anatomical features provide clues, we've learned that physical and anatomical appearances can be deceiving when trying to determine which organisms represent separate species. For example, Domain Archaea and Domain Bacteria share many of the same physical characteristics, and many animals have different appearances based on where they live or the time of the year. DNA analysis confirms these are two separate species of wolf.

9. You would search fossil-bearing rocks that date before the appearance of the first true whales, and after their first common terrestrial ancestor (150 million to 135 million years ago). You would look for fossil animals that lived in or near water with anatomical characteristics such as the movement of the nostrils from the front of the face to the top of head. You could also seek out those that show the development of a tail fluke and side fins from legs and arms.

10. The fact that it is single-celled does not help, since there are single-celled organisms in all three domains. You could examine the organism to determine if it had internal organelles such as a nucleus (if so, it would be a eukaryote). Since it lives on the sea floor near a thermal vent, it suggests that the microbe might be an archaea. If so, you would look for DNA information that would place it closer to eukarya than bacteria. If the genetic clues proved negative and it had no internal organelles, you would then conclude that it belonged to Domain Bacteria.

11. Theropods.
12. Horned dinosaurs.
13. The hip structure is different among these lineages.

14. Modern birds are lizard-hipped and armored dinosaurs were bird-hipped.
15. You cannot say a lot about the hip structure of crocodiles. About all you can say is that they have a different structure than dinosaurs and modern birds.
16. Be careful not to misinterpret the tree by selecting the closest branch. Always trace the branch back to the node. In this case Sauropods are more closely related to birds than armored dinosaurs.

Potential Student-Generated Questions

1. Would molecular studies help to clarify the evolutionary relationships among these three species of plants?
2. Are fossils of these plants found in lake or pond sediments?
3. Are the trapping mechanisms homologous structures between the waterwheel and the Venus flytrap?
4. Do the snap-trap mechanisms of the waterwheel and Venus flytrap represent convergent evolution?

CHAPTER 3

3.2 Check + Apply Your Understanding

1. Genetic variation is the raw material upon which natural selection acts (eliminates less adapted variants) and results in evolutionary change (changes in the genetic composition of the population).
2. First, meiosis is a cellular process that reduces and mixes genetic information in spores, sperm, and egg cells. Second, fertilization combines two unique sex cells to produce a new combination of traits in offspring. The third way is through mutations (the rearrangement, duplication, or deletion of genes).
3. One reason is that mutations can create new genetic combinations and the second reason is that it can activate or inactivate genes and thus shape the evolutionary abilities and success of organisms.
4. Surprisingly, dolphins have many of the same

genes for detecting airborne chemicals as land mammals; however, as you might predict, they have been inactivated by genetic mutations.
5. The dispersal or movement of spiders to new locations is the evolutionary process of gene flow, which will increase genetic variation in the spider populations.

3.3 Check + Apply Your Understanding

1. Chance and the natural selection of variants through the "struggle for existence."
2. Unlike natural selection, which selects the most adapted phenotypes for survival and reproduction, random events eliminate individuals regardless of their fitness—many adapted phenotypes may be eliminated along with maladapted individuals and many maladapted individuals will survive.
3. Adaptive traits increase an organism's fitness by increasing its ability to survive and reproduce.
4. This scenario shows several evolutionary processes taking place. First, the population of fiddler crabs is marked by individual variation in the size of the male's claw and the ability to build a structure. Second, sexual selection is operating—the enlarged claw is a sexually adaptive trait that is being selected for—males with the large claws outcompete other males and mate more often, passing the genetic information for large claws and other traits to the next generation. The advantages of reproduction outweigh the disadvantages of making and carrying a large claw and attracting predators with the waving movements.
5. Selection pressures include predation by other birds and animals and food delivered by the adult bird to feed them. Traits selected **for** include long necks to project their mouth to be first in line, a rapid growth rate to develop more rapidly, and a large mouth to feed on larger prey brought

by the mother. Traits selected **against** include small body size and a slow development time.

3.4 Check + Apply Your Understanding

1. Speciation is the origin of new species and new kinds of organisms. It is driven by the divergence of populations that eventually become reproductively isolated from each other.

2. (1) geographical, (2) temporal, (3) mechanical, and (4) behavioral.

3. Reproductive isolation prevents or reduces gene flow into a population and thus reduces genetic diversity. Isolating barriers keep species different from other species.

4. This is an example of a post-mating barrier called **hybrid inviability.**

5. This female preference for mating with its own population selects for reproductive isolation. The reproductive barrier is a behavioral barrier because it involves communication before mating (pre-mating barrier). Yes, this barrier is sexual selection in which the female selects the male to mate with based on its call rhythm and duration.

3.5 Check + Apply Your Understanding

1. A living fossil is a group of closely related species that have changed very little in millions of years as shown by a similarity between living and fossilized specimens.

2. (1) Strong stabilizing selection, where variations are selected against; (2) effective reproductive barriers to speciation; and (3) a developmental constraint that "boxes in" phenotypic change. A fourth factor is a highly stable, constant environment that promotes stabilizing selection.

3. Yes, I agree with the statement. Speciation is a process that isolates and keeps species from interbreeding. It is a diverging process. On the other hand, the process of forming hybrids is a mixing or breeding of two species and is

the opposite, or reverse, of speciation.

4. (A) A species with little genetic diversity is generally more susceptible because it lacks the raw material for natural selection to bring about adaptations compared to a species with greater diversity and resilience or response to change.

5. Purple loosestrife has found an opportunity in the United States and is outcompeting the native wetland flora for space and other resources. The success of purple loosestrife is driving the local wetland plants to local extinction by eliminating them from the area. This has a ripple effect by severely disrupting the food chain or other parts of the wetland ecosystem such as refuges for fish. If they are unable to respond to this change, fish and other wildlife may also become locally extinct.

Chapter 3
End of Chapter Questions

1.
Fertilization   c.
Gene flow   d.
Meiosis   b.
Mutation   a.

2.
Allele   a., e.
Genome   c.
Gene   b., d.

3. c.

4. d.

5.
Coevolution   d.
Natural selection   a.
Selection pressure   b.
Sexual selection   c.

6. c.

7. c.

8.
Adaptive radiation   c.
Local extinction   e.
Mass extinction   a.
Reproductive isolation   b.
Speciation   d.

9. d.

10. b.

11. c.

12. Very thick bony skulls would be heavy and be a disadvantage to flying fast and would throw the bird off balance.

13. Bdelloid rotifers must rely on mutation or random events.

14. Zone lines represent a pre-mating behavioral barrier because the two fungi are exchanging communication signals that prevent the fusion of the two fungal bodies. This inhibits genetic variation and promotes speciation.

15. The graph shows a gradual increase in cranial capacity over the last 3.5 million years.

16. No, it remained about 400 to 500 mL.

17. Selected **for.** Clearly cranial capacity is an example of a trait under strong selection pressure to increase.

18. Living humans have an average cranial capacity of about 1400 mL; half this capacity occurs about 1 to 1.5 million years ago.

19. The sample of skulls may include (1) several different populations and (2) individual variation, and (3) differences between children, males and females, and geological dates may not be that precise.

20. The farther back in time you go, the fewer the fossils.

21. Not necessarily; a larger brain does not necessarily mean that the brain has a more complex "wiring" or reflects greater intelligence. Note: This concept was used in the late nineteenth century to show the alleged racial superiority of Caucasians. Of course, Caucasian scientists generated the data.

Potential Student-Generated Questions

1. How does a robber know where the nectar is placed?

2. Do nectar robbers, by virtue of their landing and probing, cause self-pollination?

3. Is nectar robbing an instinctive or a learned activity?

4. Are some flower shapes easier to rob than others?

5. Does the flower produce an overabundance of nectar so that the robbers do not hinder pollination?

6. Is nectar robbing always detrimental to a plant's fitness?

CHAPTER 4
4.2 Check + Apply Your Understanding

1. An organism's macrohabitat represents the regional environment in which it lives. The microhabitat is the set of conditions that the organism experiences and that influence its life activities.

2. Life zones in aquatic habitats can be classified based on depth of light penetration and distance from shore, whereas life zones on land typically are described based on the dominant vegetation that grows there.

3. Abiotic factors include the chemical and physical aspects of the environment that influence an organism's survival, growth, and reproduction.

4. The effects of a vertical orientation include (1) amount of direct sunlight is reduced, (2) cliffs are sheltered from winds, and (3) gravity acts on cliff plants to dislodge them. Other factors that may be mentioned include (4) cliff habitats have a reduced amount of soil, and (5) cliff rocks contain less moisture and (6) less minerals than soil.

5. Douglas fir generally occupies north-facing slopes because they are cooler and retain moisture; south-facing slopes are warmer and drier.

4.3 Check + Apply Your Understanding

1. A resource is any substance or energy that an organism requires for its life activities. For example, organic compounds are resources for consumers such as animals and fungi. They obtain and break down organic compounds as a source of carbon.

2. Under normal conditions, the elements in fertilizers are considered environmental constraints. When excess nutrients enter the aquatic habitat, algae quickly absorb them and channel this essential resource into reproduction and growth; hence the population increases rapidly.

3. A species' ecological niche is the specific range

of conditions in which it lives and carries out its activities. The conditions included in a niche are abiotic and biotic components of its habitat. Niche also includes the description of its general ecological role.

4. The smooth blue aster likely lives in the shaded, moist habitat. As a general rule, plants that live in light-limited habitats have wider, larger leaves. When light is limited, plants often make the tradeoff from root growth to leaf growth. Open sunny habitats generally contain plants with narrow short leaves, like the tansy aster.

5. Bog spicebush occupies a narrow ecological niche because it is restricted to a relatively narrow range of soil conditions—wet, acidic pH, and high in organic matter. Its narrow niche on a local level translates into its spotty distribution in the southeastern United States.

4.4 Check + Apply Your Understanding

1. These include dormancy, hibernation, and migration.

2. Tardigrades enter dormancy when their aquatic habitat dries up (desiccates), freezes, loses its oxygen, or receives an influx of salts.

3. Although both animals become inactive, a true hibernator such as the hummingbird shows a marked decrease in its body temperature and metabolic rate. This includes breathing rate and heart rate. Although black bears are inactive for much of the winter and their heart rate slows dramatically, their body temperature does not drop as significantly.

4. Following disturbance, opportunistic species are able to rapidly colonize sites because they typically possess broad ecological tolerances, they are able to withstand increased light and changing moisture and temperatures, and their seeds germinate quickly.

5. The alders are less tolerant of shady conditions than the trees that grew previously on the site—which would have had to survive under competing conditions

from existing overstory trees shading them. The plants growing under the mature alder must be shade tolerant since the alders intercept most sunlight.

Chapter 4
End of Chapter Questions

1.
niche   b.
life zone   d.
habitat   c.
resource   f.
constraint   a.
tradeoff   e.

2. e.
3. c.
4. a.
5. b.
6.
dormancy   a.
hibernation   c.
migration   d.
opportunism   b.

7. c.
8. d.
9. e.
10. c.

11. Students should use Figure 4.32 to list factors found in the habitat where they live.

12. Even a small increase of 5 degrees would have a dramatic effect on the extent of tundra life zones by converting them to grassland/deserts, woodland/shrublands, or more likely taiga life zones. The thawed soils would also release more carbon back into the atmosphere—impacting climate change patterns and making the life zone available to many organisms.

13. Nitrogen is a critical nutrient used by all living organisms. It is used in many important biomolecules, such as DNA and proteins. A lack of nitrogen in a habitat represents a potential constraint.

14. The term habitat includes a broad-scale description of the living and environmental factors found where an organism lives. The term niche is more specific and describes both the habitat where an organism lives and the ecological role that an organism is adapted to.

15. Over time the acid rain would alter niches by shifting nutrient availability. You would then expect to see a shift toward plants that are tolerant of acidic

soil as the acid rains continue to lower the pH of the soils and shift mineral nutrient availability.

16. Habitats describe locations that share a range of resource availability and environmental conditions, which combine to support biological communities. A niche describes a specific range of resource needs and environmental conditions that an organism is adapted to. Thus a lake habitat can support more than one species of fish, but each species of fish is specifically adapted to a specific niche within the habitat.

17. Prior to entering hibernation the bat must build up its energy reserves by spending more of its time finding food. This may mean moving to habitats that have higher populations of insects and suitable places to roost during hibernation. Hibernation is characterized by lowering of cellular metabolic rate in order to lower energy use as well as decreasing the core body temperature.

18. The graph shows a series of sharp increases and decreases between 1957 and about 1980. After 1980 the population has a less dramatic upward/downward series of changes.

19. Before, in about 1960 the salmon count peaked at 1,800.

20. No. With the exception of one upward spike in the late 1970s, the population trends have continued to decrease or stay low. Although the fish ladders provide some means for fish to get around the dam barriers, they do not seem to be the complete answer to the problem of low fish populations.

21. The Ice Harbor Dam represents the first major physical barrier to salmon migration relative to fish migration upstream to the Lemhi River. Because this is the first barrier depicted on the graph, it appears that the vast majority of salmon making the return migration were unable to bypass the dam following its construction, and subsequent

generations were accordingly reduced.

22. Both the population spike that occurred about 1979 and the low-level maintenance of fish counts after that time may be related to fish hatchery activities.

23. The hatchery release and the investment in dam improvements for downstream migration both are likely influencing ongoing fish counts in the Lemhi River. There is a chance that the hatchery program is the only reason salmon counts are remaining relatively steady since the late 1970s.

Potential Student-Generated Questions

1. What role does rainfall play in the reproduction of the kingfisher?
2. In a drought year, how might that condition influence the habitat location of the bird's breeding cycle?
3. How would climate change that produces longer, dryer periods for the year influence the movements of this bird?
4. How would human activities that caused deforestation impact this species?

CHAPTER 5
5.2 Check + Apply Your Understanding

1. A population consists of a single species located in a defined geographic area. A community is a collection of populations of different species located in a defined geographic area. An ecosystem contains collections of communities as well as the abiotic conditions found there.

2. Species diversity is related to resource availability and the number of ecological niches found in a community. Periodic severe forest fires would damage and reduce the resources available to species and alter the ecological niches, likely reducing species diversity until the forest recovers.

3. Transition zones are boundary areas that possess habitat characteristics found in communities on both sides. The blending of habitat characteristics in

transition zones often means a unique mix of resources can be found there, supporting species from adjacent habitats as well as those adapted to the special characteristics of a transition zone.

4. Resource availability rarely matches the population needs from year to year. Populations will increase beyond resource availability only to decline rapidly to a point below the carrying capacity line, with the cycle repeating over time.

5. The windward, or left side in the graphic, because of the rainfall amounts and variations as you move up the mountain side. This variation will drive different vegetation and communities. The top of the mountain is an abrupt change to the leeward or dry side which is comparatively uniform in terms of numbers of community diversity.

5.3 Check + Apply Your Understanding

1. Commensalism is a type of interaction between organisms where only one of the two species involved in the relationship benefits while the other is not harmed or benefited. In a mutualistic relationship, both interacting organisms get some benefit from the relationship.

2. As the similarity in ecological niche increases, the need for similar habitat conditions and resources increases and thus the level of competition will increase.

3. Community structure describes the diversity of species, their distribution, and their relative numbers present in a defined geographic area at that point in time. Population characteristics such as age, gender, and the types of interactions experienced by these populations impact community structure.

4. The normal relationship would be characterized as commensalism, where the mite benefits and the humans are neither harmed nor benefited. When mites

cause skin problems, the relationship is best described as parasitic, where the mites benefit and humans are harmed.

5. In spite of the fact they have a very similar ecological niche, the competition between the crows and ravens is interspecific competition (between members of different species). The level of intensity will be high due to the high degree of niche overlap.

5.4 Check + Apply Your Understanding

1. Solar energy is stored in producers as biological molecules and the construction of plant tissues. The sun's energy is the base-level energy input for most ecosystems.

2. The trophic level of an organism provides information about how it obtains its energy. Understanding the feeding relationships and trophic position gives you information about how many animals might be supported in an ecosystem.

3. Both food chains and food webs illustrate feeding relationships and energy transfer within a community of organisms. Food chains illustrate a single chain of possible feeding relationships, whereas food webs show many potential feeding relationships. In both cases, they illustrate the one-way flow of energy through an ecosystem using directional arrows, only differing in the complexity of the interactions being shown.

4. Increased nutrients available in corn or similar grains reduce the amount of food a cow must eat in order to gain a specific amount of weight. The corn also likely contains more nutrients per unit of volume than pasture grasses, thus improving the nutrition to the cow.

5. The farmers are selecting cattle that show an increase in energy transfer efficiency. If the cows are fed the same diet but some get larger faster, they must have a slightly better ability to build biomass from their feed. Cows with increased

energy efficiency alter the pyramid slightly by increasing the number that can be supported by a fixed amount of primary producers (the bar is slightly wider at that trophic level).

5.5 Check + Apply Your Understanding

1. Communication involves the sending of specific signals that can be visual, acoustic, chemical, or tactile.
2. They have evolved social behavior, such as flocking, to improve protection for all members of the group. It can improve individual survival in obtaining resources and help in finding a mate.
3. Those active during the day would rely more on visual signals, while those at night might communicate more using noises or chemical cues.
4. A chemical compound, or warning pheromone, released into the atmosphere can quickly spread among the termite soldiers, causing them to quickly react to the danger.
5. The benefits would include mutual defense from predators associated with many eyes to watch for danger and the odds of one individual being taken as prey. Cooperative efforts at finding food and caring for young are other potential benefits. Intraspecific competition can be intense in large groups and a major cost. Diseases and parasites can easily be passed from one animal to another in large social groups.

Chapter 5
End of Chapter Questions
1.
transition zone d.
population   b.
community   a.
ecosystem   c.
2. d.
3. a.
4. c.
5.
keystone species   b.
trophic level   d.
food web   a.
resource partitioning c.
6.
parasitism   c.
commensalism   a.
predation   d.
mutualism   b.

7. c.
8. b.
9. c.
10. a.
11.
dominance hierarchy c.
pheromones   a.
communication   b.
12. b.
13. d.
14. c.
15. As a chemical signal, it is classified as a pheromone. Pheromone signals can be spread over a wide area as an animal moves across the landscape. The signaling animal and the receiving animal do not have to meet for the message to be transmitted. Deer can use these signals to mark territory or to signal females that might be in the area that a male is available for mating purposes.
16. Food webs provide a more realistic illustration of potential feeding relationships among the many species present in a community. Food chains are limited to showing one set of feeding relationships. Energy transfer efficiency between trophic levels limits the number that can be present in a food chain.
17. Because good vision is of limited value at night, the bats must rely more heavily on their senses of smell and hearing, both of which would allow them to sense the presence of potential threats or signals sent out by other bats.
18. As the competition for resources increases (moves right on the graph), the number of species present is reduced. This can be seen in the fewer number of lines as you move farther right. Two species are present at the highest levels of competition, whereas as many as four are present at moderate levels of competition.
19. The first relationship is an example of commensalism, where the barnacle benefits from the relationship and the whale is not harmed. The second relationship is an example of parasitism, where the nutrients needed for the barnacle are taken from the host,

depleting the amount available for it.
20. With increasing levels of plant biomass production (primary production), you have the potential for supporting larger numbers of upper trophic levels (that is, more species/greater numbers). It would likely have both a broader base and be taller—potentially including more trophic levels or more organisms per trophic level.
21. Whether or not a tool is present, the chimps prefer to use vocalization to communicate with researchers. The data indicate that use of gestures was the second most preferred form of communication, and given the variation (illustrated by the error bars on the histogram) there may be no statistically significant difference in the preference between these two forms of communication.
22. The chimpanzees recognize that the researchers are not paying attention to them, and in each case there is a much lower level of communication. Specifically, the chimpanzees recognize that gestures do not work when the subject is facing away from them.
23. Chimps seem very similar to humans in their preferred methods of communication. Clearly we share the use of gesturing and vocalizing as a means of signaling to those around us. This is not surprising given how similar chimps are to humans in other forms of behavior and in their genetics.
24. It is likely that chimpanzees have learned through repeated interactions with researchers in laboratories that they will respond to vocalized forms of communication. This is similar to observations of chimpanzees in the wild that use a wide range of vocal communication to send signals of recognition or danger, or to calm their offspring.
25. There likely would be some differences between laboratory chimps and those in the wild. This is due to the familiarity of laboratory chimps to interac-

tions with humans over the course of their captivity. Truly "wild" chimps would have to learn successful forms of communication with humans through trial and error. Given the chimps' preference for gesturing and vocalization, these forms likely would still dominate in communication with "wild chimps."

Potential Student-Generated Questions
1. How do dramatic cycles in population sizes influence the relationships shown in the block diagram?
2. In the 19th and early 20th century, many scavenger animals were actively hunted as pests. What impact does this have on other trophic levels in the food web?
3. Do animals who have genomes which favor highly efficient growth substantially alter the food web relationships?
4. What is the impact on a food chain when a habitat disturbance alters the species composition of primary producers in a community?

CHAPTER 6
6.2 Check + Apply Your Understanding
1. Epithelial tissue.
2. Connective tissue cells.
3. False. Some types of connective tissue, such as cartilage and bone, secrete a nonliving material, or matrix, in which the cells are embedded.
4. (A) Gases would move through a **single** layer of cells much faster than through a multiple layer of cells. (B) The esophagus receives food of various textures. To prevent abrasion and wear and tear, it would be composed of **many** cell layers. (C) Absorption of small nutrient molecules would be facilitated by a **single** layer of epithelial cells.
5. (A) Nervous tissue. (B) Epithelial tissue has glands in the outer layer of the tongue that secrete the sticky solution. (C) Muscle tissue, under stimulation from nervous tissue, contracts and flips the tongue outward.

6.3 Check + Apply Your Understanding
1. Mitochondrion.
2. (A) The nucleolus is the part of the nucleus that produces ribosomes, which in turn produce proteins. Nucleolus size and activity are directly related to the activity of the cell, and younger cells are more active than the older aging cells.
3. Goblet cells would have (1) ribosomes to make the protein in mucus, (2) Golgi complexes to package it, and (3) vesicles to transport it to the plasma membrane for export into the stomach cavity.
4. If adhesion proteins have become defective, the cancer cells leave their location and spread to other locations.
5. As white blood cells patrol the blood and tissues, they are sensitive to whether cells present "self" recognition marker proteins on their plasma membranes. The proteins in the cytoskeleton that allow them to move in and out of blood vessels also are involved.

6.4 Check + Apply Your Understanding
1. Covalent bond.
2. (1) Central carbon and hydrogen atoms, (2) an amine group of nitrogen and hydrogen atoms on one end of the molecule, (3) an acid group on the other end of the molecule, and (4) a side group attaches to the central carbon. The amine group of one amino acid joins the acid group of the next amino acid to form a protein chain.
3. Glycogen is constructed from thousands of different glucose molecules that are bonded together when specialized enzymes remove water between glucose molecules.
4. (B) The net movement will be from side 2 to side 1 because side 2 has the lower solute concentration and therefore the higher water concentration. Water will flow down its concentration gradient from side 2 to the lower concentration of water on side 1.

5. (A) Large proteins that are assembled inside the mammary gland cell and secreted into the milk duct are transported in vesicles by bulk transport. (B) Yes. Energy is required to secrete antibodies and other large macromolecules out of the mammary gland cell into the milk duct.

6.5 Check + Apply Your Understanding

1. A strictly controlled factor has a narrow range above and below its set point. Conditions that place it beyond this narrow range trigger action.

2. Temperature, fluid and nutrient levels, as well as concentrations of oxygen, carbon dioxide, and salts in the blood.

3. The three parts are (1) the **sensor,** which perceives the change in the factor it is monitoring, (2) the **integrator,** which compares the factor against its set point, and (3) the **effector,** which takes action to amplify or counteract the change in the factor.

4. A positive feedback system uses the same three components as a negative feedback system. The effector in negative feedback **opposes** the direction of the stimulus, whereas in positive control, the effector increases, or **amplifies,** the signal in the direction of the stimulus. Positive feedback systems are rarely used because they do not lead to balance and stability.

5. (1) The pancreas is the **sensor** of excess sugar in the blood, (2) the pancreas is also the **integrator,** comparing the concentration of blood sugar against its set point and secreting insulin into the blood, and (3) the liver is an **effector** because it brings down the blood sugar level by taking glucose and turning it into the storage molecule glycogen.

Chapter 6
End of Chapter Questions

1.

Atom	f.
Cell	e.
Molecule	d.
Organ	b.
Organ system	c.
Tissue	a.

2.

Connective tissue	b.
Epithelial tissue	d.
Muscle tissue	a.
Nervous tissue	c.

3. d.
4. b.
5.

Golgi complex	b.
Lysosome	h.
Mitochondria	g.
Nucleolus	d.
Nucleus	a.
Endoplasmic reticulum	f.
Ribosome	e.
Vesicle	c.

6.

Adhesion protein	c.
Recognition protein	b.
Transport protein	a.

7. a.
8. c.
9. b.
10.

Carbohydrate	b.
Lipids	c.
Nucleic acids	d.
Proteins	a.

11.

Active transport	a.
Bulk transport	e.
Facilitated diffusion	b.
Osmosis	c., d.
Simple diffusion	d.

12. b.
13. a.
14. c.
15.

Effector	a.
Integrator	c., d.
Sensor/receptor	b., e.

16.

Negative feedback	a.
Positive feedback	b.

17. a.
18.

a. The hairs are outgrowths of the skin, or epidermis, of the gecko. The epidermis is composed of epithelial tissue with keratin proteins. The hairs consist of proteins that are strong and linear, similar to the helix of keratin. Proteins are the most versatile of the macromolecules in terms of structure and function.

b. At the molecular level, they would look at the kinds of molecules that compose the hairs and how they are arranged. What elements do they contain that make them attach and detach so easily? What genes code for these molecules? At the cellular level, they would look at the epithelial cells and how they produce the hairs in their specific arrangement. At the tissue level, they would look at the organization of epithelial cells within the tissue—which cells function in which way in the tissue. At the organ system level, they would investigate the arrangement of muscles, bones, and ligaments that allow the gecko to run. Finally, at the organism level, they would look at the mechanics of the gecko as it runs to determine the angles and forces it exerts on the toe pads.

c. For the tail to break off, the tail's four tissues must fracture or split quickly. Epithelial tissues compose the outer skin. Connective tissues dominate the tail structure—bone in the middle must break, and blood vessels must close off quickly to prevent large loss of blood. Muscles cause the tail to wriggle and must be stimulated by nervous tissue remaining in the tail.

19. (A) This scenario describes the feeding behavior of snails. Feeding involves the digestive system and the coordinated actions of the organs of the mouth, tonguelike structure, esophagus, and stomach. (B) All four tissues are involved. Epithelial tissues line the mouth, esophagus, and stomach, and they secrete the sticky mucus. Muscles move the teeth and are connected to the tonguelike structure by connective tissue. Muscle tissue contracts to move the teeth and food particles under the activity of nervous tissue. Nervous tissue also stimulates the cilia to beat.

20. Survival of midge larvae declined with submergence time and the increase in salt concentrations.

21. Nearly all larvae survived at all concentrations for one day. After one day, survival rapidly declined with exposure to the two highest concentrations (1,500 and 2,000 mOsm kg^{-1}).

22. About one day.

23. More than 50 percent survived the highest concentration for three days.

24. 75 percent.

25. 45 percent.

26. Yes. Virtually all of them survived at 0 mOsm kg^{-1}.

27. No. Nearly all larvae exposed to these concentrations survived to day 10 of the experiment.

28. When living cells are submerged in high concentrations of salt water, osmosis will cause a net loss of water from the cell. This results in a collapse of the cell and loss of a vital component of their cells.

29. The net direction of water would be into cells. If uncontrolled, this would cause cells to swell and possibly burst.

30. The experiment showed that midge larvae have the ability to survive a wide range of salt concentrations (0–2,000 mOsm kg^{-1}) over a short period of time (one day). If inundated with seawater (1,000 mOsm kg^{-1}), most larvae can survive for about three days. The bottom line is that Antarctic midge larvae are very well adapted to tolerating wide ranges of salt concentrations.

31. The dark coloration absorbs heat energy in this cold environment.

Potential Student-Generated Questions

1. How does the fuel enter the muscle cell?

2. Is the generation of energy able to fuel the muscle activity needed to make the long-distance migration?

3. What would happen if the fuel was not able to efficiently move across the muscle cell membrane?

CHAPTER 7
7.2 Check + Apply Your Understanding

1. The process by which the cell becomes specialized in its form and function.

2. By increasing the number of cells by cell division, by increasing the size of their cells, or by secreting materials around their cells.

3. Cell division is associated with growth, whereas cell movement, cell arrangement, and cell specialization are associated with development.

4. Both lobsters and humans stop growing when they reach adulthood and maintain their size as adults; therefore, they both are determinate.

5. 1) During the tadpole stage, yolk synthesis genes would be turned off; 2) as an adult the tiger salamander uses lungs and therefore the gill development genes would be turned off; 3) during the tadpole stage the genes for limb development would be turned on.

7.3 Check + Apply Your Understanding

1. Temperature, daylight, available nutrients.

2. Hormones cross over cell membranes and bind to specific receptors. This action triggers cellular responses that turn on or off certain genes.

3. The lean mass that is of interest to the market is muscle. These particular growth hormones act on muscle cells.

4. Because the epithelial tissues could not respond to the signal, the characteristics of hair growth would not occur.

5. Because krill are arthropods, their regulation of development is similar to that of lobsters. The krill respond to the ocean's temperature as a key indicator for spawning. In summer the cue would be a temperature.

7.4 Check + Apply Your Understanding

1. Growth, development, wound/tissue repair, asexual reproduction.

2. The cell spends 90 percent of its time in interphase. At this time, it performs its specific job and prepares itself for mitosis by replicating its DNA and organelles as well as growing its cytoplasm.

3. TGGCAGC.

4. A nerve cell is active in early development and then goes dormant for the rest of the organism's life cycle. The cells in the intestine constantly divide and turn over, so they continuously go through mitosis.

5. These protein fibers are critical for moving the chromosomes and separating the sister chromatids. If the microtubules are dysfunctional, then proper cell division cannot occur.

7.5 Check + Apply Your Understanding

1. Cell communication occurs through direct physical contact between adjacent cells and through intercellular chemical messengers.

2. A promoter binds to a start region of the gene and triggers transcription.

3. Programmed cell death (apoptosis) is important to reabsorb developing tissues. For instance, the tail of a tadpole is reabsorbed through apoptosis in frog development. In humans, as the tissues around the fingers and toes develop the old tissue needs to be broken down and absorbed.

4. Transcription is the process that rewrites (or transcribes) the DNA into mRNA, whereas translation interprets (or translates) the mRNA to a protein sequence.

5. EGF is released as a response to environmental cues or signaling molecules. EGF binds to a cell membrane receptor and enters the cell. Inside of the cell, it triggers a cascade reaction that releases other signaling molecules that act as transcription factors for gene expression. The expression of various genes results in the growth and differentiation of cells.

Chapter 7
End of Chapter Questions
1.
Growth   b.
Cell differentiation   a.
Development   c.
2. c.
3. a.
4. d.
5. c.
6.
Interphase   c.
Mitosis   a.
Cytokinesis   b.
7. b.
8. c.
9. d.
10.

Transcription   c.
Translation   b.
DNA replication a.
11. a.
12. b.
13.
a. A few examples are the ears, nose, tail, and body. Other cell types that constantly divide, such as the epithelial cells of the skin and intestine, are also undergoing cell division.
b. As the puppy grows, it grows into its feet. Its ears and nose elongate, developing into their final body shape and form. The skin around the eyes grows "droopy."
c. Muscles arise from the mesoderm and the lungs arise from the endoderm.
d. Neurons arise from the ectoderm.
14. Mitosis occurs for growth, development, and tissue repair in multicellular organisms. Some animals use mitosis as a form of asexual reproduction. Growth is the increase in size of the animal; development is the formation and rearrangement of cells; and tissues need to repair or renew themselves as a part of maintaining the organism.
15. Essentially the cell cycle can be interrupted at any phase. Some examples of drug interactions include the prevention of DNA replication, the prevention of microtubule formation, and the blocking of receptors preventing growth factors from triggering the cell cycle.
16. The mRNA sequence is UUUGAGGUC. This mRNA sequence is the code for an amino acid sequence. The mRNA is made in the nucleus, modified, and transported to a ribosome in the cytoplasm. On the ribosome, enzymes, amino acids, and tRNA help to facilitate the interpretation of the mRNA code to a sequence of amino acids that make up the insulin protein.
17. See Table below.

18. Defects to the central nervous system and ears.
19. 10 percent of 4.2 million (or 420,000 births) were from mothers who used alcohol, and 1 percent were from binge-drinking mothers.
20. Development of the central nervous system, heart, and sensory organs is well under way at this point. If 50 percent of mothers drink, this can have detrimental effects on the fetus, perhaps leading to FAS or other birth defects.
21. As a result of the cells committing suicide, less brain tissue is present in babies born to mothers who drink in the third trimester. With fewer brain cells and less mass, these babies may develop other developmental issues. These babies have behavioral and learning disabilities, and some are mentally retarded.

Potential Student-Generated Questions
1. If conditions are not ideal, how long can a Monarch larva stay in a particular stage of development?
2. What is the impact of juvenile hormone being released in later stages of development?

CHAPTER 8
8.2 Check + Apply Your Understanding
1. The black bear is a generalist because it eats a wide variety of foods such as fruits to fish. The koala is a specialist feeder because it feeds exclusively on the leaves of eucalyptus trees.
2. Venoms, or toxins, paralyze the prey so that it cannot escape. Anesthetics numb nerves to prevent the host from detecting the animal and preventing feeding. Anticoagulants are blood-thinning proteins produced by blood-sucking animals to prevent blood from clotting and thickening so that it can flow more easily.

3. Spiders are fluid feeders that feed on the predigested fluids of their prey.
4. Because fruits may have a hard outer covering, fruit-eating bats have teeth adapted for piercing (canines) and crushing (large flat molars). These bats crush the pulp of ripe fruit, swallow the juice, and spit out most of the pulp and seeds. Moth-eating bats have more numerous, smaller teeth. Vampire bats have specialized incisors and canines to cut and penetrate through fur and skin.
5. The top diagram shows sharp incisors and multiple sets of molars, which are adaptive for eating grass. This is the herbivorous horse. The bottom diagram shows large canine teeth and other sharp teeth for eating meat. This is the carnivorous coyote.

8.3 Check + Apply Your Understanding
1. (1) Energy and (2) raw materials to build their tissues.
2. Water is essential for chemical reactions, metabolic processes, and regulation of body temperature. It is a chief component of digestive secretions, as well as blood and body fluids, which transport nutrients, hormones, and waste products throughout the body.
3. (A) Animal muscle contains large amounts of protein and fat, (B) growing shoots provide carbohydrates and proteins, and (C) eggs provide protein and fat. Nitrogen is supplied by all three of these foods in the form of proteins.
4. Collagen as a protein is composed of building block molecules called amino acids. To connect and pull, the best shape would be (B) a long twisted chain. In fact, collagen are three chains in a helical arrangement.
5. Glycogen is the carbohydrate storage molecule of all animals. Fat is also used by animals to store energy. Together they fuel growth and developmental changes.

8.4 Check + Apply Your Understanding
1. (1) Physical breakdown and movement of food

through the digestive system, (2) secretion of digestive enzymes and acids to chemically process food, (3) absorption of nutrients into circulation, and (4) elimination of undigested waste products from the body.
2. The shape of the enzyme determines the shape of its active site, which interacts with the substrate molecule.
3. Mucus protects the stomach and lining of the small intestine from being digested by the protein-digesting enzymes.
4. Constipation occurs when (A) a low-fiber diet causes (B) the stools to move too slowly so that (C) large amounts of water are absorbed from the stool back to the bloodstream. Thus the stools are dry and small in size, and difficult to eliminate.
5. The fox's diet consists mostly of meat and some plants, so the fox will have a shorter intestine and colon, as well as a smaller cecum, as shown on the right (Letter c). The longer intestines and larger cecum on the left (Letter b) characterize the cellulose-eating pika's digestive system.

Chapter 8
End of Chapter Questions
1.
Bulk feeder   b., d., e., f.
Filter feeder   a.
Fluid feeder   c.
2. d.
3. c.
4.
Carbohydrates   c., e., i., k.
Lipids   b., d.
Minerals   h., l.
Proteins   a., g., j.
Vitamins   f.
5.
Carbohydrates
and proteins   e.
Fats and proteins   a., b., c.
Carbohydrates, fats,
and proteins   d.
6. c.
7. c.
8. e.
9.
Gallbladder   e.
Pancreas   h., j.
Large intestine   i.
Liver   b.
Mouth   a.
Small intestine   d., g.
Stomach   c., f.
10. a.
11. d.

Group	Percent Using Any Alcohol	Percent Binge Drinking
Not pregnant	50%	10%
Pregnant	10%	1%

12. d.

13. Females have a higher percentage of fat and thus a lower percentage of water than males. This means that, in females, alcohol is dissolved into a smaller volume of water and therefore constitutes a HIGHER percentage of the blood. Females have a higher BAC than their male counterparts of the same age, body weight, and build. BAC charts build the gender factor into their calculations. The elderly male, like the female, has a higher percentage of fat and will have a higher BAC than the college-age male.

14. (A) The teeth most developed for grinding leaves are the molars, with their variable cutting and grinding surfaces. (B) The large amount of cellulose in their diet means that they probably have a rich microflora (they are, in fact, ruminants). (C) Springboks live in a hot environment, which means that a dry feces is one way for them to conserve water.

15. The rhino's milk is represented by letter B because it provides a large amount of water for the young rhino. In a hot habitat, water loss would be critical for survival. The freezing temperatures of the reindeer habitat require the young to develop a thick layer of insulating fat compared to the rhino. Thus, its milk will be high in fat, letter A. The reindeer milk contains many more calories than the rhino milk. The rhino probably will drink larger quantities of milk to acquire the water and other nutrients necessary to sustain its rapid growth.

16. A kangaroo rat is able to conserve the water it produces from metabolic reactions by reducing losses normally occurring through the lungs, urine, and feces.

17. An empty stomach may reflect (A) infrequent feeding bouts, (B) short periods of feeding followed by (C) rapid digestion. It also may mean that (D) the volume of food consumed at any time is low. (E) Porbea-

gle sharks are opportunistic feeders that consume a large diversity of prey.

18. Most obese dogs eat one meal per day.

19. 27.1 percent of obese dog owners have a strong or very strong interest compared to 39 percent of normal dog owners.

20. The obesity rates go up as the number of snacks goes up, from 30.3 percent for one snack to 42.9 percent for three snacks. The trend is not totally consistent; the obesity rate for dogs that receive two snacks per day actually is lowest, at 26.8 percent.

21. 91.7 percent of normal pet owners had normal weight, whereas 76.3% of obese pet owners had normal weight.

22. Clearly excess food consumption is the real culprit in obesity rates, but interest in dog nutrition and their own health issues, such as weight and exercise of the owner, relate to obesity in dogs. For the dog, the number of meals and snacks per day are strongly related to obesity, as well as the interpretation about whether or not the dog would trade food for a walk.

23. Yes. Small dogs with a higher metabolic rate would burn calories at a faster rate than larger dogs. In a comparison we might expect to see more small dogs in the normal category and more large dogs in the obese category.

24. Owners of obese dogs clearly have some health issues of their own: they tend to be more overweight than owners of normal dogs, they exercise less regularly, and they smoke more.

25. Unfortunately, no. The owners of obese dogs display much less interest in dog nutrition than their counterparts with normal weight dogs. Because the owners of obese pets tend to have weight control issues of their own, it's likely that because they are less concerned with their own weight that they would be less concerned with the weight of their pet.

26. No. Like their human counterparts, most domesticated dogs and cats no longer face the challenges associated with finding enough food to survive. This means that their obesity rates have risen along with those of their owners, creating health challenges that may diminish their quality of life or even shorten it through factors such as heart disease.

Potential Student-Generated Questions

1. Where are the greater sized patches of prey—near the surface or deeper?
2. Which location provides the most efficient feeding—near the surface or deeper?
3. How efficient is filter feeding by baleen whales?
4. Which swimming speed is most efficient?
5. How frequently do baleen whales lunge for food at the surface and in deeper water?
6. How is lunging frequency related to the size of their catch?

CHAPTER 9

9.2 Check + Apply Your Understanding

1. (1) Oxygen is not very soluble in water. (2) Gases diffuse 10,000 times more slowly in water than in air.
2. Oxygen is produced in water by (1) photosynthetic organisms such as aquatic plants, algae, and cyanobacteria, and (2) the diffusion of oxygen from the atmosphere (higher concentration) into the water (lower concentration).
3. Both habitats contain the SAME percentages of oxygen and carbon dioxide. As one ascends in altitude, the column of air and thus the atmospheric pressure decreases. This disperses oxygen molecules and reduces the concentration gradient.
4. (A) Diffusion is the physical process that moves oxygen into and out of the eggs. (B) Turtles that build their nest closer to the surface will have a more rapid rate of gas exchange with it. (C) The nest atmosphere will become lower in oxygen as it is consumed by the

actively digging turtles and higher in carbon dioxide.
5. Lower rainfall causes streams to run slower and therefore decreases the amount of oxygen mixed in from the atmosphere. In addition, warmer summer temperatures mean less oxygen is dissolved in the water. Together, these conditions tend to adversely affect aquatic animals.

9.3 Check + Apply Your Understanding

1. Each of their cells is in close contact with the environment and exchanges gases with the medium in which they live.
2. (1) Gas molecules will diffuse faster with increasing temperature; (2) the greater the difference in concentration or pressure, the faster the gases will diffuse; and (3) gases will diffuse faster in air than water, and faster in water than through a solid medium.
3. A whale that is blowing off gases is exhaling and this means that its diaphragm is **relaxed.** In a relaxed state the diaphragm decreases the volume of the whale's chest cavity and increases the pressure on its enormous lungs, causing it to blow.
4. 1B, 2A. The running lizard is more active and has a greater oxygen requirement and would possess larger lungs with a greater surface area. It would also have a strong diaphragm muscle to ventilate its lungs.
5. The sculpin has a large gill surface area for two reasons: (1) it lives in a habitat that has low amounts of dissolved oxygen and is adapted with a large gill surface area to extract larger quantities of oxygen, and (2) the sculpin is a bottom feeder and is not very active; thus it does not move or ventilate its gills as an adaptation to extract more oxygen from the water.

9.4 Check + Apply Your Understanding

1. First, closed circulatory systems have blood with cells that contain oxygen-binding proteins. Second,

they have a distribution network of blood vessels that delivers oxygen to body cells. Third, animals have a pumping mechanism that keeps blood flowing and circulating.
2. A capillary has very thin walls that reduce the distance that gases have to travel to cross into tissue fluids or the lung.
3. The heart is a double pump that pumps blood with two different pressures. The right side of the heart is a low-pressure pump that supplies blood to the nearby lungs through the pulmonary circuit. The left side of the heart is a high-pressure pump that pushes blood to distant parts of the body through the systemic circuit.
4. Bats have higher energy demands for oxygen because of their flying activities and thus have a THINNER blood-gas barrier than nonflying mammals such as a mouse.
5. (A) Oxygen is transported as dissolved oxygen in the blood plasma, which can occur when the water is very cold and oxygen rich. (B) Ice fishes compensate for a lack of hemoglobin and red cells by having (1) a larger heart, (2) wider blood vessels, and (3) a higher cardiac output. All of these choices increase the transportation of oxygen to body cells. (C) Not likely, because warmer water holds lower levels of dissolved oxygen. Oxygen will also be less soluble in the ice crocodile fish's warmer blood.

9.5 Check + Apply Your Understanding

1. ADP captures energy through the bonding of a phosphate group to become ATP.
2. The higher the energy demand, the higher the rate of metabolism, which in turn requires a larger oxygen supply.
3. (A) With the exception of digestive enzymes inside lysosomes, digestive enzymes operate outside of cells in the digestive tract, whereas metabolic enzymes operate inside cells. (B) Digestive enzymes

break down nutrient molecules into building block molecules, while metabolic enzymes can do this plus build new larger macromolecules from their building blocks. (C) Both enzymes act very specifically on particular types of bonds in particular types of molecules.

4. To contract a muscle, food nutrients, a form of chemical energy, are transformed into electrical energy in nerve cells. The chemical energy is transformed into mechanical energy in muscle cells. The fourth kind of energy is thermal energy or heat. Yes, it is a vigorous and sustained activity and must be fueled by a large supply of energy.

5. Order of rate of metabolism: (2) Flapping wings to take off. (1) Gliding–soaring uses less energy, but still requires muscle effort to hold the wings out. (3) Resting in a stationary position is the least active or demanding and poses the least energy demand on albatrosses. Since heart rate is directly related to metabolism, the sequence of activities is the same: 2-1-3.

9.6 Check + Apply Your Understanding

1. Glycolysis converts glucose into three products: (1) pyruvate, (2) ATP, and (3) hydrogen atoms and their carrier.

2. The carbon in the CO_2 originates with the carbon in the glucose, fatty acids, or amino acids that are metabolized in the citric acid cycle.

3. Without mitochondria, red blood cells must use anaerobic metabolism by glycolysis and fermentation.

4. This long-term sustained activity used by the dogs is aerobic metabolism. The amount of calories they need can be produced from energy-dense foods that are high in **fats.**

5. If the flow of electrons in the electron transport chain is blocked, the cell cannot make the large amounts of ATP to meet the energy needs of the animal. The

cyanogens would be toxic and potentially fatal to the animal eating them.

Chapter 9
End of Chapter Questions

1.
Terrestrial habitat   c.
Aquatic habitat   a., b., d.
2. d.
3. b.
4.
Gills   b.
Lungs   d.
Skin   c.
Tracheal system   a.
5. c.
6. d.
7.
Arteries   a., e.
Capillaries   c.
Veins   b., d., f.
8. a., b., and c. are all false.
9. a.
10.
Citric acid cycle   b.
Electron
transport chain   a., e., f., g.
Fermentation   d.
Glycolysis   c.
11. c.
12. b.
13. The insect is growing and developing and has a high oxygen requirement. By penetrating the water's surface, the insect is tapping into a greater medium to obtain oxygen—the air. Its spiracle is an opening to deliver air directly into its tracheal system. The air physically goes from the atmosphere directly to its cells.
14. The breeding season places extra demands on the male frogs to court a mate and produce sperm. They need more energy and oxygen from their environment. The hairy extensions greatly increase the surface area of the skin and have a rich blood supply for increased diffusion of oxygen during the warm summer season when oxygen supplies are lower.
15. The book lung brings **air** directly into contact with blood. Its many "pages" increases the **surface area** between the air and the blood. The thin narrow pages keep the **distance** very short for gases to travel and slow down the flow of blood to increase the **contact time.**

16. Fish kills often occur in the summer when the water temperature warms up and drives oxygen out of the water into the atmosphere through evaporation. Warm waters also speed up the decomposition of dead vegetation by decomposer microbes. By using large amounts of oxygen, these microorganisms further reduce the oxygen levels in the water. Note: "Winterkills" occur when ice prevents atmospheric oxygen from diffusing into the water.
17. The lizard's response shows a gradual decrease in the rate at which it respires or consumes oxygen from 4° to 1°C. At both ends of the graph (2° to 4°C and −3° to −4.5°C), the consumption of oxygen changes at a faster rate.
18. Graph B shows that the lizards were able to resume normal respiration after 30 hours below freezing.
19. Yes, I think you'd find differences. Lizards found in northern Spain, living in warmer habitats, would probably not have the same level of antifreeze proteins in their blood as those living in the frigid habitats above the Arctic Circle.
20. When the blood is frozen it is not delivering oxygen, so the amount of oxygen available to its tissues is extremely low. I expect that its cells are generating small amounts of energy with anaerobic metabolism, using glycolysis and fermentation with lactic acid as an end product.
21. Yes. If the cells are using anaerobic metabolism virtually no oxygen is being consumed, but cellular respiration is occurring to keep the animal alive.
22. Since respiration uses 6 oxygen molecules and produces 6 molecules of CO_2, I would predict that the curves look similar. When little oxygen is consumed, little CO_2 is released from aerobic pathways.
23. The lizard's brain, along with nervous tissues, are most susceptible to injury from low oxygen levels. They have the highest rates

of energy consumption of the body's tissues.

Potential Student-Generated Questions

1. How big are Apatosaurus lungs?
2. Did Apatosaurus have a diaphragm to change the volume and pressure in its chest cavity?
3. If Apatosaurus was ectothermic like reptiles, how much air did it need to breathe?
4. If Apatosaurus was a reptile, how could its lungs support its energy demands?
5. How fast did it breathe to meet its energy demands?
6. How would bones with hollow pockets support the massive weight of Apatosaurus?

CHAPTER 10
10.2 Check + Apply Your Understanding

1. The nervous system is divided into the central nervous system and the sensory and motor neuron system. The central nervous system consists of the brain and spinal cord, locations where signals are received and processed and new signals are generated.
2. Neurons are all about communicating information from one body location to another.
3. A neuron is an individual cell of the nervous system. It may be a sensory neuron, a motor neuron, or an interneuron. Nerves are bundles of individual neuron axons, all wrapped in a protective covering of connective tissues.
4. The simplest animals, such as sponges, lack nervous systems. They have collections of sensory cells that can signal contraction of other cells on the outside of the body. As animals increased in complexity, they evolved networked nervous systems that included specializations for sensing the environment and signaling muscle contraction, culminating in the development of ganglia and brains where control and processing were centralized.
5. In order for environmental information such as pain

or temperature to be perceived and acted upon, there must be a link between the sensory neurons involved in collecting and transmitting the information to the brain, where interneurons evaluate the information and generate response signals that travel through motor neurons to move muscle groups. Without this link there is no ability to perceive or respond to the ever-changing conditions in the environment.

10.3 Check + Apply Your Understanding

1. Sensory receptors provide information about the environment as well as internal systems in the body, such as blood pressure and the fullness of our stomach.
2. Photoreceptors are stimulated by light energy; mechanoreceptors by pressure; chemoreceptors by specific types of chemicals; pain receptors by damage; and thermoreceptors by heat or cold.
3. Both are examples of mechanoreceptors that are sensitive to pressure. Those found in our inner ears are stimulated by pressure waves generated by sound, whereas those on the surface of the insect's body are stimulated by several kinds of pressure waves, which include sound and changes in air pressure.
4. Because they are active hunters during the day and night, wolves must have good color vision—or many cones for daytime and many rods for good nighttime vision. Wolves are members of the Dog family, so I expect they would have a high concentration (tens of millions) of chemoreceptors in their noses, like a bloodhound.
5. Examine Figure 10.16. The beginning of the cochlea contains hair cells sensitive to higher frequencies, such as 20,000 Hz, so you would expect to lose high-frequency hearing.

10.4 Check + Apply Your Understanding

1. A reflex arc is a link between an input to a sensory

neuron and a motor neuron that controls muscle movement, without involving an active link to the brain and central nervous system.

2. The cerebellum is responsible for coordinating fine muscle movement such as writing, along with some mental functions such as attention and language, and some emotional functions such as fear and pleasure responses.

3. The fish brain lacks a large, infolded cerebrum where complex problem solving skills are concentrated—so fish are unlikely to have these skills. Similarly, mice lack this region as well, and would not feel the emotions developed in the more developed cerebrum.

4. Because the left side of the cerebrum controls the right side of the face, I would look for a tumor on the left side of the cerebrum, which controls motor functions in the face.

5. This type of damage would be very serious. It would cause basic problems with homeostatic controls for heartbeat and breathing, endangering its life. Even if the damage was not life-threatening, the dog could have trouble sleeping or have trouble controlling which sensory nerve signals are interpreted by its brain.

10.5 Check + Apply Your Understanding

1. The three classes of neurons are sensory—those that conduct signals from the tissues in the body to the central nervous system (CNS); motor—those that conduct signals from the CNS to the muscles; and interneurons—nerve cells found in the spinal cord and brain.

2. Nerve cells have branch-like extensions of the cell membrane called dendrites attached to the cell body. A long ropelike extension called the axon extends away from the cell body and ends in small peglike projections.

3. A disease that causes an interruption in the transmission of action potentials

at the neuromuscular junctions would result in weakness of that muscle, and, if severe enough, total failure of muscles to function.

4. For an action potential to develop there must be sufficient stimulation of the nerve cell to cause changes to the permeability of the cell membrane to sodium ions. When stimulated, a resting neuron rapidly changes the electrical charge of the membrane by rapidly moving sodium ions into the cell. This creates a bioelectrical charge that moves along the length of the neuron.

5. The advantage of having one nerve signal stimulate the contraction of multiple muscle fibers at the same time is the need for fewer neurons to carry out a single task. This is a favorable adaptation for large muscle groups such as those in the legs and arms. It's a disadvantage when you need fine motor skills, where the motor neuron signals the contractions of small motor units.

10.6 Check + Apply Your Understanding

1. The ends of bones are protected by pads of cartilage; the joints contain fluid, which helps lubricate and protect bones; and ligaments wrap and stabilize joints.

2. Muscles are attached to bones by tendons. As muscles contract they shorten, pulling on the tendon. This in turn pulls against the bone, resulting in movement of a limb such as an arm.

3. Oxygen is an important element in fueling aerobic respiration, a metabolic process that produces ATP. The ATP is necessary because of the role it plays in the binding and ratcheting action of myosin molecules with actin filaments.

4. Muscles contain bundles of muscle fibers that are wrapped in connective tissue. Each muscle fiber contains a collection of sarcomeres, which are made from bundles of the contractile proteins myosin and

actin. Motor neurons terminate adjacent to muscle fibers, and when action potentials arrive they stimulate those muscle fibers to contract.

5. As you decrease the number of motor neurons you create larger motor units that have less fine muscle control. You would lose the ability to have precise control over your fingers and would have difficulty writing and performing similar activities that require this level of control.

10.7 Check + Apply Your Understanding

1. Birds have evolved strong flight muscles in their breasts that enable them to flap their wings providing lift to keep them in the air; lightweight and streamlined bodies; and lightweight but strong skeletons.

2. Swimmers need a streamlined body shape to minimize resistance as they move through the water, an energy-saving adaptation that also helps them move more quickly. Fast fish have also evolved strong, flexible muscles that have a high percentage of fast-twitch muscle fibers and high fat content for energy.

3. The length and proportion of the legs contribute to a long or short stride. Fast runners typically have long, lean legs, and the lower leg from the knee to the ankle joint is typically longer than the upper.

4. Fruit-eating bats do not echolocate. Recall that echolocation evolved as a means for bats to located flying insects. Fruit-eating bats rely primarily on their excellent sense of smell to locate their food.

5. They have a tall skinny neck with large eyes to better see potential predators. Their long powerful legs are well adapted for running; in fact, they can run faster than a horse, reaching speeds of 45 mph.

Chapter 10
End of Chapter Questions

1.

Nerves	e.
Neurons	d.
Nervous system	b.

Motor neuron	a.
Sensory neuron	c.

2.

Central nervous system	b.
Brain	c.
Ganglia	a.

3. c.
4. d.
5.

Sensory receptors	b.
Pain receptors	d.
Chemoreceptors	a.
Mechanoreceptors	c.

6. d.
7. c.
8. d.
9. d.
10.

Joint	b.
Muscle fiber	d.
Sarcomere	c.
Actin	a.

11. b.
12. a.
13.

Swimming and diving	c.
Running and jumping	a.
Digging and crawling	d.
Climbing and swinging	b.

14. b.
15. b.
16. c.
17. Based on echolocation, I would predict that dolphins rely on mechanoreceptors stimulated by the returning sound waves to help guide them to their food.

18. The inchworm likely possesses a hydrostatic skeleton similar to that of other insects and worms. To move, internal muscles contract in coordinated actions that permit this grip-and-pull movement. Each of the feet must be adapted to grip the edge of the leaf; this requires muscle control over finger-like projections on each foot.

19. Although gorillas are powerful animals, they do not use their arm strength for swinging through the trees, so their forearm and hand muscles likely are less strong than those of orangutans. They probably lack the precurved fingers found in orangutans as well because they don't require such a strong grip.

20. Rods are useful for seeing in low-light conditions, whereas cones are adapted to see different types of colors. Mutant fish with almost all rods in their retina would have excellent low-light vision but would lack

sharp color vision found in the normal fish. In the wild, the shallow coral reefs have lots of light and color, an environmental condition that favors fish with lots of cones and color vision. The mutant fish would be better adapted to deeper portions of the reef with less light.

21. As the shark gets larger, its ability to bite harder increases.

22. For each animal, the bite force is greatest in the rear of the mouth, where it is almost double the amount of bite force in the front of the mouth. This is true for both sharks and *T. rex*.

23. More force can be applied at the rear of the mouth because most of the muscle mass for biting is located there.

24. The model shows that there is a large muscle mass at the rear of the mouth, the joint where the top of the jaw joins with the bottom of the jaw. This large muscle mass is directly related to ability of the rear of the mouth to generate more bite force.

25. No, there is no apparent advantage related to bite force. A mature great white shark that weighs only 10 percent of the *T. rex* has a greater bite force with its cartilaginous jaw and skeleton.

26. To scale up muscle mass, divide the weight of the megalodon by that of the largest great white shark: 105,140/7,300, which is roughly 14. So it is likely that the megalodon had at least 14 times the muscle mass of the great white.

Potential Student-Generated Questions

1. What types of sensory neurons would you expect to find if you examined the lizard's feet?

2. What is the relationship between these sensory neurons and the lizard's thermal dance?

3. Describe the placement of the lizard's eyes and its ability to spot predators.

4. Does the brain of this tiny lizard have a large, infolded cerebrum?

CHAPTER 11

11.2 Check + Apply Your Understanding

1. Asexual reproduction creates genetically identical offspring and requires only one parent. Sexual reproduction involves two parents and creates offspring that are genetically different from one another and from the parent.

2. Sexual reproduction takes a larger investment of time and energy in the acts of finding and acquiring a mate, making gametes, and mating. However, this investment is a trade-off for increased genetic variation in the population.

3. Evolution is the genetic change in a population over time. These changes can only occur through the passing of genetic information from generation to generation, which happens through reproduction. Reproduction is a result of many years of evolutionary adaptations that help the organism to survive and reproduce.

4. Honey bees reproduce both asexually and sexually. They use sexual reproduction to increase genetic variation in their hive, but they also use asexual reproduction to produce males in a more energy- and time-efficient manner. Because finding and selecting a mate uses energy, they also mate only once in a lifetime. The female mates with several males, collecting and storing the sperm. This conserves energy but allows her to expand the colony's gene pool. The male dies after passing on his genes.

5. This is an example of parthenogenesis, which is a type of asexual reproduction.

11.3 Check + Apply Your Understanding

1. It is a strategy for reproducing in which the eggs are deposited and fertilized outside of the female's body.

2. The amniotic egg allows reproduction to take place outside of the water, providing moisture and protection to the developing embryo. Specifically, the membranes of the egg keep water close to the embryo, facilitate gas exchange, and allow the processing of nutrients and wastes.

3. Internal fertilization increases the likelihood of the sperm and egg coming together. Because the fertilized egg is better protected inside the female, it is more likely to survive the initial stages of development. Internal fertilization also eliminates the need to reproduce in water.

4. Land snails are hermaphrodites that reproduce sexually. They utilize courtship rituals to select their mates, they mate with multiple partners to increase genetic variation, and they utilize "love darts" to improve the efficiency of egg fertilization.

5. The male bird has a yellow head, elaborate plumage (longer feathers), brighter coloration, and is larger than the female. The male with the most elaborate feathers will be viewed by females as healthy. It will most likely attract mates, which enhances his reproductive success.

11.4 Check + Apply Your Understanding

1. Gonads, ducts, glands, and accessory glands.

2. The gonads produce the eggs (female) or sperm (male) and also the sex hormones that maintain the reproductive system.

3. The placenta provides nourishment so that the fetus can develop more fully in a protected environment before birth.

4. The sperm travels from the testes to the epididymis through the vas deferens and ejaculatory duct and out the urethra of the penis. It enters the vagina and passes through the cervix into the oviduct. In the oviduct, it will fertilize the egg, which moves down to the uterus where it implants.

5. Because the hippo is caring for one offspring and gives birth only once every couple of years, it will have a longer gestation period (240 days). The ground hog and wolf are nurturing several offspring, so they will have shorter gestation periods (30 and 63 days, respectively).

11.5 Check + Apply Your Understanding

1. Meiosis introduces two sources of genetic variation through crossing over and random assortment of chromosomes during meiosis I. The third source arises from random fertilization of gametes.

2. The sperm is powered by the midpiece, which contains mitochondria that produce ATP. To enter the egg cell, the head contains an acrosome that contains enzymes that dissolve the outer membrane of the egg. The head also contains the nucleus, which fuses with the egg's nucleus. The tail is a cellular extension known as a flagellum.

3. A pair of chromosomes that is made up of one chromosome from each parent. Each chromosome carries the same genes but perhaps different alleles (versions) of the trait.

4. Both processes produce haploid gametes in the gonads (ovaries and testes, respectively) through meiosis. Spermatogenesis starts at puberty and continues throughout the rest of a man's life, whereas oogenesis starts before birth and ends when a woman runs out of eggs (around 50 years old). At the end of spermatogenesis four functional sperm cells and one functional egg cell are formed from oogenesis.

5. Oogenesis begins before birth in women. Therefore women are born with all of their eggs. The eggs released later in life come from older cells. These cells are not able to perform as well as newer cells and are more prone to error during meiosis and gametogenesis. Because spermatogenesis happens at a fast pace, these cells are always "new." The life of a sperm cell is less than one week once it reaches maturation.

Chapter 11 End of Chapter Questions

1.
Asexual reproduction   a.
Sexual reproduction   b.

2. d.
3. b.
4.
Internal fertilization   d.
External fertilization   c.
Sexual dimorphism   b.
Amniotic egg   e.
Placenta   a.
5. d.
6. b.
7.
Ovary   c.
Testes   a.
Gonads   b.
8.
Uterine cycle   d.
Ovarian cycle   c.
Conception   b.
Pregnancy   a.
9. b.
10. b.
11.
Meiosis   b.
Gametogenesis   d.
Crossing over   a.
Nondisjunction   c.
12. c.
13. a.
14. The vagina. Both the cloaca and vagina function to receive the sperm.
15. The storage of sperm allows the female to conserve energy for laying her nest. If she has to spend energy on mating for each nest, she may not have enough energy to bury her eggs. The storage of sperm also allows her to mate with more than one mate. Sometimes this sets up a competition, improving the overall fitness of the population. A disadvantage is that if her mate selection is poor, she or her offspring may not be able to adapt to the environmental conditions.
16. LH and FSH trigger gamete production. In men, sperm production would decrease, and women would experience menstrual cycle irregularities.
17. Overall, women ages 35–44 years have a more difficult time.
18. 15–24 years old, 2.6 percent; 25–34 years old, 2.9 percent; 35–44 years old, 3.0 percent.
19. Many women are delaying the start of their family until their 30s. Because the eggs have been around for 35–45 years, they are more prone to mutation and cellular dysfunctions.

20. When you look at the data this way, there has been a 60 percent change in 15–25 year olds, a 29 percent change in 35–44 year olds, and a 25 percent change in the 35–44 year olds. This indicates that the youngest group is facing infertility problems at a faster rate than the other age groups.
21. Testosterone triggers sperm production. If testosterone levels are low, sperm production will be low.
22. A hormonal imbalance caused by high levels of estrogen disrupts both egg development and ovulation.

Potential Student-Generated Questions

1. How many fat reserves does the mother need to sustain her cubs during dormancy?
2. Usually larger animals with small litters have longer periods of development. What are the trade-offs of the strategy of delayed implantation?

CHAPTER 12

12.2 Check + Apply Your Understanding

1. Shrub.
2. A, annual herbaceous plants; B, there are no woody annuals.
3. Both annuals and biennials have a single burst of flowering or reproduction. For annuals this burst occurs within 12 months, whereas biennials grow and store energy for a year before they flower in the second and final year of their life.
4. All 10 life processes are related to photosynthesis. Students may ask how reproduction and dispersal are related to photosynthesis, and the answer is that without sugar, the ability of a plant to produce flowers or reproductive structures and to have a successful seed survive are very limited.
5. Mistletoes are green and can make most of their own sugars. Therefore they are parasites that steal (b) water and minerals from their host.

12.3 Check + Apply Your Understanding

1. (1) cell walls, (2) central vacuole, (3) plastids (chloroplasts, chromoplasts, leucoplasts).

2. Plant cells that store substances generally have thin walls made of cellulose. Plant cells specialized to transport water lay down additional walls, making their walls thicker with a rigid compound called lignin.

3. Leucoplasts within storage cells with a thin cell wall.

4. The chloroplasts produce sugars, and the central vacuole stores toxins that defend Eucalyptus against herbivores (except koalas).

5. The purple flower petals contain red and blue pigments in their vacuoles, and the yellow pollen sacs contain carotenoid pigments in chromoplasts.

12.4 Check + Apply Your Understanding

1. Cell division.

2. One component of growth is cell expansion, which occurs when water within a vacuole pushes or exerts pressure against the growing cell wall.

3. (a) cork cambium, (b) vascular cambium, (c) shoot apical meristem.

4. Cell division requires that microtubules of the cytoskeleton form, attach, and pull chromosomes apart.

5. The red circles (#1) are transport tissues: xylem and phloem. The tender tissue (#2) that appears white is storage cells composing parenchyma tissue.

12.5 Check + Apply Your Understanding

1. Root hairs are located just behind the root tips of the root system.

2. (1) transport and (2) energy storage.

3. Xylem dominates the structure and functions to transport water and support the plant. An annual ring represents one year's growth; thus, the older part of the tree next to your knees would have more growth rings than the part next to your head.

4. Sun-grown coffee has higher rates of photosyn-

thesis. Therefore (a) sun-grown coffee uses more water, (b) sun-grown coffee has higher concentrations of stomata on its leaves to supply greater amounts of CO_2 to the process, and, because it produces more sugars by photosynthesis, (c) sun-grown produces more coffee beans.

5. (a) The cuticle would be very thick to prevent the loss of water to the dry environment. (b) Stomata would not be on the heated upper epidermis; rather, they would be clustered on the shaded lower surface. (c) The leaves are exposed to the sun and would have characteristics of sun leaves, with thick, densely packed mesophyll. (d) The leaves in the figure look stout and self supporting, so the plant would have large amounts of fibers supporting the leaves. (e) With lots of fibers and support tissues, the leaves are costly to build and would be long-lived.

12.6 Check + Apply Your Understanding

1. Plant disease occurs when three factors interact: (1) a capable pathogen, (2) a susceptible host plant, and (3) the right environmental conditions.

2. (1) structural barriers and (2) chemical barriers.

3. (1) penetrating directly through the cell walls of the epidermis, (2) natural plant openings such as stomata, and (3) through wounds.

4. Alkaloids repel herbivores with their bitter taste and stimulate their nervous systems. Terpenoids deter feeding by herbivores. Cyanogens poison the animal's mitochondria.

5. The environment, in this case drought, causes the tree to be more susceptible to attack. The fungus has not changed; it simply is present. A stressed plant has less energy to make chemical or structural barriers to withstand the attack.

Chapter 12
End of Chapter Questions
1.
Annual plant   b., e.
Biennial plant   a.
Perennial plant   c., d.

2.
Cork cambium   a.
Root apical meristem   d., e.
Shoot apical meristem b.
Vascular cambium   c.
3. d.
4. b.
5.
Epidermis   b., c.
Parenchyma   a., g.
Support tissue e.
Phloem   d.
Xylem   f.
6.
Cell wall   b.
Chloroplast   f.
Chromoplast   g.
Leucoplast   a., d.
Vacuole   c., e.
7. a.
8. d.
9.
Leaf   d., f.
Root   a., b.
Stem   c., e.
10. d.
11. a.
12.
Alkaloids   a., d.
Cyanogens   c.
Terpenoids b., e.
13. c.
14. b.
15. (a) All three have mitochondria. (b) Leaf mesophyll has chloroplasts. (c) Root parenchyma has leucoplasts. (d) All three have ribosomes. (e) All three have Golgi complexes (Golgi secrete cell wall components in plant cells. (f) All three have a nucleus containing chromosomes.

16. By tunneling through the mesophyll the insect destroys tissue involved in photosynthesis and making sugar for the aspen tree. This affects all parts of the plant that require sugar for respiration and energy production. With less sugar to fuel cell division, the vascular cambium will produce fewer and smaller cells of both the springwood and summerwood. Therefore, the wood will show a narrower annual growth ring.

17. Girdling a tree removes the phloem tissue that transports sugars from the leaves to the roots. With no way to get sugars to the roots, the roots eventually starve to death, which in turn blocks water from being transported to the

leaves. The tree dies a slow death.

18. Buttresses help anchor the tree. It is predicted that buttresses are dominated by thick-walled support fibers. Although deep taproots are not predictable because the buttresses support the tree, experimental evidence shows that many buttressed trees have deep taproots. Buttressed trees tend to be tall and form the upper canopy of the rainforest.

19. The pumpkin starts out growing slowly in the early stage (days 0–30) before it rapidly adds weight through the mid stage (days 30–70) and then slows down in the late stage (days 70–90). As the pumpkin gets larger, it slows down its growth in the later stages of late September–early October.

20. No, its growth rate varies throughout its 90-day growing period. It begins slowly only adding 3 lbs per day for the first 30 days. Over its mid stage (30–70 days) it grew at over 17 lbs per day. The last 20 days (70–90) it added less than 2 lbs per day.

21. Its growth rate was really fast during the second six weeks from August to late September (days 30–70). From days 40–48 the pumpkin added about 220 pounds total, for 27.5 pounds per day. From days 70–80, the pumpkin added about 3.4 pounds per day.

22. The large central vacuole in the fruit flesh cells expanded with inputs of water.

23. Cell division dominates the early phase as the pumpkin adds more cells; this is shown by the relatively low weight gain for the first 30 days. With millions of new cells that gained water and expanded, the pumpkin's weight rapidly increased from days 45 to 55.

24. Yes, a pumpkin cannot get this huge without generating lots (millions and billions) of new cells over a short period of time. The genetic composition of the seed is the starting point for this rapid growth rate.

25. The warm temperatures of August and the absorption of water really pumped up the metabolism for cell division, and the water expanded the cells to grow that fast and big. Water is also necessary for photosynthesis, which produced sugars to build the pumpkin cell walls and cell structures. Minerals play an important role in the growth of cell structures and in functioning.

26. Pumpkins and animals both show a determinate growth pattern and actually are very similar, with a rapid initial growth spurt followed by a leveling of the growth rate when they reach maturity.

27. Flowers and fruits draw water and sugars, so by removing the flowers and fruits the growers help the plant direct all the sugars, water, and nutrients to a single growing point—the growing pumpkin.

28. In general, yes, I would expect them to show a determinate growth pattern in the S-shaped growth curve. Of course the weights and the scale would be much lower. Some of these other fruits may show some other growth curves and ripen sooner or later than pumpkins and thus flatten the curve or show a steeper increase.

Potential Student-Generated Questions

1. How is growth rate affected by early emergence of leaves?

2. Do annuals store enough energy to emerge early in cold temperatures?

3. How does early leaf emergence affect growth rate?

4. How would a warm spring affect the time available for growth?

5. Which lifespan is favored by plants in the forest floor environment?

6. Do plants that emerge early have a competitive advantage over plants that emerge later?

7. Are plants with a fast or slow growth rate favored to grow on the forest floor in early spring?

CHAPTER 13

13.2 Check + Apply Your Understanding

1. Adaptations happen over many generations as a function of natural selection and result in a change in the types of certain genes within the DNA. The changes to the DNA occur because plants with these genes have higher levels of survival and reproduction—thus passing along this new collection of DNA to future generations. A response is a short-term redirection of resources or alteration of cell chemistry to survive changing environmental conditions—like seasonal changes.

2. A lack of water creates a severe metabolic stress that the plant attempts to adjust to by using all internal water available, causing it to wilt at first and then finally to die from its inability to photosynthesize, conduct normal metabolic chemical reactions, and support plant structures dependent on water.

3. Plants will respond to resource limitations by slowing growth, development, or reproductive activities that rely on the availability of that resource. In severe cases, like a prolonged drought, the limitation can exceed a plant's ability to respond and the plant dies.

4. A late spring frost would create problems for a plant because the new, young tissues have not had time to acclimate to cold weather. A late spring frost would likely kill a plant.

5. This change in the plant as a function of water stress is both a response and an adaptation. The plant **responds** to the change in moisture conditions by drying out its aboveground tissues. The plant's ability to respond to and survive this dramatic change in moisture is a function of past natural selection—yielding a highly **adaptive** plant.

13.3 Check + Apply Your Understanding

1. The two stages are the light stage and the Calvin cycle. The products of the light reactions (ATP and a hydrogen carrier molecule) are used in the Calvin cycle, and the altered ADP and hydrogen carriers cycle back for reuse in the light stage.

2. The wavelengths of light we can see with our eyes approximate the light that chlorophyll molecules are also sensitive to. Light energy raises the electron energy level in chlorophyll molecules, which is concentrated, resulting in one electron being stripped away to be used during the light stage.

3. The three products are an electron, hydrogen, and oxygen. The electron enters an electron transport chain, and the hydrogen is first used to create a chemiosmotic concentration difference, which ultimately powers the production of ATP and the creation of a hydrogen carrier molecule. Oxygen diffuses out of the cells as a waste product.

4. The evidence would indicate that this plant has one of the two photosynthetic adaptations evolved to help plants survive in hot, dry climates. In this case, the light stage and the Calvin cycle are separated in time.

5. Enhanced CO_2 concentrations will likely improve growth in these trees, assisting the carbon fixation process.

13.4 Check + Apply Your Understanding

1. Both sugars and oxygen must be present for aerobic metabolism to occur.

2. Sugar loading at the source refers to the active transport of sugars where they were produced, and the unloading at the sink refers to the active transport out of phloem tissues into cells where sugar can be used in aerobic metabolism, stored, or converted to other types of organic molecules.

3. Water is vital to many plant metabolic processes, and oxygen is required to support aerobic metabolism in rapidly growing root tissues. When these two resources are present in insufficient amounts, plant growth must slow down or stop. Too much water displaces oxygen and shuts down the aerobic metabolism.

4. There is a direct relationship between higher temperatures, high metabolic rate, and tissues that have a high level of activity: when each increases, the rate of metabolism also increases.

5. It is likely the plants would be less healthy, grow less, and not be as desirable to purchase as those grown in the greenhouse. The soils in your neighborhood likely contain adequate amounts of many mineral elements, but not all that would be needed to support the full growth and development of the plant. Similarly, the environmental conditions would be less than optimal—also interfering with plant growth and health.

13.5 Check + Apply Your Understanding

1. Organic matter is the decaying remains of plants, animals, and other organisms; it contains nutrients that are made available to growing plants through the decomposition process. Organic matter also provides pore space for oxygen to permeate into soils where active root tissues grow. It also acts as a sponge, holding water and helping soils to keep from drying out too quickly.

2. Spaces between soil particles permit water to percolate between them and dissolve mineral nutrients—thus making both available for plants to absorb. Pore space also provides a location for oxygen to permeate the soil and support aerobic metabolism by the metabolically active tissues.

3. Phosphorous is an important mineral element in meristems, leaves, and other metabolically active sites in the plant. It is used in many critical molecules, such as ATP, phospholipids in cell membranes, sugars, and DNA.

4. Trimming roots removes the active site on plants for water and mineral uptake. These components are critical to photosynthesis and growth. By reducing the amount of roots, you reduce the ability of the plant to get larger.

5. The two missing mineral elements would be nitrogen and magnesium, used in the production of the green pigment chlorophyll.

Chapter 13
End of Chapter Questions

1.
ecological niche   b.
stress   c.
disturbance   a.

2. b.
3. b.
4. a.
5.
chloroplast   c.
chlorophyll   a.
light reaction   b.
Calvin cycle   d.
6. c.
7. b.
8. d.
9. a.
10. b.
11.
soil texture   f.
mineral availability   c.
compost   d.
macronutrient   a.
micronutrient   b.
transpiration   e.
12. b.
13. a.
14. c.
15. a.
16. Phosphorus is a critical macronutrient used to construct the plasma membrane for every cell in the plant. If you eliminate this nutrient from fertilizer you can limit new cell construction since phosphorus is a key component of membranes, thus decreasing overall grass growth in the lawn. Phosphorus is also critical to the building of ATP—the energy currency of the cells.

17. The plant on the left is successfully infected with the mutualistic rhizobium: the leaves are green, indicating the development of healthy chlorophyll molecules, which requires the presence of nitrogen. The leaves at right are not green; they lack chlorophyll.

18. Acid rain over a number of decades will ultimately affect the pH of soils in plant communities downwind from the power plant. A change in pH alters the molecular structure of mineral nutrients in solution—altering a plant's ability to gain access to these resources. The change in resource availability will change the mix of plants growing there, favoring new groups of plants better able to compete in the new resource availability environment.

19. During the daylight when the plant is photosynthesizing, your meter would tell you that carbon dioxide was moving from the air into the leaf—different from at night, when the meter would show no movement of carbon dioxide into the leaf. During the day, oxygen diffuses out of the leaf and into the atmosphere; at night there is no light reaction, so no oxygen is diffusing out.

20. The century plant used a warm, dry climate–adapted photosynthetic pathway—indicated by its adaptation to desert regions of the southwestern United States—and the clue is that the stomata are closed during daylight hours. The light reaction occurs in the same way as in all other plants, inside the chloroplast at the thylakoid membrane, as the lead-in set of reactions to the Calvin cycle, which occur during the nighttime hours. The agave has a competitive advantage in that this adaptation minimizes water loss and allows it to maintain higher levels of metabolism.

21. The most growth occurred in May, early in the growing season for temperate climates.

22. In May, peak growth occurs at peak rainfall, which was over 100 mm. Temperatures were 8° to 20°C, with an average of 12°C, approximately 10°C below the peak temperatures experienced during the summer months.

23. Yes, there are spikes in light reception on the angled leaves that occur in both March and November.

24. Yes, the plant's adaptation to cooler weather accounted for a significant portion of its total yearly photosynthesis as indicated by Fig. 13.36. The yucca's cold tolerance allows it to photosynthesize during the times of the year when other plants are likely to be dormant.

25. Yes, most of the growth occurs during a time of the year when temperatures are milder; less growth occurs during the hotter summer months.

26. The peak interception of light by the angled leaves corresponds with the highest periods of growth for the yucca.

27. Yucca's cold tolerant photosynthetic advantage over the warm weather photosynthetic adapted plants will be limited. Those plants with the warm, dry weather adapted photosynthetic pathways will potentially be favored and potentially exclude the yucca.

28. Since yucca plants possess photosynthetic pathways adapted to warm, dry climates there would be no need for yucca to have a cool weather adaptation. If yucca possessed the alternate photosynthetic pathway I would expect growth to shift away from spring and fall and occur during the warmer summer months of June to September.

Potential Student-Generated Questions

1. As C3 plants, how are seagrass communities potentially affected by increased water temperatures?

2. Seagrasses often grow in sandy soils. Would you expect these to contain many available nutrients?

3. What effect does government planning that closely controls land use close to the coastline have on potential seagrass growth?

4. What effect would land conversion to crop farming in coastal areas potentially have on seagrasses?

5. As cool weather–adapted plants, would you expect seagrasses to have a high

level of water loss from transpiration?

CHAPTER 14

14.2 Check + Apply Your Understanding

1. Sexual reproduction creates a genetically diverse population.

2. The plant life cycle includes a sporophyte and gametophyte generation. Meiosis and genetic variation are introduced by the sporophyte as it makes spores. The sporophyte generation results from fertilization.

3. Aspens can quickly clone their surviving underground structures to invade the open habitat following a fire. Conifers reproduce by seed and must progress through the many stages of sexual reproduction and build new trees "from scratch."

4. Even before flowers begin their development, vegetative organs absorb water, make sugar, and store starch. Once the flower begins to form, roots and stems deliver water, sugars, and raw materials to the growing flower. Water and cell expansion keep the flower parts supported during the period of blooming.

5. The photo shows aspen trees that are separate, as if they were scattered as seeds. The trees also show variations in color. This most likely is an example of sexual reproduction. Aspen clones that form from asexual reproduction appear as a clustered dense stand of trees.

14.3 Check + Apply Your Understanding

1. Long-day plants flower when the hours of daylight exceed a genetically determined day length and therefore flower in the spring and early summer.

2. The sepals, petals, stamens, and pistil develop in this sequence.

3. Meiosis, spores, mitosis, egg cell development.

4. Within a single pollen grain, the two nuclei are genetically identical because they both arose by mitosis. However, the genetic information between the two pollen grains arose by meio-

sis and are genetically different. Because the pollen grain is the site of meiosis and produces gametes, it is referred to as the gametophyte generation of the life cycle.

5. Egg cells contained in each ovule = 1; egg cells contained in each flower = 100; considering only the carpels, number of times meiosis occurred in each flower = 100; number of haploid spores initially produced in each flower = 400; number of spores that proceeded through mitosis in each flower = 100.

14.4 Check + Apply Your Understanding

1. Pollination, pollen germination, sperm formation, pollen tube growth, fertilization, embryo development.

2. Animals visit flowers for food to eat, materials to build their nests, sex, and laying their eggs. Flowers offer rewards to attract animals to return with pollen from other flowers. Some flowers defend themselves by rewarding ant guards that defend them or have nectar with small amounts of toxins and defensive chemicals to repel herbivores, pathogens, and thieves.

3. Because the relationship between pollen and seeds is 1:1, the pumpkin required at least 500 pollen grains. Because each pollen grain produces one pollen tube, 500 pollen tubes are required. Each pollen grain produces two sperm; thus, 1,000 sperm were required to produce the 500 seeds.

4. Short-tongued flies are restricted to flowers that have an open, bowl shape with readily accessible pollen and nectar—flower b. Long-tongued flies are able to gather nectar at the end of a long floral tube—flower a.

5. (a) This species must be self-compatible if the closed flowers produce viable seeds. (b) The closed flowers never receive pollen other than their own and therefore produce seeds similar to the parent. (c) No. Because the egg cells and the sperm cells are geneti-

cally unique (due to meiosis), each seed will be genetically similar but not identical.

14.5 Check + Apply Your Understanding

1. A fruit is a ripened ovary that comes from the female portion of a flower.

2. Dispersing seeds beyond the seed shadow prevents competition from the parent plant, avoids seed predators, and places a seed into a position to compete in a new environment.

3. Plants that live in seasonal habitats often require a specific period of cold temperature before germination so that they can accurately perceive the season is spring and conditions are safe to germinate and grow. Without this requirement, many seeds might begin to germinate during a spell of warm autumn weather and then be killed by the cold winter frost.

4. By absorbing the ethylene, the chemical will slow the ripening of the fruit.

5. Regarding similarity, both seeds and spores are "packages" that function to move genes (alleles) to new locations and start to form a new plant. Structurally they are different in that the spore is a single cell and the seed is composed of a multicellular embryo and nutritive tissues. Seeds are generally much larger and contain more food reserves.

Chapter 14
End of Chapter Questions

1.
Asexual   a., b.
Sexual   c., d., e., f.
2.
Egg cell   b.
Sperm cell   a., d.
Spore   c., e., f.
3. a.
4. b.
5.
Petal   a.
Carpel   c., f., g.
Sepal   b.
Stamen   d., e.
6.
Animal   b., f.
Wind   a., c., d., e., g.
7.
Self-compatible   b., d.
Self-incompatible   a., c.
8. c.
9. d.

10.
Embryo   a., d., f.
Endosperm   c., e.
Seed coat   b.
11. d.
12. a.
13.
Animal   a., b., c.
Water   d.
Wind   e.
14. a.
15. b.
16. b.
17. The poinsettia flowers when the days are shorter than 11.5 hours and is a short-day plant. It enters its photoperiod right around the autumnal equinox—September 21. Poinsettias require about six weeks of its photoperiod to flower around Thanksgiving. Its red "leaves" (not leaves, they are bracts) attract hummingbirds.

18. Plowing affects at least three environmental factors for weed seeds pulled to the surface: (1) light, (2) oxygen, and (3) temperature. Seeds can only germinate when the temperature is within its range (above a minimum and below a maximum) and oxygen is available for respiration. Of all three factors, light probably is the factor that is most changed by plowing. Weed seeds with a light requirement are now favored and will germinate and emerge. By stimulating germination and emergence, the farmer can remove as many weeds as possible from the soil.

19. The outer **receptacle** wall forms the hooks that catch onto animal fur. The burdock receptacle, fruit (an achene), and its many seeds are dispersed as a unit. The tiny crescent-shaped fruits/seeds fall to the ground when the burr is removed. A tall stem will only catch the fur of larger, taller mammals. A smaller-sized burr would more likely escape detection than a larger one and therefore be carried farther from the seed shadow.

20. The gas-filled wings help the pine pollen stay aloft (3) and not settle out of the air before they are carried long distances.

21. The duration of feeding increased with increasing sucrose concentration. Fe-

males remained longer at concentrations greater than 10 percent, and at 55 percent females remained three times longer on the flower than did males.

22. Skippers increase the volume of nectar up to 25 percent sucrose and then sharply reduce the volume as the nectar solution becomes more concentrated. The volume decreased at high concentrations.

23. Female butterflies have greater energy requirements because they must nourish and lay eggs.

24. No. Butterflies drink fairly concentrated sugar solutions. A 40 percent sugar solution is very concentrated.

25. One reason may be that the skipper meets its sugar requirement with a lower volume of highly concentrated sugars. At some point the sugar solution would be so syrupy that the amount of force needed to feed would be too great to feed.

26. The highly concentrated sugar solution is more costly for the plant to produce. By preferentially feeding on and pollinating flowers with concentrated nectars, the butterfly places selective pressure on plants in this direction.

27. A hummingbird has a greater energy requirement than a butterfly. Birds also have shorter duration feeding, so I can predict that they feed on more highly concentrated sugars than do butterflies. **Note:** This is not necessarily true; most hummingbird flowers produce a 25 percent sugar solution.

Potential Student-Generated Questions

1. Do bats or birds eat larger fruit?

2. How does the fig's color affect the kind of animal disperser attracted?

3. How does the distance traveled affect the number of seeds dispersed?

4. How does feeding behavior affect the seed germination rate?

CHAPTER 15

15.2 Check + Apply Your Understanding

1. (a) the filamentous growth form, (b) the yeast growth form, (c) filamentous, (d) filamentous.

2. The fungus needs to absorb nitrogen in order to build proteins and chitin, which both contain nitrogen.

3. Growth and building a large interconnected network of hyphae is expensive; it requires a continual supply of raw materials such as chitin, sugars, amino acids, and phospholipids. These supplies need to be transported to the tip of the growing hypha, where they are assembled into new cell walls, plasma membranes, and cellular structures. Growth requires a continuous supply of energy in the form of ATP.

4. Your friend is not correct. Yeasts do not have the ability to break down starch—only simple sugars. Beer is brewed from **malted** barley, which is the conversion of starch to simpler sugars when the grain germinates.

5. The photo shows that bioluminescence is coming from the fruiting body, which is only a small portion of its body. The light glow is generated by chemical reactions that would become more intense with increased moisture and oxygen.

15.3 Check + Apply Your Understanding

1. Fungi harm their living hosts as parasites, pathogens, and predators. Two mutualistic relationships are (1) mycorrhizae, which are fungi that colonize plant roots; and (2) lichen fungi, which partner with populations of green alga and/or cyanobacteria.

2. Digestive enzymes are secreted directly onto a substrate where they break it down so that the fungus can absorb the small building block molecules. Without them the fungus would starve and be unable to acquire energy or nutrients to live. Like enzymes, water is required to break bonds of large macromolecules to digest them. Water is also needed to dissolve the breakdown products and absorb them into their body.

3. (a) produces lots of spores that can find or fall onto the strawberries; (b) has genes that produce enzymes that break down sucrose; (c) *Rhizopus* grows fast because it can break down the relatively simple molecule sucrose relatively quickly.

4. (1) Spore numbers reach their peak in the fall when there is a flush of fallen leaves into the stream. (2) A fast-moving stream has higher levels of dissolved oxygen than a stagnant pond and thus supports greater metabolism and larger populations of aquatic fungi.

5. The toenail fungus is a pathogen that lives off living tissues under the toenail and then causes the characteristic symptoms. If it were to degrade only the nonliving proteins of the nail, then it would act as a decomposer. No; treating only the nail will not work because the fungus lives in the nail bed.

15.4 Check + Apply Your Understanding

1. The fungal spore functions in (1) reproduction of the fungal body and (2) dispersal to new locations. A third function is dormancy—survival through unfavorable environmental conditions.

2. Spores are dispersed by wind, rain, and insects. *Pilobolus* uses water pressure to disperse its spores.

3. (a) Mitosis occurs as the hypha grows out of a spore; (b) meiosis occurs within the spore-producing cell to shuffle genetic information and reduce the number of chromosomes in the production of four unique spores; (c) the fruiting body grows through mitosis; (d) the growth of the mycelium is through mitosis of cells at the tip of the hyphae.

4. (a) 1—sexual spores are haploid; (b) 2—the hyphal cells of a fruiting body contain two separate nuclei, each with a set of chromosomes; (c) 1—cells formed by mitosis from a haploid spore are also haploid.

5. Spores of the puffball are dispersed upward, so they have a greater chance of catching air currents. The pore mushroom on the left disperses its spores in a downward direction, where they have a smaller chance of wind dispersal.

15.5 Check + Apply Your Understanding

1. The death cap toxins cause the liver and kidneys to fail. Psilocybin affects the central nervous system, which leads to hallucinations and visions. Ergot constricts blood vessels, leading to muscle spasms and seizures.

2. Antibiotics such as penicillin disrupt the building of the bacterial cell wall. Other antibiotics inhibit bacterial ribosomes and their gene expression, or block a step in the synthesis of critical enzymes. These compounds inhibit or kill bacteria that compete with fungi for the substrate molecules they require.

3. (1) Cellulose is a large macromolecule, such as starch, which is broken down to glucose, so it should be placed in A. (2) Carbon dioxide is a by-product of fermentation and is represented by C. (3) Pyruvate is the molecule converted to ethanol by fermentation, so it is in B.

4. The packages deny oxygen and water to fungi and therefore deny them two essential resources needed to grow and degrade the foods. Without water, the fungi cannot hydrolyze the substrates and absorb the building block molecules. Without oxygen, the molds cannot metabolize the sugars to produce the energy needed to grow.

5. (a) Yes. An herbivore eats plant material and is the source of energy for rumen microbes such as *Neocallimastix*. (b) No. The anaerobic pathways do not include the mitochondria. (c) Yes. It ferments the carbohydrates to produce ethanol or lactic acid. (d) Yes. As a member of the kingdom Fungi, the cell walls contain chitin.

Chapter 15
End of Chapter Questions

1.

Fruiting body	b.
Hypha	a.
Mycelium	c.
Yeast	d., e.

2. d.
3. d.
4. b.
5.

Decomposer	d., e.
Lichen	h.
Mycorrhiza	b., g.
Parasite	a.
Pathogen	c.
Predator	f.

6. c.
7. d.
8.

Asexual spores	b., e., f.
Sexual spores	a., c., d.

9. d.
10. d.
11.

Antibiotic	b., d.
Ethanol	a.
Toxin	c., e.

12. b.
13. b.
14. To form a fruiting body, a sexual spore must germinate, grow, encounter, and then merge with a spore of the same species with a **different** mating type. Sexual spores are produced by meiosis and will carry different mating type alleles. They are more likely to encounter another inky cap spore and its hyphae in the **inky fluid** than those dispersed into the air and dispersed away from one another.

15. (1) The fungus infects bats that live in moist cool conditions with fairly few air currents. (2) The fungus probably is not adapted to infecting bats that roost outside of caves where the temperature and humidity are very different. (3) Because there are few wind or air currents in the caves, the fungus probably is spread by bats that pass spores through contact. (4) Cavers disinfect their gear and clothing to prevent spreading spores from one cave to another in their explorations.

16. (a) Tremella, (b) Dacrymes, (c) Tremella.

17. *Dacrymyces capitatus* decayed the hardwood blocks, which lost more than a third of its weight in 12 weeks.

18. *Dacrymyces capitatus* causes a white rot (degrades both lignin and cellulose) on both hardwoods and conifers. *Dacrymyces stillatus* causes white rot on hardwoods only, and *Calocera* is a brown rotter (only cellulose) on conifer wood only.

19. Yes, the data supports this generalization. The data show that *Calocera* is a brown rotter that decayed the conifer and *Dacrymyces stillatus* is a white rotter that degraded the hardwood.

20. Springwood consists of thinner-walled cells than the thick-walled summerwood. A thin wall is easier to break through than a thick wall, so the thin-walled springwood is easier for the decay fungi to breach than the thick-walled summerwood.

21. Because the metabolic process is aerobic cellular respiration, one of the end products is water. The carbon in the cellulose is converted to CO_2 (plus energy and water).

22. The jelly fungi require oxygen to degrade the wood, which is provided by a coarse soil rather than a packed soil. The lack of oxygen would reduce the metabolism of the fungi, so the choice of a loose soil allows the wood decay fungi to metabolize and make enzymes to decompose the wood.

23. Yes. You are alarmed because these jelly fungi cause significant wood decay to the structural integrity of the window frame. They indicate a water problem, and the window should be replaced and the source of moisture located.

Potential Student-Generated Questions

1. How many spores does a single fruiting body produce?

2. How do animals find them when they fruit underground?

3. Which animals eat deer truffles as part of their diet?

4. How much nutrition do truffles provide to an animal?

5. How are spore structured so that they can withstand the digestive system of animals?

6. Are spores stimulated by passage through the digestive tracts of mammals?

CHAPTER 16

16.2 Check + Apply Your Understanding

1. A typical diagram should include most of the portions of the water cycle depicted in Figure 16.6. Students may not consider the movement of water below ground because it is out of sight.

2. A watershed is a geographic region where all precipitation drains into specific rivers or lakes.

3. Residence time refers to the time water spends in any one reservoir. For instance, the average time water resides in the atmosphere is about nine days, whereas water may remain in the ocean for more than a thousand years.

4. Gravity plays a role in moving surface water downhill, through watersheds, and eventually to lakes or the ocean, where evaporation may take over the process of moving water back into the atmosphere.

5. Higher amounts of solar energy would mean greater heat on Earth—resulting in greater evaporation rates and less liquid water and more water vapor in the atmosphere.

16.3 Check + Apply Your Understanding

1. Soil microbes (bacteria and archaea) convert nitrogen into forms that are usable by plants.

2. They both contain large inorganic reservoirs (carbon in ocean water and nitrogen in the atmosphere), and both are critical to the living world. Each exists as a gas and in molecular forms found in biomolecules.

3. Volcanoes move carbon out of very long-term reservoirs stored in Earth's crust and onto the surface or into the atmosphere; therefore, there would be a substantial increase in the amount of carbon moved each year. Essentially carbon would be moved from very long-term storage into shorter-term storage in the atmosphere where it can contribute to atmospheric warming.

4. They will increase because of the large inputs of ammonia and other protein-based materials.

5. Nitrogen would likely be the most affected by the contamination, which

would kill the microbes that are responsible for converting the element into a form usable by plants, perhaps limiting the ability to revegetate the site.

16.4 Check + Apply Your Understanding

1. Proxy data provide climate scientists with clues about the changes in climate over long periods of time prior to humans having the equipment to make such measurements.

2. Climate refers to long-term statistically measurable variations in factors, such as temperature and precipitation. Weather refers to short-term climatic conditions, such as those predicted by weather station reporters on TV.

3. Organisms with a narrow ecological niche are most vulnerable to climate change. They are well adapted to a very narrow range of environmental conditions, which likely would be in jeopardy during climatic change, thus putting them at risk for extinction if they are unable to migrate to suitable new habitats.

4. A rise in sea level is tied to the temperature of the water. As water warms it expands, thus contributing to higher ocean levels. Higher temperatures associated with global warming also melt large reservoirs of ice in polar regions; this melt water runs into the oceans and increases the volume of water, causing a rise in level.

5. There would likely be no immediate influence on global warming. Water vapor accounts for up to 76% of the greenhouse gas effect, and the residence time for other gases is years or decades. This means that it would take a few years for the influence of human contributions to the greenhouse effect to begin to be realized.

Chapter 16
End of Chapter Questions

1.

Reservoir	d.
Watershed	c.
Carbon budget	a.
Nitrogen fixation	b.

2. c.
3. a.
4. c.
5. a.
6.

Weather	d.
Global warming	c.
Climate	b.
Greenhouse gas effect	a.

7. a.
8. d.
9. c.
10. d.
11. Like carbon dioxide, methane is a greenhouse gas that absorbs heat energy and causes the atmosphere to warm. The extinction of large herds of methane-producing herbivores would eliminate a natural source of this greenhouse gas, causing the atmosphere to retain less heat and resulting in a cooling period.

12. Increased evaporation rates combined with higher transpiration rates put more water vapor into the atmosphere, adding to the global warming process. Because water vapor has a short residence time, it's likely this phenomenon has a more limited impact compared to greenhouse gases with their greater influence on trapping heat.

13. The greenhouse gases emitted by Mt. Tambora would be expected to have a long-term impact because of the long residence times of the gases and their influence in trapping solar energy and warming the atmosphere. Ash and certain gases, which reflect sunlight, would have an opposite effect. Based on the severe winters and cold summers that followed, the major influence on global climate was the ash and, despite their shorter residence times, gases that reflect solar radiation.

14. Based on the diagram, as the moisture-laden air is pushed up against the mountains it rises and cools. The cooler air is able to hold less water, so precipitation in the form of rain or snow occurs. The land on the far side of the mountain receives less moisture as the air continues over the mountains because the amount of water being held by the

cooler air has been reduced, creating the rain shadow effect. This is illustrated on a map of the western Cascade mountains showing annual precipitation patterns in the Cascades. The west side of the mountains receives more moisture than the east side, and this effect carries out several hundred miles into Eastern Washington and Oregon. The side where more rain falls has denser plant life that includes trees. The rain shadow side does not receive enough moisture to support trees, and the landscape is dominated by grasses.

15. Woody plants such as trees capture and hold carbon for decades or longer, an important biological reservoir in the overall carbon cycle. By cutting down and burning the trees, carbon is moved from the biological reservoir into the atmosphere or into the soil as ash. The water cycle is impacted by the fact that the trees and other water transpiring plants are no longer present. and soil moisture would cycle at slower rates than normal.

16. There have been five peaks (including the one we are currently in) along with four extended cold periods during this time period.

17. They appear to mirror one another. As the temperature climbs and falls, so does the CO_2 level.

18. Dust tends to be at low levels, with a series of short-lived peaks. When dust levels are at their highest, more solar radiation is reflected back into space, cooling Earth during these times. You can see this in the corresponding temperature graph: when dust is high, temperatures are correspondingly low.

19. Any nutrient increase in oceans results in more photosynthesis, which drives a reduction in atmospheric carbon dioxide through the Calvin cycle. Thus, an increase in ocean productivity means less potential global warming from greenhouse gases.

20. There may be evidence to support this claim. The graph clearly illustrates the rise and fall of temperatures associated with ice ages and warm periods. We understand that greenhouse gases act to warm our atmosphere—a process critical to life on earth. Additional warming may offset some effects from long-term or cyclical cooling processes associated with the orientation and energy output from our sun that might favor a cooling period.

Potential Student-Generated Questions

1. How can farmer fertilization practices be altered to help reduce this problem?
2. Do you think fertilizer production will alter the global atmospheric nitrogen reservoir in a large-scale fashion?
3. How would further hybridization research aimed at increasing corn yields from the same quantity of resources alter the relationship between ethanol production and aquatic ecosystems?
4. Would altering the corn genome to promote symbiotic relationships with nitrogen-fixing bacteria solve this problem?
5. Speculate about the role soil microbes play in decreasing or increasing the impact of this process.

CHAPTER 17

17.2 Check + Apply Your Understanding

1. The designation allows governments to take specific action to protect habitats and species from further harm caused by human activities.
2. Endangered species are in imminent danger of extinction, whereas threatened species are likely to become endangered in the near term throughout all or a significant portion of their range.
3. Habitats with intact levels of biodiversity represent working ecosystems that can carry out natural biogeochemical processes, such as water and carbon cycling, as well as having working community relationships, such as food webs that keep populations in check.

4. You need to recognize which species are native or invasive, whether or not functioning food webs still exist, and, if not, which species are missing and need to be reintroduced, and whether or not there is sufficient habitat and resources available to support the organisms. You also need to repair damage caused by humans that fragmented the landscape or damaged physical processes, such as drainages or soils critical to supporting a healthy ecosystem.
5. Yes, the human activity associated with the dam posed an imminent danger to the survival of the snail darter, and under the Endangered Species Act the court was able to stop this activity and prevent the destruction of this species.

17.3 Check + Apply Your Understanding

1. Zoos are reservoirs for breeding stock for threatened and endangered species. These animals can be deliberately bred to preserve the species and their genetic diversity.
2. Humans can capture wild animals and (1) regulate which other animals they mate with or sequester eggs or young away from the parents to improve hatching and survival rates for infants, and (2) use enhanced reproduction techniques such as artificial insemination to increase the fertilization rate for animals.
3. Organisms that humans believe are charismatic receive greater attention than others, and with greater attention comes more emphasis on preservation or protection.
4. In extreme cases, we could capture the remaining bats and treat them for the disease or isolate them from diseased bats. Once we improve the health of these bats, we could begin a controlled breeding program at zoos, eventually releasing bats back into the wild. If research helps us better understand the cause, we could alter the genome of captured bats and insert

disease-resistant genes, making the bats less susceptible to the effects of the disease.
5. The effort is not likely to succeed. Humans have moved into and developed much of the Great Plains, building cities, farms, and roads throughout the region. Thus, the pristine habitat that once supported more than 100 million animals no longer exists. Bison are no longer able to roam from Canada to northern Mexico as they did 200 years ago. Smaller efforts to restore bison to refuges and large tracts of land, as on western ranches, can succeed, providing suitably large habitats with sufficient resources where the bison can live year-round.

17.4 Check + Apply Your Understanding

1. Diversity increases as you move toward the equator. Year-round temperatures increase along with available precipitation.
2. Each species possesses an intrinsic value and serves a role in ecosystem functioning, such as providing ecosystem services or filling a role within a food web. As we discover new species, we learn more about their role and relationship to healthy, functioning ecosystems.
3. Because biodiversity hotspots harbor large concentrations of biodiversity per unit area, preserving one acre might provide critical habitat for many more species than would preserving an acre in another less-diverse habitat.
4. In general, human activities related to development have a dramatic, negative impact on levels of biodiversity. This is due to the disturbance or destruction of habitat and the alteration of resource availability for the species native to that area.
5. Although separated by a great distance, they must share similar characteristics that support abundant biodiversity such as rainfall amounts, mild abundant resources, and varying landcapes to support many eco-

logical niches. While the Pacific Northwest has mild temperatures throughout the year, it does not have high year-round temperatures.

Chapter 17
End of Chapter Questions

1.
Endangered species   b.
Threatened species   a.
2. b.
3. d.
4. c.
5. c.
6. c.
7. d.
8. a.
9. a.
10. c.
11. c.
12. b.
13. b.
14. A sunken ship provides a range of niches that marine species can colonize and exploit as habitat. Surfaces are available for sponges and corals to grow on. Fish use the structure of the ship as hiding cover to protect them against predation. Herbivores can feed on plants growing on the ship, and larger fish can feed on the smaller ones.
15. The biodiversity in the wilderness area presumably would be intact, with habitats in near pristine condition. Elements such as food webs and ecosystem services would function normally. In the adjacent national forest where human activities would disturb the natural landscape, varying levels of habitat fragmentation and disturbance to ecosystem services would be occurring. Some habitats would be sufficiently fragmented, for example from road or timber harvesting disturbances, that the native organisms would be displaced or isolated.
16. I would expect the American white pelican to be more susceptible to problems with habitat fragmentation because of the broad range of land it relies upon for habitat. As the bird migrates between summer and winter range, any habitat fragmentation that occurs along the way impacts the species. Much of

its northern habitat is impacted by development, agriculture, and ranching activities, which alter the habitat structure to a point where it may be unsuitable. You could make the argument that the brown pelican is in greater danger because of the rapid development of coastal habitat by humans. Because the brown pelican is confined to these rapidly developing coastal communities, its only habitat choice is being rapidly depleted.
17. Income generated from ecotourism provides an ecologically friendly tax base to the government. This promotes habitat preservation in two significant ways: (1) governments want to preserve the natural beauty and biological characteristics of a habitat to ensure people continue to visit and spend money, and (2) ecotourism provides a revenue stream that can be used to purchase additional natural lands, which would preserve the habitat and biodiversity for the future.
18. The number of seeds produced per mustard plant is cut in half from 40 to about 20.
19. There is a distinct negative relationship between the number of seeds produced per radish fruit and the amount of distance the bees must travel. At the grassland boundary you obtain about five seeds per plant, and by the time you are 400 meters away that number has been cut in half.
20. The number of seeds produced per plant increases from less than 30 to greater than 40.
21. The grassland should be located less than 100 meters from the plants to achieve this level of fruit set.
22. It would increase the farmers' desire to preserve native grasslands that harbor native bee populations near their crops. This would help ensure that the crops get the highest levels of pollination and fruit production.
23. Placing the portable hives does increase crop production, but it removes the incentive farmers would

have to conserve native grasslands near their fields. In addition, because there is no incentive to preserve the native grasslands, the native bee populations would certainly be diminished because they would be displaced. Also, as habitat fragmentation increases, a larger number of portable hives must be placed around fields to ensure sufficient pollination.

24. Small isolated populations of native plants that are dependent upon specific types of bees for pollination are in grave danger of localized extinction. If they are unable to be pollinated with the proper pollen grains from distant plants, they will not produce seeds. For annual plants that must produce seeds each year to continue the species, it means that they will potentially disappear from the landscape in a few seasons. For perennial plants, you would expect to see a general decline as their numbers dwindle from lack of pollination and seed set.

Potential Student-Generated Questions
1. How might increasing temperatures associated with climate change ultimately influence mussel biodiversity?
2. How would shifts in mussel biodiversity impact ecosystem services?
3. How do local climate patterns influence stream flow in a watershed?
4. What is the relationship between changes in the abundance of mussels and mussel metabolism?
5. How might the placement of a dam influence ecosystem services?

CHAPTER 18
18.2 Check + Apply Your Understanding
1. Asexual. Sexual reproduction takes more energy—energy and resources are used for producing eggs, attracting a mate, and, in some animals, nurturing the developing embryo and young.

2. It provided for a controlled experiment. Mendel was able to examine one variable at a time.
3. The genetic makeup of the rose will dictate its color.
4. Good model organisms, like the pea plant, have a short generation time and allow the researcher to easily control reproduction. They also have genome information available and are easily manipulated. Mendel was successful due to his scientific approach; other plants or animal could have been used with the same success. While mice have a relatively quick reproductive cycle, since they have a larger genome, they would not be as easy to work with.
5. See Table below.
In the first cross, 75% will be dominant with one hump and 25% will be recessive with two humps (3:1 chance). In the second cross, there is a 100% chance that all will be heterozygous and express the dominant single hump trait (4:0 chance).

18.3 Check + Apply Your Understanding
1. If present in the genotype (as one copy or two), a dominant trait is expressed. A recessive trait is expressed only if there are two copies. The forms are not always observed. Some traits are not viable and the offspring does not survive.
2. In pleiotropy (such as sickle-cell anemia or albinism), multiple traits are affected by one gene. In polygenic patterns, the expression one trait (such as height in humans) is an in-

Heterozygous	H	h
H	HH	Hh
H	Hh	hh
Homozygous	H	H
h	Hh	Hh
h	Hh	Hh

teraction of multiple genes.
3. Incomplete dominance. The mother is EE, the father is ee, and the child is Ee.
4. The amount of sun influences plant height and leaf size; water and mineral levels also effect plant growth and health. In some animals, such as humans, sunlight influences skin color. Growth, weight, and health are influenced by mineral levels in humans and other animals.
5. Because there is spectrum, or wide distribution of phenotypes, it is polygenic. Multiple genes are involved in the expression of eye color.

18.4 Check + Apply Your Understanding
1. 22 pairs, or 22 chromosomes, are autosomes, meaning they carry information available for both sexes. Autosomes are not involved in sex determination.
2. Through sex chromosomes. Many animals use the XX/XY system.
3. Because men only need one copy of a recessive allele to express it and women need two.
4. Chromosome 9 is larger and carries more genes. (It is estimated that chromosome 9 carries 800–1,300 genes, and chromosome 13 carries 600–700.)
5. There are several indicators that this disease is passed on through autosomal dominant inheritance. You know it is passed on the autosome because it is passed from the father (II-3) to the son (III-5) (this does not happen with X-linked traits). Also, males

and females are affected, with roughly the same probability. You know it is dominant because it occurs in three consecutive generations (this does not happens with recessive traits). A final indicator of a dominant inheritance is that the individual II-1 has the allele for the disease but he does not express it.

18.5 Check + Apply Your Understanding
1. Artificial selection is the intentional breeding of selected traits (also referred to as selective breeding). Domestication is accomplished through many generations of artificial selection.
2. It can amplify masked recessive traits that are undesired.
3. Inbred; outcrossed.
4. The smaller population has less genetic variation and fewer mates to select from; therefore, it is more susceptible to inbreeding depression.
5. Belyaev and his colleagues showed that domestication can occur rather quickly in animals. More importantly, they showed a connection between behavior and genetics. The experiments also provided insight on the process of how wolves were domesticated into tame dogs. Because these experiments connected genetics and behavior, they opened doors to studies on the impact of genes on behavioral and social traits. By understanding what genetic changes occur to domesticate a wild animal and what traits cause aggressive behavior, researchers may be able to better understand autism, aggression, fear, and other social disorders in humans.

18.6 Check + Apply Your Understanding
1. All cells are derived from the zygote. The genotype is replicated during embryonic development.
2. 100% because he carries two recessive alleles. All sperm will carry one copy.
3. Proteins.

4. Structural.
5. As functional proteins, enzymes catalyze chemicals reaction. In fat metabolism, they catalyze the breakdown of fat into smaller molecules.

Chapter 18
End of Chapter Questions
1.
Gene	f.
Allele	b.
Heterozygous	a.
Homozygous	e.
Dominant allele	d.
Recessive allele	c.

2.
Meiosis	a.
Independent assortment	c.
Law of segregation	b.

3. c.
4. e.
5.
Incomplete dominance	c.
Codominance	d.
Pleiotropy	a.
Polygenic	b.

6. b.
7. b.
8.
Linked	b.
X-linked	c.
Sex chromosomes	d.
Autosomes	a.

9. e.
10. e.
11. a.
12.
Genotype	c.
Phenotype	d.
Structural protein	b.
Functional protein	a.

13. The parents are both heterozygous for the trait. If F represents freckles, each parent is Ff.
14. Mr. Jones is ii and Mrs. Jones could be I^Ai or I^AI^A. The baby will have type A or type O blood.

Parents	I^A	i
i	I^Ai	ii
i	I^Ai	ii

OR

Parents	I^A	I^A
i	I^Ai	I^Ai
i	I^Ai	I^Ai

15. The pedigree on the left shows a dominant trait that appears in each generation. The pedigree on the right is a recessive trait. In the first generation, neither parent shows the trait, yet the trait appears in their offspring.

16.
a. Genotypes are listed in the table below.

1	X^rX^r	9	X^rY
2	X^RY	10	$X^RX^?$
3	X^RY	11	X^RY
4	X^RX^r	12	$X^RX^?$
5	X^rY	13	X^rY
6	X^RX^r	14	X^rX^r
7	X^rY	15	X^rY
8	X^rX^r		

b. Because this is an X-linked recessive trait, males pass on a Y chromosome but never an X chromosome. Because the X chromosome is also inherited from the mother, males cannot pass the trait on to male offspring.
c. If the mother has the trait, 100% of the time she will pass it on to a male offspring because it is an X-linked recessive trait and the X chromosome comes from her.
17. See Table below.
18. Autosomal recessive.

19.

Family Member	Genotype	Phenotype
I-1	Cc	Carrier
I-2	Cc	Carrier
II-1	CC (if Cc, likely an offspring would be affected)	Normal
II-2	cc	Cystic fibrosis
III-1	Cc	Normal

20. There is a 1:4 chance that with each reproduction a child will have cystic fibrosis.
21. III-3 is Cc and III-4 is either CC or Cc. If both are Cc, there is a 1:4 chance the child will have cystic fibrosis. If III-4 is CC, then there is not a chance that the child will have cystic fibrosis.
22.
a. No. She carries the gene.
b. Yes, especially if there is a family history of the disease in his family.
c. 1:4.
23. Zero chance.

Potential Student-Generated Questions
1. If I am homozygous recessive what are my chances of developing breast cancer or ovarian cancer? Is it 100%?
2. Is the likelihood of developing breast cancer or ovarian cancer equal?
3. How many other genes are involved in these cancers?

4. How do these mechanisms compare in males and females? In different ethnicities?

CHAPTER 19
19.2 Check + Apply Your Understanding
1. The production of yogurt, cheese, bread, beer, and wine all rely on biological systems (microbes) and therefore are examples of biotechnology.
2. Genomics provides the foundational information to design efficient biological systems. In order to manipulate the genes of animals, plants, and bacteria, scientists must first understand the genes of the targeted organisms. When developing new technologies, the more that is understood about the genes and interactions of the genes, the better the system can be designed. Genomics will also enable scientists to more quickly manipulate genomes to make useful products.

3. Some ethical concerns of genetic testing include confidentiality, privacy, insurance coverage, and employment discrimination. In addition, what to do with the information is a topic to consider. For instance, if the genetic test was performed on the fetus, is it ethical to terminate the pregnancy?
4. Many diseases are cheaper to treat when diagnosed early. Preventive measures might save insurance companies money in the end treating diseases such as cancer and diabetes.
5. By identifying and targeting the genes that make enzymes in bacteria or other organisms that break down these hazardous substances, biotechnologists could design a robust biological system that is specialized for breaking down wastes after an oil spill or at nuclear waste sites.

19.3 Check + Apply Your Understanding
1. "Trans" means to cross. Transgenic organisms have genes transferred from one organism into another.
2. Bacteria and yeast do not have the mechanisms needed to make many complex proteins properly folded and assembled in the correct format for eukaryotes.
3. In genetic engineering techniques, a plasmid is the vector that is used to carry new genetic material into a new organism. Plasmids, found in bacteria, are naturally occurring, double-

stranded, circular DNA molecules that are separate from the bacterial chromosome.
4. Viruses target specific cells. In gene therapy, viruses that target unintended cells could have negative effects on the patient. Also, the replication mechanisms of viruses are deactivated for gene therapy. If these mechanisms are not fully disabled, the viruses can disrupt gene regulation, leading to cancer.
5. If the genes for rust immunity could be transferred to these other cereal crops, they too would be resistant to this agricultural pest. First, the gene for rust immunity would need to be identified and isolated. It then could be inserted into a vector that would next be inserted into the plant genome.

19.4 Check + Apply Your Understanding
1. First, they are unspecialized cells capable of renewing themselves through cell division. Second, under the proper conditions, they can be induced to become tissue- or organ-specific cells with specialized functions.
2. Personalized medicine allows doctors to treat patients as individuals. By considering an individual's genetic markers, genetic profile, family genetics, and the results from other tests, doctors can provide the best treatment for the individual rather than just practicing the norm.
3. Genetic markers are sequences of DNA that are known to be risk factors for certain diseases. By analyzing the genetic markers of a disease compared to the DNA of a patient, medical researchers gain insight into the genetic makeup of the patient and the likelihood for disease to develop.
4. Although the cloned pet likely will look the same as the original pet, it will not behave the same. From a genetic point of view, the cloned pet will be identical to the original pet; however, genes interact with the environment (nutrition, positive reinforcement, dominance),

Pattern	Number of Allele Copies Needed for Expression	Are Parents of Affected Child Usually Affected (Y/N)?	Equally Likely To Be Expressed in Both Sexes (Y/N)?	Transmitted Father to Son (Y/N)?	Is Expressed Trait Likely to Skip a Generation (Y/N)?
Autosomal dominant	1	Y	Y	Y	N
Autosomal recessive	2	N	Y	Y	Y
X-linked dominant	1	Y	Y	N	N
X-linked recessive	2 if female, 1 if male	N	N	N	N

so the cloned pet will not be the same. Behavioral aspects affect the development and personality of the pet as well. Because these will be different, the cloned pet will not be the same.

5. Following the model of therapeutic cloning for diabetes research, DNA from the patient's skin could be isolated and placed into a donor egg cell. This cell then could be grown in culture until embryonic stem cells can be harvested. The stem cells could be triggered to produce nondefective cells that could be injected back into the patient to reverse the muscular dystrophy disorders.

19.5 Check + Apply Your Understanding
1. Genomics, genetic profiling, and DNA markers can be used to investigate genetic relatedness.
2. Hair, blood, semen, saliva, or skin each can be used as DNA evidence.
3. Like a regular fingerprint, a DNA fingerprint is unique to an individual (unless you are comparing twins). It is a DNA profile that can identify an individual based on unique features of his or her genome.
4. Statistically, a profile comparing 20 loci provides more evidence, but it likely will take longer and cost more money to analyze. An analysis of only eight loci may provide enough data to make a case beyond a reasonable doubt.
5. The DNA profiles of two closely related individuals will share more similarity than those of two unrelated individuals. Comparing the DNA from a child to the DNA of the potential father will yield a higher percent of matches. Specifically, because the Y chromosome comes from the father, genetic markers from the Y chromosome can be used to determine genetic relatedness.

Chapter 19
End of Chapter Questions
1.
Biotechnology   d.
Genomics   b.
Genetic engineering   a.
Gene therapy   c.

2. d.
3. c.
4.
Genetic markers   c.
Stem cells   a.
DNA profiling   d.
Therapeutic cloning   b.
5. a.
6. b.
7.
Embryonic stem cells   b.
Adult stem cells   a.
8. d.
9. d.
10. As in a criminal investigation, DNA profiling was used to identify victims of the 9/11 attacks looking at least 13 loci. DNA samples of remains, such as bone and tissue fragments, were compared to DNA from family members, previously stored medical samples, and other personal items from each victim. Due to the nature of the attack, some of the DNA evidence was damaged through harsh conditions such as fire and heat. Also, with more than 2,500 victims and 21,000 remains, the sheer number of samples to analyze has been daunting.
11. Rainfall shortages, infertile soil, lack of productive land, pests, and diseases all negatively impact food production in Africa. The use of genetically modified organisms can improve food production by addressing some of these issues. Today, food distribution is one of the issues facing developing nations. The use of GMOs can help local food production and, therefore, local access. Concerns over biotechnology include access to technology, human safety, and long-term effects on the continent's biodiversity.
12. A positive diagnosis for a fatal disease is difficult to deal with. This information affects the individual, family members, and close friends. He or she may be treated as a social outcast because people don't know what to say to someone in this situation. In some cases, early diagnosis may prevent the individual from reproducing and passing the defective gene onto another generation. In disorders such as familial

hypercholesteremia, early treatment can delay the onset of cardiovascular disease, and it is effective to diagnose early. In terms of discrimination, both life and health insurance could be denied, and employers might discriminate as well.
13. Suspects 1 and 2 and the victim have a different DNA profile than the DNA from the blood collected at the crime scene.
14. The DNA profile of Suspect 3 matches the DNA from the blood at the crime scene.
15. See Table 19.3.
16. Each number represents an allele. Because the genotype of an individual is made of two alleles (one inherited from each parent), there are two values.
17. Sources of error include contamination of the sample, sample handling (mix-up), proper storage, interpretation of results, and calibration of the equipment analyzing the DNA. Bias is another source of error. If the data are accompanied with a note indicating the innocence or guilt of a suspect, the data may be analyzed with a "preconceived notion."
18. The data only indicate the analysis of three loci. Standard procedure is to examine 13 loci. The more loci analyzed, the stronger the evidence. This evidence should be used in conjunction with other forensic evidence, such as hair samples, ballistics, eyewitness accounts, etc.
19. Understanding allele frequencies in a given population helps to determine the strength of the evidence shown by a DNA profile. Crime labs compute the frequency of each allele in a sample population and then compound the individual frequencies by multiplication. For example, if 10 percent of Caucasian Americans are known to exhibit the 15 allele at the first locus (D3S1358) and 20 percent (1:5) are known to have the 16 allele, then the frequency of the pair of alleles would be estimated as $2 \times 0.10 \times 0.20 = 0.04$, or 4 percent among Cauca-

sian Americans. Frequency estimates for the overall profile can be staggeringly small: one in a billion or even less. These statistics make DNA profiles powerful evidence.

Potential Student-Generated Questions
1. What kind of virus do you use? Will the virus be safe? Will it deliver the right number of genes?
2. How will the virus deliver the genes to the neurons? Will the genes insert into the neuron genome?
3. If the cells of the brain are nondividing, how long does the effect of gene therapy last?

CHAPTER 20
20.2 Check + Apply Your Understanding
1. The wall determines (1) the cell's shape and (2) protects the bacteria from expanding and bursting when they are in freshwater or shrinking when they are in saltwater.
2. Plasmids are small pieces of circular DNA found in bacterial cells in addition to their larger circular chromosome. During cell division, plasmids replicate independently and are inherited by the daughter cells. Plasmids carry genes that are important to pathogens such as toxins or proteins, which bind and halt antibiotics from acting and thus give antibiotic resistance.
3. Nutritionally, bacteria require a source of energy along with carbon, hydrogen, oxygen, and nitrogen to build cellular structures such as phospholipid membranes, DNA molecules, and proteins. They need minerals for smooth functioning of enzymes and metabolic pathways. Yes, phosphorus is needed to build ATP, phospholipid membranes, and the backbone of DNA.
4. A carrier who does not show symptoms is a person who is infected and not diseased. Food poisoning by either *Staphylococcus* or *Clostridium* is an example of a toxin that is ingested that leads to disease symptoms without the person being infected by the pathogen.

5. (a) During the incubation period, bacteria attach, evade the body defenses, multiply, spread, and begin to damage body cells. As damage increases, the body is in the diseased state or illness period. (b) A condom blocks the portal of entry from the reservoir, which is an infected sex partner. (c) Antibiotics inhibit the multiplication stage of the disease cycle.

20.3 Check + Apply Your Understanding
1. Viruses do not have the characteristics of life. They cannot survive on their own. They obtain their energy and raw materials to reproduce by using a host cell. In this sense, they are nonliving parasites. They are intracellular because they only exist inside of cells. Outside of the host cell, they are particles with no means of reproducing.
2. All viruses have a nucleic acid core (made of either DNA or RNA) protected by a protein capsid. Most animal viruses have an outer envelope made of lipids and rich in proteins.
3. When the virus incorporates part of the host cell, the host will recognize the virus as part of itself. Using the cell membrane helps the virus to elude immune responses as well as gain easy entry into other cells.
4. Some bacteria viruses can transform benign bacteria into toxin-producing, disease-causing agents.
5. It is important to administer this vaccine before females become sexually active. Once a person is infected, the vaccine is not effective at preventing cancer. Because this virus shows stability, it does not mutate quickly. Therefore, it shows characteristics of a DNA virus.

20.4 Check + Apply Your Understanding
1. These fungi are adapted to penetrate animal cells to gain access to nutrients in the outer skin. They utilize keratin protein as a nitrogen and energy source.
2. If the worm kills its host, it has killed both its habitat and food source.

3. Human cells are also eukaryotic cells; therefore, it is more difficult to distinguish the parasite from the host.

4. This attractant is released in stages 5 and 6, when the gamete cells are released. At this stage, the mosquito is attracted to the host so that the life cycle can be completed inside of the mosquito.

5. Recall that intermediate hosts contain the larval stages and the final host is the reproductive stage that produces eggs. In this life cycle, both the snail and the fish are intermediate hosts, and the human is the final host.

20.5 Check + Apply Your Understanding

1. An antigen is any foreign substance that when introduced into the body provokes an immune response. Two antigens that commonly challenge the body are foreign molecules (proteins, polysaccharides) composing virus capsids and bacterial cell walls.

2. The second line of defense (1) is preformed, (2) is general, and (3) has no memory of past encounters. The third line of defense (1) is acquired or adapted by experience, (2) is specific, and (3) remembers previous encounters.

3. See Table below.

4. This blood test means that your friend has been exposed to antigens of the herpes simplex virus and not to HIV.

5. The thymus is where T-cells are educated during early life to distinguish between self and nonself markers. The consequence may be that the T-cells mistakenly identify the viral antigens as self and not recognize them as foreign. This would allow the virus to establish an infection that lasts the entire life of the person.

Chapter 20
End of Chapter Questions

1.
Capsule   b.
Cell wall   d.
Toxin   a.
Plasmid   c.

2. 2, 7, 1, 4, 3, 6, 5

3. a., c.

4.
Capsid   b., d.
Envelope   a., d.
Nucleic acid core c.

5. c.

6. d.

7.
Protozoan   a., c.
Fungus   a., b.
Worm   b., d., e.

8. d.

9. a., c., d.

10.
First line of
defense   a., b., c.
Second line
of defense   d., f., h.
Adaptive
immunity   e., g., i.

11. d.

12. a.

13. d.

14. (a) macrophages, (c) B-cells, and (d) antimicrobial proteins are associated with inflammation and bacterial infection. (b) Killer T-cells are defensive responses to virus-infected cells.

15. (a) The biofilm protects the bacteria from salivary antibacterial enzymes and prevents them from being washed down the throat into the acidic stomach. (b) The bacteria acquire the sugar for metabolism by colonizing food left in the mouth between teeth and gums. (c) The mouth is an environment conducive for bacterial growth. It is moist, warm, and frequently filled with food. (d) Because the bacteria produce lactic acid, they must ferment sugars in the low oxygen pockets near the teeth and gums. (e) Brushing and flossing remove the energy supply and the biofilm necessary for oral bacteria to produce lactic acid and break down the tooth enamel characteristic of tooth decay.

16. (a) B, (b) A, (c) A, (d) A, (e) C, (f) D.

17. All three viral infections peak during the winter months.

18. Yes, remarkably so, which makes this disease very predictable.

19. The graph shows that the influenza season starts in November, peaks in January–February, and declines in March. So the flu season lasts about three months during the coldest part of the year.

20. Rubella shows an epidemic in 1964, when a greater than fivefold increase in the number of cases occurred. During this outbreak more than 100,000 people were infected, whereas about 20,000 people contracted the virus during a normal, or smaller, outbreak.

21. No, in the southern hemisphere, winter occurs from June to September. This fact is important because people in the northern hemisphere can infect or become infected by people in the southern hemisphere and keep the seasonal cycles going.

22. Because outbreaks occur in winter and not in summer, temperature may play a role in the disease cycle. Correlations have been established for cycles of humidity, rains, and winds.

23. The explanation would have to include the life cycle, the environmental influences on the mosquito, and its complex interaction with human behaviors and biology.

24. Mutations cause the dominant strain of the virus to change from year to year. Yes, the flu shot you receive contains and protects you against a few specific flu strains, but there are many different strains actively infecting people in the world during any flu season.

Potential Student-Generated Questions

1. Are all the sugar molecules on the pathogenic strain of bacteria the same?

2. Do lysosomes fuse to vesicles containing tuberculosis bacteria?

3. How do bacteria look when they are in a lysosome? Do they have walls?

4. Do tuberculosis bacteria add layers to resist killing?

CHAPTER 21
21.2 Check + Apply Your Understanding

1. A disease-causing variable that can be changed. For example, drinking alcohol, physical inactivity, and smoking all are modifiable behaviors that are risk factors for disease.

2. Smoking, overweight/obesity, high blood pressure, high blood glucose (sugar), and physical inactivity.

3. The cells of adipose tissue release leptin, a hormone (chemical signal) that adjusts appetite to balance energy intake and expenditure.

4. Because chronic illnesses are multifactorial and each person's health is truly individualized, it is difficult to predict the outcome of a particular illness.

5. Yes. I would consider bronchiectasis a chronic disease. Because scar tissue will build up, the disease is long-term and affects a person's health for more than three months. As scar tissue builds up, the body will respond. These responses will trigger other symptoms, which will perpetuate the disease. In this sense the disease cannot be cured and will not go away.

21.3 Check + Apply Your Understanding

1. A diet low in sodium and high in fruits and vegetables helps to reduce blood pressure. Also, increasing physical activity and reducing stress and smoking improve the cardiovascular circuit.

2. The three are directly related. Atherosclerosis causes plaques that can break loose and lodge in blood vessels in the brain. If a plaque blocks the blood flow to the brain, a stroke can occur.

3. Atherosclerosis narrows the arteries, which in turn increases blood pressure. As a result, the intense pressure of the blood forced against the artery wall damages the artery and increases inflammation, which in turn increases the rate of atherosclerosis.

4. I would emphasize that his blood pressure needs to be monitored and, if controlled to normal levels, could prevent a heart attack, atherosclerosis, or a stroke. I also would make sure he understood that several lifestyle factors (listed in Question 1) could lower blood pressure.

5. (a) At rest in bed, the heart needs to work less hard. As a result, the blood pressure is the lowest. (b) While standing, the heart needs to work harder, and the blood pressure is higher. (c) Blood pressure would decrease in response to blood loss after a car accident. Under these conditions, the loss of blood volume lowers the blood pressure and increases the heart rate.

21.4 Check + Apply Your Understanding

1. Cancer cells grow and divide uncontrollably as a result of DNA mutations that disrupt normal cell cycle and function.

2. Cancer, regardless of the tissue of origin, spreads through the blood and lymph vessels to other parts of the body.

3. Through heredity they have received a mu-

	Recognize and Attack Pathogens Outside of Cells	Recognize and Attack Pathogens Inside Cells
Molecules	Histamine, enzymes, mucus, gastric juice	—
	Complement proteins	—
	Antibodies	—
Cells	B-cells	Helper T-cells
	Antibody-producing cell	Killer T-cells
	Memory cells	—

tated copy of the gene. Fewer mutations need to accumulate for cancer to develop.

4. In normal cell division, cell regulator genes encourage cell growth and differentiation. Tumor suppressor genes keep cell growth and differentiation in check by regulating transcription and the activity of DNA repair genes. DNA repair genes add another layer of control and trigger either cell death or repair when DNA damage has occurred. In cancerous cell division, oncogenes accelerate cell growth and differentiation. Mutations in tumor suppressor genes and DNA repair genes allow the cell cycle to continue in an unregulated fashion, leading to tumors.

5. Tumors acquire nutrients and resources from blood vessels. If the blood vessels and blood supply are cut off, the tumor cannot grow.

Chapter 21
End of Chapter Questions

1.
Chronic disease	d.
Maladaptation	c.
Unmodifiable risk factor	a.
Modifiable risk factor	e.
Risk factor	b.

2. d.

3. d.

4.
Atherosclerosis	a., c., d.
High blood pressure	a., b., d.

5. a.

6. c.

7.
Hereditary cancer	c.
Tumor	a.
Malignant tumor	b.

8. b.

9. d.

10. (a) This person is at high risk for developing heart disease since he has two high risk factors—age and diet. While he is physically active in his job, his diet consists of fast foods that are high in saturated fats, sodium, and calories,

raising his risk for cardiovascular disease. From the information provided, I would assess the risk for diabetes to be low. (b) This person is at high risk for developing heart disease and diabetes since she has two high risk factors—physical inactivity and diet. This person has a high-stress, sedentary job and a diet high in sugar. (c) Long-term, if this student's behavior does not change, he is at high risk for developing heart disease since he has one high risk factor—diet. Though this person is young, he is at risk to develop cardiovascular disease since his diet is high in saturated fats, sodium, and calories. From this information provided, I would assess the risk for diabetes to be low.

11. See Table below.

12. Because the lung cancer has metastasized, surgery cannot treat the cancer. Chemotherapy is the best treatment option because it is the only systemic approach. Your aunt should discuss any additional benefits that would result from shrinking any of the tumors with radiation. Shrinking the tumors with radiation may improve her breathing, comfort, and quality of life.

13. About 2,500 per year, which is about seven cigarettes per day, or one-third of a pack.

14. It essentially follows the same trends. Three thousand cigarettes per year is about eight per day, or slightly less than half a pack per day.

15. Those smoking more cigarettes get cancer sooner than do those smoking fewer cigarettes.

16. Overall, from 1993 to 1997, there was a 27.8 percent increase in college smokers. Both females and African-Americans had the largest rise in the number of smokers. When comparing males and females, female smokers increased 31 percent, whereas males increased 23 percent. The trends among the various classes are inconsistent. Freshman and juniors have a higher rate of increase than sophomores and seniors.

17. Because cancer is a disease characterized by mutations in several genes that accumulate over time, cancer does not develop overnight. Lung cancer develops after years of exposure to carcinogens.

18. I would expect the lag time to be similar between men and women, but both lines on the graph shifted to the right.

19. Joe Camel is a cartoon character that appeals to young children. His demeanor is a smooth, cool character that will also appeal to teenagers looking to be "hip and cool" like Joe Camel. The second ad targets women. The slogan "You've come a long way, baby" is a phrase of empowerment. The ad reminisces about the days when women could not vote and belonged in the kitchen. Both messages are misleading. Smoking cigarettes does not make a person cool, hip, smooth, or empowered.

20. Many students smoke as a response to stress or peer pressure, or as a way to lose weight. Many teens see smoking as an adult activity that they are forbidden to do. Media campaigns and following people they admire who smoke are appeals to teens.

Potential Student-Generated Questions

1. Will drugs that block estrogen receptors reduce the growth of tumors?

2. Will these drugs have other side effects on bone growth?

3. What are other modifiable factors from the diet or environment that mimic estrogen?

TABLE 21.4 Breast Cancer Genes

Gene Mutation	Function	Oncogene or TSG?
BRCA-1	Encodes a protein that is directly involved in repairing damaged DNA	TSG—acts like stoplight, inactivating/repairing DNA
PTEN	An enzyme that acts as part of a chemical pathway that signals cells to stop dividing and triggers cells to undergo a form of programmed cell death	TSG—acts like stoplight, triggering cell death
ERBB-2	One member of a family of genes that provide instructions for producing growth factor receptors (growth factors are proteins that stimulate cell growth and division)	Oncogene—acts like green light, activating a cellular response

Glossary

Action potential Nerve signals begin when one or more neurons respond to a stimulus and generate an action potential—a rapid and temporarily irreversible change in the charged ion concentration across the cellular membrane of all cells. The electric charge created by the movement of charged ions isn't very strong.

Active transport A process that moves molecules uphill against the concentration gradient with the expenditure of energy.

Adaptation A change in a trait, behavior, or structure of an organism that allows the organism to be suited to its current environment; feature of an organism that enables it to survive and reproduce in its environment.

Adaptive immune system A powerful third line of defense for the body that includes a wide range of specialized white blood defensive cells and molecules, which have the ability to recognize antigens at a very fine level of detail.

Adaptive radiation Species have branched or radiated into numerous lineages, each with many species.

Adenosine diphosphate (ADP) A large nitrogen-containing macromolecule with two phosphate groups. Captures energy from energy-releasing pathways by adding a third phosphate group to become ATP.

Adenosine triphosphate (ATP) A large nitrogen-containing macromolecule with three phosphate groups that carries chemical energy to the sites where energy is used. Releases energy to energy-using pathways by splitting off a phosphate group to become ADP.

Adult stem cells Undifferentiated cells derived from nonembryonic cells. Through cell division, these cells continue to generate replacements for cells that are lost through normal wear and tear, injury, or disease.

Aerobic metabolism Energy-releasing metabolic pathways that use oxygen to completely break down amino acids, fatty acids, and glucose to make large numbers of ATP molecules; also called *cellular respiration.*

Alcoholic fermentation Metabolic pathway used by yeast to derive energy from sugars without oxygen; produces ethanol (alcohol) and carbon dioxide.

Alleles Different forms of a gene that bring about particular traits; the different forms of a gene.

Amino acids Building block molecules for proteins; each amino acid consists of four parts: (1) central carbon and hydrogen atoms, (2) a group of nitrogen and hydrogen atoms called an amino group on one end of the molecule, (3) an acid group on the other end of the molecule, and (4) a side group that attaches to the central carbon; there are about 20 different types of amino acids.

Amniotic egg An egg found in reptiles, birds, and mammals that has a food source and protective membranes to protect and aid the developing embryo; often called a "self-contained pond" of fluid that supports the growth and development of the embryo.

Anaerobic metabolism Energy-releasing metabolic pathways that do not use oxygen; they incompletely break down glucose to make a small number of ATP molecules.

Annual growth ring A ring of cells produced by woody plants each growing season; consists of a light-colored ring of larger, wider xylem cells (springwood) followed by a dark ring of narrower, thicker-walled cells (summerwood).

Annuals Herbaceous plants that grow, flower, and produce seeds within a single growing-season.

Anther The top part of the stamen (male) that produces pollen grains.

Antibiotics Chemicals that interfere with the metabolism of bacteria and other fungi and thus inhibit their growth or kill them.

Antibodies A group of Y-shaped proteins that circulate in the blood and bind specifically to antigens. This recognition occurs when the antigen binds to the specific binding site on the tip of the antibody molecule.

Antigens Specific molecules on the surface of microbes that our body is able to detect as being "out of place." Examples of antigens are proteins that make up the viral capsid and envelope, bacterial cell walls, and the outer cuticle of hookworm parasites.

Apical meristem Growing points at the root and shoot tips and in buds; in roots it is the part that perceives and responds to gravity and grows downward; shoot apical meristem grows upward, increasing the height of the plant.

Arteries Strong, muscular blood vessels that carry blood away from the heart.

Asexual reproduction The process that creates offspring without the fertilization of the egg. In most cases, the offspring are genetically identical to the parent and develop through the process of mitosis; forms clones.

Atherosclerosis A medical condition of reduced blood flow caused by buildup of plaques on the inner walls of the arteries. This condition is associated with high blood pressure and leads to hardening of the arteries. Most heart attacks and strokes are caused by atherosclerosis.

Atria (plural of *atrium*) Chambers of the heart that receive blood from veins.

Autosome A chromosome that does not determine sex gender; in humans, chromosome pairs 1–22.

Autotrophs Organisms, like plants, algae, and some microbes, that can "feed themselves" through photosynthesis.

Bark Consists of all tissues outside of the vascular cambium.

B-cells A specialized group of lymphocytes that remain in the bone marrow or migrate to the spleen, where they mature and learn to recognize specific antigens. B-cells specialize in eliminating parasites and pathogens such as staph and strep bacteria, which attack cells in the blood and tissues.

Behavioral ecology A branch of biology that studies the adaptive behaviors of social animals to their environment that provide them a selective advantage over solitary animals.

Biennials Herbaceous plants that have a single period of reproduction in the second and final year of their life; lifespan spread out over two years.

Bile A yellowish green solution made by the liver that acts like a detergent, separating fat globules into many small droplets; the liver makes half a quart of bile each day.

Biodiversity (or *biological diversity*) The number and variety of organisms found within a given region; refers to the richness of living systems, which includes the diversity of all organisms, the diversity within and among species and populations, and the diversity of ecosystems.

Bioethics Field of study that examines legal, social, and moral standards as they apply to biotechnology and medicine.

Biotechnology A scientific field that uses biological agents, such as microorganisms or cellular components, in a controlled manner to solve problems or to make products that have beneficial uses.

Blood pressure The force of blood pushing against artery walls created by contraction of the heart muscles. Your blood pressure is recorded as two numbers, such as 120/80. The top number is the *systolic pressure,* which is the maximum pressure in the arteries when the heart beats. The bottom number is the *diastolic pressure,* which represents the lowest pressure in the blood vessels when the heart is at rest between beats.

Brain In animals, a large concentration of nerve cells that act as the information-processing center, controlling the other organ systems of the body either by activating muscles or by causing secretions of chemicals such as hormones.

Brain stem A protected area located underneath the brain that is dedicated to vital housekeeping functions in the body, such as blood pressure, respiratory rate, heart rate, coughing, and reflex reactions such as vomiting.

Bud An undeveloped part of a plant that grows into a new stem, leaf, or flower.

Budding Form of asexual reproduction that produces a daughter cell that is genetically identical to parent organism; common in yeasts.

Bulk transport A type of active transport that moves large macromolecules by enclosing them in a vesicle before moving them into the cell or secreting them out of the cell.

Calorie A measure of energy contained in food.

Calvin cycle The second stage of photosynthesis, a set of linked chemical reactions that do not directly require sunlight but instead rely on the products of the light reaction.

Cancer A family of diseases that have one thing in common: the uncontrolled growth of abnormal cells. One feature that makes many cancers deadly is their ability to spread to various parts of the body through the blood or lymph fluids. Because different types of cells perform specific functions in the body, each cancer has a different set of characteristics.

Canine Type of tooth that occurs on either side of the incisors and is specialized for piercing and tearing food; well developed in carnivorous mammals.

Capillaries Narrow, thin-walled blood vessels that serve as the sites for exchange between the blood, tissue fluid, and body cells.

Capsid A protective protein coat that covers an individual virus.

Capsule Outer layer of carbohydrates and proteins secreted by some bacteria that helps them stick to surfaces. For example, strep bacteria have a capsule that helps them stick to the epithelial cells of the throat and prevents them from being washed away into the stomach during swallowing.

Carbon cycle The global biogeochemical cycle in which carbon is exchanged among the living and nonliving portions of Earth. It is one of the most important cycles of Earth and allows for carbon to be recycled and reused throughout Earth and all of its organisms.

Carcinogens Agents that cause genetic mutations leading to uncontrolled cell division and cancer. Tobacco smoke, ultraviolet light, and many pesticides are familiar carcinogens.

Carnivore Animal that feeds on other animals or their eggs.

Carpel Female reproductive organ of a flower located in the central portion of the flower. The carpel consists of three regions: the top is the stigma, a surface that receives pollen; the middle is the style, which elevates the stigma; and the bottom is an enlarged ovary.

Carrying capacity The ability of a habitat to sustain a particular population size based on the resources available over a specific period of time.

Cecum A pouch located between the small intestine and large intestines in many cellulose-eating animals; houses large populations of microbes that further digest cellulose and release nutrients. In humans, the cecum is the first part of the large intestine.

Cell cycle The distinct stages that a cell goes through to duplicate and divide. The cell cycle provides growth and development in organisms, as well as replacement cells. It includes interphase, mitosis, and cytokinesis.

Cell differentiation The process that changes each cell into its final form—a specialized cell type with a clearly defined structure and function.

Cell wall Provides rigidity, protection, and support to plant, fungi, and bacteria cells; encloses the contents of the cell and determines its size and shape; in fungi, composed of chitin, proteins, and carbohydrates; composed of molecules that are highly cross-linked to provide mechanical strength and support along with flexibility; works with the plasma membrane to control the exchange of substances into and out of the cell; also prevents the cell from bursting if a large amount of water enters by osmosis.

Cellulose A structural carbohydrate used by plants to build their cell walls and seed coats; macromolecule that forms the framework of the plant cell wall; a complex carbohydrate that consists of several thousands of glucose molecules bonded end to end as long fibers that are twisted and wound together to form a very strong material; fiber.

Central nervous system (CNS) The nervous system in vertebrate animals comprising the brain and spinal cord, which receive and integrate information delivered from various types of sensory neurons.

Cerebellum A large, bell-shaped outgrowth of the pons that plays an important role in muscle control, some mental functions such as attention and language, some emotional functions such as fear and pleasure responses, and fine motor skill control of muscles.

Cerebrum The large, infolded region at the top/front of the brain. This region functions in behavior, movement, sensory processing, communication, problem solving, and memory.

Chemoreceptors Sensory receptors that, when stimulated by chemicals in the environment, allow for the perception of taste and smell in vertebrates.

Chemotherapy A medical treatment for cancer that involves treating patients with drugs that destroy cancer cells by interfering with the cell growth and multiplication phases of the cell cycle. Chemotherapy is delivered through the blood to reach cancer cells wherever they have spread. Because these drugs are most effective against rapidly dividing cells, normal cells such as those of the intestinal tract, inside of the mouth, and hair follicles also are often harmed.

Chitin A modified (contains nitrogen) complex carbohydrate composed of long chains of glucose; a very strong and durable structural molecule that forms the outer shell, or exoskeleton, of insects, crabs, and other arthropods. It is a strong compound that forms the cell walls of fungi.

Chlorophyll Green pigments embedded in the surface of chloroplasts. A chlorophyll molecule is shaped like a tennis racket with its fatty acid chain. The chlorophyll molecules perform two functions: (1) they absorb light, and (2) they transfer that energy to other compounds.

Chloroplasts An organelle in a plant cell that is specialized for photosynthesis; a type of plant plastid that contains the green pigment chlorophyll; the organelles where the chemical reactions of photosynthesis take place.

Chromatin A macromolecule found in the nucleus; composed of long strands of DNA wrapped around proteins; packaged into chromosomes.

Chromosomes The structures in the cell that store the DNA; "genetic packages" in the nucleus of a cell; consist of long strands of chromatin (DNA and protein) that store and express genetic information.

Chronic disease An illness or medical condition that lasts over a long period and sometimes causes a long-term change in the body that affect a person's function for three or more months. Examples are heart disease, diabetes, cancer, Alzheimer disease, obesity, and arthritis.

Circulatory system The animal organ system that transports and distributes oxygen to body cells and carries carbon dioxide away from body cells. Closed systems consist of a pump and a network of blood vessels that carry blood; open systems do not have blood vessels but rather bathe each of the body cells in the body cavity.

Citric acid cycle Stage of aerobic metabolism that follows glycolysis; a series of eight reactions that yield two ATP molecules, hydrogen atoms attached to carriers, and carbon dioxide. Fatty acids and many amino acids enter metabolism in the citric acid cycle. Also referred to as the Krebs cycle.

Classify (or *classification*) The process of placing organisms into categories based on similar characteristics or traits. This includes body form, anatomical structures, developmental events that occur, and the biochemistry of the organisms.

Climate Generally describes weather patterns that occur in a location or region over decades or longer.

Clone A group of organisms that are genetically identical copies of a single parent; offspring from asexual reproduction; a stand of aspen trees that are genetically identical.

Codominance The expression of both the dominant and recessive traits in the phenotype of the organism.

Codons Sets of three nucleotides that comprise the genetic code. Each of the 20 amino acids used in protein synthesis is coded for by one or more codons. The codon base pairs with an anticodon on the tRNA that carries a specific amino acid.

Coevolution Process that occurs when two or more organisms interact and exert selection pressures on each other to produce adaptations.

Commensalism A relationship between species where only one of the two species involved receives a benefit from the relationship. The other organism involved is often a host that provides a home or transportation to another species but is neither harmed nor benefited by the interaction.

Communication Animal behavior that involves a signal or stimulus transmitted by one animal and received by another. In general, as social complexity increases in a population, so does the complexity of the signals the group uses.

Communities Collections of populations of different species located in the same habitat.

Complex carbohydrates Large macromolecules that consist of long chains of sugars bonded together; functions as energy-storage molecules; in plants, structural molecules that compose cell walls; chemically, consists of long chains of glucose molecules.

Conception The stage of reproduction that is marked by fertilization and the formation of a zygote.

Connective tissue Tissue involved in support, storage, and protection; makes and secretes a nonliving layer, or matrix, in which cells are embedded; widely distributed throughout the body.

Conservation biology The field of biology that deals with the protection of biological diversity.

Constraint A resource available in less than optimal amounts that causes an organism to alter its behavior and/or metabolic processes in order to survive.

Consumers Cannot make their own food and get energy and nutrients indirectly by eating producers and other organisms. Animals are consumers; organisms such as animals and fungi that consume and convert the energy stored in producers, perhaps in the form of sugars, into fuel to run their life activities.

Cork Protective plant tissue on the outside of roots and stems of woody plants; produced by cork cambium; functional and dead at maturity.

Cork cambium A lateral meristem found in woody plants; divides to form a protective outer layer of dead cells (cork) and an inner layer of living storage cells (parenchyma).

Cortex The outer region of the root that specializes in storing starch.

Covalent bond Strong bonds occur between atoms when a pair of electrons orbits both nuclei; serves as the backbone of macromolecules found in cells and determines a molecule's basic structure.

Crossing over Source of genetic variation that occurs in meiosis during prophase I, when the homologous chromosomes align and exchange segments of their chromosomes.

Cyst A dormant stage of the life cycle of some parasites that consists of a structure which has a hard outer surface that protects it from desiccation and helps with dispersal.

Cytokinesis The process that occurs after mitosis that splits the organelles, cell membrane, and cytoplasm into two identical daughter cells.

Cytoplasm Complex mixture of water and dissolved substances that is located between the plasma membrane and the nucleus in eukaryotic cells. In prokaryotic cells, it is the fluid inside of the cell.

Cytoskeleton Forms the cell's internal "skeleton" through a network of protein tubules and filaments; supports and maintains the cell's shape, moves chromosomes during cell division, and moves different structures within the cell.

Decomposers Organisms that use and obtain chemical energy from wastes or the remains of dead organisms, breaking down these remains into smaller molecules that can be used again by producers or be recycled in an ecosystem.

Deforestation The clearing of forests by humans for the development of agriculture, for animal grazing, and for settlement and logging. Deforestation is a primary cause of habitat fragmentation and reduction of biodiversity.

Deoxyribonucleic acid (DNA) A long double-stranded molecule that contains the genetic information to direct life's activities and the instructions for each cell and organism to reproduce. DNA contains a deoxyribose sugar-phosphate backbone and the nitrogenous bases

adenine, guanine, cytosine, and thymine. In eukaryotic cells, DNA is found in the nucleus.

Development The changes in shape, form, function, and structure organisms undergo as they mature from embryos to adults.

Diaphragm A large sheet of muscle used by mammals to ventilate their lungs.

Diffusion The movement of a chemical substance from an area of high concentration (or pressure) to an area of lower concentration (or pressure).

Digestive enzyme Speeds up chemical reactions that break down large macronutrient molecules into smaller molecules outside of cells; breaks a specific chemical bond of a specific molecule under specific ranges of temperature and pH.

Diploid The number of chromosomes in most cells in the body, representing a full genetic set of chromosomes with contributions from both parents.

Disease A condition affecting an organism, most often associated with specific symptoms such as physical ailments. Diseases are often caused by infectious agents, which include viruses, bacteria, fungi, protists, and multicellular parasites.

Disease cycle Infectious diseases caused by bacteria and other pathogens follow a series of steps called the disease cycle. Not all diseases progress through the steps in the same order. The disease cycle provides a framework for understanding how pathogens cause infection and disease.

Dissolved oxygen The amount of oxygen dissolved in water; the oxygen content of water.

Disturbance A range of physical events, such as a fire, hurricane, flood, or volcanic eruption, which alters the existing environmental conditions along with the ability of organisms to survive in that habitat. Disturbance includes biological events such as widespread disease or insect infestations that alter the plant and animal community structure.

DNA profile A genetic fingerprint established by analyzing several highly variable sites in the genome.

DNA repair genes A gene family associated with cancers, which as their name implies, function to detect and repair DNA damage from mutations or other damage to DNA.

DNA replication The process of copying the DNA molecule from an existing DNA strand as a template.

Dominance hierarchy A type of social interaction among members of a group that ranks members and creates a strict social order that often arises from the physical differences among individuals in relation to their access to resources based on certain characteristics.

Dominant In reference to alleles, the allele that is preferentially expressed; only one copy of the allele is needed for expression of the phenotype.

Dormancy A survival-strategy adaptation of many organisms where they temporarily suspend growth, development, or reproduction in response to stressful environmental conditions. Dormancy is a temporary condition that is reversed when conditions improve; a state of arrested growth; an adaptation with great survival value for plants, fungi, some protists, and bacteria.

Double fertilization A distinguishing process in angiosperms (flowering plants) where the two sperm nuclei from the pollen grain unite with two cells in the ovule. The first sperm cell fertilizes the egg cell to form a zygote, and the second sperm cell fertilizes the central cell to form the endosperm.

Echolocation A form of biological sonar that uses high-frequency sound waves to locate food and to navigate.

Ecosystems Geographic areas that contain collections of communities and habitats. Often defined by the dominant plant communities or environmental conditions found there.

Ecosystem services The economic valuation of the collection of physical resources found in a functioning ecosystem combined with the contribution of each species found there. Ecosystem services are grouped into broad categories, such as nutrient and water cycling, carbon sequestration from photosynthesis, food production, and recreational benefits.

Ectothermic Animals that do not *internally* regulate their body temperature and rely almost exclusively on environmental sources of heat. Formerly called cold-blooded animals, ectothermic animals include invertebrates, as well as fish, amphibians, and reptiles.

Egg cell Haploid female reproductive cell (gamete); after fertilization it becomes a diploid zygote.

Electron transport chain The third stage of aerobic metabolism; releases the energy in the hydrogen atoms from glycolysis and the citric acid cycle to yield a large number of ATP molecules.

Elements Substances that cannot be broken down into simpler substances. Examples include carbon, oxygen, and hydrogen.

Embryo A young plant or animal that grows from a zygote (fertilized egg); in seed plants it grows inside an ovule and then a seed.

Embryonic stem cells Undifferentiated cells derived from the inner cell mass during the first week of embryonic development; develop into all of the various types of cells and organs such as the heart, lung, skin, sperm, eggs, and other tissues.

Endangered species An organism that is near extinction throughout all or a significant portion of its biogeographic range.

Endodermis The innermost layer of cells of the cortex; controls the passage of minerals and other molecules into the xylem.

Endomembrane system A system that includes endoplasmic reticulum, Golgi complex, and vesicles that work together to make, sort, and transport substances within or out of the cell.

Endosperm Food-reserve tissue of the seeds of flowering plants that surrounds the embryo; stored energy is absorbed as the embryo develops.

Endothermic Animals that maintain a constant body temperature considerably above the temperature of the environment. Formerly called warm-blooded animals, endothermic animals include birds and mammals.

Envelope Another form of protective outer covering found on some viruses. The envelope is comprised of a glycolipid coat rich in proteins.

Enzyme A protein (or sometimes an RNA) that speeds up chemical reactions in living things; protein that speeds up a chemical reaction, such as digestion of food and conversion of macronutrient molecules to energy.

Epidermis Plant tissue on the outside edge of the plant body that is in direct contact with the environment; generally a single layer of cells; its functions depend on its location; remains throughout the life of herbaceous plants; is replaced in roots and stems in the first year in woody plants. In animals, the upper or outer layer of the two main layers of cells that make up the skin. In animals, the epidermis is mostly made up of flat, scale-like cells. The deepest part of the epidermis also contains melanocytes that produce melanin, which gives the skin its color.

Epithelial tissue Tissue that covers the inner and outer surfaces of the body; lines internal passageways of the body (digestive, urinary, reproductive); forms specialized glands that excrete waste products or secrete substances.

Esophagus The digestive tube that moves food from the back of the mouth to the stomach; lined with many glands that secrete mucus.

Essential nutrients Nutrients the body cells can't make and must be obtained through the diet.

Estrogen Hormone that acts on the various tissues and organs in the female reproductive tract, causing the breasts, ovaries, uterus, and vagina to mature.

Ethylene A gaseous hormone that causes some fruits to ripen.

Eukaryotic cells Large, complex cells consisting of many specialized compartments. Animals, plants, fungi, and protists are composed of these types of cells.

Evolution The process that scientifically explains the unity and diversity of living things; process that causes changes to the genetic composition of a population from generation to generation.

Evolutionary tree A tool that provides a visual summary of a complex set of scientific data, linking taxonomy with evolutionary relationships.

Experiment A carefully controlled test designed to make a discovery or to determine the validity of a hypothesis.

External fertilization Union of sperm and eggs outside the body, usually in water or a moist environment.

Extinction Elimination of a species, which may be local or global as in the total elimination of a species from Earth.

Fatty acids Insoluble in water; building block molecules of fats (triglycerides); long chains of carbon, hydrogen, and oxygen.

Feces Mass of undigested solid matter, or the stool, which is eliminated from the body; substantial amounts of water may be lost from the body as a component of feces.

Fermentation Second stage in anaerobic metabolism; a metabolic pathway that converts pyruvate to lactic acid or ethanol with the release of a small number of ATP molecules.

Fertilization A process that involves the union of two sex cells (gametes) and is marked by a sequence of events (contact and recognition and fusion of nuclei).

Fiber The indigestible portion of plant foods; cellulose, a complex carbohydrate that composes plant cell walls.

Fibers Extremely long, narrow plant cells with thick, lignified walls; provide structural support to plant body; generally dead at maturity; abundant underneath the epidermis of stems and help the plant bear bending stresses. In broad-leaved woody plants, fibers lend rigid support to strengthen wood.

Fibrous root system A network of many similar-sized roots that all grow from the base of the stem.

Filamentous A multicellular growth form of fungi composed of a network of hyphae; in contrast to the single-celled yeast growth form.

Fitness The ability of an organism to survive and reproduce in its environment.

Follicle-stimulating hormone (FSH) A reproductive hormone present in both males and females that stimulates the growth of follicles in the ovary and induces the formation of sperm in the testis.

Food chain An illustration that maps the specific feeding arrangements between organisms in an ecosystem.

Food web An illustration showing the range of potential feeding relationships among many organisms in an ecosystem.

Forensic science Field that uses scientific techniques to analyze crime scenes, identify accident victims, and establish paternity in child support cases.

Fossils The physical evidence of organisms that lived in the past and includes mineralized bones, teeth, casts, molds, shells, and wood. Trace fossils are evidence of an organism's activity such as footprints, burrows, impressions, and preserved feces.

Freshwater Naturally occurring water stored in ice, lakes, rivers, and underground as groundwater in aquifers. Freshwater excludes saltwater found in oceans and is generally low in concentrations of dissolved salts.

Fruit A ripened ovary that develops from flowers; contains seeds and other parts of the flower.

Fruiting body The structure of a fungus that produces spores; a mushroom is an example of a fruiting body.

Functional proteins Proteins that protect, communicate, transport gases, or catalyze chemical reactions; examples are antibodies, hormones, hemoglobin, and enzymes.

Gallbladder Muscular digestive sac that stores and concentrates excess bile; contracts and empties bile into small intestine when stimulated by fats in the intestine.

Gametes Reproductive cells, such as egg cells in females and sperm cells in males, that fuse during fertilization to form a zygote.

Gametogenesis Process of forming and developing haploid gametes; in animals occurs in gonads; involves the coordinated events of meiosis, mitosis, and cell differentiation.

Ganglia (singular, *ganglion*) Clusters of neurons that perform basic functions of integrating sensory inputs and controlling limb movement in the absence of a true brain.

Genes The functional units of inheritance that pass on traits from generation to generation. Genes are individual segments of DNA coding for specific molecules, usually proteins.

Gene flow The movement (also called migration or dispersal) of alleles between different populations.

Gene therapy A technique that inserts a normal gene into the genome, replacing the defective gene that causes disease.

Genetic engineering The genetic manipulation of taking a gene from one organism and inserting that sequence into the DNA of another species.

Genetic markers DNA sequences that indicate an increased risk of developing a specific disease or disorder. Genetic markers have been identified for many human diseases, including colon, prostate, and breast cancer, heart disease, Alzheimer disease, type 1 diabetes, and Parkinson disease.

Genetically modified (GM) food Crops that are genetically altered with genes from another organism.

Genome All of the genetic content (DNA, genes, and chromosomes) of the organism.

Genomics The study of all the genes of an organism, examining both the DNA sequences and the interactions of these genes with each other and with the organism's environment.

Germ layer Any of the three cellular layers (ectoderm, endoderm, or mesoderm) that give rise to the various tissues and organs of the animal body.

Germination The emergence of a plant or fungus from a spore or seed; growth of an embryo within a seed that breaks through the seed coat; often occurs after a period of dormancy.

Gills Respiratory structures used by many groups of aquatic animals that function to increase the surface area in contact between water and the body fluids for gas exchange.

Global climate change Refers to global changes in the statistical patterns of weather over periods that range from decades to millions of years; in modern times it refers to those human activities that have had a direct impact on global factors such as temperature.

Global warming Refers to a long-term increase in temperatures near Earth's surface; in typical use it refers to climate changes associated with humans.

Glucose A six-carbon ring-shaped molecule; sugar that fuels life activities; the most abundant sugar molecule in human diets; the main source of energy for cells; the sugar flowing through human blood vessels (blood sugar).

Glycogen Complex carbohydrate used to store energy in animals and fungi; a long, branched chain of thousands of glucose molecules.

Glycolysis Transfers energy in glucose to pyruvate (a three-carbon compound) through a series of 10 different chemical reactions. Some of the hydrogen atoms are stripped from glucose and attached to hydrogen-carrier molecules; generates two molecules of ATP by enzymes in the cytoplasm.

Golgi complex Organelle that functions to process substances for secretion out of the cell; receives and modifies proteins and lipids from the endoplasmic reticulum for transport.

Gonads Primary animal reproductive organs (ovaries and testes) that are responsible for producing eggs, sperm, and hormones.

Grasslike Herbaceous plant growth form marked by slender, green stems that bear several long narrow leaves.

Greenhouse gases Atmospheric gases such as water vapor and carbon dioxide, which absorb incoming solar radiation and trap heat in the atmosphere.

Growth An increase in size as a result of three different but closely related processes: increasing cell numbers, increasing individual cell size, and in animals, the secretion of various proteins and other materials around the cells.

Growth factors A group of proteins or hormones that stimulate cells to divide by signaling the cell to remain in the cell cycle, to exit the cycle, or to continue to differentiate and become specialized.

Growth form A plant's shape, structure, and appearance.

Habitat The environment in which an organism lives, which includes all of the living and nonliving factors that surround and influence its life processes. At a large scale the organism's macro-habitat includes regional environmental characteristics; its microhabitat is the smaller-scale environment—the conditions in its immediate vicinity that influence its physiology.

Haploid A single set of chromosomes; the number of chromosomes in a gamete, representing half the number of chromosomes after fertilization.

Helper T-cells A specialized class of white blood cells that can work in concert with macrophages to destroy pathogens. As the name implies, helper T-cells help in the function and control of virtually all immune system activities.

Hemoglobin A red-pigmented protein found in red blood cells that transports and delivers oxygen to body cells.

Herbaceous plant A plant that does not actively grow in diameter by adding wood; it grows only through apical meristems.

Herbivores Animals that feed exclusively on plants or parts of plants (roots, stems, leaves, flowers, fruits, and seeds).

Heterotrophs Organisms that cannot make their own food and must obtain energy and nutrients from other organisms such as plants, fungi, or other animals.

Heterozygous Individual with two different alleles for a single trait.

Hibernation A survival strategy characterized by a relatively long, seasonal period of inactivity during which the organism's body alters its normal metabolic patterns by lowering temperatures and heart rates and overall slowing of other bodily functions.

Hierarchy (refers to the *hierarchy of life*) Larger units of life are composed of smaller, nested units. Each level includes and builds on the level below it.

From the smallest to the largest, the organization is atoms, molecules, cells, tissues, organs, organ systems, populations, communities, ecosystem, and finally the biosphere.

High blood pressure (hypertension) A physical condition in which the blood pressure in arteries is regularly above the normal range. Healthcare professionals define high blood pressure as a consistently elevated pressure of 140 or higher systolic and/or 90 or higher diastolic.

Homeostasis The process in which an organism senses, adjusts, and maintains conditions in its internal environment within a range that favors survival and reproduction; process that keeps conditions within the internal body environment within their normal ranges.

Homologous chromosomes The matched pairs of chromosomes (one from each parent) found in diploid cells.

Homologous structures The similarities in anatomical traits that result from common ancestry.

Homozygous Individual with two similar alleles for a single trait.

Hormones Highly specialized chemical messengers that bind to cellular receptors on specific target tissues, causing metabolic, physiological, or behavioral responses in organisms.

Host An organism that is infected by or fed upon by another living organism; a living source of nutrients and energy.

Hybrid The offspring of mating between two different species.

Hydrogen bond A type of weak bond where a hydrogen atom with a positive charge is attracted to another negatively charged atom, often an oxygen or nitrogen atom.

Hydrolysis The splitting of long-chained macro-molecules into their building block molecules through the addition of water and enzymes.

Hyphae Thin, microscopic cellular filaments that make up the main body of a multicellular fungus; grow into a highly branched network (mycelium).

Hypothesis A testable, plausible explanation for a natural phenomenon.

Inbreeding depression Reduced health and vigor as a result of breeding between closely-related individuals.

Incisors Type of teeth at the front of the mouth, which are used for cutting, clipping, or gnawing food; incisors determine the size and shape of the bite and thus the food available to an animal.

Incomplete dominance Occurs when two alleles are needed to express a trait, but the dominant trait does not completely mask the recessive trait.

Independent assortment The random distribution of maternal and paternal chromosomes into gametes; the independent movement of each pair of homologous chromosomes during meiosis.

Infection (infect) The entry and multiplication of pathogens in an animal or plant body.

Inflammatory response A fundamental, nonspe-cific type of response by the body to disease and

injury characterized by the classic signs of pain, heat (localized warmth), redness, and swelling.

Inflorescence A group of flowers in a definite arrangement on a plant stem.

Intermediate hosts Required host necessary for the completion of particular life stages of some parasitic organisms such as tapeworms. Most parasitic worms involve one host for the develop-ment of their larval stages and a second host for the adult stages of life.

Internal fertilization Union of the sperm and egg inside the female's body. This strategy consumes more energy and results in fewer offspring, but it enhances the organism's reproductive success by protecting the gametes and maintaining a moist environment for the developing young.

Interphase The phase of cell division in which the cell spends most of its time. During this phase the cell performs its job, grows its cytoplasm, replicates its organelles, and duplicates its chromosomes located in the nucleus.

Interspecific competition Competition between members of different species.

Intraspecific competition An intense form of competition between members of the same species that occurs when there is complete niche overlap.

Joint The location where two bones involved in movement come together. Many joints are filled with fluid, which helps lubricate and cushion bone-to-bone connections.

Keystone species A term used to describe organisms that have a disproportionally large effect on community structure by helping to regulate the types and numbers of various other species in a community.

Killer T-cells A class of white blood cells that work in concert with helper T-cells. Killer T-cells directly kill infected cells before the pathogen has had a chance to replicate. Some T-cells become memory cells that circulate in the blood for years, waiting for the pathogen or antigen to reappear.

Large intestine Also called the colon; the last segment of the digestive system that functions to absorb water and consolidate feces for elimination through the anus.

Law of segregation Each parent contributes one of the two alleles that are passed down to the offspring. The members of each pair of alleles that code for a specific trait separate during meiosis.

Lichens A mutualistic partnership between fungi and populations of green algae or cyanobacteria; commonly found on trees or fallen logs, in soil, or encrusted on rocks and gravestones.

Life history Series of changes an organism undergoes throughout its lifetime; the time a plant spends in each of its stages of growth and reproduction.

Life zone Broad geographic areas with similar physical characteristics that have comparable types of plants and animals living there.

Ligaments Straplike connective tissues that hold bones in place and stabilize joints. Most joints, including those in the knee, have multiple ligaments.

Light reactions The first of two stages in photosynthesis characterized by a complex set of reactions in which the absorption of light energy is used to transform water and other resources into ATP, and hydrogen carrier molecules, which are used during the second stage.

Lipids A class of organic macromolecules that do not dissolve in water; include phospholipids, cholesterol, oils and fats.

Lungs Inflatable internal respiratory structures used by vertebrates for gas exchange.

Luteinizing hormone (LH) A reproductive hormone present in both males and females. In males, LH stimulates cells in the testes to produce testosterone. In females, LH controls the ovarian and uterine cycles and acts on cells in the ovaries to produce estrogen and progesterone.

Lymph Tissue fluid that flows through a system of vessels and collects in lymph nodes.

Lymphocytes A group of white blood cells that always patrol the blood and lymphatic fluids of the body. There are many different types of lymphocytes, each of which perceives different antigens and performs different immune system functions.

Lysosome Special kind of vesicle that contains powerful digestive enzymes; if broken open, capable of digesting the entire cell.

Macromolecules Large, complex molecules that compose living things; also referred to as biomolecules. Examples include nucleic acids, proteins, carbohydrates, and lipids.

Macronutrients Organic molecules required in large amounts by the body; large, complex molecules containing energy stored in their chemical bonds that, when broken, release energy.

Macrophage One type of phagocyte that is able to capture and ingest and destroy as many as 100 bacteria at a time.

Malignant tumor A tumor that spreads throughout the body (metastasizes); the most dangerous type of tumor that can cause deaths.

Mass extinctions The death of all individuals comprising a species; elimination of many species in a relatively short period in geologic time.

Mechanoreceptors Sensory receptors located in skin and ears that, when stimulated by the environment, allow for the perception of sound and physical touching along with contributing to balance. The strength of the sensation is directly associated with the number of sensory cells stimulated.

Meiosis Reproductive process in specialized cells that reduces the number of chromosomes in half, rearranges its information to create new combinations of alleles, and then splits twice to produce four cells; a special type of cell division in which a diploid cell divides to produce four genetically unique haploid gametes.

Memory cells Cells made from the initial contact between B-cells and an antigen. Memory cells express receptors for the antigen, but they circulate in the body for years and even decades. If the same antigen is encountered in the future, memory cells respond rapidly to overwhelm the bacterium before it has a chance to cause disease.

Menopause The period in a woman's life when ovulation and the ovarian cycle cease. Estrogen and progesterone levels decline, creating physical and emotional changes that can be uncomfortable.

Meristem A growing point of a plant; grows by increasing the number of cells (cell division) and the size of cells (cell expansion).

Mesophyll The middle portion, or region, of a leaf; dominated by large numbers of parenchyma cells that carry out photosynthesis; also is composed of support and vascular tissues.

Metabolic enzymes Enzymes that speed up chemical reactions in metabolic pathways. Operate inside cells (as opposed to digestive enzymes that generally work outside of cells). Function by breaking bonds or making bonds.

Metabolic pathways Chemical reactions that occur as a series of steps in energy-releasing and energy-using pathways.

Metabolism Chemical reactions that release and use energy; the process of transferring energy from food nutrients to molecules that bring about cellular work; chemical process of using energy to build macromolecules from building block molecules.

Metastasize The shedding of cancerous cells from tumors that move into the blood and lymph system to spread throughout the body.

Microbes Single-celled organisms, like bacteria and some protists and fungi.

Micronutrients Nutrients that are required in smaller amounts to support normal body functioning.

Migration The seasonal movement between habitats in response to changing habitat conditions—chiefly environmental extremes of temperature, moisture, or other resources. The physical expense involved must be exceeded by the availability of resources in the new habitat.

Minerals Inorganic substances; animals need a variety of minerals, which usually are ingested as salts.

Mitochondrion Organelle that is the "power generator" of the cell where the chemical energy from food nutrients combines with oxygen to be transformed into fuel that drives the cell's many activities.

Mitosis Also called cell division; the part of the cell cycle where the cell separates its duplicated chromosomes into two identical daughter cells. This process is common for tissue growth or repair, or for asexual reproduction.

Molars Type of teeth used for chopping up and grinding food; are especially well developed in herbivores; located toward the back of the mouth.

Molecules Two or more atoms interacting, or bonding, together.

Mucous membranes Epithelial tissues and their underlying connective tissue that line the body cavities that are open to the external environment. Lines the mouth, nose, eyes, ears, anus, urethra, and vagina. These linings represent an enormous surface area where pathogens frequently make their first contact with the body.

Mucus A thick secretion composed mainly of water, complex carbohydrates, and proteins; binds food

particles together and lubricates their passage through the digestive tract; protects the stomach from digesting itself; helps create a barrier to infection.

Multicellular Organisms made of many cells that work together.

Muscle fibers Long, cylindrical cells found in skeletal muscles composed of bundles of specialized contracting protein fibers wrapped in a covering of tough connective tissue. Larger, more powerful muscles contain more bundled muscle fibers.

Muscle tissue Tissue that contracts to produce movement; an effector in a feedback loop.

Mutation A change in a DNA sequence.

Mutualism A relationship between different species in which both species benefit from the interaction, which develops because it results in an increased level of fitness for both species.

Mycelium An interconnected network of hyphae that branch throughout a host or substrate to find and absorb nutrients; vegetative part of the fungus.

Mycorrhiza A symbiotic (mutually beneficial) partnership between fungi and the roots of plants; the fungus receives sugars and the plant receives water and minerals.

Natural selection One mechanism of evolution that causes differential survival and reproduction among individuals of a population. Natural selection acts over time on variation; a process of favoring adapted individuals for survival and successful reproduction.

Negative feedback A control system that detects changes in a factor and counteracts those changes to maintain the steady state.

Nerve nets The simplest nervous systems that evolved in radial animals such as jellyfishes and hydras, consisting of a mesh of nerves spread throughout their bodies that detect stimuli and control cells that contract, allowing the animals to change shape or direction.

Nerves Bundles of neurons that provide a signal pathway to and from an animal's brain. Each nerve is a cablelike structure that contains the signal-input regions from many adjacent neurons.

Nervous system A collection of tissues and specialized cells that provides information about the environment, signaling pathways throughout the body, and the ability to integrate information input and generate responses.

Nervous tissue Tissue that senses and communicates information throughout the body; tissue that causes muscle tissue to contract.

Neurons Cells that receive information and communicate with other neurons in a complex network of connections; structural and functional units of the nervous system.

Neurotransmitter A chemical signal that crosses the gap between adjacent neurons and stimulates the downstream cell. In effect, the neurotransmitter "carries" the nerve signal across the synapse, allowing the communication to continue on its pathway.

Niche An organism's range of habitat conditions and resources needed, along with the variety of interactions a species engages in for survival and reproduction.

Nitrogen cycle The global movement of nitrogen, where nitrogen is converted between its various chemical forms. The movement between reservoirs can be carried out via both biological and nonbiological processes. Microbes are a critical living resource in the cycling of this element.

Nondisjunction A lack of chromosome separation that occurs in meiosis I or meiosis II, resulting in a gamete with either an extra or a missing chromosome.

Nuclear envelope Membrane that encloses the nucleus and forms a boundary between the nucleus and cytoplasm.

Nucleolus Darkened region of the nucleus that contains instructions (genetic information) for making ribosomes.

Nucleus A large structure in eukaryotic cells that houses the majority of the DNA in the cell; control center of the cell.

Nutrients Chemical substances essential for survival because they provide energy and raw materials and/or support body processes such as growth, maintenance, or repair of tissues.

Nutrition A specialization in biology that examines how nutrients are digested, broken down, absorbed, and used in the body.

Oncogenes A family of genes that transform normal cells into tumor cells. Mutated forms of "green light" genes found in many forms of cancer that cause cells to always move forward in the cell cycle and divide.

Oogenesis Process occurring in the ovaries, producing ova (or eggs).

Opportunistic organisms Pioneering organisms that are able to quickly become established in recently disturbed habitats and exploit the changed conditions.

Organelles The nucleus and other membrane-bound structures that perform specialized functions in eukaryotic cells.

Organic molecules Compounds composed of more than one type of element produced by living organisms, which contain carbon–hydrogen bonds. The four major classes of organic molecules include carbohydrates, proteins, lipids, and nucleic acids.

Organs A group of two or more types of tissues that work together to perform a specific function; a group of tissues that work together to perform specialized functions; they form the structure and function of organ systems.

Organ system A group of several different organs that work together to perform specialized functions, such as gas exchange, digestion, or excretion.

Osmosis Passive movement of water across a cell membrane; water moves down its concentration gradient toward the compartment with the highest concentration of dissolved materials.

Ovarian cycle Series of cyclic events in the ovaries of mammals which results in the release of eggs.

Ovary In female animals, the structure that produces eggs and reproductive hormones. In flowering plants, the structure at the base of the

carpel that contains ovules. After pollination and fertilization, it swells to become fruit.

Pain receptors Sensory receptors that, when stimulated, allow for the perception of pain.

Pancreas Organ that secretes digestive enzymes, substances that neutralize stomach acids and also insulin and glucagon critical to regulating blood sugar levels.

Parasitism An interaction between organisms where one organism, the parasite, benefits at the expense of another organism, the host, which supplies resources.

Parenchyma Plant tissue consisting of large thin-walled cells; stores water and starch in leucoplasts in roots and stems; performs photosynthesis in leaves.

Passive transport Movement of substances "downhill," or down its concentration gradient.

Pedigree A type of family tree that geneticists use to track the history of a particular trait over several generations; used to track patterns of inheritance.

Perennials Plants that live for more than two years and have repeated episodes of reproduction; may be herbaceous or woody.

Petals Modified leaves of a flower that form the whorl just inside the sepals, protecting the reproductive organs; often colorful, with a sweet fragrance to attract pollinators.

Phagocytes A special group of white blood cells able to recognize, ingest, and kill microbes that display nonself antigens on their surface.

Phenotype The observable characteristics of a trait.

Pheromones Specialized chemical signals produced by animals to communicate information.

Phloem Plant tissue that carries sugars throughout the plant; flows from sites where sugars are made or stored to places where it is used or stored; occurs and functions with xylem in vascular bundles.

Phospholipid Insoluble molecule that forms the membranes of the cell; an individual phospholipid molecule consists of a head (phosphate groups) and two fatty acid tails.

Photoperiod The daily cycle of light and darkness that influences flowering in many plants.

Photoreceptors Specialized cells that are sensitive to light energy.

Photosynthesis The metabolic process that transforms sunlight energy and carbon dioxide into larger biological molecules, such as sugar.

Placenta Organ in the uterus of mammals that provides nutrient and waste exchange to the developing fetus.

Plasma membrane A thin, structured lipid bilayer that separates cells from other cells and their environment. This membrane encloses the cytoplasm and acts as a physical envelope. Regulates movement of molecules in and out of the cell.

Plasmids Small pieces of circular DNA found in many prokaryotes that often code for proteins important to infection or survival.

Plastids Specialized plant organelles that include chloroplasts, chromoplasts, and leucoplasts; found in

all vegetative and reproductive plant structures; contain a small circular piece of DNA and can divide and multiply within a cell.

Pleiotropy Genetic condition in which one gene controls two or more phenotypic traits.

Pollen grains In seed-producing plants, the structure produced in the male sex organs of a flower that carries the male genome to female sex organs of a flower and produces sperm cells.

Pollen tube A slender single-celled tube that develops from a pollen grain; grows from the stigma and down the style to penetrate the ovule and deliver sperm (male gamete) to unite with an egg cell (female gamete).

Pollination The transfer of pollen from male anthers to a female stigma.

Polygenic inheritance Genetic condition in which one phenotypic characteristic is controlled by many genes.

Population A single species living in one geographic area at the same time.

Positive feedback Control system that increases, or amplifies, the signal in the direction of the stimulus.

Predation An interaction in which one animal kills and eats another to obtain its food energy.

Pregnancy The period in mammalian development in which the embryo develops in the uterus of the mother; begins when the zygote implants into the uterine wall.

Primary producers Autotrophic photosynthetic or chemosynthetic organisms able to transform light or chemical energy into biological molecules. Members of the first trophic level.

Primary structure The sequence of amino acids in a protein.

Producers Organisms in the first trophic level that use photosynthesis or chemosynthesis to produce sugars and other biomolecules.

Prokaryotic cells Small, simple cells that lack organized compartments; includes bacteria and archaea.

Promoter A specific start sequence for a gene.

Proteins Large, complex, specifically shaped molecules that carry out many of the functions of life including the synthesis of other biological molecules.

Protozoans Unicellular, heterotrophic protists, many of which are mobile and act as parasitic pathogens.

Punnett square A graphic tool geneticists use to show and predict the odds of potential allelic combinations in gametes based on the genotype of the parents.

Races Two different populations of the same species that can interbreed, but display different phenotypes.

Radiation therapy A cancer treatment using targeted ionizing radiation internally or externally to kill cancer cells and shrink tumors.

Recessive An allele that is not preferentially expressed and that requires two copies for expression of that phenotypic characteristic.

Red blood cells Blood cells that pick up and transport oxygen to body cells.

Reflex arcs Localized, involuntary loops between nerves and muscle groups that, when stimulated, produce actions without signal processing in the brain.

Reproduction The process that creates new individuals of the same kind from previously existing individuals.

Reproductive barriers Conditions that prevent two organisms or populations from breeding.

Reproductive isolation Condition in which populations are unable to interbreed and can lead to separate species; drives speciation.

Resources Substances required for an organism to survive that are provided by its habitat; see Table 4.2.

Resource competition Competition within a community for limited resources such as food that occurs in all habitats. May be between members of the same or different species.

Resource partitioning A process of subdividing the limited resources within a habitat that ultimately reduces competition intensity between species.

Respiratory system The animal organ system that supplies body cells with oxygen and excretes carbon dioxide from the body; involves moving air in and out of the lungs and the exchange of gases with cells in the blood.

Retroviruses Class of RNA viruses like HIV that use their unique enzymes to direct the replication of their genome by a host nucleus.

Ribosomes Organelles, found free in the cytoplasm or attached to endoplasmic reticulum, where translation (protein synthesis) takes place.

Risk factors Behaviors and genetic factors that increase the chance of developing a disease. Modifiable risk factors are those that involve diet or other behaviors, which individuals can influence. Examples include smoking, diet, and certain forms of high blood pressure.

RNA (ribonucleic acid) Single stranded nucleic acid found in the nucleus and cytoplasm of a cell that is involved in protein production. Contains a ribose sugar backbone and nitrogenous bases (adenine, guanine, cytosine, and uracil). There are three classes: messenger (mRNA), transfer (tRNA), and ribosomal (rRNA).

Root cap Dome of epidermal cells that protects the growing root as it grows through the soil; secretes a slimy substance that allows the growing tip to navigate around soil particles in search of resources.

Root hairs Delicate, short-lived narrow extensions of root epidermal cells that contact soil particles and absorb water and minerals.

Root system A branched network of roots of a plant.

Salivary glands Glands that open into the mouth; produce mucus and enzymes that begin the digestion of starch.

Science The systematic investigation of the natural world.

Scientific method A proper and objective systematic investigation that includes appropriate experimental design and rigorous review of the results by other scientists. Designed to reduce human bias in the perception and interpretation of data.

Seed A multicellular structure that consists of an embryo and food reserves (endosperm) and is enclosed by a protective seed coat; originates as an ovule that contains a fertilized egg.

Selection pressures Conditions that exert natural selection on an organism at its particular stage of life.

Semen Fluid that protects, nourishes, and helps to deliver the sperm.

Sepals Outermost parts of a flower; often green and having the same number as petals. Sometimes colorful to attract pollinators.

Sex chromosomes Chromosomes that determine gender; in humans, the X and Y chromosomes.

Sex-linked Any gene or trait located on a sex chromosome.

Sexual dimorphism A difference in physical characteristics between the males and females of a species; examples are differences in size, coloration, or body structure.

Sexual reproduction Creation of genetically unique offspring from the fusion of male and female gametes. In this mode, half of the offspring's genetic material comes from each parent.

Sexual selection Pressures that favor particular traits that attract a mate.

Shared characteristics Evolutionary novelties, like anatomical features, that groups of organisms have in common. These traits or characteristics can include physical attributes such as the presence of a bony skeleton or behavioral characteristics such as singing to attract a mate. Shared characteristics are used to separate organisms into groups and establish relationships.

Shrubs Woody plants that do not have a main trunk but have many similar-sized branches that grow near the base of the plant; shorter in height than trees.

Simple carbohydrates Sugars containing one or two sugar units bonded together that provide the major fuel source to power cellular activities.

Simple diffusion Movement of substances such as gases down their concentration gradient from an area of higher concentration to an area of lower concentration.

Sister chromatids Two identical copies of the DNA strands produced during chromosome replication at the end of the S phase of cell division.

Skeletal muscle One of three types of muscle tissue in the animal body that contract and result in movement.

Speciation The process that produces new species.

Species Members of a population that actually or potentially interbreed and produce viable offspring in nature.

Species diversity A measure of the number and relative density of different species found in a geographic area; a type of biological census, tallying the species present, their numbers, and their density in comparison to other species present.

Spermatogenesis Process occurring in the testes that produces sperm.

Spinal cord Bundles of nerves located in the spine that serve as the primary signal pathway between the brain and the rest of the body, as well as a location where many reflex reactions are controlled.

Spore In plants, a reproductive cell that is genetically unique; it is dispersed into new habitats and grows into a new individual; in fungi, a reproductive cell that arises by mitosis or meiosis and that disperses, germinates, divides, and grows into a new fungus.

Stamens Fertile male flower parts that produce pollen; consist of a filament that supports the pollen-producing anther.

Starch The energy-storage carbohydrate made by plants; abundant in plant roots, bulbs, and seeds; a long chain of thousands of glucose molecules bonded together.

Stem cells Unspecialized cells that have the potential to develop into many different cell types and tissues during early life and growth.

Stomata (singular form, **stoma**) Openings in the epidermis of leaves and herbaceous stems controlled by guard cells that respond to environmental cues. Permit the exchange of gases with the atmosphere for photosynthesis.

Stroke A blockage of a blood vessel that supplies resources to the brain. When oxygen supply is limited, the part of the body controlled by that area of the brain can no longer function.

Structural proteins Proteins that give body parts form and structure, including hair, muscle, skin, tendons, cartilage, and connective tissues.

Substrate In metabolism, a molecule acted upon by a specially shaped enzyme. For fungi, the nonliving source of nutrients and energy; examples include cow dung, hair, wood, and fallen leaves.

Succulent Plants such as cactuses that store large amounts of water in their roots, stems, and leaves, which gives them a swollen or bulging appearance; often found in dry, arid habitats.

Sucrose Chemical name for table sugar; composed of two sugar units (glucose and fructose molecules) bonded together.

Sugars Small carbohydrate molecules that contain five or six carbon atoms that form a single or double ring; building block molecules for complex carbohydrates.

Surface-to-volume ratio The ratio, or relationship, that determines how much surface area is available to supply a given volume of cells with oxygen and nutrients.

Tactile communication Physical touching between animals used for communication. Often used to establish bonds between members or to send important signals.

Taproot A thick primary root specialized for storage.

Taxonomy A field of biology that classifies, identifies, and names organisms. Biologists that specialize in this field are referred to as taxonomists.

T-cells A group of lymphocytes, which are "educated" in the thymus gland and learn to make the sharp distinction between molecular markers that say "self" from antigens that say "nonself," or are foreign to the body. Unlike B-cells, T-cells probe and detect markers on cell surfaces and specialize in identifying and eliminating virus-infected cells.

Tendons Straplike connective tissues that link muscles to bones. As muscle tissues contract, they shorten and pull against tendons attached to bones, producing movement.

Testosterone Important developmental and reproductive hormone responsible for many activities in the male reproductive system. Stimulates development of male secondary sexual characteristics, such as increased bone and muscle mass, and growth of body hair.

Theory An explanation based on a large body of scientific observations, experiments, and reasoning; it is tested and confirmed as a general principle helping to explain and predict natural phenomena.

Therapeutic cloning A technique that produces stem cells genetically matched to the patient by making embryonic stem cells from his or her own adult tissues; aims to generate cloned embryos, which provide embryonic stem cells for repair of damaged or defective tissues.

Thermoreceptors Sensory receptors located in the skin that, when stimulated, allow for the perception of hot and cold.

Threatened species An organism likely to become endangered in the near future.

Tissues Groups of cells that associate and work together to form body organs and perform specialized functions such as movement or reproduction.

Tissue fluid Fluid surrounding cells and tissues that allows for the exchange of materials with both the cells and the blood.

Tradeoff A behavioral or metabolic shift that occurs in response to a resource constraint. All habitats have some resource limitations that influence the reproduction and survival of the organisms that live there.

Transcription The gene expression process that transfers the code in the DNA to intermediate messenger RNA (mRNA) molecules.

Transition zone A location where organisms and environmental conditions of adjacent habitats blend together. These blended conditions support species and contain ecological niches found in habitats on both sides of the zone.

Translation The process of gene expression that converts messenger RNA (mRNA) to a protein sequence made from amino acids.

Transpiration The evaporation of water out of a plant into the atmosphere through open stomata. As water evaporates from the leaf, a thin column of water is pulled all the way through the plant from the roots up the stem and to the leaves.

Trees Woody plants that consist of a large main vertical, woody trunk that supports numerous smaller branches and a canopy of leaves; generally tall plants.

Triglyceride Chemical name for fats and oils; a type of fat that contains three fatty acid chains attached to a glycerol molecule; macromolecule that stores energy in fat cells and tissues.

Trophic level One position or level within a food chain or food web that describes the feeding or energy relationships in an ecosystem.

Tumor An abnormal mass of cells resulting from uncontrolled cell division. Not all tumors are cancerous, but even noncancerous ones grow and ultimately interfere with the function of normal cells around them.

Tumor suppressor genes A family of genes involved in regulating transcription, DNA repair, and cell-to-cell communication. These genes serve two functions: to prevent cell division and to lead to cell death.

Turgor pressure The outward push exerted by water against the cell wall; plays a key role in cell expansion during growth and in flexible support of herbaceous plants and their parts.

Uterine cycle Series of changes in mammals that prepare the uterus to receive the egg, often called the menstrual cycle.

Vaccination Typically, an injection that contains materials specifically designed to activate the immune system in advance of coming into contact with a known pathogen.

Vaccines A biological preparation that introduces a killed or inactivated virus or bacterial antigen into the body to "educate" the body's immune system to recognize and eliminate a particular threat based on its specific molecular structure.

Vacuole Plant cell organelle that stores water, pigments, and toxins; a large central vacuole occupies the majority of the cell volume.

Vascular cambium A lateral plant meristem found in woody plants; produces xylem (wood) to the inside and phloem to the outside of roots and stems.

Vascular system Plant tissue system that is composed of conducting tissues that transport water and minerals (xylem) and sugars (phloem); runs throughout the plant and connects roots, stems, and leaves.

Vector A carrier. In biotechnology, a carrier of genetic information to another organism (for example, a plasmid). In infectious diseases, a carrier of an infectious agent to another organism.

Vegetable A root, stem, or leaf of a plant that is eaten by humans.

Veins Flexible blood vessels that carry blood back to the heart.

Ventricles Chambers of the heart that pump blood into arteries.

Vesicles Small round organelles that transport substances within the cell or to the plasma membrane for secretion.

Villi Fingerlike projections that greatly extend the absorptive surface area of the small intestine; contain microvilli projections that further increase surface area.

Vines Weak-stemmed plants that use the support of the ground, woody plants, or other sturdy objects to grow toward the sunlight.

Viruses Intracellular parasites composed of a nucleic acid core (DNA or RNA) wrapped in a protein coat.

Vitamins A diverse group of organic molecules generally required in small amounts that cannot be synthesized by the organism, but play a vital role by assisting enzymes in metabolic reactions.

Water cycle The continual movement of water on, above, and below the surface of Earth, including water contained in living organisms. Energy from the sun powers much of the movement of water.

Woody plant Plant that adds layers of woody tissues to its stems and roots each year so that it increases in diameter; it grows in length through apical meristems and in diameter through lateral meristems (vascular and cork cambiums).

Xylem Plant tissue that transports water and minerals throughout a plant; composed mostly of rigid, dead cells and dominates the structure of wood; occurs and functions with phloem in vascular bundles.

Yeast A general term for a single-cell fungus; most reproduce by budding and are capable of metabolizing sugar by alcoholic fermentation.

Zygote A fertilized egg that grows and develops first into an immature juvenile and then into an adult organism.

Index

The letter f designates figure; t designates table; **bold** designates key terms.